Functional assessment of wetlands

Functional assessment of wetlands

Towards evaluation of ecosystem services

Edited by
Edward Maltby

CRC Press
Boca Raton Boston New York Washington, DC

WOODHEAD PUBLISHING LIMITED
Oxford Cambridge New Delhi

Published by Woodhead Publishing Limited, Abington Hall, Granta Park
Great Abington, Cambridge CB21 6AH, UK
www.woodheadpublishing.com

Woodhead Publishing India Private Limited, G-2, Vardaan House, 7/28 Ansari Road,
Daryaganj, New Delhi – 110002, India

Published in North America by CRC Press LLC, 6000 Broken Sound Parkway, NW
Boca Raton FL 33487, USA

First published 2009, Woodhead Publishing Limited and CRC Press LLC

British Library Cataloguing in Publication Data
A catalogue record for this book is available from the British Library.

Library of Congress Cataloging in Publication Data
A catalog record for this book is available from the Library of Congress.

Woodhead Publishing ISBN 978-1-85573-834-8 (book)
Woodhead Publishing ISBN 978-1-84569-516-3 (e-book)
CRC Press ISBN 978-0-8493-2600-4
CRC Press order number WP2600

The publishers' policy is to use permanent paper from mills that operate a
sustainable forestry policy, and which has been manufactured from pulp
which is processed using acid-free and elemental chlorine-free practices.
Furthermore, the publishers ensure that the text paper and cover board used
have met acceptable environmental accreditation standards.

Typeset by Replika Press Pvt Ltd, India
Printed by CPI Antony Rowe, Chippenham, Wiltshire, UK

Contents

Section 4 Scoring and appendices

Appraisal sheets and Look-up tables

Appraisal sheets are used to keep a record of answers to process-specific questions about the wetland area under study. The procedures select the appropriate description of wetland process performance using look-up tables. Wetland interactions are complex, sometimes involving more than one process to perform a single function.

Introduction to SWIMMER

The Institute for Sustainable Water, Integrated Management and Ecosystem Research (SWIMMER) is a hub for water-related research within the University of Liverpool, UK. The Institute was established in 2005 to encourage integrated thinking among the water and environmental sciences, to promote interdisciplinary collaboration within the university and with other organisations, and to facilitate both fundamental and applied research through active partnerships with end-users.

Led by Professor Edward Maltby, the Institute has extensive experience in the science and management of wetlands, including scientific understanding of the hydrological, biogeochemical and ecological processes that support sediment and ecosystem functioning, and the consequent provision of ecosystem services and goods to human communities. SWIMMER's international outlook provides a focus for research into climate change, the water cycle, ecosystem functioning, and for pioneering work on the 'ecosystem approach', a methodological framework for achieving sustainability.

Commissioned research on behalf of government, supra-government and non-governmental organisations focuses on the development of tools such as functional assessment and decision-support systems. These facilitate more intelligent environmental management, and help manage urgent societal needs such as poverty alleviation, maintenance of essential ecosystem services and mitigation of climate change impacts. SWIMMER's website can be accessed through www.liv.ac.uk/swimmer/.

Registration details

It is anticipated that *Functional assessment of wetlands* will be both updated and expanded in the future. If you would like to know more about updates, please register with SWIMMER by contacting us at: swimmer@liv.ac.uk. Updating will be greatly enhanced by feedback from readers using the functional assessment procedures for themselves. We look forward to hearing your comments and suggestions.

Foreword

Wetlands perform functions that deliver benefits to society, often referred to as *ecosystem services*, as a result of the interactions that take place among natural processes within the structure of these ecosystems. However, wetlands do not all perform the same functions, nor is any one function carried out to the same extent or intensity among different wetland types. A functional approach to wetland assessment enables a holistic view to be taken of the wide range of services wetlands can provide. Some of these may be mutually exclusive, or else one service may be optimised only by reducing the effectiveness of another. The functional assessment procedures (FAPs) can be used to determine the likelihood of particular functions occurring, and how this may vary with changing environmental conditions and management. Questions can be addressed such as 'what are the implications for biodiversity or water quality regulation if water levels are raised by a given level?' The method can be used to indicate the potential and priorities for management options, and at what point further studies are required to improve the knowledge necessary for strategic land-use or site-specific management decisions to be made with more confidence.

The FAPs provide a methodology, which can be used by both experts and non-experts to assess wetland functioning relatively rapidly. Primarily this is achieved by the identification of key characteristics or predictors, which can be related to functions without the need for detailed studies. The most effective performance of some key functions, such as nitrate removal by denitrification, is sometimes focused on small discrete areas of the landscape. To cater for this, assessments are made of distinctive, often small areas of the wetland, characterised by features of hydrology and landform. Better understanding of such variation can lead to improved management and more satisfactory resolution of planning and land-use conflicts, offering options for alternative uses and addressing development pressures.

Previous development of sufficiently robust, science-based tools for wetland functional assessment has been impeded by the lack of integration of the wide range of specialised scientific and geographical perspectives necessary. However, a sequence of EC-funded research projects undertaken since 1991 was targeted largely to overcome this impediment through international and inter-disciplinary collaboration.

It is anticipated that the FAPs will be used by a range of individuals or organisations concerned with wetland management who wish to gain a better understanding of the processes, functions, services or benefits and potential of the wetlands for which they have responsibility or interests. The FAPs are a first step in providing guidance for on-site management and decision making by presenting information on how a wetland is likely to be functioning and the key processes operating within it. On a different level, the FAPs can assist strategic planners at local and catchment scales to make decisions on wetland management and land-use through assessment of overall wetland functioning and the contribution of wetlands to ecosystem services such as water quality, water quantity and biodiversity. Knowledge of functioning can be used to assist with the implementation of national and wider policy, such as the Water Framework Directive, where they can assist with decisions such as the location, type, number and area of wetland buffer zones required to deliver desired water quality benefits, or the risks to the integrity of water bodies and contiguous wetlands resulting from activities such as abstraction that affect water resources.

The modular structure of the package facilitates its expansion to include additional functions to meet the requirements of other specific users, such as the development of a module for wetland archaeologists. The heritage value of wetlands derives mainly from the ability of waterlogged anoxic environments to preserve organic and associated archaeological evidence, which is otherwise rapidly destroyed by biological decay processes under the oxidised conditions that prevail when drying out takes place.

Assessment outcomes from the FAPs can be used in conjunction with other information relating to, for example, societal priorities, costings and policy limitations to assist and support environmental decision making. The FAPs translate best available scientific knowledge into reasonable predictions of how component parts of wetlands function in different landscape contexts. FAP outcomes are linked to societal priorities such as flood control, pollution reduction and biodiversity conservation. The FAPs recognise and emphasise the 'natural capital' of healthy wetland ecosystems, and reinforce the fact that wetland management is a question of choice: for individuals as well as society.

There have been at least three major developments over the gestation period of the FAPs, which underlie their potential significance and points of relevance to operational application:

1. elaboration of the Ecosystem Approach as the primary instrument for delivery of the objectives of the UN Convention on Biological Diversity (http://www.biodiv.org);
2. the UN Millennium Ecosystem Assessment (MA: http://www.maweb.org), which assesses the links between ecosystems and human wellbeing, governmental support through the Millennium Development Goals and the Plan of Implementation following the World Summit of Sustainable Development (http://www.un.org/events/wssd/);
3. in Europe, the development of a more integrated and holistic approach to water management through implementation of the Water Framework Directive (2000/60/EC).

The common philosophy underpinning the rationale behind these initiatives is the recognition that ecosystem functioning is directly or indirectly responsible for the delivery of goods and services that are essential to human welfare. The value of such benefits is sometimes recognised as the 'natural capital' of the world's ecosystems. There is considerable interest from the policy community in the development of methodologies for the assessment of ecosystem services as support to improved decision making in the management of natural resources. Wetlands deliver a wide range of ecosystem services that contribute to human wellbeing, such as water supply, flood regulation, water purification, climate regulation, biodiversity and productivity (e.g. of fish and grazing land) and amenity.

Functional assessment enables the user to predict the functioning of a wetland area without the need for comprehensive empirical research. This provides indicators for the quality and degree of delivery of specified ecosystem services. The link to specific ecosystem services has been developed subsequently through a geographical information system (GIS) mounted decision support tool called the Wetland Ecosystem Decision Support System (WEDSS; http://www.liv.ac.uk/swimmer/research/past_projects.html).

Application of the FAPs enables the user to make more informed and rational decisions about wetlands, based on the better understanding of how they work and what they are contributing, or could contribute, to various sectoral interest groups and to society as a whole.

Edward Maltby

Key contributors

Many individuals in institutions across Europe have contributed material to these procedures, both as scientific researchers and editorial contributors.

Core editorial team:
C. J. Baker
T. Barker
U. Digby
D. V. Hogan
E. Maltby
R. J. McInnes

Lead scientific authors:
K. H. Bishop
M. S. A. Blackwell
B. Clément
E. Maltby
D. Papadimos
M. Scholz
C. Schulz-Zunkel
M. Seferlis
J. Tellam
J. T. A. Verhoeven
M. L. Verhoeven
The Functional Assessment Procedures CD-ROM was programmed by A. Aidoud.

Contributing research personnel:

Wetland Ecosystems Research Group (now located at SWIMMER: Institute for Sustainable Water, Integrated Management and Ecosystem Research, University of Liverpool)
Principal Investigator: E. Maltby
Researchers: C. J. Baker, T. Barker, P. Cheesman, U. Digby, D. V. Hogan, A. Hooijer, M. Hutchins, C. P. Immirzi, E. C. Jewkes, I. Jones, J. N. Kellett, C. Linstead, R. J. McInnes, N. Melville, O. Paramour, F. Sgouridis, R. Thorne, J. E. Tollervey, M. D. Wood
Research students: M. S. A. Blackwell, M. J. Everington, N. J. Mockler, G. A. Oliver, M. A. Russell, M. Simpson

Free University of Amsterdam
Principal Investigator: B. Engelen
Researcher: A. Hooijer

University of Birmingham
Principal Investigator: J. W. Lloyd
Researchers: J. H. Tellam, K. S. J. Massey
Research student: K. Papatolios

University of Glasgow
Principal Investigator: K. J. Murphy
Researchers: C. Allan, T. H. Flowers, I. D. Pulford
Research students: V. Abernethy, J. M. Hills

Complutense University, Madrid
Principal Investigator; J. J. Sanz Donaire
Researchers: M. R. Llamas Madurga, J. Cruces de Abia, D. Díaz Alvarez, A. Sánchez Pérez de Evora
Research students: J. M. Fornés Azcoiti, M. García Rodríguez, E. Perez Gonzalez

Automonous University, Madrid
Researchers: C. Montes del Olmo, F. G. Bernáldez, C. Levassor, C. Guerrero,
Research student: M. Florín Beltrán

University of Rennes
Principal Investigator: B. Clément
Researchers: A. Aidoud, C. Faribault,
Research students: I. Gilles, B. Touzard, S. David

University of Utrecht
Principal Investigator: J. T. A. Verhoeven
Researchers: M. L. Verhoeven, A. Spink, R. Aerts
Research students: M. M. P. van Oorschot, M. J. van der Peijl, I. J. J. van den Wyngaert

Auen Institute, Rastatt
Principal Investigator: P. Obrdlik
Researchers: E. Castella, E. Schneider

The National Parks and Wildlife Service, Office of Public Works, Dublin
Principal Investigator: J. B. Ryan
Researchers: M. C. D. Speight, B. Healy, P. Galvin.

Swedish University of Agricultural Sciences
Principal Investigator: K. H. Bishop
Researcher: M. Nilsson
Research student: S. J. Köhler

Trinity College, Dublin
Principal Investigator: P. Johnson
Research student: O. Dubois

Station Biologique de la Tour du Valat, France
Researcher: P. Grillas

UFZ – Helmholtz Centre for Environmental Research
Principal Investigator: K. Henle
Researchers: C. Schulz-Zunkel, V. Eisenmann, O. Brandt, K. Henle, S. Geyer, H. Potesta, M. Scholz

Greek Biotope Wetland Centre (EKBY)
Researcher: M. Seferlis
Researchers: D. Papadimos, M. Anastasiadis

University of Bucharest
Principal Investigator: A. Vadineanu
Researcher: S. Christofor

Acknowledgements

We would like to acknowledge the following funding sources:

FAEWE I (Functional Analysis of European Wetland Ecosystems I), Contract STEP CT90 0084

FAEWE II (Functional Analysis of European Wetland Ecosystems II), Contract EV5V-CT95-0060

PROTOWET (Procedures for the Operationalisation of Techniques for the Functional Analysis of European Wetland Ecosystems), Contract ENV4-CT95-0060

EVALUWET (European Valuation and Assessment Tools supporting Wetland Ecosystem Legislation), Contract EVK1-CT-2000-00070

TECWET (Techniques and Procedures for the Functional Analysis of Wetland Ecosystems), Contract EVK1-2001-80001

ECOFLOOD (Towards Natural Flood Reduction Strategies), Contract EVK1-2002-00579

Euro-limpacs (Integrated Project to evaluate the Impacts of Global Change on European Freshwater Ecosystems), Contract GOCE-CT-2003-505540

Special acknowledgement is due to Dr Hartmut Barth, Scientific Programme Officer at the Directorate General Research of the European Commission, for his sustained support of the interdisciplinary scientific group contributing to this effort. Lastly to my wife, Rosemary Maltby, for forbearance and taking on many other unreasonable tasks whilst I indulged in the numerous international meetings leading eventually to this publication.

For
Rosemary, Alistair, Peter, Geoffrey and Penny

Preface

This document is a guidance manual and explanatory methodology for the procedures for functional assessment of wetlands, known commonly through its development as the FAPs. The FAPs process information entered by the user and transform the data into outputs that can assist in management decisions based on best available knowledge.

The paper version contains full instructions and supplementary guidance to enable a user to make an effective functional assessment of a wetland. The Functional Assessment Procedures CD-ROM takes much of the more tedious aspects out of the assessment once the initial recording stage is completed. It provides a full record of wetland processes, the functioning resulting from them, and their assessment. This information is stored electronically for future reference.

The essence of the FAPs is the conversion of documented and observable information into an interpretive picture of functioning in the wetland. It begins with desk-based studies, local knowledge and field observations of the characteristics and variables controlling processes that determine the functional characteristics of the wetland. This information constitutes a database against which to measure supplementary information obtained by examination of key variables that have important influences on specific aspects of wetland functioning.

The outputs of the assessment exercises can be used directly in order to evaluate a specific process or function in a defined part of the wetland, or can be combined to assess the overall performance of functions in part or all of the wetland area. This capability can be used to compare wetland areas for functional performance, for monitoring performance before and after management activities, or predicting the possible outcomes of different scenarios related to policy or environmental (such as climate) change. Outputs can be fed directly into the GIS-based computer programme Wetland Evaluation Decision Support System (WEDSS, downloadable free from the WEDSS page on the SWIMMER website http://www.liv.ac.uk/swimmer/research/past_projects.htm), which offers the option of asking 'what if' type questions about potential management strategies.

Both the FAPs and WEDSS provide practical methods of assessing and managing wetlands for optimal delivery of one or more of the important ecosystem services that they provide. The FAPs is an on-going project and future editions will greatly benefit from user feedback. The authors welcome your comments and suggestions.

Section 1
Introduction

1.0

Introduction to wetlands and functional assessment

1.1 Introduction

Human perception of and relationships with wetlands have evolved over time, together with the transformations that have taken place in the organisation and aspirations of society, technological innovations and the level of scientific understanding. Early human communities were strongly and directly dependent on the wetlands of river, lake and coastal margins for water, food, shelter, fuel, security, tradable products and transport. Understanding of how the ecosystem worked, in terms of hydrology, biogeochemical cycling and ecological dynamics, was not defined, but was simply part of everyday life, and this conditioned the community's relationship with the flood cycle, provision of potable water and harvesting of fish and wildlife. Whilst such knowledge was rarely codified into forms that would eventually become a distinct scientific discipline, its understanding has been highly influential in the development of civilisation. The early Sumerians, and subsequently the Babylonians, were as well versed in the ecology of the Tigris–Euphrates marshes as they were farmers, architects and politicians (Kramer, 1963).

There are still high levels of direct dependency on wetland resources by human communities in many parts of the world. This is due to the ability of wetlands to perform certain functions such as flood control and storm protection, and to supply products such as wildlife, fisheries and forest resources. There are many examples of wetlands that have retained their functions and are considered as important resources for the local communities. These include lake Megali Prespa in Greece that performs groundwater recharge, lakes Volvi and Langada in Greece that provide effective flood control, the Odiel Marshes in Spain that perform sediment/toxicant retention, the Akgol Lagoon in Turkey and the Danube delta in Romania and Ukraine that provide local villages with fish and thatching materials, and the Camargue in France that is used for hunting and a wide range of other amenities, including village tourism (Hollis and Finlayson, 1996). Human decisions currently being taken to exploit wetlands result more from a response to economic, social and political pressures than any appreciation of the sustainable benefits that can derive from natural ecosystem functioning (Maltby, 1991). Despite considerable improvements in public awareness, there is still a failure to appreciate that human wellbeing and economic development can be consistent with, and indeed enhanced by, the maintenance of fully functioning wetland ecosystems.

Historically, human endeavour has concentrated on the conversion of wetland ecosystems such as swamps, bogs, marshes and floodplains to non-wetland conditions, allowing their more intensive or direct economic utilisation for farming, forestry or urban development. Wetland degradation and loss has continued throughout Europe, while the conservation community and, increasingly, environmental scientists, resource planners and economists claim growing evidence for their importance for biodiversity, and for the maintenance or enhancement of various aspects of environmental quality. Thus, while society seems to have become increasingly detached from wetlands, this independence may be more apparent than real. The structures and investments necessary to replace the functions previously carried out by wetlands may be both more expensive as well as unsustainable.

The professional challenge to the wetland scientist is to interpret the wetland ecosystem as a functioning unit within the complex human and often dynamic natural environment, to evaluate their tolerance of various uses, and advise on optimum management strategies to maintain functional integrity. His or her more difficult task is to collaborate with other professionals and lobbyists to bring about major changes in the institutional structures required at local, regional and international levels to ensure the future wise stewardship of some of the globe's most precious yet fragile environmental resources (Maltby, 1991).

A wide range of factors have promoted and continue to promote wetland degradation in Europe. These include science and information deficiencies, imbalances in the distribution of costs and benefits of wetland conservation, inadequate planning systems and instruments, policy conflicts and institutional weakness (Dugan, 1990; Maltby, 1991; Maltby *et al.*, 1994; see also Table 1.1.1).

Wetlands in Europe are widespread and exhibit enormous variety across the continent (Finlayson and Moser, 1991). This diversity of wetland types reflects a wide range in climate arising from latitudinal, longitudinal and altitudinal gradients, in geological and geomorphological landscapes and the distinctive pattern of human, animal and plant life within which these ecosystems occur.

In common with other parts of the world, the European wetland resource has become progressively depleted and continues to degrade under a range of pressures. Population pressures and the concomitant factors of agricultural development, urbanisation and industrialisation are the overriding factors that have led to alteration or destruction of wetlands. The historic widespread perception among the public and decision makers that wetlands are dangerous, disease-ridden places and that they present obstacles to economic development and other social gains, has contributed to this process of progressive wetland alteration and loss. In recent times, only rarely, if at all, has there been general appreciation of the wider benefits that may accrue from intact wetland ecosystems.

More recently, the opposition to wetland loss has tended to come from an increasing array of conservation-based non-governmental organisations (NGOs) and other stakeholder groups such as wildfowlers and fishing organisations, rather than from the governmental statutory conservation agencies. Such interests have been more or less successful in conserving wetlands because of their value as habitat for rare and endangered species as well as areas for recreational pursuits

Table 1.1.1 Some overriding factors contributing to wetland loss and degradation (after Maltby *et al.*, 1994)

Science and information base	Science demonstrating the true value of wetlands is often lacking or inadequately communicated to decision makers. This means that the goods and services arising from wetland functioning are not adequately taken into account when considering the relative merits of preserving a wetland or, for example, draining it for agricultural purposes.
Imbalance of costs and benefits	While the benefits of wetland degradation (e.g. drainage) may be apparent locally, the costs, such as increased sedimentation or deterioration in water quality, may be manifested in another part of the catchment. This imbalance means that the costs of wetland degradation are not adequately taken into account in local land planning.
Policy conflicts	Governmental departments are normally organised along sectoral boundaries. It may be the case that agricultural policy is contrary to environmental policy.
Institutional deficiencies	Decision making bodies may not have adequate integration or appropriate systems and staff to develop and enforce appropriate land and water management strategies.

such as bird watching, hunting and fishing. Five factors highlight the inadequacy of a traditional conservation perspective as the only approach to the wise use of wetland resources.

1. Limited finance will mean that only a small proportion of the remaining wetland resource can ever become part of a formal protected area network.
2. Increasing evidence is emerging of the significant contributions wetlands can make to wider socio-economic values that arise from their functioning, but which are rarely taken into account in decisions affecting them.
3. Even though many wetlands appear intact there is circumstantial evidence, such as declining bird populations and fisheries, that ecosystem functioning has been progressively impaired.
4. The greater part of the wetland resource is at the mercy of decision making that takes no account of the larger functional roles of these ecosystems.
5. Lack of effective policy instruments or legislation, which specifically links wetland ecosystem functioning to environmental protection (other than meeting nature conservation objectives). However, the new Water Framework Directive (WFD) gives implicit recognition to the role that wetlands can play in meeting the objectives of water quality and 'good ecological condition' (EC, 2000). This importance has been recognised in the Horizontal Guidance on Wetlands and the WFD recently produced by the WFD Common Implementation Strategy Working Group on Wetlands (www.wfd.mop.gov.si/navodila/guidancedoc12wetlands.pdf).

Until recently the value of wetlands in flood control, water quality improvement, regulation of greenhouse gas emissions, support of food chains and biodiversity maintenance has rarely featured in political or environmental decision making that has subsequently had impact on wetlands. If wetlands are to be protected from further damage and the resource used to optimal extent, then there is a clear need for this gap to be filled in Europe. This situation is in stark contrast to that in North America and particularly in the USA, where a rich array of assessment techniques has been developed based on an assessment of functioning and the benefits that wetland functions can deliver (Bartoldus, 1999).

1.2 What is a wetland?

Wetlands occupy the transitional zone between aquatic and terrestrial environments, but the issue of their formal definition has long been a source of controversy among experts and non-experts alike. The Convention of Wetlands of International Importance (UNESCO, 1971, http://www.ramsar.org/key_conv_e.htm) includes rivers, lakes and estuarine and coastal shallow water bodies within the definition of wetlands (see Section 1.2.2). Their transitional character and the associated varied nature of their positions in the landscape produce a plethora of wetland types (Fig. 1.2.1). Much scientific confusion has arisen from the lumping together of highly diverse ecosystems and landscape units under the single wetland banner.

Whatever the scientific conflicts, there are certain features of wetlands that can be generally regarded as diagnostic:

1. Wetlands are distinguished by the predominant presence and dynamics of water, either at the surface or within the root-zone.
2. Wetlands often have unique soil or sediment conditions that differ from adjacent non-wetland areas.
3. Wetlands support vegetation (and generally animals) specifically adapted to permanently or seasonally wet conditions. (Figs 1.2.2 and 1.2.3).

1.2.1 *Search for clarity in definition*

Although general characteristics of many wetland types are relatively easy to identify, agreeing the exact definition of a wetland has been problematic, and the interpretation of the term has been subject to considerable debate. Cowardin *et al.* (1979) acknowledged that

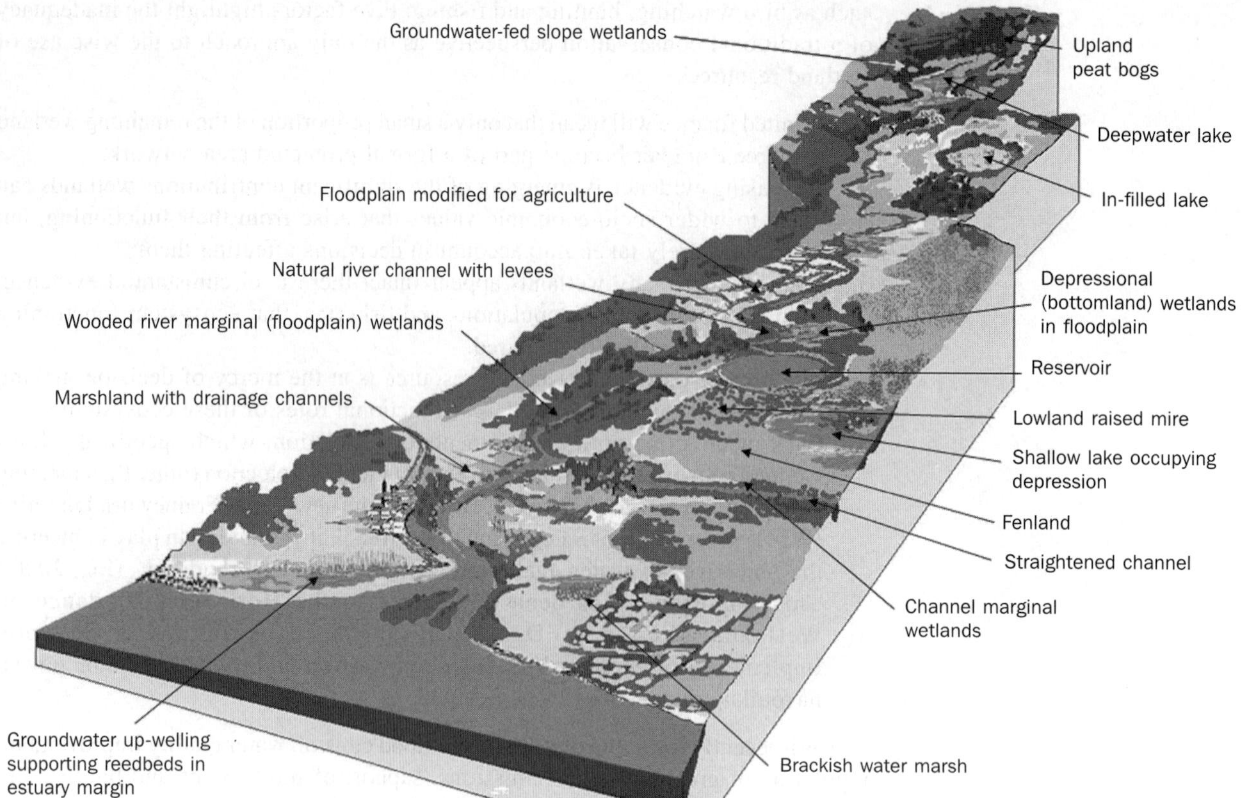

1.2.1 Elements in the catchment of a small river system, showing channel, associated lakes, estuary and wetlands (illustration by A. Aidoud).

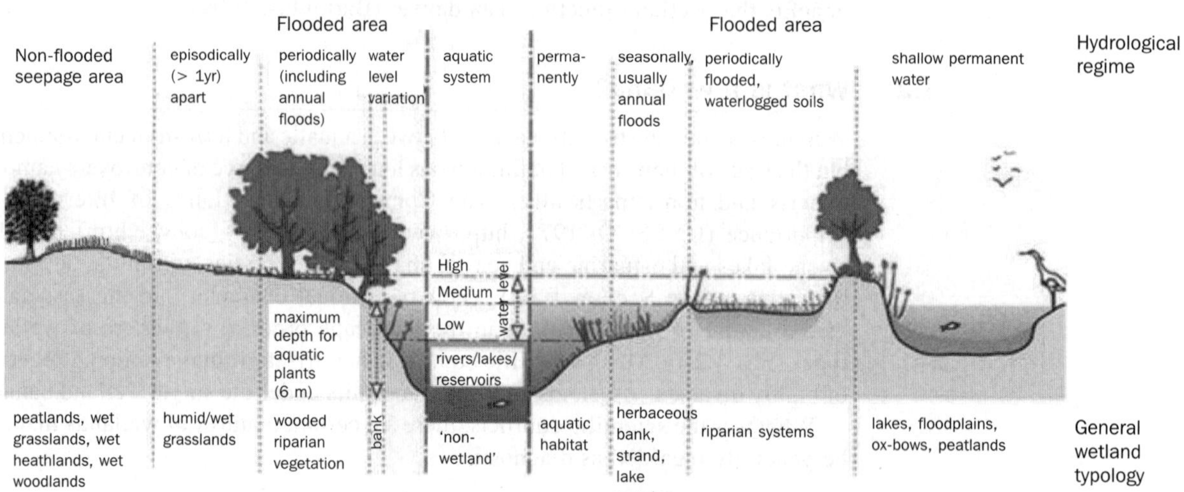

1.2.2 River and lake marginal wetland types (adapted from Agence de l'Eau Seine Normandie).

> 'there is no single, correct, indisputable definition for wetlands, primarily because of the diversity of wetlands and because the demarcation between dry and wet environments lies along a continuum.'

Different wetland areas may exhibit certain properties, which are absent in others, and the combination of characteristics in a given wetland may be unique. For this reason, some proposed definitions have been found to be too restrictive. Further, there have been areas of land that have remained dry for many years, yet, following

1.2.3 Varied wetland features and landscapes in Europe: (a) waterfowl; (b) river in flood; (c) peat burning, North York Moors, UK; (d) Lake Vistonis, Northern Greece; (e) fishing boats, Northern Greece; (f) view of the Donana Wetlands, Spain; (g) cotton grass (*Eriophorum angustifolium*); (h) a wetland in the Danube Delta (photos: E. Maltby).

a period of rainfall, have supported wetlands of international importance (Scott and Jones, 1995). Again, under some definitions, these areas would not have been classified as wetlands despite their importance arising from the (temporary) presence of water.

An agreed definition of wetlands in Europe would be useful for a variety of reasons:

1. There is currently no widely accepted definition and the term 'wetlands' is interpreted differently between countries and sectors within countries.
2. A widely accepted definition would assist interpretation and implementation of the Water Framework Directive (WFD) (EC, 2000) and other relevant European-wide legislation with respect to wetlands.
3. The term wetland is relatively new and has no easily translated equivalent in some European languages. Therefore, a definition is required to facilitate communication within Europe.

1.2.2 *Alternative definitions*

A variety of existing definitions are given in Table 1.2.1. One widely used and internationally accepted definition is that of the 'Ramsar' Convention: The Convention on Wetlands of International Importance, Especially as Waterfowl Habitat (UNESCO, 1971, http://www.ramsar.org/key_conv_e.htm).

Table 1.2.1 Selected definitions

Date	Source	Definition/Description
1956	Shaw and Fredine	'The term "wetlands" refers to lowlands covered with shallow or sometimes temporary or intermittent waters. They are referred to by such names as marshes, swamps, bogs, wet meadows, potholes, sloughs, and river-overflow lands. Shallow lakes and ponds, usually with emergent vegetation as a conspicuous feature, are included in the definition, but the permanent waters of streams, reservoirs, and deep lakes are not included. Neither are water areas that are so temporary as to have little or no effect on the development of moist-soil vegetation.'
1971	Ramsar Convention on Wetlands	'Areas of marsh, fen, peatland or water, whether natural or artificial, permanent or temporary, with water that is static or flowing, fresh, brackish or salt, including areas of marine water, the depth of which at low tide does not exceed six meters.'
1972	Clean Water Act (USA)	For regulatory purposes the term wetlands means 'those areas that are inundated or saturated by surface or ground water at a frequency and duration sufficient to support, and that under normal circumstances do support, a prevalence of vegetation typically adapted for life in saturated soil conditions. Wetlands generally include swamps, marshes, bogs, and similar areas.'
1979	Cowardin *et al.*	'Wetlands are lands transitional between terrestrial and aquatic systems where the water table is usually at or near the surface or the land is covered by shallow water. Wetlands must have one or more of the following three attributes: • at least periodically, the land supports predominantly hydrophytes (water-loving plants); • the substrate is predominantly undrained hydric soil (wet soils); • the substrate is non-soil and is saturated with water or covered by shallow water at some time during the growing season of each year.'
1986	Swedish Environmental Protection Agency	In Sweden, "wetlands" are considered to have two key features: • they do not have permanent surface waters, but are flooded by surface waters or groundwater for large parts of the year (Leonardson, 1994); • they have vegetation which is at least 50 % hydro-phile (SEPA, 1996). This definition includes all different types of mires and wet forestrand, shore-pastures, and small waters and shallow waters along the shores (SEPA, 1986). However, this definition would not be appropriate for inclusion of the extensive lake and river systems considered as wetlands in some countries and especially in the Mediterranean region.
1987	US Army Corps of Engineers	'The term "wetland" means those areas that are inundated or saturated by surface water or ground water at a frequency and duration sufficient to support, and that under normal circumstances do support, a prevalence of vegetation typically adapted for life in saturated conditions. Wetlands generally include swamps, marshes, bogs and similar areas.'
2000	Mitsch and Gosselink	'Wetlands have many distinguishing features, the most notable of which are the presence of standing water, unique wetland soils, and vegetation adapted to or tolerant of saturated soils.'

Table 1.2.1 cont'd

Date	Source	Definition/Description
1995	Clymo *et al.*	Wetland systems have three features in common: 1. Porous solid matrices partially or completely filled with water that may be moving or stagnant; 2. Water present near or above the surface of the matrix for at least part of the year, thus emphasising the transitional state of the water rather than its permanence; 3. Microbiological activity on substrates within the system, usually creating anoxic conditions in at least part of the matrix.
1999	Lewis *et al.*	'The generic term "wetland" is now used worldwide and includes specific ecosystems known regionally as bogs, bottomlands, carr, fens, floodplains, mangroves, marshes, mires, moors, muskegs, playas, peatlands, pocosins, potholes, reed swamps, sloughs, swamps, wet meadows, and wet prairies.'

The Ramsar Convention currently has over 140 contracting parties, including the European Commission. Its definition of wetlands is extremely broad and is commonly criticised for a lack of scientific consistency as it lumps together a range of different ecosystems with little apparent underpinning rationale other than a general reference to 'wetness'. This is exacerbated further by the recent addition of coral reefs and cave systems. The implementation of wetland-specific legislation and policies is made difficult to develop and implement because they cover an extremely wide range of habitats and affect many stakeholder groups.

1.2.3 A new, functional definition of wetlands

Wetlands are particularly valuable in sustainable environmental management because of the functions that they can perform, such as groundwater recharge, floodwater retention, removal of nutrients and ecosystem support. A new definition is proposed here, which recognises their functional aspects rather than their descriptive characteristics. It is intended also to help those non-specialists who are seeking to interpret the new Water Framework Directive (EC, 2000) and the particular challenges of river basin management, where wetland functioning provides key contributions to meeting new water policy objectives.

A suggested functional definition is:

> Wetlands are heterogeneous but distinctive ecosystems in which special ecological, biogeochemical and hydrological functions arise from the dominance of water, in particular sources, chemistry and periodicity of inundation or saturation. They occur in a wide range of landscapes and may support permanent shallow (generally < 2 m) or temporary standing water. They have soils, substrates and biota adapted to flooding and/or waterlogging and associated conditions of restricted aeration.

Notes:

1. Permanent standing waters greater than 2 m in depth are not considered to be wetlands under this definition as this represents the common approximate depth limit for rooted aquatic vegetation, but the existence of deeper water bodies such as lakes and rivers is often essential to the integrity of connected wetlands.
2. While rivers and streams could be included in a strict interpretation of this definition, permanent running waters within distinct channels are significantly different from wetlands in functional terms and, therefore, should not be considered as wetlands.
3. However, temporary or seasonal streams should be considered as wetlands as the periodic nature of their inundation is a key factor determining the nature

of these ecosystems, which are distinct from otherwise similar permanent waters.

4. The common European wetland types covered by this definition are described in Table 1.2.2.

Notwithstanding the importance of naming different types of wetland for reasons of consistency and understanding, the rationale of the functional assessment procedures (FAPs) is built around the characterisation of the basic and more generic building blocks of wetlands. These landscape units, defined later in terms of hydrogeomorphology, are not specific to any established classification of wetland type, but enable the ecosystem to be interpreted without preconceived ideas of how previously classified or named wetlands might work.

1.3 The importance of the wetland resource

1.3.1 *Wetland processes and functions*

Wetlands are extremely diverse not only in their physical characteristics and geographical distribution but also within particular landscape units such as floodplains, mires and marshes. This variation supports biodiversity and provides goods and services of economic and societal benefit (Fig. 1.3.1; Table 1.3.1). Despite this, wetlands remain a threatened asset. They have been degraded by direct and indirect human actions arising from an often mistaken belief of their low intrinsic value and their general consideration as wastelands.

Since the 1970s, this view has gradually changed with increased knowledge of the valuable habitats available within wetlands for specialised and often rare threatened or endemic plant and animal species and with the use of the areas for recreational pursuits such as hunting, fishing and bird watching. This has afforded wetlands with some acknowledgement of their conservation values; however, this has often been insufficient to achieve protection or preference for maintenance in competition with the perceived higher economic value of land conversion. Wetlands can actually contribute immediate and wider socio-economic benefits that arise from their functioning quite apart from nature conservation values. These include flood control, water quality improvement, preservation of palaeohistorical remains (such as pollen and wooden artefacts), regulation of greenhouse gas emissions and the production of food and natural building materials. Rarely have such values figured in the economics of decision making, largely through lack of sound scientific data. The consequences of this have had serious implications and may continue to do so if wetlands are not sufficiently incorporated into catchment-scale water protection initiatives.

Whilst many wetlands influence flood-flow dynamics, there are numerous factors that determine whether the function and benefit of floodwater control is performed. Depending on factors such as their position in the landscape, size, spatial distribution (Mitsch and Gosselink, 2000), proximity to floodwaters, 'roughness' of vegetation cover and consequent microtopography, and the presence or absence of other floodwater interceptors, wetlands can stem floodwaters and provide a socio-economic value by protecting adjacent and downstream areas from the damage caused by floods. Those that do retain floodwaters may act as a temporary store, holding water during periods of increased run-off, or following inundation from a river or lake, and allowing water to return to the water body relatively slowly. The effect may be to modify the river flood pulse, reducing its amplitude while extending its duration (Fig. 1.3.2). Floodwater may be discharged over a longer time period or removed completely through evapotranspiration and percolation to groundwater (Davies and Claridge, 1993). All such circumstances can reduce considerably the damage caused by floodwaters.

Not all wetlands are valuable in this regard. Analysis of published studies between 1930 and 2002 showed that five out of 28 wetlands situated in floodplains and, more significantly, 27 out of 66 wetlands in headwaters may actually be 'flood generating' (Bullock and Acreman, 2003). Nevertheless, there are wetlands that do control flood waters and their conversion creates areas more prone to flooding,

Table 1.2.2 Some common wetland types with local names in various European countries

English	German	Dutch	Greek	Romanian	Swedish	French	Description
Carr*	Bruchwald, Sumpfwald	Broekbossen	Eli me xilodi vlastisi	Mlastina, Padure umeda/ aluviala	Sumpskog	Bois humide Forêt humide	Wet woodland often dominated by *Alnus* and *Salix* and frequently associated with organic topsoil. Generally synonymous with swamp in U.S. terminology.
River marginal wetlands*	Auen, Flussauen, historische und rezente Aue	Rivier-begeleidende wetlands	Parapotamioi ygrotopoi	Zona inundabila	Långgrundå-stränder, Kärr	Zone humide alluviales Zone humide riveraines	All currently active and historical floodplain ecosystems in which inundation by surface water and/or the prevalence of an elevated water table are regular phenomena. Also included may be adjacent valley slopes, which are regularly affected by overland flow, groundwater seepage or otherwise have an elevated water table.
Temporary streams**	Temporäre Fließgewässer	Tijdelijke beken	Ryakia asynechous rois	Cursuri de apa temporare	Tillfälliga vattendrag	Chenaux temporaires	Streams or parts of streams, which flow for part of the time only with some periodicity. They may be fed primarily by groundwater or surface run-off.
Floodplains*	Auen (rezent), Überflutungs-aue	Overstromings-vlakten	Kataklisomenes ektaseis	Lunci	Älv-eller åstränder	Plaines alluviales inondables	Areas subject to regular over-flooding by adjacent rivers.
Wet meadows*	Feucht- und Nassgrünland	Vochtige graslanden	Ygroi leimones	Pajisti umede (cu "exces de umiditate")	Fuktängar, Fukthedar	Prairies humides	Grassland with waterlogged soil but without surface water for majority of year.
Marsh*	Marsch	Moeras	Eli	Mlastina Japsa Balta	Mader, Källor	Marais Marécage	Maintained by water sources other than rainfall and dominated by herbaceous vegetation. They do not accumulate peat.
Freshwater marsh*	Süßwasser-marsch	Zoetwater-moeras	Eli glikon neron	Mlastina Japsa Balta	Strandäng	Marais d'eau douce	Marshes maintained by freshwater from springs, streams, run-off or groundwater and dominated by grasses and sedges.
Tidal salt marsh**	Salzwiesen, Salzmarsch	Schor, kwelder	Paliroiko aloelos	Melea Lacuri litorale (littoral lakes)*/[2]	Havsstrand-äng	Marais salé (saline) Marais saumâtre (brackish)	Marshes maintained by inputs of saline or brackish water due to tides and characterised by salt-tolerant species. Often associated with complex of tidal creeks.
Tidal freshwater marshes**	Tidesüßwass-ermarsch	Zoetwater-getijden-moeras	Paliroiko elos glykon neron	Lacuri litorale*[1]/*[2]	Strandäng	Lagune d'eau douce (n.b. with indirect tidal influence)	Marshes influenced by tidal cycles but not exposed to salt water stress. More diverse vegetation than salt marsh.

Table 1.2.2 cont'd

English	German	Dutch	Greek	Romanian	Swedish	French	Description
Peatlands**	Moor	Veengebieden	Omvrodietoi tyrfones	Turbarii	Myrar, Mosse	Marais tourbeux	Generic term for wetland landscapes, which have accumulated organic material.
Bogs**	Hochmoor	Hoogvenen	Tyrfones	Turbarii	Mossar	Tourbières à sphaignes	Peat-forming wetland in which precipitation is the dominant water input. They are low in nutrients and often acidic.
Fens**	Niedermoor, Flachmoor, Durchströ-mungsmoor	Laagvenen	Elos tyrfones mi apoklistikos omvrodieto	Turbarii	Rikkärr	Bas marais (base rich) Bas marais acides (acidic)	Peat-forming freshwater wetlands primarily supported by groundwater sources and related nutrients, generally non-acidic, with a marsh-like vegetation often dominated by reeds.
Mires**	Moor	Venen	Eli tyrfonon	Turbarii Mlastini	Myrar, Kärr	Marais tourbeux	All active peat-forming wetland ecosystems of variable nutrient status (rich, intermediate and poor).
Blanket bog**	Hochmoor, Regenmoor	Spreihoogvenen	Epipedoi tirfones	Turbarii	Mosse	Tourbière de couverture	Extensive peat deposits generally occurring in high rainfall upland and oceanic regions. Generally 1–3 m deep. Develop independently of details of relief.
Raised bog**	Hochmoor, Regenmoor	Hoogveen	Ypsiloi tyrfones	Mlastina bombata / inalta Bahna Tinov Molhas	Högmosse	Tourbière bombée Haut marais	Dome-shaped peat bodies sustained by high rainfall and an internal mound of water, generally more than 5 m deep.
Quaking bogs**	Schwingmoor	Trilveen	Tremodeis tyrfones	Sfagnet Bahna Tinov Molhas	Gungfly på myr/mosse	Radeau tourbeux Tourbière flottante	Similar characteristics to raised bog and formed when a floating mat of peat-forming vegetation progressively covers a small lake.
Shallow lakes/ lake margins*	Flachwasser-See, Verlandungs-zone	Ondiepe meren/ meeroevers	Riches limnes/ perithoria limnon	Lacuri putin adanci/malul lacurilor Ghioluri/japse (in the Danube Delta)	Grunda sjöar	Zone palustres Pelouses/ Prairies amphibies (lake margins) Herbiers aquatiques (shallow lakes)	Lakes/lake margins under 2 m in depth.
Swamps*	Sumpf, Verlandungs-moor, Flachmoor	Rietlanden	Eli me xilodi vlastisi	Japse Plaur Balta	Träsk	Roselières Magnocariçaies	Wetlands forming in or around still waters (e.g. lakes, oxbows) with surface more or less permanently saturated. Often with extensive growth of reeds (*Phragmites*).

Coastal lagoons**	Lagune, Baltic See: Bodden	Kustlagunes	Limnothalasses	Lagune litorale	Avsnörda vikar	Lagunes salées (saline) Lagunes saumâtre (brackish)	Shallow area of permanent saline or brackish water cut off from sea.
Mud flats**	Watt	Slikken, wadden	Diapalliroiki laspotopoi	Intinderi/plaje maloase[3]	Strand	Vasières	Intertidal areas of mud and other fine substrates.

*Habitat suitable for evaluation with functional assessment procedures.

**Habitat not suitable for evaluation with functional assessment procedures.

[1] There are no real tides on the Black Sea coast (tidal variation is limited to few centimetres). Seawater can influence littoral lakes and wetlands as a result of easterly winds. The term melea refers to the shallow water zone behind littoral sand barriers.

[2] Not applicable in Romanian specific conditions; the specified Romanian terms could show relative similar circumstances.

[3] Not applicable in Romanian specific conditions; the specified Romanian terms are just translations from English.

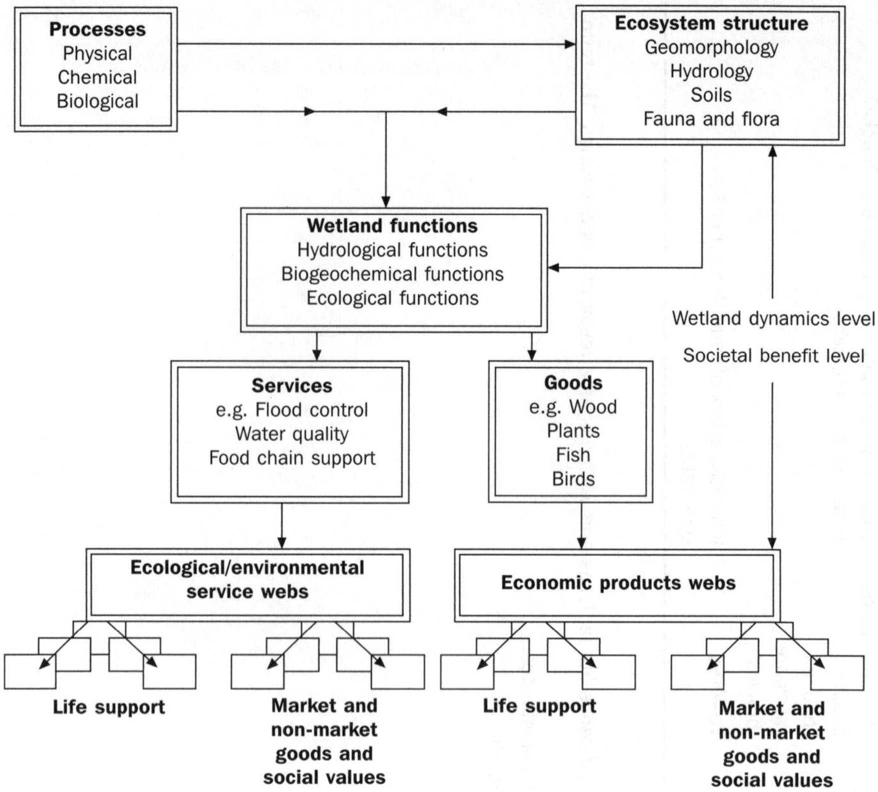

1.3.1 Wetland ecosystem processes, functions and structure sustain numerous goods and services that are environmentally and economically important. Adapted from Maltby *et al.*, 1996.

causing serious problems for human populations. The separation of floodplain wetlands from their river, usually by the construction of flood defence embankments alongside the river channels, is a major concern because of the implications for flood risk management, together with the loss of other functional benefits (see also Section 1.4). Recent guidelines have been prepared to advise on the potential values of reconnecting rivers with their natural floodplains (Blackwell and Maltby 2006; http://levis.sggw.waw.pl/ecoflood/).

Groundwater recharge is an important socio-economic benefit derived from wetlands. Water moves via the wetland to underground aquifers that hold 97 % of the world's potable unfrozen water (Bouwer, 1978). During the process of infiltration, water undergoes various physical and chemical processes, which ameliorate the potable quality, resulting in a reduced treatment cost if ultimately used for human consumption. The Environment Agency of England and Wales estimates that about one third of public water supply originates from groundwater sources.

There are many biogeochemical processes that wetlands perform, which result in significant benefit to humans as well as supporting overall ecological quality. These include the immobilisation and transformation of excess nutrients, heavy metals and other toxicants generally in forms that are tightly bound to sediments or soil particles (Gambrell and Patrick, 1978). Eutrophication, which may threaten potable water supplies as well as impairing ecological quality, can be reduced or prevented by wetland processes. Phosphorus may be inactivated by chemical bonding to inorganic ions (Richardson, 1985) or through storage in plant biomass. Nitrate can also influence eutrophication (Horan, 1993; Fisher and Acreman 2004), though is of particular interest due to its additional toxicity to human as well as aquatic life (Bruning-Fann and Kaneene, 1993; James *et al.*, 2005). It has been estimated in Sweden that a wetland area of 2 km^2 can be sufficient to reduce the input of nitrogen into adjacent surface waters by 2000t per year, primarily through plant storage (Fleischer, 1990). Further, it has been shown that in certain river marginal and estuarine soils, denitrification may account for the permanent annual

Table 1.3.1 The values of wetland functions (adapted from Maltby *et al.*, 1996)

Wetland function	Wetland process	Benefits/comments
Hydrological:		
Floodwater detention	Floodwater retention	Floodwater detention decreases peak flow in rivers, reducing flood damage downstream. Provides wildlife habitat especially important for fisheries support.
Groundwater recharge	Groundwater recharge	Replenishment of groundwater resources. Maintenance of dependent ecosystems in discharge areas. Maintenance of base flow in rivers fed by discharge elsewhere.
Groundwater discharge	Groundwater discharge	Emergence at springs or seepage zones. Maintenance of river base flow. Maintenance of ecosystems dependent on soil-water regimes.
Sediment retention	Sediment retention	Improvement of river water quality due to the reduced input of suspended sediments and associated sediment nutrients.
Biogeochemical:		
Nutrient retention	Long-term retention of nutrients (N and P) through plant uptake Storage of nutrients (N and P) in soil organic matter Adsorption of N as ammonium Adsorption and precipitation of P in the soil Retention of particulate nutrients (N and P)	Retaining or delaying the release of the nutrients N and P into water bodies may prevent potentially deleterious effects on watercourses. Eutrophication changes the ecosystem composition and character and adversely affects potable water quality.
Nutrient export	Gaseous export of N by: denitrification ammonia volatilisation Export of nutrients through vegetation management Export of nutrients via wind- and water-mediated processes	In addition to gaseous removal of N, nutrients may also be exported at varying rates from the wetland due to remobilisation and delayed transportation.
In-situ carbon retention	Organic matter accumulation	Carbon may be stored for long periods, often as peat, mitigating acceleration of climate change. Peat supports rare and valuable biological communities. Peat can be used as fuel, a growth medium and a source of pharmaceuticals (though not necessarily sustainably).
Trace element storage	Physical retention of trace elements Biogeochemical retention of trace elements	Improvement of river water quality due to the reduction of suspended or dissolved trace element loads. Trace element loads can have toxic effects on the stream ecosystem. Prevention of groundwater contamination and uncontrolled translocation of trace elements within the river marginal wetland.

Table 1.3.1 cont'd

Wetland function	Wetland process	Benefits/comments
Trace element export	Plant uptake of trace elements Physical remobilisation of trace elements Biogeochemical remobilisation of trace elements	When soil/sediment storage capacity is exceeded, there can be consequent risks for food production (through remobilisation by plant uptake). Possibility of controlled removal of trace elements, but danger of toxic effects on plants and recontamination of river water or groundwater.
Organic carbon concentration control	Organic carbon export into surface waters	Wetlands strongly influence the concentration of dissolved organic carbon in run-off water, key features of water quality and the aquatic ecosystem in areas with DOC-rich water (over ca 5 mg L^{-1} DOC.)
Ecological:		
Ecosystem maintenance	Provision of overall habitat structural diversity Provision of micro-sites for: macroinvertebrates fish herptiles birds mammals	The physical, chemical and biological processes occurring in wetlands all contribute to the provision of unique habitats, which support a variety of adapted organisms. This contributes to global biodiversity and supports recreation and quality of life.
Food-web support	Provision of plant and habitat diversity Biomass production Biomass import via physical processes: watercourses overland flow wind transport Biomass import via biological processes: via fauna via anthropogenic means Biomass export via physical processes Biomass export by biological processes	The food web (interlinked network of organisms feeding on one another) of the wetland may be supported by biomass production on-site or by detritus and/or organisms externally transported into the system. The production of biomass by the wetland can also be responsible for supporting food webs at other sites.

removal of some 30 kg of nitrate per hectare (Maltby *et al.*, 1991; Koch *et al.*, 1992).

Wetlands provide a variety of often specialised habitats, which provide the structure for maintaining biodiversity on a number of scales. Many wetland areas support threatened, rare, endemic and endangered species, both those that have become adapted to the wetland environment and those that rely on the wetland to support part or all of their life cycle. In the USA, 43 % of all federally listed threatened and endangered species use wetlands (5 % of the land area) to support some or all of their life cycle (Feierabend, 1992). In the UK, the decline and disappearance of the bittern (*Botaurus stellaris*) had been attributed to loss of reedbed habitat, the restoration of which has now begun to see a recovery of species numbers.

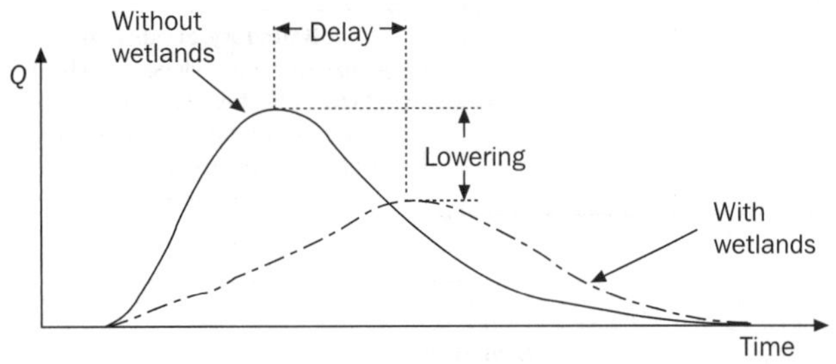

1.3.2 A common representation of the potential effect of wetlands on river flow (*Q*).

1.3.3 Sustainable wetland product use in Romania (photo: E. Maltby).

The nature conservation value of wetlands alone can be a controversial and insufficient argument for protection, giving rise to conflicts between development and conservation. Difficulties arise when attempts are made to separate people from the environment they rely on to sustain their livelihoods. However, the sustainable use of wetlands allows wetland products to be utilised whilst protecting biodiversity and natural hydrological, biogeochemical and ecological functioning (Fig. 1.3.3).

1.3.2 The economic context

The following sections describe the kinds of approaches that can be adopted to enable an economic evaluation of wetlands to be undertaken, linked to functioning, (Fig. 1.3.4), though this is not part of the formal system described in this volume.

Economics of wetlands

The essence of an overall socio-economic evaluation is to determine how society is affected by the functions an ecosystem might perform, and by changes in such functioning. The key to valuing a change in an ecosystem function is establishing the link between that function and some resulting service valued by people. If that link can be established, then the concept of derived demand can be applied through the change in value of the ecosystem service it supports. In the case of wetlands, such changes can occur readily as a result of drainage and other impacts.

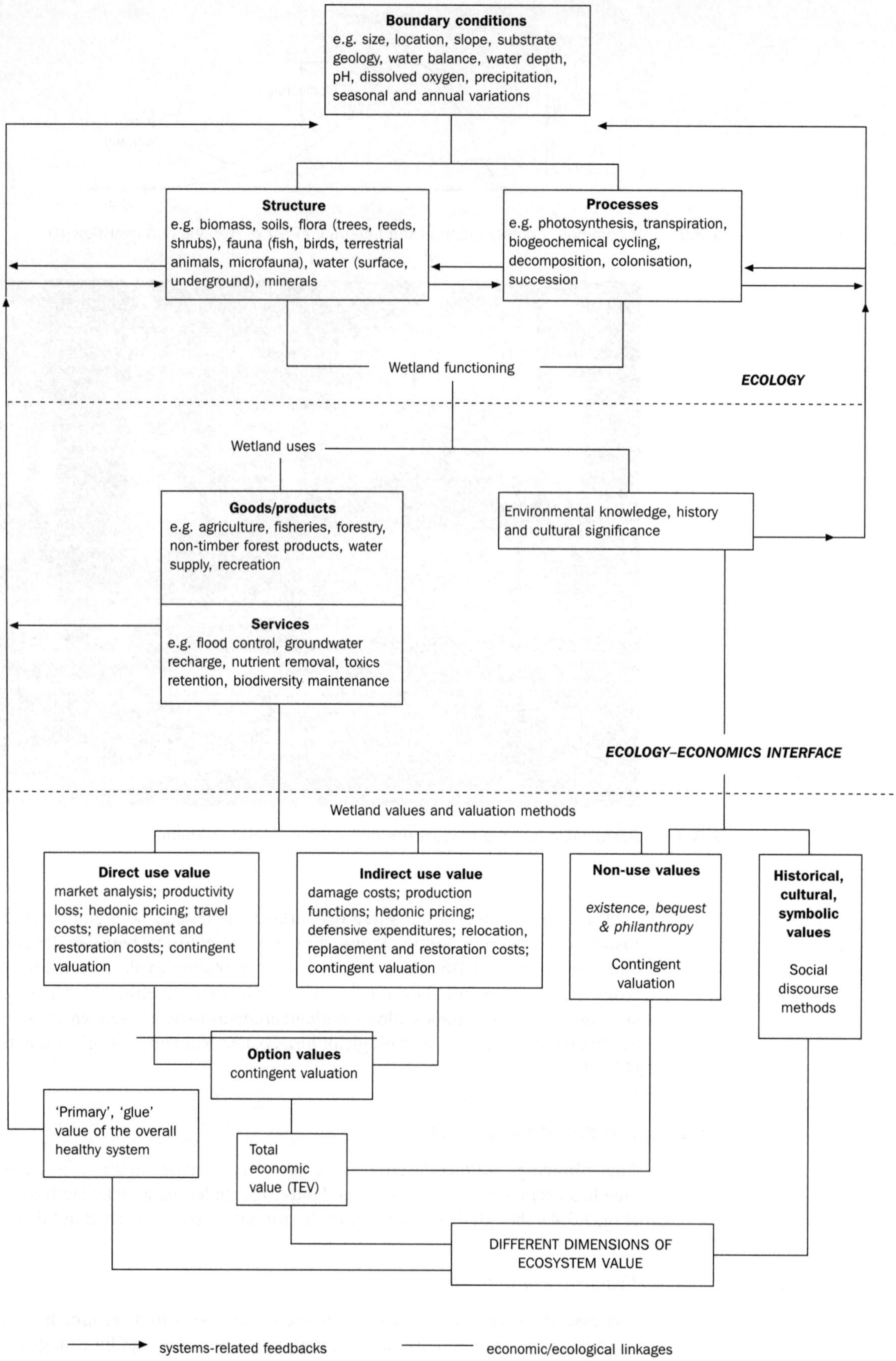

1.3.4 Wetland functions, uses and values (after Turner *et al.*, 2000, modified from Maltby *et al.*, 1996).

In assigning values to the ecosystem structure and functions, the key issues and ecological principles that must be considered include:

- the spatial and temporal scale of ecological processes;
- the structure, complexity and diversity that underlie ecosystem functions;
- the dynamic (in space and time) nature of ecosystems;
- the uncertainty associated with ecosystems.

The total economic value of wetlands

Ecologists may use the term 'value' to imply 'that which is desirable or worthy of esteem for its own sake; something or some quality having intrinsic worth'. Economists use the term to describe 'a fair or proper equivalent in money or commodities', where 'equivalent in money' represents that sum of money that would have an equivalent effect on the welfare or utilities of individuals.

A number of ecosystem goods and services can be valued in economic terms, while others cannot because of uncertainty and complexity of conditions. Wetlands provide a wide range of goods and services of significant value to society, such as storm and pollution buffering, flood alleviation, recreation and aesthetic services. 'Valuing' a wetland ecosystem is essentially the same as valuing the characteristics of other systems. Nevertheless, since it is the case that the component parts of a system are contingent on the existence and continued proper functioning of the whole, then putting an aggregate value on wetland ecosystems is complicated.

In valuing a resource such as a wetland ecosystem, the use of the total economic value (TEV) classification enables the values to be usefully broken down into categories (Table 1.3.2). The initial distinction is between *use value* and *non-use value*, with use values involving some interaction with the resource, whether direct or indirect:

Direct use value: involves interaction with the wetland itself rather than via the services it provides. It may be consumptive use, such as fisheries or timber, or it may be non-consumptive, as with some recreational and educational activities.

Indirect use value: derives from services provided by the wetland. This might, for example, include the removal of nutrients, providing cleaner water to those downstream, or the prevention of downstream flooding.

Table 1.3.2 Components of economic value

<table>
<tr><td>Aesthetic/ educational use value</td><td rowspan="2">Non-consumptive use value</td><td rowspan="3">Direct use value</td><td rowspan="6">Use value</td><td rowspan="9">TOTAL ECONOMIC VALUE (TEV)</td></tr>
<tr><td>Distant use value</td></tr>
<tr><td>Consumptive use value</td><td></td></tr>
<tr><td>Indirect use value</td><td></td><td></td></tr>
<tr><td>Option value</td><td></td><td></td></tr>
<tr><td>Quasi-option value</td><td></td><td></td></tr>
<tr><td>Existence value</td><td></td><td></td><td rowspan="3">Non-use value</td></tr>
<tr><td>Bequest value</td><td></td><td></td></tr>
<tr><td>Philanthropic value</td><td></td><td></td></tr>
</table>

There is also the possibility of deriving value from 'distant use' through media such as television or magazines, although it is unclear whether this type of value is actually a use value, and to what extent it can be attributed to the ecosystem involved.

Non-use value is associated with benefits derived simply from the knowledge that the wetland ecosystem is maintained. It is by definition not associated with any use of the resource or tangible benefit derived from it, although users of a resource might also attribute non-use value to it. Non-use value is closely linked to ethical concerns, often being linked to altruistic preferences, although according to some analysts it stems ultimately from self-interest. It can be split into three basic components, although these may overlap depending upon exact definitions.

Existence value: derived simply from the satisfaction of knowing that the wetland continues to exist, whether or not this might also benefit others. This value notion has been interpreted in a number of ways and seems to straddle the instrumental/intrinsic value divide.

Bequest value: associated with the knowledge that a resource will be passed on to descendants to maintain the opportunity for them to enjoy it in the future.

Philanthropic value: associated with the satisfaction derived from ensuring resources are available to contemporaries of the current generation.

Finally, there are two categories not associated with the initial distinction between use values and non-use value:

Option value: an individual derives benefit from ensuring that a resource will be available for *use* in the future. In this sense, it is a form of use value, although it can be regarded as a form of insurance to provide for possible future but not current use.

Quasi-option value: associated with the potential benefits of awaiting improved information before giving up the option to preserve a resource for future use. It suggests a value in particular of avoiding irreversible damage that might prove to have been unwarranted in the light of further information.

These various elements of total economic value are assessed using economic valuation methods, and some of these elements are more easily valued than others, especially those with easily identifiable uses (usually the use type values). Non-use values are usually more difficult to assess. The main problem when including the full range of catchment environmental services in economic choices is that many of these services are not valued in markets. There is a gap between market valuation and the economic value of many catchment functions such as flood control or water quality improvement. To fill these gaps, the non-marketed services must first be identified and then where possible monetised. In the case of many catchment functions, the identification of economically relevant services is of special importance, since over time those benefits not allocated by the market have continuously gained importance.

Why the value of ecosystems can be greater than total economic value

The social value of a wetland ecosystem may in fact be more than the sum of the aggregate total economic value of that same system's components, for a number of reasons. Firstly the full complexity and coverage of the underpinning 'life-support' functions of healthy evolving ecosystems is currently not precisely known in scientific terms. Second, total economic value may fail to capture the redundancy reserve value of the ecosystem. This includes a pool of latent keystone species or processes, which are required for system maintenance in the face of stress and shock. Finally, because the range of secondary values (use and non-use) that can be derived from an ecosystem is contingent on the prior existence of such a healthy and evolving system, there is, in a philosophical sense, a 'prior value' that could be ascribed to the system itself. There is a sense in which the operating system yields or possesses 'glue' value, i.e. value related to the structure and functioning properties of the system, which hold everything together.

Included here are those services of the catchment that act to stabilise the ecological system and perform a protective and supportive function for the economic system. The ecological functions of natural systems include the following (somewhat overlapping) categories of values under this umbrella:

Inherent value: describes those services without which there would not be the goods and services provided by the system.

Contributory value: considers the economic–ecological importance of species diversity, such that even species not useful for human use are important since they contribute to increases in diversity which contribute to the generation of more species.

Indirect use value: the support and protection provided to economic activity by regulatory environmental services.

Primary value: the existence of the catchment structure prior to the range of function or service values (Turner and Pearce, 1993).

Infrastructure value: relates to a minimum level of ecosystem 'infrastructure' as a contributor to its total value (Costanza *et al.*, 1997).

A framework for the ecological–economic analysis and evaluation of the functions and values of wetlands is presented in Fig. 1.3.4. Further guidance on the economic approaches to wetland assessment can be found in Barbier (1994), Barbier *et al.* (1996), Costanza *et al.* (1997), Turner *et al.* (2000), Balmford *et al.* (2002) and Turner *et al.* (2003).

1.4 Realising the benefits of wetland functioning

1.4.1 *Wetlands and flood defence*

Floods give rise to some of Europe's most common and widespread disasters. Between 1980 and 2002, the greatest number of floods occurred in France (22 %), Italy (17 %) and the UK (12 %), with the highest number of flood-related fatalities occurring in Italy (38 %), Spain (20 %) and France (17 %). The estimated costs of flooding during this period are €11 billion in both Germany and Italy, and approximately €6 billion in both Spain and the UK. In response to this, the EU has launched around 50 research projects in recent years, with a total budget of €58 million, in areas such as flood risk assessment, flood hazard and risk mapping, flood forecasting and preventative land-use planning (WWF, 2002). A major aspect of this research has been investigation of the role of wetlands and floodplains in relation to flooding and flood prevention. While climate change is widely considered to be one of the main causes of increased flooding, the increased isolation of wetlands and floodplains from rivers by the construction of flood defences has exacerbated the problem, and in many cases simply moved flooding downstream to areas where increased damage occurs. EU funded projects such as the ECOFLOOD Project (2003–2005) and the Wise Use of Floodplains Project (1999–2002) have investigated the role of naturally functioning floodplains and wetlands in flood alleviation.

Concern has widened recently as questions about the future of home and business insurance in flood-prone areas further threatens the livelihoods of those living in or near impounded areas (Fig. 1.4.1). Insurance companies are understandably becoming concerned about offering coverage to businesses and people located behind ineffective flood defence structures. The numbers involved are huge: The International Commission for the Protection of the Rhine (ICPR) estimates that properties and businesses worth €1500 billion are located on potentially floodable areas along the Rhine alone (WWF, 2002).

During the 1990s, serious flooding occurred along many European rivers (Fig. 1.4.2), as well as numerous smaller streams in both urban and rural areas.

1.4.1 Emergency measures against flooding along the River Mulde, Germany (photo: F. Dziock).

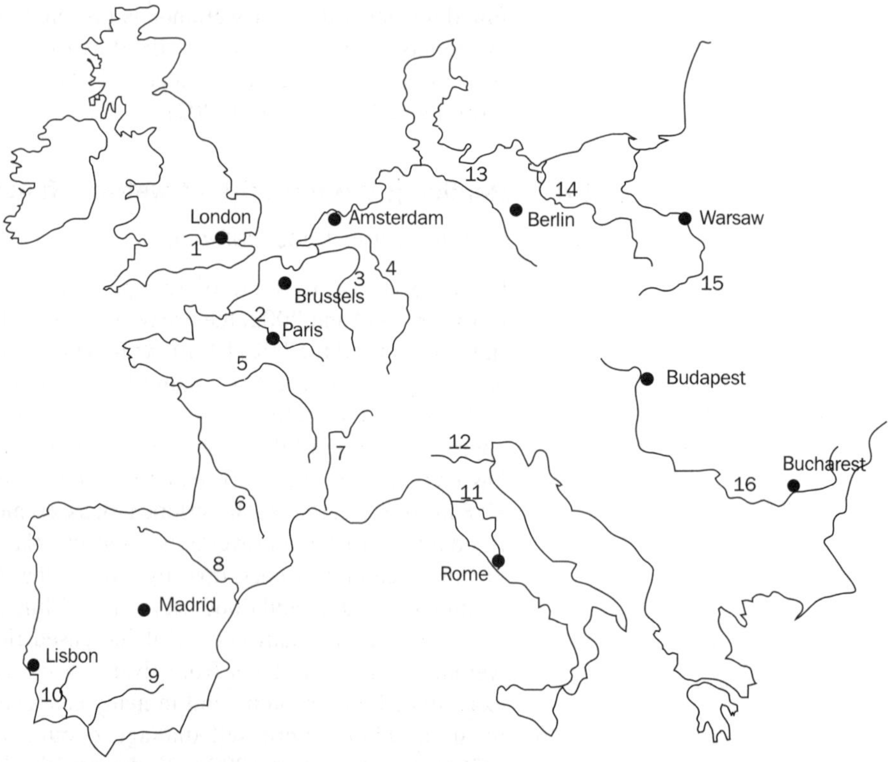

1.4.2 Map of major European rivers where severe floods occurred in the 1990s. Key to rivers: 1. Thames, 2. Seine, 3. Meuse, 4. Rhine, 5. Loire, 6. Garonne, 7. Rhone, 8. Ebro, 9. Guadalquivir, 10. Gudiana, 11. Arno, 12. Adda, 13. Elbe, 14. Oder, 15. Vistula, 16. Danube.

Particularly significant events in terms of damage and policy impact occurred along the River Oder in the former East Germany and Poland, in areas of northern France, Belgium and southern Netherlands, and in southern England along the Thames, particularly around Oxford and areas to the north. In addition, southern France, northern Spain and parts of Italy suffered from devastating flash flooding and heavy death tolls (Handmer, 2001). In 2006, the European Commission proposed an 'ethic of prevention, protection and preparedness' for flooding in Europe. This was followed in November 2007 with a new EU Floods Directive (EC, 2007). The

Directive requires governments to assess flooding risks, construct maps of vulnerable areas and make flood risk management plans for river basins and coastal areas in their territory. The FAPs can contribute significant additional insight into the probable hydrological functioning of marginal wetlands relevant to the implementation of flood risk policies.

In August 2002, flooding in the Elbe river basin exceeded all previous capacity limits. The River Mulde, a tributary of the Elbe, re-established its former course and inundated extensive opencast brown coal workings now occupying the former floodplain. The cost of damage in the Elbe catchment was calculated to be at least €11.33 billion (IKSE, 2004). Of a previous 6172 km^2 floodwater retention area, only 838 km^2 (= 13.6 %) now remain on floodplains still connected directly to the river channel (Fig. 1.4.3), the remainder now being isolated by flood defence embankments. Historic straightening of the riverbed has shortened the watercourse in the Czech Republic by 55 km and in Germany by 60 km. This engineering work has increased the flow rate of the river. However, the European Environment Agency (EEA, 2003) noted that the Elbe floods of 2002 did not reach Hamburg as feared, because the large and relatively undeveloped floodplain in the lower course of the river absorbed a substantial amount of the flows.

The flooding caused significant damage, amounting to some €8.9 billion in Germany and €2.43 billion in the Czech part of the Elbe catchment. More than 337 000 people were affected by the flood and more than 100 000 people had to leave their homes in Germany. Some 180 bridges and 740 km streets, as well as 94 railway bridges and 400 km railway lines were damaged. The overall cost to agriculture alone is estimated at €287 million (Bundesministerium des Inneren, 2002). The money for the rebuilding, provided by the EU, the German federation, the counties and the communes, has amounted to around €10 billion for all damage caused by floods in Germany in 2002 (Schwarze and Wagner, 2003).

Between 1998 and 2002, Europe suffered over 100 major damaging floods, including the catastrophic floods along the Danube and Elbe rivers in 2002. Since 1998, floods have caused some 700 fatalities, the displacement of about half a

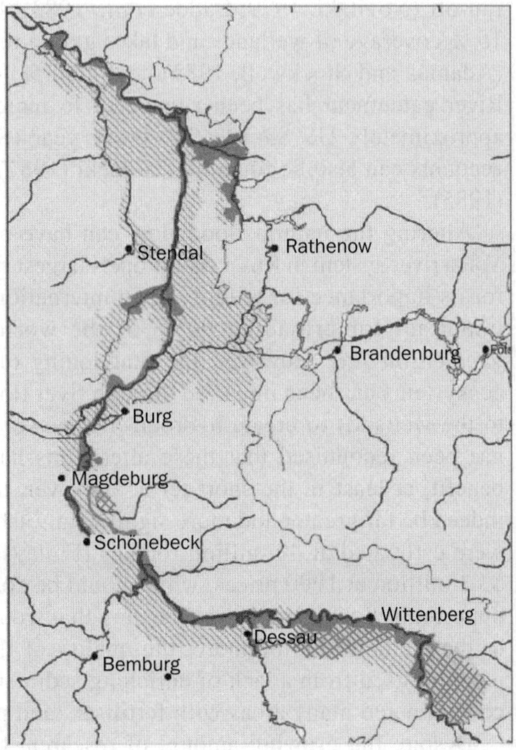

1.4.3 Retention area of the River Elbe in Saxony–Anhalt, past (dark lines), present (shaded areas) and in 2002 as a result of broken dykes (cross-hatching) (source: Bundesanstalt für Gewässerkunde, 2002).

million people and at least €25 billion in insured economic losses (EEA, 2003). The assets at risk of flooding can be enormous. Coastal areas are also at risk of flooding. The total value of economic assets located within 500 m of the European coastline, including beaches, agricultural land and industrial facilities, is currently estimated at €500–1000 billion (Commission of the European Communities, 2004).

In the UK, 1.8 million residences and 140 000 commercial properties, representing the interests of 4–5 million people could be affected by floods. The recent National Assessment of Defence Needs and Costs estimated the capital value of assets at risk to be approximately £250 billion (at 2004 prices) and the average damages with the defences now in place at just over £1 billion per year (DEFRA, 2001).

Similar economic concerns exist also in the Netherlands where about 70 % of the properties are at flood risk because they lie either below sea-level or below the river water-level (Kok *et al.*, 2002). About 10 million people live in these areas, and large industries have been developed. The estimated present value of the potential flood damage is currently at €0.5 billion (Vis *et al.*, 2003). Other estimations carried out by insurance companies raise the potential losses by flooding in Central-Holland, for example, up to €300 billion. In this case, it should be noted that the gross national product of the Netherlands is €400 million per year and, in order to compensate flood damage of €300 billion, each inhabitant would have to pay some €20 000 per year, which is the average salary of an employee (Kok *et al.*, 2002).

In Romania a total of 1.3 million ha and 500 000 people are at serious risk of being flooded every year. The risk of flood damage is presently estimated at about $100 million per year, whereas the expected annual property loss is estimated at around US$ 400 million (including earthquakes and floods) (Government of Romania, 2002).

If suitable wetlands are protected or restored and floodwaters are detained then downstream areas can be protected through the alteration of the timing and magnitude of peak flow in rivers. Studies have shown that in the North Dakota prairies, the potholes reduce peak stream flow by storing close to 75 % of total run-off (Novitzki, 1979; Ludde *et al.*, 1983). In Wisconsin, catchments with just 15 % coverage of wetlands and lakes gave a reduction in flood peaks of 60–65 % (Adamus and Stockwell, 1983), and a 40 % loss of wetland area in the Charles River catchment has been calculated to increase the cost of flood damage by approximately US $3 193 000 every year (Sather and Smith, 1984). Detailed accounts can also be found in Bardecki (1987), Koerselman (1989) and Novitzki (1985).

Altering the natural flood flow can have other economic consequences. The Volga river system in Russia is Europe's largest river system and its delta is renowned for its importance for wildlife at an international level. The area is economically important for providing 90 % of the world's Russian sturgeon (*Acipenser gueldenstaedtii*). However, the functioning of the river and associated wetland ecosystems has been impaired through river flow alteration (and flow contribution to the wetland) to create hydroelectric power stations and storage reservoirs. It has been recognised that these alterations have undoubtedly had an economic benefit, at least in the short term. However, the longer-term consequences may indeed be far greater and more significant. Fishery losses between 1959 and 1990 were estimated at 40 million roubles (Finlayson, 1992), equivalent to about US $3.3 million at 1990 prices, which could be conservative considering the reduced opportunity for spawning and feeding that would result from a lesser spring flood to the wetland. Furthermore, the quality of the soil in the catchment has been compromised from a lack of enriching sediment, and reclamation for polders has contaminated many areas with fertilisers and pesticides.

Despite the growing amount of research, there is currently no legislation in Europe that specifically targets the protection or restoration of wetlands for the purpose of reducing flood risk and there has been reluctance amongst governments to move from hard to soft engineering solutions for flood control. Ironically, it is often the construction of flood defences such as dykes and dams that results in the degradation and loss of wetlands and their ability to deliver a flood defence function. This is exemplified in the UK where the Department of Environment, Food and Rural Affairs (DEFRA) is responsible for both inland and coastal defence work, typically comprising river impoundment and diversion. There is a requirement that Natural England and other conservation organisations are consulted before any scheme can be implemented, so that impacts on wetlands can be assessed. Generally, no consideration is given to the loss of flood defence function. Only now is the role of wetlands as part of an ecological solution to flood defence being considered in river basin management strategies. The WFD provides an opportunity to restore the flood defence function of wetlands, but for this to happen the objectives of the WFD must be implemented appropriately. Mainstream development pressures from agriculture, urbanisation, transport and navigation continue to drive wetland loss, and governments appear reluctant to promote environmentally appropriate approaches, despite existing knowledge and models that indicate that better wetland management could provide a large part of the solution to flood defence.

1.4.2 *Water purification: wetlands as buffer zones for protecting or enhancing water quality*

A buffer zone is a vegetated area situated between agricultural land and a surface water body, and acting to protect the water body from harmful impacts such as high nutrient, pesticide or sediment loadings resulting from land-use practices. Many buffer zones are wetlands, although not exclusively, but they all offer protection to a water body through a combination of physical, chemical and biological processes. The degree to which this protection is provided depends on a number of factors including the size, location, hydrology, vegetation and soil type of the buffer zone (Dosskey *et al.*, 1997; Leeds-Harrison *et al.*, 1996), as well as the nature of the impacts that threaten the water body. Functions typically operating within buffer zones include nutrient, heavy metal, pesticide/herbicide and sediment removal from run-off via processes such as denitrification, plant uptake, adsorption and sediment retention (Blackwell and Maltby, 1998; Blackwell *et al.*, 1999; Maltby *et al.*, 2000; Maltby and Blackwell, 2005). A buffer zone may provide additional benefits such as bank stabilisation, floodwater detention and wildlife habitat (Fig. 1.4.4), though appropriate location of a buffer zone is essential for optimal performance of a desired function.

Research has indicated that the establishment of buffer zones alongside rivers and streams can be an effective measure in preventing nitrate from reaching water bodies (Cooper, 1990; Ambus and Christensen, 1993; Haycock and Burt, 1993; Weller *et al.*, 1994). However, an example is given later of the effectiveness of buffer zones located away from the river channel. Comparisons of the efficacy of

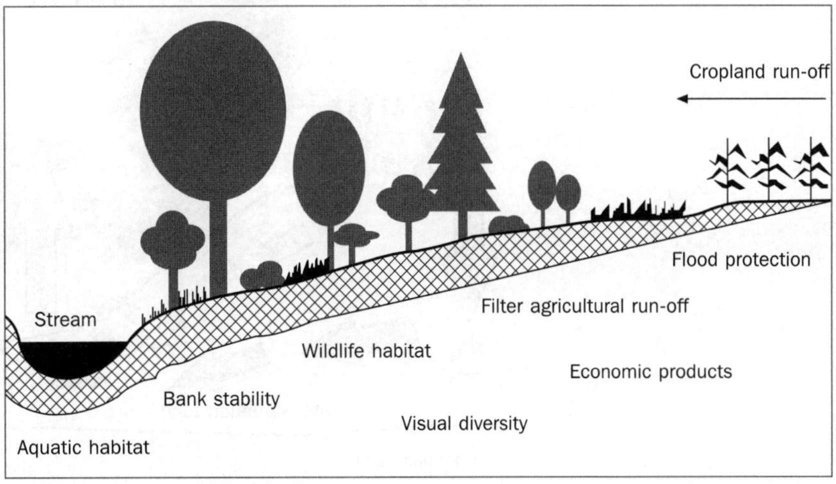

1.4.4 Benefits that a riparian buffer can provide (after Dosskey *et al.*, 1997).

such zones and their cost effectiveness have shown that, as a solution to the problem of excess nitrate, buffer zones are favoured above other preventative measures such as the operation of Nitrate Sensitive Areas (Burt and Trudgill, 1993; Haycock *et al.*, 1993; Johnes and Burt, 1993). Reduction of nitrogen in waters passing through wetlands has been reported as being regularly in excess of 80 % (Haycock and Burt, 1993). This is well demonstrated in a study by Hefting *et al.* (2003) in the Netherlands (Fig. 1.4.5). Mean nitrate–nitrogen concentrations in soil water were seen to decline from 15.29 mgl^{-1} at the upslope edge of a riparian wetland buffer zone adjacent to agricultural land, to 0.11 mgl^{-1} approximately 20 m downslope, near the stream the buffer zone was protecting. However, most of this nitrate removal occurred near the upslope edge of the buffer zone as, mid-way between the upslope edge and the stream, mean concentrations were already observed to have declined to 0.12 mgl^{-1}. The reductions in nitrate concentrations were attributed to a combination of dilution by groundwater, denitrification and plant uptake.

In the United Kingdom the current protocol for the implementation of buffer zones is described by both the Environment Agency (Environment Agency, 1996) and DEFRA. Generally, it is recommended that buffer zones should be located in riparian areas adjacent to main watercourses, and extend to between 5 and 30 m in width, depending on the type of pollution they are attempting to ameliorate.

Models for buffer zone designs such as REMM (Riparian Ecosystem Management Model) (Lowrance *et al.*, 1998) and RiMS (Riparian Management Systems) (Isenhart *et al.*, 1995) also focus on riparian buffers, although these models have been developed specifically for application in the United States. Buffer zones can range from large areas of land such as that shown in Fig. 1.4.6, sometimes complex in design and management, to simple, fenced, narrow strips of land adjacent to streams (Fig. 1.4.7). The recommended width of buffer zones for the effective delivery of different functions is shown in Fig. 1.4.8.

It has been argued that riparian wetlands contribute in a significant way if they remove at least 30 % of the total nitrogen load (Thoren *et al.*, 2005). To achieve this level of reduction, studies in the USA (Mitsch *et al.*, 2001), Sweden (Arheimer *et al.*, 2005) and China (Yin and Shan, 2001) suggest that, as a global rule, wetlands can contribute significantly to water quality improvement at the catchment level if they account for at least 2–7 % of the catchment area (Verhoeven *et al.*, 2006).

While riparian buffer zones are sometimes highly effective at removing pollutants, especially nitrate, from diffuse sources such as shallow groundwater or surface run-off, they are often bypassed where natural hydrological flows are intercepted by ditches or drains, which are common features in the UK and wider European landscapes (Goudie, 1986). In this way a riparian buffer zone can be rendered ineffective for pollutant removal from water draining agricultural land

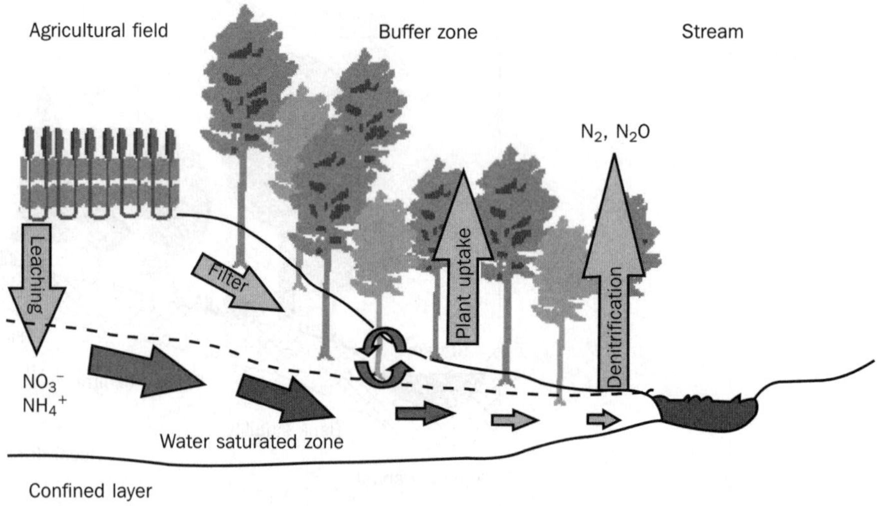

1.4.5 Nitrogen transformation and retention in riparian buffer zones (after Hefting, 2003).

1.4.6 A river marginal wetland area in West Cornwall, UK, comprising wet woodland, wet grassland and fen communities (photo: Environment Agency).

1.4.7 An example of poaching by cattle along the River Clyst, Devon, UK. Suitable fencing could be used to deny stock access, creating a riparian buffer zone that would intercept nutrient-rich run-off (photo: E. Maltby).

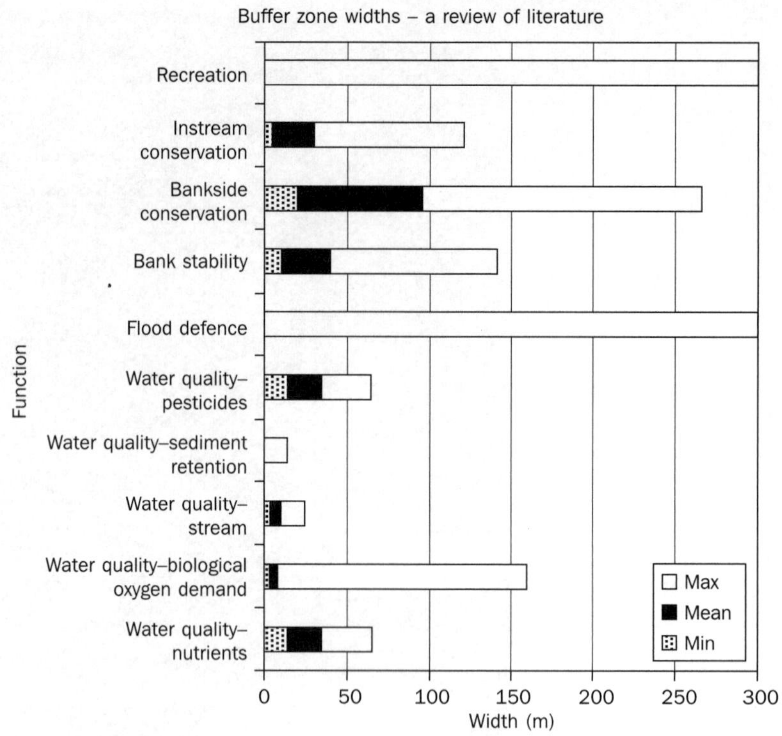

1.4.8 The widths of buffer zones necessary to achieve prescribed functions (based on Haycock and Muscutt, 1995).

upslope (Leeds-Harrison *et al.*, 1996). In such cases, it may prove more effective and cost efficient to establish buffer zones in association with ditches or areas of discharge that may be acting as zones of enhanced denitrification (Blackwell and Maltby, 1998). Similar conclusions have been drawn from work by Haycock and Muscutt (1995) who reported that while 85 % of some sub-catchments of the River Avon in Hampshire in the UK were served by effective riparian buffer strips, 60 % of polluting material in the river was delivered by roads and drains that effectively bypassed them.

An example of an alternatively located, non-riparian buffer zone for nitrate removal (Fig. 1.4.9). was studied in the catchment of the River Torridge, south-west England. A ditch draining improved agricultural land and passing through the wetland became disrupted at several locations along its course, causing areas of overland flow and the development of locally wetter patches in the wetland (Fig. 1.4.10). Monitoring of nitrate concentrations in the ditches indicated that more than 90 % of the nitrate in the ditch water is removed regularly in these zones (Fig. 1.4.9), with the lowest observed removal efficiency being 60 % (Blackwell, 1997). These areas, which occupy only a small percentage of the total wetland area, are therefore acting as effective buffer zones for the removal of nitrate. The establishment of a conventional riparian buffer zone along the banks of the River Torridge in accordance with current policy recommendations would have little or no impact on the quality of the water in the ditch as it by-passes any potential conventional buffer zone, discharging directly into the river. In such situations, targeting buffer zones away from the riparian margins may be most effective for water quality maintenance and improvement. Generally, it is accepted that for the delivery of most water quality enhancing functions, buffer zones should be established on low-order streams or ditches, where the majority of diffuse pollutants generally enter streams, and buffer zones are less likely to be by-passed.

Successful riparian zone management requires an understanding of the effects of buffer strips on all nitrate sources, including remote sources, such as deeper groundwater (Martin *et al.*, 1999). In areas with intensive land use and very heavy use of fertiliser, nitrate concentrations in the groundwater are so high that high N_2O emissions in the riparian wetlands are to be expected, contributing to the concentration of greenhouse gases in the atmosphere. Therefore restoration of

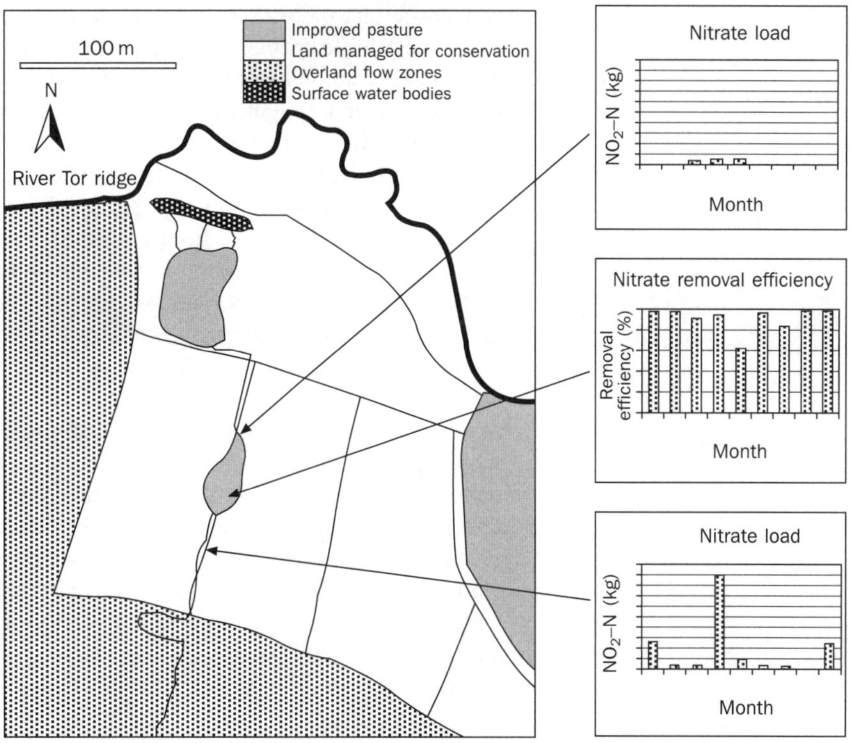

1.4.9 Kismeldon Meadows: an example of alternative wetland buffer zone locations for the removal of nitrate from agricultural run-off (from Blackwell *et al.*, 1999).

1.4.10 A buffer zone for arable run-off along the River Clyst, Devon, UK. (photo: E. Maltby).

wetlands in riparian zones should be accompanied by lower fertiliser applications to avoid switching the adverse effects on the environment from the reduction in water quality to the amplification of global warming (Verhoeven *et al.*, 2006).

1.4.3 Wetlands and biodiversity

Wetland ecosystems provide the water and primary productivity upon which diverse species of plants and animals depend for survival. They support high concentrations of birds (especially waterfowl), mammals, reptiles, amphibians, fish, invertebrate species and plants. Levels of species diversity do, however, vary considerably among wetlands: for example, while some lakes display high levels of diversity and endemism, others may support few species and little biomass. Compared with terrestrial ecosystems, understanding about freshwater biodiversity remains relatively poor. Identification and classification of wetland species is hampered by the fact that many species share their life cycle between freshwater and marine environments (Ramsar, 1996).

The wetland environment can be physiologically harsh at both ends of the hydrological spectrum. Aquatic organisms are not adapted to deal with periodic drying that occurs in many wetlands, while terrestrial organisms are stressed by long periods of flooding. In shallow water conditions, temperature extremes at the wetland surface tend to be greater than those of aquatic environments. The most severe stress, however, is probably the absence of oxygen in flooded wetland soils, which prevents organisms from respiring through normal aerobic metabolic pathways. In the absence of oxygen, the supply of nutrients available to plants is also modified, and concentrations of certain elements and organic compounds can reach toxic levels. In consequence, those plants and animals regularly found in wetlands have evolved functional mechanisms to deal with these stresses, thus increasing the species and genetic diversity of many wetland ecosystems (Mitsch and Gosselink, 2007).

Plants have developed a wide range of adaptations in order to survive and exploit their wetland environments. The most obvious developments are structural, concerned especially with the problem of supplying oxygen to roots growing in anaerobic (oxygen-deficient) soils or sediments. Many aquatic plants, such as water lilies, are extremely porous and contain special tissue, called aerenchyma, which has large, air-filled intercellular spaces. Most flood-tolerant wetland species respond to flooding by increased root aerenchyma development (Armstrong, 1979; Burdick and Mendelssohn, 1990; Pezeshki *et al.*, 1991). Since oxygen diffuses 10 000 times faster in air than in water, the aerenchyma can facilitate the movement of oxygen from the leaves to the rhizomes (Greenwood, 1967).

Other plants have developed biochemical adaptations, thought to be particularly significant in tolerating flood or waterlogged conditions. Some species oxidise elements such as iron, which is toxic in the reduced ferrous (Fe^{2+}) form. Other species excrete the toxic products of anaerobic respiration. Some, for example, diffuse acetaldehyde and alcohol through the large surface area provided by finely divided roots. Additionally, many wetland plants appear to have a much higher than average metabolic tolerance for toxic ions (Ernst, 1990).

In rainfall-dependent wetlands, such as peat bogs, the structure of bog mosses (*Sphagnum* spp.) enables them to hold enormous quantities of water, sometimes as much as 15 times the weight of the plant itself. In this way, a layer of *Sphagnum* moss can maintain high levels of ground saturation even during prolonged dry periods. The carnivorous habit of the sundew (*Drosera* spp.) (Fig. 1.4.11) is an adaptation to low levels of nutrients in some wetlands. Insects are trapped by sticky droplets on leaf protruberances, and enzymes break down the prey, allowing nitrogenous compounds to be absorbed into the plant.

Many animals, like plants, are adapted to waterlogging, and particularly breeding and feeding in a wetland environment. However, although it is possible to identify specific response mechanisms to individual kinds of environmental stresses, in reality an organism responds to a complex of environmental factors, and it is the success of this integrated response that determines its fate (Mitsch and Gosselink, 2007; Begon *et al.*, 2006).

Some insects, such as dragonflies and damselflies, lay their eggs on submerged plants. Water is also essential for amphibians, which can spend much of their time on land but return to the water to breed in order to provide their eggs with protection from drying out. Frogs and toads are found in wetlands throughout the world, and frequent a wide range of altitudes, water salinities and qualities, as well as vegetation types.

Fish are found in enormous diversity in rivers, lakes and coastal waters, though

1.4.11 Common sundew (*Drosera rotundifolia*) (photo: E. Maltby).

most of them are dependent to some degree on the marginal wetlands, whether for food, spawning, nursery or other habitat requirements. Some species, such as the European carp (*Cyprinus carpio*), migrate from rivers to seasonally inundated floodplain forest or grassland for spawning, while others move from deepwater lakes to vegetated shallows, or from the open sea to coastal mud flats, lagoons and marshes. In the case of migratory species such as the salmon (*Salmo salar*), the distances covered may be thousands of kilometres from ocean feeding grounds to shallow gravel-bed spawning grounds of catchment headwaters.

There are a number of mammals which show both physical and behavioural adaptations to life in and around water. One of the best examples is the beaver (*Castor fiber*) with its notable behavioural adaptation of building dams to create the pools in which lodges are built. The otter (*Lutra lutra*) has many of the physical adaptations of the beaver but with the added advantage of having a slim body, which aids swimming.

The high productivity of many wetlands also provides support for large numbers of birds, many of which depend on a linked network of wetland sites during long seasonal migrations, or as breeding or over-wintering grounds. One of the most impressive concentrations of wetland fauna is the number of migrating waterfowl that visit wetland sites during annual migrations to and from feeding and breeding grounds. Seven million shorebirds breeding in northern Eurasia winter along the Atlantic shores of Africa. Individual wetlands, such as those of the Inner Niger or Senegal deltas, are especially important as these are some of the first freshwater bodies encountered south of the Sahara: More than 3 million migrants, mainly ducks and waders, use these deltas every year. European wetlands also attract large numbers of birds. In eastern Europe, the 6500 km^2 of the Volga Delta provide 7 million waterbirds with a migratory stop-over. Within the delta of the Rhine and Meuse on the North Sea coast, the Eastern Schelde provides another element along the Western Palearctic migration route, hosting some 47 000 ducks, geese and swans and 160 000 shorebirds (Finlayson and Moser, 1991). Along the Mediterranean, Black Sea and Iberian coasts are many deltas of major rivers, which are important in attracting large numbers of birds for breeding and feeding. Notable examples are those of the Danube, Evros, Po, Rhône (Camargue) and Ebro, together with coastal wetlands such as Doñana and coastal lagoon complexes like those of the northern Adriatic and Greece.

However, the importance of wetlands is not always related to numbers of species. Some sites are vital for rare or endemic species (Maltby, 1986). Amongst endemic species found in European wetlands of international importance is the fish *Barbus prespensis*, found only at Lake Prespa in Greece. Because of the pressures on natural ecosystems in Europe, a very high proportion of wetland species are considered to be under threat; around a quarter of all fish species and nearly half of all amphibians in Europe are considered threatened (Halls, 1997). They depend on habitats that have become highly fragmented and degraded, A recent BirdLife International study of population trends in European bird species concluded that a quarter of all declining species had been adversely affected by wetland drainage. The European otter has been extirpated from much of western and central Europe, though recent conservation measures have enabled some recovery to take place, such as in south-west England. Intensive agricultural practices have led to many species of wetland plants, such as orchids, becoming increasingly confined to nature reserves and other protected areas. In the Netherlands, it has been estimated that 30 % of all species require a high water table. However, in many parts of the country, water tables have fallen substantially as a result of groundwater extraction for use in industry and drinking water supply, and because of more efficient drainage and irrigation (Halls, 1997).

Europe's long history of settlement, permanent agriculture and industrialisation has brought major changes to the natural ecosystems of the region and consequently the species of flora and fauna which are an integral part of them. Whilst it can be argued that human activity in the past gradually increased the biodiversity of Europe over the course of many centuries by creating an intricate, artificial mosaic of habitats, more recent activities have seen a reversal in that trend. Biodiversity loss in wetlands is a fact that has to be tackled sooner rather than later. For this to happen, the biodiversity value of wetlands must be included on the agenda of decision makers and managers who can decide the fate of these valuable but threatened resources.

1.4.4 *Cultural heritage*

The significance of losses of wetland habitats and dependent species is matched, some would argue exceeded, by a parallel loss in heritage value. A review of 'monuments' (i.e. archaeological and palaeoenvironmental remains preserved in peat and waterlogged soils) at risk in England's wetlands (Van de Noort *et al.*, 2002) indicates that 78 % of wetland monuments have been destroyed or damaged in the last 50 years, mainly as a result of agricultural land drainage and ploughing. The anoxic conditions associated with wetland deposits enable preservation of otherwise fragile organic remains which provide evidence of past environmental conditions and human culture. Nevertheless, while degraded wetland habitats can frequently be restored and certain types created, together with their associated species, once archaeological or palaeoenvironmental evidence is destroyed, the heritage value is lost for ever.

Organic remains can be regarded broadly as natural or cultural (Coles, 2001). Natural remains are of plant or animal origin, the former most evidently preserved in peat deposits, but with plant fragments frequently observable in waterlogged mineral deposits of silt and clay or mud. Plant remains, including pollen grains and spores, together with animal remains, particularly invertebrate exoskeletons and shells, represent an important record of conditions during the development of the wetland. Cultural remains comprise the artefacts and structures left by humans in and around the wetland, which form such an important archaeological record for interpreting past aspects of human culture. In exceptional cases, complete human bodies have been recovered from wetland sites (Turner and Scaife, 1995). In the past, human communities have often been dependent on wetlands for natural resources to provide the basic requirements of food, fuel and protection. Archaeological discoveries of settlements, commonly referred to as lake villages, associated with standing water bodies, began in the mid-19th century in Switzerland, following which many examples have been reported across Europe. The dynamics of communities living in the vicinity of wetlands were moulded by the conditions found there, including seasonal variations in water level, plant growth cycles and animal behaviour (Coles and Coles, 1989).

A number of laws operating at the global, European and national levels are available to protect the heritage value of wetlands. Some have been specifically developed to enable designation of important sites, while others target the needs of wildlife conservation, but in so doing also protect cultural heritage values. Marsden (2001) has reviewed the legislation developed to designate and protect the archaeological heritage value of wetlands in England and Wales. The study concluded that there remains a lack of integration within the legislative provision, and that the different objectives and requirements of scale for nature conservation (ecosystems) and archaeology (sites) have mitigated against developing co-ordinated policies and prescriptions for integrated site management.

In the 1990s, it was noted that a holistic approach was beginning to develop for addressing issues of nature conservation and the protection of archaeological evidence (Carman, 1994). However, it is now widely recognised that the common objectives for wildlife and cultural heritage management of wetlands should be promoted with practical steps to enable this to take place on the ground. It is a major focus of English Heritage's strategy for wetlands (http://www.english-heritage.org.uk/server/show/conWebDoc.3208), while the Ramsar Convention, at the Conference of Parties in Valencia, November 2002, adopted Resolution VIII 19 providing guiding principles for taking into account the cultural values of wetlands for the effective management of sites (http://www.ramsar.org/index_cop8.htm). Guiding principles include safeguarding cultural landscapes, management planning, monitoring, environmental impact assessment, research and communications.

The development of the procedures for functional assessment has focused on producing a modular package in order to enable the technique to be readily extended, as required, to cover additional functions to meet the needs of a wider range of potential users. One example currently underway, in collaboration with English Heritage, is to develop a module dealing with the functioning of wetlands to preserve organic archaeological remains, palaeoenvironmental evidence and palaeolandscapes.

Sutton Common, South Yorkshire, UK contains two Iron Age enclosures of uncertain purpose. The archaeological evidence has been seriously damaged by land management operations. One of the enclosures has been ploughed out, and timber remains are decomposing as a result of drying out following land drainage. Integrated management plans are now under development to protect the remaining cultural heritage, promote wildlife conservation and provide public access for leisure and recreation by the local community. Details of the archaeological investigations are summarised in Smith (2004).

A good example of wetland management for wildlife conservation and archaeology is provided by the Sweet Track (Coles and Coles, 1986), a Neolithic timber trackway discovered in 1970 in fen peat of the Somerset Levels and Moors, UK (Fig. 1.4.12, Fig. 3.2.8). Where it passes through a National Nature Reserve (Shapwick Heath), a raised water level regime has been maintained to meet both objectives, including a period when it was necessary to pump water onto the site to prevent the water table falling below the level of the timbers. The success of the strategy to maintain anoxic conditions has been demonstrated by a re-examination of the condition of the remains in conjunction with a programme of monitoring environmental conditions in the burial horizon (Brunning *et al.*, 2000).

1.4.5 *Wetlands and landscape*

Wetlands often provide distinctive landscapes, as a result not only of their inherent properties, such as the occurrence of deep peat or having an extensive flooding regime, but also because of the modifications and management that may have been developed to enable their exploitation by local communities, often over a considerable period of time. Peat may have been dug for fuel and perhaps more recently drained and mined for horticultural products. If subsequently abandoned, a site can leave a legacy of flooded depressions and dry banks, remnants of a former ground surface. Woodland has often become established on former peat workings. The flooded former mediaeval peat workings of the Norfolk Broads form a distinctive landscape in East Anglia, UK. The Broads have an international reputation for wetland nature conservation, yet also provide for important water-

1.4.12 Excavation of part of the Sweet Track running through Shapwick Heath National Nature Reserve, Somerset Levels and Moors, UK (photo: Somerset Levels Project).

based leisure activities which generate considerable income to the local economy (Moss, 2001).

Where wetland has been reclaimed for agricultural use, a system of ditches and drains is often created, forming a distinctive broad open landscape of fields surrounded by water-filled ditches, often of high nature conservation value. These landscapes of fields and drains can form distinctive landscapes, such as the Somerset Levels and Moors in South West England, where pollarded withy (willow; *Salix* spp.) beds have provided the material for a range of products including hurdles, fencing, fuel and, particularly nowadays, basketry. Depending on the climatic limitations of the area, land-use history and cultural context, the land may be reclaimed and drained to varying degrees for agricultural use. Traditional management practices such as summer grazing of wet grassland with local breeds of cattle or horses, the mowing of fens and marshes or the cutting of reed for thatching have been critical in maintaining and conserving increasingly rare habitats and dependent species.

The extensive Trebon fish ponds of the Czech Republic, dug by settlers from the 12th century onwards, are the dominant feature in a landscape that was formerly characterised by woodland, rivers and wetlands. During the middle ages, fish became an important food resource, and brought prosperity to the region (Fig. 1.4.13).

In the semi-arid regions of southern Europe, salt pans form distinctive landscape features, which provide economic returns but also important breeding and feeding sites for flamingos (Verhoeven *et al.*, 2006). In the Camargue of southern France, for instance, flamingos, together with white horses, are icons of the area that attract many thousands of tourists each year.

Within wider landscapes, wetlands often support vegetation that contrasts strongly with adjacent dry land, such as fens occupying valley bottoms, or peaty seepage zones on footslopes. Alternatively, wetlands may indicate cultural boundaries such as where the artificial drainage of surrounding areas was precluded

(a)

(b)

1.4.13 The fish ponds of Trebon brought wealth to the area (photos: E. Maltby).

owing to excessively wet conditions, or to circumstances of land ownership. The wetland boundary may be distinguished by a distinctive feature such as a ditch or drain, embankment, stone wall or elevated road or track, or there may be simply a change in vegetation type.

1.5 Threats and impacts on the wetland resource

1.5.1 Wetland loss

Wetlands have been subjected to major changes for centuries. Population pressure, agricultural development and industrialisation have resulted in the alteration or destruction of many of these areas, driven by the widespread perception of wetlands as being impediments to development.

Over the last 200 years, wetland loss has been substantial. It is estimated that, across the lower (conterminous) 48 states of North America, 53 % of wetlands have been lost due to human activities (Tiner, 1984), some 80 % of this resulting from agriculture. Following a longer history of economic development in Europe, wetland loss had been generally a slower process, though some more recent losses have been equally rapid, resulting in a significant overall decline in both quantity and quality. Some estimates show wetland loss on a global scale to have been at least 50 % since 1900, mainly in northern countries in the first half of the 20th century, but with increasing pressures now focused on tropical and sub-tropical wetlands. Jones and Hughes (1993) in the most recent pan-European study reported more than 50 % of the original wetland cover lost from many countries. Négrel *et al.* (2005) provide some specific examples: Spain lost half of its wetlands (around 500 000 ha) between 1950 and 1970; 70 % of wetlands in Portugal have been converted to cultivated or industrial land; 66 % of wetlands in Italy disappeared between 1938 and 1984; while 62 % were lost in Greece in the period 1920–1999. In France, wetlands of the Camargue reduced from 67 % in 1942 to 39 % in 1984, representing a loss of 33 000 ha. In the United Kingdom, between the middle of the 19th century and 1978, 84 % of lowland raised bog was lost through afforestation, agricultural reclamation and commercial peat cutting. Much of the remainder has been severely damaged by burning and draining, leaving only 6 % of the original area intact (Nature Conservancy Council, 1986).

1.5.2 Kinds of impact

Wetlands are ecosystems vulnerable to many threats because of their dependence on very specific environmental conditions. In nature conservation terms, small changes to the water table depth may have magnified biological consequences owing to the development of unique communities (Etherington, 1983). However, the relationships between hydrological variables and their effects on plant growth and distribution are complex and imperfectly understood (Wheeler and Shaw, 1995).

Through degradation, the many important socio-economic benefits offered by wetlands can also be lost. Table 1.5.1 indicates the major causes of wetland degradation and loss, which may be due to direct on-site impacts leading to complete replacement of the ecosystem (e.g. land drainage, construction work or peat mining – Fig. 1.5.1), or to indirect impacts such as diffuse pollution from nearby agricultural sources, leading to undesirable alteration of the wetland. Some examples of these impacts are discussed below.

Agricultural intensification

This includes both the direct loss of a wetland that is reclaimed for agricultural use, and the impacts a wetland may experience when intensive agriculture is carried out nearby. Depending on the intended land-use, wetland reclamation may involve drainage, cultivation and drilling a new crop, and a programme of fertiliser and pesticide use. Indications of former wetland may continue to exist in ditches and drains created as part of the water management system. In arable systems (such as the Fens of eastern England), lowering of the water table and annual cultivation have lowered the ground surface by peat wastage amounting to about 2 cm a year (Seale, 1975). Eventually, ploughing turns up mineral substrate, and agricultural potential is reduced. Bare soil exposed at critical times of the year enhances the risk of erosion taking place, with consequent deposition of sediment elsewhere, sometimes within adjacent wetland systems.

Diffuse pollution

This involves the unintentional introduction of nutrients, heavy metals, pesticides and persistent organic pollutants from surrounding land, most usually as a result of agricultural activities. The pollution is 'diffuse' because it does not derive from identifiable point sources, but from the results of general land management. In consequence it can be difficult to address. While it is often possible to reduce the impact on water bodies by the use of buffer strips (see Section 1.4.2), a more sustainable solution is to focus on 'best management practice' to reduce the risk at source, supplemented by the use of buffer strips to catch pollutants 'leaking'

Table 1.5.1 Causes of wetland degradation and loss (after Gosselink and Maltby, 1990)

Source of impact	Examples of specific impact or purpose of activity
Natural processes	Vegetational succession Erosion Accretion Storm events
Drainage	Agriculture (crop production) Forestry (timber production) Mosquito control
Stream dredging and channelisation	Navigation Flood protection Housing and development Reservoir maintenance
Filling	Solid waste disposal Road construction Commercial, residential and industrial development
Land-use change	For agriculture For aquaculture
Construction	Dykes, dams and levees Sea walls for flood control Water supply Storm protection Irrigation
Additions to land/crops	Pesticides, herbicides and nutrients (domestic sewage, run-off, sediments from dredging and filling) Atmospheric deposition
Mining and quarrying	e.g. peat, coal, gravel, sand and clay
Water abstraction from aquifers, reservoirs and rivers	Potable water supply Industrial processes Crop irrigation
Civil engineering works (e.g. canals, roads and other structures)	Alterations to hydrological flow paths
Subsidence due to mining or abstraction	Wetter ground conditions, but often with nutrient-rich waters

1.5.1 Large-scale peat extraction for power generation in Finland (photo: E. Maltby).

from the system. The Common Agricultural Policy (CAP) of the EU has been responsible for promoting agricultural intensification by making payments available for crop production and stocking. It is only now that the focus for support payments is on land management rather than production.

Many wetland types are particularly sensitive, not only to land drainage but also to changes in water chemistry. Specialised wetland plant communities can be lost as more competitive plants with high nutrient requirements become established at the expense of species typical of infertile conditions. Increase of nutrients in peatlands is also likely to enhance rates of organic matter decomposition. This causes wastage of the peat with loss of any archaeological evidence (see Section 1.4.4) or palaeoenvironmental record it contains, together with release to the atmosphere of carbon dioxide, a greenhouse gas (Mander *et al.*, 2005)

Isolating river channels from their floodplains

Wetlands developed on floodplains have often become isolated from river channels and their flows, due to the effects of a wide variety of flood defence measures including land drainage, flood embankments, the straightening of river channels and other in-channel works. Where natural flooding no longer takes place, flood-related functions are not performed (see Section 1.4.1), including loss of flood storage capacity, with consequent enhanced risk of flooding taking place elsewhere along the river corridor. For example, 80 % of the Elbe catchment has been disconnected from its floodplain and, in 2002, peak flows of 5000 m^3 s^{-1} were recorded, approximately 10 times the average (Fig. 1.5.2). This was due to an exceptional combination of meteorological factors, including more than 300 mm of rain falling in 24 hours in the mountains at the German–Czech border (Rudolf and Rapp, 2003). Rudolf and Rapp (2003) pointed out the likelihood of increased flood risks in the future and the need to take precautions, including restoring or creating areas of floodable land, and the prevention of new developments on floodplains, alongside improved transnational flood warning and disaster control systems.

Where floodwaters are able to be detained on the floodplain, this enables water quality to be improved through processes such as denitrification to remove nitrate from the water column (Maltby *et al.*, 1998b), and the deposition of sediments to which are adsorbed nutrients (especially P) and pesticides.

1.5.2 Overbank flooding of a town on the River Elbe, Germany in 2002. Pumps can be seen operating towards the bottom of the picture (photo: André Künzelmann).

1.5.3 *Wetland degradation: the example of water supply to the Tablas de Daimiel, central Spain*

An example of a wetland degraded by a variety of impacts is the Tablas de Daimiel in central Spain. It occupies some 1928 ha at the confluence of the rivers Cigüela and Guadiana in south-west Castilla-La Mancha, and comprises a series of shallow lakes and surrounding marshes important for a range of breeding and wintering waterbirds. In the past, the wetland has been maintained by groundwater discharge from a large underlying aquifer, together with periodic flooding from the River Cigüela. However, despite its designations as a Ramsar site, a National Park, a UNESCO Man and Biosphere Reserve and an EU Special Protection Area, it has suffered serious desiccation. This results mainly from depletion of the underlying aquifer by illegal abstraction for crop irrigation, which increased substantially during the period 1970–2000, and as a result of which the Tablas have changed from an area of groundwater discharge to one of recharge (Fig. 1.5.3). Bromley *et al.* (2001) report a lowering of the groundwater table by as much as 50 m, which has also affected flows in the adjacent river systems as groundwater and surfacewater systems have become decoupled. Although there is a designated buffer zone around the Tablas, this focuses on elimination of hunting, so irrigated crops continue to be grown up to the boundary of the site.

Maintaining adequate wetness in the Tablas is complicated by the substantial natural variation in rainfall (and consequent aquifer recharge) from year to year. From time to time, dry conditions have led to the site becoming completely dry, followed by periodic flooding resulting from wet periods. During dry periods in 1986 and 1987, serious fires occurred at the site, and peat burned underground for many months. The amounts of water being abstracted from the La Mancha aquifers have led to conflicts among farmers, regulators and nature conservationists in the area. In order to address the situation, three actions were undertaken:

1. Passing of legislation in 1985 to prevent any further expansion in the amount of irrigated land, and to reduce irrigation in existing areas.
2. In 1993, a five-year programme of subsidies to encourage farmers to take land out of production.
3. From 1988–1996, a series of water transfers from the Tagus–Segura aqueduct to the River Cigüela as a means of ensuring water supply to the Tablas.

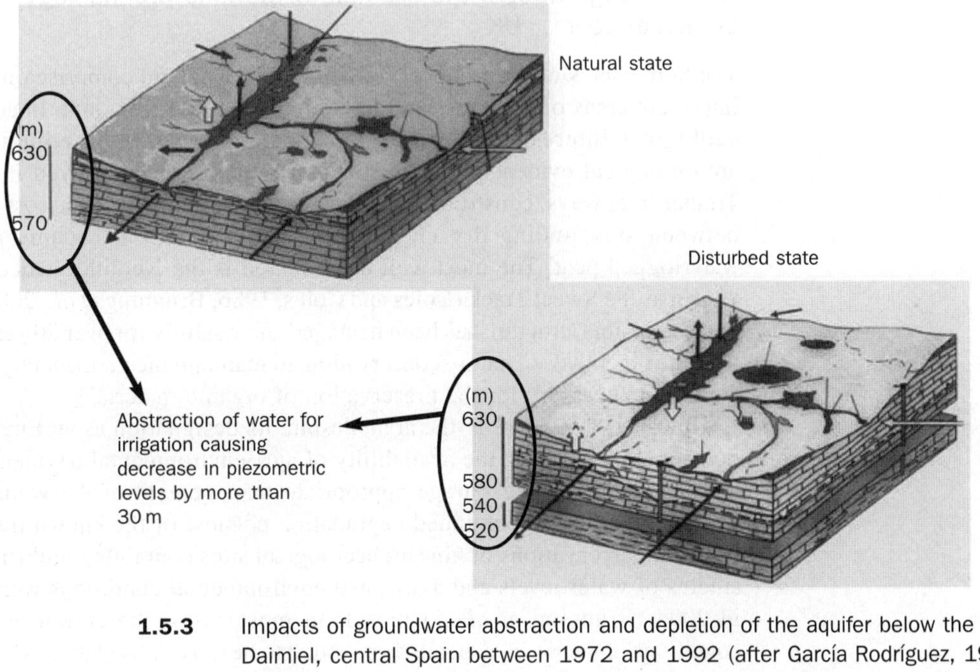

1.5.3 Impacts of groundwater abstraction and depletion of the aquifer below the Tablas de Daimiel, central Spain between 1972 and 1992 (after García Rodríguez, 1996). Location of the groundwater table and surface water indicated by shading.

1.5.4 Channel of the River Cigüela, La Mancha, Spain, after straightening and reprofiling (photo: E. Maltby).

The Cigüela was also cleaned to improve its supply capacity (Fig. 1.5.4). However, groundwater levels continued to decline. In addition, water was lost from the river channel mainly by leakage, together with degradation of river marginal wetland habitats, including the La Mancha lakes upstream on the original floodplain of the river.

Groundwater over-exploitation is a complex issue involving an understanding of economic costs and benefits in addition to environmental and social issues. Long-term variations in rainfall in La Mancha make it difficult to interpret the significance of a decline in groundwater levels in terms of 'over-exploitation' without careful examination of a large amount of data. Conflict resolution between stakeholders requires effective programmes to provide information and education, and ultimately local control is seen as the best solution to management and use of groundwater resources (Llamas and Martinez-Santos, 2005).

1.5.4 Wetland degradation: the example of wetland archaeology in the Somerset Levels, UK

The Somerset Moors and Levels of south-west England comprise one of the most important areas of wet grassland found in the United Kingdom. In addition to the ecological interest of the area, fen peat deposits there contain important archaeological evidence of prehistoric communities which lived in the vicinity. Timber trackways, constructed to allow access to and passage across the wetlands between surrounding dry elevated sites of habitation, became preserved in waterlogged peat. The most well documented is the Neolithic wooden trackway known as the Sweet Track (Coles and Coles, 1986, Brunning *et al.*, 2000). Uniquely in the area, this structure has been managed successfully for over 20 years in tandem with the objectives of nature conservation, to maintain the permanently waterlogged condition necessary for the preservation of organic materials.

However, elsewhere in the area, despite its designation as an Environmentally Sensitive Area (ESA), the availability of agri-environmental payments to farmers and landowners to encourage appropriate management of the wetlands has not generally prevented continued degradation of most of the known trackway sites. Following excavations of nine archaeological sites containing prehistoric timbers, studies of water levels and associated environmental conditions were carried out in 2004 (Brunning *et al.*, in press). In most cases, the current environmental conditions of the burial deposits were found to be too oxidised to expect preservation to take place. Some sites were completely oxidised while others experienced prolonged seasonal periods of water table draw-down and consequent oxidation. In most cases, on-going degradation of archaeological evidence was likely to be taking place together with continued loss by wastage of the peat resources of the area.

1.6 Challenges for the future of wetlands and their management

1.6.1 International agreements: the Ramsar Convention

The Ramsar Convention (The Convention on Wetlands of International Importance, especially as Waterfowl Habitat) is perhaps the most significant inter-governmental agreement on wetlands. Conservation was the original motive in the only international agreement to cover a specific single group of ecosystem types. The Convention recognised the vital importance of protecting not just single sites but networks, linked chain-like, often over long distances, by the essential requirements of migratory birds for breeding, over-wintering, resting and feeding sites. Its goal is the conservation and wise use of wetlands through local, regional and national actions and international cooperation, as a contribution towards achieving sustainable development (Ramsar COP8, 2002, http://www.ramsar.org/index_cop8.htm). By the end of 2006, 154 countries had signed the Convention since its adoption in 1971, designating 1651 sites (totalling 150 million ha) for inclusion on the Ramsar List of Wetlands of International Importance (see www.ramsar.org/ for more recent updates). Signatories are obliged to take wetland conservation into account in planning and to encourage the 'wise use' of all wetlands. Furthermore, they are required to establish protected wetland reserves and designate Wetlands of International Importance. The Ramsar Conference of Parties (COP 9, 2005, http://www.ramsar.org/index_cop9_e.htm) updated the definition of 'wise use' of wetlands as:

> the maintenance of their ecological character, achieved through the implementation of ecosystem approaches, within the context of sustainable development.

A wetland qualifies for the designation of 'international importance' if it meets stated criteria relating to functioning, such as hydrology, ecology and support of human as well as wildlife populations. This is a significant development, since the original emphasis was on the populations of waterfowl that characterised the early history of the Convention (Williams, 1990). In Europe, 46 contracting parties have signed the Convention (Fig. 1.6.1) and the total surface area of Ramsar sites in Europe (April 2007) is around 22 million ha (Table 1.6.1).

1.6.1 Ramsar sites in continental Europe and the Russian Federation (courtesy of Wetlands International). The large dots on this *Web Mapper* map mark the sites of all the Ramsar sites in Europe and are interactive. By clicking on a dot, the Web Mapper takes the user to a map with links to detailed information about the site. Associated, searchable maps are global and country-specific. Go to http://www.wetlands.org/rsis/

Table 1.6.1 European Ramsar Sites (adapted from Ramsar Sites Information Service, http://www.wetlands.org/rsis/ Ramsar Secretariat, April 2007)

Country	Convention came into force (date)	No. of sites	(hectares)
Albania	29.03.96	3	83,062
Armenia	06.11.93	2	492,239
Austria	16.04.83	19	122,277
Azerbaijan	21.05.01	2	99,560
Belarus	25.08.91	8	283,107
Belgium	04.07.86	9	42,938
Bosnia and Herzegovina	01.03.92	2	10,911
Bulgaria	24.01.76	10	20,306
Croatia	25.06.91	4	80,455
Cyprus	11.11.01	1	1,585
Czech Republic	01.01.93	12	54,656
Denmark	02.01.78	38	2,078,823
Estonia	29.07.94	11	218,344
Finland	21.12.75	49	799,518
France	01.12.86	24	828,803
Georgia	07.06.97	2	34,223
Germany	26.06.76	32	839,327
Greece	21.12.75	10	163,501
Hungary	11.08.79	26	206,379
Iceland	02.04.78	3	58,970
Ireland	15.03.85	45	66,994
Italy	14.04.77	50	59,796
Latvia	25.11.95	6	149,158
Liechtenstein	06.12.91	1	101
Lithuania	20.12.93	5	50,451
Luxembourg	15.08.98	2	313
Malta	30.01.89	2	16
Monaco	20.12.97	1	10
[Montenegro] [awaiting official notification of legal status & name]	[27.04.92]	1	20,000
Netherlands	23.09.80	49	818,908
Norway	21.12.75	37	116,369
Poland	22.03.78	13	125,760
Portugal	24.03.81	17	73,784
Republic of Moldova	20.10.00	3	94,705
Romania	21.09.91	5	683,628
Russian Federation	11.02.77	35	10,323,767
Serbia	27.04.92	6	28,025
Slovakia	01.01.93	14	40,414
Slovenia	25.06.91	3	8,205
Spain	04.09.82	63	281,768
Sweden	21.12.75	51	514,506
Switzerland	16.05.76	11	8,676
The FYR of Macedonia	08.09.91	1	18,920
Turkey	13.11.94	12	179,482
Ukraine	01.12.91	33	744,651
United Kingdom	05.05.76	165	895,144

1.6.2 *European directives*

Although individual nations within the EU have their own legislation affecting wetland management, there are a number of wider EU Directives which influence wetland conservation and management. A network of protected sites, known as Natura 2000, exists across Europe and consists of Special Protection Areas (SPAs), designated under the Birds Directive (EC, 1979) and Special Areas of Conservation (SACs) designated under the Habitats Directive (EC, 1992).

Birds Directive

The Birds Directive aims to protect rare and migratory bird species and requires member states to designate SPAs to this end. To qualify, a site must meet one of the following criteria:

1. An area is used regularly by 1 % or more of the national population of a species listed in Annex I of the Birds Directive (EC, 1979, as amended) in any season.
2. An area is used regularly by 1 % or more of the biogeographical population of a regularly occurring migratory species (other than those listed in Annex I) in any season.
3. An area is used regularly by over 20 000 waterfowl (waterfowl as defined by the Ramsar Convention) or 20 000 seabirds in any season.
4. An area which meets the requirements of one or more of the Stage 2 guidelines in any season, where the application of Stage 1 guidelines 1, 2 or 3 for a species does not identify an adequate suite of most suitable sites for the conservation of that species.

Habitats Directive

The Habitats Directive aims 'to maintain or restore the extent and quality of rare habitat types and to ensure that rare species can survive and maintain their populations and natural range on a long term basis'. The directive lists 168 habitat types and 632 animal and plant species that require protection through the designation of SACs.

Water Framework Directive

Wetlands are not a primary emphasis within the Water Framework Directive (WFD) (EC, 2000), although their water requirements are addressed where they are dependent on surface or groundwaters. Despite limited attention to wetlands, it is widely recognised by the Commission, NGOs and the agencies responsible for implementation that wetlands will be very important in delivering the aims of the WFD (e.g. Environment Agency, 2002; Defra 2003; European Commission, 2003).

The European Parliament and Council agreed on the WFD in 2000. It will have a major impact on water management in Europe, and some sections are particularly relevant to wetland management, notably:

- Article 1: Establishment of a wetland protection framework.
- Article 4a: An emphasis on ecological quality as well as water quality.
- Article 5: The characterisation of wetlands in the context of their river basins.
- Article 8: The assessment of waters moving in and out of wetlands in terms of flow rates, chemical quality and ecological potential.
- Article 13: The development of river basin management plans.

The emphasis on river basin management and the assessment of ecological quality will encourage more complete assessments to be made of wetland functions and the benefits they provide, in the context of an entire catchment.

The WFD is the most significant piece of European legislation affecting water resources. It provides both a vision for sustainable water management in Europe and a framework for its delivery based on river basin management plans.

The protection of waters that support wetlands is one of the aims of the Directive (Article 1), and the restoration and re-creation of wetlands is listed as a Supplementary Measure which may be implemented in River Basin Districts (Annex VI). With increasing knowledge of the economic and social benefits provided by wetland functions, it might be argued that effective wetland management is one of the most cost-effective tools for the delivery of the aims of the WFD. As an integral part of the water cycle, wetlands can contribute to sustainable catchment management through benefits such as pollution control and the protection of ground and surfacewater quality, flood regulation and control, recharge and discharge of groundwaters, maintenance of biodiversity and provision of direct as well as indirect economic benefits.

The need for catchment-wide solutions to environmental problems has been emphasised by the recent floods in several European countries, exacerbated by the functional separation of rivers from the floodplain (see also Sections 1.4.1 and 'Isolating river channels from their floodplains', in Section 1.5.2). The social and economic costs of this flooding have been considerable. This increased understanding of the role of wetlands culminated in the production of the horizontal guidance on wetlands and the WFD for member states, produced by the WFD Common Implementation Strategy Working Group on Wetlands (European Commission, 2003).

On the one hand, water management needs to be appropriate for wetland protection whilst, on the other hand, the maintenance of wetlands is an effective means to deliver WFD objectives. For wetlands to be integrated into the management of river basins under the WFD, and for the decision making process in general, innovative tools are required to provide assessments of wetland functioning within catchments and of the full social and economic valuation of these functions. The assessment procedures presented here are intended to assist a decision making framework by enabling functioning of wetlands to be taken more fully into account when making decisions regarding water management and the implementation of the WFD.

1.6.3 European Common Agricultural Policy

Under the Treaty of Rome (1957), the CAP was established in 1962 with the objective of securing food supplies to the member states through increasing agricultural production, protection of the domestic market and provision of subsidies for producers. As such it has been the largest influence on agricultural land-use for all member states since their accession to the European Union. The subsequent intensification in land management not only produced agricultural surpluses and associated costs, but led to environmental degradation.

In July 1992, reforms were carried out to the CAP, including artificial constraints and price cuts to deal with over-production, incentives (agri-environment schemes) for environmentally-beneficial forms of farming, and reductions in support for environmentally-damaging investments such as afforestation of bogs. More recently, the Agenda 2000 Agreement (http://ec.europa.eu/agenda2000/index_en.htm), agreed at the Berlin Council in 1999, included the requirement for member states to produce Rural Development Plans for the period 2000–2006, which should include the provision of rural development and agri-environment support to meet identifiable needs. This aspect of the CAP reform aimed to develop an integrated EU policy shift from production support to payments that promote and enhance environmental and rural development measures. While the primary objectives remain food quality, safety and competitiveness, protection of the environment is viewed as a non-negotiable aspect of quality. In July 2002, the European Commission, while respecting the objectives of Agenda 2000, tabled a mid-term review of the CAP, expressing concerns for improvements, among other things, in food quality and the protection of the environment including landscapes and cultural heritage. Government responses are focusing on the whole farm approach, and are now incorporating environmental management into farmers' business planning.

With specific regard to wetlands, a major consequence of implementing the original CAP objectives has been the widespread drainage and reclamation of wetlands to increase agricultural production. Because of the extensive impact of this, most substantive wetlands now remaining are likely to be those for which a nature conservation designation has been established, i.e. 'the jewels in the crown'. Wetlands lacking designation are often unprotected by the planning process; for example, where an Environmental Impact Assessment may be required, consultation will provide information only for designated sites. The drive for agricultural production has led to wetland loss and degradation, notably in areas where water resources are in limited supply and where decisions have favoured agricultural production at the expense of wildlife and habitat conservation. This is particularly evident in the more arid regions of southern Europe.

Though the management of wetlands tends to be targeted towards biodiversity objectives, other issues can often be addressed, such as the *in situ* preservation of organic archaeological remains, a water quality problem or a flood storage requirement. However, since optimal requirements for different target objectives may be more or less mutually exclusive, it is important to be able to identify and prioritise potential benefits. This requires multidisciplinary collaboration among key elements of the policy, research and end-user communities.

The importance of having flexibility in the depth and breadth of agri-environment schemes is now recognised. Whereas to date the schemes available (deep and narrow) have been targeted to achieve specific environmental objectives, often within relatively small areas of land, new entry schemes (broad and shallow) are being piloted to enable a much more extensive uptake by farmers and landowners, in order to deliver overall more general environmental benefits. In the UK, for example, the findings from selected pilot areas (to represent arable/horticultural, mixed, pastoral and upland farming regimes) are currently under review. Rewarding farmers and landowners who engage in positive countryside management not only addresses the questions of declining biodiversity, but can also help to retain distinctive landscape features, protect soil, water and air from degradation and safeguard archaeological sites.

Contributions of farming to the landscape are often currently not rewarded in the marketplace. For example, farmers and landowners do not receive compensatory payments for managing wetlands as washlands (allowing riparian areas to flood naturally in order to reduce flood risk to properties and premises), or buffer zones (to reduce the impacts of diffuse pollution on aquatic ecosystems), which deliver not only environmental but also social and economic benefits. With specific regard to the future role of washlands, given the generally accepted view that climate change is likely to lead to increased risks of flooding, serious consideration is being given to developing a policy framework that encourages appropriate washland management to minimise flood risk while at the same time delivering a wider range of environmental and public benefits (English Nature, 2001).

Surveys by many organisations across Europe frequently indicate that rural environments are valued by the public, and that wetlands often comprise a key element in the local landscape, supporting distinctive and specialised habitats and species, which have become increasingly rare following wetland loss and degradation. Supporting farmers can achieve many important objectives, including regeneration of threatened landscapes or habitats, attracting tourism, improving the quality of life for rural communities and encouraging traditional forms of land management, which can maintain traditional rural skills and employment.

The CAP and related environmental policies of the EU will have a critical role to play in the protection and rehabilitation of many ecosystems, including wetlands, in new member states of the EU, and those seeking membership in the future. An example of important wetlands here are the alluvial meadows of central Europe, such as the Lužnice River floodplain in the Czech Republic, where research has indicated ease of restoration after widespread abandonment of traditional mowing regimes (Prach and Straškrabová, 1996). However, without economic incentives, local farmers are unlikely to be persuaded to adopt appropriate management to reinstate these productive and ecologically important habitats.

1.6.4 The North American experience

In the USA, wetlands are subject to numerous Federal and State laws, the most important being those associated with the '404 Program', authorised under Section 404 of the Clean Water Act (33 US Code 1344). This requires the Army Corps of Engineers to regulate dredging and filling of 'waters' through a permit system. Although wetlands are not mentioned in Section 404, legal precedent has demanded their inclusion in the definition of 'waters' (Smith, 2007). An additional significant step was taken in 1988 when the report of the National Wetland Policy Forum was published (Conservation Foundation, 1988). This stated that there should be 'no overall net loss of the nation's remaining wetlands'. This principle has been accepted by all government agencies and has become a major factor in the protection and restoration of wetlands.

Wetland functions are singled out for particular attention during permit review. Section 320.4 (4) of the above legislation states: 'No permit will be granted which involves the alteration of wetlands identified as important by paragraph (b)(2)...' which states:

> Wetlands considered to perform functions important to the public interest include:
>
> (i) Wetlands which serve significant natural biological functions, including food chain production, general habitat and nesting, spawning, rearing and resting sites for aquatic or land species;
> (ii) Wetlands set aside for study of the aquatic environment or as sanctuaries or refuges;
> (iii) Wetlands the destruction or alteration of which would affect detrimentally natural drainage characteristics, sedimentation patterns, salinity distribution, flushing characteristics, current patterns, or other environmental characteristics;
> (iv) Wetlands which are significant in shielding other areas from wave action, erosion, or storm damage. Such wetlands are often associated with barrier beaches, islands, reefs and bars;
> (v) Wetlands which serve as valuable storage areas for storm and flood waters;
> (vi) Wetlands which are ground water discharge areas that maintain minimum baseflows important to aquatic resources and those which are prime natural recharge areas;
> (vii) Wetlands which serve significant water purification functions; and
> (viii) Wetlands which are unique in nature or scarce in quantity to the region or local area.

The need to assess wetlands in the context of the 404 Program, coupled with the 'no net wetland loss' policy, has stimulated the progressive development of functional assessment methodologies in the USA. In particular, it has led to progression of a hydrogeomorphic (HGM) approach, discussed further in Section 1.8.2 (Smith *et al.*, 1995; Brinson, 1996; Brinson in press), which parallels the methodology developed for use in Europe, but meets the requirements of the regulatory framework of the USA.

Canada supports 24 % of the world's wetlands, covering an area of over 127 million ha (Rubec, 1994). Despite this, approximately 20 million ha of the original resource has been lost, most significantly in the heavily populated parts of the country, such as southern Ontario, where there has been a decline of 68 %. The Federal Policy on Wetland Conservation (FPWC), adopted in 1992, has spear-headed Canadian policy on wetlands. This policy (http://www.wetlands.ca/wetcentre/wetcanada/federal-policy/fedpol.html) was conceived to deliver wetland protection and meet the Canadian commitments to the Ramsar Convention and other international agreements. It has adopted the Ramsar Convention principle of *wise use* of wetlands, together with the concept of *no net loss*. The FPWC applies to the 29 % of Canadian wetlands under direct federal control, and has some influence on the remainder, which are the responsibility of provincial governments, organisations or private individuals.

1.6.5 Climate change

It is now generally accepted that climate change is occurring beyond the range of expected natural variation (Mann *et al.*, 1998). The evidence for this has recently been reviewed by the Intergovernmental Panel on Climate Change (Houghton *et al.*, 2001; IPCC, 2007). Since 1900, global mean air temperature has increased by 0.6 °C and future changes are likely to be more rapid. Climate models predict a temperature rise in Europe of 2.0–6.3 °C by the end of the current century (EEA, 2004), with pronounced warming in the south, and warmer summers and winters in the north. Patterns of precipitation are also expected to change, with rainfall reduced in the south but increased in the north. Furthermore, extreme events, such as floods, droughts and storms, are likely to become more common.

Although there is some uncertainty regarding the pattern of these changes, it is recognised that they will affect the hydrology, biogeochemistry and ecology of water bodies and wetlands (Schindler, 1997; Murdoch *et al.*, 2000), which are intimately linked within catchments, as recognised in the WFD. These changes will present challenges for wetland decision makers and managers by altering the distribution, nature and functioning of wetlands in Europe.

1.6.6 *Societal change*

In addition to climate change and the developments in environmental policy, changes in human society are also likely to impact wetlands and their management. The way in which society views wetlands is changing, as wetland values become more widely understood and are seen as having a role in addressing a range of environmental issues. There may also be more indirect pressures on wetlands resulting from societal changes, such as movements of population, increasing *per capita* water demand and development of infrastructure, often focused on specific geographical areas such as south-east England and parts of the Mediterranean coasts of France and Spain. Societal changes may also have impacts on wetlands through changes in agricultural practice, by preferences for particular crops or methods of cultivation (e.g. organic production). The relationships among societal choices, policy and climate change are complex, but all can influence the future management of wetlands.

1.7 What is a functional approach?

1.7.1 *Background*

The historical view of wetlands has been dominated largely by sectoral viewpoints, for example the need to drain them to enable agriculture development or the need to retain their water-dominated character to conserve particular species and habitats represent the extremes of a wide range of common perspectives. The traditional conservation approach has been characterised by features such as rarity, uniqueness and representativeness of the organisms and habitats found in wetlands. It has relied largely on the establishment of protected area networks to safeguard at least a proportion of the total wetland resource. There has been increasing recognition, however, of the need to strengthen this more traditional conservation-based approach in ways that recognise the wider importance of the functioning of wetlands. Perversely, the European wetland resource has been and continues to be at the mercy of decision making that takes little or no account of these functional dimensions. Specific individual and societal interests in economic welfare, employment and other aspects of social wellbeing have often, but misleadingly, been viewed as contradictory to wetland conservation.

1.7.2 *Why a functional methodology?*

A functional approach to wetland assessment is one that acknowledges that wetlands can perform work at a variety of scales in the landscape, which may result in significant direct and indirect benefits to people, wildlife and the environment. However, it should also be acknowledged that there may be disbenefits for human society, such as impacts on health from disease vectors, and restriction on mobility, access and certain economic activities. Adopting a functional approach to wetland evaluation broadens the consideration of wetlands from one of solely viewing them as icons of conservation with an ecological emphasis, to one which attempts to interpret and assess the dynamics of the wetland as a complex and often functionally varied ecosystem in both space and time.

Some key operational challenges for the FAPs are:

- making better decisions for the long-term benefit of society concerning wetlands outside the protected area network;
- avoiding the often prohibitive costs and time constraints of empirical research;
- preventing gross misinterpretation of wetland functioning based on generalisations;
- recognising that different parts of a single wetland may function differently;
- supporting practical wetland restoration targets and priorities;
- translating fundamental scientific understanding into transferable knowledge.

There is still very limited awareness of the nature and significance of wetland functioning in civil society, including among the various policy makers and resource managers whose decisions may have far-reaching effects on wetlands. Recent research has done much to alert the scientific community to the multiple functions performed by wetlands and the resulting values that can be attributed to wetland ecosystems. Nevertheless, the variation among and within ecosystems demands the development of innovative tools for wetland functional assessment, so that the retention of wetlands and their management needs can more effectively compete with the other land-use and sectoral demands of society. It is argued that public support for wetlands is more likely if, in addition to their traditionally recognised role as habitat for wildlife and rare, endangered or threatened species, they can also be demonstrated to be key contributors to sustainable economic development and quality of human life. Thus their roles in water quality, flood control, groundwater protection and support of economically important resources such as fisheries or grazing land may be more appealing to individuals less concerned with a commitment to biodiversity than the immediate needs of day-to-day living. This balance of priorities also tends to be reflected in the political process and associated decision making.

The argument for adopting a functional approach to wetland assessment is based on:

1. increasing evidence of the significant contribution wetlands can make to wider socio-economic values, which arise from their functioning;
2. more convincing arguments, which can be considered by decision makers (such as planning authorities, public inquiries, and stakeholder representatives) faced with the apparent conflict between economic and ecological needs;
3. the limited financial resources available for inclusion of anything more than the most outstanding wetland sites within protected areas networks (the 'jewels in the crown' of the overall resource);
4. an intimate and often interdependent relationship between wetlands, rural activities, especially agriculture, and water resource management throughout Europe;
5. the fact that not all wetlands perform the same functions or any single function to the same extent, which makes essential a methodology enabling discrimination among and within wetlands;
6. an implicit need for understanding and managing wetlands according to functional criteria in order to effectively implement the new European WFD and the accommodation of related policy areas such as agriculture, fisheries and biodiversity.

By way of comparison, the approach to wetland assessment in the USA is conditioned by distinct programmes regulating wetlands at both the State and Federal levels (Bartoldus, 1999). The methods do not easily translate into the European context for three reasons:

1. There is no comparable regulatory programme in Europe and, notwithstanding the new WFD, which makes provision for protected habitats, wetlands *per se* are not subject to specific planning or other controls.

2. There is a strong level of human intervention through direct management in Europe, which tends to be reduced or absent in the USA.
3. The extreme regional and geographical diversity of wetlands, and the inadequacy of the science base, limit our understanding of how they function (Maltby *et al.*, 1994; Bullock and Acreman, 2003).

1.7.3 *The functional approach and a changing policy environment*

The WFD is the most substantial piece of European water legislation to date. It requires all inland (surface- and groundwaters), transitional and coastal waters to reach 'good ecological status' by 2015. This is to be achieved through the establishment of integrated river basin management plans with the aim of meeting environmental objectives, including ecological targets for surface waters. Table 1.7.1 indicates how the functions assessed by the FAPs link to specific parts of the WFD. The full text of the WFD can be found at: http://europa.eu.int/comm/environment/water/water-framework/index_en.html).

1.8 Introduction to the functional assessment procedures (FAPs)

1.8.1 *Underlying rationale*

It is recognised that wetlands are capable of performing functions of varied significance to society as a result of interactions among natural processes and ecosystem structure. It should be noted, however, that not all wetlands perform the same functions, nor is any one function carried out to the same extent in different wetlands.

A method of evaluation based on the functions a wetland is likely to perform can assist in making appropriate decisions on the management and protection of wetland sites. Wetland functions may yield benefits to local or more distant human populations. The detailed evaluation of functions can be a time-consuming and resource-intensive activity, but the pressures on wetlands are such that management decisions need to be made within a much shorter timescale. The FAPs can be used by experts and non-experts alike to assess wetland functions relatively rapidly. Primarily, this is achieved by the identification of key elements or predictors, which can be related to functions without the need for detailed empirical research or monitoring studies.

1.8.2 *Scope of the FAPs*

The FAPs currently address two broad groups of wetlands: river marginal and lake marginal wetlands. Initially, work focused on river marginal systems because of their importance as biological corridors, the possibilities for restoration under EU policy, their high potential environmental benefits and high historical rates of loss. This was expanded subsequently to include lake marginal wetlands. Further work will be required to incorporate estuarine and other wetlands into the FAPs.

River marginal wetlands include all currently active floodplains and historical floodplain ecosystems in which inundation by surfacewater or the prevalence of an elevated water table are regular occurrences. Also included may be adjacent valley slopes, which are regularly affected by overland flow or groundwater seepage, or otherwise have an elevated water table. Waterlogging normally should be of sufficient duration to support hydrophytic vegetation and enable the development of hydromorphic soils. There may or may not be hydrological connections between the floodplain and the valley slopes.

Lake marginal wetlands occupy fringing areas or topographic depressions adjacent to bodies of standing water, which may be permanent or temporary. Inundation from the water body or the presence of an elevated water table are characteristic features. There may be active hydrological connection with the water body, or this may have become disrupted or severed by human actions. Spatial variability at a variety of scales is important in controlling functional performance of a wetland. Some examples are given in Table 1.8.1.

Table 1.7.1 Wetland functions relevant to delivery of the objectives of the Water Framework Directive

Function	Article 1 – Purpose	Article 4 – Environmental Objectives
Floodwater detention	"mitigating the effects of floods and droughts" "water needs, (of) terrestrial and wetlands"	1ai "to prevent deterioration ... of all bodies of surface water" 1aii "protect, enhance and restore all bodies of surface water" 1aiii "to protect and enhance all artificial and heavily modified bodies of water"
Groundwater recharge	"mitigating the effects of floods and droughts" "water needs, (of) terrestrial and wetlands"	1ai "to prevent deterioration ... of all bodies of surface water" 1aii "protect, enhance and restore all bodies of surface water" 1aiii "to protect and enhance all artificial and heavily modified bodies of water" 1bii "balance between recharge and discharge of groundwater"
Groundwater discharge	"mitigating the effects of floods and droughts" "water needs, (of) terrestrial and wetlands"	1ai "to prevent deterioration ... of all bodies of surface water" 1aii "protect, enhance and restore all bodies of surface water" 1aiii "to protect and enhance all artificial and heavily modified bodies of water" 1bii "balance between recharge and discharge of groundwater"
Sediment retention	"protects and enhances the status of aquatic systems"	1ai "to prevent deterioration ... of all bodies of surface water" 1aii "protect, enhance and restore all bodies of surface water" 1aiii "to protect and enhance all artificial and heavily modified bodies of water"
Nutrient retention	"protects and enhances the status of aquatic systems" "reduction of pollution of groundwater"	1ai "to prevent deterioration ... of all bodies of surface water" 1aii "protect, enhance and restore all bodies of surface water" 1aiii "to protect and enhance all artificial and heavily modified bodies of water" 1bi "limit the inputs of pollutants into groundwater and prevent the deterioration ... of all bodies of groundwater" 1bii "enhance and restore all bodies of groundwater"
Nutrient export	"protects and enhances the status of aquatic systems" "reduction of pollution of groundwater"	1ai "to prevent deterioration ... of all bodies of surface water" 1aii "protect, enhance and restore all bodies of surface water" 1aiii "to protect and enhance all artificial and heavily modified bodies of water" 1bi "limit the inputs of pollutants into groundwater and prevent the deterioration ... of all bodies of groundwater" 1bii "enhance and restore all bodies of groundwater"
In-situ carbon retention	"protects and enhances the status of aquatic systems" "reduction of pollution of groundwater"	1ai "to prevent deterioration ... of all bodies of surface water" 1aii "protect, enhance and restore all bodies of surface water" 1aiii "to protect and enhance all artificial and heavily modified bodies of water" 1bi "limit the inputs of pollutants into groundwater and prevent the deterioration ... of all bodies of groundwater" 1bii "enhance and restore all bodies of groundwater"

Table 1.7.1 cont'd

Function	Article 1 – Purpose	Article 4 – Environmental Objectives
Trace element storage	"protects and enhances the status of aquatic systems" "reduction of pollution of groundwater"	1ai "to prevent deterioration ... of all bodies of surface water" 1aii "protect, enhance and restore all bodies of surface water" 1aiii "to protect and enhance all artificial and heavily modified bodies of water" 1bi "limit the inputs of pollutants into groundwater and prevent the deterioration ... of all bodies of groundwater" 1bii "enhance and restore all bodies of groundwater"
Organic carbon concentration control	"protects and enhances the status of aquatic systems" "reduction of pollution of groundwater"	1ai "to prevent deterioration ... of all bodies of surface water" 1aii "protect, enhance and restore all bodies of surface water" 1aiii "to protect and enhance all artificial and heavily modified bodies of water" 1bi "limit the inputs of pollutants into groundwater and prevent the deterioration ... of all bodies of groundwater" 1bii "enhance and restore all bodies of groundwater"
Ecosystem maintenance	"protects and enhances the status of aquatic systems"	1ai "to prevent deterioration ... of all bodies of surface water" 1aii "protect, enhance and restore all bodies of surface water" 1aiii "to protect and enhance all artificial and heavily modified bodies of water"
Food-web support	"protects and enhances the status of aquatic systems"	1ai "to prevent deterioration ... of all bodies of surface water" 1aii "protect, enhance and restore all bodies of surface water" 1aiii "to protect and enhance all artificial and heavily modified bodies of water"

The fundamental functional unit

Within the FAPs, the fundamental unit of land on which functional assessment is performed is a sub-area of the wetland called a hydrogeomorphic unit (HGMU). The HGMU approach was developed from an underlying hydrogeomorphic concept and application in the USA (Brinson, 1993), and further developed for specific application in Europe (Maltby *et al.*, 1994). The FAPs rely on the successful identification and delineation of HGMUs, which are defined as:

> Areas of homogeneous geomorphology, hydrology and/or hydrogeology, and under normal conditions, homogeneous soil/sediment (Maltby *et al.*, 1998a).

Vegetation is not used as a defining characteristic owing to the significant influence of historic and current land-use, particularly in Europe. However, vegetation is described for each HGMU as part of the procedures, and can give clues to the hydrological and geomorphological dynamics.

Previous attempts to develop tools for the rapid assessment of wetland functioning have tended to treat a wetland site as a single functional unit. Pioneering work on assessing wetland functions and values in the USA (e.g. Adamus, 1983; Adamus and Stockwell, 1983; Lonard and Clairain, 1985) led to the development of the Wetland Evaluation Technique (WET) (Adamus *et al.*, 1987), which was based fundamentally on the assessment of an 'entire' wetland area. However, it was recognised that many wetlands were too complex to be evaluated as a single unit, and so guidelines were published for the evaluation of multiple 'assessment areas' (Adamus *et al.*, 1987). These were normally envisaged as necessary for 'larger' wetlands and were spatially distinct rather than forming more or less subtly

Table 1.8.1 Processes and functions associated with properties of the wetland landscape

Landscape properties	Examples of possible process controls	Examples of functional significance
Geology and landform (geomorphology)	Discharging groundwater affects waterlogging and nutrient pathways, which control denitrification, mineralisation and adsorption; it also controls the type and distribution of vegetation, which may form a mosaic of types	Discharge of groundwater, removal of nutrients by retention and export; habitat diversity optimised by complexity of vegetation mosaic
Floodplains and their relationship to river dynamics; microrelief	Floodplain configuration and river discharge affect the pattern and duration of flooding and soil water regimes	Floodwater retained on the floodplain reduces peak discharge and consequent flood risk; sediment, nutrient and pollutant removal from floodwaters by storage on the floodplain; detention of floodwater enables nutrient export (primarily N-removal by denitrification) to take place
Surface water flows	Flows of water determine areas where nutrient transformations are possible	Even small areas can be very effective in improving water quality; important information for selecting locations for buffer zones
Soil variability; particle-size often reflects position on the floodplain or slope	Particle-size class and organic matter content affect nutrient and pollutant adsorption and release	Soil–water interactions influence resulting water quality, retention of nutrients and contaminants
Small-scale variations in hydrology and vegetation	Distribution of plants, habitats for invertebrates, distribution and type of food supply and consequent feeding activities of birds	Maintenance of biodiversity
Management impacts (e.g. grazing, burning, drainage, cultivation)	Most rapidly affects vegetation and hydrology, though soil properties can also be changed; many fundamental effects on hydrological, biogeochemical and ecological processes	Appearance of a wetland can change rapidly and substantially; changes in structure can enhance, reduce or have little effect on functioning

different parts of a single complex system. Outputs as simple ratings were recognised as being inadequate alternatives to quantitative or more discriminatory methods required in decision making (Adamus *et al.*, 1991). In the 1990s, in response to the perceived shortcomings of WET, Smith *et al.* (1995) presented an approach based on a HGM classification of wetlands. Brinson (1996) elaborated the advantages of the HGM approach based on a standard, hydrologically-biased classification, the use of 'reference' wetlands and the distinction between wetland functions and values (often previously lumped together). The methodology, however, is based on the assessment of 'whole' wetlands, rather than subdivisions with spatial and temporal variations in functional behaviour. This has been necessary partly because of the limited extent of precise scientific information about process variation at smaller scales, and partly because of the limited need for application at any other scale. There has been considerable sharing between researchers in the USA and Europe (Brinson, 1996) of the rationale and logic of the HGM approach to wetland functional assessment. The methodologies evolved should be viewed as complementary, and individually or collectively appropriate to their particular contexts and purposes of application.

While the wetland assessment techniques developed in the USA have been driven largely by the requirements of legislation, such as the US Federal Clean Water Act, and the need to avoid, minimise or mitigate the adverse effects of road construction (Larson and Mazzarese, 1994), in Europe, the primary emphasis has, until recently, been on biodiversity and conservation. Legislation has required assessment of species and habitat but not of particular wetland functional characteristics or their variation. Most European assessment methods associated with wetlands focus on the river channel, though some techniques also include the adjacent riparian areas and floodplain. Methods such as the River Habitat Survey and the System for Evaluating Rivers for Conservation (Boon *et al.*, 1997; 1998; Raven *et al.*, 1998) require only broad categorisation of floodplain features, while other approaches such as the Riparian Channel and Environmental Inventory (Petersen, 1992) and the Riparian Forest Quality (Qualitat del Bosc de Ribera) (Munné *et al.*, 2003) do not extend beyond riparian areas, and exclude wetland types such as lake margins or peatlands. Techniques available tend to focus on either species and habitat integrity, or the condition of the river channel and its banks, rather than hydrological, biogeochemical and ecological functioning. Yet it is clear from increasing scientific knowledge that functional properties may vary considerably within single contiguous wetlands. Better understanding of such variation may lead to improved management and more satisfactory resolution of conflicts, offering options for alternative uses and addressing development pressures. Such understanding has been sought by adopting a distinctively European-style HGM approach to wetland analysis (Maltby *et al.*, 1994).

1.8.3 Development of the functional assessment concept

Previously the development of sufficiently robust, science-based tools for wetland functional assessment has been impeded by the lack of integration of the wide range of specialised scientific and geographical perspectives necessary. However, the sequence of EC research projects investigating wetland functioning in Europe, summarised below, was targeted largely to overcome this impediment and enable production of the FAPs to take place. Figure 1.8.1 shows the range of studies contributing to the development of the procedures.

Four major research projects (FAEWE phase I, 1991–4; FAEWE phase II 1994–6; PROTOWET 1996–9; and EVALUWET 2001–4; see Acknowledgements), which used field and laboratory studies, were carried out by a variety of European institutes. Additionally, the theoretical and desk-based TECWET (2001–4) further developed the procedures. A reference list of published outputs is given in Appendix 4. Details of these projects are provided by the EU Community Research and Development Information Service (CORDIS) via the website www.cordis.europa.eu/en/home.html. Together, the projects explored and refined the functional analysis approach:

- FAEWE (phase I, 1991–4; **F**unctional **A**nalysis of **E**uropean **W**etland **E**cosystems). This field- and laboratory-based study identified and measured key processes operating in river marginal wetland ecosystems, assessed the effects on functioning of alterations to wetland ecosystem processes, and began development of the system of wetland functional analysis.
- FAEWE (phase II, 1994–6) improved the predictive capacity of the functional assessment procedures, developed socio-economic evaluations and investigated the extension of their application to the catchment scale.
- PROTOWET (1996–9; **Pr**ocedures for the **O**perationalisation of **T**echniques for the Functional Analysis of European **W**etland **E**cosys**t**ems) extended the procedures for application to estuarine and lake marginal wetlands and for

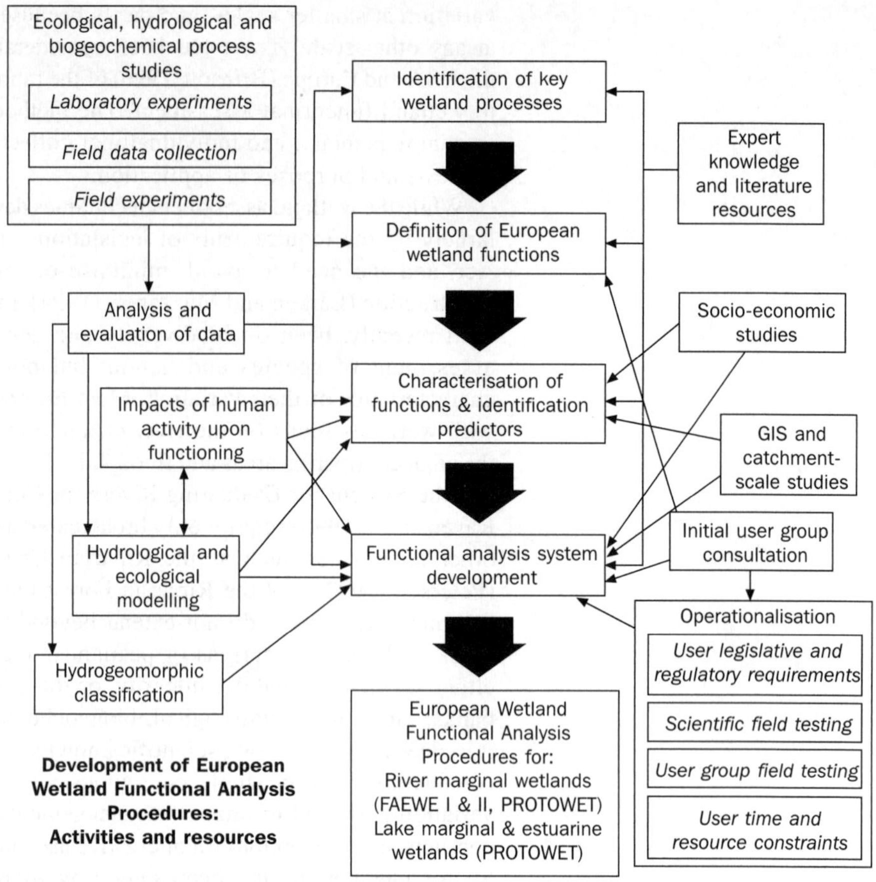

1.8.1 Activities and resources used in the development of the European Wetland Functional Analysis Programme.

high latitude and continental systems. It also developed a generic HGM classification of European wetlands.

- EVALUWET (2001–4; **E**uropean **valu**ation and assessment tools supporting **wet**land ecosystem legislation) linked the procedures with European legislation (particularly the WFD) and developed the concept of a Wetland Evaluation Decision Support System (WEDSS), which integrates functional assessment and socio-economic valuation within a geographical information system (GIS) mapping system.
- TECWET (2001–4; **Tec**hniques and Procedures for the Functional Analysis of **Wet**land Ecosystems) was a refining project, which improved the user-friendliness of the procedures, introduced quantitative and semi-quantitative outputs, developed a computer-based version and ensured compatibility with current European legislation. This document is the final output of the four major research programmes, based on the further development and testing done in TECWET.

1.8.4 The study sites

Details of the study sites used to support development of the FAPs are shown in Table 1.8.2. They were chosen to represent a broad range of climatic, geological, soil and vegetation types, together with large variations in hydrological regime. One key focus for study had been to examine the impact on functioning of human disturbance. To address this, sites in FAEWE 1 were selected in contrasting pairs, with and without the following impacts – silt and peat sedimentation, past agricultural reclamation, river regulation and water abstraction (Maltby *et al.*, 1996). Later stages expanded the studies of impacts to include the creation of polders, afforestation and drainage, grazing, nutrient enrichment and the presence of heavy metals. The sites also provided initial testing locations for the development of the FAPs.

Table 1.8.2 Details of study sites used in FAPs development

Name and location	Project	Description	Purpose of study
River Torridge/ Walden, UK	FAEWE	River marginal wetlands (wet grassland) on floodplains and footslopes	Nutrient enrichment impact
River Shannon/ Little Brosna, Ireland	FAEWE	River marginal wetlands (wet grassland – callows) on floodplains	Sediment deposition impact
River Loire/ Allier, France	FAEWE	River marginal wetlands (dry and wet grassland, scrub) on floodplains	River flow regulation impact
River Guadiana headwaters, Spain	FAEWE	River marginal wetlands (sedge and reed beds and saltmarsh)	Groundwater abstraction impact
River Danube, Romania	FAEWE II	River marginal wetlands (floodplain grassland and forest) and polders (rice)	Flooding regimes and polders impacts
Lac de Grande Lieu, France	PROTOWET	Lake marginal wetlands (reed beds and humid grassland)	Grazing impact
Tour du Vala, France	PROTOWET	Swamps, wet grassland and lakes of the Camargue	Grazing impact
River Svartberget headwaters, Sweden	PROTOWET	River marginal wetlands (coniferous forest)	Afforestation and drainage impacts
Oostvaarder-plassen, The Netherlands	PROTOWET	Lake marginal wetlands (reed beds)	Grazing and nutrient enrichment impacts
River Mulde/Elbe, Germany	PROTOWET	River marginal wetlands (floodplain grassland, reed and sedge beds)	Heavy metals impact
River Tamar/ Carey, UK (Tamar catchment)	EVALUWET	River marginal wetlands (floodplain wet grassland and footslope seepage zones of wet grassland and woodland)	FAP testing
Waterland catchment, The Netherlands	EVALUWET	Fen meadows, lakes	FAP testing
Hovran catchment, Sweden	EVALUWET	Wet pasture, forests and lakes	FAP testing
Paríž Creek catchment, Slovakia	EVALUWET	Reed beds, wet grassland	FAP testing
Neajlov catchment, Romania	EVALUWET	Marshes, lakes and artificial ponds	FAP testing
Lake Cheimaditida catchment, Greece	EVALUWET	Lake marginal reed beds, wet meadows and calcareous fens	FAP testing
Elbe catchment, Germany	EVALUWET	Floodplain hardwood forest, floodplain wet grassland, oxbow lakes with marginal swamps	FAP testing

1.8.5 *Application of the FAPs*

The FAPs are designed to be used by both experts and non-experts in wetlands, and mostly all that is required is a general knowledge of the natural environment, a few basic pieces of equipment and time dedicated to carrying out the assessment. Users are guided through the procedures in a step-by-step way, and unfamiliar skills and techniques can be learnt along the way. In just a few places, an indication is given of where more specialist advice or expertise may be required.

It is anticipated that, primarily, the FAPs will be used by a range of individuals or organisations concerned with wetland management who wish to gain a better understanding of the processes, functions, benefits and potential of the wetlands for which they have responsibility or interests. The FAPs can assist with on-site management and decision making by providing better information on how a wetland is functioning, and the key processes operating within it. These types of user are most likely to be statutory government agencies and NGOs such as wildlife trusts. Private landowners and those guiding decision makers will also be able to use the procedures to provide more informed decisions.

Thus, the FAPs will assist strategic planners at local and catchment scales to make decisions on wetland management and land-use through assessment of overall wetland functioning and contribution to factors such as water quality, water quantity and biodiversity. Knowledge of this functioning can be used to help with the implementation of national, international and supranational policy, such as the WFD. In strategic decision making, wetlands may be used as basic or supplementary measures for achieving policy goals. For example, decisions on the location, type, number and area of wetland buffer zones required to deliver desired benefits can be assisted by the information obtained from a functional assessment.

At the regional, national and supranational scale, the information provided by the FAPs on wetland functioning can contribute to the further development of policy by providing greater information on the importance and value of wetland functions in both environmental and economic terms. Indeed, economists can also benefit from the output of the FAPs, as the information provided enables more accurate economic valuation of wetlands, which in turn can feed into policy development. Some examples of potential users and benefits are given in Table 1.8.3.

Across Europe the package has been developed in consultation with potential users of the system, notably in the UK and the Netherlands, where consultative groups were established. From these consultations, the need for a flexible, modular package was identified to address the following requirements:

- for planners to resolve decision making dilemmas in relation to development control and catchment management planning;
- for guidance on appropriate conditions and the management necessary to optimise wetland functioning;
- for identification of levels of impact that alter functioning;
- where appropriate, for quantification, semi-quantification or rapid quantification of functioning;
- for provision of guidelines to assist non-experts to conduct assessments;
- for a scientific rationale for the protection and appropriate management of wetlands;
- for the identification of potential areas for restoration and the means for achieving this.

Functional assessment outcomes can be used in conjunction with other information relating to, for example, societal priorities, costings and policy limitations to assist and support relevant decision making. The FAPs translate best available scientific knowledge into reasonable predictions of how different parts of wetlands function in different landscape contexts. The complexity of individual wetland ecosystems can be taken into account by applying the FAPs at the 'field' scale for individual site management purposes. FAP outcomes are linked to socially relevant priorities such as flood control, pollution reduction and biodiversity conservation. The FAPs recognise and emphasise the 'natural capital' of healthy

Table 1.8.3 Functional assessment procedures: examples of users and benefits

Examples of anticipated users	Benefits and applications
EC environment	Assisting integrated management and conservation of wetland ecosystems in the EU through policy innovation and support of legislation, specifically the WFD
Water regulators and industry	Implementation of WFD
Planning authorities	Improving the efficiency of wetland management decision making for national or regional planning authorities and implementation of the Ramsar Convention and Convention on Biological Diversity
Government agencies	Assisting implementation of EC Directives aimed at improving environmental protection and wetland conservation
Non-government agencies	Aiding site managers and implementation of the Convention of Wetlands of International Importance (Ramsar) 'wise use' requirements
Scientific research community	Investigation of wetland functioning
Organisation for Economic Co-operation and Development	Providing assistance in the economic evaluation of wetland management options and interpreting ecosystem functioning in terms of its social and economic benefits

wetland ecosystems (Costanza *et al.*, 1997), and reinforce the fact that wetland management is a question of choice; for individuals as well as society.

It should be borne in mind, however, that the FAPs do not deal with all possible processes and functions carried out by wetlands. It is anticipated that more can be added over time as the procedures themselves are more thoroughly and extensively tested.

How should they be used?

The FAPs should be used as a first step for land managers wishing to assess the functional value of a wetland. An indication is given in Fig. 1.8.2 (see colour section) of the type of output and interpretation possible from FAPs. This takes as an example the function of nutrient removal, assessed for HGMUs in a study site in the catchment of the River Tamar, UK. The outcomes provide important information on the potential sources and likelihood of nutrient loss from identifiable parts of the wetland.

It is important to remember that management decisions should not be based on the FAPs outcomes alone, which should instead be used as a guide. This also relates to sustainability issues: the procedures assess the wetland in its present management state and conditions, which may not be stable. In these cases, further analysis is recommended using models and longer-term monitoring techniques to determine the effects on functioning of alternative environmental or socio-economic scenarios.

How do the FAPs work?

The procedures are separated into two distinct stages, the establishment of a database of site-specific characteristics and the functional assessment section (Fig. 1.8.3). They are designed around the identification and characterisation of HGMUs using field and desk-based information, and provide simple but detailed explanations of how to do this. The information is structured into a powerful database. The functional assessment process then interrogates this database and, through detailed decision trees for each function, derives an output.

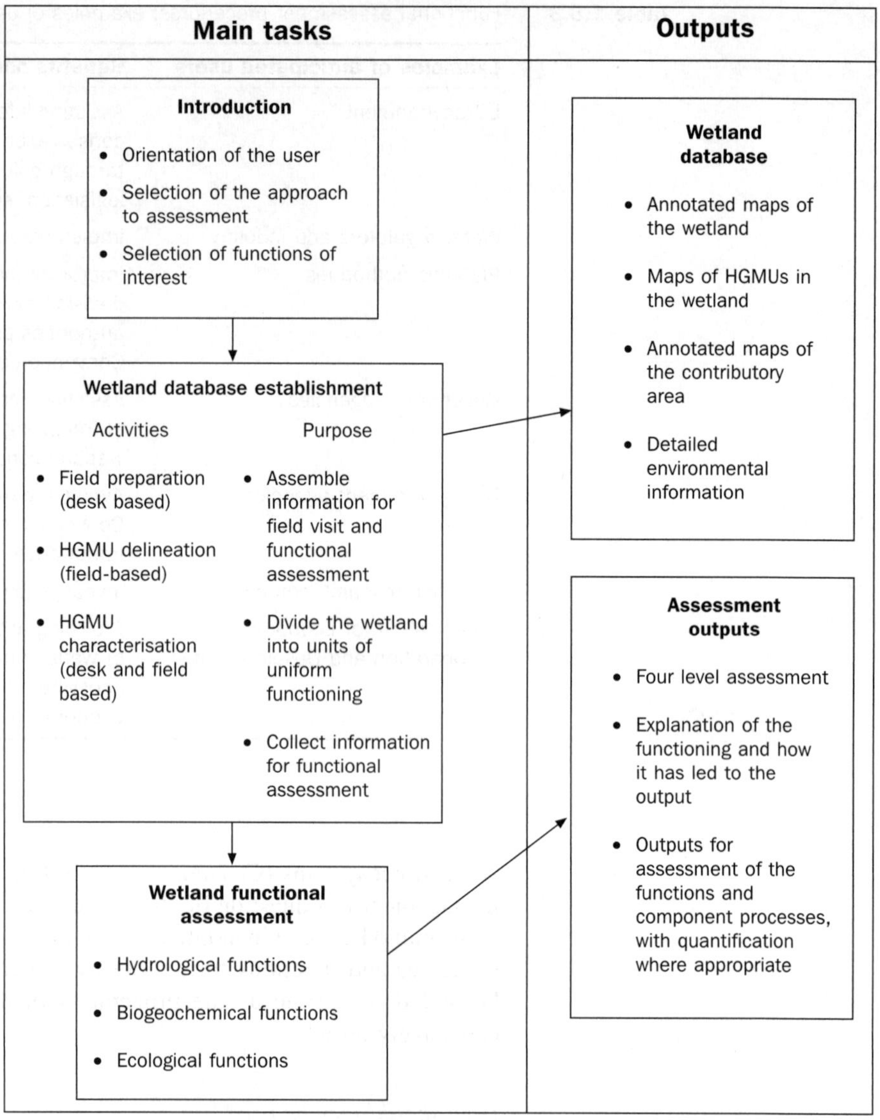

1.8.3 Basic pathway through wetland functional assessment procedures.

The database

The database is compiled within the Wetland Database Establishment section of the procedures (Fig. 1.8.3). The collection of data is within the context of HGMUs (areas of predictable functioning). Their identification and delineation are an essential step in undertaking the FAPs, and clear instructions are given to users to enable this task to be completed accurately. The data for each HGMU are obtained from field observations and archive information and, again, instructions are written to ensure that the appropriate answers are obtained.

Assessment of processes and functions

Determination of process performance is based upon the degree of operation of controlling variables (CV), and determination of function performance is based upon the performance of one or more processes underlying them. Assessment for each HGMU therefore has two main steps: determination of the occurrence and performance of relevant processes within each function; and, where applicable, the combination of groups of process outputs to give an estimate of functioning.

Assessment outcomes are reached through interrogation of the wetland database previously established, and determination of appropriate outputs from the answers gained. The user conducts the interrogation by answering questions for each process, which are structured in decision trees. The answers are combined to give information about the operation of processes in the form of a rationale or explanatory statement coupled with a code. Each output code corresponds to one of four general statements:

- The process is definitely being performed.
  Code **1**
- The process is being performed but probably only to a small degree, and there are constraining factors or uncertainties.
  Code **1/x**
- The process is not (significantly) being performed.
  Code **2**
- The process is definitely not being performed.
  Code **4**

The extent of process performance will vary according to operation of the CV that lie behind each process. This means that the process may be performed fully (100 %) or at any level below this. The FAPs accommodate this by the inclusion of a qualifying sub-code that indicates the level of performance. Where quantification of values is possible, for example in the assessment of nutrient concentrations, the qualifier is a letter (**1/a**, **1/b**, **1/c**) accompanied by an appropriate range of values. Where quantification is not possible, the qualifier is in the form **1/1**, **1/2**, ..., **1/*n***.

The first statement therefore is modified into one of two forms, i.e.

- ***The process is definitely being performed*** (quantification is often possible)
  Codes **1/a**, **1/b** or **1/c** (**1/a** being full performance, **1/b** being performance at less than maximum, and so on) plus a range of expected values;
- ***The process is definitely being performed*** (quantification is not possible but outcomes are ranked)
  Codes, **1/1**, **1/2**, ..., **1/*n*** (**1/1** being full performance, **1/2** being performance at less than maximum, and so on).

The remaining three codes are unaffected.

Some functions comprise just one process, but others result from the sum of more than one process. In these cases, the outputs from all relevant processes are combined to give an assessment output for the entire function. Additionally, where quantification is possible, an overall functional statement can be given, resulting from addition of the values estimated for individual processes. These outputs are given in the same format as those for the processes. Management decisions can then be founded upon a process-based assessment of functioning.

The rationale statement that forms part of the assessment output tells the user what is happening in the HGMU for each process; the accompanying code gives an indication of the degree of function performance within the broad definitions of the four general statements. Where a function depends on a single process, this is enough for a user to gain an understanding of current functioning. For functions that result from the combination of several processes, the rationale statements are valid for the component processes, but the codes that accompany the rationale statements must be combined in order to evaluate the degree of performance of the function as a whole. Accordingly, a functional summary stage follows each set of processes. Steps are summarised in Fig. 1.8.4.

In the electronic version of the FAPs on the enclosed CD, all function performance summaries are expressed as numerical values between 0 and 1, reflecting the degree of performance for each function. There is provision in Section 4.0 to do this manually if required. The electronic version, together with information on electronic data handling for wetland assessment, is outlined in Section 1.9.

Flexibility of approach

The approach to the assessment and evaluation of wetland functions can vary considerably depending on the target objectives of the user. If the user is interested

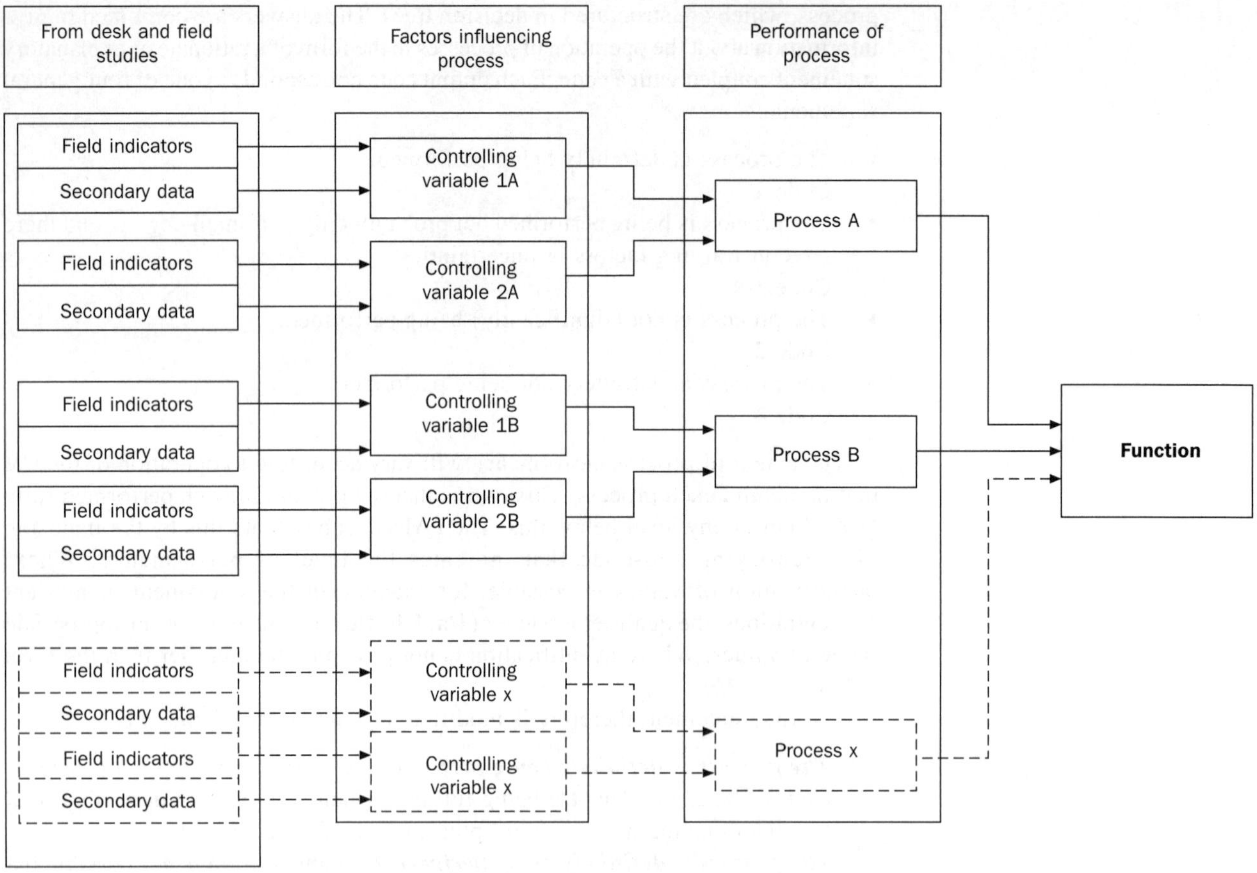

1.8.4 Procedural steps in functional assessment.

in assessing all the functions listed in the FAPs, then all sections of the wetland database establishment and functional assessment need to be completed; however, the user may wish to assess only a single function or a selection of functions. The structure of the FAPs permits this, so that only information appropriate for assessing a particular function needs to be collected. To carry out the assessment the user simply completes the wetland database establishment and functional assessment sections identified as being assigned to the selected function.

Functions are delivered through the operation of natural processes, one or more of which may be of specific interest to the user, rather than the overall function. For example, the user may wish to know how much nitrate is lost by denitrification, rather than the overall export of N from the system. The FAPs enable the identification of the appropriate process(es), and indicate to the user the need to complete the entire wetland database establishment section together with those functional assessment sections relevant to the selected processes.

1.8.6 Further development

The FAPs outcomes indicate the likelihood of a function being performed on a scale from *definitely* to *definitely not* (Section 1.8.5). However, where function performance is dependent on the effectiveness and combination of CV that drive the processes responsible for functioning, there will be differences in the strength or possibility of that functional performance, as discussed above. In these cases, it will be useful to be able to express these outcomes in terms of the degree to which the function is performed.

On-going developmental work stemming from EVALUWET (for details see Technical Implementation Plan at www.cordis.europa.eu/en/home.html) has indicated how all outcomes of the FAPs could be summarised on a single scale according to their codes. This is based on the possible outcomes of assessing individual processes and functions, using derived standardised outputs generated for a multi-criteria analysis-based model, DEFINITE (**DE**cision on a **FINITE** set of alternatives) (Janssen *et al.*, 2001). The method was developed to improve the quality of environmental decision making and has been used in designing a decision support system for wetlands, now in development, which incorporates socio-economic valuations within a GIS environment (Janssen *et al.*, 2001).

For a given HGMU, a functional value can be derived (based on Basso *et al.*, 2000) from the average of the scores allocated to the individual processes or functions ranging from 1 (for the lowest value or condition) to *n* (for the highest in a series of 1 to *n*). No value is given to processes or functions not operating or not significantly performed, as shown in an example in Table 1.8.4. This approach can be adopted independently of factors such as the number of outcome categories or absolute values. Nevertheless, the following points should be borne in mind. A zero value assigned to an unknown situation might not be a strictly correct interpretation, but would be appropriate where the process is operating at low intensity and functioning could be disregarded. The score codes also imply a linear scale, which may not be the case across the spectrum of process operation or functional outputs.

The functional value for an HGMU is the average value of the scores for all processes or functions operating. In the case of the example shown in Table 1.8.5, this would be $(4 + 2 + 2 + 3 + 1) \div 5 = 2.4$. Similarly, the values for each HGMU could be considered to give an overall functional value to the entire wetland assessment area. In this way, it is possible to indicate how effective a particular wetland can be, either as a whole or through one or more of its component elements, in delivering a wide range of environmental benefits, together with consequent social and economic benefits.

Development of additional functional assessment modules

The modular structure of the package facilitates its expansion to include additional functions to meet the requirements of other specific users such as wetland archaeologists. The heritage value of wetlands derives mainly from the ability of waterlogged anoxic environments to preserve organic and associated archaeological evidence, which is otherwise rapidly destroyed by biological decay processes under oxidised conditions when drying out takes place. A protocol is currently under development using the FAPs as a basis to enable archaeologists to assess three functions relating to cultural heritage, i.e. the preservation of buried organic archaeological remains, palaeoenvironmental evidence and palaeolandscapes.

1.9 The Functional Assessment Procedures CD-ROM

The FAPs have been translated into CD-ROM form to facilitate easier data entry and handling, and to provide automatic processing. Computerisation will facilitate updates of software and further improvements. Here, we discuss the general principles and distinctive features of the CD-ROM version. Issues specific to updated

Table 1.8.4 Example of assessment outcomes and performance codes

Assessment outcome	**1/a**	**1/b**	**1/c**	**1/x**	**2** or **4**
Score code	4	3	2	1	N/A

Table 1.8.5 Deriving functional value for an HGMU from assessment outputs

	Process 1	Process 2	Process 3	Process 4	Process 5	Process 6
FAP outcome code	**1/a**	**1/c**	**1/c**	**1/b**	**1/x**	**2**
Score	4	2	2	3	1	Not considered

versions, e.g. technical questions and installation procedures, are included in a technical note to be found on the CD-ROM.

1.9.1 *The general principles and algorithm*

Computerisation of the FAPs has two essential objectives: first, to provide a structure for management of data, i.e. entering and modifying the database or deleting from it; second, to automate the procedures, thereby skipping the appraisal sheet and assessment chart stages. In the paper version, these stages provide the user with fundamental understanding of the rationale of the procedures.

A general schematic algorithm is represented in Fig. 1.9.1. For a given process and question about the controlling variables (CV; see Section 1.8), the program selects the answer as a modality (or class) of the relevant variable within the database. Sets of answers are combined according to the requirements of the assessment table, and a functional code (score) is given to the process. The program assigns a functional score that expresses the functional performance of the HGMU (see Section 1.8) for the process or function under assessment. A detailed algorithm is given in Fig. 1.9.2, which shows the procedure for a given process (among 39;

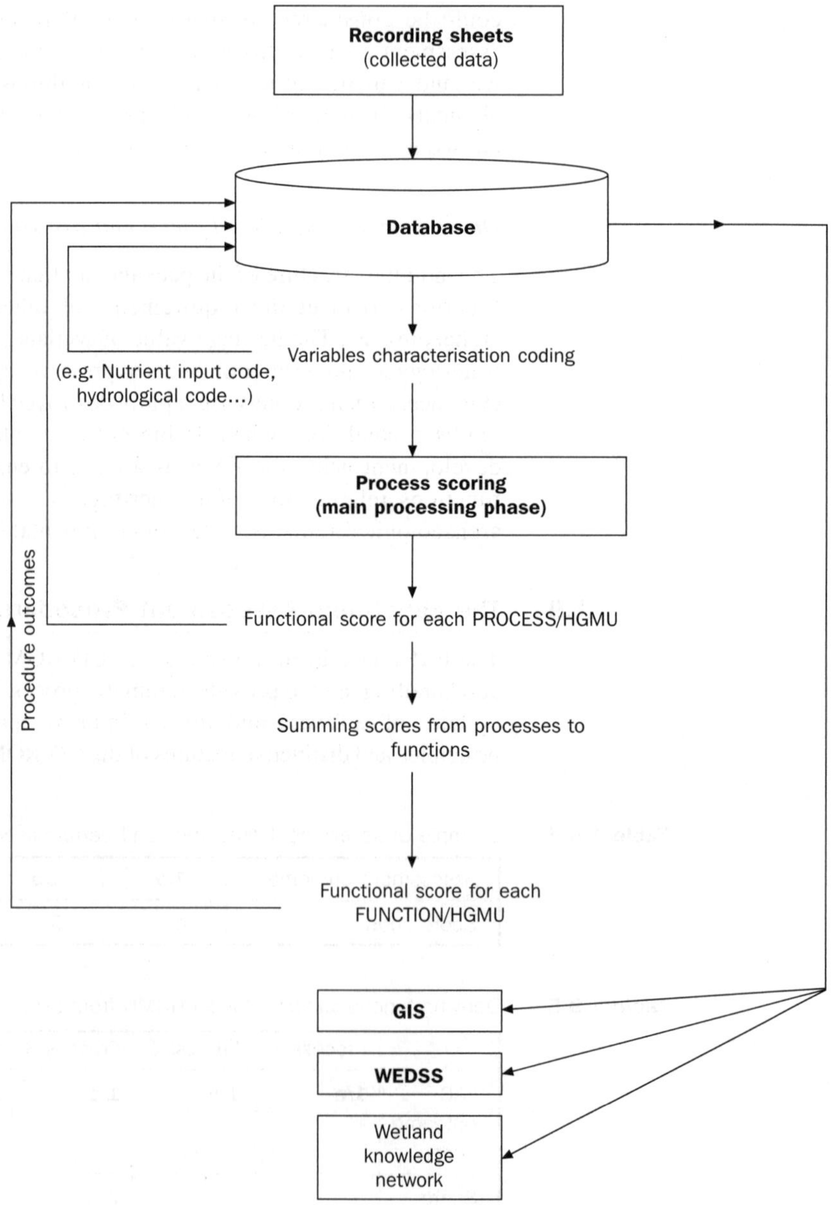

1.9.1 General algorithm of the software.

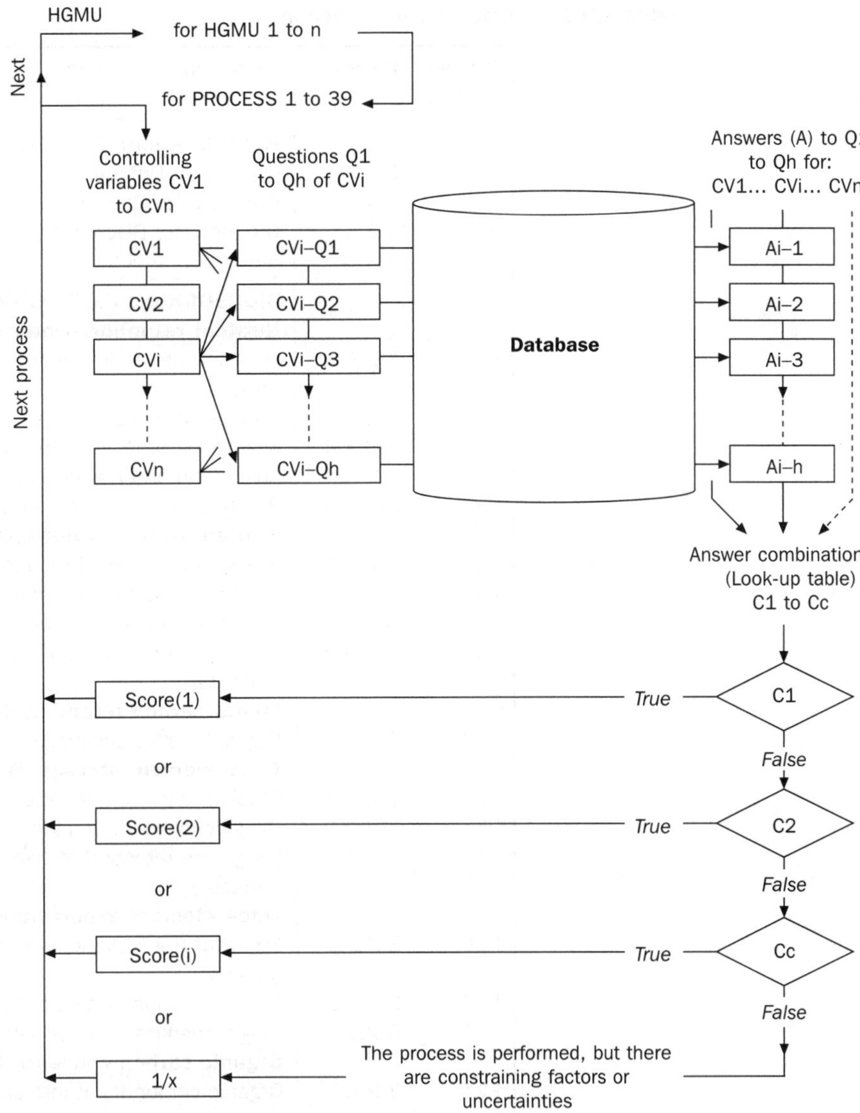

1.9.2 Scoring procedure algorithm.

see Table 1.9.1) and an HGMU among *n* HGMUs of a given assessment area (AA; see Section 2.1.A).

All outcomes are saved automatically and can be accessed if necessary for use in future developments relating to the FAPs, e.g. WEDSS, other GIS frameworks or any ecosystem knowledge network.

1.9.2 Features of the electronic database

Recorded data are entered into the database using the keyboard and, with only slight differences, using the same data order and structure as on the recording sheets (RS) in the paper version of the FAPs. In the electronic version, the 'recording sheets' (RS) of the paper version are referred to as 'recording forms' and are here abbreviated 'e-RS'.

Recording sheet 1 is divided into five forms (or sub-e-RSs) because of the quantity of data needed. In e-RS2, which corresponds to RS2a (and in some cases RS4), soil data to be entered represent only average values. No data entries are needed for RS2b–e, because the *Nutrient Input* and *Trace Element* input codes are computerised. Recording Sheet 3 is represented by one form.

There were numerous potential problems to overcome in the development of the computer program, not least the requirement for information from the HGMU under assessment, and from neighbouring areas, which may be adjacent HGMUs (i.e. within the AA) or a part of the contributory area (CA; see Section 2.1.B).

Table 1.9.1 Process and function list

Process number	Process code	Process/function name
		HYDROLOGICAL FUNCTIONS
P1	3.1.A	Floodwater Detention
P2	3.1.B	Groundwater Recharge
P3	3.1.C	Groundwater Discharge
P4	3.1.D	Sediment Retention
		BIOGEOCHEMICAL FUNCTIONS
		Nutrient retention (water quality function)
P5	3.2.A.1	Long-term retention of nutrients (N & P) through plant uptake (process)
P6	3.2.A.2	Storage of nutrients (N and P) in soil organic matter (process)
P7	3.2.A.3	Adsorption of N as ammonium (process)
P8	3.2.A.4	Adsorption and precipitation of P in the soil (process)
P9	3.2.A.5	Retention of particulate nutrients (N and P) (process)
		Nutrient export (water quality function)
P10	3.2.B.1.1	Gaseous export of N (denitrification) (process)
P11	3.2.B.1.2	Gaseous export of N (ammonia volatilisation) (process)
P12	3.2.B.2	Export of nutrients through vegetation management (process)
P13	3.2.B.3	Export of nutrients via water and wind mediated processes (process)
		***In-situ* carbon retention (function)**
P14	3.2.C.1	Organic matter accumulation
		Trace element storage (function)
P15	3.2.D.1	Physical retention of trace elements (process)
P16	3.2.D.2	Biogeochemical retention of trace elements (process)
P17	3.2.D.3	Long-term retention of trace elements through plant uptake (process)
		Trace element export (function)
P18	3.2.E.1	Export of trace elements through vegetation management (process)
P19	3.2.E.2	Physical remobilisation of trace elements (process)
P20	3.2.E.3	Biogeochemical remobilisation of trace elements (process)
		Organic carbon concentration control (function)
P21	3.2.F.1	Organic carbon input into surface waters (through leaf fall) (process)
P22	3.2.F.2	Dissolved organic input into surface waters (through soil) (process)
		ECOLOGICAL FUNCTIONS
		Ecosystem maintenance (function)
P23	3.3.A.1	Provision of overall structural habitat structural diversity
P24	3.3.A.2.1	Provision of micro-sites for macroinvertebrates
P25	3.3.A.2.2	Provision of micro-sites for fish
P26	3.3.A.2.3	Provision of micro-sites for herptiles
P27	3.3.A.2.4	Provision of micro-sites for birds
P28	3.3.A.2.5	Provision of micro-sites for mammals
P29	3.3.A.3	Provision of plant and habitat diversity
		Food-web support (function)
P30	3.3.B.1	Productivity/biomass support
		Biomass import via physical processes
P31	3.3.B.2.1	Biomass import via watercourses
P32	3.3.B.2.2	Biomass import via overland flow
P33	3.3.B.2.3	Biomass import via wind transport
P34	3.3.B.3	Biomass import via biological processes
		Biomass export via physical processes
P35	3.3.B.4.1	Biomass export via watercourses
P36	3.3.B.4.2	Biomass export via overland flow
P37	3.3.B.4.3	Biomass export via wind transport
		Biomass export via biological processes
P38	3.3.B.5.1	Biomass export via fauna
P39	3.3.B.5.2	Biomass export via anthropogenic means

This kind of information is not always contained in the recording sheets, and relies on map and data manipulation that it is not easy to computerise. To overcome these constraints, a special recording form (NAMI) containing 'Neighbouring Area and Map Information' was added to the electronic forms to accommodate complementary information.

Typically, these concern hydrological and biogeochemical processes and functions that deal with important variables upslope or downslope of the assessed HGMU, e.g.:

- Is the HGMU located in an area recorded as 'RECRE' within AA Map2? (y/n)
- Does the HGMU share a boundary with an HGMU assigned one of the following geomorphic codes: TB, FC, DT, LK, OX, BS, IL, CL or CN? (y/n)
- Is there a clear flow path from the neighbouring area ? (y/n)
- Is there a gradient of >15° in neighbouring HGMUs or CA? (y/n)
- Are there springs in upslope areas (HGMUs or CA)? (y/n)

Some variables will require compilation. Calculations (e.g. mean annual rainfall, mean annual evapotranspiration, number of different habitat codes in the AA) or procedures (e.g. nutrient input or hydrological codes) are performed by the program, and the results for the computerised variables are displayed separately.

The user's understanding of the processes and functioning occurring in the wetland will be greatly enhanced by going through the various stages of the manual FAPs, especially when new to the assessment process. Once this stage is completed, use of the electronic version will be found to be quicker, and may enable considerable flexibility in use and interpretation. This will be developed further by combination with the decision support facility offered by the WEDSS.

1.9.3 *Displays of outputs*

Once all HGMU data are recorded, the program will perform the assessments and give the user a choice of five display options. These range from a simple report that can be pasted into any word processor, to evaluations of process outputs displayed to fulfil different user requirements. For example, Figure 1.9.3 shows the 'computerised variables' screen which comprises a bar chart indicating the assessment outcomes of all processes in one screen, with rationale statements viewable by mouse selection. The comparative contributions of an HGMU to a range of ecosystem services can be seen by selecting the 'provision of ecosystem services' screen (Figure 1.9.4). In this way, the user can make comparative assessments of HGMU performance and pinpoint candidate areas for management attention. The display options are particularly useful when making comparative assessments of future management options (see Section 1.10).

1.10 The application of functional analysis

1.10.1 *Overall objectives*

A functional approach to wetland assessment enables a holistic view to be taken of the wide range of benefits wetlands can provide. However, some of these may be mutually exclusive, or else one may be optimised whilst reducing the effectiveness of another. The FAPs can be used to determine to what degree a selection of (or all) functions listed operates under current conditions and management. It is also possible to test 'what if' scenarios by changing the inputs for appropriate environmental variables (such as water regime) to see the effects on functional performance of potential changes in management. The case studies described here illustrate application of the FAPs at both individual site and broader landscape scales, and the kinds of interpretation of outcomes that can be made. Outcomes highlight the potential and possible priorities for management options to meet not only environmental but also social and economic target objectives. They can indicate where further studies would be beneficial to improve the knowledge required for strategic land-use or site-specific management decisions to be made with greater confidence.

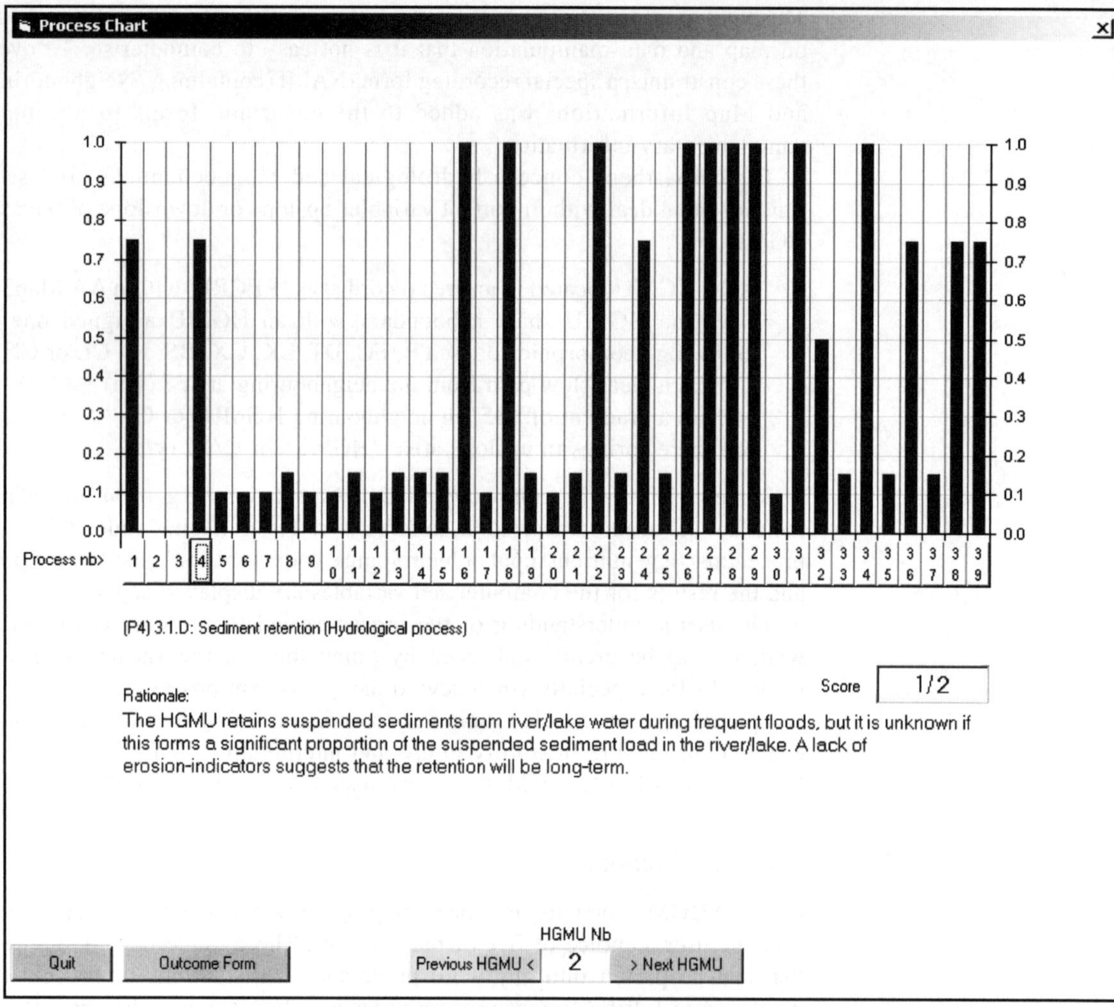

1.9.3 Process scores display screen with annotation accessible by mouse. Scores are quantified according to the output scoring in the FAPs (see Section 1.8). Scores are: 0 for 4 (not functioning), 0.1 for 2, 0.15 for 1/x, 0.25 for 1/d, 0.5 for 1/c or 1/3, 0.75 for 1/b or 1/2, and 1 for 1/a or 1/1 (full performance of function).

1.10.2 Application of FAPs at the site scale

The FAPs can be used as a first step for land managers wishing to assess the functional value of a wetland. An example is given in Section 1.8.5 (Fig. 1.8.2) of the type of output and interpretation possible from FAPs using the example of the nutrient removal function assessed for HGMUs identified at a study site (Tetcott) in the catchment of the River Tamar, UK. The outcomes provide important information on the potential sources and likelihood of nutrient losses from identifiable parts of the wetland. The approach required mapping of all the HGMUs at the site, using the methods described in Section 2.2 (HGMU delineation). Initial differentiation was made between land on the floodplain of the River Tamar and the sloping land of adjacent valley sides. Most of the floodplain was identified as a flat area, while minor levees alongside the river channel were mapped as elevations, and former channels were described as depressions. A number of slope units were separated by gradient category, the steepest land being dry backslopes furthest from the floodplain, and the gentle footslopes being the wettest, representing seepage areas at the base of the backslopes.

Once the basic functional units have been identified and delineated, probable functional performance can be assessed for any function of interest, under both present circumstances and also for any chosen alternative land-use and management scenario. By way of example, Fig. 1.10.1 (see colour section) is a GIS-based display (from the WEDSS) of the outcomes for the function of nutrient retention under

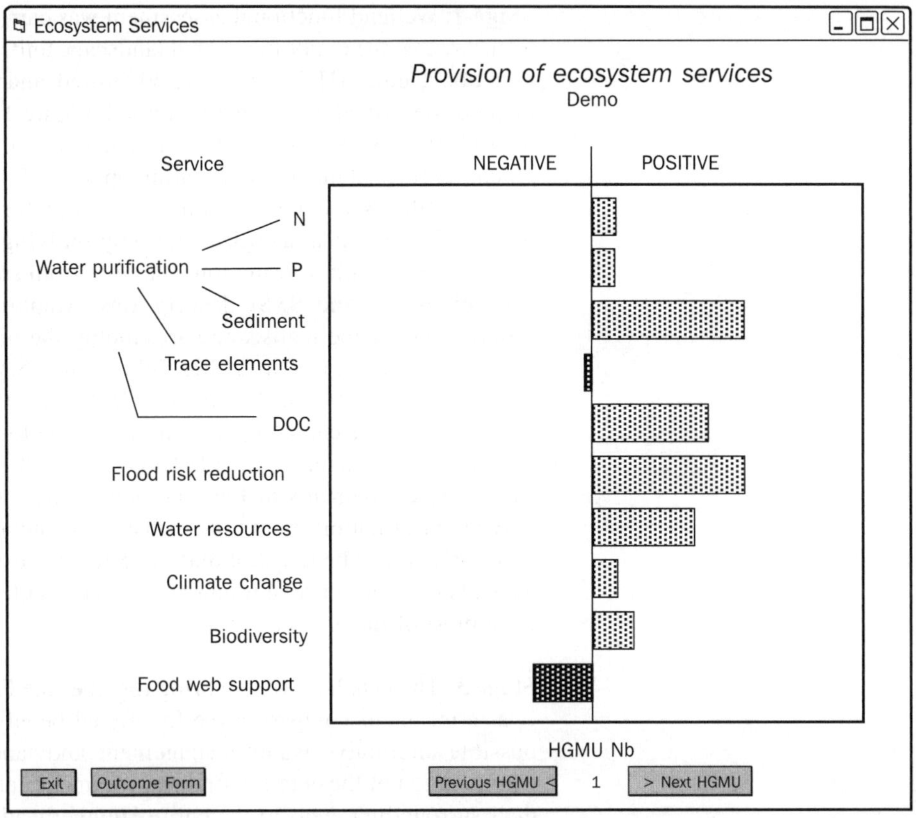

1.9.4 The output display screen showing the assessment of provision of ecosystem services by the HGMU.

the three potential options of increased intensity of management, compliance with minimum environmental management standards and an environmentally focused (green) scenario.

1.10.3 *Application of the FAPs at the landscape scale*

Though the FAPs have been developed for individual site-based assessment, it is possible to adopt an approach that enables them to be applied at the wider landscape scale. A possible way of doing this was tested in the Humberhead Levels, UK, as part of a land management initiative to address economic, social and environmental issues through a programme of actions to set objectives for sustainable water and land management in an arable area (see www.naturalengland.org and search for *Humberhead Levels Land Management Initiative*).

The Humberhead Levels comprise some 2275 km^2 of flat land to the south of York at the head of the Humber Estuary, an area of former wetland now largely drained for intensive arable farming. A number of important remnant wetlands survive, notably one of the largest areas of raised bog in the UK: the Thorne and Hatfield Moors. The object of the assessment was to examine the opportunities for raising water levels, and to demonstrate the wider public benefits of changes in hydrological management. Wetland assessment was undertaken in three stages:

Stage 1: Surviving wetlands of the Humberhead Levels were identified and their status determined in terms of their present condition and their restorability to a more pristine condition. The majority of the Humberhead Levels, former wetlands now mostly in intensive agricultural use, were considered to qualify as *degraded* and *non-restorable* under current socio-economic circumstances, though it was acknowledged that future policy changes could enable these areas to become more realistic candidates for re-wetting and consequent restoration.

Stage 2: Wetland functional assessment was carried out at nine pilot sites, chosen to ensure that the major functional landscape units of the area were all represented. Functional units (HGMUs) were identified and mapped at each site using the methods described below in Section 2.2. Figure 1.10.2 (see colour section) shows the HGMUs mapped at Sutton Common, one of the pilot sites. The maps and FAP outcomes formed the basis of scaling up to 5 × 5-km areas surrounding each pilot. Figure 1.10.3 (see colour section) shows this for the Sutton Common area. This was achieved partly as a desk exercise by applying patterns established at the study sites, together with information available from sources such as Ordnance Survey and soil maps, and SSSI descriptions. Mapping was also assisted by using observations of the landscape surrounding the pilot study sites. Resources were not available to carry out further field studies beyond the boundaries of the pilot sites. At this smaller scale, some HGMUs identified at pilot sites could not be mapped, either because separations would be too small to delineate, or because data were insufficiently detailed or precise. These HGMUs were included as subsidiary components of the main landscape functional units. It was important to retain information about these subsidiary components at the smaller scale, since it can sometimes be the case that the effectiveness of functional performance can depend on the presence of minor components with distinctively different properties from most of the area concerned.

Stage 3: The task here was to bring together the FAP outcomes from Stages 1 and 2, as a means of comparing the functional benefits that could be delivered from possible alternative water management and land-use scenarios. This required extrapolation of the detailed findings of functional assessment from pilot sites and their surrounding areas to the entire Humberhead Levels by a scaling-up process similar to that applied in Stage 2. This uses information such as soil and geology maps to indicate the broad extent of different landscape types. The overall landscape-scale functional units identified were mapped (Fig. 1.10.4, see colour section) and compiled into a database using a GIS. These units formed the basis for wetland functional assessment of the Humberhead Levels. In applying this approach of scaling-up, it is crucially important to select pilot sites representative of the wider areas to which detailed functional assessment outcomes are to be extrapolated. Where uncertainty about this exists, or patterns of wetland type are complex, it is wise to undertake some reconnaissance survey work beyond the pilot site boundaries in order to improve the reliability of extrapolation of the results in the scaling-up exercise.

For each functional landscape type across the Humberhead Levels, assessment of functions was made for the three, increasingly wet, land-use and management scenarios listed below. Further details of a selection of these, in which the greatest scope exists for variation among scenarios, are shown in Table 1.10.1.

1. ***Degraded***, in which present drainage is maintained to enable largely intensive agricultural use to continue.
2. ***Policy compliance*** under which agri-environment schemes are fully implemented.
3. ***Green***, under which drainage is abandoned and land is allowed to return to its natural wet condition.

1.10.4 *Wetland functional assessment for the Humberhead Levels*

An example is given below of the main findings of applying the FAPs to the landscape units of the entire Humberhead Levels assessment area, based on FAPs outcomes from HGMUs of individual study sites (Table 1.10.2). This uses the main outputs only, and not the subtleties of interpretation and degree of performance that are given in the individual process outputs. This method gives a general indication of the way in which the landscape works to provide environmental benefits. Many of these are enhanced with increasing degrees of wetness, as processes supporting wetland functions are enabled to operate more effectively.

Table 1.10.1 Re-wetting the Humberhead Levels – example land use scenarios

Landscape type	Degraded	Policy compliance	Green
River alluvial flats	Permanent pasture on alluvial gley soils, some sections embanked, otherwise flood pastures; occasional spring cereals where drained	Drains and ditches blocked to raise water levels and convert any arable land to permanent pasture	Embankments removed or set back to enable flood pastures to extend across the floodplain
Sands (land affected by ground-water)	Pump-drained, arable farming	Altered pumping regime to allow the water table to rise and provide groundwater irrigation for damp grassland; some blocked ditches and aquatic ecosystems established	Surface water penned in ditches, and drains blocked; wet grassland and fen established
Clay lands (on former lake clay deposits)	Drained for dry/ damp grassland; clayey stagnogley soils	Drains blocked for wet grassland	Drains blocked and surface water penned in ditches; wet grassland and fen with aquatics in ditches
Valley fen peat	Wetland nature reserves where land disturbed or subsided – otherwise drained for arable; organic soils, peat locally stabilised but mostly continuing rapid degradation	Wetland nature reserves where land disturbed or subsided – otherwise less intensively drained, mainly wet grassland; organic soils, peat locally stabilised, otherwise continued slow degradation of peat	Wetland nature reserves where land disturbed or subsided – otherwise surface water penned in ditches and drains blocked, fen or fen-carr established; organic soils, peat stabilised with some growth locally

Degraded

Most hydrological functions are not currently performed because of a lack of opportunity for interaction between the wetland and surface or groundwaters, due to the impacts of artificial drainage. Where flow rates in ditches are reduced by the growth of vegetation, there is some possibility that the flood detention function can be performed, but not where an efficient ditch clearance programme is carried out. The exceptions are depressions (natural landform or artificial features such as ditches and drains), which may allow groundwater to discharge where peat is developed over permeable materials. Many water quality functions are carried out to some degree even in arable areas, though key functions of N-removal by denitrification and nutrient storage in soil organic matter are precluded owing to dry conditions. Nutrient retention in soil organic matter is limited when peaty soil is cultivated for arable cropping, since the oxidised conditions promote decomposition of the peat. Adsorption and precipitation of phosphorus in the soil is definitely performed where drainage is in operation, giving predominantly oxidised conditions. P inputs on arable land and associated ditch systems are from inorganic fertiliser additions. Lime added to neutralise naturally acidic soils under arable production will provide additional adsorption sites for phosphorus. Nutrients

Table 1.10.2 A compilation of the Humberhead Levels FAPs outputs for groundwater impacts on areas of sand lands based on the three scenarios of Degraded, Policy Compliance and Green. D = the function is definitely being performed; P = the function is probably being performed, but there are constraining factors or uncertainties; N = the function is not, or not significantly, being performed. Where a function is definitely or probably being performed, it is also shaded in the table of outcomes, so any patterns in functional performance can be more easily seen (adapted from Hogan and Maltby, 2005b)

Land management scenarios	Degraded						Policy compliance					Green				
Functional units / Process/function	Flat areas	Uniform slopes	Undifferentiated depressions	Ditches	Lakes	River alluvial flats	Flat areas	Uniform slopes	Undifferentiated depressions	Ditches	Narrow floodplains	Flat areas	Uniform slopes	Undifferentiated depressions	Ditches	Narrow floodplains
Hydrological																
3.1.A Floodwater detention	N	N	N	D	N	P	N	N	N	D	D	N	N	N	D	D
3.1.B Groundwater recharge	N	N	N	P	N	N	N	N	N	N	N	N	N	P	N	N
3.1.C Groundwater discharge	N	N	N	P	N	N	N	N	N	D	N	N	N	P	D	N
3.1.D Sediment retention	N	N	N	N	N	N	N	N	N	N	D	N	N	N	N	D
Biogeochemical																
3.2.A.1 Long-term retention of nutrients by plant uptake	NA	NA	NA	NA	NA	NA	P	P	P	P	P	P	P	P	P	P
3.2.A.2 Storage of nutrients in soil organic matter	N	N	N	N	N	N	N	N	N	P	N	P	N	D	P	N
3.2.A.3 Adsorption of N as ammonium	P	P	P	N	N	P	P	P	P	N	P	P	P	P	N	P
3.2.A.4 Adsorption and precipitation of P in soil	D	D	D	D	N	D	D	D	D	N	D	D	D	D	N	N
3.2.A.5 Retention of particulate nutrients (N and P)	N	N	N	P	N	N	N	N	N	N	D	N	N	N	N	D
3.2.B.1.1 Gaseous export of N by denitrification	D	D	D	P	D	D	D	D	D	P	D	D	D	D	P	D
3.2.B.1.2 Gaseous export of N by ammonia volatilisation	N	N	N	N	N	N	D	D	D	P	D	D	D	D	P	D
3.2.B.2 Export of nutrients (N and P) by vegetation management	P	P	P	N	N	P	P	P	P	P	P	D	D	P	P	D
3.2.B.3 Export of nutrients by wind and water	N	N	N	N	N	N	N	N	N	N	D	N	N	N	N	N
3.2.C.1 Organic matter accumulation	N	N	N	N	N	N	P	P	P	P	P	P	N	D	P	N
Ecological																
3.3.A.1 Provision of overall habitat structural diversity	N	N	N	P	P	N	P	P	P	D	P	N	N	N	D	P
3.3.A.2.1 Micro-sites for macroinvertebrates	N	N	N	P	N	N	N	N	N	D	N	P	P	P	D	D
3.3.A.2.2 Micro-sites for fish	N	N	N	D	D	N	N	N	N	P	N	P	P	P	P	P
3.3.A.2.3 Micro-sites for herptiles	N	N	N	P	P	N	N	N	N	P	N	P	P	P	P	D
3.3.A.2.4 Micro-sites for birds	N	N	N	N	N	N	P	P	P	P	P	P	P	P	P	D
3.3.A.2.5 Micro-sites for mammals	P	P	P	P	P	P	D	D	D	D	D	D	D	D	D	D

Table 1.10.2 cont'd

Land management scenarios	Degraded						Policy compliance					Green				
Functional units / **Process/function**	Flat areas	Uniform slopes	Undifferentiated depressions	Ditches	Lakes	River alluvial flats	Flat areas	Uniform slopes	Undifferentiated depressions	Ditches	Narrow floodplains	Flat areas	Uniform slopes	Undifferentiated depressions	Ditches	Narrow floodplains
3.3.A.3 Provision of plant and habitat diversity	N	N	N	N	N	N	D	D	D	D	D	D	D	D	D	D
3.3.B.1 Productivity/biomass production	N	N	N	D	D	N	D	D	D	D	D	D	D	D	D	D
3.3.B.2.1 Biomass import via water courses	N	N	N	P	N	N	N	N	N	N	D	P	P	P	N	P
3.3.B.2.2 Biomass import via overland flow	N	N	N	N	N	N	N	N	N	N	N	P	P	P	N	D
3.3.B.2.3 Biomass import via wind transport	P	P	P	P	P	P	P	P	P	P	P	P	P	P	P	P
3.3.B.3 Biomass import via biological processes	P	P	P	P	P	P	P	P	P	P	P	P	P	P	P	D
3.3.B.4.1 Biomass export via water courses		N	N	D	D	P	P	P	P	N	D	P	P	P	N	P
3.3.B.4.2 Biomass export via overland flow		N	N	P	N	P	D	D	N	N	D	N	N	N	N	P
3.3.B.4.3 Biomass export via wind transport		P	P	P	P	P	P	P	P	P	P	P	P	P	P	P
3.3.B.5.1 Biomass export via fauna		N	N	P	P	N	N	N	N	P	N		N	N	P	D
3.3.B.5.2 Biomass export via anthropogenic means		N	N	N	N	N	D	D	D	P	D	D	D	D	P	D

in particulate form can be removed only where surface water movement takes place, i.e. in flowing drains and ditches. Ammonium loss by volatilisation takes place widely, with the N supply pool maintained mainly from fertilisers, together with some atmospheric input. Organic matter is not considered to accumulate under drained conditions, but rather peat degradation and decomposition continue to take place. The ecological function of habitat structural diversity is not significantly performed on arable land. Micro-sites are provided for most ecological groups, assessments applying in general to all the component features of the functional landscape type. There are generally limitations on the production, import and export of biomass.

Policy compliance

Hydrological functions remain unaltered from the more intensive agricultural use, unless ditches are blocked (see 'green' scenario, below). Water quality functions are enhanced by raising water levels in a number of ways. Wetter conditions and a cessation of ploughing will reduce peat wastage and enable the storage function of nutrients in soil organic matter to be performed. Although uncertain, some organic matter accumulation may take place. While more reduced conditions make adsorption and precipitation of P less likely, they do promote export of N by denitrification. Raising water levels for wet grassland habitat management improves the functioning for overall habitat structural diversity. Biomass export is also promoted by a change to grassland farming, when the crop is removed by cutting for hay.

Flood detention is definitely performed by the ditch systems where discharge has been blocked to retain water on the site, largely precluding contributions of flow to the surrounding drainage network. Water stored in ditches can also enable some recharge of groundwater to take place. Under fen and other herbaceous plant communities, much of the vegetative material is recycled at the end of the growing season, precluding significant long-term retention of nutrients. Similarly, in ditches blocked for re-wetting purposes, much of the plant material dies back in the winter, recycling nutrients within the ditch system. However, the process is definitely performed where the land is under woodland, because nutrients are stored in timber. Biomass import and export through the agency of water are not significant, as water tends to be held within the systems rather than flowing in and out. Habitat structure and a variety of faunal micro-sites are provided with the development of fen or fencarr habitats. Biomass production is likely to be important, especially in the wettest areas. Bird movements are likely to be important agents of nutrient export though, where agricultural management is limited, there is little likelihood of biomass being exported by anthropogenic means.

1.11 Further reading and references

Adamus, P.R. (1983). *A Method for Wetland Functional Assessment, Volume II*, FHWA Assessment Method, Rep. No. FHWA-IP-82-24, Federal Highway Administration, US Department of Transportation, Washington, DC.

Adamus, P.R. and Stockwell, L.T. (1983). *A Method for Wetland Functional Assessment: Volume I, Critical Review and Evaluation Concepts*, Rep No. FHWA-IP-82-23, Federal Highway Administration, US Department of Transportation, Washington, DC.

Adamus, P.R., Stockwell, L.T., Clairain, E.J., Smith, R.D. and Young, R.E. (1987). *Wetland Evaluation Technique (WET) Volume II*, Operational Draft TRY-87, US Army Corps of Engineers, Waterways Experiment Station, Vicksburg, MS.

Adamus, P.R., Stockwell, L.T., Clairain, E.J., Morrow, M.E., Rozas, L.P. and Smith, R.D. (1991). *Wetland Evaluation Technique (WET), Volume I, Literature Review and Evaluation Rationale*, Technical Report WRP-DE-2, US Army Corps of Engineers, Waterways Experiment Station, Vicksburg, MS.

Ambus, P. and Christensen, S. (1993). Denitrification variability and control in a riparian fen irrigated with agricultural drainage water, *Soil Biology and Biochemistry* **25**, 915–923.

Arheimer, B., Löwgren, M., Pers, B.C. and Rosberg, J. (2005). Integrated catchment modelling for nutrient reduction: scenarios showing impacts, potential and cost of various measures, *Ambio* **34**(7), 513–520.

Armstrong, W. (1979). Aeration in higher plants, *Advances in Botanical Research* **7**, 225–332.

Balmford, A., Brouer, A., Cooper, P., Costanza, R. *et al.* (2002). Economic reasons for conserving wild nature, *Science* **297**, 950–953.

Barbier, E.B. (1994). Valuing environmental functions: tropical wetlands, *Land Economics* **70**, 155–173.

Barbier, E.B., Acreman, M.C. and Knowler, D. (1996). *Economic Valuation of Wetlands: a Guide for Policy Makers and Planners*, Ramsar Convention Bureau, Gland, Switzerland.

Bardecki, M. (1987). Cumulative impacts of agricultural land drainage on watershed hydrology, *Proceedings of the National Wetland Symposium on 'Wetland Hydrology'*, Association of Wetland Managers Inc., Chicago, IL, 124–129.

Bartoldus, C.C. (1999). *A Comprehensive Review of Wetland Assessment Procedures: A Guide for Wetland Practitioners*, Environmental Concern Inc., St. Michaels, MD.

Basso, F., Bove, E., Dumontet, S., Ferrara, A., Pisante, M., Quaranta, G. and Taberner, M. (2000). Evaluating environmental sensitivity at the basin scale through the use of geographic information systems and remotely sensed data: an example covering the Agri basin Southern Italy, *Catena* **40**, 19–35.

Begon M., Townsend, C.R. and Harper, J.R. (2006). *Ecology, From Individuals to Ecosystems*, Blackwell Publishing, Oxford.

Blackwell, M.S.A. (1997). *Zones of Enhanced Denitrification in River Marginal Wetlands*, Unpublished PhD Thesis, Department of Geography, Royal Holloway University of London, UK.

Blackwell, M.S.A. and Maltby, E. (1998). Identification and field testing of zones of enhanced denitrification (ZEDs) in the wetland landscape. In: A.J. McComb and J.A.

Davis (eds), *Wetlands for the Future*, Proceedings of INTECOL's International Wetlands Conference, Gleneagles Press, Adelaide, Australia, 443–455.

Blackwell, M.S.A. and Maltby, E. (2006). *Ecoflood Guidelines. How to Use Floodplains for Flood Risk Reduction*, European Communities, Brussels, Belgium.

Blackwell, M.S.A., Hogan, D.V. and Maltby, E. (1999). The use of conventionally and alternatively located buffer zones for the removal of nitrate from diffuse agricultural run-off, *Water Science and Technology* **399**(12), 157–164.

Blackwell, M.S.A., Hogan, D.V. and Maltby, E. (2002). Wetlands as Regulators of Pollutant Transport. In: P.M. Haygarth and S.C. Jarvis (eds), *Agriculture, Hydrology and Water Quality*, CABI, Wallingford, UK, 321–339.

Boon, P.J., Holmes, N.T.H., Maitland, P.S., Rowall, T.A. and Davies, J. (1997). A system for evaluating rivers for conservation (SERCON): development structure and function. In: Boon P.J. and Howell D.L. (eds), *Freshwater Quality: Denying the Undeniable*? HMSO, Edinburgh, UK 299–326.

Boon, P.J., Wilkinson, J. and Martin, J. (1998). The application of SERCON (System for Evaluating Rivers for Conservation) to a selection of rivers in Britain, *Aquatic Conservation, Marine and Freshwater Ecosystems* **8**, 597–616.

Bouwer, H. (1978). *Ground Water Hydrology*, McGraw Hill, New York.

Brinson, M.M. (1993). *A Hydrogeomorphic Classification for Wetlands*, Wetlands Research Program Technical Report WRP-DE-4, US Army Corps of Engineers, Waterways Experiment Station, Vicksburg, MS.

Brinson, M.M. (1996). Assessing wetland functions using HGM, *National Wetlands Newsletter* **18**, 10–16.

Brinson, M.M. (in press). The United States HGM (hydrogeomorphic) approach. In: Maltby, E. and Barker, T. (eds), *The Wetlands Handbook*, Wiley-Blackwell, Oxford, UK.

Bromley, J., Cruces, J., Acreman, M., Martinez, L. and Llamas, M.R. (2001). Problems of sustainable groundwater management in an area of over-exploitation: the Upper Guadiana catchment, Central Spain, *Water Resources Development* **17**(3), 379–396.

Bruning-Fann, C.S. and Kaneene, J.B. (1993). The effects of nitrate, nitrite and N-nitroso compounds on human health: a review, *Veterinary and Human Toxicology* **35**(6), 521–538.

Brunning, R., Hogan, D.V., Jones, J., Jones, M.J., Maltby, E., Robinson, M. and Straker, V. (2000). Saving the Sweet Track: The *in-situ* preservation of a Neolithic wooden trackway, Somerset, UK, *Conservation and Management of Archaeological Sites* **4**, 3–20.

Brunning, R., Hogan, D.V., Jones, J., Tinsley, H., Jones, M., Kenward, H. and Tyres, I. (in press), *A Fragile Past – Monuments at Risk in Somerset's Peatlands*, English Heritage, Oxbow Books, Oxford, UK.

Bullock, A. and Acreman, M. (2003). The role of wetlands in the hydrological cycle, *Hydrology and Earth System Sciences* **7**(3), 358–389.

Bundesanstalt für Gewässerkunde – BfG (2002). *Das Augusthochwasser im Elbeeinzugsgebiet 2002* (The August flood in the Elbe catchment 2002), http://elise.bafg.de/?3967

Bundesministerium des Inneren – Federal Ministery of the Interior of the Federal Republic of Germany 6.11 (2002), Press release, http://www.bmi.bund.de/top/dokumente/Pressemitteilung/ix_90912.htm?nodeID=3735

Burdick, D.M. and Mendelssohn, I.A. (1990). Relationship between anatomical and metabolic responses to soil waterlogging in the coastal grass *Spartina patens*, *Journal of Experimental Botany* **41**, 223–228.

Burt, T.P. and Trudgill, S.T. (1993). Nitrate in groundwater. In: Burt, T.P., Heathwaite, A.L. and Trudgill, S.T. (eds), *Nitrate. Processes, Patterns and Management*, Wiley, Chichester, UK, 213–238.

Carman, J. (1994). Date for the future or an amenity for the present? The values of the natural and historic wetland environment. In: Cox, M., Straker, S. and Taylor, D. (eds), *Wetlands: Archaeology and Nature Conservation*, HMSO, London.

Clean Water Act (1972). Federal Water Pollution Control Amendments of 1972, enacted by the 92nd United States Congress Effective October 18, 1972.

Clymo, R.S., Dawson, F.H., Bertram, B.C.R., Burt, T.P., Gilman, K., Ingram, H.A.P., James, R., Kirkby, M.J., Lee, J.A., Maltby, E., Wheeler, B.D. and Wilcock, D. (1995). Conclusion: directions for research on wetlands in Britain. In: Hughes, J.M.R. and Heathwaite, A.L. (eds), *Hydrology and Hydrochemistry of British Wetlands*, Wiley, Chichester, 467–478.

Coles, B.J. (2001). A Past Less Foreign: wetland archaeology and its survival in European perspective. In: Coles, B. and Olivier, A. (eds), *The Heritage Management of Wetlands in Europe*, Europae Archaeologiae Consilium and WARP, Royal Library, Brussels, Belgium, 1–6.

Coles, B.J. and Coles, J.M. (1986). *Sweet Track to Glastonbury. The Somerset Levels in Prehistory*, Thames and Hudson, London, UK.

Coles, B.J. and Coles, J.M. (1989). *People of the Wetlands: Bogs, Bodies and Lake-Dwellers*, Thames and Hudson, London, UK.

Commission of The European Communities (2004). *Flood Risk Management, Flood Prevention, Protection and Mitigation*, Communication from the Commission to the Council, the European Parliament, the European Economic and Social Committee and the Committee of the Regions, Brussels, Belgium, COM(2004)472 Final.

Conservation Foundation (1988). *Protecting America's Wetlands: An action agenda.* Final report of the National Wetlands Policy Forum, Conservation Foundation, Washington, DC.

Cooper, A.B. (1990). Nitrate depletion in the riparian zone and stream channel of a small headwater catchment, *Hydrobiologia* **202**, 13–26.

Costanza, R., d'Arge, R., de Groot, R., Farber, S., Grasso, M., Hannon, B., Limburg, K., Naeem, S., O'Neill, R.V., Paruelo, J., Raskin, R.J., Sutton, P. and van den Belt, M. (1997). The value of the world's ecosystem services and natural capital, *Nature* **387**, 253–260.

Cowardin L.M, Carter V, Golet F.C & LaRoe E.T. (1979). *Classification of wetlands and deepwater habitats of the United States*, US Fish & Wildlife Service Pub. FWS/OBS-79/31, Washington, DC.

Davies, J. and Claridge, C.F. (eds) (1993). *Wetland Benefits. The Potential for Wetlands to Support and Maintain Development*, Asian Wetland Bureau Publication No. 87; IWRB Special Publication No. 27; Wetlands for the Americas Publication No. 11.

Davis, R., Cunningham, R. and Avis, C. (2002). *The role of wetlands in the Water Framework Directive.* Unpublished paper to the Strategic Coordination Group of the Common Implementation Strategy of the Water Framework Directive, Brussels, Belgium.

DEFRA (2001). *National Appraisal of Assets at Risk from Flooding and Coastal Erosion, including the Potential Impact of Climate Change*, Final report, http://www.defra.gov.uk/environ/fcd/policy/NAAR1101.pdf

DEFRA (2003). *Third consultation paper on the implementation of the EC Water Framework Directive (2000/60/EC)*, London, Department for Environment, Food and Rural Affairs.

Dosskey, M., Schultz, D. and Isenhart, T. (1997). *How to design a riparian buffer for agricultural land*, Agroforestry Note 4, USDA-FS, USDA-NRCS, http://waterhome.brc.tamus.edu/projects/afnote4.htm

Dugan, P.J. (1990). *Wetland Conservation: A Review of Current Issues and Required Action*, IUCN, Gland, Switzerland.

Dugan, P. (ed.) (1993). *Wetlands in Danger*. Mitchell Beazley–IUCN, Gland, Switzerland.

EC (1979). Council Directive 79/409/EEC of 2 April 1979 on the conservation of wild birds, *OJ* **L103**, 25.04.1979, 1–18.

EC (1992). Council Directive 92/43/EEC of 21 May 1992 on the conservation of natural habitats and of wild fauna and flora, *OJ* **L206**, 22.07.1992, 7–50.

EC (2000). Directive 2000/60/EC of the European Parliament and the Council of 23 October 2000 establishing a framework for Community action in the field of water policy, *OJ* **L327**, 22.12.2000, 1–72.

EC (2007) Directive 2007/60/EC of the European Parliament and of the Council of 23 October 2007 on the assessment and management of flood risks, *OJ* **L288**, 6.11.2007, 27–34.

EEA (2003). *Mapping the impacts of recent natural disasters and technological accidents in Europe*, Environmental issue report no. 35, European Environment Agency, Copenhagen, Denmark.

EEA (2004). *Impact of Europe's Changing Climate: an indicator-based assessment*, EEA Report No. 2, Office for Official Publications of the European Communities, Luxembourg.

English Nature (2001). *Sustainable Flood Defence: The Case for Washlands*, English Nature Research Report No. 406, Peterborough, UK.

Environment Agency (1996). *Understanding Buffer Strips*, An information booklet, Environment Agency, Bristol, UK.

Environment Agency (2002). *The Water Framework Directive: Guiding Principles on the Technical Requirements*, Environment Agency, Bristol, UK.

Environmental Laboratory (1987). *Corps of Engineers wetlands delineation manual*, Technical Report Y-87-1, US Army Corps of Engineers, Waterways Experiment Station, Vicksburg, MS.

Ernst, W.H.O. (1990). Ecophysiology of plants in waterlogged and flooded environments, *Aquatic Botany* **38**, 73–90.

Etherington, J.R. (1983). *Wetland Ecology*, Edward Arnold, London, UK.

European Commission (2003). *Horizontal Guidance on the Role of Wetlands in the Water Framework Directive*, Directorate General Environment of the European Commission (DGXI), Brussels, Belgium.

Feierabend, J.S. (1992). *Endangered Species, Endangered Wetlands: Life on the Edge*, National Wildlife Federation, Washington, DC.

Finlayson, C.M. (ed.) (1992). *Strategy and Action Plan to Conserve the Wetlands of the Lower Volga*, IWRB, Slimbridge, UK.

Finlayson, M. and Moser, M. (eds) (1991) *Wetlands*, International Waterfowl and Wetlands Research Bureau IWRB, Facts on File Limited, Oxford, UK.

Fisher, J. and Acreman, M.C. (2004). Wetland nutrient removal: a review of the evidence, *Hydrology and Earth System Sciences* **8**(4), 673–685.

Fleischer, S. (1990). Wetlands – a nitrogen sink, *Acid Environment* **9**, 26–28.

Gambrell, R.P. and Patrick, Jr, W.H. (1978). Chemical and microbiological properties of anaerobic soils and sediments. In: Hook, D.D. and Crawford, R.M.M. (eds), *Plant Life in Anaerobic Environments*, 1st edn, Vol. 1, Ann Arbor Science, Michigan, USA, 375–423.

García Rodríguez, M. (1996). *Hydrogeology of the Tablas de Daimiel and the Ojos del Guadiana (Hydrological basis for a functional classification of riparian wetlands)*, Unpublished PhD Thesis, Faculty of Geological Sciences, Department of Geodynamics, Complutense University of Madrid.

Gosselink, J.G. and Maltby, E. (1990). Wetland losses and gains. In Williams, M. (ed.), *Wetlands: A Threatened Landscape*, Blackwell, Oxford, UK, 296–322.

Goudie, A. (1986). *The Human Impact on the Natural Environment*, The MIT Press, Cambridge, MA.

Government of Romania (2002). *Romania–Hazard Risk Mitigation and Emergency Preparedness Project*, Report No. PID11109, ROPE75163.

Greenwood, D.J. (1967). Studies on the transport of oxygen through the stems and roots of vegetable seedlings, *New Phytologist* **66**, 337–347.

Halls, A.J. (ed.) (1997). *Wetlands, Biodiversity and the Ramsar Convention: The Role of the Convention on Wetlands in the Conservation and Wise Use of Biodiversity*, Ramsar Convention Bureau, Gland, Switzerland.

Handmer, J. (2001). Improving flood warnings in Europe: a research and policy agenda, *Environmental Hazards* **3**, 19–28.

Haycock, N.E. and Burt, T.P. (1993). Role of floodplain sediments in reducing the nitrate concentration of subsurface run-off: a case study in the Cotswolds, UK, *Hydrological Processes* **7**(3), 287–295.

Haycock, N.E., Pinay, G. and Walker, C. (1993). Nitrogen retention in river corridors – European perspective, *Ambio* **22**, 340–346.

Haycock, N.E. and Muscutt, A.D. (1995). Landscape management strategies for the control of diffuse pollution, *Landscape and Urban Planning* **31**, 313–321.

Hefting, M.M. (2003), *Nitrogen Transformation and Retention in Riparian Buffer Zones*, PhD Thesis, Utrecht University, The Netherlands.

Hefting, M.M., Bobbink, R. and De Caluwe, H. (2003). Nitrous oxide emission and denitrification in chronically nitrate-loaded riparian buffer zones, *Journal of Environmental Quality* **32**, 1194–1203.

Hogan, D.V. and Maltby, E. (2005a). *The social, economic and environmental benefits of wetlands in the Humberhead Levels: Stage 3 Report Part 1*, Unpublished report to the Countryside Agency, Royal Holloway Institute for Environmental Research, Royal Holloway University of London.

Hogan, D.V. and Maltby, E. (2005b). *The social, economic and environmental benefits of wetlands in the Humberhead Levels. Stage 3 Report, Part 2*, Unpublished report to the Countryside Agency, Royal Holloway Institute for Environmental Research, Royal Holloway University of London.

Hollis, G.E. and Finlayson C.M. (1996). Ecological change in Mediterranean wetlands. In: Vives, T.P. (ed.), *Monitoring Mediterranean Wetlands: A Methodological Guide*, MedWet Publication, Wetlands International, Slimbridge, UK and ICN, Lisbon, Portugal.

Horan, N.J. (1993). *Biological Wastewater Treatment Systems*. Wiley, Chichester, UK.

Houghton, J.E.T., Ding, Y., Griggs, D.J., Nouger, M., van der Linden, P.J. Dai, X., Maskell, K. and Johnson, C.A. (2001). *Climate Change 2001: the Scientific Basis*, Cambridge University Press, Cambridge, UK.

Hughes, J. and Heathwaite, L.J. (eds) (1995). *Hydrology and hydrochemistry of British wetlands*, Wiley, Chichester, UK.

IKSE (2004). Dokumentation des Hochwassers vom August 2002 im Einzugsgebiet der Elbe, Internationale Kommission zum Schutz der Elbe, Magdeburg, http://elise.bafg.de/servlet/is/6889/

IPCC (2007). *Climate Change 2007: Impacts, Adaptation and Vulnerability*, Working Group

II Contribution to the Intergovernmental Panel on Climate Change Fourth Assessment Report, IPCC Secretariat, Geneva, Switzerland.

Isenhart, T.M., Schultz, R.C., Colletti, J.P. and Rodrigues, C.A. (1995). Design, function, and management of integrated riparian management systems, *Proc. National Symposium on Using Ecological Restoration to Meet Clean Water Act Goals*, USEPA, Chicago, IL, 93–102.

James, C., Fisher, J., Russell, V., Collings, S. & Moss, B. (2005) Nitrate availability and hydrophyte species richness in shallow lakes, *Freshwater Biology* **50**(6), 1049–1063.

Janssen, R., Herwijnen, M.V. and Beinat, E. (2001). *DEFINITE: Cases and User Manual*, Report R-01/01, Institute for Environmental Studies, Vrije Universiteit Amsterdam, Amsterdam, The Netherlands.

Johnes, P.J. and Burt, T.P. (1993). Nitrate in surface waters. In: Burt, T.P., Heathwaite, A.L. and Trudgill, S.T. (eds) *Nitrate: Processes, Patterns and Management*. Wiley, Chichester, UK, 269–321.

Jones, T.A. and Hughes, J.M.R. (1993). Wetland inventories and wetland loss studies: a European perspective. In: Moser, M., Prentice, R.C. and van Vessem, J. (eds) *Waterfowl and Wetland Conservation in the 1990s: a Global Perspective*. Proceedings of the IWRB symposium, St Petersburg, Florida/IWRB Special Publication 26, IWRB, Wagemigen, The Netherlands.

Koch, M.S., Maltby, E., Oliver, G.A. and Bakker, S.A. (1992). Factors controlling denitrification rates of tidal mudflats and fringing saltmarshes in South-west England, *Estuarine Coastal Shelf Science* **34**, 471–485.

Koerselman, W. (1989). Groundwater and surface-water hydrology of a small groundwater-fed fen, *Wetland Ecol. Manage.* **1**, 33–43.

Kok, M., Vrijling, J.K., van Gelder, P.H.A.J.M. and Vogelsang, M.P. (2002). Risk of flooding and insurance in the Netherlands, *The Second International Symposium on Flood Defence (ISFD 2002)*, Beijing, September 10–13.

Kramer, S.N. (1963). *The Sumerians*, The University of Chicago Press, Chicago, IL.

Larson, J.S. and Mazzarese, D.B. (1994). Rapid assessment of wetlands: history and application to management. In: Mitsch, W.J. (ed.) *Global Wetlands: Old World and New*, Elsevier Science, Amsterdam, The Netherlands, 625–636.

Leeds-Harrison, P.B., Quinton, J.N., Walker, M., Harrison, K.S., Gowing, D.G., Tyrrel, S.F., Morris, J., Mills, J. and Harrod, T.R. (1996). *Buffer zones in headwater catchments*, MAFF–English Nature Buffer Zone project – Final Report – CSA 2285, Cranfield University, Sitsoe, UK.

Leonardson, L. (1994) *Wetlands as nitrogen sinks: Swedish and international experience*, SEPA Report 4176, Swedish Environmental Protection Agency, Stockholm, Sweden.

Lewis, M.A., Mayer, F.L., Powell, R.L., Nelson, M.K., Klaine, S.J., Henry, M.G., Dickson, G.W. (eds) (1999). *Ecotoxicology and Risk Assessment for Wetlands. Proceedings from the Pellston Workshop on Ecotoxicology and Risk Assessment for Wetlands* (30 July–3 August 1995, Anaconda, MT), SETAC Press, Pensacola, FL.

Llamas, M.R. and Martinez-Santos, P. (2005). Intensive groundwater use: a silent revolution that cannot be ignored, *Water Science and Technology* **51**(8) 167–174.

Lonard, R.I. and Clairain, E.J. (1985). Identification of methodologies and the assessment of wetland functions and values. In: Kusler, J.A. and Riexinger, P. (eds) *Proceedings of the National Wetland Assessment Symposium*, Portland, ME, 66–72.

Lowrance, R., Altier, L.S., Williams, R.G., Inamdar, S.P., Bosch, D.D., Sheridian, J.M., Thomas, D.L. and Hubbard, R.K. (1998). The riparian ecosystem management model: simulator for ecological processes in riparian zones, *Proceedings of the First Federal Interagency Hydrologic Modelling Conference*, Las Vegas, NV, 1.81–1.88.

Ludde, A.P., Frank, A.L. and Johnson, D.H. (1983). Water storage capacity of natural wetland depressions in the Devil's Lake Basin of North Dakota, *Journal of Soil and Water Conservation* **38**, 45–48.

Maltby, E. (1986). *Waterlogged Wealth, Why waste the world's wet places?* Earthscan, London, UK and Washington, DC.

Maltby, E. (1991). The world's wetlands under threat – developing wise use and international stewardship. In: Hansen J.A. (ed.) *Environmental Concerns. An Inter-disciplinary Exercise*, Elsevier, London, 109–136.

Maltby, E. and Blackwell, M.S.A. (2005). Managing riverine environments in the context of new water policies in Europe, *International Journal of River Basin Management* **3**(2), 133–141.

Maltby, E., Hogan, D.V., Immirzi, C.P and Oliver, G.A. (1991). *Nutrient Dynamics and Functional Assessment of European Wetlands*, Report to EC DG XI, Wetland Ecosystems Research Group, Report 4, Department of Geography, University of Exeter, UK.

Maltby, E., Hogan, D.V., Immirzi, C.P., Tellam, J.H. and van der Peijl, M.J. (1994) Building a new approach to the investigation and assessment of wetland ecosystem functioning. In: Mitsch, W.J. (ed.) *Global Wetlands: Old World and New*, Elsevier, Amsterdam, The Netherlands, 637–658.

Maltby, E., Hogan, D.V. and McInnes, R.J. (1996). *Functional Analysis of European Wetland Ecosystems – Phase 1 (FAEWE)*, European Commission, Ecosystems Research Report No. 18, European Commission, Brussels.

Maltby, E., McInnes, R.J., Hutchins, M.G. and Hogan, D.V. (1998a). Assessment of wetland functioning in the river marginal environment: Recent scientific progress and future priorities. In: Bailey, R.G., José, P.V. and Sherwood, B.R. (eds) *United Kingdom Floodplains*, Westbury Publishing, Otley, UK, 154–168.

Maltby, E., Jewkes, E.C., Baker, C.J., Mockler, N.J., Flynn, N. and Gardner, P.J. (1998b). A simple method for estimating nitrate removal from floodwaters in river marginal wetlands, *Soil Use and Management* **14**, 90–95.

Maltby, E., Blackwell, M.S.A. and Baker, C.J. (2000). Linking wetland science to policy: meeting the challenge with special reference to water quality issues. In: Balazs, E., Galante, E., Lynch, J.M., Schepers, J.S., Toutant, J-P., Werner, D. and Werry, P.A.T.J. (eds) *Biological Resource Management: Connecting Science and Policy*, Springer-Verlag, Berlin, Heidelberg, New York, 291–308.

Mander, Ü., Teiter, S. and Augustin, J. (2005). Emission of greenhouse gases from constructed wetlands for wastewater treatment and from riparian buffer zones, *Water Science & Technology* **52**(10–11), 167–176.

Mann, M.E., Bradley, R.S. and Hughes, M.K. (1998). Global-scale temperature patterns and climate forcing over the past six centuries, *Nature* **392**, 779–787.

Marsden, S. (2001). The heritage management of wetlands: legislative designation and protection, a viewpoint from England and Wales. In: Coles, B.J. and Olivier, A. (eds) *The Heritage Management of Wetlands in Europe*, European Archaeological Council and WARP, Royal Library, Brussels Belgium, 7–15.

Martin, T.L., Kaushik, N.K., Trevors, J.T. and Whiteley, H.R. (1999). Review: Denitrification in temperate climate riparian zone, *Water, Air, Soil Pollution* **111**, 171–186.

Mitsch, W.J. and Gosselink, J.G., (2000). The value of wetlands: importance of scale and landscape setting, *Ecological Economics* **35**, 25–33.

Mitsch, W.J. and Gosselink, J.G. (2007). *Wetlands*, 4th edn. Wiley, New York.

Mitsch, W.J., Day, J.W., Gilliam, J.W., Groffman, P.M., Hey, D.L., Randall, G.W. and Wang, N. (2001). Reducing nitrogen loading to the Gulf of Mexico from the Mississippi River Basin: Strategies to counter a persistent ecological problem, *Bioscience* **51**, 373–388.

Moss, B. (2001). *The Broads: the peoples' wetland*, New Naturalist, Harper Collins, London, UK.

Munné, A., Prat, N., Solà, C., Bonada, N. and Rieradevall, M. (2003). A simple field method for assessing the ecological quality of riparian habitat in rivers and streams: QBR index, *Aquatic Conservation: Marine and Freshwater Ecosystems* **13**, 147–163.

Murdoch, P.S., Baron, J.S. and Miller, T.L. (2000). Potential effects of climate change on surface-water quality in North America, *Journal of the American Water Resources Association* **36**, 347–366.

Nature Conservancy Council (1986). *Nature Conservation and Afforestation in Britain*. NCC, Peterborough.

Négrel, P., Petelet-Giraud, E., Sgouridis, F. (2005). *EUROWET, Integration of European Wetland Research in Sustainable Management of the Water Cycle*, Final Report, Volume 2, Synthesis of the current wetland research in Europe, EU Contract GOCE-CT-2003-505586, Royal Holloway Institute for Environmental Research, University of London.

Novitzki, R.P. (1979). Hydrological characteristics of Wisconsin's wetlands and their influence on floods, stream flow and sediment, In: Greeson, P.E. and Clark, J.E. (eds) Wetland Functions and Values: the state of our understanding, American Water Research Association, Minneapolis, MN, 377–388.

Novitzki, R.P. (1985). The effects of lakes and wetlands on flood flows and base flows in selected northern and eastern states, *Proceedings of Conference on Wetlands of the Chesapeake*, Easton, Maryland, Environmental Law Institute, 143–154.

Petersen, R.C. (1992). The RCE: a riparian, channel, and environmental inventory for small streams in the agricultural landscape, *Freshwater Biology* **27**, 295–306.

Pezeshki, S.R., Matthews, S.W. and Delaune, R.D. (1991). Root cortex structure and metabolic responses of *Spartina patens* to soil redox conditions, *Environmental and Experimental Bot.* **31**, 91–97.

Prach, K. and Straškrabová, J. (1996). Restoration of degraded meadows: an experimental approach. In: Prach, K., Jenik, J. and Large, A.R.G. (eds) *Floodplain ecology and management. The Lužnice River in the Trebon Biosphere Reserve, Central Europe*, SPB Academic Publishing, Amsterdam, The Netherlands, 87–93.

Ramsar Convention Bureau (1996). *Wetlands and Biological Diversity*. Third Meeting of the Conference of the Parties to the Convention on Biological Diversity, Buenos Aires, Argentina, 4–15 November 1996, document UNEP/CBD/COP/3/Inf.21.

Richardson, C.J. (1985). Mechanisms controlling phosphorus retention capacity in freshwater wetlands, *Science* **228**, 1424–1427.

Raven, P.J., Holmes, N.T.H., Dawson, F.H. and Everard, M. (1998). Quality assessment using River Habitat Survey data, *Aquatic Conservation: Marine and Freshwater Ecosystems* **8**, 477–499.

RHIER (2004). *European Evaluation and Assessment Tools supporting Wetland Ecosystem Legislation (Evaluwet)*, EU Contract EVK1-CT-2000-00070, Final Report, Royal Holloway Institute for Environmental Research, Royal Holloway University of London.

Rubec, C.D.A. (1994). Canada's federal policy on wetland conservation: a global model. In: Mitsch, W.J. (ed.). *Global Wetlands: Old World and New*, Elsevier, Amsterdam, 909–917.

Rudolf, B. and Rapp, J. (2003). The century flood of the River Elbe in August 2002: synoptic weather development and climatological aspects, *Quarterly Report of the Operational NWP-Model of the Deutscher Wetterdienst, Special Topic July 2003*, German Meteorological Service, Offenbach/Main, Germany.

Sather, J.H. and Smith, R.D. (1984). *An Overview of Major Wetland Functions and Values*, Western Energy and Land use Team, US Fish and Wildlife Service, FWS/OBS-84/18, Washington, DC.

Schindler, D.W. (1997). Widespread effects of climatic warming on freshwater ecosystems in North America, *Hydrological Processes* **11**, 1043–1067.

Schwarze, R. and Wagner G. (2003). *Marktkonforme Versicherungspflicht für Naturkatastrophen – Bausteine einer Elementarschadenversicherung*. Wochenbericht des DIW Berlin 12/03 http://www.diw.de/deutsch/produkte/publikationen/wochenberichte/docs/03-12-1.html

Scott, D.A. and Jones, T.A. (1995). Classification and inventory of wetlands: a global overview, *Vegetatio* **118**, 3–16.

Seale, R.S. (1975). *Soils of the Chatteris District of Cambridgeshire (Sheet TL 38)*, Soil Survey Special Survey No. 9 Harpenden, Soil Survey. Soil Survey and Land Research Centre, Sitsoe, UK.

SEPA (1986), *Environmental Effects of Draining Wetland and Forest*, Final report no 3270, Solna: Statens naturvårdsverk.

SEPA (1996). *Draining the Landscape: Application of the Relevant Environmental Laws* (in Swedish), SEPA General Guidelines 96:3, Swedish Environmental Protection Agency, Stockholm, Sweden.

Shaw, S.P. and Fredine, C.G. (1956). *Wetlands of the United States*, Circular 39, US Fish and Wildlife Service, Washington, DC.

Smith, R. (ed.) (2004). *The Marsh of Time: Saving Sutton Common*, Halsgrove, Tiverton, UK.

Smith, R.D. (in press). Wetlands assessment in practice: development and application in the United States regulatory context. In: Maltby, E. and Barker, T. (ed), *The Wetlands Handbook*, Wiley-Blackwell, Oxford, UK.

Smith, R.D., Ammann, A. Bartoldus, C. and Brinson, M.M. (1995). An Approach for Assessing Wetland Functions Using Hydrogeomorphic Classification, Reference Wetlands, and Functional Indices, Technical Report TR WRP-DE-10, US Army Corps of Engineers, Waterways Experiment Station, Vicksburg, MS.

Thoren, A.K., Legrand, C. and Tonderski, K.S. (2005). Temporal export of nitrogen from a constructed wetland: influence of hydrology and senescing submerged plants, *Ecological Engineering* **23**, 283–290.

Tiner, R.W. Jr (1984). *Wetlands of the United States: currents status and recent trends*, US Fish and Wildlife Service, Newton Corner, MA.

Turner, R.K. and Pearce, D.W. (1993). 'Sustainable economic development: economic and ethical principles'. In Barbier, E.B. (ed.), *Economics and ecology: New Frontiers and Sustainable Development*, Chapman and Hall, London, 177–194.

Turner, R.C. and Scaife, R.G. (eds) (1995). *Bog bodies. New discoveries and new perspectives*, British Museum Press, London, UK.

Turner, R.K., van den Bergh, J.C.J.M., Soderqvist, T., Barendregt, A., van der Straaten, J., Maltby, E. and van Ierland, E.C. (2000). Ecological-economic analysis of wetlands: scientific integration for management and policy, *Ecological Economics* **35**, 7–23.

Turner, R.K., van den Bergh, J.C.J.M. and Brouwer, R. (2003), *Managing wetlands: an ecological economics approach*, Edward Elgar, Cheltenham, UK.

United Nations Educational, Scientific and Cultural Organization (UNESCO) (1971). *Convention on Wetlands of International Importance especially as Waterfowl Habitat*, Ramsar, Iran, 2.2.1971, http://www.ramsar.org/key_conv_e.htm

Van de Noort, R., Fletcher, W., Thomas, G. Carstairs, I. and Patrick, D. (2002). *Monuments at risk in England's Wetlands*, University of Exeter, UK, http://www.ex.ac.uk/marew

Verhoeven, J.T.A., Arheimer, B., Yin, C. and Hefting, M.M. (2006). Regional and global concerns over wetlands and water quality, *Trends in Ecology and Evolution* **21**, 96–103.

Vis, M., Klijn, F., de Bruijn, K.M. and van Buuren, M., (2003). Resilience strategies for flood risk management in the Netherlands, *International Journal of River Basin Management* **1**(1), 33–40.

Weller, D.E., Correll, D.L. and Jordan, T.E. (1994). Denitrification in riparian forests receiving agricultural discharges. In: Mitsch, W.J. (ed.), *Global Wetlands: Old World and New*, Elsevier, New York, 117–131.

Wheeler, B.D. and Shaw, S.C. (1995). Plants as hydrologists? An assessment of the value of plants as indicators of water conditions in fens. In: Hughes, J.M.R. and Heathwaite, A.L. (eds) *Hydrology and Hydrochemistry of British Wetlands*, Wiley, Chichester, UK, 63–82.

Williams, M. (ed.) (1990). *Wetlands. A Threatened Landscape*, The Institute of British Geographers, Blackwell, Oxford, UK.

Wise Use of Flood Plains Project (1999–2002). *Wise use of floodplains – a demonstration of techniques to evaluate and plan floodplain restoration,* Final report for contract LIFE99 ENV/UK/000203.

WWF (2002). *Managing floods in Europe: the answers already exist*, Background briefing paper, WWF Danube–Carpathian Programme, WWF Living Waters Programme – Europe.

Yin, C. and Shan, B. (2001). Multipond systems: A sustainable way to control diffuse phosphorus pollution, *Ambio* **30**, 369–375.

Section 2

Wetland database establishment

2.0 Introduction

Objectives

To perform a full *functional analysis* of a wetland site it is necessary to assess the characteristics of the wetland, the structure of the site, the activities that occur within the site and the features of the surrounding area which may influence the site. A full and detailed study of the site is required to establish a database that can be used to assess these characteristics. This is the purpose of the *wetland database establishment*.

The primary objective of the database establishment is to divide the assessment area (AA) and contributory area (CA) landscape into discrete functional hydrogeomorphic units (HGMUs), which can then be characterised to determine what functions these units are performing.

- Each AA is the river or lake marginal wetland area being assessed, defined by field boundaries, administrative boundaries, ownership boundaries or wetland extent (Section 2.1.A).
- Each CA is the area that drains directly into the AA, either through surface flow or through groundwater flow (Section 2.1.B).
- Each discrete HGMU is a landscape unit of uniform geomorphology and hydrological regime, with soils uniform in as much as they are a reflection of the hydrology (Section 2.2.C).

Method

The wetland database establishment is divided into three sections, central to which is a field visit to the wetland. Activities take place in the field or office.

I Preparation
Determination of information that is required before entering the field in order to aid in the accurate interpretation of field characteristics. Conducted in the office.

II HGMU delineation
Division of the landscape into discrete functional HGMUs. The construction of a field map that outlines the structure of the site and the coded HGMUs that comprise it. Conducted in the field.

III HGMU characterisation
Identification of the characteristics of the site that are important to its functioning. Construction of a database characterising the individual HGMUs that make up the site. Conducted in the field and office.

Follow through the three sections of the wetland database establishment carefully. Instructions are given for each stage, and each of these should be followed in the order in which they are presented. In some sections you may be able to omit sections depending upon either the functions you are assessing or information previously collected in the wetland database establishment.

The instructions are written as clearly as possible to aid the user. They assume minimal expert knowledge. However, conducting the wetland database establishment is a skill that will develop with experience. As familiarity develops, the process will become quicker and users can develop their own shortcuts according to their individual requirements.

Output from database establishment

The output from this exercise is a database detailing the characteristics of the wetland area being assessed, and its surroundings. This database is held on a series of maps and recording sheets. The maps give details of the whole site and the position of the HGMUs within the site; the recording sheets give details on site management and characteristics of individual HGMUs.

Data recording and references

The functional assessment procedures (FAPs) work by interrogating the data collected in this section through a series of questions and decision trees. Each entry must therefore be recorded with a reference that can be used to find it easily. Maps are referred to by use of map numbers (i.e. AA Map 3). Data on recording sheets are referenced by the recording sheet number (e.g. Recording Sheet 1) and a hierarchical number, which is derived from the section requested in the wetland database establishment section. Data are requested by either an ACTION, a question under an ACTION or an INSTRUCTION under an ACTION.

Where you are instructed to record data upon a recording sheet, the number of the ACTION (e.g. C.5 or F.3) is used as the point of reference for this entry on the recording sheet and the FAPs (during assessment). For example, in the Field Preparation section, Question C.3.3.1 has an entry space for the answer on Recording Sheet 1, in a box labelled C.3.3.1. During the assessment, if this information is required, the question asked would refer to this number in the recording sheet, where the answer will be found.

Where questions beneath an ACTION request data, the question number becomes part of the data reference. For instance, if you are answering Question 13 under ACTION C.5, the reference becomes C.5.13. When an INSTRUCTION under an ACTION requests data, the ACTION reference number is used. This type of hierarchy is used to refer to all data recorded on the recording sheets.

The final database comprises:

AA Map 1 (base map)
- Working copy for use in the field, containing
  - Archive information, site name, worker and date
  - Assessment area boundary
  - Boundaries of major landscape components
  - Boundaries of hydrogeomorphic units
  - Codes identifying hydrogeomorphic units
  - Transects walked

AA Map 2 (support map)
- Working copy produced from the desk study and updated in the field, containing
  - Archive information, site name, worker and date
  - Assessment area boundary
  - Land-use in assessment area
  - Underlying geology
  - Extent of flooding

AA Map 3 (final map)
- A final, neat copy of AA Map 1 updated after the field visit, containing the final version of all the information gathered on AA Map 1.

AA Map 4
- A spare copy in case of damage to AA Maps 1, 2 or 3

CA Map 1
- Final copy produced from desk work, containing
  - Archive information, site name, worker and date
  - Assessment area boundary

Contributory area boundary
Land-use in contributory area
Underlying geology

CA Map 2
A spare copy of the CA Map in case of damage to CA Map 1.

Recording Sheet 1
Details of the land-use and management of the AA obtained from land owners, land managers, etc. (double sided).

Recording Sheet 2a
Containing a hydrogeomorphic code unique to each HGMU, and comprising four sections

SL	G1	FA	H6
Landscape code	Gradient code	Geomorphic code	Hydrological code

Also soil descriptions for each HGMU in the AA (single sided).

Recording Sheet 2b
Contains an appraisal sheet for the section: Nutrient Input Characterisation (H1 and H2) for each HGMU in the AA (single sided).

Recording Sheet 2c
Contains an assessment chart for the assignment of nutrient inputs for each HGMU in the AA using the results of Recording sheet 2b (single sided).

Recording Sheet 2d
Contains an appraisal sheet for the section: Trace Element Input Characterisation (I1 and I2) for each HGMU in the AA (single sided).

Recording Sheet 2e
Contains an assessment chart for the assignment of trace element inputs for each HGMU in the AA using the results of Recording sheet 2d (single sided).

Recording Sheet 3
Contains geomorphological, hydrological, ecological indicators, vegetation codes and HGMU area for each HGMU in the AA (single sided).

Recording Sheet 4
Contains soil descriptions for a single HGMU in an AA when more than one profile is described (HGMU delineation, Section 2.2.C). Not always used (double sided).

Further reading and references

Lloyd J.W., Tellam J.H., Rukin N. and Lerner D.N. (1993). Wetland vulnerability in East Anglia: A possible conceptual framework and generalised approach, *Journal of Environmental Management* **37**, 87–102.

Maltby E., Hogan D.V. and McInnes R.J. (1996). *Functional Analysis of European Wetland Ecosystems – Phase 1 (FAEWE)*, European Commission, Ecosystems Research Report No. 18. Office for Official Publications of the European Communities, Luxembourg.

Mitsch, W.J. and Gosselink, J.G. (2007). *Wetlands*, 4th edn, Wiley, New York.

2.1 Fieldwork preparation

Introduction

Fieldwork preparation is the first stage in the development of the wetland database and has several purposes. One of the key aims is to produce maps to be used in the field, which will illustrate the wetland area of interest called the assessment area (AA) and the wider area that can influence its functioning, called the contributory area (CA).

Other information that can assist the user in the field is gathered. This information may relate to climate, land-use, management, conservation status and history of flooding. Local experts, landowners, environmental agencies and conservation bodies may be useful sources of such information. A range of possible sources is given in Appendix 2.

- Work through all sections (2.1.A–2.1.E) of the Fieldwork Preparation section regardless of the functions you are assessing and in the order presented.
- Complete all the following tasks irrespective of the chosen assessment approach or functions of interest. A prerequisite of this volume of the functional assessment procedures (FAPs) is that the user already knows the boundaries of the AA under assessment. The FAPs are not intended to operate as wetland delineation procedures.

2.1.A Assessment area (AA) delineation and map preparation

The purpose of this section is to define the river or lake marginal wetland area to be assessed (hereafter referred to as the assessment area, AA) on a map and to prepare a baseline AA map using archive or secondary information. This will form the basis for all further data collection and functional analysis.

2.1.A.1 *ACTION: Define AA*

Consider the exact purpose of the planned functional assessment, and therefore which precise area you should be looking at. The hydrogeomorphic unit (HGMU) approach allows the user to assess both wetlands and associated areas, including those that may possess non-wetland characteristics.

The AA will fall into one or more of the following categories in which delineation is described by:

1. field boundaries;
2. administrative boundaries;
3. ownership boundaries;
4. wetland extent.

For categories 1–3 delineation is relatively straightforward and can be easily transferred to a map. In category 4, it is up to the user to set the boundary according to a personal or institutional definition of a wetland. The FAPs have provided a conceptual definition of both river and lake marginal wetlands (Section 1.8.2). A practical, more general definition of wetlands has also been provided (Section 1.4.3). However, to ascribe a single definition here would make assumptions regarding the nature of your assessment. Therefore if your site falls in category 4 you must define your own wetland area.

Notes:

- The AA boundaries can be drawn on any available map.
- There is no minimum or maximum size for an AA.

2.1.A.2 *ACTION: Collect archived/secondary source data*

Collect all potentially valuable archived information about the site. The following list provides guidelines for suitable sources.

- It is desirable to have either detailed up-to-date topographic maps of the site and surrounding area (e.g. Fig. 2.1.1, see colour section), or aerial photographs to improve a less recent or less detailed map.
- Older maps and aerial photographs (Fig. 2.1.2, see colour section) can be useful for identifying changes in landcover and management.
- Geological maps, land-use maps and soil maps.
- Vegetation maps.
- Records of site management.
- Records of site flooding regime.
- Records or maps of land cover in the upstream river catchment.
- Details of land owners at and around the site, land managers, persons and institutions that have studied the site, persons or organisations that regularly use the site for recreational purposes such as nature conservation, bird watching, angling and hunting.

Notes:

- The process of constructing a HGMU map is demonstrated sequentially throughout this document through a series of photos and maps of Kismeldon Meadows, an extensively studied wetland site (see Table 1.8.2 and Section 1.8.5). Located in Devon, UK, it is a typical river marginal wetland comprising wet grassland, and demonstrates features commonly found on wetland sites. The maps and photographs used throughout the document show how these features should be represented on an HGMU map.
- This information is used throughout the AA Map Preparation section. It is most sensibly collected at the same time to prevent undue replication of effort and time wasting. Full details of potential sources of this information can be found in Appendix 2.
- Care should be taken in this section to ensure copyright laws are not breached.

2.1.A.3 *ACTION: Prepare map of AA*

Follow the guidelines below for the preparation of this map. The guidelines are not stepwise so it is advisable to read through them before starting AA map preparation:

1. Select a topographic map that details features such as watercourses, field boundaries, contours and spot heights of the area in which the AA is located.
2. Copy, scan or print the map to create at least four copies, following the guidelines given below for scale and size. Three of these are AA maps to be used during Delineation and Characterisation, the fourth is a backup map to act as spare in case of accident or damage in the field. An extra map might be needed for the wider area of the CA. Ensure that copyright laws are not breached if published maps are photocopied.
3. The scale of the available maps will vary from case to case. If necessary, enlarge or reduce the map by scanning it and printing it out, or by photocopying it. Note that this will change the map-scale and that most photocopies will distort the original map slightly, especially towards the edges. Copy the scale bar onto the map prior to photocopying. Note that digital images may also be available for the area of interest.
4. Copying and magnifying/reducing a topographic map might result in map

lines becoming unclear or too wide or thin. In such cases it is advisable to enhance or redraw the map after copying and use copies of the redrawn map as the base map for further work.

5. The copied map should be no smaller than one piece of A4 paper (297 × 210 mm) or larger than A3 (420 × 297 mm). Since these maps are intended for use in the field, their size should be appropriately manageable.
6. Recommended scales of the copied map are 1:1250 or 1:2500, depending on the size of the AA.
7. The AA should cover about half the map, leaving a border (10–20 % of the map width) around the sides to cover areas adjacent to the AA, which may be annotated later. Also, leave space for the information box shown in Action A.6 at the end of this section (below).
8. Draw or copy an 'information box' on the base map, similar to the one shown in Action A.6.
9. If the AA is too large to give the required level of detail (1:5000 map on A3 format) subdivide it along boundaries clearly identifiable in the field and produce a map for each division with approximately 10 % overlap. Name the divisions accordingly. Also prepare a smaller scale map covering the entire AA, with the positions of the larger scale maps noted on them.
10. Label the maps **AA Map 1–AA Map 4**, respectively. Use suffixes for partial AA maps if the AA is subdivided (e.g. **AA Map 1a**, **AA Map 1b**). N.B. Replicate maps might be needed to keep the information recorded on the maps clear, for instance, if there is too much information on one map an additional one will be needed.
11. Retain the original map for reference and protect it with a transparent plastic sleeve. Use the copies in the field.

2.1.A.4 ACTION: Delineate AA boundary

Record information: AA Map 1, AA Map 2, AA Map 3 + spare(s)
Draw the AA boundary in red on the map. Make sure the defined assessment area forms a practical unit on the map following topographic features such as streams, ditches and roads. It helps when the exact position of the AA boundary is clearly identifiable in the field. Using a pencil, draw the AA boundary on the AA map. The precise nature and application of the assessment purpose (and not just a matter of convenience) might dictate delineation.

Note: The use of a soft pencil can help to differentiate between the AA boundary and topographic features printed on the map.

2.1.A.5 ACTION: Enhance AA map

Record information: AA Map 1, AA Map 2
As a general rule in the wetland database establishment, to reduce the amount of written information needed on the map, use the following colours to draw or redraw certain key features on your AA maps:

- **Grey**: preliminary notes (for all objects below); boundaries outside the AA.
- **Blue**: streams, drains, ditches, springs and lentic water, e.g. ponds, reservoirs.
- **Green**: embankments.
- **Red**: AA and HGMU boundaries (the latter are explained in Section 2.2).
- **Yellow**: delineation transects (used in Action C.1 in Section 2.2).
- **Grey**: all other objects, such as roads, paths, land-use boundaries, etc. (added to the maps in Sections 2.2 and 2.3).

In this action:

- Enhance **AA Map 1** for all features listed above (as far as possible at this stage) except for the delineation transect and HGMU boundary.
- Use all other available resources such as topographic maps and aerial photographs to check if the pattern of ditches, field drains, streams, canals,

roads, paths, embankments are correct according to your best information. Modify in the appropriate colour where necessary.

2.1.A.6 *ACTION: Put archive information on maps*

Record information: AA Map 1, AA Map 2, AA Map 3 + spare(s)
For later interpretation and database purposes it is important to note the following information on each map:

- **AA name**: If the AA has a given or assigned name, this should be included on the AA map. If the AA does not possess a known name then the user should assign a name to it. The user can devise a personal identifier, so long as the adopted name does not duplicate the name of nearby wetland, or in any way cause confusion. Some AAs, for example, might be named after a river or lake that they are associated with, e.g. River Thames Site 1, River Thames Site 2. Alternatively, they can be named after a nearby landmark, building or village, e.g. Bridge Farm Site 1, Newtown Wetland Site 1. Whatever system is adopted, the objective is for the AA to be identified and the name to be clearly displayed on the AA map.
- **Date of visit**: The date of the field visit must be included on the AA map, on the day of the field visit.
- **Field recorder**: Both the person responsible for the field recording and the person responsible for the preparation of the AA Map must include their name on it. This will serve as an important item of reference information if the field worker is not the same user that is completing the remainder of the functional assessment.
- **North arrow**: An arrow indicating magnetic north must be drawn on the AA map.
- **Reference co-ordinates**: Either nationally acceptable reference co-ordinates based on a geometric grid, or geographic co-ordinates representing lines of latitude and longitude.
- **Scale**: It is very important that the field recorder enters the scale on the AA map. This can be of the form x cm = y m, or as a ratio such as 1:5000. If the map is enlarged or reduced by scanning or by photocopying, ensure that a scale-line (e.g. 1 km or 1 mile) is placed on the map before this is done.
- **Weather conditions**: It is worth adding a few brief notes about the weather conditions on the day of the field visit, such as general conditions and antecedent conditions (for example, 'sunny after rainy spell').

An example of a suggested method for representing the pertinent information on the AA map can be found in Table 2.1.1 and an AA map with annotations to this stage is illustrated in Fig. 2.1.3 (see colour Section).

2.1.B Contributory area (CA) delineation and map preparation

The purpose of this section is to identify and delineate the area of land affecting the wetland under assessment (the Contributory Area, CA). This sets the AA into a wider landscape context, which is essential for certain functions.

Note: It is possible that the data necessary for delineation and subdivision of the CA are not readily available. In this case carry out Actions B.1–B.4 as far as possible and see if more information can be gathered during the delineation and characterisation phases.

2.1.B.1 *ACTION: Define the contributory area (CA)*

Define as well as possible the expected location and limits of the CA from available maps and aerial photographs. The CA is the area that drains directly into the AA; either through surface flow, subsurface flow or groundwater flow but not through direct flooding from rivers or lakes. The CA is therefore the area between the AA

Table 2.1.1 Example of pertinent map information

Name: Bridge Farm Site 1	Date: 01 August 2004	Field recorder: A. Fieldworker	North ↑
Reference co-ordinates: *ST924882 (Ordnance Survey)*	**Scale:** *1 cm = 50 m; (1:5000)*	**Weather conditions:** *Fine, sunny, clear skies, max. Temp 25 °C*	

boundary (upslope from the water body) and the first topographic watershed divide. Artificial structures outside the AA such as canals and major roads are also CA boundaries if they intercept run-off from slopes and land drains further up and lead it to a stream or ditch bypassing the AA.

The CA may be defined as the area in which if a drop of water were to fall, the water would enter the AA before entering the river or lake (assuming all rainfall runs overland). In other words the CA is the hydrological catchment of the AA.

Note: In highly modified, flat landscapes with artificial, often pump drainage, the CA may not easily be defined or might have been separated. This means that the AA might not have a CA. If this is the case, you are still able to carry out assessments, but this section no longer applies and should be omitted.

2.1.B.2 ACTION: Prepare contributory area (CA) map

1. The CA map should cover the AA as well as the CA, including a 200 m zone (for the purposes of site influence) around the CA and AA.
2. Choose a base map for delineation. The scale of this map will usually be much smaller than the AA map, possibly 1:50 000 or smaller – depending on the expected size of the CA.
3. The requirements for these maps are otherwise the same as for the AA map, see Action A.3.
4. Enhance the CA map as was previously done for the AA map (Action A.5).
5. Make two copies of the CA map. Label them **CA Map 1** and **CA map 2**. **CA Map 2** is a spare in case of damage or accident.

2.1.B.3 ACTION: Delineate contributory area (CA)

Record information: CA Map 1 + spare
Draw the CA boundary – in red – on both maps following the watershed boundary and artificial structures that would intercept run-off from slopes. Also draw the AA boundary on these maps.

2.1.B.4 ACTION: Subdivide the CA by gradient

Record information: CA Map 1
On **CA Map 1**, divide the CA into units of uniform gradient and label each unit with the appropriate *gradient code* as detailed below. As this task is meant to get an impression of the overall gradient of the CA, no great detail is needed. The gradient codes can be more easily identified in the field, and therefore it is advisable to only delineate areas with distinctive gradients from the map (which can be identified from the contour spacing), and determine the codes for the main units when in the field.

0–2°	***Level–Gentle***, commonly describes floodplain components	**G0**
>2–5°	***Gentle–Moderate***, often describes foot-slopes	**G1**
>5–11°	***Moderate–Steep***	**G2**
> 11°	***Steep***	**G3**

Where more than one CA unit has been assigned the same gradient code, add a numerical suffix to ensure that each unit has its own distinct code, e.g. $\mathbf{G2}_1$ and $\mathbf{G2}_2$ for two moderate slope units.

Figure 2.1.4 (see colour section) illustrates how the CA map might look at this stage.

2.1.C Collection of desk information on assessment area (AA) and contributory area (CA)

The purpose of this section is to collect as much information about the site as possible in advance of the field visit. This information will be used to help HGMU delineation and as a database for reference by the functional assessment procedures.

2.1.C.1 ACTION: Contact local experts for information

Key resources for the wetland database establishment are people with knowledge of the site. They can often provide information that is not available from maps, field surveys or archives. This is especially true for AA and CA management practices and determination of flooding regime. Suitable organisations, people and institutions to contact are listed in Appendix 2.

Contact as many people as necessary or relevant (landowners are the best starting point) in order for them to:

- Assist in the collection of information under Sections 2.1.C and 2.1.D (wetland database establishment).
- Assist in Section 2.2, HGMU delineation, and Section 2.3, HGMU characterisation, which are also part of the wetland database establishment. It is advisable to read through the requirements of these sections before contacting experts (refer to Appendix 2 for suggestions). Ensure information is gained about any areas in the AA where extra caution is needed during field surveys – e.g. swampy areas or areas that can flood very quickly.

2.1.C.2 ACTION: Determine land-use in and adjacent to the AA

Record information: AA Map 2 and CA Map 1

If relatively recent and accurate land-use data for the AA are available they can greatly assist orientation in the field. Sources of data can be topographic maps (note that different scales/types can carry different information), aerial photographs, satellite images and information recorded about sites and properties. The land-use practice in the areas adjacent to the AA may also influence the functioning of the site.

Using the information available from maps and the resources collected in Action A.2, delineate and annotate the areas of the following land types using the codes provided within the AA and within the CA. All boundaries should be drawn with a grey pencil, on **AA Map 2** and **CA Map 1**:

Sea/estuary S/E – general undifferentiated unit for marine and brackish areas (though the FAP methodology is devised for freshwater ecosystems, there may be tidal areas adjacent to the AA which might need to be identified on the map).
Inland (non-marine) water IW – standing and flowing freshwater ecosystems, both natural and artificial.
Beach/mudflat/cliffs B/M/C – coastal and estuarine habitats.
Salt marsh SM – inter-tidal herbaceous plant communities.
Improved grassland IG – often intensively managed grassland dominated by a few fast-growing species.
Wet grassland WG – unimproved or lightly managed grassland having a high water table for much of the year and containing species adapted to wet conditions.
Dwarf shrub heath DSH – vegetation with more than 25 % ericaceous or *Ulex* species, usually on impoverished acid soils.

Broadleaved/mixed woodland BMW – woodland dominated by broadleaved species or a mixture of broadleaved and coniferous trees.
Permanent crops PC – farmland with crops other than grass not sown on a seasonal basis.
Coniferous woodland CW – dominated by conifers but also includes areas recently felled but intended for return to similar land cover.
Bog BG – rain-fed peat-forming ecosystems.
Fen, marsh and swamp FMS – seasonally or permanently waterlogged or flooded by groundwater or run-off, developed on organic or mineral soils.
Arable and horticultural land AH – farmland seasonally cultivated for crops such as cereals and vegetables, also includes annual grass leys, set-aside and fallow.
Suburban/rural development SRD – built structures and associated development often in rural areas or intimately mixed with one or more of the above habitat types.
Urban/industrial development UID – intensely developed, often long-established urban areas and associated infrastructure (e.g. roads and railways), including current industry and business parks and post-industrial derelict land (unless other habitat types have subsequently developed).
Bare ground/rocks and screes BG/R+S – Unvegetated land (excluding sand and mudflats) and exposed rock surfaces.
Other grassland OG – semi-natural or lightly improved grasslands not included above (e.g. acid or calcareous grasslands).

2.1.C.3 ACTION: Answer the following questions about the site and its management

Record information: Recording Sheet 1 unless otherwise indicated

Notes

- Not all of the information in Action C.3 can always be collected in the Field Preparation phase of the wetland database establishment because the data might not be available from secondary data sources, or contact with local experts might not (yet) have been possible. These questions are repeated in Action E.1, Section 2.3 but re-worded so that they can be filled out using observable field indicators rather than secondary data sources. If a local expert is able to help in the field or the user has sufficient knowledge of the site, then the questions in this Action, C.3, should be used. If there is no expert available or the user does not have sufficient site knowledge then the question in Action E.1, Section 2.3 should be used.
- The hierarchical reference code for data in this section has four levels, e.g. C.3.16.1. This would be explained as follows: C.3.16.1 refers to question 1 in subsection 16 ('Conservation and protected status: designated areas') of Action C.3.
- At this stage, if there is no longer enough room on the AA map, move onto a fresh map, remembering to label it clearly as an additional sheet.
- Throughout this section, requests are made to record information on the AA and CA maps using *codes*. Codes are used to reduce the amount of writing on the maps, which would otherwise become crowded and confusing. A reference list of what each of the codes refers to can be found in Appendix 3.
- Scientific notation has been used generally. This means that e.g. 15 kilogrammes of nitrogen per hectare per year is noted as 15 kg N ha^{-1} a^{-1}.

C.3.1 Climatic and atmospheric information

C.3.1.1 Record the mean total monthly precipitation for each month of the year and also record the annual precipitation (mm), in the spaces provided on **Recording Sheet 1** (**C.3.1.1i** for mean monthly and **C.3.1.1ii** for mean annual). For information, consult your nearest meteorological recording site.

C.3.1.2 Record the mean rainfall in the 4-week period previous to the field visit. Record this information on **Recording Sheet 1** (**C.3.1.2**). For information, consult your nearest meteorological recording site.

Note: The information for QC.3.1.2 might not be available immediately, in which case, remember to come back to this point at a suitable time.

C.3.1.3 Record the mean total annual and monthly potential evapotranspiration (mm) from the nearest meteorological recording site in the spaces provided on **Recording Sheet 1** (**C.3.1.3i** for mean monthly and **C.3.1.3ii** for mean annual).

C.3.1.4 Calculate and record the mean annual wind speed from the nearest meteorological recording site. Then classify the values into the following and record the category:

a. ≤ 4.8 m s^{-1}
b. > 4.8 and ≤ 6.6 m s^{-1}
c. > 6.6 m s^{-1}

C.3.1.5 Classify the mean annual soil temperature (°C) from data recorded at the nearest meteorological recording station into one of the following categories:

a. Pergelic < 0 °C
b. Cryic 0 °C – 8 °C
c. Mesic > 8 °C – 15 °C
d. Thermic > 15 °C – 22 °C

C.3.1.6 Record the mean date of the last spring frost recorded at the nearest meteorological recording station.

C.3.1.7 Record the mean date of the first autumn frost recorded at the nearest meteorological recording station.

C.3.1.8 Calculate in days the approximate duration of the last growing season.

Note: Growing season = number of days between the mean date of the last spring frost and the mean date of the first autumn frost.

C.3.1.9 Is the nitrogen concentration in atmospheric deposition (estimated from Fig. 2.1.5, see colour Section):

a. < 15 kg N ha^{-1} a^{-1}
b. ≥ 15 kg N ha^{-1} a^{-1}
c. Unknown

Note: National or more local information might be available and should be used in preference to Fig. 2.1.5.

C.3.1.10 What is the mean annual temperature (°C) of the water column of surface water (if present) in the AA?

a. < 8 °C
b. ≥ 8 °C
c. Unknown/not applicable

C.3.1.11 What amount of the mean annual precipitation accumulates as snow?

a. < 20%
b. ≥ 20%
c. Unknown/not applicable

C.3.1.12 Is the area assessed situated in an arid or a semi arid region? Refer to a suitable climate map (Fig. 2.1.6, see colour section).

C.3.2 Land-use and management: application of fertilisers and organic wastes

C.3.2.1a How often is inorganic fertiliser applied within the AA?

a. No history of application in the last 20 years (go to QC.3.2.4a)

b. None currently applied but previously applied within the last 20 years or applied less than once a year
c. Currently applied once or more than once a year
d. Unknown (go to QC.3.2.4a)

C.3.2.2a If the answer to QC.3.2.1a is b or c, does this fertiliser contain:
a. Both nitrogen and phosphorus (go to QC.3.2.3a (i))
b. Nitrogen but no phosphorus (go to QC.3.2.3a (i))
c. Phosphorus but no nitrogen (go to QC.3.2.3a (ii))
d. Unknown

C.3.2.3a (i) What is the rate of nitrogen applied to the AA?
a. < 15 kg N ha^{-1} a^{-1}
b. 15–50 kg N ha^{-1} a^{-1}
c. > 50 kg N ha^{-1} a^{-1}
d. Unknown

C.3.2.3a (ii) What is the rate of phosphorus applied to the AA?
a. < 5 kg P ha^{-1} a^{-1}
b. 5–10 kg P ha^{-1} a^{-1}
c. > 10 kg P ha^{-1} a^{-1}
d. Unknown

C.3.2.4a How often is organic fertiliser applied by spraying or spreading (includes slurry, animal manure, stockyard washout, parlour washings, septic tank disposal; treated domestic sewage; sewage sludge, fishmeal, bloodmeal, bonemeal) to the surface of the AA?
a. No history of application in the last 20 years
b. None currently applied but previously applied within the last 20 years or applied but less than once a year
c. Currently applied once or more than once a year
d. Unknown

C.3.2.5a How often is organic fertiliser applied by injection (includes slurry, animal manure) within the AA?
a. No history of application in the last 20 years
b. None currently applied but previously applied within the last 20 years or applied less than once a year
c. Currently applied once or more than once a year
d. Unknown

C.3.2.6a How often is organic industrial waste (includes industrial organic waste; municipal waste; food processing waste) applied within the AA?
a. No history of application in the last 20 years
b. None currently applied but previously applied within the last 20 years or applied less than once a year
c. Currently applied once or more than once a year
d. Unknown

C.3.2.7a How often is lime applied to (parts of) the soil of the AA?
a. No history of application of lime in the last 10 years
b. None currently applied but previously applied within the last 10 years or applied less than once a year
c. Currently applied once or more than once a year
d. Unknown

C.3.2.8 How often is inorganic fertiliser applied within the CA?
a. No history of application in the last 20 years (go to QC.3.2.10)
b. None currently applied but previously applied within the last 20 years or applied less than once a year
c. Currently applied once or more than once a year
d. Unknown (go to QC.3.2.10)

C.3.2.9 If the answer to QC.3.2.8 is b or c, does this fertiliser contain:
a. Both nitrogen and phosphorus
b. Nitrogen but no phosphorus
c. Phosphorus but no nitrogen
d. Unknown

C.3.2.10 How often is organic fertiliser applied by spraying or spreading (includes slurry, animal manure, stockyard washout, parlour washings, septic tank disposal; treated domestic sewage; sewage sludge, fishmeal, bloodmeal, bonemeal) applied within the CA?

a. No history of application in the last 20 years
b. None currently applied but previously applied within the last 20 years or applied but less than once a year
c. Currently applied once or more than once a year
d. Unknown

C.3.2.11 How often is organic fertiliser applied by injection (including slurry, animal manure) within the CA?

a. No history of application in the last 20 years
b. None currently applied but previously applied within the last 20 years or applied less than once a year
c. Currently applied once or more than once a year
d. Unknown

C.3.2.12 How often is organic industrial waste (including industrial organic waste; municipal waste; food processing waste) applied within the CA?

a. No history of application in the last 20 years
b. None currently applied but previously applied within the last 20 years or applied less than once a year
c. Currently applied once or more than once a year
d. Unknown

C.3.2.13 How often is lime added to (parts of) the soil of the CA?

a. No history of application in the last 10 years
b. None currently applied but previously applied within the last 10 years or applied less than once a year
c. Currently applied once or more than once a year
d. Unknown

C.3.2.14 Record on **AA Map 2** the areas within the AA where fertiliser (using the code **FERT**), organic waste (using the code **OWASTE**) or lime (using the code **LIME**) are applied and indicate the type, rate and frequency of fertiliser applied.

C.3.2.15 Record on **CA Map 1** the areas within the CA where fertiliser (using the code **FERT**), organic waste (using the code **OWASTE**) or lime (using the code **LIME**) are applied and indicate the type, rate and frequency of fertiliser applied.

C.3.3 Land-use and management: burning of vegetation

C.3.3.1a How often is vegetation within the AA burnt as part of land management practices?

a. No history of burning in the last 20 years
b. Not burnt within the last 5 years but previously burnt between 5 and 20 years ago
c. Currently burnt once or more than once in the last five years
d. Unknown

C.3.3.2 Record on **AA Map 2** the areas within the AA that are subjected to burning, using the code **BURN**.

C.3.4 Land-use and management: harvesting of herbaceous vegetation

Includes: mowing and removal of grass; hay production; silage; harvesting of arable crops; harvesting of cereals or vegetables; market gardening; production and removal of lawn turf; cutting of reeds.

C.3.4.1a How often is vegetation within the AA harvested?

a. No history of harvesting of herbaceous vegetation in the last 20 years
b. Not currently harvested but previously harvested within the last 20 years or harvested less than once a year

c. Once or more than once a year
d. Unknown

C.3.4.2a For the crops where herbaceous vegetation is harvested, which of the following categories would best describe the harvesting practices?
a. Predominantly mechanised practices
b. Predominantly manual practices

C.3.4.3a How often and when does harvesting take place:
a. Once, only in spring (between April–June)
b. Once, only in summer (between July–August)
c. Once, only in autumn (between September–November)
d. Once, only in winter (between December–March)
e. Twice or more than twice a year

C.3.4.4 Record on **AA Map 2** the areas within the AA where harvesting of herbaceous vegetation occurs (using the code **HARV**). Where different harvesting practices are used (from QC.3.4.2a above) also indicate these on the map.

C.3.5 *Land-use and management: ploughing of land*

C.3.5.1a How often is any land within the AA ploughed?
a. No history of ploughing in the last 20 years
b. Not currently ploughed but previously ploughed within the last 20 years or ploughed but less than once a year
c. Currently ploughed (every year)
d. Unknown

C.3.5.2 How often is any land within the CA ploughed?
a. No history of ploughing in the last 20 years
b. Not currently ploughed but previously ploughed within the last 20 years or ploughed but less than once a year
c. Currently ploughed once or more than once a year
d. Unknown

C.3.5.3 Record on **AA Map 2** the areas within the AA that are ploughed, using the code **PLOU**.

C.3.5.4 Record on **CA Map 1** the areas within the CA that are ploughed, using the code **PLOU**.

C.3.6 *Land-use and management: woodland management and forestry*

Includes: felling; coppicing; pollarding; clearing; planting.

C.3.6.1a How often does felling and removal of timber in woodland areas occur within the AA?
a. Not in the last 20 years
b. No current felling and removal but previously practised within the last 20 years or felling and removal occurs but less than once in 5 years
c. Current felling and removal or previously practised within the last 5 years
d. Unknown

C.3.6.2 Record on **AA Map 2** the woodland areas within the AA where felling and removal of trees, coppicing or pollarding occurs (using the code **FOREST**).

C.3.7 *Land-use and management: livestock density*

Proportions of livestock animals relate approximately to those of the livestock unit (LSU) – see glossary in Appendix 1. More details about LSUs can be found on DEFRA's Rural Development Service Technical Advice Note 33.

C.3.7.1a Are any of the following forms of livestock kept within part/all of the AA?

- Cattle or horses or donkeys $\geq$ 1 ha^{-1}
- Pigs $\geq$ 3 ha^{-1}
- Sheep or goats $\geq$ 6 ha^{-1}
- Turkeys or geese $\geq$ 30 ha^{-1}
- Chickens or ducks $\geq$ 100 ha^{-1}

a. No history of livestock kept in the last 20 years
b. None currently kept but previously kept within the last 20 years or currently kept but at a density lower than the figure indicated above
c. Currently kept at a density equal to or higher than the figure indicated above
d. Unknown

C.3.7.2a Within the AA, are any of the livestock listed in QC.3.7.1a kept for the purpose of meat, milk or wool production?

C.3.7.3a Within the AA, do the animals feed/graze in the same area as where they ruminate/rest/sleep?

C.3.7.4 Record on **AA Map 2** the areas within the AA where the above forms of livestock are kept at densities above those listed in QC.3.7.1a, using the code **LV** and locate the areas used for feeding, grazing, resting and sleeping by shading.

C.3.7.5 Are any of the following forms of livestock kept within part/all of the CA?

- Cattle or horses or donkeys $\geq$ 1 ha^{-1}
- Pigs $\geq$ 3 ha^{-1}
- Sheep or goats $\geq$ 6 ha^{-1}
- Turkeys or geese $\geq$ 30 ha^{-1}
- Chickens or ducks $\geq$ 100 ha^{-1}

a. No history of livestock kept in the last 20 years
b. None currently kept but previously kept within the last 20 years or currently kept but at a density lower than the figure indicated above
c. Currently kept at a density equal to or higher than the figure indicated above
d. Unknown

C.3.7.6 Record on **CA Map 1** the areas within the CA where the above forms of livestock are kept at densities above those listed in QC.3.7.5, using the code **LV**.

C.3.8 *Land-use and management: bird populations*

Includes breeding colonies; nesting colonies; feeding colonies; roosting colonies; moulting colonies.

C.3.8.1a Have any large populations of birds (greater than 1000 ha^{-1}) been recorded within the AA?

C.3.8.2 Record on **AA Map 2** (using the code **LPB**), the areas where large populations of birds occur within the AA.

C.3.8.3a Record the approximate density (below) of the large populations of birds present within the AA.

a. 1000–2000 birds ha^{-1}
b. > 2000 but < 5000 birds ha^{-1}
c. 5000–10 000 birds ha^{-1}
d. > 10 000 birds ha^{-1}
e. < 1000 birds or not applicable

C.3.9 *Land-use and management: drainage*

Surface artificial drainage: includes: arterial drainage; drainage ditches.

C.3.9.1 Are any areas within the AA drained through the use of surface drainage?

C.3.9.2 Record on **AA Map 2** any currently active surface drains within the AA, using the code **ACTSD**.

C.3.9.3 Are any areas within the CA drained through the use of surface drainage?

C.3.9.4 Record on **CA Map 1** any currently active surface drains within the CA, using the code **ACTSD**.

Subsurface artificial drainage: includes pipe drains, secondary drainage (moling).

C.3.9.5a Are any areas within the AA drained through the use of subsurface drainage?

C.3.9.6 Record on **AA Map 2** any currently active subsurface drains within the AA, using the code **ACTSSD**.

C.3.9.7 Are any areas within the CA drained through the use of subsurface drainage?

C.3.9.8 Record on **CA Map 1** any currently active subsurface drains within the CA, using the code **ACTSSD**.

C.3.10 *Land-use and management: peat cutting and extraction*

C.3.10.1a How frequently is peat cut and removed from the AA?

a. No peat cutting or extraction occurs
b. Hand cutting or extraction of peat occurs at a local or traditional scale
c. Mechanised cutting or extraction of peat occurs at an industrial or commercial scale
d. Unknown

C.3.10.2 How frequently is peat cut and removed from the CA?

a. No peat cutting or extraction occurs
b. Hand cutting or extraction of peat occurs at a local or traditional scale
c. Mechanised cutting or extraction of peat occurs at an industrial or commercial scale
d. Unknown

C.3.10.3 Record on **AA Map 2** the location and extent (by shading with a pencil) of current cutting of peat within the AA, using the code **PEATEX**.

C.3.10.4 Record on **CA Map 1** the location and extent (by shading with a pencil) of current cutting of peat within the CA, using the code **PEATEX**.

C.3.11 *Land-use and management: mining and extraction*

Includes: gravel extraction; mineral extraction; quarrying.

C.3.11.1a Are any areas within the AA currently undergoing mining, quarrying or extraction of any form (other than peat mining)?

a. No indication of mining, quarrying or extraction
b. Historical areas of mining, quarrying or extraction, currently inactive
c. Currently active areas of mining, quarrying or extraction.
d. Unknown

C.3.11.2 Are any areas within the CA currently undergoing mining, quarrying or extraction of any form (other than peat mining)?

a. No indication of mining, quarrying or extraction
b. Historical areas of mining, quarrying or extraction, currently inactive
c. Currently active areas of mining, quarrying or extraction.
d. Unknown

C.3.11.3 Record on **AA Map 2** the location and extent (by shading with a pencil) of mining, quarrying or extraction activities within the AA, using the code **MINEX**.

C.3.11.4 Record on **CA Map 1** the location and extent (by shading with a pencil) of mining, quarrying or extraction activities within the CA, using the code **MINEX**.

C.3.12 *Land-use and management: application of industrial inorganic waste*

Includes: mine and quarry spoil, mineral washings, slag heaps, dredgings.

C.3.12.1a How often is industrial inorganic waste applied within the AA?
a. No history of application in the last 20 years
b. None currently applied but previously applied within the last 20 years or applied less than once a year
c. Currently applied once or more than once a year
d. Unknown

C.3.12.2 How often is industrial inorganic waste applied within the CA?
a. No history of application in the last 20 years
b. None currently applied but previously applied within the last 20 years or applied less than once a year
c. Currently applied once or more than once a year
d. Unknown

C.3.12.3 Record on **AA Map 2** the areas within the AA subjected to industrial inorganic waste disposal, using the code **IIWASTE**.

C.3.12.4 Record on **CA Map 1** the areas within the CA subjected to industrial inorganic waste disposal, using the code **IIWASTE**.

C.3.13 *Land-use and management: construction*

Includes: road construction, railway construction, building construction; other construction work.

C.3.13.1a Are any areas within the AA currently undergoing construction work?

C.3.13.2 Are any areas within the CA currently undergoing construction work?

C.3.13.3 Record on **AA Map 2** the location and extent of construction activities within the AA, using the code **CONSTR**.

C.3.13.4 Record on **CA Map 1** the location and extent of any construction activities within the CA, using the code **CONSTR**.

C.3.14 *Land-use and management: hunting, shooting, fishing*

C.3.14.1 Are any areas within the AA used for hunting, shooting or fishing?
a. Not used for hunting, shooting or fishing
b. Used for hunting, shooting or fishing at a recreational scale
c. Used for hunting, shooting or fishing at a commercial scale
d. Unknown

C.3.14.2 Record on **AA Map 2** the location and extent of hunting, shooting or fishing activities within the AA, using the code **HUNSF**.

C.3.15 *Land-use and management: recreation*

Includes: footpaths; bridleways; playing fields; cycle paths; off-road vehicle access; camping sites; riverside watersport access.

C.3.15.1 Do any areas within the AA possess some form of direct recreational use?
a. No direct recreational use
b. Light recreational use
c. Heavy recreational use
d. Unknown

C.3.15.2 Record on **AA Map 2** the location and extent of recreation activities within the AA, using the code **RECRE**.

C.3.16 *Conservation and protected status: designated areas*

Includes: internationally important/protected areas; nationally important/protected areas; locally important/protected areas; Ramsar sites; national nature reserves; sites of special scientific interest; important bird areas; conservation 'red flags'; other designations, local or national (see Appendix 2).

C.3.16.1 Are there any areas within the AA that have been assigned an international conservation status?

C.3.16.2 Are there any areas within the AA that have been assigned a national conservation status?

C.3.16.3 Are there any areas within the AA that have been assigned a local conservation status?

C.3.16.4 Record on **AA Map 2** the location and extent of conservation areas within the AA of

a. International importance (use the code **ConII**)
b. National importance (use the code **ConNI**)
c. Local importance (use the code **ConLI**)

C.3.17 *Conservation and protected status: species of conservation status*

Includes: The Habitats Directive; The Bern Convention; The Bonn Convention; internationally important/protected species; nationally important/protected species; The Wildlife and Countryside Act (amended 2000) (UK) and the Countryside and Rights of Way Act (2000) (UK); Red Data Book species; nationally scarce species; locally important species (see Appendix 2).

C.3.17.1 Has the AA been surveyed for its flora and/or fauna? (Check local conservation bodies)

C.3.17.2 Are other records of the flora and/or fauna of the AA available? (Check local conservation bodies, and local floras and fauna lists)
[No to QC.3.17.1 and QC.3.17.2, move on to QC.3.19]

C.3.17.3 If the answer to either QC.3.17.1 or QC.3.17.2 is yes, has the presence of any of the following species been recorded previously of international conservation importance?
Tick all appropriate boxes on the recording sheet.

a. Macroinvertebrates
b. Fish
c. Herptiles
d. Birds
e. Mammals
f. Plants

C.3.17.4 If the answer to either QC.3.17.1 or QC.3.17.2 is yes, has the presence of any of the following species been recorded previously of national conservation importance?
Tick all appropriate boxes on the recording sheet.

a. Macroinvertebrates
b. Fish
c. Herptiles
d. Birds
e. Mammals
f. Plants

C.3.17.5 If the answer to either QC.3.17.1 or QC.3.17.2 is yes, has the presence of any of the following groups been recorded previously as locally scarce?
Tick all appropriate boxes on the recording sheet.

a. Macroinvertebrates
b. Fish
c. Herptiles
d. Birds

e. Mammals
f. Plants

C.3.17.6 If the answer to QC.3.17.3, QC.3.17.4 or QC.3.17.5 is yes (boxes a, b, c, d, e or f have been ticked), is there any reason to believe that the group might be no longer present?
Tick all appropriate boxes on the recording sheet.
a. Macroinvertebrates might no longer be present
b. Fish might no longer be present
c. Herptiles might no longer be present
d. Birds might no longer be present
e. Mammals might no longer be present
f. Plants might no longer be present

C.3.18 *Conservation and protected status: migratory species*

C.3.18.1 Is there any record of invertebrates that migrate in large numbers frequenting the AA?
C.3.18.2 Is there any record of migratory fish species in the river channel or tributaries?
C.3.18.3 Is there any record of birds that migrate in large numbers frequenting the AA?
C.3.18.4 Is there any record of mammals that migrate in large numbers frequenting the AA?

C.3.19 *River and lake water information: river and lake flooding regime*

C.3.19.1 Reference level: Identify a point or feature in the AA or its surroundings with a known elevation relative to the national reference level (e.g. Ordnance Datum (m.O.D.)). Mark the location of the reference point and its level on **AA Map 2**.
C.3.19.2 Flood level: For the AA, determine the following as accurately as possible from inlet hydrographs, spot measurements, anecdotal records and local evidence:
a. Water level (m.O.D.) for frequent (at least once a year) flood events
b. Water level (m.O.D.) for occasional (once every 2–10 years) flood events
c. Water level (m.O.D.) for rare (once every 10–50 years) flood events
d. River bank height (m.O.D.) or mean maximum level of lake

C.3.19.3 Flood extent: If possible, outline the following in grey on **AA Map 2**, from information gained from flood event maps, aerial photographs, historical records of flood events and local knowledge: [Flood limits may also be estimated from water level information in combination with topographic (contours, spot heights) information for the AA, if these are of good quality].
a. Statistically determined flood extent lines for frequent (once at least every 2 years), occasional (once every 2–10 years) and rare (once every 10 or more years) floods
b. Statistically determined flood levels for frequent, occasional and rare floods. These can be converted into flood extent lines if a good topographical map is available and if it is evident that full hydrological connectivity between the river or lake and the AA exists – i.e. if floodwater can enter the site freely
c. Flood extent lines for riverbank full height (the height of the river bank before it breaches during flooding)
d. Maximum flood extent lines, which may have been determined earlier from a range of historical and field sources
e. Historical flood extent lines, mapped during a particularly extreme flood in the past or represented on an aerial photograph
f. Indications of areas that are 'liable to flooding'

C.3.19.4 Flood discharges: Based on the inlet hydrograph and spot flow measurements, determine the following peak discharges in m^3/day

a. Maximum daily discharge during frequent (once at least every 2 years) flood events
b. Maximum daily discharge during occasional (once every 2–10 years) flood events
c. Maximum daily discharge during rare (once every 10–50 years) flood events

C.3.19.5 Duration of floods: Based on the inlet hydrograph, historical flood event records and local knowledge, classify flood durations into three categories: (1) < 1 week, (2) 1 week–1 month, (3) > 1 month:

a. Duration of frequent (once at least every 2 years) flood events
b. Duration of occasional (once every 2–10 years) flood events
c. Duration of rare (once every 10–50 years) flood events

C.3.19.6 From inlet hydrograph, historical flood event records and local knowledge estimate the months in which flooding would normally occur in a typical year.

C.3.19.7a Non-overbank flooding interactions: Is it possible that water from the (main) water body (river/stream/ditch/lake) has contact with the AA?

a. No never/not applicable
b. Yes, the AA borders the (main) water body and horizontal exchange (either via surface or via subsurface flow) can occur freely between the AA and the water body
c. Yes, although the AA does not border the water body, the AA has contact with the water from the water body during flood events
d. Contact between water in the water body and the AA does take place once or more than once a year, but in another way than mentioned in answers b and c (e.g. indirectly, via drainage ditches)
e. Contact between water in the water body and the AA does take place, but in another way than mentioned in answers b and c (e.g. indirectly, via drainage ditches) and less than once a year
f. Unknown

C.3.19.8a Flooding frequency: How often is the AA (or parts of the AA) flooded with water from the (main) water body?

a. Frequently (once or more than once a year)
b. Occasionally (once every 2–10 years)
c. Rarely (once every 10–50 years)
d. Unknown

C.3.20 *River/lake water information: river/lake water quality*

Note: C.3.20–C.3.23 asks questions of specific European Directives where the Water Framework Directive (WFD) uses this information to denote chemical/biological quality. Standards and recommended limits can be found in the Annexes of these Directives see also http://ec.europa.eu/environment/water/water-framework/index_en.html).

C.3.20.1 Does the water in the stream, river, pond or lake passing through or adjacent to the AA fail to reach any of the standards set in the EC Freshwater Fish Directive (EC, 1978)?

C.3.20.2 Does the water in the stream, river, pond or lake passing through or adjacent to the AA fail to reach any of the standards set in the EC Dangerous Substances Directive (EC, 1976)?

C.3.20.3 Does the water in the stream, river, pond or lake passing through or adjacent to the AA fail to reach any of the standards set in the EC Urban Wastewater Directive (EC, 1991a)?

C.3.20.4 Does the water in the stream, river, pond or lake passing through or adjacent to the AA fail to reach any of the standards set in the EC Nitrate Directive (EC, 1991b)?

C.3.20.5 Does the water in the stream, river, pond or lake passing through or adjacent to the AA fail to reach any of the standards set in the EC Surface Water Abstraction Directive (EC, 1975)?

C.3.20.6 What is the mean annual nitrate–nitrogen concentration of the water in the main river or lake?

a. < 1 mg l^{-1} NO_3–N
b. ≥ 1 mg l^{-1} NO_3–N
c. Unknown

C.3.20.7 What is the mean annual ammonium–nitrogen concentration of the water in the main river or lake?

a. < 1 mg l^{-1} NH_4–N
b. ≥ 1 mg l^{-1} NH_4–N
c. Unknown

C.3.20.8 What is the mean annual phosphate–phosphorus concentration of the main river or lake?

a. < 0.1 mg l^{-1} PO_4–P
b. ≥ 0.1 mg l^{-1} PO_4–P
c. Unknown

C.3.20.9 What is the mean annual total nitrogen concentration of the water in the main water body?

a. < 2.2 mg l^{-1} N
b. ≥ 2.2 mg l^{-1} N
c. Unknown

Note: Only answer QC.3.20.10–QC.3.20.13 if you wish to assess the process of dissolved organic carbon (DOC) input into surface waters through soil (Section 3.2.F.2). These questions require analytical information on the content of DOC in surface waters. To find out whether your AA of interest has this information, you could:

1. ask local experts or consult other sources of background information (Appendix 2);
2. contact university or specialist environmental research groups who might have such data;
3. carry out the analysis yourself using standard methods as described by the Swedish Environmental Protection Agency's Environmental Quality Criteria (SEPA, 2000).

C.3.20.10 Is there more than 5 mg l^{-1} of DOC in the surface water of the AA (i.e. is the water brown-coloured after filtering with a 0.45 μm filter) during baseflow?

C.3.20.11 Is there more than 5 mg l^{-1} of DOC in the surface water of the AA (i.e. is the water brown-coloured after filtering with a 0.45 μm filter) during periods of high flow?

C.3.20.12i Is there more than 5 mg l^{-1} of DOC in any nearby streams or lakes in the AA or CA (i.e. is the water brown-coloured after filtering with a 0.45 μm filter) during periods of high flow or baseflow?

C.3.20.12ii Is there less than 5 mg l^{-1} of DOC in any nearby streams or lakes in the AA or CA (i.e. is the water pale or clear after filtering with a 0.45 μm filter) during periods of high flow or baseflow?

C.3.20.13 Is there more than 2 mg l^{-1} of particulate organic carbon (i.e. that is retained by a 0.45 μm filter) at any time during the year?

a. Yes, in the stream/river/lake
b. Yes, in nearby streams/rivers/lakes
c. No

Notes:

- The trace element functions require more detailed information than generally required by these procedures. For this reason, before going any further with these particular functions you should check whether you have the capacity to collect information on the following:

- Munsell soil colour using the Munsell Soil Color Charts, Munsell Color Company Inc., Baltimore, Maryland 21218, USA (instructions in Action C.3, Section 2.2);
- analytical data on the concentration of trace elements in surface waters (instructions in QC.3.20.14);
- analytical data on the concentration of trace elements in soils (instructions in QC.3.25.1;

If you are unable to collect any of this information, you will be unable to carry out assessments for the trace element functions. Therefore you should ignore all questions labelled as specific for the assessment of Trace Element Export or Retention.

- Cadmium (Cd) is used as an example trace element because it is toxic, often abundant and easily mobilised. Cadmium is a relevant trace contaminant for many European alluvial wetlands, and this module can be expected to have wide applicability as a result. The trace element functions have been developed following extensive work in a case study site in the catchment of the Elbe in Germany, which has elevated concentrations of Cd. The concentration ranges used for classifying concentrations relative to background levels are therefore based on the Elbe data. The interpretation of the interactions between physico-chemical processes and their influence on the behaviour of Cd is based on empirical data from these studies. There are other trace elements that may need to be considered in wetlands and these include, but are not limited to, arsenic, mercury and selenium. Similar processes of assessment may be used for these, but it should be noted that background concentrations of trace elements are likely to be location-specific due to regional differences in mineralogy. In addition, the assessor will need to consider the modifying influences of physico-chemical processes within the wetland on the speciation and resultant behaviour of the trace element under consideration. As a result, prescriptive guidance based on concentration ranges may not be applicable across all regions of Europe. For this reason, the user should identify appropriate regional concentration data for the specific trace elements in their AA, then undertake a comparative assessment of measured wetland concentrations accordingly. A literature search will give guidance on processes and trace elements of relevance in European wetlands. Suitable starting references are given in Sections 3.2.D and 3.2.E.
- The trace element functions will be subject to further testing and development in future editions. At present, this set of functions should be considered as a prototype.
- QC.3.20.14 is specific for the assessment of trace element export and trace element retention. Only answer this question if you wish to carry out an assessment for these functions (Sections 3.2.D and 3.2.E). QC.3.20.14 requires analytical information on the concentration of trace elements in surface waters. To find out whether your AA of interest has this information, you could:
  1. consult local experts or other sources of background information (Appendix 2);
  2. contact university or specialist environmental research groups in the area.

C.3.20.14 What is the trace element concentration in the river or lake water?

a. ≥ normal for the area
b. < normal for the area
c. No information available

E.g. given for the Elbe catchment in Germany (see note above)

C.3.20.14 What is the trace element concentration in the river or lake water?

a ≥ normal for the area, e.g. for Cd in Elbe ≥ 2.5 mg Cd kg^{-1}
b < normal for the area e.g. for Cd in Elbe < 2.5 mg Cd kg^{-1}
c No information available

C.3.21 *Groundwater information: geological information*

Depending on the scale of the maps and of the size of the AA and CA, existing geological maps might be accurate enough to distinguish geological units. Draw the information indicated below in grey on **CA Map 1.** Also draw on **AA Map 2** if more than one geological unit underlies the AA.

C.3.21.1 If a hydrogeological map is available, outline the following areas in grey on **CA Map 1** where applicable. Also draw these areas on **AA Map 2** if more than one hydrogeological unit underlies the AA.
a. Part of a regionally important recharge area
b. Part of a locally important recharge area
c. Part of a regionally important discharge area
d. Part of a locally important discharge area

C.3.21.2 If an aquifer- or groundwater-resources map is available, draw the information indicated below in grey on **CA Map 1**. Also draw on **AA Map 2** if more than one water resources unit underlies the AA.
a. Major aquifers
b. Minor aquifers
c. Non-aquifers

C.3.21.3 If a geological map is available, delineate the areas where the rocks that comprise the solid geology underlying the AA and CA are classed as:
a. A karst landscape (characterised by features such as sinkholes, dolines, turloughs, cave systems, underground streams)
b. Thoroughly fractured sandstone (in areas with much faulting)
c. Another rock type known locally to have high permeability or make a good aquifer
d. Any other rock type

C.3.21.4 If a geological map is available, delineate the areas where the deposits that comprise the non-solid (or *drift*) geology underlying the AA and CA are classed as:
a. Gravels – possibly alluvial (rivers) or fluvioglacial (glacial rivers) in origin
b. Sand – possibly alluvial or aeolian (dunes) in origin
c. Silt – possibly alluvial or aeolian (loess) in origin
d. Clay – possibly alluvial or lacustrine (lakes) in origin
e. Peat – over 1 m deep

C.3.21.5 If a soil map is available, delineate the areas where the soils are classed as:
a. Well drained
b. Otherwise indicative of rapid infiltration of rainfall throughout the year

C.3.21.6 Is it likely that the groundwater is rich in calcium (e.g. because the groundwater flow originates in, or passes through, areas with marine sand soils or limestone)?
Record Y/N/NA in the appropriate section of Recording Sheet 1.

C.3.22 *Water table information*

C.3.22.1 Depth of groundwater table: Wherever such information is available for the AA, note down the following information on **Recording Sheet 1** and mark the location on **AA Map 2**. Keep in mind that these water levels can be highly variable throughout the AA:
a. Annual mean water level
b. Minimum groundwater level in wells or piezometers
c. Maximum groundwater level in wells or piezometers

C.3.22.2a Depth of surface water: Wherever such information is available for the AA, note down the following information and mark the location on **AA**

Map 2 as well as **Recording Sheet 1**. Keep in mind that these water levels can be highly variable throughout the AA:

a. Mean depth of surface water level in standing water, streams or ditches

b. Minimum depth of surface water level in standing water, streams or ditches

c. Maximum depth of surface water level in standing water, streams or ditches

C.3.22.3 Water table fluctuations: Is the top 20 cm of the soil prone to water table fluctuations (alternate periods of waterlogged and non-waterlogged conditions, which may be due to natural, seasonal fluctuations or water table management), and which result in dry soil conditions for at least half of the year?

C.3.22.4a Is the water level maintained at high levels in (part of) the AA at any time (e.g. for nature conservation purposes)?

C.3.22.5 Record on **AA Map 2** the areas within the AA where the circumstances described in QC.3.22.3 and QC.3.22.4a are applicable.

C.3.23 *Groundwater information: groundwater quality*

C.3.23.1 EC groundwater quality directives: If such information is available, does the groundwater contained in the aquifer underlying the AA fail to reach any of the standards set in the EC Protection of Groundwater against Pollution Directive (EC, 1980)?

C.3.24 *Catchment land-use and management: activities and features affecting sedimentation*

C.3.24.1 Erosion enhancing features: Indicate which of the following occupies over 10 % of the upstream catchment or over 50 % of the area immediately upslope (in the AA or CA) occupied by one or more of the following types of area listed below?

a. Mountainous areas

b. Sparsely vegetated areas

c. Urban areas

d. None of the above

C.3.24.2 Erosion enhancing activities: Indicate which of the activities listed below has taken place in more than 50 % of the area upslope or in more than 10 % of the upstream catchment directly along the river channel or lake shore during the last 3 years.

a. Mining

b. Urban development or road-building

c. Major channel improvements in the river and major morphological improvements (modification or extension) in the lake (e.g. dredging)

d. Logging or afforestation (including lake shore)

e. None of the above

C.3.24.3 Sediment transport reducing features: Indicate which one of the following, found within the AA, is likely to be the main factor responsible for reducing sediment transport.

a. Locks or weirs

b. Wooden bulkhead (against erosion of sediments from the lake shoreline due to wind fetch and water level)

c. Sediment basin (a temporary barrier or other suitable retention structure built across an area of water flow to intercept run-off and allow transported sediment to settle and be retained prior to discharge into river/lake waters)

d. None of the above

C.3.24.4 Sediment transport enhancing features: Are any of the following features present along significant stretches (along at least 20 % of the total channel length) of the main river channel upstream of the AA? In the case of lake marginal wetlands (LMW), the condition of 20 % of the total channel length should be applied to any inflow tributaries.

a. Major embankment schemes or river dykes reducing flood storage
b. Straightened channels bypassing natural meanders
c. None of the above

C.3.24.5 Does the river or area immediately upstream of the AA (within 20 % of the total upstream river length or lake area) have a recent history of migrating channels (during the last 20 years)?

Note: Figures 2.1.7 and 2.1.8 depict how the AA Map and CA Map (respectively) might look following the above exercise.

C.3.25 *Soil analytical data on trace element concentration*

The following questions in C.3.25 are specific for the assessment of trace element export and trace element retention (Sections 3.2.D and 3.2.E). Only answer these questions if you wish to carry out an assessment for these functions. The questions require analytical information on the concentration of trace elements in surface waters. To find out whether your AA of interest has this information, you could:

1. consult local experts or other sources of background information (Appendix 2);
2. contact university or specialist environmental research groups in the area.

C.3.25.1a Does the soil trace element concentration indicate that anthropogenically derived contamination of the AA has taken place via overbank flooding? *(Recording Sheet 1)*

a. Yes, the soil trace element concentration is slightly higher than typical background concentrations
b. Yes, the soil trace element concentration is moderately higher than typical background concentrations
c. Yes, the soil trace element concentrations are much higher than typical background concentrations
d. No, the soil trace element concentrations are not elevated
e. Unknown, no soil analytical data are available

E.g. for the Elbe catchment example site: Total soil cadmium concentration

C.3.25.1b Does the soil Cd concentration indicate that anthropogenically derived contamination of the AA has taken place via overbank flooding? *(Recording Sheet 1)*

a. Yes, the soil Cd concentration is slightly higher than typical background concentrations (> 5 mg kg^{-1} soil)
b. Yes, the soil Cd concentration is moderately higher than typical background concentrations (> 10 mg kg^{-1} soil)
c. Yes, the soil Cd concentrations are much higher than typical background concentrations (> 20 mg kg^{-1} soil)
d. No, the soil Cd concentrations are not elevated (″ 5 mg kg^{-1} soil)
e. Unknown, no soil analytical data are available

Note: This is specific to the Elbe catchment. In the UK, for example, typical background Cd concentrations are around 0.1–2 mg kg^{-1} soil

C.3.25.2a Total organic matter content: What is the soil organic matter content of the topsoil in the AA (obtained by chemical analysis or field estimate of the upper 30 cm or ploughed layer)? *(Recording Sheet 1)*

a. < 2 %
b. 2–8 %
c. > 8 but < 15 %
d. ≥ 15 %
e. Unknown, no organic matter analytical data are available

Added to the map at this stage is information collected from the desk study in Section 2.1.C., relating to the AA. The following information, if available from desk information, should appear on the map at this stage.
(NOTE: items in the list are marked with either * to indicate that this information was not available at this stage or # to indicate that this information was not applicable in this example, and therefore is not shown on the map.)
Land use in AA and adjacent areas; Areas subject to fertiliser/organic waste application; Areas subject to burning; #Areas in which vegetation is harvested; Areas subject to ploughing; #Woodland subject to felling; Livestock grazing areas; #Areas with large populations of birds; Surface drainage; Areas drained by sub-surface drains; #Areas subject to peat cutting; #Location of mining, quarrying and/or extraction activities; #Areas used for industrial organic waste disposal; #Areas subject to construction activities; #Areas used for hunting and/or fishing; #Areas used for recreation; Nature conservation designated areas; A reference level; Flood extent; *Hydrogeology; *Aquifer extents; *Ground/soil water depths; *Surface water depths.
If it is necessary to represent a large amount of the information listed above on AA Map 2, it might be necessary to use several replicates, each containing a limited amount of information, in order that the maps remain legible.

2.1.7 Example of AA Map 2, amended with environmental data from the database establishment exercise. Reproduced by permission of Ordnance Survey on behalf of HMSO. © Crown copyright 2008. All rights reserved. Ordnance Survey Licence number 100047942.

CA MAP 1

NAME	DATE	FIELD RECORDER
KISMELDON MEADOWS	16/10/02	A. FIELDWORKER
REFERENCE CO-ORDINATES	SCALE	WEATHER
SS 235117	0 100M 200M	CLOUDY, DRY MAX. TEMP. 14°C

Added to the map at this stage is information collected from the desk study in Section 2.1.C., relating to the CA. The following information, if available from desk information, should appear on the map at this stage.
(NOTE: items in the list are marked with either * to indicate that this information was not available at this stage or # to indicate that this information was not applicable in this example, and therefore is not shown on the map.)
Land use; Areas subject to fertiliser/organic waste application; Areas subject to ploughing; Livestock grazing areas; Surface drainage; Areas drained by sub-surface drains; #Location of mining, quarrying and/or extraction activities; #Areas used for industrial organic waste disposal; #Areas subject to construction activities; *Aquifer extents.
If it is necessary to represent a large amount of the information listed above on CA Map 1 it might be necessary to use several replicates, each containing a limited amount of information, in order that the maps remain legible.

2.1.8 Example of CA Map 1, after addition of information gained from the desk and field research. Reproduced by permission of Ordnance Survey on behalf of HMSO. © Crown copyright 2008. All rights reserved. Ordnance Survey Licence number 100047942.

2.1.D Confirmation of rights of access

The purpose of this section is to ensure that the field recorder acquires full rights of access to the AA prior to the field visit.

2.1.D.1 *ACTION: Confirm rights of access*

Check with the appropriate land owner(s) or land manager(s) prior to undertaking any fieldwork.

2.1.E Selection of required field equipment

The purpose of this section is to ensure that equipment appropriate to the functions being assessed and to conduct fieldwork is taken into the field.

2.1.E.1 *ACTION: Define type of field assessment and assessment area and select required equipment*

Types of assessment are:

1. hydrological
2. biogeochemical
3. ecological

The following equipment is required for each activity:

All functions:

- Site maps (see earlier in this section)
- AA Map 1, AA Map 2, CA Map 1
- Spare maps
- Recording Sheet 1
- Delineation guidelines and sufficient numbers of Recording Sheets 2, 3 and 4
- Spare paper or a notebook
- Clipboard with a transparent plastic bag as a cover
- Pencils – grey, red, blue, green, yellow, plus some spares
- Pencil sharpener
- Eraser
- Rain-proof field clothing
- Wellington boots
- Soil auger
- Compass
- Ruler
- Botanical field key (e.g. Hubbard, 1992; Rose and O'Reilly, 2006; Fitter *et al.*, 1996)
- Magnifying glass
- Soil sample bags: re-sealable plastic bags to store soil samples from augered soil samples
- Rag to clean hands

Assessment of biogeochemical functions:

- Field pH meter

Additional equipment that might aid your assessment:

- Clinometer
- Specific ion electrodes
- Conductivity meter

2.1.F Further reading and references

Bobbink, R., Hornung, M. and Roelofs, J.G.M. (1998). The effects of air-borne nitrogen pollutants on species diversity in natural and semi-natural European vegetation, *Journal of Ecology* **86**, 717–738.

EC (1975). Council Directive 75/440/EEC of 16 June 1975 concerning the quality required of surface water intended for the abstraction of drinking water in the Member States, *OJ* **L194**, 25.07.1975, 26–31.

EC (1976). Council Directive 76/464/EEC of 4 May 1976 on pollution caused by certain dangerous substances discharged into the aquatic environment of the Community *OJ* **L129**, 18.05.1976, 23–29.

EC (1978). Council Directive 78/659/EEC of 18 July 1978 on the quality of fresh waters needing protection or improvement in order to support fish life, *OJ* **L222**, 14.08.1978, 1–10.

EC (1980). Council Directive 80/68/EEC of 17 December 1979 on the protection of groundwater against pollution caused by certain dangerous substances, *OJ* **L020**, 26.01.1980, 43–48.

EC (1991a). Council Directive 91/271/EEC of 21 May 1991 concerning urban waste-water treatment, *OJ* **L135**, 30.05.1991, 40–52.

EC (1996). Council Directive 91/676/EEC of 12 December 1991 concerning the protection of waters against pollution caused by nitrates from agricultural sources, *OJ*, **L375**, 31.12.1991, 1–8

Fitter, R., Fitter, A. and Blamey, M. (1996). *Wild Flowers of Britain and Northern Europe.* Collins Pocket Guide, HarperCollins Publishers, London.

Hubbard, C.E. (1992). *Grasses*, Penguin Books, London, UK.

Maltby, E., Hogan, D.V. and McInnes, R.J. (1996). *Functional Analysis of European Wetland Ecosystems – Phase 1 (FAEWE)*, European Commission, Ecosystems Research Report No. 18. Office for Official Publications of the European Communities, Luxembourg.

Mitsch, W.J. and Gosselink, J.G. (2007) *Wetlands*, 4th edn, Wiley, New York.

Rose, F. and O'Reilly, C. (2006). *The Wild Flower Key. How to identify wild flowers, trees and shrubs in Britain and Ireland*, Frederick Warne, London, UK.

SEPA (2000) *Environmental quality criteria for lakes and watercourses*, Report 4913, Swedish Environmental Protection Agency, Stockholm, Sweden.

Tarrasón, L. *et al.* (2006) *Transboundary Acidification, Eutrophication and Ground Level Ozone in Europe since 1990 to 2004*, EMEP Status Report 1/2006, L. ISSN 1504-6109, available at http://emep.int/publ/common_publications.html

2.2

Hydrogeomorphic (HGMU) delineation

Introduction

HGMU delineation is carried out in order to divide the landscape into units, which have uniform functioning: hydrogeomorphic units (HGMUs). This is conducted in the field. At the end of this section the HGMUs will be delineated and no further alteration of their boundaries will take place. Units are characterised in more detail in Section 2.3, HGMU characterisation. A HGMU is a landscape unit of uniform geomorphology and hydrological regime; soils must also be uniform in as far as they are a reflection of the hydrology and geomorphology. Criteria for the degree of homogeneity that is required are provided in the guidelines that follow.

Delineation of HGMUs within the assessment area is carried out under the following criteria:

- hydrology (hydrological boundaries may be natural or man-made)
- geomorphology
- soil

Initial HGMUs are mapped using criteria for 'hydrology' and for 'geomorphology'. These can be amalgamated – but not usually divided – using soil information. Mapping criteria for each category are provided by the guidelines.

Instructions:

1. In addition to the equipment required in Section 2.1.E, the following maps will be used for HGMU delineation as a minimum with spares in case of error or accident:
    - **AA Map 1** – 3 copies
    - **AA Map 2** – 3 copies
    - **AA Map 3** – 3 copies
    - **CA Map 1** – 1 copy
2. In order to be consistent and avoid confusion, a strict colour coding has been defined for the annotation of maps:
    - **Grey:** preliminary notes (for all objects below); boundaries outside Floodplain HGMU Component.
    - **Blue:** streams, drains, ditches, and springs.
    - **Green:** embankments.
    - **Red:** AA and HGMU boundaries.
    - **Yellow:** delineation transects (used in Action C.1).
    - **Grey:** all other objects (added to the maps in this section and Section 2.3).
3. Mark HGMU boundaries initially on **AA Map 1** only. Transfer all information to **AA Map 3** at the end of the delineation activities.

2.2.A Check and enhance prepared AA Map

The purpose of this section is to check if the information recorded on the AA and CA maps is broadly correct and to orientate and familiarise the user with the site before conducting the field survey. This is especially important when the prepared maps are based on old information or on information of low accuracy or resolution.

This section can be conducted solely by the user but should, where possible, be conducted with the aid of a person having local knowledge. Ideally the person should be able to assist in interpretation of the field evidence during part or all of the HGMU delineation and HGMU characterisation sections. If this is not possible, then gain as much local insight as possible, preferably in advance of the field visit, by obtaining comments on the actions listed below. Information that is best collected from experts with local knowledge is listed in the List of information sources in Appendix 2.

Note: You will recognise some text that is repeated from earlier sections. This is intended to help smooth your way through the procedures and avoid the necessity of referring back to earlier pages.

2.2.A.1 *ACTION: Conduct a brief survey of the assessment area (AA)*

This can be conducted from a bridge, the slope of a nearby hill or any other point – preferably elevated – inside or outside of the AA, that allows a 'bird's eye' view of the AA. Select more than one point if needed for a full view – it may be necessary to walk around the entire AA. If more than one point is needed, repeat the following actions from each point.

Instruction 1: Compare field situation with details on AA maps

Record information: AA Map 1, AA Map 2 and CA Map 1

The following points should be checked, and enhancement should be made on **AA Map 1**:

1. ***Topography***: Is the pattern of outstanding features such as streams, ditches, intensively drained areas, embankments and roads in the AA in accordance with that on the map? Make corrections on **AA Map 1** and **CA Map 1** where relevant.
2. ***Land-use***: Sketch the patterns of the broader land-use types that are clearly discernible, on **AA Map 2** – this is especially important where access is limited or where boundaries will be hard to map in the field, such as woodland, tall reedbeds and open water. Delineate and annotate the areas of the following land types (also listed in Action C.2, Section 2.1) within the AA and within the CA. All these boundaries should be drawn with a grey pencil, on **AA Map 2** and **CA Map 1**:

Sea/estuary S/E – general undifferentiated unit for marine and brackish areas (though the FAP methodology is devised for freshwater ecosystems, there may be tidal areas adjacent to the AA which may need to be identified on the map)
Inland (non-marine) water IW – standing and flowing freshwater ecosystems, both natural and artificial
Beach/mudflat/cliffs B/M/C – coastal and estuarine habitats
Salt marsh SM – inter-tidal herbaceous plant communities
Improved grassland IG – often intensively-managed grassland dominated by a few fast-growing species
Wet grassland WG – unimproved or lightly managed grassland having a high water table for much of the year and containing species adapted to wet conditions
Dwarf shrub heath DSH – vegetation with more than 25 % ericaceous or *Ulex* species, usually on impoverished acid soils
Broadleaved/mixed woodland BMW – woodland dominated by broadleaved species or a mixture of broadleaved and coniferous trees

Permanent crops PC – farmland with crops other than grass not sown on a seasonal basis
Coniferous woodland CW – dominated by conifers but also includes areas recently felled but intended for return to similar land cover
Bog BG – rain-fed peat-forming ecosystems
Fen, marsh and swamp FMS – seasonally or permanently waterlogged or flooded by groundwater or run-off, developed on organic or mineral soils
Arable and horticultural land AH – farmland seasonally cultivated for crops such as cereals and vegetables, also includes annual grass leys, set-aside and fallow
Suburban/rural development SRD – built structures and associated development often in rural areas or intimately mixed with one or more of the above habitat types
Urban/industrial development UID – intensely developed, often long-established urban areas and associated infrastructure (e.g. roads and railways), including current industry and business parks and post-industrial derelict land (unless other habitat types have subsequently developed)
Bare ground/rocks and screes BG/R+S – unvegetated land (excluding sand and mudflats) and exposed rock surfaces
Other grassland OG – semi-natural or lightly improved grasslands not included above (e.g. acid or calcareous grasslands)

Instruction 2: Note further information on map

Record information: AA Map 1, CA Map 1 or field notebook

- The user is of course free to note any other general information that may help with the delineation and characterisation of the HGMUs and orientation in the field. Note this on **AA Map 1**.
- On **AA Map 1** and **CA Map 1**, check and modify any information in a 200 m zone around the AA and the CA.

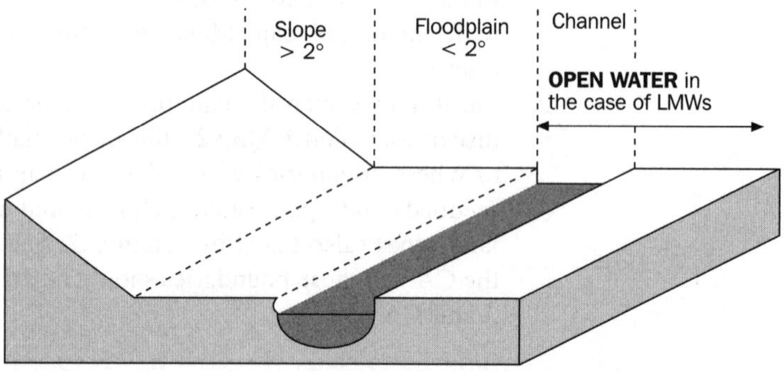

2.2.1 Schematic cross-section of an AA and its components. LMW = lake marginal wetland.

2.2.B Subdivide AA into its main components

The purpose of this section is to subdivide the AA into its main landscape components. These are later subdivided into HGMUs in the following sections. The AA is defined as the area the user wishes to assess. Potentially three main geomorphic components can occur within it: a slope, a floodplain and a channel. Before the AA can be subdivided into HGMUs, which can then be characterised and assessed, the user must delineate these components (Fig. 2.2.1).

2.2.B.1 *ACTION: Divide the AA into the major landscape components: slope, floodplain and channel*

This can often be done before entering the field or from an observation point if the entire AA can be seen. Draw component boundaries on **AA Maps 1** and **2** in red.

Definitions:

- ***Channel***: land conveying or capable of conveying water in a confined channel or complex of confined channels which, in their natural conditions, display characteristic fluviogeomorphological features, such as river bank, meander, slip-off slope, river cliff, channel bars, islands. Also included are artificial channels such as ditches and drains, which are especially characteristic of highly modified, flat landscapes.
- ***Open water***: water mass which covers an extended area (lake, pond, reservoir, etc.), not necessarily man-made.
- ***Floodplain***: land adjacent to a channel or open water component as defined above, which has an overall slope of less than 2° in a direction perpendicular to the channel or open water component.
- ***Slope***: land adjacent to a floodplain or to a channel/open water component as defined above, which has an average slope usually more than 2° in a direction perpendicular to either the floodplain or to the channel/open water landscape components.

Where there is doubt over the slope of a main component, a clinometer can be used to establish this in the field. In some cases one of these units may be absent, for instance in a headwater stream characterised by steep slopes with poor or minor floodplain development, or in association with high stream order rivers in lowland, coastal or low relief areas where the floodplain is very wide and the valley slopes are at a distance beyond the extent of the AA.

2.2.B.2 *ACTION: Note AA component boundaries on map*

Record information: AA Map 1

Draw the AA component boundaries on **AA Map 1** as far as you can see, using a red pencil. Follow the legend given below (see Fig. 2.2.2, colour section).

Legend for notation of major AA component boundaries:

- Smooth boundary (gradual without a clear point at which the slope changes): - - - - - - -
- Angular boundary (an easily identifiable change in gradient, occurring within 10 m): ————

2.2.C Subdivide AA components into HGMUs: preparation and procedure

The purpose of this section is to prepare for and give instruction upon field HGMU delineation that divides the AA into a patchwork of delineated and coded HGMUs within the three main AA landscape components. The establishment and orientation of transects is explained along with the actions to be carried out along them. The procedure for HGMU delineation and coding is described. Record the date of the field visit in the box provided on **Recording Sheet 1**.

2.2.C.1 *ACTION: Identify transects through the AA landscape components*

Record information: AA Map 1

To delineate the HGMUs, their boundaries must be mapped in the field using a combination of the information recorded on **AA Map 1** and observable natural and artificial hydrological and geomorphological features in the field. To do this, transect(s) through the AA must first be identified.

1. Identify a transect through the site, which crosses as many of the hydrological and geomorphological landscape features as possible. A single transect may be appropriate to assess the AA, but multiple transects may be established if necessary. The transects do not have to be straight, but they should meet as many of the following conditions as possible:
   - be no further than 100 m apart, though this depends on site conditions;
   - cover the main features and variations;
   - be as evenly distributed through the AA as possible;
   - be identifiable in the field;
   - be possible to follow in the field (i.e. safe and physically manageable);
   - avoid following streams or artificial structures which are likely to form HGMU boundaries;
   - include mature islands if they are included in the channel AA component – if multiple transects are used, they may intersect if necessary.
2. Transect lines should be drawn on **AA Maps 1** and **2** in a yellow pencil, and numbered with Roman numerals (e.g. I, II, X) (Figure 2.2.3, see colour Section).

Guidelines for the identification of HGMU boundaries

Hydrological boundaries (Figs 2.2.4–2.2.6). The following features are likely to relate to, or result in, changes in the hydrology of the site, and therefore represent HGMU boundaries:

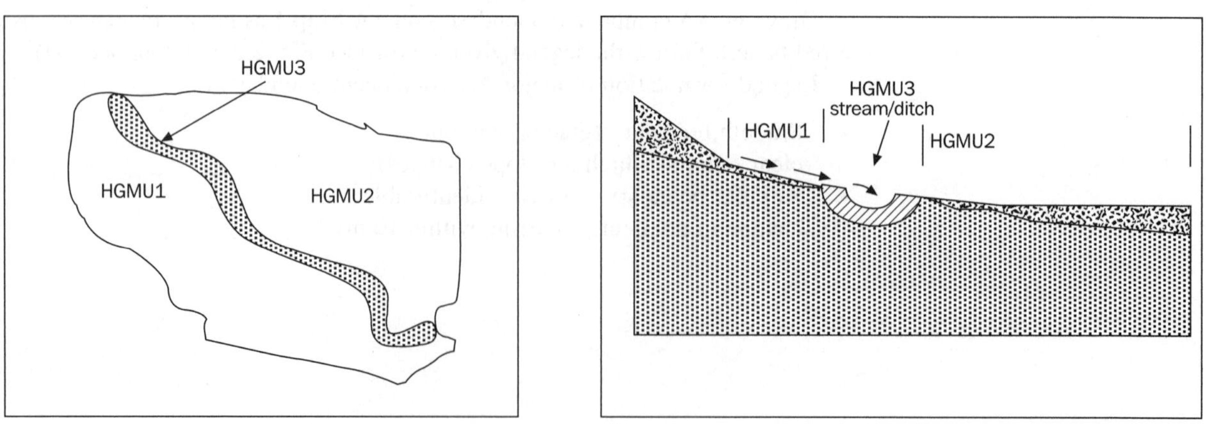

(a) (b)

2.2.4 Hydrological boundaries: (a) stream/ditch; (b) cross-section (streams/ditches may separate areas with different inundation or groundwater regimes).

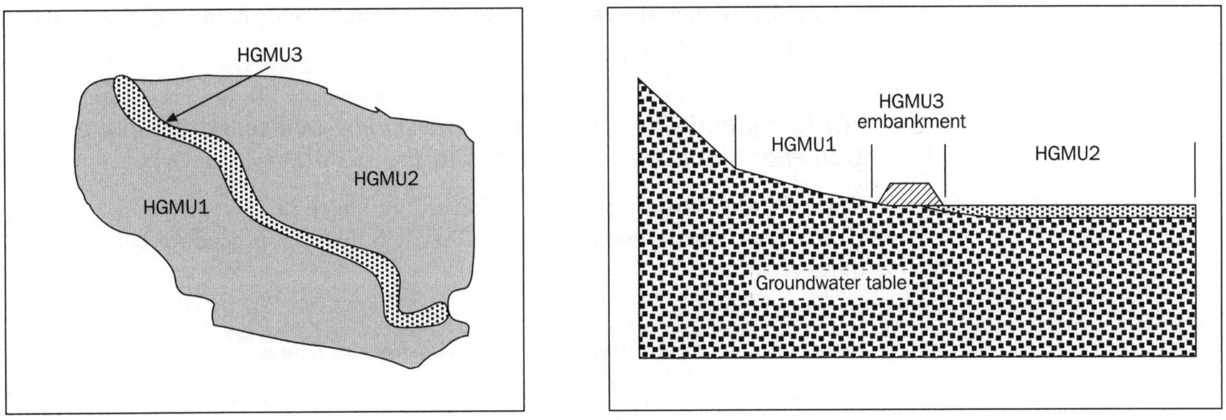

(a) (b)

2.2.5 Hydrological boundaries: embankment (HGMU3) (a) plan view; (b) cross-section.

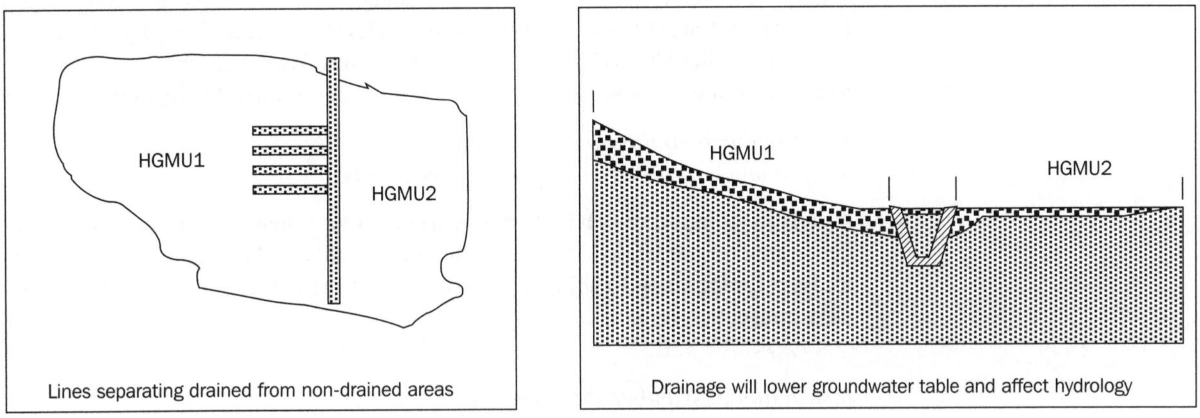

(a) (b)

2.2.6 Hydrological boundaries: subsurface drained and non-drained areas (a) plan; (b) cross-section.

- *Embankments* may separate flooded from non-flooded units (it should be held in mind that embankments are also HGMU types with a geomorphic code (Instruction 7)).
- *Streams and larger ditches* (> 1 m wide at the upper ground surface level) may separate areas with different inundation or groundwater regimes (it should be remembered that these are also HGMU types with a geomorphic code (Instruction 7)).
- *Lines separating intensively drained areas* (intervals between drains approximately < 50 m) from non-drained areas (intervals between drains > 50 m). Only consider these when drained and wet non-drained areas are clearly discernible and the drainage system is functioning fully. Drainage will lower the groundwater tables and thereby affect the hydrology.
- *Areas of overland flow originating from natural or artificial hydrological discharges.*

Geomorphological boundaries: These can be identified as significant changes in the gradient of the land surface. The following should be kept in mind:

- Sometimes the boundaries are not immediately apparent in the field, but they are often indicated by changes to the 'wetness' of the habitat, as expressed by flooded or waterlogged soils associated with specific vegetation types or structure.
- Soil wetness and, to a lesser extent, the distribution of vegetation can vary seasonally and with events.
- Vegetation characteristics and boundaries can be used as indicators only if

evidently they do not primarily result from management, i.e. they must not coincide with only a fence or a ditch.

2.2.C.2 *ACTION: Identifying, mapping and assigning landscape components, gradients and geomorphic codes to the HGMUs*

Record information: AA Map 1, Recording Sheet 2a

To identify, map and code the HGMUs, follow the procedure below for each transect.

Instruction 1: Walk through the AA following the transect

Observe any general features of interest as before. Record these on **AA Map 2**.

Instruction 2: Identify HGMU boundaries

When you reach a potential HGMU boundary, using the definition above (Action C.1 – Guidelines for the identification of HGMU boundaries), lightly draw the boundary of the HGMU you have entered on **AA Map 1** in grey pencil using the following notation; you will need to erase this later. Follow the legend given below.

- *Angular boundary between units*: ___________
- *Smooth or graded boundary between units*: _ _ _ _ _ _ _

As a rule, do not draw HGMU boundaries at more than 50 m from your position, depending on the HGMU type and viewing conditions. You may have to leave the transect temporarily to achieve a complete outline if this is not possible from the next transect.

Instruction 3: Number the HGMUs

Record information: AA Map 1 and Recording Sheet 2a (in the space marked: C.2 Number)

Number sequentially each HGMU identified and record it on **AA Map 1** and a corresponding **Recording Sheet 2a**. Use one copy of **Recording Sheet 2a** and **3** per HGMU.

Instruction 4: Code for drained areas

Record information: Recording Sheet 2a (in the space marked: C.2 Drained)

Where an HGMU has been delineated on the basis of a hydrological management boundary (Action C.1) that separates a drained from an undrained area, record **D** on **Recording Sheet 2a**.

Instruction 5: Assign a landscape component code to each HGMU

Record information: Recording Sheet 2a *(in the space marked: C.2 Landscape Code)*

Assign a landscape component code to each delineated HGMU according to the subdivision made in Action B.1 using the following notation, and record it on the corresponding **Recording Sheet 2a**:

- *slope component* **S**
- *floodplain component* **F**
- *channel component* **C**

Instruction 6: Assign a gradient code to each HGMU

Record information: Recording Sheet 2a (in the space marked: C.2 Gradient)

Assign a gradient code to each delineated HGMU using the categories below and note it on **Recording Sheet 2a**. For further information on gradients, refer to

Hodgson (1974) *Soil Survey Field Handbook: Describing and Sampling Soil Profiles*. In the floodplain AA component, gradients should be determined in the general direction of the channel or parallel to the valley slope. In a slope AA component, the steepest gradient should be determined. This should be done for all HGMUs except embankments and open water.

•	0–≤ 2°	***Level–Gentle***, commonly describes floodplain components	**G0**
•	> 2°–≤ 5°	***Gentle–Moderate***, often describes foot-slopes	**G1**
	> 5°–≤ 11°	***Moderate–Steep***	**G2**
	> 11°	***Steep***	**G3**

Where there is doubt over the slope, use a clinometer to measure slope gradient.

Instruction 7: Assign a geomorphic code to each HGMU

Record information: Recording Sheet 2a (in the space marked: C.2 Geomorphic Code)

The main landscape components (slope, floodplain, channel) comprise different types of geomorphological features, all of which need to be coded. Code each delineated HGMU using the geomorphic landform codes listed below.

Geomorphological boundaries generally follow changes in surface level/gradient, except when they are minor changes considered as microtopographical features, as indicated in the list below (see Intermediate features). However, not every boundary is marked by an abrupt geomorphological change. For example, the boundary between a floodplain and footslope is often difficult to discern, as the features may grade into each other gradually over some distance. In such cases, it is often necessary to determine the approximate boundary by examining the soil (see Action C.3, below) and marking it using a broken line (see Action C.2, Instruction 2, above). In some cases, a change in surface elevation of as little as 20 cm over tens of metres will constitute a boundary.

All units will be classified as either *elevated, intermediate* or *depressional*, and a prefix code (E, I or D, respectively) will be assigned, with the help of the definitions below. Thereafter, a geomorphic code will be assigned to the HGMU, using the descriptions of geomorphic landscape features below. Record the geomorphic code on **Recording Sheet 2a**.

Notes:

- The following geomorphic and artificial units can be found commonly in river and lake marginal wetlands but not all may be present in the AA. Some are found only in specific main landscape components and these have been indicated accordingly.
- Mature islands, when they occur within the AA channel or lake component, must be delineated using the same codes as for floodplain components. A mature island is defined as:

  a permanent mid-channel or lake feature, prone to flooding at high flows or lake levels and capable of supporting perennial vegetation
- Other channel features that differ from this description should not be mapped as HGMUs.

Elevated (prefix '**E**'):

Levee	**LV**	a raised natural riverbank formed by the deposition of sediment during periods of floods (floodplain only)
Terrace	**TC**	an elevated shelf adjacent to the valley slopes at higher elevation than the floodplain and only inundated during exceptionally high flood events
Embankment	**EM**	a man-made embankment, built adjacent to the channel, set back from the channel, or built up on top of natural levees (floodplain only)

Ditch embankment	**DE**	a man-made embankment found in association with ditches on slopes or floodplains
River dune	**RD**	sand dunes on floodplain, between 1 and 10 m high (floodplain only)
Peat dome	**PD**	a smooth convex elevated feature comprising a peat soil deposit developed on the floodplain or slope, or bordering a slope or terrace.
Unidentified elevation	**UE**	any elevation that does not fall in any of the above categories.

Intermediate (prefix **'I'**):

Flat area	**FA**	a smooth plain (floodplain component only) lacking distinct microrelief (features generally less than 0.5 m in elevation and 10 m across)
Uniform slope	**US**	a smooth slope (slope component only) lacking surface microrelief (features generally less than 0.5 m in elevation and 10 m across)
Riverbank	**RB**	floodplain component only, with distinct microrelief (within 0.5 m elevation and 10 m across)

Depressional (prefix **'D'**):

Backswamp	**BS**	an area of low elevation and seasonal or permanent waterlogging at the back of the floodplain and adjacent to the valley slope or terrace wall; not as low-lying or well-defined as a depression unit (floodplain only)
Cut-off channel	**CO**	a former river channel, currently cutoff, containing no permanent water but seasonally flooded and may be partially filled with sediment (floodplain only)
Ox-bow lake	**OX**	a former river channel, currently cutoff, containing permanent standing water (floodplain only)
Former channel	**FC**	a sinuous depression, seasonally flooded, still connected to the river channel (floodplain only)
Point bar	**PB**	an area of sediment deposited on the inside of a meander loop, often occurring as concentric ridges and swales; following inundation, sloughs of stagnant water sometimes form in the swales; can be vegetated or unvegetated
Tributary	**TB**	a stream or channel traversing the slope or floodplain, which enters the river channel
Cave	**CV**	a natural cave, deep undercut, or disused mine
Lake or pond	**LK**	a natural or man-made lake or pond, including fishponds, millponds, gravel pits
Ditch	**DT**	a man-made agricultural drainage ditch that may contain permanently flowing water, intermittent flows (mark the flow direction), or stagnant water
Canal	**CN**	an artificial navigational waterway
Artificial depression	**AD**	any dry artificial depression
Unidentified depression	**UD**	any depression that does not fall in the above categories
Artificial channel	**CA**	a linear feature designed to hold or conduct water

Figures 2.2.7 and 2.2.8 (see colour section) indicate ways in which these categories can be depicted on a map. In Fig. 2.2.7, a lake marginal wetland, different HGMUs with the same code are distinguished on the map by colour and a unique numerical suffix (see note, below) to clarify their identity in the assessment process. Descriptors are given in the key. The HGMUs with the same code in the river marginal wetland shown in Fig. 2.2.8 would also need to be distinguished by a different numerical suffix in the same way. Figure 2.2.9 (see colour section) gives

an illustrated example of transect lines, delineated HGMUs and their tabulated codes on an AA map.

Where more than one HGMU has been assigned the same code, add a numerical suffix to ensure each unit has its own distinct code, e.g. **FA1** and **FA2** for two flat areas. If these are further subdivided (e.g. on the basis of soil type), these would then become (**FA1/1** and **FA1/2**, **FA2/1** and **FA2/2**).

Instruction 8: Repeat Action C.2, steps 1–7 for each transect.

Record information: Recording Sheet 2a

2.2.C.3 *ACTION: Verify and refine HGMU boundaries using soil information*

Record information: AA Map 1 and Recording Sheet 2a (for a single soil sample) or Recording Sheet 4 (for multiple soil samples)

The purpose of this section is to collect soils data that can be used to verify and refine provisional HGMU delineation and characterise HGMU soil profiles. The subdivision of the AA is into units with different soils. Soil information can be used to test if the delineation of an HGMU, based on criteria of hydrology and geomorphology, has been correct. Soils are a reflection of hydrological and geomorphological processes, which should be uniform within HGMUs. Soil information is used primarily to confirm or invalidate the existing boundaries but is, in some cases, used to delineate new boundaries.

Notes:

- The information required is of a relatively simple level of detail, and guidelines are provided to aid the interpretation of soil features. However, some expert knowledge may still be required; therefore store part of each sample in a sealed plastic sample bag if this is thought necessary.
- Only conduct the soil survey of the HGMUs once they have been provisionally delineated using hydrological management and geomorphological criteria according to the procedure outlined above in Action C.2.

Instruction 1: Select soil-sampling point(s)

The HGMUs delineated at this stage are provisional, but are uniform and distinct according to geomorphic and hydrological criteria. Use the following guidelines to select soil-sampling points:

- In most cases only one soil sample description is needed for each provisional HGMU (use **Recording Sheet 2a**). However for larger provisional HGMUs it is better to have more than one sample (use **Recording Sheet 4**). Estimate the maximum HGMU length in any direction from **AA Map 1** and select the minimum number of soil samples from the table below (more samples can be taken if there are indications of changes in soil composition):

HGMU length (m)	Number of soil profile descriptions
0–100	1
100–200	2
200–500	3
500–1000	4
> 1000	5

- When a single sample is being taken from an HGMU, sample in the middle of the HGMU area. If more than one sample is taken, either attempt to locate them in different vegetation types or attempt to space the sampling locations at regular intervals.
- Mark the sampling points with a cross on **AA Map 1** with a grey pencil. If more than one soil sample has been taken in a single HGMU, number the samples sequentially.

Note: Exercise due care and apply safe practice if sampling from areas with surface water.

Instruction 2: Record the moisture status of the topsoil

Record information: Recording Sheet 2a or 4 (in the space marked: C.3.2, Topsoil Moisture)

The soil can be described in terms of four different moisture conditions:

- **Flooded**: The soil is submerged beneath standing or flowing water.
- **Wet**: The soil is waterlogged but is not submerged. Squeezing a soil sample with one hand will yield some water or (in the case of peat or clay) saturated soil material squirting out between fingers.
- **Moist**: The soil is moist but squeezing will not yield water.
- **Dry**: The soil lacks any signs of moisture, is pale in colour and tends to be hard or brittle.

Instruction 3: Collect a soil sample

Using a soil auger (screw type is preferable) sample the soil to a total depth of 0.5 m using the method below.

1. Remove any litter, such as leaves and twigs, from the surface of the soil.
2. Measure the soil auger and carefully screw the auger into the soil to the full depth of the screw and then withdraw, ensuring that the soil sample is retained on the auger.
3. Lay out the soil sample along a ruler.
4. Reinsert the soil auger in the hole and continue removing cores of sediment until a depth of 0.5 m is reached or until penetration is no longer possible.
5. At the following intervals conduct Instructions 4–9, recording all information on **Recording Sheet 2a** or **4** (one of these sheets is needed for each provisional HGMU): 0–5 cm; 5–10 cm; 10–15 cm; 15–20 cm; 20–30 cm; 30–40 cm; 40–50 cm.

Instruction 4: Record soil colour

Record information: Recording Sheet 2a or 4 (in the space marked: C.3.4 Matrix colour)

To examine the soil colour, carefully remove an intact lump of soil from the auger and break it open. Examine the 'fresh' face of soil. The colour of the soil can be described in terms of the overall colour of the soil material, or matrix (supplemented by information on the colour of mottles that may be present within the profile – see Instruction 5). The colour of the matrix (and mottles) should be selected from the following:

- **Brownish**
- **Yellowish**
- **Reddish**
- **Greenish**
- **Greyish**
- **Orange (ochreous)**
- **Rusty (ferruginous)**
- **Black**

If parts of the soil are black, distinguish between material that is hard to crush between the fingers (termed manganiferous) and material that is soft and peaty or similar to charcoal (termed organic). Example soil profiles are shown in Figs 2.2.10a and b (see colour Section).

If you wish to carry out an assessment of the functions of trace element storage or trace element export, another sample is required. A 30 cm sample of the topsoil should be mixed and the Munsell Colour Values should be recorded in the space marked **C.3.4ii** on **Recording Sheet 2a** (Munsell Soil Color Charts, Munsell Color Company Inc., Baltimore, Maryland 21218, USA).

Instruction 5: Record the colour of mottles

Record information: Recording Sheet 2a or 4 *(in the space marked: C.3.5 Mottle colour)*

To examine the soil for mottles, carefully remove an intact lump of soil from the auger and break it open. Examine the 'fresh' face of soil. Mottles can appear within the soil profile as spots or blotches of different colour or shades interspersed within the dominant colour of the soil matrix. Their colour can be described using the categories detailed above in Instruction 4. Sometimes mottling is not easily detected or described – as a rule of thumb it should be noted if at least 5 % of the exposed soil section is of a clearly different colour than the remaining 95 %.

Instruction 6: Identify the type of soil material

Record information: Recording Sheet 2a or 4 (*in the space marked: C.3.6 Type of soil material*)

Determine whether the soil is predominantly inorganic or organic before proceeding with Instructions 7 and 8. Inorganic soil material consists of a mixture of clay, silt, sand and gravel in varying proportions, whilst organic soils contain much greater proportions of organic matter in various stages of decay. Where the organic matter content exceeds 50 %, the soil material is described as peat.

A soil profile (sequence of layers or horizons) is classed as organic generally if more than half of the upper 80 cm comprises organic soil material. However, it should be noted that an inorganic soil profile class (soil type) can have a peaty topsoil or subsoil layers provided that inorganic horizons predominate in the profile. If soil material is inorganic then proceed with Instruction 7 followed by Instruction 9. If it is organic, go to Instruction 8.

Instruction 7: Record the particle size of mineral material

Record information: Recording Sheet 2a or 4 *(in the space marked: C.3.7 Mineral particle size)*

Mineral material can be described in terms of its particle-size class (or texture), estimated in the field by rubbing a moist sample of material (moisture can be adjusted by adding water (saliva) between the thumb and forefinger to examine the texture. The following categories should be used for describing mineral material:

Class	Description
Fine	clayey; sticky when wet, hard when dry
Medium	silty; feels smooth and soapy; or loamy; intermediate characteristics; comprising sand, silt and clay
Coarse	sandy; feels coarse and abrasive; visible sand grains
Gravel	small stones or rock fragments

Record the soil texture class for every 5 cm over a depth of 50 cm. If more than one description is required within a particular 5 cm depth range (e.g. where an organic topsoil is less than 5 cm thick), record the texture classes separately in boxes a and b.

If you wish to carry out an assessment on the functions of trace element storage or trace element export, it is necessary to record an average of the mineral particle size in the top 30 cm of the topsoil. Record this in the space marked **C.3.7ii** on **Recording Sheet 2a**.

Instruction 8: Record the characteristics of organic material

Record information: Recording Sheet 2a or 4 *(in the space marked: C.3.8 Organic matter characteristics)*

The organic material present within the soil should be described using the following terms:

- **Fibrous peat**: yellowish or reddish brown organic matter with prominent identifiable plant remains.
- **Amorphous peat**: dark brown or black organic matter with little or no recognisable plant remains except wood.
- **Semi-fibrous peat**: intermediate character between fibrous and amorphous peat.
- **Humus**: dark surface layer of decomposed litter without mineral material.
- **Humose**: predominantly dark brown or black organic-rich layer in mineral soil; especially prevalent in topsoils; can feel smooth in parts.

Instruction 9: Note presence and depth of hard layers

Record information: Recording Sheet 2a or 4 *(in the space marked: C.3 General comments)*

If the soil over-lies hard rock or an indurated (hardened) layer within 0.5 m (or at greater depth), note the depth and characteristics of the hard material on **Recording Sheet 2a** or **4**.

Instruction 10: Note salt layers

Record information: Recording Sheet 2a or 4 *(in the space marked: C.3 General comments)*

If the soil contains high concentrations of salts, note the depth and characteristics in the **General comments** box on **Recording Sheet 2a** or **4**.

Instruction 11: Compare soil profiles in the same provisional HGMU

Record information: Recording Sheet 2a

Where more than one soil profile description is made for an HGMU (in this case they will have been recorded on **Recording Sheet 4**), produce one composite, average description of the soil profile representative of the complete HGMU and record this on **Recording Sheet 2a**. If the profiles are radically different, then further examination of the HGMU is necessary.

Soil characteristics within an HGMU must be sufficiently, but not necessarily entirely, uniform. Use the following guidelines to test this:

- A soil profile composed predominantly of mineral (sandy or silty or clayey) components cannot occur in the same HGMU as a predominantly organic (peaty) soil.
- Two soil profiles composed predominantly of sandy and clayey components respectively cannot occur in the same HGMU.
- Soil profiles composed of predominantly silty material can occur with profiles that are predominantly either sandy or clayey in the same HGMU.
- Two soils with each profile characterised by a mixture of clay, silt, sand or peat horizons can occur together in the same HGMU.

Where soils are radically different, the provisional HGMU should be re-surveyed for any geomorphological or hydrological management features that may have been overlooked, and then the HGMU should be subdivided. Differences in vegetation management may help to locate the new boundary. If this does not help, then intensify the soil survey to establish the boundary of this unit and draw it onto **AA Map 1**.

Where provisional HGMUs have been subdivided using soil information, add a numerical suffix to the geomorphic code and create a new **Recording Sheet 2a** for the new HGMU, i.e. a provisional HGMU that is a uniform slope and therefore coded **US** could be divided into two units on the basis of slope angle giving **US/G1**$_1$ and **US/G1**$_2$ each with their own **Recording Sheet 2a**.

2.2.C.4 *ACTION: Final delineation of the HGMUs*

Record information: AA Map 1 and 3 and Recording Sheet 2a

The purpose of this section is to finalise the boundaries of the provisional HGMUs and to produce a map of coded HGMUs with a **Recording Sheet 2a** corresponding to each. This provides the structure for subsequent characterisation of the HGMUs and functional analysis.

By this stage you should have an **AA Map 1** with HGMUs delineated from hydrological management, geomorphology and soils information. The HGMUs should be numbered, and copies of **Recording Sheet 2a** for each of these should have geomorphic codes for landforms and slope gradient/position (where applicable) and information from at least one soil profile.

At this stage it is useful to check the HGMUs you have delineated, redraw HGMU boundaries in red, and ensure that HGMU numbering and corresponding recording sheets match up (Figure 2.2.11, see colour Section).

Instruction 1: Refine HGMU boundaries using soils information

Record information: AA Map 1

The objective of this final step is to ensure uniformity within HGMUs and verifiable delineation and distinctions between HGMUs, using the description of the *upper 0.5 m* of the soil column.

- Soil characteristics within an HGMU must be similar although not necessarily entirely uniform.
- A soil composed predominantly of mineral (sandy or silty or clayey) components cannot occur in the same HGMU as a predominantly organic (peaty) soil.
- Two soils composed predominantly of sandy and clayey components respectively cannot occur in the same HGMU.
- Soils composed of predominantly silty material can occur with soils that are predominantly either sandy or clayey in the same HGMU.
- Two soils with profiles characterised by similar mixtures of clay, silt, sand or peat horizons can occur together in the same HGMU.

Instruction 2: Draw final HGMU boundaries

Record information: AA Map 3

Once it is certain that the boundaries are correct according to the information available, draw them on **AA Map 3** in red. Code all HGMUs and ensure that this corresponds with the appropriate **Recording Sheet 2a**.

Instruction 3: Check if the following items and information is now available, and correct any gaps before starting the HGMU characterisation phase

1. On **AA Map 1,** check that you have outlined the rough boundaries for HGMUs from field delineation.
2. On **AA Map 3,** check that you have outlined the final boundaries of all HGMUs in slope, floodplain and channel AA components.
3. Check that you have subdivided **CA Map 1** on the basis of slope gradient with each unit coded.
4. The partially completed **Recording Sheet 2a** and **1** for each HGMU in the AA should contain the following information at this stage:
   - archive information: date, surveyor name and site name;
   - geomorphic codes;
   - hydrological management information;
   - slope gradient and position in the landscape information (where appropriate);
   - soil description.
5. Possibly soil samples.

2.2.D Further reading and references

Brinson, M.M. (1993). *A Hydrogeomorphic Classification for Wetlands*, Wetlands Research Program Technical Report WRP-DE-4, US Army Corps of Engineers, Waterways Experiment Station, Vicksburg, MS.

Gilvear, D.J., Tellam, J.H., Lloyd, J.W. and Lerner, D.N. (1990). *The Hydrodynamics of East Anglian Fen Systems*, Nature Conservancy Council Contract Survey Report No. 88, Peterborough.

Goode, D.A. and Ratcliffe, D.A. (1977). Peatlands. In: Ratcliffe, D.A. (ed.), *A Nature Conservation Review*, Volume 1, Cambridge University Press, Cambridge, UK, 249–287.

Gosselink, J.C. and Turner, R.E. (1978). The role of hydrology in freshwater wetland ecosystems. In: Good, R.E., Whigham, D.F. and Simpson, R.L. (eds) *Freshwater wetlands: Ecological processes and management potential*, Academic Press, New York, 63–78.

Gregory, S.V., Swanson, F.J., McKee, W.A. and Cummings, K.W. (1991). An ecosystem perspective of riparian zones, *Bioscience* **4**, 540–551.

Hodgson, J.M. (ed.) (1974). *Soil Survey Field Handbook*, Soil Survey Technical Monograph No. 5, Rothamsted Experimental Station, Harpenden, UK.

Hollands, G.G (1987). Hydrogeologic classification of wetlands and their management. *In:* Gerakis, P.A. (ed.) *Wetland Hydrology, Proceedings from a national wetland symposium*, Association of State Wetland Managers, Inc., Berne, New York.

Kieselev, P.A. (1975). Study of the water balance of stratified aquifers based on the analysis of their regime during hydrogeological investigations of marshy lands and adjacent areas. In: *Proceedings of Minsk Symposium, Hydrology of Marsh-ridden Areas*, UNESCO, IAHS, 35–46.

Lloyd, J.W., Tellam, J.H., Rukin, N. and Lerner, D.N. (1993). Wetland vulnerability in East Anglia: A possible conceptual framework and generalised approach, *Journal of Environmental Management* **37**, 87–102.

Lugo, A.E., Brinson, M.M. and Brown, S. (eds) (1990). *Forested Wetlands*, Elsevier, Amsterdam.

Maltby, E., Hogan, D.V. and McInnes, R.J. (1996). *Functional Analysis of European Wetland Ecosystems – Phase 1 (FAEWE)*, European Commission, Ecosystems Research Report No. 18. Office for Official Publications of the European Communities, Luxembourg.

Maltby, E., Hogan, D.V., Immirzi, C.P., Tellam, J.H. and Peijl, M.J. van der. (1994). Building a new approach to the investigation and assessment of wetland ecosystem functioning. In: Mitsch, W.J. (ed.), *Global Wetlands: Old World and New*, Elsevier, Amsterdam, 637–658.

Mitsch, W.J. and Gosselink, J.G. (2007). *Wetlands*, 4th edn, Wiley, New York.

Munsell Soil Color Charts, Munsell Color Company Inc., Baltimore, MD 21218, USA.

Novitzki, R.P. (1979). Hydrological characteristics of Wisconsin's wetlands and their influence on floods, stream flow and sediment. In: Greeson, P.E., Clark, J.R. and Clark, J.E. (eds) *Wetland Functions and Values: The state of our understanding*, American Water Resources Association, Minneapolis, USA, 377–388.

Wheeler, B.D. (1984). British fens: A review. In: Moore P.D. (ed.), *European Mires*, Academic Press, London, 237–282.

2.3

HGMU characterisation

Introduction

Following the identification of the component HGMUs, it is necessary to record a range of features that influence functioning. These features may be hydrogeomorphic, hydrological or ecological and act as indicators for the potential functions. Each indicator is relevant to a different number of functions and, therefore, if only a number of the functions are of interest, not all of the actions detailed in the following sections need to be carried out.

Characterisation of the HGMUs is carried out in the field using the following criteria:

2.3.A Recording of geomorphic indicators
2.3.B Recording of hydrological indicators
2.3.C Recording of ecological indicators
2.3.D Habitat and vegetation identification

Details are recorded on **Recording Sheet 3** in Section 2.4. Most features provide information which is used in all the functional assessments; others are function specific and a selective approach can be used by referring to the list of functions on the right hand side of the page for each indicator.

2.3.E Part 1: Land use and site management, field indicators

This section enables answers to be found for some of the questions in Action C.3, Section 2.1 through the use of field indicators rather than maps, archives and local experts. However, the level of detail possible is lower than in Action C.3, therefore 2.3.E Part 1 should be completed only if the information required was previously unavailable.

2.3.E Part 2: Collection of HGMU information

Eventually the field data will be compiled with the data collected from the desk study and a further four tasks completed:

2.3.F Assignment of hydrological codes
2.3.G Estimation of HGMU area
2.3.H Nutrient input characterisation
2.3.I Trace element input characterisation

2.3.A Recording of geomorphic indicators

The purpose of this section is to record the geomorphic features that predict functioning. Read through prior to undertaking field studies. Follow the guidelines through in the field. On completion, record the geomorphic information for each HGMU on **Recording Sheet 3**. An example of a completed recording sheet is given in Fig. 2.3.1.

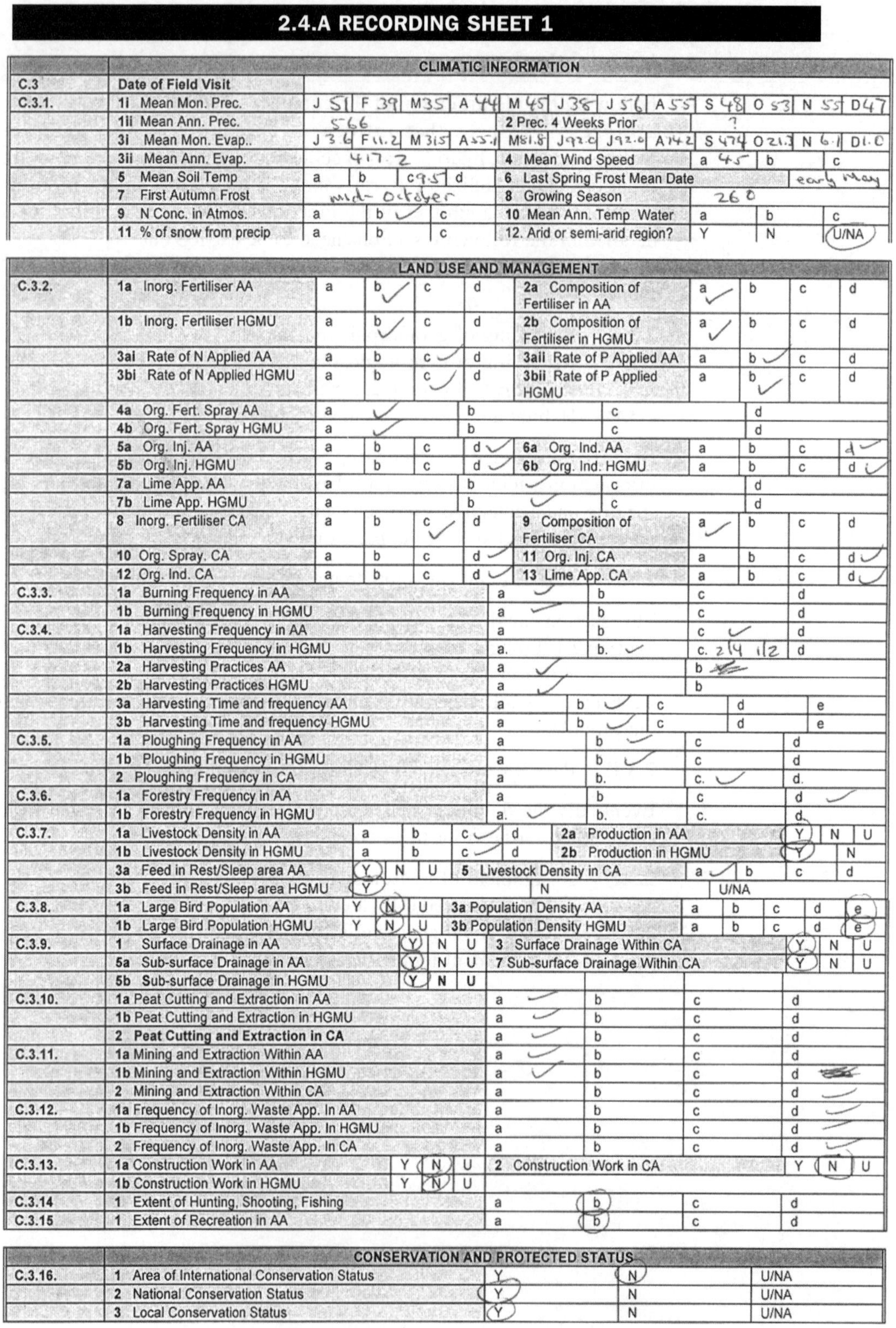

2.4.A RECORDING SHEET 1

CLIMATIC INFORMATION			
C.3	Date of Field Visit		
C.3.1.	1i Mean Mon. Prec.	J 51 F 39 M 35 A 44 M 45 J 38 J 56 A 55 S 48 O 53 N 55 D 47	
	1ii Mean Ann. Prec.	566	2 Prec. 4 Weeks Prior: ?
	3i Mean Mon. Evap.	J 3.6 F 11.2 M 31.5 A 55.1 M 81.8 J 92.0 J 92.0 A 74.2 S 47.4 O 21.3 N 6.1 D 1.0	
	3ii Mean Ann. Evap.	417.2	4 Mean Wind Speed: a 4.5 / b / c
	5 Mean Soil Temp	a / b / c 9.5 / d	6 Last Spring Frost Mean Date: early May
	7 First Autumn Frost	mid-October	8 Growing Season: 260
	9 N Conc. in Atmos.	a / b ✓ / c	10 Mean Ann. Temp Water: a / b / c
	11 % of snow from precip	a / b / c	12. Arid or semi-arid region? Y / N / (U/NA)

LAND USE AND MANAGEMENT			
C.3.2.	1a Inorg. Fertiliser AA: a / b ✓ / c / d	2a Composition of Fertiliser in AA: a ✓ / b / c / d	
	1b Inorg. Fertiliser HGMU: a / b ✓ / c / d	2b Composition of Fertiliser in HGMU: a ✓ / b / c / d	
	3ai Rate of N Applied AA: a / b / c ✓ / d	3aii Rate of P Applied AA: a / b ✓ / c / d	
	3bi Rate of N Applied HGMU: a / b / c ✓ / d	3bii Rate of P Applied HGMU: a / b ✓ / c / d	
	4a Org. Fert. Spray AA: a ✓ / b / c / d		
	4b Org. Fert. Spray HGMU: a ✓ / b / c / d		
	5a Org. Inj. AA: a / b / c / d ✓	6a Org. Ind. AA: a / b / c / d ✓	
	5b Org. Inj. HGMU: a / b / c / d ✓	6b Org. Ind. HGMU: a / b / c / d ✓	
	7a Lime App. AA: a / b ✓ / c / d		
	7b Lime App. HGMU: a / b ✓ / c / d		
	8 Inorg. Fertiliser CA: a / b / c ✓ / d	9 Composition of Fertiliser CA: a ✓ / b / c / d	
	10 Org. Spray. CA: a / b / c / d ✓	11 Org. Inj. CA: a / b / c / d ✓	
	12 Org. Ind. CA: a / b / c / d ✓	13 Lime App. CA: a / b / c / d ✓	
C.3.3.	1a Burning Frequency in AA: a ✓ / b / c / d		
	1b Burning Frequency in HGMU: a ✓ / b / c / d		
C.3.4.	1a Harvesting Frequency in AA: a / b / c ✓ / d		
	1b Harvesting Frequency in HGMU: a. / b. ✓ / c. 2/4 1/2 / d		
	2a Harvesting Practices AA: a ✓ / b		
	2b Harvesting Practices HGMU: a ✓ / b		
	3a Harvesting Time and frequency AA: a / b ✓ / c / d / e		
	3b Harvesting Time and frequency HGMU: a / b ✓ / c / d / e		
C.3.5.	1a Ploughing Frequency in AA: a / b ✓ / c / d		
	1b Ploughing Frequency in HGMU: a / b ✓ / c / d		
	2 Ploughing Frequency in CA: a / b. / c. ✓ / d.		
C.3.6.	1a Forestry Frequency in AA: a / b / c / d ✓		
	1b Forestry Frequency in HGMU: a. ✓ / b. / c. / d.		
C.3.7.	1a Livestock Density in AA: a / b / c ✓ / d	2a Production in AA: (Y) / N / U	
	1b Livestock Density in HGMU: a / b / c ✓ / d	2b Production in HGMU: (Y) / N	
	3a Feed in Rest/Sleep area AA: (Y) / N / U	5 Livestock Density in CA: a ✓ / b / c / d	
	3b Feed in Rest/Sleep area HGMU: (Y) / N / U/NA		
C.3.8.	1a Large Bird Population AA: Y / (N) / U	3a Population Density AA: a / b / c / d / (e)	
	1b Large Bird Population HGMU: Y / (N) / U	3b Population Density HGMU: a / b / c / d / (e)	
C.3.9.	1 Surface Drainage in AA: (Y) / N / U	3 Surface Drainage Within CA: (Y) / N / U	
	5a Sub-surface Drainage in AA: (Y) / N / U	7 Sub-surface Drainage Within CA: (Y) / N / U	
	5b Sub-surface Drainage in HGMU: (Y) / N / U		
C.3.10.	1a Peat Cutting and Extraction in AA: a ✓ / b / c / d		
	1b Peat Cutting and Extraction in HGMU: a ✓ / b / c / d		
	2 Peat Cutting and Extraction in CA: a ✓ / b / c / d		
C.3.11.	1a Mining and Extraction Within AA: a ✓ / b / c / d		
	1b Mining and Extraction Within HGMU: a ✓ / b / c / d		
	2 Mining and Extraction Within CA: a / b / c / d ✓		
C.3.12.	1a Frequency of Inorg. Waste App. In AA: a / b / c / d ✓		
	1b Frequency of Inorg. Waste App. In HGMU: a / b / c / d ✓		
	2 Frequency of Inorg. Waste App. In CA: a / b / c / d ✓		
C.3.13.	1a Construction Work in AA: Y / (N) / U	2 Construction Work in CA: Y / (N) / U	
	1b Construction Work in HGMU: Y / (N) / U		
C.3.14	1 Extent of Hunting, Shooting, Fishing: a / (b) / c / d		
C.3.15	1 Extent of Recreation in AA: a / (b) / c / d		

CONSERVATION AND PROTECTED STATUS				
C.3.16.	1 Area of International Conservation Status	Y	(N)	U/NA
	2 National Conservation Status	(Y)	N	U/NA
	3 Local Conservation Status	(Y)	N	U/NA

2.3.1 A completed recording sheet for an HGMU in Sutton Common, UK.

2.3.A.1 *ACTION: Measure soil pH*

Record information: Recording Sheet 3

In the space provided in **A.1i**, record the pH of the top 10 cm of topsoil using a field probe. This is most easily done by collecting a 10 cm deep soil core, removing stones and plant material, and mixing it thoroughly in a container with a volume of de-ionised water usually less than half the volume of the soil core, in order to create a thick slurry. The soil pH is measured by inserting into the slurry a pH probe attached to a calibrated pH meter and recording the measurement direct from the meter display.

If you wish to carry out an assessment of the functions of trace element storage or trace element export, a 30 cm sample of the topsoil is needed. This sample should be mixed and the pH should be recorded in the space marked **A.1ii** on **Recording Sheet 3**.

2.3.A.2 *ACTION: Record the composition of the land surface*

Record information: Recording Sheet 3

For HGMUs where there has been little or no soil development, record the composition of the land surface using the categories below and record the information on **Recording Sheet 3**. If this is not applicable to the HGMU then circle '**g**' in the box on **Recording Sheet 3**.

The following categories are used to identify the surface of the HGMU where encrusting lichens are likely to represent the only vegetation:

a. bare boulders (stones > 10 cm diameter)
b. bare gravel (stones 0.2 cm–10 cm diameter)
c. bare sand (grains < 0.2 cm diameter)
d. bare silt or mud
e. bare peat
f. bare rock
g. NA (not applicable)

2.3.A.3 *ACTION: Identify microtopography*

Record information: Recording Sheet 3

Identify whether the microtopography of the HGMU is negligible or significant, as defined below. Record the results on **Recording Sheet 3**.

Microtopography or microrelief (see Figs 2.3.2a, b and c) refers to undulations and irregularities at the ground surface *of less than 0.5 m amplitude*, and which can be described in the following terms:

1. negligible (surface is virtually flat) ______________
2. significant (noticeable surface variations, including grass tussocks, and irregularities in the soil surface resulting from poaching and soil creep)

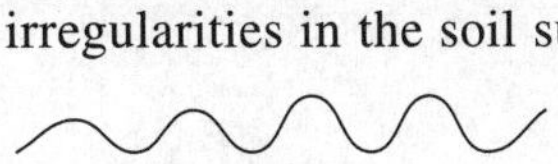 or 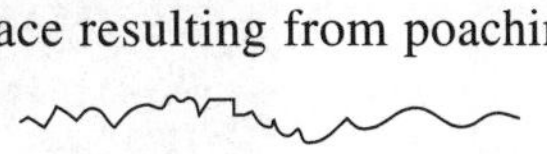

2.3.A.4 *ACTION: Record the topographical exposure to wind*

Record information: Recording Sheet 3

Estimate the topographical exposure or openness as defined below. Record results on **Recording Sheet 3**. The degree of topographical exposure of the HGMU is assessed by estimating the position of the site relative to the horizon, using the topex system. Estimate the angle of inclination between the site and the horizon (which can also be woodland tree tops) at the eight compass points (every 45° around 360°), using a clinometer if necessary. Add these angles together to give

2.3.2 (a) Negligible microrelief at a fen area near Utrecht, The Netherlands (photo: J. Verhoeven); (b) significant microrelief in south-west England (photo: E. Maltby); (c) significant microrelief created by *Molinea caerulea* on blanket bog (photo: E. Maltby).

the topex value, an index corresponding to the following categories of topographical exposure:

Total value	
0–30°	**1** (exposed)
31–100°	**2** (moderate)
101–160° +	**3** (sheltered)

2.3.A.5 *ACTION: Indicate the presence of wind cropped trees*

Record information: Recording Sheet 3.

The growth of trees often gives a further indication of the exposure of a site. Strong winds tend to kill new growth on the windward side of the tree, and the tree therefore extends towards the leeward side. This results in the tree having a somewhat lopsided shape, with large numbers of dead twigs on one side and a lot of growth on the other. Trees of this shape are referred to as wind-cropped. Record the presence of these trees within the HGMU on **Recording Sheet 3**.

2.3.B Recording of hydrological indicators

The purpose of this section is to record the hydrological characteristics of each HGMU. The movement of water around and through a wetland is one of the primary determinants of how it functions hydrologically, ecologically and biogeochemically. As far as possible, the hydrological functioning of river marginal and lake marginal wetlands can be quantified in terms of floodwater control through flood response mechanisms, groundwater recharge and discharge and surfacewater generation. In addition, hydrological processes and other characteristics (such as water resources, flow velocities, quantities and flow paths) dictate much of the nutrient movement in the ecosystem that is important in the assessment process. The results may also be used to predict the potential impairment of functioning arising from vulnerability to hydrological impacts.

The following section characterises hydrology following a list of questions. The answers to these questions are entered into **Recording Sheet 3**. Once the sheet has been completed as fully as possible, the result can be used to (i) assign a hydrological code to *each* HGMU (Section 2.3.F) for use in all functional assessment procedures (FAPs), and (ii) assess specific hydrological functions for *specific* HGMUs.

Some of the questions below can be answered only when specific conditions occur in the HGMU (e.g. flooding, summer). It will usually be impossible to answer all questions during a single site visit. In order to answer as many questions as possible it is advisable to consult a local expert – preferably while visiting the HGMU. Not all of the questions need to be answered. If hydrological functions (Sections 3.1.A–3.1.D) are not being assessed then only answer Section 2.3.F (Assignment of hydrological codes). It may be useful to go through Section 2.3.F first to avoid answering unnecessary questions.

Unless specified otherwise, the questions in this section can be answered with:

- **Y** (yes)
- **N** (no)
- **U** (unknown)
- **NA** (not applicable).

NA applies, for example, when a question asks 'is there water in the channel?' but a channel is not present.

Read through prior to undertaking field studies. Follow the guidelines through in the field. On completion, record the hydrological information for each HGMU on **Recording Sheet 3**. Use a new copy of the recording sheet for each HGMU.

2.3.B.1 ACTION: Record HGMU elevation: estimate or confirm reference level

Record information: Recording Sheet 3 *(B.1 Record HGMU elevation)*

If an assessment of the Floodwater Detention function is planned, the elevation of the HGMU must be determined, preferably with reference to a national level such as the ordnance datum level (m.O.D.), otherwise the elevation relative to another HGMU or other feature in the AA may be used. A reference point in the AA may already have been determined and recorded on **AA Map 2 C.3.19.1**). If water bodies in the AA are in open connection with each other or with the river channel or lake, they can provide very good reference, as the water surface will be horizontal in the (near-) absence of flow. If you have not yet identified a reference level, you should do so now:

B.1.1 Identify the reference point by name.

B.1.2 Identify the absolute elevation of this point relative to a local, regional or national reference level of choice (such as m.O.D.). This point represents the local relative elevation level 'zero metres'.

B.1.3 Estimate the mean elevation of the HGMU (in metres) relative to the chosen reference level. The accuracy required varies with AA topography, but estimates should generally be within 1 m, and preferably 0.5 m. *All HGMU* *levels must relate to the same reference level.* If it is not possible to directly estimate the elevation relative to the reference level, this must be done via the elevation of another HGMU. Apart from rough estimates, detailed topographic maps may provide this information.

B.1.4 Determine the absolute elevation of the HGMU by combining information from QB.1.2 and QB.1.3.

2.3.B.2 *ACTION: Record characteristics of surface water flow*

Record information: AA Map 2 and Recording Sheet 3 *(B.2 Surface water flows)*

Identify surface water flow direction and flow features of the HGMU, answer Y/N/U/NA unless specified otherwise. Watercourses have been mapped in the delineation phase: ditches, stream channels and canals, either wet or dry. However, their position relative to other HGMUs is important. Mark flow directions in ditches and stream channels in and around the HGMU (only in the main ones if there are many) on **AA Map 2**.

Answer QB.2.1–QB.2.8 only if there is any sign of water flow or water channels (ditches or streams) in the HGMU or leading into or out of the HGMU. Otherwise fill in NO for all questions and go to Action B.3.

B.2.1 Has water flow been observed over the soil surface or through rivulets into the HGMU from upslope areas?

B.2.2 Has water flow been observed over the soil surface or through rivulets or drains from the HGMU into downslope areas?

B.2.3 Are there any watercourses (including dry beds, ditches and underground drains) originating in the HGMU and leading to the river channel or lake (either directly or indirectly), or to a downslope HGMU?

B.2.4 Are there any watercourses (including dry beds, ditches and underground drains) that lead into the HGMU from a higher area and terminate within the HGMU?

B.2.5 If there are any watercourses originating in the HGMU (including ditches over 0.5 m deep), do they all hold water for only part of the year? Indications are: submerged terrestrial vegetation, exposed aquatic vegetation, or the absence of water when there are clear indications of recent inundation, such as deposits of fine matter, flood marks (strand lines) or erosion features (such as under-cut banks, channelling in the bed, scour pots).

B.2.6 If there are watercourses originating in the HGMU (including ditches) do some of them hold water throughout the year? Indicators for such permanent streams include submerged aquatic plants in the dry season and populations of non-migrant fish.

B.2.7 If there are watercourses originating in the HGMU (including ditches), do any of them still carry flowing water at least two weeks after significant rainfall in summer?

B.2.8 If there are water courses originating in the HGMU (including ditches), do any of them still carry sufficient flow to make a noticeable contribution to flow in the main river or lake, or to water levels in an adjoining HGMU, at least two weeks after significant rainfall in summer?

2.3.B.3 *ACTION: Record springs and seepage zones and characteristics of standing water*

Record information: Recording Sheet 3 *(B.3 Springs and seepage)*

Identify springs and seepage features of the HGMU, answer Y/N/U/NA unless specified otherwise.

B.3.1 Have active springs (flowing water emerging from a very localised source area) been observed within the HGMU?

B.3.2 At least two weeks after significant rainfall in summer, is the soil in the HGMU still waterlogged without any obvious surface water inputs?

B.3.3 After prolonged frost, which has resulted in freezing of standing water nearby, is there standing water within the HGMU that is not frozen?

B.3.4 When snowfall has covered all surrounding areas, are there any wet (but not inundated), snow-free areas?

B.3.5 Is the HGMU situated immediately downslope of a unit with a gradient > 5° and is the gradient of the HGMU surface less than that of the upslope unit; i.e. is the surface profile concave?

B.3.6 If the HGMU is in an arid or semi-arid region where water shortage may be a limiting factor to plant growth, is the vegetation within the HGMU considerably more 'lush' (higher, greener or more dense) than that in surrounding areas?

Answer QB.3.7–QB.3.12 only if accompanied by a local expert. Otherwise fill in NA for all questions and go to QB.3.13.

B.3.7 What is the mean (in time) surface water table depth (relative to the soil surface) in watercourses in the HGMU?

B.3.8 What is the lowest (in time) surface water table depth (relative to the soil surface) in watercourses in the HGMU?

B.3.9 What is the highest (in time) surface water table depth (relative to the soil surface) in watercourses in the HGMU?

Answer QB.3.10–QB.3.12 only if there are groundwater wells in the HGMU. Otherwise fill in NA for all questions and go to QB.3.13.

B.3.10 What is the mean (in time) groundwater table depth (relative to the soil surface) in wells in the HGMU?

B.3.11 What is the lowest (in time) groundwater table depth (relative to the soil surface) in wells in the HGMU?

B.3.12 What is the highest (in time) groundwater table depth (relative to the soil surface) in wells in the HGMU?

B.3.13 Is the HGMU situated directly downslope from an artificial reservoir, pond or lake where water levels are kept higher than they would be under natural conditions?

B.3.14 If present, is standing water brown or rust coloured?

B.3.15 If present, is standing water covered with a small, oily layer (iron bacteria)? N.B. The layer breaks apart when touched.

B.3.16 Do the major flow pathways through the HGMU exit directly into a river or lake?
a. Yes, at low flow
b. Yes, at high flow
c. Unknown/not applicable

B.3.17 Do the major flow pathways through the HGMU traverse organic rich soils?
a. Yes, at low flow
b. Yes, at high flow
c. Unknown/not applicable

2.3.B.4 *ACTION: Record inundation indicators*

Record information: Recording Sheet 3 *(B.4 Inundation)*

Identify inundation features of the HGMU, answer Y/N/U/NA unless specified otherwise.

B.4.1 Has (part of) the HGMU ever been inundated?
If NO move on directly to ACTION B.5.

B.4.2 Has flooding (inundation) been absent from the HGMU for at least 5 years?

B.4.3 Is (part of) the HGMU permanently inundated with standing water?

B.4.4 Has the river or lake ever been observed to flow into the HGMU as overbank flooding or through channels?

B.4.5 If the HGMU is not immediately adjacent to the main river or lake, are the connections between the main river or lake and the HGMU ever inundated by standing water or by water flowing in the direction of the HGMU?

B.4.6 What is the maximum observed flow velocity of floodwater in the HGMU (in m s^{-1}; estimate by following a floating object for 5 seconds)?

Note: Do not carry out this exercise during bankfull height where it is not possible to gain access.

B.4.7 If the flood in the river channel or lake has receded more than a week ago, at the time of the field visit, what is the mean water depth in the HGMU (in metres)?

B.4.8 Do any of the following buoyant materials of local origin occur in the HGMU?

- A well-developed litter layer (partly decomposed leaves) under trees or bushes.
- Old dead wood of local origin (such as branches lying around a dead tree).

B.4.9 Has buoyant material, which does not appear to have originated in the HGMU, been deposited within it so as to form a flood strand line? (see Fig. 2.3.3).

B.4.10 Are there old structures (buildings, wooden fences, wooden walkways, etc.) in the HGMU that would clearly be damaged by flooding but without any specific measures being taken to avoid flooding?

B.4.11 If there are buildings within the HGMU, are there any marks on them which would indicate previous flooding?

B.4.12 Is the HGMU used for orchards or other tree crops that would not tolerate flooding?

B.4.13 If the HGMU is at the foot of the AA 'Slope' component, are there sometimes indications of standing water but only directly after rainfall events and not caused by overbank flooding?

2.3.3 Stranded vegetation and flotsam in the riparian area (photo: E. Maltby).

B.4.14 Is the HGMU cut off from the river channel or lake by a major continuous embankment, clearly constructed to prevent any flooding?

B.4.15 Are there other obvious field indicators of frequent flooding (i.e. a flood at least once every 2 years), not listed above?

B.4.16 What is the mean annual water level in the main river channel or lake (in metres, relative to a reference level)?

B.4.17 What is the mean water level (in metres, relative to a reference level) in the main river channel or lake during frequent floods (i.e. at least 1 flood every 2 years)?

B.4.18 What is the mean water level (in metres, relative to a reference level) in the main river channel or lake during occasional floods (i.e. 1 flood every 2–10 years)?

B.4.19 What is the mean water level (in metres, relative to a reference level) in the main river channel or lake during rare floods (i.e. 1 flood every 10 or more years)?

B.4.20 Is the HGMU mostly covered with 'rough' vegetation such as dense reed, bushes, shrubs or forest, which will slow down (flood) waters considerably?

B.4.21 If the HGMU has only one (potential) outlet for floodwater, is the width of the outlet less than 10 % of the mean width of the area usually flooded in the HGMU, i.e. is the outflow from the HGMU restricted?

B.4.22 From inlet hydrographs, historical flood event records and local knowledge, estimate the months in which waterlogging or flooding would occur in the HGMU in a normal year (to be recorded as first and last month numbers e.g. '1103' for Nov to March – if duration is less than one-month code this as '0000').

B.4.23 At the time of the site visit, was the HGMU inundated regardless of the water source?

2.3.B.5 *ACTION: Record erosion/sedimentation indicators*

Record information: Recording Sheet 3 *(B.5 Erosion/sedimentation)*

Record erosion/sedimentation features of each HGMU, answer Y/N/U/NA unless specified otherwise.

B.5.1 Are there are any indications of erosion within the area immediately upslope from the HGMU (in the AA or CA): e.g. scouring in steeper areas; rills or gullies in cultivated land; local exposure of bare rock; sediment accumulation at lower slopes?

B.5.2 Are there are any indications of erosion within the HGMU: e.g. scouring in steeper areas; rills or gullies in cultivated land; local exposure of bare rock; steep exposed banks adjacent to the river or lake; sediment accumulation at lower slopes?

B.5.3 Has the river or lake that floods the HGMU been observed to carry high concentrations of suspended sediment during peak discharges? N.B. This can be observed as turbid or muddy water with a yellowish or brownish opaque colour.

B.5.4 Is there evidence of recently deposited mineral (sand, silt, clay) or organic (grass, leaves, branches) material at the soil surface? (Fig. 2.3.4)

B.5.5 Is there evidence of recently deposited mineral material on the surfaces of growing plants?

B.5.6 Has surface run-off from the area immediately upslope (in AA and CA) been observed to carry high concentrations of suspended sediment?

B.5.7 Is the HGMU unvegetated or sparsely vegetated?

B.5.8 Is the HGMU adjacent to a river that is known to shift its course locally, and situated on an outer bend?

B.5.9 Are there any substantial patches of bare or eroded land within the HGMU? An example is shown in Fig. 2.3.5.

2.3.4 High rates of flow (spates) can carry gravel and silt in the water column, which is deposited in areas when flow is reduced (photo: E. Maltby).

2.3.5 Erosion channels in peatland (photo: E. Maltby).

2.3.C Recording of ecological indicators

The purpose of this section is to identify and record the ecological characteristics of the HGMUs and AA to provide predictors for the FAPs. Characteristics such as the community composition are used as predictors of wetland functioning. For instance, they are used as part of the assessment of habitat and habitat diversity (for ecosystem maintenance) and biomass import and export (for food web support).

The answers given to the questions asked in this section will be entered onto **Recording Sheet 3** and will provide a basis for determining the types of habitat available for species colonisation. Read through prior to undertaking field studies. Follow the guidelines through in the field. On completion, record the ecological information for each HGMU on **Recording Sheet 3.**

2.3.C.1 ACTION: Estimate the architectural diversity of the vegetation as the total number of layers, as defined below

Record information: Recording Sheet 3

For each HGMU, the physical diversity of structure within the habitat (*architectural diversity*) needs to be recorded. The architectural diversity is described in terms of the number of layers present. Examples of some communities of differing architectural diversity and the number of layers in each are given in Fig. 2.3.6.

(i) 1 layer composed of herbaceous plants, grasses, mosses, etc.
(ii) 1 layer composed of shrubs, shrub canopies, branches, leaves, etc.
(iii) 2 layers: one composed of shrubs, the other of trees.
(iv) 4 layers: one composed of herbaceous plants, one of shrubs, and two layers of trees, with some trees much taller than the others.
(v) 2 layers: one composed of herbaceous plants, one of trees; no shrubs present.

Often community layers are missing or, alternatively, they can be further divided into more layers (for example some trees that protrude far above the others). Very complex communities may have up to eight layers or more.

C.1 Count the number of vegetation layers present in the HGMU and record this number on **Recording Sheet 3** in **'Architectural Diversity'**.

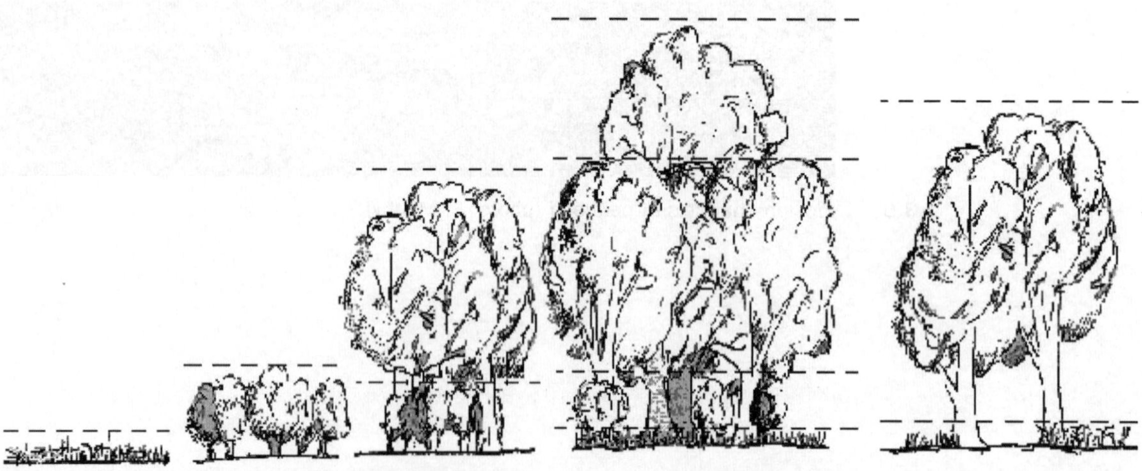

(i) 1 layer (ii) 1 layer (iii) 2 layers (iv) 4 layers (v) 2 layers

2.3.6 Architectural diversity.

2.3.C.2 *ACTION: Identify the type of litter layer present in the HGMU*

Record information: Recording Sheet 3

The type of *litter layer* present will usually reflect the nature of the *standing biomass*, i.e. there will usually be herbaceous litter under grasses, and forest litter of leaves and woody debris under woodlands.

C.2 If a litter layer is present, record the litter layer present from the list of four basic types described below:

a. Negligible – virtually no litter present
b. Herbaceous litter layer – distinct litter layer present comprising grasses and herbaceous matter
c. Forest debris litter layer – distinct litter layer present comprising leaves and woody debris
d. Combination litter layer – a combination of herbaceous and forest litter
e. Not applicable (NA)

2.3.C.3 *ACTION: Record if the litter layer reflects the vegetation present in the HGMU*

Record information: Recording Sheet 3

C.3 If a litter layer is present, does it reflect the living vegetation present, i.e. do the species of dead leaves match the trees or herbaceous vegetation present? Record Y/N or NA.

2.3.C.4 *ACTION: Record the presence of any of the features of the vegetation listed below*

Record information: Recording Sheet 3

Certain features of the vegetation are important for the organisms that live there. Record if the following are present within the HGMU:

a. Canopy forming trees
b. Under-storey trees, including juvenile canopy species
c. Shrubs or bushes
d. Tall herbaceous vegetation
e. Medium height herbaceous vegetation
f. Low herbaceous vegetation
g. Dense ground vegetation
h. Sparse ground vegetation
i. Solitary trees or windbreaks

2.3.C.5 *ACTION: Record the presence of any of the habitat features listed below*

Record information: Recording Sheet 3

Animal species have specific requirements. Hence, different habitats are colonised by different species, and a complex habitat will have many species.

Record the presence of any of the habitats listed below:

a. Very old, mature trees
b. Standing dead trees, generally lacking bark
c. Standing dead trees with a substantial covering of bark
d. Fallen trees, generally lacking bark

e. Fallen trees with a substantial covering of bark
f. Tree stumps
g. Tree wounds or sap runs
h. Rotted holes in trees
i. Hollow trees
j. Trees overhanging the water
k. Gravel banks
l. Herbivorous animal dung
m. Ant nest mounds (obvious mound of exposed fine soil, with entrance holes and ants)
n. Grass tussocks
o. Wasp nests (paper nests in trees, holes or in the ground, with wasps)
p. Dry, open, sunny areas of SE aspect suitable as basking sites for reptiles (any such area, which can vary from a whole hillside to individual rocks, or be artificial, such as a sheet of metal or a road)
q. Damp, shady areas
r. Areas of abundant algae
s. River and lake cliffs (Figs 2.3.7a and b)
t. Rocky outcrops
u. Derelict or uninhabited buildings
v. Caves, disused tunnels and mines
w. Old bridges, aqueducts or viaducts
x. Bat boxes
y. Burrows, dens, lairs, holts, holes or other domiciles of mammals
z. Animal tunnels or runways through the vegetation

2.3.7 Features of lake and river edge habitat: (a) lake cliff (code 's'); (b) river cliff (code 's'); (c) river boulder (code 'bb') (all photos: photo archive EKBY/Miltiadis Seferlis).

aa Beaver or otter constructions (dams across a river, or shelters on lake shores built out of trees or logs with characteristic gnaw marks). Dams as such often lead to the formation of a lake upstream of the dam. In lake marginal wetlands beavers and otters often collect woody debris or gnaw logs to build a lair or den on the lake shore

bb Boulders or rocks within or adjacent to the river channel (Fig. 2.3.7c)

cc Rocks on lake shores

ddi Deciduous vegetation that overhangs the watercourse

ddii Deciduous vegetation that has the potential to overhang the watercourse

2.3.D Habitat and vegetation identification

The purpose of this section is to identify accurately the habitat types present in each HGMU. It is mainly a vegetation-based identification. The vegetation will provide useful information about the environment and management of the HGMU, and therefore the processes that occur.

A key is provided to help identify the vegetation types present within the HGMU. The user requires a degree of botanical knowledge, as accurate identification of plant species will provide evidence of certain ecological conditions.

A good field key (e.g. Hubbard, 1992; Rose and O'Reilly, 2006; Fitter *et al.*, 1996) will be required for the accurate field identification of plant species. Attempts to match plants with pictures often lead to misidentification and an inaccurate assessment of wetland functioning. Always use an appropriate key.

Despite their often limited distribution, natural and semi-natural communities are more indicative of the environmental conditions that regulate wetland functioning than some of the broader categories, such as cropland, which may cover larger areas of the landscape.
Note: Some habitats may be unvegetated or sparsely vegetated.

2.3.D.1 *ACTION: Identify the vegetation/habitat type in each HGMU*

Record information: Recording Sheet 3

Read through the vegetation/habitat key prior to undertaking field studies. Follow the key in the field, identifying the plant species as required. On completion, record the vegetation/habitat code given for each HGMU on **Recording Sheet 3** and estimate the percentage of cover of this vegetation type (record this on **Recording Sheet 3** and **AA Map** 2). If more than one distinct vegetation/habitat type is present in different parts of the HGMU, determine a code for each type separately, then estimate the percentage of the HGMU that is covered by each type. Record the information on **Recording Sheet 3**. Do not attempt to identify vegetation/habitat covering the same area separately, for example, trees and understorey, separately.

If the vegetation/habitat has already been identified and given a classification type under some other system, for example the CORINE in Europe or NVC in the UK, then use Table 2.3.1 at the end of the key (p.168) to convert this to the vegetation/habitat types used in the FAPs.

Note: The key provided is more comprehensive than that generally required by the procedures.

Background to vegetation/habitat key

The vegetation/habitat types identified in this section are intended to cover most of those found within Europe, except restricted island floras. The system followed is the EUNIS Habitat Classification (http://eunis.eea.eu.int/habitats.jsp; formerly CORINE), which is used by the European Union to designate habitats that are of special interest within the context of the Habitats Directive 1992 (see NATURA 2000: http://dataservice.eea.europa.eu/dataservice/metadetails.asp?id=774). This is a Europe-wide system of habitat classification based on vegetation.

Since individual plants cannot migrate, their presence in a biotope indicates their ability to grow and survive in that environment. The evolutionary strategies of plant species are the result of trade-offs between physiological requirements and tolerance of constraints and disturbances in their environment (see e.g. Ellenberg *et al.*, 1992; Grime, 2002). This means that the plant species assemblage can be used as a bio-indicator of major ecological factors, and acts as a surrogate for these when making habitat assessments. We have used the CORINE method of plant assemblage classification because it accommodates more or less all the trade-offs in the species combinations illustrated and, when compared with wetland systems in FAEWE and PROTOWET, has been found to be a good indicator of ecological factors in the habitat.

Vegetation codes: In some situations, such as levees, there may be highly distinctive vegetation communities occupying the same HGMU. A common example would be the presence of riparian woodland occupying the river fringe, with herbaceous grassland present at the landward side of the levee. The user should apply the procedures in a way that is appropriate to the application, which could include running two assessments for the same HGMU, changing only the vegetation code.

The vegetation/habitat types identified can be one of three categories critical to the application of FAPs to river marginal wetlands and lake marginal wetlands:

- those mainly humid types characterising *river marginal and lake marginal wetland ecosystems*, as defined within this document;
- other wetland vegetation/habitat types;
- non-wetland vegetation/habitat types occupying *contributory areas* within the catchment of specified river marginal and lake marginal wetlands.

Many of the vegetation/habitat types included in the key are not typical of river and lake marginal wetland ecosystems. These vegetation/habitat types have been included for three reasons:

- Non-wetland vegetation/habitat types may occur within the floodplain, particularly if the flooding regime has been altered.
- Non-wetland vegetation/habitat types are likely to occur in the area surrounding the floodplain and give valuable information on the impacts that these areas are having on the wetland.
- Non-wetland vegetation/habitat types in the area surrounding the floodplain may be performing important functions, which are important for the overall functional assessment of the site.

The key leads the user to vegetation/habitat types, which have been grouped into six main categories. Each of these main categories is briefly described below:

Unvegetated and sparsely vegetated	**UX**	Either as a result of poor soil conditions or frequent disturbance, this category has little or no vegetation/habitat.
Forest and woodland	**WX**	Trees are dominant, and may be natural or planted. These vegetation/habitat types can occur on wet, flooded or dry soils.
Aquatic	**AX**	Vegetation/habitat adapted to life in water (Fig. 2.3.8)

2.3.8 An example of a wetland aquatic habitat with floating-leaved and emergent plant species (photo: E. Maltby).

Grassland	**GX**	Dominated by species of grass, sometimes with substantial amounts of sedges, rushes or broad-leaved plants. These vegetation/habitat types can occur on wet, flooded or dry soils. (Fig. 2.3.9)
Mires	**MX**	These are vegetation/habitat types composed of plants that are adapted to life in waterlogged organic soils with actively-accumulating peat.
Scrub and dwarf shrubs	**SX**	Dominated by woody shrub species. These vegetation/habitat types can occur on wet, flooded or dry soil.

Definition of terms used in vegetation/habitat key

Vegetation canopy the tallest continuous cover of plants (occupying more than 50 % of the surface area).

Height of the canopy the average heights of the tallest continuous plant cover. Isolated plants or reproductive structures may protrude above the canopy but these should be ignored when estimating canopy height.

Dominant species the species that comprise the largest proportion of the vegetation/habitat canopy. Dominance is defined in terms of the area covered by the plants, not the number of individuals. This is most easily perceived by imagining yourself looking down from above the vegetation/habitat, including trees, and then attempting to estimate the proportion of the surface you can see that is covered by each species.

Woody plants perennial plants that produce a hard, fibrous thickening to their stems, laid down in annual rings, and with an outer layer of darker, softer bark.

Herbaceous plants perennial or annual plants, which lack a hard, fibrous thickening. Although the stems of some can be very fibrous and last for a long time when dead, they tend to have a softer texture.

Bryophytes mosses and liverworts; small, delicate plants without flowers, which lack the developed tissues (xylem) that transfer water within the stem.

2.3.9 A dry grass-dominated vegetation/habitat (photo: E. Maltby).

Macroalgae mostly aquatic plants that are individually distinguishable with the naked eye (i.e. not microscopic).

Key to vegetation/habitat types

APPEARANCE OF THE HABITAT

1. There is some vegetation (include mosses, ferns, liverworts and macro-algae) 2
   or
   There is no apparent vegetation **Key A**
2. The vegetation is largely continuous *without* substantial areas (total < 30 %) of exposed rock, boulders or soil (if the vegetation is tall, imagine yourself up above the vegetation, including trees, looking down, and try to estimate the proportion of the surface you can see that is not covered by plants) 3
   or
   The vegetation is largely discontinuous *with* substantial areas (total > 70 %) of exposed rock, boulders or soil **Key B**
3. The vegetation is dominated by woody plants 4
   or
   The vegetation is dominated by herbaceous plants. Include also vegetation dominated by herbaceous plants totally submerged in water, or plants with leaves that float on the water surface **Key C**
4. The vegetation is a canopy dominated by plants over 4 m tall **Key D**
   or
   The vegetation is a canopy dominated by plants less than 4 m tall **Key E**

Key A UNVEGETATED HABITATS (no vegetation)

1. The habitat is predominantly open water with no plants (care must be taken as plants may be present but submerged, if so return to main key Q1) 2
   or
   The habitat is terrestrial with no plants (without a column of water above the ground surface) 3
2. The water is obviously flowing
   River, stream **U17**
   or
   The water is standing with no obvious direction of flow
   Lake, pond **U18**
3. The habitat has been recently subjected to human activities such as ploughing or excavation
   Agriculture, building **U1**
   or
   The habitat appears natural or semi-natural 4
4. The habitat is predominantly snow or ice
   Snow packs, glaciers **U19**
   or
   The habitat is predominantly volcanic ash, lava fields, lava tubes, etc.
   Volcanic features **U16**
   or
   The habitat is predominantly sand
   Sand dunes, beaches **U2**
   or
   The habitat is predominantly mud, gravel or shingle
   Mud flats, gravel banks **U3**

or
The habitat is predominantly rock, boulders or cliffs (can be with encrusting lichens)

Bare cliffs **U10**

or
The habitat is a cave

Caves **U15**

Key B SCREES, VEGETATED CLIFFS, COLONISING VEGETATION (< 30 % vegetation)

1. The habitat has been recently subjected to human activities such as ploughing or excavation

   Agriculture, building, etc. **U1**

   or
   The habitat appears natural or semi-natural 2

2. The vegetation is almost entirely mosses, liverworts and ferns 3

   or
   The vegetation is dominated by plants other than mosses, liverworts and ferns 6

3. The vegetation is only apparent in cracks in otherwise bare, flat rock exposed at the ground surface

   Pavements **U11**

   or
   The vegetation is growing on the ground or rock surface 4

4. The vegetation is predominantly composed of mosses, liverworts, filmy-ferns (Hymenophyllaceae) or *Adiantum capillus-veneris*

   Wet inland cliffs **U14**

   or
   The vegetation is predominantly composed of other ferns 5

5. The fern *Cryptogramma crispa*, and the moss *Rhacomitrium lanuginosum* are present

   Alpine and northern siliceous screes **U5**

   or
   The fern *Gymnocarpium robertianum* is present

   Western Mediterranean and thermophilous screes **U7**

   or
   Any of the ferns *Asplenium viride*, *A. scolopendrium*, *A. trichomanes*, *A. petrachae* or *Polypodium cambricum* are present

   Vegetated calcareous inland cliffs **U12**

   or
   Either of the ferns *Asplenium septentrionale or A. adiantum-nigrum* are present. Flowering plants, particularly *Saxifraga* spp., may also be present

   Vegetated siliceous inland cliffs **U13**

6. *Rhynchospora alba*, *R. fusca*, *Drosera intermedia*, *D. rotundifolia* or *Lycopodiella inundata* are present

   White beak-sedge communities **M4**

   or
   The above plants are lacking 7

7. *Sempervivum*, *Sedum* or *Scleranthus* are present

   Alpine weathered rock and outcrop communities and pioneer swards **U9**

   or
   The above plants are lacking 8

8. The vegetation has low cushion-forming, often spiny shrubs, such as *Acantholimon, Astragalus, Erinacea, Bupleurum, Ptilotrichum, Genista* and *Echinospartum*
Hedgehog-heaths **S11**
or
The vegetation is not like the above 9

9. The substrate is gravel, mud or sand 10
or
The substrate is unconsolidated stones, boulders, or rock 11

10. *Isoetes, Littorella uniflora, Lobelia dortmana, Eleocharis acicularis, Juncus bufonius, Juncus bulbosus, Potamogeton, Nuphar* or mats of drying macroalgae are present
Amphibious communities **A4**
or
The trees or shrubs *Myricaria germanica, Salix, Hippophae, Alnus, Betula* or the herbaceous plants *Oenothera biennis, Filipendula, Petasites, Epilobium fleischeri* are present
Vegetated gravel banks **S2**
or
Bidens, Rorippa, Polygonum or *Paspalum* are present
River mud banks **A5**
or
Corynephorus canescens, Spergula morisonii or *Salix arenaria* are present with carpets of fruticose lichens
Fluvio-glacial dunes **U4**

11. *Achillea nana, Oxyria digyna* or *Saxifraga bryoides* are present
Alpine and northern siliceous screes **U5**
or
Thalaspi rotundifolium or *Draba hoppeana* are present
Alpine calcareous screes **U6**
or
Gymnocarpium robertianum, Galeopsis angustifolia, Leontodon hyoseroides, Linaria spp. or *Saxifraga* spp. are present
Western Mediterranean and thermophilous screes **U7**
or
Drypis spinosa is present
Eastern Mediterranean screes **U8**

Key C HERBACEOUS VEGETATION (with small shrubs < 1 m)

1. The vegetation is a produce-bearing plant grown as an arable crop 2
or
The vegetation appears natural or semi-natural, or is a grassland other than a cereal crop 3

2. The vegetation is a largely monospecific stand with few weeds, often grown in large fields
Intensive cropland **G34**
or
The vegetation is an arable crop, particularly a cereal, with a high number of weeds such as *Agrostema githago, Centaurea cyanus, Papaver* spp., *Chrysanthemum segetum*, or *Adonis* spp.
Extensive cultivation **G33**

3. *Sphagnum* is frequent or abundant (cover ≥ 10 %) 4
or
Sphagnum is absent or rare (cover < 10 %) 10

4. Presence of *Betula pendula*, *B. pubescens*, *Pinus sylvestris*, *P. rotundata* or *Picea abies* trees (cover > 10 %)

Birch and conifer swamp woods **W4**

or

There are no trees present or < 10 % tree cover 5

5. The dwarf shrubs *Calluna*, *Erica* or *Vaccinium* are abundant 6

or

The above dwarf shrubs are absent or rare (cover < 10 %) 8

6. *Erica tetralix* is equal to or more abundant than *Calluna*, *Vaccinium* or other *Erica* spp., the vegetation dominated by *Molinia caerulea* and *Sphagnum* with *Narthecium ossifragum*

Wet heath **M5**

or

Calluna, *Vaccinium* or other *Erica* spp. are more abundant than *Erica tetralix* 7

7. The community is characterised by a high number of hummocks and hollows of colourful (red, yellow, green and brown) *Sphagnum* spp. (*S. cuspidatum*, *S. recurvum*, *S. papillosum*, *S. magellanicum*) with other mosses. Open pools are present and characterised by *Eriophorum vaginatum*, *Menyanthes trifoliata*, *Utricularia* spp. and *Drosera* spp., other parts with *Narthecium ossifragum* and *Vaccinium oxycoccus*, *Erica tetralix* and other ericaceous shrubs.

Near natural raised bogs **M1**

or

The community is characterised by a patchy mix of *Eriophorum vaginatum*, *E. angustifolium*, *Scirpus cespitosus*, *Molinia caerulea*, *Schoenus nigricans*, *Rhynchospora alba*, *Polygala serpyllifolia*, *Narthecium ossifragum*, and *Pinguicula lusitanica* together with *Calluna vulgaris*, *Erica tetralix*, and *Myrica gale*. Abundant mosses including *Sphagnum* spp. (*S. subnitens*, *S. tenellum*, *S. papillosum*, *S. magellanicum*) and *Rhacomitrium lanuginosum*

Lowland blanket bog **M2**

or

The community is characterised by a patchy mix of *Eriophorum vaginatum*, *Calluna vulgaris*, *Erica tetralix*, *Scirpus cespitosus*, *Narthecium ossifragum*, *Rubus chamaemorus* and *Pinguicula lusitanica*. Abundant mosses including *Sphagnum* spp. (*S. subnitens*, *S. tenellum*, *S. papillosum*, *S. magellanicum*) and *Rhacomitrium lanuginosum*

Upland blanket bog **M3**

8. The vegetation is dominated by bryophytes, with *Montia fontana*, *Stellaria alsine*, *Saxifraga spp.*, *Chrysosplenium alternifolium*, *C. oppositifolium*, *Cardamine amara*, *C. flexuosa*, *C. raphanifolia*, *Equisetum telmateai* or *Pinguicula vulgaris*, with *Sphagnum auriculatum*

Springs **M7**

or

The vegetation is dominated by plants other than bryophytes 9

9. The vegetation is a mix of small sedges or rushes growing over a carpet of mosses, particularly *Sphagnum* (*S. recurvum*, *S. auriculatum*, *S. russowii*, *S. cuspidatum*, *S. subsecundum*, *S. papillosum*), with *Carex echinata*, *C. curta*, *C. nigra*, *Eriophorum angustifolium*, *Agrostis canina*, *Juncus filiformis*, *Viola palustris* and *Ranunculus flammula*

Acidic fens **M6**

or

The vegetation is a colonist community of swaying swards, floating carpets or quaking mires, composed of medium-tall sedges *Carex rostrata*, *C. lasiocarpa*, *C. limosa*, *C. chordorrhiza*, *Eriophorum* spp., *Scheuchzeria palustris*, and herbs *Menyanthes trifoliata*, *Potentilla palustris*, *Liparis loeselii*, *Calla palustris*, with brown mosses and *Sphagnum*. Often open water communities occur with *Potamogeton polygonifolius* or *P. coloratus*.

Transition mires **M11**

10. The vegetation is dominated by ferns or bryophytes 11
or
The vegetation is dominated by plants other than ferns or bryophytes 13

11. The vegetation is dominated by *Pteridium aquilinum*, often mixed with plants such as *Rubus* and *Ulex*

Thickets **S18**

or
The vegetation is dominated by bryophytes, with *Montia fontana*, *Stellaria alsine*, *Saxifraga spp.*, *Chrysosplenium alternifolium*, *C. oppositifolium*, *Cardamine amara*, *C. flexuosa*, *C. raphanifolia*, *Equisetum telmateia* or *Pinguicula vulgaris*

Springs **M7**

or
The vegetation is predominantly composed of mosses, liverworts, filmy-ferns (Hymenophyllaceae) or *Adiantum capillus-veneris*

Wet inland cliffs **U14**

or
The vegetation is predominantly composed of other ferns 12

12. The fern *Cryptogramma crispa* and the moss *Rhacomitrium lanuginosum* are present, frequently growing on unconsolidated areas of stones, boulders or rock

Alpine and northern silicious screes **U5**

or
The fern *Gymnocarpium robertianum* is present, often with *Galeopsis angustifolia*, *Leontodon hyoseroides*, *Linaria* spp. or *Saxifraga* spp. frequently growing on unconsolidated areas of stones, boulders or rock

Western Mediterranean and thermophilous screes **U7**

or
Any of the ferns *Asplenium viride*, *A. scolopendrium*, *A. trichomanes*, *A. petrachae* or *Polypodium cambricum* are present

Vegetated calcareous inland cliffs **U12**

or
Either of the ferns *Asplenium septentrionale* or *A. adiantum-nigrum* is present. Flowering plants, particularly *Saxifraga* spp., I may also be present

Vegetated siliceous inland cliffs **U13**

13. The vegetation (> 10–20 cm) is dominated by grasses, sedges, rushes or reeds (also for rich mixes of broad-leaved herbs and grasses, sedges, rushes or reeds, where no overall dominance is obvious) 23
or
The vegetation (< 10–20 cm) as above without reeds or non-dominant reeds 14

14. *Isoetes*, *Littorella uniflora*, *Lobelia dortmanna*, *Eleocharis*, *Pilularia*, *Marsilea*, *Juncus bulbosus* or *J. bufonius* are present

Amphibious communities **A4**

or
Broad-leaved herbs dominate the vegetation: pondweeds, water lilies, free-floating plants or other aquatic plants. Typical genera include *Potamogeton*, *Myriophyllum*, *Elodea*, *Callitriche*, *Nuphar*, *Nymphaea*, *Lemna*, *Hydrocharis*, *Azolla*, *Salvinia*, *Utricularia*, *Chara* or *Nitella* 46
or
The above species are not present 15

15. The vegetation is characterised by frequent patches of bare soil, stone, boulders or rock 16

or

The vegetation covers the soil entirely or almost entirely 18

16. The vegetation is dominated by a mixture of plants including *Oenothera biennis*, *Petasites*, *Filipendula* or *Epilobium*

Gravel banks **S2**

or

Bidens, *Rorippa*, *Polygonum* or *Paspalum* dominates the vegetation

River mud banks **A5**

or

The vegetation is not as above, and typically growing on unconsolidated areas of stones, boulders or rock 17

17. *Sedum*, *Sempervivium*, *Scleranthus* or *Jovibarba* are present

Middle European pioneer swards and Alpine weathered rock and outcrop communities **G20**

or

The vegetation is dominated by *Achillea nana*, *Oxyria digyna* and *Saxifraga bryoides*, frequently growing on unconsolidated areas of stones, boulders or rock

Alpine and northern siliceous screes **U5**

or

The vegetation is dominated by *Thalaspi rotundifolium* and *Draba hoppeana*, frequently growing on unconsolidated areas of stones, boulders or rock

Alpine calcareous screes **U6**

or

The vegetation is dominated by *Galeopsis angustifolia*, *Leontodon hyoseroides*, *Linaria* spp. or *Saxifraga* spp., with *Gymnocarpium robertianum* frequently growing on unconsolidated areas of stones, boulders or rock

Western Mediterranean and therophilous screes **U7**

or

The vegetation is dominated by *Thlaspi rotundifolium* and *Drypis spinosa*, frequently growing on unconsolidated areas of stones, boulders or rock

Eastern Mediterranean screes **U8**

18. The vegetation occurs as largely monospecific stands, or with very few species 19

or

The vegetation appears to be diverse mix with several species 20

19. The vegetation is composed of monospecific stands, or mixes of a few species, with *Sagittaria sagittifolia*, *Acorus calamus*, *Butomus umbellatus*, *Oenanthe aquatica*, *Rorippa amphibia*, *Equisetum fluviatile*, *Sium latifolium* or *Hippurus vulgaris*, often occurring next to stands of, or with, *Phragmites australis*, *Scirpus lacustris*, *Typha* spp., *Sparganium* spp., *Eleocharis palustris*, *Glyceria maxima*, *Phalaris arundinacea* or *Scirpus* spp. Include all these stands together

Reed beds **M13**

or

The vegetation is composed of monospecific stands, or mixes of a few species, with *Rorippa nasturtium-aquaticum*, *Rorippa microphylla*, *Veronica beccabunga*, *V. anagallis-aquatica*, *Apium nodiflorum* or *Sium erectum*, often occurring next to stands of, or with, *Glyceria fluitans*, other small *Glyceria* spp., small *Sparganium* spp. or *Catabrosa aquatica.* Include all these stands together

Small reed beds of fast-flowing waters **M15**

or

The vegetation is composed of stands of tall perrenial herbs, either native or introduced, such as *Alliaria petiolata*, *Lamium album*, *Impatiens* spp.,

Reynoutria spp., *Solidago* spp., *Rudbeckia* spp., *Aster* spp., *Eupatorium canibum*, *Sonchus palustris* or *Urtica* spp., often with *Galium aparine*, *Glechoma hederacea*, *Calystegia sepium* or *Geum urbanum*. Include all stands together

Humid tall herb fringes **G11**

20. The vegetation is a mixture of *Trifolium repens*, *Leontodon autumnalis*, *Bellis perennis*, *Ranunculus acris*, *R. repens* and *Cardamine pratensis* together with the grasses *Lolium perenne*, *Poa* spp., *Festuca* spp. and *Cynosurus cristatus*

Mesophile grasslands **G10**

or

The vegetation is not as above 21

21. The vegetation is a mixture of the tall herbs *Filipendula ulmaria*, *Achillea ptarmica*, *Angelica sylvestris*, *Iris pseudacorus*, *Sparganium erectum* or *Valeriana officinalis* together with rushes such as *Juncus effusus* and *J. acutiflorus*

Meadowsweet stands and related communities **M14**

or

The vegetation is composed of tall perennial herbs, either native or introduced, such as *Alliaria petiolata*, *Lamium album*, *Impatiens* spp., *Reynoutria* spp., *Solidago* spp., *Rudbeckia* spp., *Aster* spp., *Eupatorium canibum*, *Sonchus palustris* or *Urtica* spp., often with *Gallium aparine*, *Glechoma hederacea*, *Calystegia sepium* or *Geum urbanum*. Include all stands together

Humid tall herb fringes **G11**

or

The vegetation is not as above 22

22. The vegetation is dominated by annual herbs, with *Malcolmia* spp. and *Erodium* spp.

Mediterranean annual deep-sand communities **G19**

or

The vegetation includes *Geranium sanguineum*, *Bupleurum* spp., *Origanum vulgare*, *Trifolium rubens*, *T. medium* or *T. ochroleucum* and is typically on the edge of a wood

Thermophile forest fringes **G29**

or

The vegetation includes *Dryas octopetala*, *Draba azoides*, *Gentiana* spp. and *Oxytropis* spp. together with grasses and sedges such as *Festuca* spp., *Sesleria* spp., *Carex ferruginea* and *C. sempervirens*

Alpine and sub-alpine calciphilous grasslands **G28**

or

The vegetation contains a mixture of *Cicerbita alpina*, *C. plumieri*, *Cirsium heleniodes*, *C. spinosissimum*, *C. flavispina*, *Aconitum* spp., *Trollius europaeus*, *Pedicularis foliosa* and *Calamagrostis arundinacea*

Sub-alpine and alpine tall herb communities **G13**

23. The vegetation is composed of monospecific stands, or mixes of a few species, with *Phragmites australis*, *Scirpus lacustris*, *Typha* spp., *Sparganium* spp., *Eleocharis palustris*, *Glyceria maxima*, *Phalaris arundinacea* or *Scirpus* spp., often occurring next to stands of, or with, *Sagittaria sagittifolia*, *Acorus calamus*, *Butomus umbellatus*, *Oenanthe aquatica*, *Rorippa amphibia*, *Equisetum fluviatile*, *Sium latifolium*, *Hippurus vulgaris*. Include all these stands together

Reed beds **M13**

or

The vegetation is composed of monospecific stands, or mixes of a few species, with *Glyceria fluitans*, other small *Glyceria* spp., small *Sparganium* spp. or *Catabrosa aquatica*, often occurring next to stands of, or with, *Rorippa nasturtium-aquaticum*, *Rorippa microphylla*, *Veronica beccabunga*,

V. anagallis-aquatica, *Apium nodiflorum* or *Sium erectum*. Include all these stands together

Small reed beds of fast-flowing waters **M15**

or

The vegetation is not as above 24

24. The vegetation is dominated by sedges 25

or

The vegetation is dominated by grasses or rushes 29

25. The vegetation is a mixture of *Scirpus holoscheonus*, *Cyperus longus* and *Schoenus nigrans*, with *Agrostis stolonifera*, *Molinia caerulea* and various tall herbs

Mediterranean tall humid grassland **G7**

or

The vegetation is not as above 26

26. The vegetation is dominated by *Cladium mariscus*

Fen-sedge beds **M10**

or

The vegetation is composed of largely monospecific stands of tussock- or bed-forming sedges, such as *Carex acuta*, *C. acutiformis*, *C. oenensis*, *C. buekii*, *C. aquatilis*, *C. riparia*, *C. vesicaria*, *C. rostrata*, *C. paniculata*, *C. pseudocyperus*, *C. otrubae*, *C. vulpina* or *Cyperus* spp.

Large sedge communities **M9**

or

The vegetation is not as above 27

27. The community is composed of *Rynchospora alba* and *R. fusca*, with *Drosera intermedia*, *D. rotundifolia* and *Lycopodiella inundata*, often growing on exposed peat

White beak-sedge communities **M4**

or

The vegetation is not as above 28

28. The vegetation is a very diverse community often of rare or restricted species, characterised by a layer of brown mosses, with *Schoenus nigricans*, *S. ferrugineus*, *Eriophorum latifolium*, *Carex davalliana*, *C. dioica*, *C. flacca*, *C. flava*, *C. viridula*, *C. hostiana*, *C. pulicaris*, *C. panicea*, *Juncus subnodulosus*, *Scirpus cespitosus*, *Eleocharis quinqueflora* together with many orchid species (*Dactylorhiza* spp., *Epipactis palustris*), and often *Parnassia palustris*, *Pinguicula vulgaris*, *Linum catharticum* or *Tofieldia* spp. This community can also have sedge, fen-sedge or reed beds present

Basic rich fens **M12**

or

The vegetation includes the sedges *Carex bicolor*, *C. microglochin*, *C. maritima*, *C. atrofusca*, *C. vaginata*, *Kobresia simpliciuscula*, *Scirpus pumilus*, and rushes *Juncus arcticus*, *J. alpinoacticulatus*, *J. castaneus*, *J. triglumis*, sometimes with reeds *Typha minima*, *T. lugdunensis*, *T. shuttleworthii*, *Toefieldia pusilla* or *Saxifraga aizoides*

Arcto-alpine riverine swards **G3**

or

The vegetation is a community of small sedges or rushes which dominate over a carpet of mosses, particularly *Sphagnum* (*S. recurvum*, *S. auriculatum*, *S. russowii*, *S. cuspidatum*, *S. subsecundum*, *S. papillosum*), with *Carex echinata*, *C. curta*, *C. nigra*, *Eriophorum angustifolium*, *Agrostis canina*, *Juncus filiformis*, *Viola palustris* and *Ranunculus flammula*

Acidic fens **M6**

or

The vegetation is a colonist community of swaying swards, floating carpets or quaking mires of medium-tall sedges *Carex rostrata*, *C. lasiocarpa*, *C.*

limosa, *C. chordorrhiza*, *Eriophorum* spp., *Schechzeria palustris*, and herbs *Menyanthes trifoliata*, *Potentilla palustris*, *Liparis loeselii*, *Calla palustris*, with brown mosses and *Sphagnum*. This vegetation is often associated with open water communities with *Potamogeton polygonifolius* or *P. coloratus*
Transition mires **M11**

29. *Juncus* spp. are frequent or abundant (cover ≥ 10 %) 30
or
Juncus spp. are absent or rare (cover < 10 %) 31

30. There is an abundance of tall, broad-leaved herbs, such as *Filipendula ulmaria*, *Achillea ptarmica*, *Angelica sylvestris*, *Iris pseudacorus*, *Sparganium erectum*, *Valeriana officinalis* together with rushes such as *Juncus effusus* and *J. acutiflorus*
Meadowsweet stands and related communities **M14**
or
The vegetation is a mix of rushes, particularly *Juncus effusus*, *J. inflexus*, *J. subnodulosus*, *J. acutiflorus*, and grasses such as *Deschampsia caespitosa*, *Holcus lanatus*, *Agrostis stolonifera* or *Festuca rubra*
Eutrophic humid grassland **G9**
or
The vegetation is a mixture of *Juncus squarrosus*, *Molinia caerulea*, *Festuca ovina*, *Nardus stricta*, *Deschampsia flexuosa* and *Potentilla erecta*
Oligotrophic humid grassland **G6**
or
The vegetation is a mixture of *Juncus trifidus*, *Nardus stricta* and *Carex biglowii*, *C. curvula*, *Festuca paniculata*, *F. eskiae* or *F. varia*, with the moss *Rhacomitrium lanuginosum*, sometimes with dwarf shrubs
Alpine and sub-alpine acidophilous grassland **G5**

31. *Agrostis capillaris*, *Festuca ovina* or *Nardus stricta* or *Molinia caerulea* are present or abundant 32
or
Lolium perenne or *L. multiflorum* are present or abundant 36
or
Bromus spp. or *Brachypodium* spp. are present 38
or
Stipa spp. are present 39
or
The above grasses are either absent or rare 40

32. *Nardus stricta* is abundant (cover ≥ 10 %) 33
or
Nardus stricta is absent or rare (cover < 10 %) 34

33. The vegetation is a mixture of *Nardus stricta* and *Carex biglowii*, *C. curvula*, *Juncus trifidus*, *Festuca paniculata*, *F. eskiae* or *F. varia*, with the moss *Rhacomitrium lanuginosum*, sometimes with dwarf shrubs
Alpine and sub-alpine acidophilous grassland **G5**
or
The vegetation is dominated by *Nardus stricta* and *Luzula multiflora*, with *Crocus vernus*
Mediterraneo-montane mat-grass swards **G15**
or
Nardus stricta, *Festuca ovina*, *Agrostis capillaris* or *Galium saxatile* dominate the vegetation
Atlantic mat-grass swards **G12**

34. *Nardus stricta* is absent; the vegetation is a mix of rushes, particularly *Juncus effusus*, *J. inflexeus*, *J. subnodulosus*, *J. acutiflorus*, and grasses such as *Holcus lanatus*, *Agrostis stolonifera* or *Festuca rubra*
Eutrophic humid grassland **G9**

or

The vegetation is different from the above 35

35. *Molinia caerulea* and *Potentilla erecta* are abundant, together with *Festuca ovina, Nardus stricta, Carex panicea, C. nigra, Succisa pratensis*

Oligotrophic humid grassland G6

or

Agrostis capillaris and *Anthoxanthum odoratum* are present or abundant, together with *Festuca ovina, Nardus stricta, Deschampsia flexuosa, Galium saxatile* or *Agrostis curtisii, Calluna vulgaris, Vaccinium myrtillus*

Atlantic matgrass swards and related communities G12

36. *Lolium perenne, Holcus lanatus, Dactylis glomerata, Trifolium repens, Bellis perennis, Taraxacum officinale* are present and *Juncus* spp. are absent 37

or

Agrostis stolonifera, Juncus effusus, J. acutiflorus, Rumex obtusifolius, R. crispus are present with some herbs above

Humid improved grasslands G8

37. *Phleum pratense, Poa trivialis, Lolium multiflorum, Festuca pratensis* are present

Dry improved grassland G32

or

Festuca rubra, Cynosurus cristatus, Lotus corniculatus, Centaurea nigra, Anthoxanthum odoratum, Cerastium fontanum or *Lathyrus pratensis* are present

Mesophile grasslands G10

38. *Bromus erectus* and *Brachypodium pinnatum* are present with *Festuca ovina, F. valesiaca, Sesleria albicans* and *Thymus praecox*

Dense perennial grasslands and middle European Steppes G27

or

Bromus fasciculatus, B. intermedius, B. alopecuros and *Brachypodium retusum* are present

Mediterranean xeric grasslands G26

or

Bromus fasciculatus, B. madritensis, B. intermedius, B. alopecuros are present in a community of mostly annuals, with *Aegilops (A. neglecta, A. geniculata), Avena (A. sterilis, A. barbata), Vulpia (V. ciliata, V. bromioides, V. geniculata)*, with *Trifolium* spp., *Medicago* spp., *Astragulus* spp., *Brassica* spp.

Mediterranean subnitrophilous grassland G31

39. *Stipa tenacissima, S. gigantea, S. lagascae* or *S. offneri* are present

Mediterranean tall grass steppes G25

or

Stipa pennata, Koeleria vallesiana or *Ononis striata* are present

Mediterraneo-montane grasslands G24

40. *Arrhenatherum elatius* is the dominant grass, with *Trisetum flavescens* 41

or

The dominant grass is some other species 42

41. The vegetation has abundant herbs such as *Anthriscus sylvestris, Daucus carota, Knautia arvensis, Filipendula ulmaria, Leucanthemum vulgare, Heracleum sphondylium* or *Geranium pratense* present

Lowland hay meadows G17

or

The vegetation has abundant herbs such as *Astrantia major, Carum carvi, Crepis mollis, C. Pyrenaica, Crocus albiflorus, Heracleum sphondylium* or *Trollius europaeus* present

Mountain hay meadows G16

or

Agrostis alpina and *Phleum alpinum* are abundant

Alpine and sub-alpine fertilised grasslands **G30**

42. The vegetation is rich in mosses and lichens, with *Minuartia verna*

Lowland heavy metal grassland **G18**

or

The vegetation is a mixture of *Agrostis capillaris*, *Festuca ovina* and *Vulpia bromoides*, with *Rumex acetosella* and *Teesdalea nudicaulis*

Medio-european open siliceous grassland **G23**

or

Dryas octopetala, *Draba azoides*, *Gentiana* spp. and *Oxytropis* spp. occur in a grassland of *Festuca* spp. or *Sesleria* spp.

Alpine and sub-alpine calciphilous grasslands **G28**

or

The vegetation is not as above 43

43. The vegetation is dominated by *Festuca* spp. 44

or

The vegetation is dominated by some other grass species 45

44. The vegetation is dominated by *Festuca costei*, *F. indigesta* or *F. simmilusitana*, with cushion-forming rosette plants such as *Jasione crispa* and *Scleranthus perennis*

Iberian fescue-plantain swards **G21**

or

The vegetation is dominated by *Festuca elegans*

Iberian tall fescue grasslands **G22**

or

Festuca ovina, *F. valesiaca* or *Sesleria albicans* is present and abundant, with *Briza* spp. and *Thymus praecox*

Dense perennial grasslands and middle European Steppes **G27**

45. The vegetation is dominated by tall cane species, such as *Imperata cylindrica*, *Saccharum ravennae*, *S. strictum*, *Arundo plinii* or *A. donax*

Riparian cane formations **G4**

or

The vegetation is a mixture of *Agrostis castellana*, *A. pouretii*, *Gaudinia fragilis*, *Vulpia ciliata*, *V. Myuros* and *V. bromoides*

Iberian vallicares **G2**

or

The vegetation is dominated by *Deschampsia media*, with *Centaurium pulchellum* and *Plantago serpentina*

Mediterranean short humid grassland **G1**

or

The vegetation is dominated by *Alopecurus utriculatus* and *Hordeum murinum*, with *Trifolium resupinatum* and several other clover species

Eastern supra-Mediterranean humid meadows **G14**

46. The vegetation is of *Rorippa nasturtium-aquaticum*, *R. microphylla*, *Oryza sativa* or some other species, and evidently managed as a commercial crop

Flooded crops **A8**

or

The vegetation is not as above 47

47. The vegetation is floating on the water surface, without roots penetrating the sediment. Typical species include *Lemna* spp., *Hydrocharis* morsus-ranae, *Azolla* spp., *Salvinia* spp. *and Utricularia* spp. 48

or

The vegetation is rooted in the sediment 49

48. *Lemna* spp., *Hydrocharis* morsus-ranae, *Azolla* spp. or *Salvinia* spp. are present

Free-floating vegetation **A6**

or

Utricularia spp. is present

Peatmoss-bladderwort bog pools **A1**

49. *Potamogeton pectinatus, P. nodosus* or *Zannichellia palustris* are abundant

Eutrophic vegetation **A7**

or

Potamogeton polygonifolius, Myriophyllum alterniflorum, or *Callitriche hamulata* are abundant

Acid oligotrophic vegetation **A3**

or

The above species are absent 50

50. The plants have leaves floating on the surface of the water; typical species include *Nymphaea* spp., *Nuphar* spp., *Nymphoides* spp., *Trapa* spp., or *Polygonum* spp. or species with both submerged and floating leaves such as *Potamogeton natans*, *Ranunculus* spp., *Callitriche* spp. or *Sparganium* spp. 51

or

The plants have only submerged leaves, which never fully break the surface of the water 52

51. *Bidens* spp., *Rorippa* spp., *Chenopodium* spp. or *Paspalum* spp. are present

River mud banks **A5**

or

The above species are absent

Rooted floating vegetation **A10**

52. *Chara* spp. or *Nitella* spp. are abundant

Chandelier algae submerged carpets **A2**

or

The above are either absent or not abundant

Rooted submerged vegetation **A9**

Key D FORESTS AND WOODLAND (> 4 m, woody)

1. The trees are arranged in straight lines (care must be taken here as regeneration in some old plantations can give them a semi-natural appearance; check if the older, mature trees are arranged in straight lines) 2

or

The trees are arranged in a random fashion, giving the area a natural or semi-natural appearance 3

2. The dominant trees are olives, walnuts, chestnuts, fruit trees, palms or some other produce-bearing trees

High stem orchards **W3**

or

The trees are timber trees, conifers or broad-leaved trees that do not produce commercial fruit or nut crops

Plantations **W1**

or

The trees are arranged in a narrow line in such a way as to delineate a boundary

Treelines and hedgerows **W2**

As many treelines, hedgerows and plantations are of considerable age and are, in essence, highly managed, semi-natural vegetation, they can be used to provide additional information on the environmental conditions and functioning. Continue with Key C from question 3, but in the final vegetation code indicate that the vegetation is either part of a hedge or treeline (code

W2) or a plantation (code **W3**). Care must be taken when exotic species are encountered as the dominant species, as these will not key out correctly.

3. The vegetation is dominated by coniferous trees, with needle-like or scale-like leaves 4

   or

   The vegetation is dominated by broad-leaved trees 13

4. *Pinus* spp. are the dominant trees 7

   or

   Abies spp. are the dominant trees

   Fir forests **W23**

   or

   Picea abies is the dominant tree 5

   or

   Other tree species are dominant 6

5. *Betula pendula* or *B. pubescens* are present, with a ground cover of ericaceous shrubs, *Vaccinium myrtillus*, *V. uliginosum*, *Empetrum nigrum*, with *Molinia caerulea* and *Sphagnum* spp.

   Birch and conifer swamp woods **W4**

   or

   Betula pendula or *B. pubescens* are not present and the ground cover is different from above.

   Spruce forests **W21**

6. *Larix decidua* is present, either as the dominant tree or mixed with *Pinus cembra*

   Larch–arolla forests **W34**

   or

   Cupressus sempervirens, *Juniperus* spp. or *Taxus baccata* (often with *Ilex aquilifolium*) are the dominant tree species

   Cypress, juniper and yew forest **W31**

7. *Pinus cembra* is present, either as the dominant tree or mixed with *Larix decidua*

   Larch–arolla forests **W34**

   or

   There is a rather open canopy of *Pinus uncinata* with a well developed understorey of shrubs

   Mountain pine forests **W29**

   or

   The dominant tree is some other *Pinus* spp. 8

8. *Pinus sylvestris* is dominant 9

   or

   The dominant tree is some other *Pinus* spp. 10

9. *Betula pendula* or *B. pubescens* are present, with a ground cover of ericaceous shrubs, *Vaccinium myrtillus*, *V. uliginosum*, *Empetrum nigrum*, with *Molinia caerulea* and *Sphagnum* spp.

   Birch and conifer swamp woods **W4**

   or

   Betula pendula or *B. pubescens* are not present in large numbers, *Pinus sylvestris* var. *scotica* or *Pinus sylvestris* is the dominant tree, often with a ground cover of *Vaccinium myrtillus*, *Juniperus communis*, *Luzula* spp. or mosses

   Scots pine forest **W26**

10. The dominant tree is *Pinus rotundata* with *Betula pendula* or *B. pubescens* present, with a ground cover of ericaceous shrubs, *Vaccinium myrtillus*, *V. uliginosum*, *Empetrum nigrum*, with *Molinia caerulea* and *Sphagnum* spp.

    Birch and conifer swamp woods **W4**

or

The dominant tree is some other *Pinus* spp. 11

11. The dominant tree is *Pinus nigra*, *P. salzmannii*, *P. laricio* or *P pallasiana*

Black pine forests **W24**

or

The dominant tree is *Pinus heldreichii*, *P. leucodermis* or *P. peuce*

High oro-Mediterranean pine forests **W2**

or

The dominant tree is *Pinus pinaste*, *P. pinea*, *P. halepensis* or *P. brutia* 12

12. The tree canopy is very open, dotted with trees, and the ground cover is shrubby evergreen bushes

Matorral **W33**

or

The tree canopy is closed

Mediterranean pine woods **W30**

13. The vegetation is dominated by deciduous trees; leaves tend to be soft to the touch and lack a highly glossy, waxy surface 19

or

The vegetation is domiated by evergreen trees; leaves with a highly glossy, waxy surface 14

14. The dominant is tree an oak, *Quercus* spp. 15

or

The dominant is tree some other evergreen broad-leaved species 17

15. The dominant trees are *Quercus suber*, *Q. rotundifolia* or *Q. faginea* with a fairly closed to open canopy with an agricultural ground cover of crops, pasture or scrub, grown in juxtaposition or rotation

Dehesa **W43**

or

The dominant trees are *Quercus suber*, *Q. rotundifolia* or *Q. faginea* with a fairly closed to open canopy with a ground cover of shrubby evergreen bushes

Matorral **W33**

or

The vegetation is not as above 16

16. *Quercus suber* is the dominant tree species

Cork oak forests **W37**

or

Quercus ilex or *Q. rotundifolia* are the dominant tree species

Meso- and supra-Mediterranean holm-oak forests **W32**

or

Quercus coccifera is the dominant tree species

Kermes oak forests **W35**

17. The dominant tree species are *Olea europaea*, *Ceratonia siliqua*, *Pistaca lenticus* 18

or

Laurus nobilis is the dominant tree species

Continental laurel–oak forests **W39**

or

Ilex aquilifolium is the dominant tree species

Holly woods **W19**

18. The canopy is open with a ground cover of shrubby evergreen bushes

Matorral **W33**

or

The canopy is closed

Olive–carob forests **W28**

19. The dominant trees are *Quercus* spp., *Salix* spp., *Betula* spp., *Alnus* spp. or *Populus* spp. 20
or
The dominant trees are of some other species 44

20. *Quercus robur* or *Q. petraea* are present 21
or
Other *Quercus* spp. are present 24
or
There are no *Quercus* spp. present 26

21. The dominant trees are *Quercus petraea* or *Q. robur* 22
or
The canopy is dominated by a mixture of trees, with *Quercus petraea, Q. robur, Fraxinus excelsior, Acer campestre, A. pseudoplatanus, A. platanoides, Ulmus galbra, Fagus sylvatica* or *Tilia* spp. 23

22. The dominant trees are a mixture of *Quercus petraea, Q. robur, Betula pendula* or *B. pubescens*, with field layer of *Pteridium aquilinum, Deschamspia flexuosa* or *Vaccinium myrtillus*
Acidophilous oak forests **W16**
or
Quercus robur is the dominant tree species, but with other trees present, particularly *Carpinus betulus, Tilia* spp., *Aesculus* spp., *Castanea* spp., *Acer pseudoplatanus, A. platanoides, A. campestre* and *Ilex* spp., and a field layer often of *Pteridium aquilinum* and *Hyacinthoides non-scripta*
Oak–hornbeam forests **W20**
or
Quercus petraea is the dominant tree species
Thermophilous and supra-Mediterranean oak woods **W25**

23. The community is of a relatively simple structure (maximum 3 layers; tree, shrub, ground), the canopy dominated by a mixture of trees, with *Quercus petraea, Q. robur, Fraxinus excelsior, Acer campestre, A. pseudoplatanus, A. platanoides, Ulmus galbra, Fagus sylvatica* or *Tilia* spp., which develop on steep or relatively unconsolidated ground
Mixed ravine and slope forests **W42**
or
The community is dominated by a mixture of *Fraxinus excelsior, Quercus petraea* and *Q. robur*, but with *Sorbus aucuparia, Acer campestre, Ulmus glabra*, understorey of *Corylus avenula, Cretaegus* spp., *Sambucus nigra*, field layer of ferns, grasses, *Mercurialis perennis, Hyacinthoides non-scripta, Circaea lutetiana, Arum maculatum*
Ash forests **W17**
or
The community is very tall, multilayered (up to 8 layers), and highly complex, consisting of *Quercus robur, Fraxinus excelsior, Ulmus minor, U. leavis, U. glabra, Populus alba, P. canescens, P. nigra, Acer pseudoplatanus, A. platanoides, Salix alba, Alnus glutinosa, Prunus avium* in the upper stratum, *Malus sylvestris, Tilia cordata* in the lower stratum, and a sub-arborescent layer of *Alnus incana, Prunus padus, Cretaegus monogyna*. There are various high and low shrubs and numerous lianas, *Clematis vitalba, Tammus communis, Humulus lupus, Hedera helix, Vitis vinifera*. There are only a few scattered examples of this community remaining, located in the great river systems of Europe
Mixed oak–elm–ash forests of great rivers **W9**

24. *Quercus pubescens* or other deciduous oaks, e.g. *Q. cerris, Q. faginea*, are the dominant tree species
Thermophilous and supra-Mediterranean oak woods **W25**
or
Quercus pyrenaica is the dominant tree species 25

25. The canopy is open with a ground cover of shrubby evergreen bushes
Matorral **W33**

or

The canopy is closed
Quercus pyrenaica forests **W18**

26. *Betula* spp. are the dominant trees 27

or

Populus spp. are the dominant trees 31

or

Salix spp. are the dominant trees 33

or

Alnus spp. are the dominant trees 35

27. Trees other than *Betula* spp. are present 28

or

Betula spp. occur almost exclusively, with very few other tree species 30

28. *Sorbus aucuparia* is abundant
Rowan woods **W27**

or

Sorbus aucuparia is not present or scarce 29

29. *Alnus glutinosa* is present, with *Osmunda regalis* in the field layer
Southern alder and birch galleries **W5**

or

Pinus sylvestris, *Pinus rotundata* or *Picea abies* are present, with a ground cover of ericaceous shrubs, *Vaccinium myrtillus*, *V. uliginosum*, *Empetrum nigrum*, with *Molinia caerulea* and *Sphagnum* spp.
Birch and conifer swamp woods **W4**

30. The ground cover is made up of ericaceous shrubs, *Vaccinium myrtillus*, *V. uliginosum*, *Empetrum nigrum*, with *Molinia caerulea* and *Sphagnum* spp.
Birch and conifer swamp woods **W4**

or

The ground cover is not as above
Birch woods **W14**

31. The dominant tree species is *Populus tremula*
Aspen woods **W15**

or

The dominant tree species is either *Populus alba* or *P. nigra* 32

32. *Fraxinus angustifolia* and *Ulmus minor* are abundant, producing a multilayered community with *Populus alba*, *P. nigra*, *Fraxinus angustifolia*, *Ulmus minor*, *Salix alba* and other *Salix* spp., *Alnus*, with many lianas. *P. alba* is usually the tallest tree present
Mediterranean poplar–elm–ash forests **W10**

or

Fraxinus angustifolia and *Ulmus minor* are either not present or not abundant, *Salix* spp. present in high numbers, and the vegetation is not as above
Riparian willow formations **W7**

33. Either *Populus alba* or *P. nigra* are present 34

or

Populus alba and *P. nigra* are absent 35

34. *Fraxinus angustifolia* and *Ulmus minor* are abundant, producing a multilayered community with *Populus alba*, *P. nigra*, *Fraxinus angustifolia*, *Ulmus minor*, *Salix alba* and other *Salix* spp., *Alnus* spp., with many lianas. *P. alba* is usually the tallest tree present
Mediterranean poplar–elm–ash forests **W10**

or
Fraxinus angustifolia and *Ulmus minor* are either not present or not abundant, *Salix* spp. present in high numbers, and the vegetation is not as above
Riparian willow formations **W7**

35. The ground cover is mainly tall reeds and sedges (e.g. *Phragmites australis, Typha latifolia, Carex pendula*) with some tall herbs, and the dominant trees *Salix cinerea, S. pentandra, S. fragilis, S. aurita. Alnus glutinosa* can also be present in high numbers
Alder, willow and bog-myrtle swamp woods **W8**
or
The ground cover is mainly tall herbs, particularly *Urtica dioica* 36

36. *Alnus glutinosa* is present and abundant
Alder woods **W12**
or
Alnus glutinosa is either infrequent or not present
Riparian willow formations **W7**

37. *Fraxinus* spp. are present 38
or
Fraxinus spp. are absent 39

38. Either *Populus alba* or *P. nigra* are present, together with *Fraxinus angustifolia, Ulmus minor, Salix alba* and other *Salix* spp., producing a multilayered community with many lianas. *P. alba* usually the tallest tree present
Mediterranean poplar–elm–ash forests **W10**
or
Populus alba and *P. nigra* are absent, and the ground cover is *Carex* spp. (*C. pedula, C. remota, C. riparia, C. acutiformis*) or tall herbs (*Lysimachia nemorum, Eupatorium cannabinum, Urtica dioica, Cirsium oleraceum, Iris pseudacorus, Equisetum telmateia, Osmunda regalis*)
Medio-European stream ash–alder woods **W11**

39. *Alnus incana* is the dominant tree species
Grey alder galleries **W6**
or
Alnus glutinosa is the dominant tree species 40

40. *Alnus glutinosa* occurs almost exclusively, with very few other tree species 41
or
Trees other than *Alnus glutinosa* are present 42

41. *Osmunda* regalis is present. If other trees are present are they *Betula* spp.
Southern alder and birch galleries **W5**
or
Urtica dioica is present. If other trees are present they are *Sambucus nigra* or *Salix* spp.
Alder woods **W12**

42. *Sambucus nigra* is present, with *Urtica dioica* present in the ground cover. If other trees are present they are *Salix* spp.
Alder woods **W12**
or
Sambucus nigra is absent 43

43. The ground cover is mainly tall reeds and sedges (e.g. *Phragmites australis, Typha latifolia, Carex pendula*) with some tall herbs, and the trees *Salix cinerea, S. pentandra, S fragilis, S. aurita* abundant
Alder, willow and bog-myrtle swamp woods **W8**

or

The ground cover is not tall reeds and sedges. The trees *S. viminalis*, *S. purpurea*, *S. triandra* or *S. alba* are abundant, sometimes with *Populus nigra* or *P. alba*

Riparian willow formations **W7**

44. *Fraxinus excelsior* is abundant 45

or

Fraxinus angustifolia or *F. ornus* are abundant 48

or

Fraxinus spp. are absent or infrequent 49

45. *Alnus* spp. are present 46

or

Alnus spp. are absent 47

46. The community is very tall, multilayered and highly complex, of *Quercus robur*, *Fraxinus excelsior*, *Ulmus minor*, *U. leavis*, *U. glabra*, *Populus alba*, *P. canescens*, *P. nigra*, *Acer pseudoplatanus*, *A. platanoides*, *Salix alba*, *Alnus glutinosa*, *Prunus avium* in the upper stratum, *Malus sylvestris*, *Tilia cordata* in the lower stratum, and a sub-arborescent layer of *Alnus incana, Prunus padus*, *Cretaegus monogyna*. There are various high and low shrubs and numerous lianas, *Clematis vitalba*, *Tammus communis*, *Humulus lupus*, *Hedera helix*, *Vitis vinifera.* Only a few scattered examples remain of this community, located in the great river systems of Europe.

Mixed oak–elm–ash forests of great rivers **W9**

or

The community is fairly simple, of *Fraxinus excelsior* and *Alnus glutinosa* arranged in few layers, and the ground layer composed of *Carex* spp. (*C. pedula*, *C. remota*, *C. riparia*, *C. acutiformis*) or tall herbs (*Lysimachia nemorum*, *Eupatorium cannabinum*, *Urtica dioica*, *Cirsium oleraceum*, *Iris pseudacorus*, *Equisetum telmateia*, *Osmunda regalis*)

Medio-European stream ash–alder woods **W11**

47. The canopy is dominated by a mixture of trees, *Fraxinus excelsior* with any of *Quercus petraea*, *Q. robur*, *Acer campestre*, *A. pseudoplatanus*, *A. platanoides*, *Ulmus galbra*, *Fagus sylvatica* or *Tilia* spp. present, which develop on steep or relatively unconsolidated ground.

Mixed ravine and slope forests **W9**

or

The community is dominated by a mixture of *Fraxinus excelsior*, *Quercus petraea*, *Q. robur*, *Acer campestre* and *Ulmus glabra* but with *Sorbus aucuparia*, and an understorey of *Corylus avenula*, *Cretaegus* spp., *Sambucus nigra*, and field layer of ferns, grasses and *Mercurialis perennis*, *Hyacinthoides non-scripta*, *Circaea lutetiana*, *Arum maculatum*

Ash forests **W17**

48. Either *Populus alba* or *P. nigra* are present, together with *Fraxinus angustifolia*, *Ulmus minor*, *Salix alba* and other *Salix* spp., producing a multilayered community with many lianas. *P. alba* usually the tallest trees present

Mediterranean poplar–elm–ash forests **W10**

or

Populus alba or *P. nigra* are absent, the dominant trees either *Fraxinus angustifolia* or *Fraxinus ornus*

Hop–hornbeam, oriental hornbeam and mixed thermophilous woods **W36**

49. The dominant trees are *Fagus* spp. 50

or

The dominant trees are *Acer* or *Ulmus* spp. 51

or

The dominant trees are of other species 53

50. The community is dominated by *Fagus sylvatica* (or *F. orientalis*, *F. moesiaca* in Greece), if other trees are present they are only in the understorey, and a field layer of either *Deschampsia flexuosa*, *Rubus fruticosus*, *Mercuralis perennis*, *Luzula luzuloides* or *Scilla lilio-hyasinthus*

Beech forests **W44**

or

The community is dominated by a mixture of trees, with *Fagus sylvatica*, *Fraxinus excelsior*, *Acer campestre*, *A. pseudoplatanus*, *A. platanoides*, *Ulmus galbra*, *Quercus* or *Tilia* spp. present, which develop on steep or relatively unconsolidated ground

Mixed ravine and slope forests **W42**

51. The canopy is dominated by a mixture of trees 52

or

The canopy is dominated by *Ulmus minor* (*U. carpinifolia*, *U. procera*), *U. glabra* or *U. laevis*

Small-leaved elm woods **W41**

52. The canopy is dominated by a mixture of trees, with *Acer campestre*, *A. pseudoplatanus*, *A. platanoides*, *Ulmus galbra*, *Fraxinus excelsior*, *Quercus petraea*, *Q. robur*, *Fagus sylvatica* or *Tilia* spp. present, which develop on steep or relatively unconsolidated ground

Mixed ravine and slope forests **W42**

or

The community is very tall, multilayered and highly complex, of *Ulmus minor*, *U. leavis*, *U. glabra*, *Acer pseudoplatanus*, *A. platanoides*, *Quercus robur*, *Fraxinus excelsior*, *Populus alba*, *P. canescens*, *P. nigra*, *Salix alba*, *Alnus glutinosa*, *Prunus avium* in the upper stratum, *Malus sylvestris*, *Tilia cordata* in the lower stratum, and a sub-arborescent layer of *Alnus incana*, *Prunus padus*, *Cretaegus monogyna*. There are various high and low shrubs and numerous lianas, *Clematis vitalba*, *Tamus communis*, *Humulus lupulus*, *Hedera helix*, *Vitis vinifera*. Only a few scattered examples remain of this community, located in the great river systems of Europe

Mixed oak–elm–ash forests of great rivers **W9**

53. The dominant trees are *Ostrya carpinifolia*, *Carpinus orientalis*, *Tilia* spp. or *Celtis australis*

Hops–hornbeam, oriental hornbeam and mixed thermophilous forests **W36**

or

The dominant trees are *Castanaea sativa*

Chestnut woods **W36**

or

The dominant trees are *Sorbus aucuparia*

Rowan woods **W27**

or

The dominant trees are *Tilia* spp.

Lime woods **W40**

or

The dominant trees are *Platanus orientalis* or *Liquidamber orientalis*

Oriental plane and sweetgum woods **W13**

Key E HEATHS AND SCRUB (< 4 m, woody)

1. There is evidence that the vegetation has been cut recently (e.g. trash, exposed wood which is still fresh looking, stumps, cut stools)

If possible use Key C to identify the vegetation, which was previously present; if not return to main key and ignore cut vegetation unless it is obviously regenerating

or

There is no evidence that the area has been cut recently 2

2. The plants are arranged in straight lines (care must be taken here as regeneration in some old plantations can give them a semi-natural appearance, check if the older, mature plants are arranged in straight lines) 3
or
The plants are arranged in a random fashion, giving the area a natural or semi-natural appearance 4

3. The dominant plants are *Vitus vinifera*, espaliers of Rosaceae, or other commercial produce-bearing plants
Shrub orchards **S19**
or
The plants are arranged in a narrow line in such a way as to delineate a field boundary
Treelines and hedgerows **W2**
As many treelines and hedgerows are of considerable age and are, in essence, highly managed, semi-natural vegetation, they can be used to provide additional information on the environmental conditions and functioning. Continue with Key D from question 4, but in the final vegetation code indicate that the vegetation is part of a hedge or treeline. Care must be taken when exotic species are encountered as the dominant species, as these will not key out correctly; give these vegetation types a code of **W2**.

4. *Salix* spp. are present 5
or
No *Salix* spp. are present 8

5. Those *Salix* spp. present are any of the dwarf species *Salix lapponum*, *S. lanata*, *S. arbuscula*, *S. reticulata*, *S. alpina*, *S. repens*, *S. arenaria*, *S. herbaceae* 6
or
Those *Salix* spp. present are arborescent or shrub species 7

6. The vegetation is composed of the dwarf willow species *Salix lapponum*, *S. lanata*, *S. arbuscula*, *S. reticulata*, *S. alpina* or *S. repens*, sometimes with other shrubs, such as *Alnus viridis*
Sub-alpine bush and tall herb **G13**
or
The vegetation is composed of the dwarf willow species *Salix herbaceae*, together with mosses such as *Polytrichum* and *Racomitrium*
Snow patch communities **S1**
or
The vegetation is composed of the dwarf willow species *Salix arenaria*, with *Corynephorus canescens*, *Spergula morisonii* and carpets of fruticose lichens
Fluvio-glacial dunes **U4**

7. The vegetation is dominated by *Salix cinerea*, *S. pentandra*, *S. fragilis* or *S. aurita* sometimes together with *Alnus glutinosa*, *Frangula alnus* and some other trees, ground vegetation typified by tall reeds and sedges, *Phragmites australis*, *Typha latifolia*, *Carex pendula* and tall herbs
Alder, willow and bog myrtle swamp woods **W8**
or
The vegetation is dominated by almost entirely by *Salix* spp., particularly *S. viminalis*, *S. purpurea*, *S. triandra*, *S. alba*, sometimes with *Alnus glutinosa*, *Populus nigra* or *P. alba*
Riparian willow formations **W7**
or
There is no overall dominant species, but a mixture of *Salix* spp., *Hippophae* spp., *Alnus* spp., *Betula* spp. or *Myricaria germanica* present, together with tall herbs such as *Oenothera biennis*, *Filipendula Petasites*, *Epilobium fleischeri*
Gravel banks **S2**

8. The vegetation is dominated by *Erica* spp., *Calluna* spp. or *Vaccinium* spp. 9
or
The vegetation is dominated by some other species 16

9. *Sphagnum* spp. mosses are present 10
or
Sphagnum spp. mosses are absent 13

10. Occasional *Betula pendula*, *B. pubescens*, *Pinus sylvestris*, *P. rotundata* or *Picea abies* trees are present
Birch and conifer swamp woods **W4**
or
There are no trees present 11

11. The vegetation is dominated by *Erica tetralix*, together with *Molinia caerula* and *Sphagnum*, with *Narthecium ossifragum*
Wet heath **M5**
or
The vegetation is dominated by species other than *Erica tetralix* 12

12. The community is characterised by a high number of hummocks and hollows of colourful (red, yellow, green and brown) *Sphagnum* spp. (*S. cuspidatum*, *S. recurvum*, *S. papillosum*, *S. magellanicum*) with other mosses. Open pools are present and characterised by *Eriophorum vaginatum*, *Menyanthes trifoliata*, *Utricularia* spp. and *Drosera* spp., other parts with *Narthecium ossifragum*, and *Vaccinium oxycoccus*, *Erica tetralix* and other ericaceous shrubs
Near natural raised bogs **M1**
or
The community is characterised by a patchy mix of *Eriophorum vaginatum*, *E. angustifolium*, *Scirpus cespitosus*, *Molinia caerulea*, *Schoenus nigricans*, *Rhynchospora alba*, *Polygala serpyllifolia*, *Narthecium ossifragum* and *Pinguicula lusitanica*, together with *Calluna vulgaris*, *Erica tetralix*, *Myrica gale*. Abundant mosses including *Sphagnum* spp. (*S. subnitens*, *S. tenellum*, *S. papillosum*, *S. magellanicum*) and *Rhacomitrium lanuginosum*
Lowland blanket bog **M2**
or
The community is characterised by a patchy mix of *Eriophorum vaginatum*, *Calluna vulgaris*, *Erica tetralix*, *Scirpus cespitosus*, *Narthecium ossifragum*, *Rubus chamaemorus* and *Pinguicula lusitanica*. Abundant mosses including *Sphagnum* spp. (*S. subnitens*, *S. tenellum*, *S. papillosum*, *S. magellanicum*) and *Rhacomitrium lanuginosum*
Upland blanket bog **M3**

13. The vegetation is dominated by tall *Erica* spp., such as *Erica arborea*, *E. lusitanica* or *E. scoparia*, often with *Cistus* spp. and *Lavendula stoechas*
Meso-Mediterranean silicicolous maquis **S10**
or
The vegetation is not as above 14

14. *Drayas octopetala*, *Arctostphylos alpinus*, *A. uva-ursi* or *Loiseleuria procumbens* are present
Alpine and boreal heaths **S9**
or
The above species are absent 15

15. The community is a mixture of *Vaccinium myrtillus*, *Rubus* spp. and tall herbs, sometimes with *Salix* spp., *Sorbus* spp. or *Rhamnus* spp.
Sub-alpine bush and tall herb communities **S4**
or
The community is a mixture of shrubs, particularly *Calluna vulgaris*, but often with *Vaccinium myrtillus*, *V. vitis-idea*, *V. uliginosum*, *Empetrum*

nigrum, *Ulex* spp. and *Erica* spp. other than *E. tetralix* (e.g. *E. cinerea*, *E. aragonensis*, *E. umbellata*). *Genista* spp. are often present

Dry heaths **S8**

16. The vegetation is characterised by hemispherical, cushion-forming, spiny vegetation, typically of the genera *Acantholimon*, *Astragalus*, *Erinacea*, *Bupleurum*, *Ptilotrichum*, *Genista* and *Echinospartum* 17
or
The vegetation is not as above 18

17. The species *Astragalus massiliensis* and *Anthyllis hermanniae* are present

Western Mediterranean clifftop phrygana **S7**

or
The above species are not present

Hedgehog heaths **S11**

18. The vegetation is dominated by *Cistus* spp., *Rosmarinus* spp., *Lavendula* spp., *Euphorbia* spp., *Genista* spp. or *Salvia* spp. 19
or
The vegetation is dominated by species other than described above 23

19. There are occassional trees, such as *Quercus* spp., *Juniperus* spp., *Carpinus* spp. or *Ilex* spp.

Pseudomaquis **S17**

or
There are no apparent trees 20

20. The vegetation is characterised by short labiate species such as *Teucrium* spp., *Thymus* spp. and *Salvia* spp., often with *Lavendula* spp., *Buxus* spp. and *Genista* spp.

Supra-Mediterranean garrigues **S16**

or
The vegetation is not as above 21

21. *Rosmarinus* is abundant 22
or
Rosmarinus is absent, the vegetation dominated by *Lavendula stoechas* or *Cistus* spp.

Meso-Mediterranean shrub formations **S15**

22. *Lavendula latifolia* is present together with *Euphorbia spinosa* and *Genista* spp.

Western meso-Mediterranean calcicolous garrigues **S14**

or
Lavendula latifolia is absent. *Euphorbia* spp. other than *E. spinosa*, and *Genista* spp. are present

Eastern garrigues **S13**

23. The ferns *Asplenium viride*, *A. scolopendrium*, *A. trichomanes*, *A. petrarchae* or *Polypodium cambricum* are present

Vegetated calcareous inland cliffs **U12**

or
The above ferns are absent 24

24. The vegetation is dominated by *Ulex* spp. 25
or
The vegetation is dominated by species other than *Ulex* spp. 26

25. The dwarf shrubs *Calluna* spp., *Erica* spp., *Vaccinium* spp. or *Empetrum* spp. are present

Dry heaths **S8**

or
Olea europea, *Pistacia lentiscus*, *Asparagus* spp., *Rhamnus* spp. or *Genista* spp. are present

Thermo-Mediterranean shrub formations **S12**

or

Ulex europaeus occurs as the sole shrub, or with *Prunus spinosa*, *Rosa*, *Rubus*, *Cornus* spp., *Cretaegus* spp., *Lonicera* spp., *Pteridium aquilinum*, *Juniperus communis* or *Buxus sempervirens*

Thickets **S18**

26. The vegetation is dominated by small trees or bushes 27

or

The vegetation is dominated by dwarf shrubs 28

27. The vegetation is dominated by a combination of *Olea europea*, *Pistacia lentiscus*, *Asparagus* spp., *Rhamnus* spp., *Ulex* spp., *Bupleurum* spp. and *Genista* spp.

Thermo-Mediterranean shrub formations **S12**

or

The vegetation is dominated by *Prunus spinosa*, *Cretaegus* spp., *Rosa* spp., *Rubus* spp., *Cornus* spp., *Lonicera* spp., *Pteridium aquilinum*, *Ulex europaeus*, *Juniperus communis* or *Buxus sempervirens* either, alone or in combination

Thickets **S18**

or

The vegetation is dominated by *Pinus mungo*, *Rhododendron hirsutum* and *Rhodothamnus chamaecistus*

Dwarf mountain pine scrub **S6**

or

The vegetation is dominated by *Spartium junceum*

Spanish-broom fields **S5**

or

The vegetation is dominated by *Nerium oleander*, *Vitex agnus-castus* or *Tamarix* spp., mostly occurring in costal areas

Southern riparian galleries and thickets **S3**

or

The vegetation is a mixture of *Betula* spp., *Hippophae* spp., *Alnus* spp. and *Myricaria germanica*, with tall herbs such as *Oenothera biennis*, *Filipendula* spp., *Petasites* spp. or *Epilobium fleischeri*

Gravel banks **S2**

28. The vegetation is dominated by *Myrica gale*

Alder, willow and bog myrtle swamp woods **M8**

or

The vegetation is dominated by *Drayas octopetala*, dwarf *Juniperus* spp., *Arctostphylos alpinus*, *A. uva-ursi* or *Loiseleuria procumbens*

Alpine and boreal heaths **S9**

REFERENCE TABLE OF VEGETATION CODES

The table starting on page 168 has been provided as a quick reference guide to the vegetation codes used here, together with the corresponding vegetation types and CORINE codes. Also, as a further aid, the appropriate National Vegetation Classification codes are provided for vegetation types found in Britain. Those vegetation types that do not occur in Britain are marked Not British; vegetation types where the species are planted or have become naturalised in Britain are marked Not British Native; those vegetation types that occur in Britain but do not have a National Vegetation Classification code are left blank.

Table 2.3.1 Reference table of vegetation codes

FAP codes	CORINE vegetation type	CORINE codes	National Vegetation Classification codes
A 1	Peatmoss-bladderwort bog pools	C22.45	A24, A24
A 2	Chandelier algae submerged carpets	C22.44	
A 3	Acid oligotrophic vegetation	C22.433, C24.41	A7b, A13, A14, A22, A23, A24
A 4	Amphibious communities	C22.3	A22, A23, S19, M29
A 5	River mud banks	C24.5	
A 6	Free-floating vegetation	C22.41	A1, A2, A3, A4
A 7	Eutrophic vegetation	C24.44	A11, A12
A 8	Flooded crops	C82.4	
A 9	Rooted submerged vegetation	C22.42	A5, A6, A11, A12, A13, A14, A15
A 10	Rooted floating vegetation	C22.43	A7, A8, A9, A10, A16, A17, A19, A20
G 1	Mediterranean short humid grassland	C37.5	Not British
G 2	Iberian vallicares	C38.4	Not British
G 3	Arcto-alpine riverine swards	C54.3	M11
G 4	Riparian cane formations	C53.6	Not British
G 5	Alpine and sub-alpine acidophilous grassland	C36.3	U7, U8, U9, U10
G 6	Oligotrophic humid grassland	C37.3	M24, M25, M26, U6
G 7	Mediterranean tall humid grassland	C37.4	Not British
G 8	Humid improved grassland	C81.2	MG7c, MG7d
G 9	Eutrophic humid grassland	C37.2	M22, M23, MG9, MG10, MG11, MG12, MG13
G 10	Mesophile grasslands	C38.1	MG4, MG5, MG6, MG7, MG7c, MG7d, MG8
G 11	Humid tall herb fringes	C37.7	
G 12	Atlantic mat-grass swards and related communities	C35.1	CG11, U2, U3, U4, U5
G 13	Sub-alpine and alpine tall herb communities	C37.8	cf U16, U17, U19
G 14	Eastern supra-mediterranean humid meadows	C37.6	Not British
G 15	Mediterraneo-montane mat-grass swards	C35.7	Not British
G 16	Mountain hay meadows	C38.3	Not British
G 17	Lowland hay meadows	C38.2	MG1, MG2, MG3
G 18	Lowland heavy metal grassland	C34.2	CG9, U4
G 19	Mediterranean annual deep-sand communities	C35.4	Not British
G 20	Middle-European pioneer swards and alpine weathered rock and outcrop communities	C34.1, C36.2	
G 21	Iberian fescue-plantain swards	C35.5	Not British
G 22	Iberian tall fescue grasslands	C35.6	Not British
G 23	Medio-European open siliceous grassland	C35.2	U1
G 24	Mediterraneo-montane grasslands	C34.7	Not British
G 25	Mediterranean tall grass steppes	C34.6	Not British
G 26	Mediterranean xeric grasslands	C34.5	Not British
G 27	Dense perennial grasslands and Middle-European steppes	C34.3	CG1, CG2, CG3, CG4, CG5, CG6, CG7, CG8, CG9
G 28	Alpine and sub-alpine calciphilous grasslands	C36.4	Not British
G 29	Thermophile forest fringes	C34.4	
G 30	Alpine and sub-alpine fertilised grasslands	C36.5	Not British
G 31	Mediterranean subnitrophilous grassland	C34.8	Not British
G 32	Dry improved grassland	C81.1	MG7a, MG7b, MG7e, MG7f
G 33	Extensive cultivation	C82.3	
G 34	Intensive cropland	C82.1	
M 1	Near natural raised bogs	C51.1	M1, M2, M3, M18, M21

Table 2.3.1 *cont'd*

FAP codes	CORINE vegetation type	CORINE codes	National Vegetation Classification codes
M 2	Lowland blanket bog	C52.1	M17
M 3	Upland blanket bog	C52.2	M19, M20
M 4	White beak-sedge communities	C54.6	M1, M2, M3, M21
M 5	Wet heath	C31.1	M14, M15, M16
M 6	Acidic fens	C54.4	M6, M6d, M7
M 7	Springs	C54.1	M31, M32, M33, M34, M35, M37, M38
M 8	Alder, willow and bog myrtle swamp woods	C44.9	S24g, M15a, M25a
M 9	Large sedge communities	C53.2	S1, S3, S6, S7, S9, S11, S17, S18, S27
M 10	Fen-sedge beds	C53.3	S2
M 11	Transition mires	C54.5	M4, M5, M8, M9
M 12	Rich fens	C54.2	M10, M11, M12, M13, S24, S25
M 13	Reed beds	C53.1	S4, S5, S8, S10, S12, S13, S14, S15, S16, S19a, S20, S21, S26, S28
M 14	Meadowsweet stands and related communities	C37.1	M27, M28
M 15	Small reed beds of fast-flowing waters	C53.4	S22, S23
S 1	Snow patch communities	C36.1	U11, U12, U13, U14, U18, CG12
S 2	Gravel banks	C24.2	
S 3	Southern riparian galleries and thickets	C44.8	Not British
S 4	Sub-alpine bush and tall herb	C31.6	U16, U17, U19, W20
S 5	Spanish-broom fields	C32.A	Not British native
S 6	Dwarf mountain pine scrub	C31.5	Not British
S 7	Western mediterranean clifftop phrygana	C33.1	Not British
S 8	Dry heaths	C31.2	H1, H2, H3, H4, H5, H6, H7, H8, H9, H10, H12, H13, H14, H16, H18, H19, H20, H21, H22
S 9	Alpine and boreal heaths	C31.4	CG13, CG14, H15, H17
S 10	Meso-mediterranean silicicolous maquis	C32.3	Not British
S 11	Hedgehog-heaths	C31.7	Not British
S 12	Thermo-mediterranean shrub formations	C32.2	Not British
S 13	Eastern garrigues	C32.5	Not British
S 14	Western meso-mediterranean calcicolous garrigues	C32.4	Not British
S 15	Meso-mediterranean shrub formations	C32.3	Not British
S 16	Supra-mediterranean garrigues	C32.6	Not British
S 17	Pseudomaquis	C32.7	Not British
S 18	Thickets	C31.8	U20, W19, W21, W22, W23, W24, W25
S 19	Shrub orchards	C83.2	
U 1	Agriculture, building, etc.	C82.1, C82.2, C82.3, C87, C88	
U 2	Sand dunes, beaches	C24.3, C64	
U 3	Gravel banks	C22.2	
U 4	Fluvio-glacial dunes	C64.1	
U 5	Alpine and northern siliceous screes	C61.1	U21
U 6	Alpine calcareous screes	C61.2	Not British
U 7	Western mediterranean and thermophilous screes	C61.3	U24

Table 2.3.1 *cont'd*

FAP codes	CORINE vegetation type	CORINE codes	National Vegetation Classification codes
U 8	Eastern mediterranean screes	C61.4	Not British
U 9	Alpine weathered rock and outcrop communities and pioneer swards	C36.2, C34.1	
U 10	Bare cliffs	C62.4	
U 11	Pavements	C62.3	
U 12	Vegetated calcareous inland cliffs	C62.1	U22, U23
U 13	Vegetated siliceous inland cliffs	C62.2	
U 14	Wet inland cliffs	C62.5	U15
U 15	Caves	C65	
U 16	Volcanic features	C66	Not British
U 17	River, stream, etc.	C24.1	
U 18	Lake, pond, etc.	C22.1	
U 19	Snow packs, glaciers	C63	
W 1	Plantations	C83.3	
W 2	Treelines and hedgerows	C84.1, C84.2	
W 3	High stem orchards	C83.1	
W 4	Birch and conifer swamp woods	C44.A	W4
W 5	Southern alder and birch galleries	C44.5	Not British native
W 6	Grey alder galleries	C44.2	Not British native
W 7	Riparian willow formations	C44.1	W6b, W6c
W 8	Alder, willow and bog-myrtle swamp woods	C44.9	W1, W2, W3, W5, W6, W6a
W 9	Mixed oak-elm-ash forests of great rivers	C44.4	Not British
W 10	Mediterranean poplar-elm-ash forests	C44.6	Not British
W 11	Medio-european stream ash-alder woods	C44.3	W6a, W6d, W6e, W7
W 12	Alder woods	C41.C	W6d, W6e
W 13	Oriental plane and sweetgum woods	C44.7	Not British
W 14	Birch woods	C41.B	W4, W16a
W 15	Aspen woods	C41.D	
W 16	Acidophilous oak forests	C41.5	W11, W16, W17
W 17	Ash forests	C41.3	W8, W9, W10
W 18	Quercus pyrenaica forests	C41.6	Not British
W 19	Holly woods	C45.8	
W 20	Oak-hornbeam forests	C41.2	W8a, W8b, W8f, W10
W 21	Spruce forests	C42.2	Not British native
W 22	High oro-mediterranean pine forests	C42.7	Not British native
W 23	Fir forests	C42.1	Not British native
W 24	Black pine forests	C42.6	Not British native
W 25	Thermophilous and supra-mediterranean oak woods	C41.7	Not British native
W 26	Scots pine forest	C42.5	W18
W 27	Rowan woods	C41.E	W11, W17
W 28	Olive-carob forests	C45.1	Not British
W 29	Mountain pine forests	C42.4	Not British native
W 30	Mediterranean pine woods	C42.8	Not British native
W 31	Cypress, juniper and yew forest	C42.A	W13
W 32	Meso- and supra-mediterranean holm-oak forests	C45.3	Not British native
W 33	Matorral	C32.1	Not British
W 34	Larch-arolla forests	C42.3	Not British native
W 35	Kermes oak forests	C45.4	Not British native
W 36	Hop-hornbeam, oriental hornbeam and mixed thermophilous woods	C41.8	Not British
W 37	Cork oak forests	C45.2	Not British
W 38	Chestnut woods	C41.9	
W 39	Continental laurel-oak forests	C45.5	Not British native

Table 2.3.1 *cont'd*

FAP codes	CORINE vegetation type	CORINE codes	National Vegetation Classification codes
W 40	Lime woods	C41.G	W8, W10
W 41	Small-leaved elm woods	C41.F	W8, W8e, W8f, W8g, W9
W 42	Mixed ravine and slope forests	C41.4	W8e, W8g
W 43	Dehesa	C84.6	Not British
W 44	Beech forests	C41.1	W12, W14, W15

2.3.E Part 1: Land-use and site management: field indicators

The questions asked in this section are equivalent to those in Action C.3, Section 2.1 but have been designed to obtain answers using only observable field indicators rather than the desk or anecdotal sources used in Action C.3. Answers are entered on the same recording sheet (**Recording Sheet 1**) and in the same spaces as answers from Action C.3 and are given the same codes. This means that the normal reference coding system does not apply in this section; the Action number is not used as the data reference code. Instead the equivalent reference codes from Action C.3 are used.

Not all the questions or parts of questions in Action C.3 can be answered using field indicators. Therefore some questions or parts of questions in Section 2.3.E Part 1 have been omitted, with the consequence that numbering sometimes may not follow the same sequence.

The questions in this section should be used *only* to:

- provide information that *could not be established* in Action C.3;
- provide information when desk sources and communication with local experts has not given answers.

If site information is to be obtained from local experts at this stage in the wetland database establishment, then use the questions provided in Action C.3 not those in this section (Part 1).

Note: The characterisation in this section applies to the entire AA rather than a single HGMU.

2.3.E.1 *ACTION: Record site management and land use information for the entire AA, using observable field indicators*

Record information: Recording Sheet 1 *(unless otherwise indicated)*

Use the statements after each possible answer to determine, from observable field indicators, which answer is true. Enter the date of the field visit in the space marked **C.3** at the top of **Recording Sheet 1**.

C.3.3 *Land-use and management: burning of vegetation (Fig. 2.3.10)*

C.3.3.1a How often is vegetation within the AA subjected to burning as part of land management practices?

2.3.10 Vegetation management by burning: Kismeldon Meadows, Devon, UK (also showing significant microrelief – see Action A.3, Section 2.3) (photo: E. Maltby).

a. No history of burning in the last 20 years
This statement cannot be confirmed by field indicators. Evidence of burning typically disappears relatively soon after it has occurred. Only information obtained from local knowledge at the time of the field visit or from the earlier desk study can be used to confirm this statement.

b. Not burnt within the last 5 years but previously burnt between 5 and 20 years ago
This statement cannot be confirmed by field indicators. Evidence of burning typically disappears relatively soon after it has occurred. Only information obtained from local knowledge at the time of the field visit or from the earlier desk study can be used to confirm this statement.

c. Burnt once or more than once in the last 5 years
This statement is true if there is evidence of burnt or charred vegetation of either a woody or herbaceous nature. Woody vegetation (e.g. heather or gorse) can retain evidence of burning in the form of blackened or charred stems for up to five years, while evidence of blackened or charred herbaceous vegetation usually lasts only for one growing season.

d. Unknown
If there is no evidence of burnt or charred vegetation, then this statement must be accepted. The short-term duration of evidence of burning means that a lack of field evidence does not mean it has not taken place recently.

C.3.3.2 Record on **AA Map 2** the areas within the AA that are subjected to burning (using the code **BURN**).

C.3.4 *Land-use and management: harvesting of herbaceous vegetation*

Harvesting includes mowing and removal of grass; hay production; silage; harvesting of arable crops; harvesting of cereals or vegetables; market gardening; production and removal of lawn turf; cutting of reeds (Fig. 2.3.11).

C.3.4.1a How often is vegetation within the AA harvested?

a. No history of harvesting of herbaceous vegetation in the last 20 years
This statement cannot be confirmed by field indicators. Evidence

2.3.11 Harvesting of reeds (photo: E. Maltby).

of harvesting of herbaceous vegetation typically disappears relatively soon after it has occurred. Only information obtained from local knowledge at the time of the field visit or from the earlier desk study can be used to confirm this statement.

b. Not currently harvested but previously harvested within last 20 years or harvested less than once a year
This statement cannot be confirmed by field indicators. Evidence of harvesting of herbaceous vegetation typically disappears relatively soon after it has occurred. Only information obtained from local knowledge at the time of the field visit or from the earlier desk study can be used to confirm this statement.

c. Once or more than once a year
This statement is true if there is evidence of harvesting. This might include the presence of stubble, bales or sheaves of the crop, uniformly cut vegetation or the remains of vegetation that has obviously been cut mechanically, but not removed from the HGMU.

d. Unknown
If there is no evidence of harvesting of herbaceous vegetation, then this statement must be accepted. The short-term duration of evidence of harvesting means that a lack of field evidence does not mean it has not taken place recently.

C.3.4.2a For the crops that are subjected to harvesting of herbaceous vegetation, which of the following categories would best describe the harvesting practices?

a. Fully mechanised practices
This method is probably used if mechanised harvesting machinery is present nearby, or there are indications, such as deep tyre tracks, of movement of heavy machinery on the land.

b. Traditional, manual practices
This method is probably used if there is evidence of scythes or other manual implements nearby, or small sheaves are present. Also, partial and irregular patterns of vegetation harvesting could indicate traditional, manual methods of harvesting.

C.3.4.3a How often and when does harvesting take place?

a. Once a year, only in spring (between April–June)
b. Once a year, only in summer (between July–August)
c. Once a year, only in autumn (between September–November)
d. Once a year, only in winter (between December–March)
e. Twice or more than twice a year
To answer this question from field indicators, harvesting of vegetation must be witnessed. It is most likely that this question can only be answered using information obtained during the desk study.

C.3.4.4 Record on **AA Map 2** the areas within the AA that are subjected to harvesting of herbaceous vegetation (using the code **HARV**). Where different harvesting practices are used (from QC.3.4.2 above) also indicate these on the map, along with areas where harvesting at different times of year occurs (from QC.3.4.3 above).

C.3.5 *Land-use and management: ploughing of land*

C.3.5.1a How often is any land within the AA ploughed?

a. No history of ploughing in the last 20 years
This statement cannot be confirmed by field indicators. Evidence of ploughing can disappear relatively quickly after it has occurred. Only information obtained from local knowledge at the time of the field visit or from the earlier desk study can be used to confirm this statement.

b. Not currently ploughed but previously ploughed within the last 20 years, or ploughed but less than once a year

This statement cannot be confirmed by field indicators. Evidence of ploughing can disappear relatively quickly after it has occurred. Only information obtained from local knowledge at the time of the field visit or from the earlier desk study can be used to confirm this statement.

c. Currently ploughed
Observation of freshly turned soil or recently seeded crops provides evidence that this statement is true. However, whether the soil is ploughed every year cannot be determined without desk study information.

d. Unknown
If there is no evidence of ploughing, then this statement must be accepted. The short-term duration of evidence of ploughing means that a lack of field evidence does not mean it has not taken place recently.

C.3.5.2a How often is any land within the CA subjected to ploughing?

a. No history of ploughing in the last 20 years
This statement cannot be confirmed by field indicators. Evidence of ploughing can disappear relatively quickly after it has occurred. Only information obtained during the desk study can be used to confirm this statement.

b. Not currently ploughed but previously ploughed within last 20 years, or ploughed but less than once a year
This statement cannot be confirmed by field indicators. Evidence of ploughing can disappear relatively quickly after it has occurred. Only information obtained during the desk study can be used to confirm this statement.

c. Currently ploughed once or more than once a year
Observation of freshly turned soil or recently seeded crops provides evidence that this statement is true. However, whether the soil is ploughed every year cannot be determined without desk study information.

d. Unknown
If there is no evidence of ploughing, then this statement must be accepted. The short-term duration of evidence of ploughing means that a lack of field evidence does not mean it has not taken place recently.

C.3.5.3 If there is evidence that areas within the AA are ploughed, mark these on **AA Map** 2, using the code **PLOU**.

C.3.5.4 If there is evidence that areas within the CA are ploughed, mark these on **CA Map** 2, using the code **PLOU**.

C.3.6 *Land-use and management: woodland management and forestry*

Includes: felling; coppicing; pollarding; clearing; planting.

C.3.6.1a How often does felling and removal of timber in woodland areas occur within the AA?

a. Not in the last 20 years
If there is no evidence of sawn-off tree stumps, log piles, sawn-off branches or regular gaps within woodland, and the trees are obviously more than 20 years old, then this statement is true.

b. No current felling and removal but previously practised within the last 20 years, or felling and removal occurs but less than once in 5 years
Evidence of sawn-off tree stumps, log piles, sawn-off branches or regular gaps within a woodland indicate felling has occurred, but this evidence must show considerable signs of weathering. Usually desk study information will be required to accept this statement.

c. Current felling and removal, or previously practised within the last 5 years

This statement is true if there is evidence of felling within the woodland. This could be indicated by the presence of stumps within the woodland that have been cut by a saw, or remains of branches that have obviously been removed from felled trunks. Also, regular gaps within a plantation indicate felling has occurred. The presence of log piles within a wood also indicate felling has occurred. However, care must be taken that these features are not historic, and that these practices have actually occurred within the last five years.

d. Unknown

If there is no evidence that felling has occurred within the last 20 years, but the age of the trees is uncertain, then this statement must be accepted.

C.3.6.2 Record on **AA Map 2** the woodland areas within the AA that are subjected to felling, thinning and removal of trees, coppicing or pollarding (using the code **FOREST**).

Note: Suitable species of trees can be cut regularly to ground level (coppiced) or above grazing height at the the top of the stem (pollarded) to encourage regrowth of wood for a variety of cropping purposes (Fig. 2.3.12).

C.3.7 *Land-use and management: livestock density*

C.3.7.1a Are any of the following forms of livestock kept within part or all of the AA at the stocking densities indicated?,

- Cattle or horses or donkeys ≥ 1 ha^{-1}
- Pigs ≥ 3 ha^{-1}
- Sheep or goats ≥ 6 ha^{-1}
- Turkeys or geese ≥ 30 ha^{-1}
- Chickens or ducks ≥ 100 ha^{-1}

a. No history of livestock kept in the last 20 years

This statement cannot be confirmed by field indicators. Only information obtained from local knowledge at the time of the field

2.3.12 Pollarding of willow: Somerset Levels, UK (photo: E. Maltby).

visit or from the earlier desk study can be used to confirm this statement.

b. None currently kept but previously kept within the last 20 years or currently kept but at a density lower than the figure indicated above Evidence of previous stocking can be indicated by the presence of feeding or watering troughs, animal hair caught on fences and hedges, poaching of soil (Fig. 2.3.13), footprints or faecal remains. Care must be taken not to confuse indicators of livestock with similar evidence that might be left by wild animals. It is unlikely that evidence would survive for more than 5 years, and care must be taken if troughs are the sole indicator.

c. Currently kept at a density equal to or higher than the figure indicated above

d. Unknown

If there is no evidence of livestock, then this statement must be answered positively. A lack of evidence cannot be assumed to mean that no livestock have been present.

C.3.7.2a If there are livestock in the AA, is there evidence that it is provided with supplementary feed; troughs, feeders, scattered feed?

C.3.7.3a Is there any evidence or indication that the animals feed or graze in the same area as where they ruminate, rest or sleep?

C.3.7.4 Record on **AA Map 2** the areas within the AA where the above forms of livestock are kept at densities above those listed in QC.3.7.1a (using the code **LV**) and locate the areas used for feeding, grazing, resting and sleeping, indicating whether the stocking is current or historic.

C.3.7.5 Are any of the following forms of livestock kept within part or all of the CA?

- Cattle or horses or donkeys ≥ 1 ha^{-1}
- Pigs ≥ 3 ha^{-1}
- Sheep or goats ≥ 6 ha^{-1}
- Turkeys or geese ≥ 30 ha^{-1}
- Chickens or ducks ≥ 100 ha^{-1}

2.3.13 Poaching of soil by livestock (photo: E. Maltby).

a. No history of livestock kept in the last 20 years
This statement cannot be confirmed by field indicators. Only information obtained from local knowledge at the time of the field visit or from the earlier desk study can be used to confirm this statement.

b. None currently kept but previously kept within the last 20 years, or currently kept but at a density lower than the figure indicated above
Evidence of previous stocking can be indicated by the presence of feeding or watering troughs, animal hair caught on fences and hedges, poaching of soil (Fig. 2.3.13), footprints or faecal remains. Care must be taken not to confuse indicators of livestock with similar evidence that might be left by wild animals. It is unlikely that evidence would survive for more than 20 years, and care must be taken if troughs are the sole indicator.

c. Currently kept at a density equal to or higher than the figure indicated above

d. Unknown
If there is no evidence of livestock, then this statement must be answered positively. A lack of evidence cannot be assumed to mean that no livestock have been present

C.3.7.6 Record on **CA Map 1** the areas within the CA where the above forms of livestock are kept at densities above those listed in QC.3.7.4 (using the code **LV**).

C.3.8 *Land-use and management: bird populations*

Includes: breeding colonies; nesting colonies; feeding colonies; roosting colonies; moulting colonies.

C.3.8.1a On the day of visit were large populations of birds (greater than 1000) present within the AA? These may be breeding, nesting, feeding, roosting, moulting or over-wintering colonies. Evidence of bird colonies can be obtained from the presence of bird droppings. This might also help estimate the population and density.

C.3.8.2 Record on **AA Map 2** the areas where large populations of birds occur within the AA, using the code **LBP**.

C.3.8.3 Record the approximate density of the large populations of birds present within the AA.

a. 1000–2000 birds ha^{-1}
b. > 2000 but < 5000 birds ha^{-1}
c. 5000–10 000 birds ha^{-1}
d. > 10 000 birds ha^{-1}
e. < 1000 birds or not applicable

C.3.10 *Land-use and management: peat cutting and extraction*

C.3.10.1a How frequently is peat cut and removed from the AA?

a. No peat cutting or mining occurs
This statement is true if the peat surface is intact, or historic mining has occurred but is not carried out currently. In the latter case, it is likely that re-vegetation of old workings would have commenced (Fig. 2.3.14).

b. Hand cutting or mining of peat occurs at a local or traditional scale
Peat cuttings are present with turfing spade marks along the wall of the cut (10–15 cm wide, slightly concave cut giving a scalloped appearance), or piles of cut peat turf are present. These workings are usually shallow (2 m or less) and surrounding areas are often intact or other cut areas are grown over (Fig. 2.3.14).

2.3.14 Revegetation of abandoned peat workings: Somerset Levels, UK. Note remnant of former ground surface (baulk) seen as a raised strip curving away from the foreground to the middle distance (photo: E. Maltby).

2.3.15 Land-use and management; quarrying in an Irish catchment (photo: E. Maltby).

c. Mechanised cutting or mining of peat occurs at an industrial or commercial scale
   This is indicated by the presence of heavy machinery, peat cuttings containing wide scrapes from machine buckets or wide level areas of dry or disturbed peat with no vegetation.

d. Unknown

C.3.10.2 If there is any evidence or indication of current cutting of peat and extraction within the AA, mark the affected locations and record the extent on **AA Map 2**, using the code **PEATEX**.

C.3.11 *Land-use and management: mining and extraction*

Includes: gravel extraction; mineral extraction; quarrying (Fig. 2.3.15).

C.3.11.1a Is there any evidence that areas within the AA have been subject to mining, quarrying or extraction of any form (other than peat mining)?

a. No evidence of mining, quarrying or extraction
b. Evidence of current mining, quarrying or extraction; heavy machinery, gravel dredgers, quarrying equipment, recently cut, unweathered open rock faces, landscape disturbed by mining, quarrying or extraction activities
b. Evidence of mining, quarrying or extraction in the past; re-vegetated gravel pits, re-vegetated quarry faces, evidence of landscaping
c. Unknown

C.3.11.2 Is there any evidence that areas within the CA have been subject to mining, quarrying or extraction of any form (other than peat mining)?

a. No evidence of mining, quarrying or extraction
b. Evidence of current mining, quarrying or extraction; heavy machinery, gravel dredgers, quarrying equipment, recently cut, unweathered open rock faces, landscape disturbed by mining, quarrying or extraction activities (Fig. 2.3.15)
c. Evidence of mining, quarrying or extraction in the past; re-vegetated gravel pits, re-vegetated quarry faces, evidence of landscaping
d. Unknown

C.3.11.3 If there is any evidence or indication of mining, quarrying or extraction activities within the AA, record the location and extent on **AA Map 2**, using the code **MINEX**.

C.3.11.4 If there is any evidence or indication of mining, quarrying or extraction activities within the CA, record the location and extent on **CA Map 1**, using the code **MINEX**.

C.3.12 Land-use and management: application of industrial inorganic waste

Includes: mine and quarry spoil, mineral washings, slag heaps, dredgings.

C.3.12.1a How often is industrial inorganic waste applied within the AA?

a. No evidence of application in the last 20 years
   This statement is true if there is no evidence of piles of slag, washings or dredgings within the AA, or if piles of slag, washings or dredgings are present and they support vegetation of greater than 20 years of age (identifiable by observation of tree growth rings).
b. None currently applied but previously applied within last 20 years or applied less than once a year
   This statement is true if piles of slag, washings or dredgings are present and they show signs of colonisation by vegetation or signs of erosion.
c. Currently applied once or more than once a year
   This statement is true if fresh, unvegetated, uneroded piles of slag, washings or dredgings are present.
d. Unknown
   If none of the above statements can be confirmed, this statement must be answered positively.

C.3.12.2 How often is industrial inorganic waste applied within the CA?

a. No evidence of application in the last 20 years
   This statement is true if there is no evidence of piles of slag, washings or dredgings within the CA, or if piles of slag, washings or dredgings are present and they support vegetation of greater than 20 years of age (identifiable by observation of tree growth rings).
b. None currently applied but previously applied within last 20 years, or applied less than once a year
   This statement is true if piles of slag, washings or dredgings are present and they show signs of colonisation by vegetation or signs of erosion.

c. Currently applied once or more than once a year
This statement is true if fresh, unvegetated, uneroded piles of slag, washings or dredgings are present.

d. Unknown
If none of the above statements can be confirmed, this statement must be answered positively.

C.3.12.3 If there is any evidence or indication of areas within the AA where industrial inorganic waste disposal occurs, record these and the frequency of application on **AA Map 2**, using the code **IIWASTE**.

C.3.12.4 If there is any evidence or indication of areas within the CA where industrial inorganic waste disposal occurs, record these and the frequency of application on **CA Map 1**, using the code **IIWASTE**.

C.3.13 *Land-use and management: construction*

Includes: road construction, railway construction, building construction; other construction work.

C.3.13.1a Is there any evidence or indication that areas within the AA are currently undergoing construction work?

C.3.13.2 Is there any evidence or indication that areas within the CA are currently undergoing construction work?

C.3.13.3 If there is any evidence or indication of construction activities within the AA, record the location and extent on a copy of **AA Map 2**, using the code **CONSTR**.

C.3.13.4 If there is any evidence or indication of construction activities within the CA, record the location and extent on a copy of **CA Map 1**, using the code **CONSTR**.

C.3.14 *Land-use and management: hunting, shooting, fishing*

C.3.14.1 Is there any evidence or indication of hunting, shooting or fishing in areas of the AA?

a. No evidence of hunting, shooting or fishing
b. Evidence of hunting, shooting or fishing at a recreational scale; fishing pegs, gamekeepers' gibbets, spent cartridges, recreational club signs (Figs 2.3.16 and 2.3.17)
c. Evidence of hunting, shooting or fishing at a commercial scale; fish traps, fish nets, fishing boats, bird traps, bird nets, game traps, shooting hides, game rearing
d. Unknown

C.3.14.2 If there is any evidence or indication of hunting, shooting or fishing activities within the AA, record the location and extent on **AA Map 2**, using the code **HUNSF**.

C.3.15 *Land-use and management: recreation*

Includes: footpaths; bridleways; playing fields; cycle paths; off-road vehicle access; camping sites; riverside watersport access.

C.3.15.1 Is there any evidence or indication that areas within the AA possess some form of direct recreational use?

a. No evidence of direct recreation use
b. Evidence of light recreation use; lightly used footpaths, bridleways or cycling paths
c. Evidence of heavy recreation use; heavily used footpaths, bridleways, cycling paths, or off-road vehicle access, playing fields, camping sites, riverside watersport access
d. Unknown

2.3.16 A boat used for wildfowling, France (photo: E. Maltby).

2.3.17 Fly fishing on the mid-reaches of a UK river (photo E. Maltby).

C.3.15.2 If there is evidence or indication of recreation activities occurring within the AA, record their extent and location on **AA Map 2**, using the code **RECRE**.

2.3.E Part 2: Collection of HGMU information

The questions asked in this section are equivalent to those asked of the AA in Action C.3, Section 2.1. In this section however, the questions are asked of the HGMU. To complete this section, use local expert knowledge and desk information to answer each question. Where this is not available and information cannot be found from desk studies, use observable field indicators. Answers should be recorded on **Recording Sheet 1** (see Section 2.4) in the appropriate spaces provided (for instance, the answer to QC.3.2.1.b should be recorded on **Recording Sheet 1** in the space labelled **C.3.2.1b.**)

Not all the questions or parts of questions in Action C.3 need to be answered for the HGMU. Therefore some questions or parts of questions in Section 2.3.E Part 2 have been omitted, with the consequence that numbering sometimes may not follow the same sequence.

The normal reference coding system does not apply in this section, the Action number acting as the data reference number is not used. The equivalent reference codes from section Action C.3 are used.

Note: The characterisation in this section applies to the HGMU. Fill out a new Recording Sheet for each HGMU.

2.3.E.2 ACTION: Record site management and land use information for the HGMU

Record information: Recording Sheet 1 *(unless otherwise indicated)*

Use desk information, expert knowledge and observable field indicators.

C.3.2.1b How often is inorganic fertiliser applied within the HGMU?
- a. No history of application in the last 20 years
- b. None currently applied but previously applied within the last 20 years or applied less than once a year
- c. Currently applied once or more than once a year
- d. Unknown

C.3.2.2b If the answer to QC.3.2.1b is a, b or c, does this fertiliser contain:
- a. Both nitrogen and phosphorus (go to QC.3.2.3b (i))
- b. Nitrogen but no phosphorus (go to QC.3.2.3b (i))
- c. Phosphorus but no nitrogen (go to QC.3.2.3b (ii))
- d. Unknown

C.3.2.3b (i) What is the rate of nitrogen applied to the HGMU?
- a. < 15 kg N $ha^{-1}a^{-1}$
- b. 15–50 kg N $ha^{-1}a^{-1}$
- c. > 50 kg N $ha^{-1}a^{-1}$
- d. Unknown

C.3.2.3b (ii) What is the rate of phosphorus applied to the HGMU?
- a. < 5 kg P $ha^{-1}a^{-1}$
- b. 5–10 kg P $ha^{-1}a^{-1}$
- c. > 10 kg P $ha^{-1}a^{-1}$
- d. Unknown

C.3.2.4b How often is organic fertiliser applied by spraying or spreading (includes slurry, animal manure, stockyard washout, parlour washings, septic tank disposal; treated domestic sewage; sewage sludge, fishmeal, bloodmeal, bonemeal) to the surface of the HGMU?
- a. No history of application in the last 20 years
- b. None currently applied but previously applied within the last 20 years, or applied but less than once a year.
- c. Currently applied once or more than once a year
- d. Unknown

C.3.2.5b How often is organic fertiliser applied by injection (includes slurry, animal manure) within the HGMU?

a. No history of application in the last 20 years
b. None currently applied but previously applied within the last 20 years, or applied less than once a year
c. Currently applied once or more than once a year
d. Unknown

C.3.2.6b How often is organic industrial waste (includes industrial organic waste; municipal waste; food processing waste) applied within the HGMU?
a. No history of application in the last 20 years
b. None currently applied but previously applied within the last 20 years, or applied less than once a year
c. Currently applied once or more than once a year
d. Unknown

C.3.2.7b How often is lime applied to (parts of) the soil of the HGMU?
a. No history of application of lime in the last 10 years
b. None currently applied but previously applied within the last 10 years, or applied less than once a year
c. Currently applied once or more than once a year
d. Unknown

C.3.3 *Land-use and management: burning of vegetation*

C.3.3.1b How often is vegetation within the HGMU burnt as part of land management practices?
a. No history of burning in the last 20 years
b. Not burnt within the last 5 years but previously burnt between 5 and 20 years ago
c. Currently burnt once or more than once in the last 5 years
d. Unknown

C.3.4 *Land-use and management: harvesting of herbaceous vegetation*

Includes: mowing and removal of grass; hay production; silage; harvesting of arable crops; harvesting of cereals or vegetables; market gardening; production and removal of lawn turf; cutting of reeds.

C.3.4.1b How often is vegetation within the HGMU harvested?
a. No history of harvesting of herbaceous vegetation in the last 20 years
b. Not currently harvested but previously harvested within last 20 years or harvested less than once a year
c. Once or more than once a year
d. Unknown

C.3.4.2b For the crops where herbaceous vegetation is harvested in the HGMU, which of the following categories would best describe the harvesting practices?
a. Predominately fully mechanised practices
b. Predominately traditional, manual practices

C.3.4.3b How often and when does harvesting take place in the HGMU:
a. Once, only in spring (between April–June)
b. Once, only in summer (between July–August)
c. Once, only in autumn (between September–November)
d. Once, only in winter (between December–March)
e. Twice or more than twice a year

C.3.5 *Land-use and management: ploughing of land*

C.3.5.1b How often is any land within the HGMU ploughed?
a. No history of ploughing in the last 20 years
b. Not currently ploughed but previously ploughed within last 20 years, or ploughed but less than once a year

c. Currently ploughed (every year)
d. Unknown

C.3.6 *Land-use and management: woodland management and forestry*

Includes: felling; coppicing; pollarding; clearing; planting.

C.3.6.1b How often does felling and removal of timber in woodland areas occur within the HGMU?
a. Not in the last 20 years
b. No current felling and removal but previously practised within the last 20 years, or felling and removal occurs but less than once in 5 years
c. Current felling and removal, or previously practised within the last 5 years
d. Unknown

C.3.7 *Land-use and management: livestock density*

C.3.7.1b Are any of the following forms of livestock kept within part or all of the HGMU?
- Cattle or horses or donkeys $\geq$ 1 ha^{-1}
- Pigs $\geq$ 3 ha^{-1}
- Sheep or goats $\geq$ 10 ha^{-1}
- Turkeys or geese $\geq$ 30 ha^{-1}
- Chickens or ducks $\geq$ 100 ha^{-1}

a. No history of livestock kept in the last 20 years
b. None currently kept but previously kept within the last 20 years or currently kept but at a density lower than the figure indicated above
c. Currently kept at a density equal to or higher than the figure indicated above
d. Unknown

C.3.7.2b Are any of the livestock listed above in QC.3.7.1 kept for the purposes of meat, milk or wool production?

C.3.7.3b Do the animals feed or graze in the same area as where they ruminate, rest or sleep?

C.3.8 *Land-use and management: bird populations*

Includes: breeding colonies; nesting colonies; feeding colonies; roosting colonies; moulting colonies.

C.3.8.1b Have any large populations of birds (greater than 1000 ha^{-1}) been recorded within the HGMU?

C.3.8.3b Record the approximate density of the large populations of birds present within the HGMU.
a. 1000–2000 birds ha^{-1}
b. > 2000 but < 5000 birds ha^{-1}
c. 5000–10 000 birds ha^{-1}
d. > 10 000 birds ha^{-1}
e. < 1000 birds or not applicable

C.3.9 *Land-use and management: drainage*

Includes: pipe drains; secondary drainage (moling).

C.3.9.5b Are any areas within the HGMU drained through the use of subsurface drainage?

C.3.10 *Land-use and management: peat cutting and extraction*

C.3.10.1b How frequently is peat cut and removed from the HGMU?

a. No peat cutting or extraction occurs
b. Hand cutting or extraction of peat occurs at a local or traditional scale
c. Mechanised cutting or extraction of peat occurs at an industrial or commercial scale
d. Unknown

C.3.11 *Land-use and management: mining and extraction*

Includes: gravel extraction; mineral extraction; quarrying.

C.3.11.1b Are any areas within the HGMU currently undergoing mining, quarrying or extraction of any form (other than peat mining)?

a. No indication of mining, quarrying or extraction
b. Historical areas of mining, quarrying or extraction, currently inactive
c. Currently active areas of mining, quarrying or extraction.
d. Unknown

C.3.12 *Land-use and management: application of industrial inorganic waste*

Includes: mine and quarry spoil, mineral washings, slag heaps, dredgings.

C.3.12.1b How often is industrial inorganic waste applied within the HGMU?

a. No history of application in the last 20 years
b. None currently applied but previously applied within last 20 years, or applied less than once a year
c. Currently applied once or more than once a year
d. Unknown

C.3.13 *Land-use and management: construction*

Includes: road construction, railway construction, building construction; other construction work.

C.3.13.1b Are any areas within the HGMU currently undergoing construction work?

C.3.19 *River and lake water information: river and lake flooding regime*

C.3.19.7b Is it possible that water from the (main) water body (river, stream, ditch or lake) has contact with the HGMU?

a. No never/not applicable
b. Yes, the HGMU borders the (main) water body and horizontal exchange (either via surface or via subsurface flow) can occur freely between the HGMU and the water body
c. Yes, although the HGMU does not border the water body, the HGMU has contact with the water from the water body during flood events
d. Contact between water in the water body and the HGMU does take place once or more than once a year, but in a way other than mentioned in answers b and c (e.g. indirectly, via drainage ditches)
e. Contact between water in the water body and the HGMU does take place, but in a way other than mentioned in answers b and c (e.g. indirectly, via drainage ditches) and less than once a year
f. Unknown

C.3.19.8b How often is the HGMU flooded with water from the (main) water body?

a. Frequently (once or more than once a year)
b. Occasionally (once every 2–10 years)
c. Rarely (once every 10–50 years)
d. Unknown

C.3.22 *Water table information*

C.3.22.2b Depth of surface water: Wherever such information is available for the HGMU, note down the following information:
a. Mean depth of surface water level in standing water, streams or ditches
b. Minimum depth of surface water level in standing water, streams or ditches
c. Maximum depth of surface water level in standing water, streams or ditches

C.3.22.4b Water table fluctuations: Is the water level maintained at high levels in (at least part of) the HGMU at any time (e.g. for nature conservation purposes)? Indicate Y, N or NA in the appropriate section of **Recording Sheet 1**.

C.3.25 *Soil analytical data on trace element concentration*

The following questions are specific for the assessment of trace element export and trace element retention (Sections 3.2.D and 3.2.E). Only answer these questions if you wish to carry out an assessment for these functions. The following questions require analytical information on the concentration of trace elements in surface waters. To find out whether your AA of interest has this information, you may:

1. Consult local experts or other sources of background information (examples for the UK given in Appendix 2).
2. Contact university or specialist environmental research groups in the area.

Important: See notes on trace element concentration on p. 104.

C.3.25.1b Does the soil trace element concentration indicate that anthropogenically-derived contamination of the HGMU has taken place via overbank flooding? *(Recording Sheet 1)*
a. Yes, the soil trace element concentration is slightly higher than typical background concentration
b. Yes, the soil trace element concentration is moderately higher than typical background concentration
c. Yes, the soil trace element concentration is much higher than typical background concentration
d. No, the soil trace element concentrations are not elevated
e. Unknown, no soil analytical data are available

E.g. for the Elbe catchment example site: Total soil cadmium concentration.

C.3.25.1b Does the soil Cd concentration indicate that anthropogenically derived contamination of the HGMU has taken place via overbank flooding? *(Recording Sheet 1)*
a. Yes, the soil Cd concentration is slightly higher than typical background concentrations (> 20 mg kg^{-1} soil)
b. Yes, the soil Cd concentration is moderately higher than typical background concentrations (> 10 mg kg^{-1} soil)
c. Yes, the soil Cd concentrations is much higher than typical background concentrations (> 5 mg kg^{-1} soil)
d. No, the soil Cd concentrations are not elevated (≤ 5 mg kg^{-1} soil).
e. Unknown, no soil analytical data are available

Note: This is specific to the Elbe catchment. In the UK, for example, typical background Cd concentrations are around 0.1–2 mg kg^{-1} soil.

C.3.25.2b Total organic matter content: What is the soil organic matter content of the topsoil in the HGMU (obtained by laboratory analysis of the upper 30 cm or plough layer)? *(Recording Sheet 1)*

a. < 2 %
b. 2–8 %
c. > 8 but < 15 %
d. ≥ 15 %
e. Unknown, no organic matter analytical data are available

2.3.F Assignment of a hydrological code to the HGMU

The purpose of this section is to determine which hydrological processes are very likely to occur in each HGMU. This results in a hydrological code (**H1–H16**; see Action F.5), which is used as an input into many of the hydrological and non-hydrological FAPs. This task draws from a variety of information sources, and is best carried out on return from the field, which is why it is separated from Section 2.3.B (Recording of hydrological indicators). See also the identification of hydrological boundaries in HGMU mapping (Section 2.2.C).

It is important to realise the difference between the HGMU coding in this section and the outcomes from functional assessment in Sections 3.1.A–3.1.D. The coding indicates the likelihood of one or more selected hydrological processes occurring *to varying degrees*, whereas the functional assessment will determine whether and to what degree or likelihood hydrological functions are being performed. Some of the processes evaluated in this section (run-off input, run-off output) are not directly relevant to hydrological functions but are relevant to other functions in the procedures.

All HGMUs are characterised by combinations of inflows and outflows. The nature, timing, depth and duration of the *processes* of groundwater discharge, run-off input, run-off output and overbank flooding will control to a great extent which functions they perform, not only hydrological functions, but also ecological and biogeochemical functions. The potential occurrence of the following four main hydrological processes (Fig. 2.3.18) can be identified using information gathered during the Delineation and Characterisation tasks.

- **Groundwater discharge:** upward flow of water from a subsurface source, either a local aquifer or a regional aquifer, can occur at the soil surface as seepage or spring flow where the water table meets the surface. As this is a slow flow mechanism, inflow to the HGMU will continue for long periods after rainfall or snowmelt. It is assessed whether discharge occurs to any degree, and may thus support ecological and biogeochemical functions – regardless of whether it is an important hydrological function in itself (for functional assessment of discharge see Section 3.1.C).
- **Run-off input:** excess rainwater entering the HGMU from an adjoining HGMU or from the CA, as overland flow, or flow near the ground surface, or flow through shallow artificial drainage systems. This includes inputs from natural streams that terminate within the HGMU, but not overbank flooding from rivers that pass the HGMU (see below). As these are rapid flow mechanisms, inflow to the HGMU will cease within a few days after rainfall or snowmelt, which distinguishes it from groundwater discharge (see above).
- **Run-off output:** water leaving the HGMU towards an adjoining wetland HGMU or a channel, either as overland flow or through drains, thus potentially removing solutes from the HGMU. Water leaving the HGMU underground is

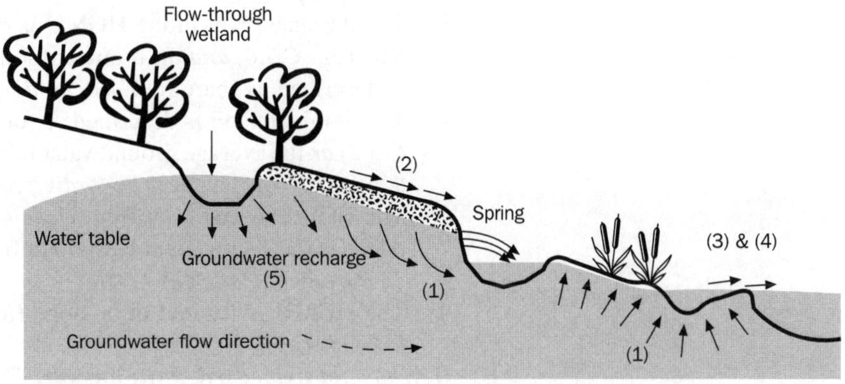

2.3.18 Landscape relationships of the four main hydrological processes: (1) groundwater discharge; (2) run-off input; (3) run-off output; (4) overbank flooding; (5) groundwater recharge.

not included. Sources of this water may include any input mechanism: rainfall, run-off input, discharge and overbank flooding.

- **Overbank flooding:** water which overtops the banks of the main river or stream passing the HGMU during a flood event, to inundate (part of) the HGMU for any period of time. Streams must originate outside the AA or the CA directly upslope from the AA, as inputs from such smaller streams are included in the run-off input process.

Note: In the assignment of the code, it is assumed that all wetland areas receive precipitation.

The following Actions (F.1–F.5) need to be performed for each HGMU upon return from the field. Use information from **Recording Sheet 3** to determine if each of the four hydrological processes occurs in the HGMU (Actions F.1–F.4). The results can then be used to assign a code to each HGMU, using the HGMU hydrological coding table (Table 2.3.2) in Action F.5. Record the code on **Recording Sheet 2a** in the box 'F. Hydrol Code'.

- If the process is occurring, place Y in box **F1/F2/F3** or **F4**.
- If the process is not occurring, place N in box **F1/F2/F3** or **F4**.
- If it is uncertain whether the process is occurring or not, place U in box **F1/F2/F3** or **F4**.

Recharge areas are zones where groundwater is replenished by infiltration through the soil of standing or flowing water or precipitation (Fig. 2.3.18). This usually occurs where the average water table surrounding the recharge area is lower than the water level in the wetland or the water table is below the soil surface. It occurs where there are no impermeable layers (e.g. clay) between the surface and the groundwater. The same area of wetland may function as a discharge or recharge area at different times as the water table rises and falls.

For guidance on Actions F.1 and F.3, the descriptions of recharge and discharge given in Sections 3.1.B (page 261) and 3.1.C (page 265) are likely to be useful when read in conjunction with Fig. 2.3.18.

2.3.F.1 ACTION: Determine occurrence of groundwater discharge

Record information: Recording Sheet 2a *(F.1)*

A. *The process of groundwater discharge is* ***unlikely*** *to occur if* ***at least two*** *of the following statements are true:*
   1. There is no area (inside or outside the AA, or both) of significant size (at least as large as the HGMU) immediately upslope of the HGMU *(AA Map 1)*.
   2. The dominant soil in the HGMU is described as 'fine' or 'medium'. *(Recording Sheet 2a: C.3.7)* ***and*** the dominant soil in the HGMU immediately upslope is described as 'fine' or 'medium' *(Recording Sheet 2a, C.3.7)*.
   3. The dominant soil in the HGMU is described as 'bare rock' (*Recording Sheet 2a: C.3.6*) ***and*** the dominant soil in the HGMU immediately upslope is described as 'bare rock' *(Recording Sheet 2a, C.3.6)*.
   4. The dominant soil is described as 'dry' at any time *(Recording Sheet 2a: C.3.2)* ***or*** the average groundwater table in the HGMU is over 50 cm depth at any time of the year *(Recording Sheet 3, B.3.10, B.3.11)*.
   5. The HGMU is flat (*Gradient Code* ***G0*** *– Recording Sheet 2a, C.2)* ***and*** the HGMU is situated at least 50 m from a sloping area (*Recording Sheet 2a, Gradient Code* ***G2–G3****)*.
   6. The HGMU is located on a slope (*Recording Sheet 2a, Gradient Code* ***G2–G3****, C.2)*.

B. ***If A. has not yielded a definite answer:*** *The process of groundwater discharge is very likely to occur if* ***at least two*** *of the following statements are true:*
   1. Springs are identifiable in the field *(Recording Sheet 3, B.2.1)*.

2. Permanent streams originate in the HGMU *(Recording Sheet 3, B.2.2, B.2.6).*
3. The particle size of the soil of the area immediately upslope of the HGMU is 'coarse' or 'gravel' *(Recording Sheet 2a, C.3.7)* or the dominant deposit in the area immediately upslope of the HGMU has a high permeability (especially sand, gravel or karstified limestone) *(AA Map 3)* or the area immediately upslope of the HGMU is located in a recharge area *(AA Map 3).*
4. **Disregard this question if the soil mineral particle size is described as 'fine' or 'medium'** *(Recording Sheet 2a, C.3.7).*
The soil in the HGMU is very 'reduced' – greyish without mottling *(Recording Sheet 2a, C.3.4, C.3.5)* or the soil is described as 'organic' *(Recording Sheet 2a, C.3.6)* or the ground- or surfacewater table remains less than 0.3 m below the soil surface throughout the year *(Recording Sheet 3, B.3.7, B.3.8, B.3.10, B.3.11) or* the soil is 'wet' or 'flooded' in the dry season *(Recording Sheet 3, B.3.2).*
5. Most of the HGMU remains snow-free or ice-free when most nearby areas are covered with snow or ice *(Recording Sheet 3, B.3.3, B.3.4).*
6. Surface water run-off is observed after at least two weeks without significant rain or flooding *(Recording Sheet 3, B.2.7).*
7. Abundant or lush vegetation (e.g. trees, reed swamp) is found in the HGMU *(Recording Sheet 3, B.2.6)* and the vegetation in nearby areas is sparse, with a large surface of exposed soil *(Recording Sheet 3 for other HGMUs, B.5.7).*

C. ***If A and B have not yielded a definite answer:*** *It is uncertain whether the process of groundwater discharge is likely to occur. The hydrological code is **H0**. It is recommended to identify the missing information, see Action F.5.*

2.3.F.2 *ACTION: Determine occurrence of run-off input*

Record information: Recording Sheet 2a (*F.2*)

Notes:

- Complete Action F.3 first.
- In answering the following questions, remember to include subsurface run-off.

A. *The process of run-off input into the HGMU is unlikely to occur if **at least one** of the following statements is true:*
1. There is no area (inside or outside the AA, or both) of significant size (at least as large as the HGMU) immediately upslope of the HGMU *(CA Map 1).*
2. The HGMU is separated from the upslope area by a feature that will intercept run-off: an embankment, canal, ditch, stream *(AA Map 3).*
3. The area immediately upslope (inside or outside the AA, or both) is unlikely to produce run-off output (*as determined under Action F.3.A*).

B. ***If A has not yielded a definite answer:*** *The process of run-off input (including subsurface run-off) into the HGMU is very likely to occur if the HGMU is located directly downslope of an area of significant size (larger than the HGMU), while **at least one** of the following statements is true:*
1. The area immediately upslope (inside or outside the AA, or both) is very likely to produce run-off output (*as determined under Action F.3.B*).
2. There are indications of standing water in the HGMU, but only directly after rainfall events and not caused by overbank flooding *(Recording Sheet 3, B.4.13).*
3. There are channels or drains originating in the area immediately upslope and terminating in the HGMU *(Recording Sheet 3, B.2.4; AA Map 3; CA Map 1).*

C. ***If A and B have not yielded a definite answer:*** *It is uncertain whether the process of run-off input is likely to occur. The hydrological code is* **H0**. *It is recommended to identify the missing information, see Action F.5.*

2.3.F.3 ACTION: Determine occurrence of run-off output

Record information: Recording Sheet 2a *(F.3)*

A. *The process of run-off output from the HGMU is unlikely to occur if* ***at least two*** *of the following statements are true:*
   1. The HGMU is located in a climate where annual potential evaporative losses exceed annual rainfall inputs *(Recording Sheet 1, C.3.1.1, C.3.1.3).*
   2. The particle size of the soil of the HGMU is 'coarse' or 'gravel' *(Recording Sheet 2a, C.3.7)* ***or*** the dominant deposit in the HGMU has a high permeability (especially sand, gravel or karstified rock) *(Recording Sheet 2a C.3.7; AA Map 3)* ***or*** the HGMU is located in a recharge area *(AA Map 3).*
   3. The average groundwater table in the HGMU always remains below 0.3 m below the soil surface *(Recording Sheet 3, B.3.11, B.3.12)* ***or*** the soil is described as 'dry' or 'moist' *(Recording Sheet 2a, C.3.2)* even after significant rainfall *(Recording Sheet 1, C.3.1.2*).
   4. The HGMU surface is near horizontal (*Gradient Code* ***G0*** *– Recording Sheet 2a, C.2)* ***and*** the HGMU is clearly a depression without any outflow channels *(Recording Sheet 2a, C.2; Recording Sheet 3, B.2.3; AA Map 2).*

B. ***If A has not yielded a definite answer:*** *The process of run-off output from the HGMU is very likely to occur if* ***at least two*** *of the following statements are true:*
   1. Surface run-off from the HGMU has been observed *(Recording Sheet 3, B.2.2).*
   2. The soil of the HGMU is described as 'fine', 'medium' or 'organic' or 'bare rock' *(Recording Sheet 2a, C.3.7, C.3.6)* ***or*** the soil is 'gleyed' or 'reduced'- greyish with or without mottling *(Recording Sheet 2a, C.3.4, C.3.5)* ***or*** the soil is described as 'organic' *(Recording Sheet 2a, C.3.6*) ***or*** the ground or surface water table remains within 0.3 m depth throughout the year *(Recording Sheet 3, B.3.8, B.3.11)* ***or*** the HGMU is described as permanently waterlogged *(Recording Sheet 3, B.3.2, B.3.8, B.3.11*) ***or*** the soil is described as 'flooded' or 'wet' *(Recording Sheet 2a, C.3.2*) in the dry season or after a dry period *(Recording Sheet 1, C.3.1.2)* ***or*** the HGMU is coded as a 'groundwater discharge' area (*as determined under Action F.1.B*).
   3. There are signs of erosion within the HGMU: e.g. scouring in steeper areas; local exposure of bare rock; sediment accumulation at lower slopes *(Recording Sheet 3, B.5.2).*
   4. There are drains or streams originating in the HGMU and leading out of it *(Recording Sheet 3, B.2.3).*
   5. The HGMU is located on a steep slope – *Recording Sheet 2a*, (*Gradient Code* **G3**, *C.2).*
   6. The HGMU receives inputs from discharge or overbank flooding (a*s determined under Action F.1.B and Action F.4.B*). *Only answer this if the other answers are negative or inconclusive.*

C. ***If A and B have not yielded a definite answer:*** *It is uncertain whether the process of run-off output is likely to occur. The hydrological code is* ***H0***. *It is recommended to identify the missing information, see Action F.5.*

2.3.F.4 ACTION: Determine occurrence of overbank flooding

Record information: Recording Sheet 2a *(F.4)*

A. *The process of overbank flooding into the HGMU is* ***unlikely*** *to occur if* ***at least two*** *of the following statements are true:*

1. The HGMU is indicated to never flood *(Recording Sheet 3, B.4.2)*.
2. The HGMU is embanked on all sides *(Recording Sheet 3, B.4.14; AA Map 3)*.
3. The HGMU has a land use that is incompatible with occasional flooding, such as orchards *(Recording Sheet 3, B.4.12)*.
4. Floatable materials of local origin are found within the HGMU *(Recording Sheet 3, B.4.8)*.
5. Buildings or fences within the HGMU indicate a lack of water flow *(Recording Sheet 3, B.4.10, B.4.11)*.
6. The HGMU is located on a steep slope – *Recording Sheet 2a*, (*Gradient Code* ***G2–G3****, C.2)*.
7. The HGMU is situated at an elevation above the probable range of water body level fluctuation *(Recording Sheet 3, B.1.2, B.4.17, B.4.18, B.4.19; Recording Sheet 1, C.3.19.2a, C.3.19.2b, C.3.19.2c, C.3.19.2d)*.

B. ***If A has not yielded a definite answer:*** *The process of overbank flooding into the HGMU is very likely to occur if* **at least two** *of the following statements are true:*

1. The HGMU has been observed to be inundated *(Recording Sheet 3, B.4.1)* to a depth over 0.2 m *(Recording Sheet 3, B.3.7, B.3.8, B.3.9, B.4.16, B.4.17, B.4.18, B.4.19)*.
2. Water on the HGMU surface has been observed to flow with significant velocity (over 0.1 m s^{-1}) *(Recording Sheet 3, B.4.6)*
3. Connections between the HGMU and the water body have been observed to be inundated *(Recording Sheet 3, B.4.5)* ***or*** overbank flow has been observed to enter the HGMU *(Recording Sheet 3, B.4.4)*.
4. The HGMU is situated at an elevation within the probable range of water body level fluctuation *(Recording Sheet 1, C.3.19.2a, C.3.19.2b, C.3.19.2c, C.3.19.2d; Recording Sheet 3, B.1.3, B.4.16, B.4.17, B.4.18, B.4.19)*.
5. Fine mineral sediment on vegetation surfaces or fine mineral or organic material on soil surfaces has been observed after floods *(Recording Sheet 3, B.5.4, B.5.5)*.
6. The HGMU is one of the following geomorphic features: a levee (geomorphic code **LV**), flat area (**FA**); backswamp (**BS**); cut-off channel (**CO**); ox-bow lake (**OX**); former channel (**FC**); point-bar (**PB**) *(Recording Sheet 2a, C.2)*.
7. The HGMU is in open connection with another HGMU (*AA Map 1*) for which it has been found that overbank flooding is very likely to occur *(Recording Sheet 3, B.4.1, B.4.4; as determined under Action F.4.B)*, with the same or lower surface elevation *(Recording Sheet 2a, B.1.3)*.
8. If the HGMU is a levee or a point-bar and erosion features – patches of exposed soil that are not caused by humans or by cattle – are widespread *(Recording Sheet 2a, C.2; Recording Sheet 3, B.5.7)*.

C. ***If A and B have not yielded a definite answer:*** *It is uncertain whether the process of overbank flooding is likely to occur. The hydrological code is* ***H0****. It is recommended to identify the missing information, see Action F.5.*

2.3.F.5 *ACTION: Assign a hydrological code to each HGMU*

Record information: Recording Sheet 2a

The hydrological codes are set out in Table 2.3.2. They are based on one or more hydrological flow processes (F1–4), assessed as described above. Select the code from the table below and enter it on **Recording Sheet 2a**.

If the resulting hydrological code is **H0**, any functional assessment will be difficult, and it is recommended to go through this action again and identify the information that is missing. It is possible that the knowledge now available on the occurrence of other hydrological processes will lead to a different outcome. Additional information can perhaps be obtained from local experts, rechecking of records or – as a last resort – an 'educated guess'.

Note: Ensure that all the information collected in Section 2.2 'HGMU delineation' has been correctly transcribed onto the field map and recording sheets, then progress to Section 3.3 'Field guidelines – function assessment specific questions'.

Table 2.3.2 Identification of hydrological codes

HYD. code	Groundwater discharge F1	Run-off input F2	Run-off output F3	Overbank flooding F4
H0	*Not enough information available*			
H0	?	–	–	–
H0	–	?	–	–
H0	–	–	?	–
H0	–	–	–	?
H1	X			
H2	X	X		
H3	X	X	X	
H4	X	X	X	X
H5		X		
H6		X	X	
H7		X	X	X
H8			X	
H9			X	X
H10				X
H11	X		X	
H12	X		X	X
H13		X		X
H14	X	X		X
H15	X			X
H16				

Key to the likelihood of process operation:
X The process is very likely to occur
Blank The process is unlikely to occur
? It is uncertain if the process occurs
– The process does not affect coding outcome

2.3.G Estimation of HGMU area

The purpose of this section is to estimate the area of the HGMUs in order to assign the relative importance of the functions, which the component HGMUs of a site performs. This must be done both in real terms and as a proportion of the whole site.

2.3.G.1 *ACTION: Estimate the area of the HGMUs using the instructions below*

Record information: Recording Sheet 3

1. Use the completed map of the AA with all the HGMUs delineated on it, together with the appropriate scale marker. If the site map has been enlarged from an existing map it will be necessary to enlarge the scale marker from that map by the same proportion.
2. Use the scale marker to draw a grid on the map dividing the site into 100 m × 100 m squares. If the site is very large then increase the size of the squares accordingly.
3. Use the grid to estimate the Total Area of the Site.
   Record the Total Area of the Site on **Recording Sheet 3 (G)**.
4. Use the grid to estimate the area of each HGMU.
   Record the HGMU Area on **Recording Sheet 3 (G)**.
5. Estimate the percentage that the area of each HGMU represents relative to the total area of the site, i.e.

$$\frac{\text{Area of HGMU}}{\text{Total area of site}} \times 100 = \text{HGMU \%}$$

6. Record the HGMU Relative Area on **Recording Sheet 3 (G)**.
7. If the mapping has been converted to digital form in a geographical information system, then the area of the AA, of each HGMU, and its proportion of the total AA, can be calculated automatically.

2.3.H Nutrient input characterisation

The purpose of this section is to assess the water quality functions (nutrient retention or export, *not* organic matter accumulation). Therefore, the nutrients entering or likely to enter an HGMU must be characterised.

A nutrient input is defined as a source of nutrient(s) (nitrogen or phosphorus) that does not naturally originate from within the wetland and is supplied by one or more of the following mechanisms:

- direct application;
- atmospheric deposition;
- groundwater discharge;
- (sub)surface run-off input;
- contact with nutrient-rich water (e.g. via overbank flooding).

An HGMU may receive several different inputs. For the purposes of functional assessment it is necessary to characterise each *nutrient source* and its *mechanism(s) of supply*.

Nutrient sources are identified in Action H.1. This section is divided into three parts:

- **Part A**: identifies nutrient sources applied within the HGMU itself (direct nutrient sources).
- **Part B**: identifies nutrient sources applied outside of the HGMU, but within either the AA or the CA (indirect nutrient sources).
- **Part C**: identifies more distant nutrient sources received by the HGMU in nutrient-rich surfacewater.

When a nutrient source originating from the area surrounding the HGMU is identified (i.e. part B above), you are also required to complete Action H.2, in order to identify the mechanism(s) of supply to the HGMU.

A potential nutrient source may have more than one mechanism of supply; therefore each different combination of nutrient source with its mechanism of supply comprises a separate nutrient input. For instance, if the CA is in receipt of inorganic fertiliser which has two mechanisms of supply to the HGMU, e.g. groundwater discharge and run-off input, then two separate nutrient inputs are recognised. This section is summarised in Fig. 2.3.19.

1. Answer the questions for each HGMU in turn in Action H.1. For each question tick the appropriate answer box on **Recording Sheet 2b**. In this section you identify the presence of one or several nutrient sources. If you identify a nutrient source in part B (in the area surrounding the HGMU) you will be instructed to:
   '...go to Action H.2 and complete Recording Sheet 2b, Section H.2'.
2. Complete Action H.2 and tick the appropriate answer boxes on **Recording sheet 2b**. (For those questions in part B that you aren't required to answer, leave the boxes blank.)
3. When all the questions in Action H.2 have been answered, return to Action H.1 and continue to answer the questions to identify other nutrient sources.
4. Once all questions in Action H.1. have been answered for the HGMU being assessed, go to Action H.3. This is the assessment of the nutrient input section in which you ascribe a nutrient input code to each identified nutrient source. Follow the instructions and enter the codes on **Recording Sheet 2c**.

Notes:

- The method of determining the nutrient input code (Action H.3) is very similar to that of functional assessment, using the same structure and syntax. Instructions and guidance on this are given at the beginning of Section 3. Some additional instructions are given at the beginning of Action H.3.
- Look-up Table 2 in Action H.3 (*A nutrient input is not (significantly) being received*) requires a slightly different approach. Instructions on its use are given in the section concerned.

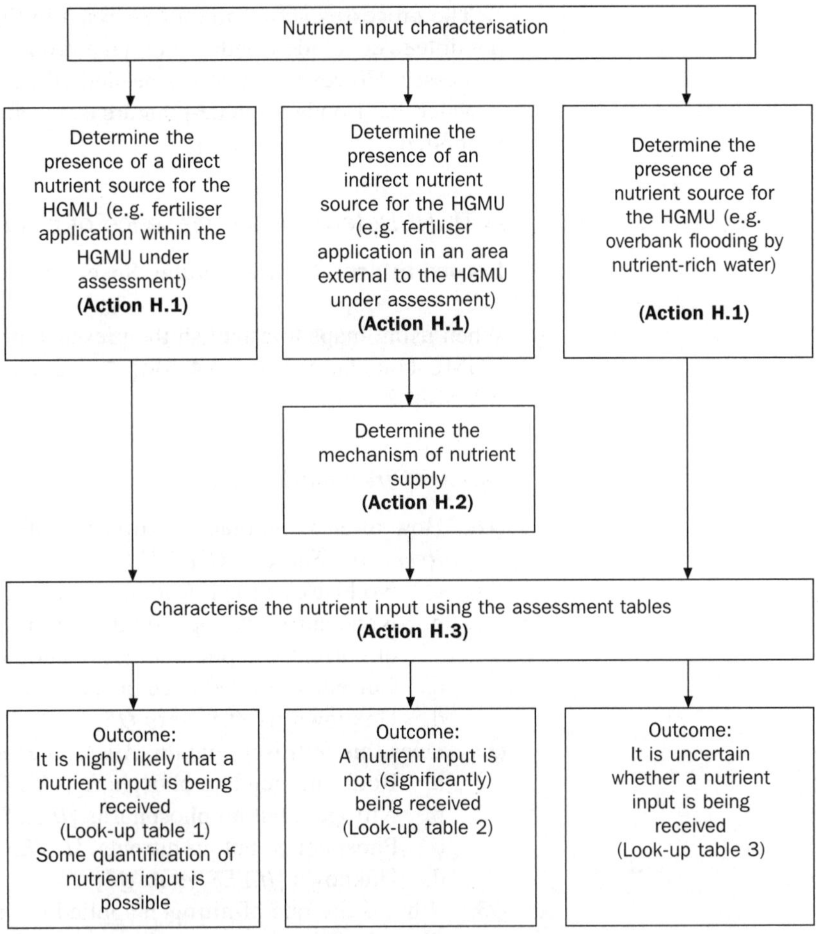

2.3.19 Summary of the nutrient input section.

- If a nutrient input cannot be characterised (because of insufficient or imprecise information) then the nutrient export and nutrient retention functions cannot be fully assessed. In this case the functional assessment section can be used to evaluate the *potential* of the wetland environment to retain or export nutrients.
- Once you have found one input code, do not stop there. You should continue the process for the whole section until you have found *all* the nutrient inputs applicable for the HGMU in question.

In this section only nutrient sources external to the wetland are considered. However, some processes can result in the mobilisation of nutrients already present in the wetland soil (naturally or due to anthropogenic activities). This is called *internal eutrophication/internal loading*.

One process that can cause internal eutrophication is flooding, or contact with water that contains sulphate. When sulphate does not occur naturally in an HGMU or when an HGMU does not frequently receive sulphate-rich water, an (occasional) inflow of sulphate-rich water (either river water or surfacewater supplied via management activity) can cause mobilisation of phosphorus; the phosphorus bound to iron in soil and sediments will be replaced by sulphide (produced by sulphate reduction in anaerobic soils). The extent to which the process is performed is dependent upon the sulphate concentration in the water, the relative concentrations of iron and phosphate in the sediments and the availability of oxygen.

Other processes that result in internal mobilisation of nutrients are, for example, drainage of peat soils and application of lime or chalk as a management technique. These actions result in conditions favourable for mineralisation of organic material. Owing to this faster mineralisation, the nutrients present in the organic material will become re-mobilised.

The processes above are not assessed in this section. Therefore it will not be possible to conclude whether internal eutrophication will occur as a result of these processes. However, caution is needed when you suspect sulphate to be present in water that floods sulphate-poor areas, when peat soils are drained or when lime is applied.

2.3.H.1 *ACTION: Determine the presence of a nutrient source*

Record information: Recording Sheet 2b (Action H.1)

When using maps to establish the presence of nutrient sources, remember to lay HGMU boundaries using **AA Map 1** or **3** over the land use boundaries shown in **AA Map 2**.

Part A: Direct nutrient sources

Q1. How often is inorganic fertiliser applied within the HGMU? *(AA Map 2; Recording Sheet 1: C.3.2.1b)*

- a. No history of application in the last 20 years [*If YES go to **Q5***]
- b. None currently applied but previously applied within the last 20 years, or currently applied but less than once a year [*If YES go to **Q2***]
- c. Currently applied once or more than once a year [*If YES go to **Q2***]
- d. Unknown [*If YES go to **Q5***]

Q2. Does this fertiliser contain (*AA Map 2; Recording Sheet 1, C.3.2.2b*):

- a. Both nitrogen and phosphorus [*If YES go to **Q3***]
- b. Nitrogen, but no phosphorus [*If YES go to **Q3** and skip **Q4***]
- c. Phosphorus, but no nitrogen [*If YES go to **Q4***]
- d. Unknown [*If YES go to **Q5***]

Q3. What is the rate of nitrogen applied to the HGMU? *(AA Map 2; Recording Sheet 1, C.3.2.3b(i))*

- a. < 15 kg N ha^{-1} a^{-1}
- b. 15–50 kg N ha^{-1} a^{-1}
- c. > 50 kg N ha^{-1} a^{-1}
- d. Unknown

Q4. What is the rate of phosphorus applied to the HGMU? *(AA Map 2; Recording Sheet 1, C.3.2.3b(ii))*

- a. < 5 kg P ha^{-1} a^{-1}
- b. 5–10 kg P ha^{-1} a^{-1}
- c. > 10 kg P ha^{-1} a^{-1}
- d. Unknown

Q5. How often is organic fertiliser applied by spraying or spreading within the HGMU? *(AA Map 2; Recording Sheet 1, C.3.2.4b)*

- a. No history of application in the last 20 years
- b. None currently applied but previously applied within the last 20 years, or currently applied but less than once a year
- c. Currently applied once or more than once a year
- d. Unknown

Q6. How often is organic fertiliser applied by injection within the HGMU? *(AA Map 2; Recording Sheet 1, C.3.2.5b)*

- a. No history of application in the last 20 years
- b. None currently applied but previously applied within the last 20 years, or currently applied but less than once a year
- c. Currently applied once or more than once a year
- d. Unknown

Q7. How often is organic industrial waste applied within the HGMU? *(AA Map 2; Recording Sheet 1, C.3.2.6b)*

- a. No history of application in the last 20 years
- b. None currently applied but previously applied within the last 20 years, or currently applied but less than once a year

c. Currently applied once or more than once a year
d. Unknown

Q8. Are any of the following forms of livestock kept within part or all of the HGMU? *(AA Map 2; Recording Sheet 1, C.3.7.1b)*

- Cattle, horses or donkeys $\geq$ 1 ha^{-1}
- Pigs $\geq$ 3 ha^{-1}
- Sheep or goats $\geq$ 6 ha^{-1}
- Turkeys or geese $\geq$ 30 ha^{-1}
- Chickens or ducks $\geq$ 100 ha^{-1}

a. No history of livestock in the last 20 years
b. None currently, but kept previously within the last 20 years, or currently kept but at a density lower than the figure indicated above
c. Currently kept at a density equal to or higher than the figure indicated above
d. Unknown

Q9. Have any large populations of birds (greater than 1000 ha^{-1}) been recorded within the HGMU? *(AA Map 2, Recording Sheet 1, C.3.8.1b, C.3.8.3a (a–d) where (and why) C.3.8.3b (a–d)*

Q10. Is the nitrogen concentration in atmospheric deposition:

a. < 1500 mg N m^{-2} a^{-1}
b. ≥ 1500 mg N m^{-2} a^{-1}
c. Unknown

(Recording Sheet 1: C.3.1.9)

Part B: Indirect nutrient sources

Q11. How often is inorganic fertiliser applied within the AA or within the CA? *(AA Map 2; CA Map 1; Recording Sheet 1, C.3.2.1a, C.3.2.8)*

a. No history of application in the last 20 years [*If YES go to **Q13***]
b. None currently applied but previously applied within last 20 years, or currently applied but less than once a year [*If YES go to **Q12***]
c. Currently applied once or more than once a year [*If YES go to **Q12***]
d. Unknown [*If YES go to **Q13***]

Q12. Does this fertiliser contain: *(AA Map 2; CA Map 1; Recording Sheet 1, C.3.2.9)*

a. Both nitrogen and phosphorus [*If YES go to Action H.2 and complete Recording Sheet 2b: Section H.2*]
b. Nitrogen, but no phosphorus [*If YES go to Action H.2 and complete Recording Sheet 2b: Section H.2*]
c. Phosphorus, but no nitrogen [*If YES go to Action H.2 and complete Recording Sheet 2b: Section H.2*]
d. Unknown [*Go to Action H.2 and complete Recording Sheet 2b: Section H.2*]

Q13. How often is organic fertiliser applied by spraying or spreading within the AA or within the CA? *(AA Map 2; CA Map 1; Recording Sheet 1, C.3.2.4a, C.3.2.10)*

a. No history of application in the last 20 years
b. None currently applied but previously applied within last 20 years, or currently applied but less than once a year [*If YES go to Action H.2 and complete Recording Sheet 2b: Section H.2*]
c. Currently applied once or more than once a year [*If YES go to Action H.2 and complete Recording Sheet 2b: Section H.2*]
d. Unknown

Q14. How often is organic fertiliser applied by injection within the AA or within the CA? *(AA Map 2; CA Map 1; Recording Sheet 1, C.3.2.5a, C.3.2.11)*

a. No history of application in the last 20 years
b. None currently applied but previously applied within the last 20 years, or currently applied but less than once a year [*If YES go to Action H.2 and complete Recording Sheet 2b: Section H.2*]
c. Currently applied once or more than once a year [*If YES go to Action H.2 and complete Recording Sheet 2b: Section H.2*]
d. Unknown

Q15. How often is organic industrial waste applied within the AA or within the CA? *(AA Map 2; CA Map 1; Recording Sheet 1, C.3.2.6a, C.3.2.12)*

a. No history of application in the last 20 years

b. None currently applied but previously applied within last 20 years, or applied but less than once a year [*If YES go to Action H.2 and complete Recording Sheet 2b: Section H.2*]

c. Currently applied once or more than once a year [*If YES go to Action H.2 and complete Recording Sheet 2b: Section H.2*]

d. Unknown

Q16. Are any of the following forms of livestock kept within the AA or within the CA? *(AA Map 2; CA Map 1; Recording Sheet 1, C.3.7.1a, C.3.7.5)*

- Cattle, horses or donkeys $\geq$ 1 ha^{-1}
- Pigs $\geq$ 3 ha^{-1}
- Sheep or goats $\geq$ 10 ha^{-1}
- Turkeys or geese $\geq$ 30 ha^{-1}
- Chickens or ducks $\geq$ 100 ha^{-1}

a. No history of livestock in the last 20 years

b. None currently, but kept previously within the last 20 years, or currently kept but at a lower density than the figure indicated above [*If YES go to Action H.2 and complete Recording Sheet 2b: Section H.2*]

c. Currently kept with a density equal to or higher than the figure indicated above [*If YES go to Action H.2 and complete Recording Sheet 2b: Section H.2*]

d. Unknown

Part C: A nutrient source reaching the HGMU via surface water

Q17. Is it possible that water from the (main) water body (river, stream, ditch or lake) has contact with the HGMU? *(AA Map 2, Recording Sheet 1: C.3.19.8b)*

a. No never/not applicable [*If YES go to Action H.3*]

b. The HGMU borders the (main) water body and horizontal exchange (either via surface or via subsurface flow) can occur freely between the HGMU and the water body [*If YES go to **Q19***]

c. Although the HGMU does not border the water body, the HGMU has contact with the water from the water body during flood events [*If YES go **Q18***]

d. Contact between water in the water body and the HGMU does take place once or more than once a year indirectly via drainage ditches, streams or (sub)surface artificial drains [*If YES go to **Q19***]

e. Contact between water in the water body and the HGMU does take place indirectly via drainage ditches, streams or (sub)surface artificial drains less than once a year [*If YES go to **Q19***]

f. Unknown [*If YES go to Action H.3*]

Q18. How often is the HGMU flooded with water from the (main) water body? *(AA Map 2; Recording Sheet 1, C.3.19.8b)*

a. Frequently (at least once every 2 years) [*If YES go to **Q19***]

b. Occasionally (once every 2–10 years) [*If YES go to **Q19***]

c. Rarely (once every 10–15 years) [*If YES go to **Q19***]

d. Unknown [*If YES go to Action H.3*]

Q19. What is the mean annual nitrate–nitrogen concentration of the water in the (main) water body? *(Recording Sheet 1, C.3.20.6)*

a. < 1 mg l^{-1} NO_3–N

b. ≥ 1 mg l^{-1} NO_3–N

c. Unknown

Q20. What is the mean annual ammonium–nitrogen concentration of the water in the (main) water body? *(Recording Sheet 1, C.3.20.7)*

a. < 1 mg l^{-1} NH_4–N

b. ≥ 1 mg l^{-1} NH_4–N

c. Unknown

Q21. What is the mean annual total nitrogen concentration of the water in the (main) water body? *(Recording Sheet 1, C.3.20.9)*

a. < 2.2 mg l^{-1} N

b. ≥ 2.2 mg l^{-1} N

c. Unknown

Q22. What is the mean annual phosphate–phosphorus concentration of the water in the (main) water body? *(Recording Sheet 1, C.3.20.8)*

a. < 0.1 mg l^{-1} PO_4–P

b. ≥ 0.1 mg l^{-1} PO_4–P

c. Unknown

[*Continue with Action H.3*]

2.3.H.2 *ACTION: Determine the mechanism of nutrient supply for each identified nutrient source outside the HGMU*

Record information: Recording Sheet 2b *(Action H.2)*

Only answer the following questions if you have identified *nutrient sources* in Part B of Action H.1. Tick the correct answers to the questions in the boxes of **Section H.2** on **Recording Sheet 2b**.

Part A: Groundwater discharge

Q1. Is the hydrology of the HGMU characterised as having a groundwater discharge (**F1**) component (hydrological code **H1**, **H2**, **H3**, **H4**, **H11**, **H12**, **H14**, **H15**)? *(Recording Sheet 2a: F)*
[*If YES go to **Q2**, if NO go to Part B: Run-off or subsurface run-off input*]

Q2. Is the nutrient source identified in Action H.1 applied to land within the AA or the CA upslope of the HGMU?
(AA Map 1; CA Map 1)
[*If YES go to **Q3**, if NO go to Part B: Run-off or subsurface run-off input*]

Q3. If the nutrient source is applied to an HGMU within the AA upslope of the HGMU being assessed, does this HGMU have a groundwater discharge component (**F1**) (hydrological code **H1**, **H2**, **H3**, **H4**, **H11**, **H12**, **H14**, **H15**)? *(Recording Sheet 2a: F; AA Map 1)*

[*Continue with Part B: Run-off or subsurface run-off input*]

Part B: Run-off or subsurface run-off input

Q4. Is the hydrology of the HGMU characterised as having a (sub)surface run-off input (**F2**) component (hydrological code **H2**, **H3**, **H4**, **H5**, **H6**, **H7**, **H13**, **H14**)? *(Recording Sheet 2a: F)*
[*If YES go to **Q5**, if NO return to Action H.1 and continue identification of nutrient sources*]

Q5. If the nutrient source identified is applied *either* elsewhere in the AA *or* in the CA, is one of the following true:

- it is applied to an area drained by either ditches, streams or (sub)surface artificial drains which flow to the HGMU *uninterrupted (AA Map 2; CA Map 1)*
- the nutrients are applied to land within the AA or the CA that is upslope of the HGMU *(AA Map 1; CA Map 1)*

[*If YES go to **Q6**, if NO return to Action H.1 and continue identification of nutrient sources*]

Q6. If the nutrient source is applied within the AA to an HGMU upslope of the HGMU being assessed, does this upslope HGMU have a (sub)surface run-off output **(F3)** component (hydrological code **H3**, **H4**, **H6**, **H8**, **H9**, **H11**, **H12**)? (*Recording Sheet 2a: F*)

[*Return to Action H.1 and continue identification of nutrient sources*]

2.3.H.3 *ACTION: Characterisation of the nutrient input*

Record information: Recording Sheet 2c *(Action H.3: nutrient input code)*

Compare results recorded for each HGMU on **Recording Sheet 2b**, Section **H.1** and **H.2** with possible outcomes below, and assign appropriate nutrient input code(s). Record code(s) on the corresponding **Recording Sheet 2c**, in Section **H.3**, in the appropriate table depending on the outcome of the assessment.

Notes:

- The method of determining the nutrient input code is very similar to that of functional assessment, using the same structure and syntax. Instructions and guidance on this are given in the Introduction to Section 3.
- In the section below (**2** A nutrient input is not (significantly) being received), a slightly different approach is required in using the look-up table, as is explained in the section concerned.
- The HGMU can receive more than one nutrient input. Therefore you have to check all the tables in turn, for each of the three levels of likely input, to determine which inputs may be taking place. Each answer combination in the tables below that fits your results on the recording sheet, is a separate nutrient input.

Look-up tables

1. IT IS HIGHLY LIKELY A NUTRIENT INPUT IS BEING RECEIVED

If all the questions listed in the first two columns (Actions H.1 and H.2) of Look-up Table 1 are answered YES, then there is a very high possibility that the HGMU is receiving a nutrient input. The outcomes are grouped into three main types of nutrient source: 'nitrogen' (**N**), 'phosphorus' (**P**) and 'nitrogen and/or phosphorus' (**Nut**). The **N**, **P** or **Nut** are followed by a subscript $_d$ or $_i$, referring respectively to direct nutrient sources (including inorganic or organic fertiliser, excretions from livestock or birds and atmospheric deposition) or indirect nutrient sources (including nutrient sources received by the HGMU in groundwater discharge or (sub)surface run-off as well as nutrient sources received by the HGMU in floodwater or managed water supplies). Finally, the codes comprise a number and a letter separated by a slash. The 1 before the slash means that a nutrient input is highly likely to be received by the HGMU. The letter after the slash gives an indication of the quantity of nutrients being received: an **a** indicating the highest range (> 50 kg N ha^{-1} a^{-1} or > 10 kg P ha^{-1} a^{-1}). An **x** after the slash means that quantification is not possible. In the quantification column (the last column of the table) quantitative information is given – if available. The figures in this column represent the amount of nutrients *at least* entering the HGMU. The figures in brackets represent an upper limit, but this information is not always known. For all assessment outcomes assigned a code containing an **x**, no quantitative information is available and therefore in these cases the last column contains a question mark (?).

For the assessment of processes, nutrient input information might be necessary. In these cases the user is referred to the codes assigned in this section. The quantitative information in the last column is not asked for in other sections but provides extra information for more insight.

The look-up table is separated into sections related to the nature of the input (i.e. 'inorganic fertiliser' is the first section). For each section record any positive answers in the corresponding box on ***Recording Sheet 2c*** (Look-up Table 1).

2. A NUTRIENT INPUT IS NOT (SIGNIFICANTLY) BEING RECEIVED

If Look-up Table 1 did not lead to an assessment, two possibilities remain: *either* the HGMU does not receive any significant nutrient inputs, *or* it is uncertain whether the HGMU receives a nutrient input. To find out if the former is the case (*no significant nutrient input*), use Look-up Table 2. If the answers to the questions agree with all the entries using the method described below, then the HGMU is

2.3.H.3 Look-up Table 1

INORGANIC FERTILISER				
H.1.ACTION	**H.2.ACTION**	**Rationale**	**Code**	**Quantification**
1c 2a 3a 4b	N/A	The HGMU is in direct receipt of nutrients from inorganic fertiliser (containing both N and P) at least annually. Amount of N and P in fertiliser is known. Because the amount of N is too low to comprise a significant nitrogen input, the nutrient input code will refer only to input of P.	**P_d 1/b**	> 5 (< 10) kg P ha^{-1} a^{-1}
1c 2a 3a 4c	N/A	The HGMU is in direct receipt of nutrients from inorganic fertiliser (containing both N and P) at least annually. Amount of N and P in fertiliser is known. Because the amount of N is too low to comprise a significant nitrogen input, the nutrient input code will refer only to input of P.	**P_d 1/a**	> 10 kg P ha^{-1} a^{-1}
1c 2a 3b 4a	N/A	The HGMU is in direct receipt of nutrients from inorganic fertiliser (containing both N and P) at least annually. Amount of N and P in fertiliser is known. Because the amount of P is too low to comprise a significant phosphorus input, the nutrient input code will refer only to input of N.	**N_d 1/b**	> 15 (< 50) kg N ha^{-1} a^{-1}
1c 2a 3b 4b	N/A	The HGMU is in direct receipt of nutrients from inorganic fertiliser (containing both N and P) at least annually. Amount of N and P in fertiliser is known.	**N_d 1/b** **P_d 1/b**	> 15 (< 50) kg N ha^{-1} a^{-1} > 5 (< 10) kg P ha^{-1} a^{-1}
1c 2a 3b 4c	N/A	The HGMU is in direct receipt of nutrients from inorganic fertiliser (containing both N and P) at least annually. Amount of N and P in fertiliser is known.	**N_d 1/b** **P_d 1/a**	> 15 (< 50) kg N ha^{-1} a^{-1} > 10 kg P ha^{-1} a^{-1}
1c 2a 3b 4d	N/A	The HGMU is in direct receipt of nutrients from inorganic fertiliser (containing both N and P) at least annually. Amount of N in fertiliser is known, but amount of P is unknown. Generally, when using inorganic fertiliser at least 5 kg P ha^{-1} a^{-1} is applied.	**N_d 1/b** **P_d 1/b**	> 15 (< 50) kg N ha^{-1} a^{-1} > 5 kg P ha^{-1} a^{-1}
1c 2a 3c 4a	N/A	The HGMU is in direct receipt of nutrients from inorganic fertiliser (containing both N and P) at least annually. Amount of N and P in fertiliser is known. Because the amount of P is too low to comprise a significant phosphorus input, the nutrient input code will refer only to input of N.	**N_d 1/a**	> 50 kg N ha^{-1} a^{-1}
1c 2a 3c 4b	N/A	The HGMU is in direct receipt of nutrients from inorganic fertiliser (containing both N and P) at least annually. Amount of N and P in fertiliser is known.	**N_d 1/a** **P_d 1/b**	> 50 kg N ha^{-1} a^{-1} > 5 (< 10) kg P ha^{-1} a^{-1}
1c 2a 3c 4c	N/A	The HGMU is in direct receipt of nutrients from inorganic fertiliser (containing both N and P) at least annually. Amount of N and P in fertiliser is known.	**N_d 1/a** **P_d 1/a**	> 50 kg N ha^{-1} a^{-1} > 10 kg P ha^{-1} a^{-1}

2.3.H.3 Look-up Table 1 cont'd

H.1.ACTION	H.2.ACTION	Rationale	Code	Quantification
1c 2a 3c 4d	N/A	The HGMU is in direct receipt of nutrients from inorganic fertiliser (containing both N and P) at least annually. Amount of N in fertiliser is known, but amount of P is unknown. Generally when using inorganic fertiliser at least 5 kg P ha^{-1} a^{-1} is applied.	**N_d 1/a** **P_d 1/b**	> 50 kg N ha^{-1} a^{-1} > 5 kg P ha^{-1} a^{-1}
1c 2a 3d 4b	N/A	The HGMU is in direct receipt of nutrients from inorganic fertiliser (containing both N and P) at least annually. Amount of P in fertiliser is known, but amount of N is unknown. Generally when using inorganic fertiliser at least 15 kg N ha^{-1} a^{-1} is applied.	**N_d 1/b** **P_d 1/b**	> 15 kg N ha^{-1} a^{-1} > 5 (< 10) kg P ha^{-1} a^{-1}
1c 2a 3d 4c	N/A	The HGMU is in direct receipt of nutrients from inorganic fertiliser (containing both N and P) at least annually. Amount of P in fertiliser is known, but the amount of N is unknown. Generally when using inorganic fertiliser at least 15 kg N ha^{-1} a^{-1} is applied.	**N_d 1/b** **P_d 1/a**	> 15 kg N ha^{-1} a^{-1} > 10 kg P ha^{-1} a^{-1}
1c 2a 3d 4d	N/A	The HGMU is in direct receipt of nutrients from inorganic fertiliser (containing both N and P) at least annually. Amount of N and P in fertiliser is unknown. Generally when using inorganic fertiliser at least 15 kg N ha^{-1} a^{-1} and 5 kg P ha^{-1} a^{-1} is applied.	**N_d 1/b** **P_d 1/b**	>15 kg N ha^{-1} a^{-1} > 5 kg P ha^{-1} a^{-1}
1c 2b 3b	N/A	The HGMU is in direct receipt of nutrients from inorganic fertiliser (containing only N) at least annually. Amount of N in fertiliser is known.	**N_d 1/b**	> 15 (< 50) kg N ha^{-1} a^{-1}
1c 2b 3c	N/A	The HGMU is in direct receipt of nutrients from inorganic fertiliser (containing only N) at least annually. Amount of N in fertiliser is known.	**N_d 1/a**	> 50 kg N ha^{-1} a^{-1}
1c 2b 3d	N/A	The HGMU is in direct receipt of nutrients from inorganic fertiliser (containing only N) at least annually. Amount of N in fertiliser is unknown. Generally when using inorganic fertiliser at least 15 kg N ha^{-1} a^{-1} is applied.	**N_d 1/b**	> 15 kg N ha^{-1} a^{-1}
1c 2c 4b	N/A	The HGMU is in direct receipt of nutrients from inorganic fertiliser (containing only P) at least annually. Amount of P in fertiliser is known.	**P_d 1/b**	>5 (< 10) kg P ha^{-1} a^{-1}
1c 2c 4c	N/A	The HGMU is in direct receipt of nutrients from inorganic fertiliser (containing only P) at least annually. Amount of P in fertiliser is known.	**P_d 1/a**	> 10 kg P ha^{-1} a^{-1}
1c 2c 4d	N/A	The HGMU is in direct receipt of nutrients from inorganic fertiliser (containing only P) at least annually.	**P_d 1/b**	> 5 kg P ha^{-1} a^{-1}

		Amount of P in fertiliser is unknown. Generally when using inorganic fertiliser at least 5 kg P ha^{-1} a^{-1} is applied.		
1c 2d	N/A	The HGMU is in direct receipt of nutrients from inorganic fertiliser at least annually. It is not known whether the fertiliser contains both N and P or only one of these nutrients. Generally when using inorganic fertiliser at least 15 kg N ha^{-1} a^{-1} or 5 kg P ha^{-1} a^{-1} is applied.	**Nut_d 1/b**	> 15 kg N ha^{-1} a^{-1} and/or > 5 kg P ha^{-1} a^{-1}
		ORGANIC FERTILISER/ORGANIC WASTE		
5c	N/A	The HGMU is in direct receipt of nutrients from the spraying or spreading of organic fertiliser at least annually.	**N_d 1/b** **P_d 1/b**	> 15 kg N ha^{-1} a^{-1} > 5 kg P ha^{-1} a^{-1}
6c	N/A	The HGMU is in direct receipt of nutrients from organic fertiliser injection at least annually.	**N_d 1/b** **P_d 1/b**	> 15 kg N ha^{-1} a^{-1} > 5 kg P ha^{-1} a^{-1}
7c	N/A	The HGMU is in direct receipt of nutrients from organic industrial waste application at least annually. It is unknown if this input contains both nitrogen and phosphorus or only one of these nutrients. Quantification is difficult because the nutrient concentration depends highly on the type of organic industrial waste.	**Nut_d 1/x**	?
		ANIMAL FAECES		
8c	N/A	The HGMU is in direct receipt of nutrients from animal faeces, arising from livestock that are kept in the HGMU at least annually.	**N_d 1/b** **P_d 1/b**	> 15 kg N ha^{-1} a^{-1} > 5 kg P ha^{-1} a^{-1}
		BIRD FAECES		
9	N/A	The HGMU is in direct receipt of nutrients from excretions from birds that are breeding, nesting, and feeding, roosting or moulting in large colonies in the HGMU. Presence of herbivore birds will also accelerate the nutrient cycling. Quantification is difficult, because the nutrient input depends on the bird species, the amount of birds and the activity of the birds.	**N_d 1/x** **P_d 1/x**	?
		ATMOSPHERIC DEPOSITION		
10b	N/A	The HGMU is in receipt of N from atmospheric deposition.	**N_d 1/b**	>15 kg N ha^{-1} a^{-1}
		DISCHARGE (INORGANIC FERTILISER)		
11c 12a	1 2 #3	It is highly likely that the HGMU is receiving a nutrient input (both N and P). The source is inorganic fertiliser that is applied at least annually outside the area of the HGMU and transported via subsurface water movement to the HGMU, where it reaches the surface by groundwater discharge. It is unlikely that the phosphate applied outside the HGMU reaches the HGMU via a groundwater flow. The high phosphorus adsorption capacity of the soil means that phosphorus will be bound to	**N_i 1/x**	?

2.3.H.3 Look-up Table 1 cont'd

H.1.ACTION	H.2.ACTION	Rationale	Code	Quantification
		soil particles during its passage through the soil. Therefore the HGMU receives a nutrient input code only for nitrogen. As nitrate is much more mobile in the soil than ammonium (also ammonium binds to soil particles), it is likely that the nitrogen in the groundwater flow comprises mostly nitrate. Quantification is difficult because various processes can alter the nutrient concentration of the groundwater flow.		
11c 12b	1 2 #3	It is highly likely that the HGMU is receiving a nitrogen input. The source is inorganic fertiliser that is applied at least annually outside the area of the HGMU and transported via subsurface water movement to the HGMU, where it is exfiltrated by groundwater discharge. Because nitrate is much more mobile in the soil than ammonium (ammonium binds to soil particles), it is likely that the nitrogen in the groundwater flow comprises mostly nitrate. Quantification is difficult because various processes can alter the nutrient concentration of the groundwater flow.	**N_i 1/x**	?
11c 12d	1 2 #3	It is highly likely that the HGMU is receiving a nutrient input. It is unknown whether this input contains both N and P or only one of these nutrients. The source is inorganic fertiliser that is applied at least annually outside the area of the HGMU and transported via subsurface water movement to the HGMU, where it reaches the surface by groundwater discharge. It is unlikely that phosphorus applied outside the HGMU reaches the HGMU via a groundwater flow. The high phosphorus adsorption capacity of the soil means that phosphorus will be bound to soil particles during its passage through the soil. As it is likely that nitrogen is present in the fertiliser, the HGMU receives a nutrient input code for nitrogen only. As nitrate is much more mobile in the soil than ammonium (also ammonium binds to soil particles), it is likely that the nitrogen in the groundwater flow comprises mostly nitrate. Quantification is difficult because various processes can alter the nutrient concentration of the groundwater flow.	**N_i 1/x**	?
		RUN–OFF (INORGANIC FERTILISER)		
11c 12a	(4 5)|(4 5 6)	It is highly likely that the HGMU is receiving nutrient input (both N and P). The source is inorganic fertiliser that is applied at least annually outside the area of the HGMU and transported via (sub)surface run-off input processes (one or a combination of ditch, tributary, overland flow and (sub)surface drainage) to the HGMU. Because nitrate is much more mobile in the soil than ammonium (ammonium binds to soil particles), it is likely that the nitrogen in the (ground)water flow comprises mostly nitrate. Quantification is difficult because various processes can alter the nutrient concentration of the (ground)water flow.	**N_i 1/x** **P_i 1/x**	?
11c 12b	(4 5)|(4 5 6)	It is highly likely that the HGMU is receiving a nitrogen input. The source is inorganic fertiliser that is applied at least annually outside the area of the HGMU and transported via (sub)surface run-off input processes (one or a combination of ditch, tributary, overland flow and (sub)surface drainage) to the HGMU.	**N_i 1/x**	?

		Because nitrate is much more mobile in the soil than ammonium (ammonium binds to soil particles), it is likely that the nitrogen in the (ground)water flow comprises mostly nitrate. Quantification is difficult because various processes can alter the nutrient concentration of the (ground)water flow.		
11c 12c	(4 5)|(4 5 6)	It is highly likely that the HGMU is receiving a phosphorus input. The source is fertiliser that is applied at least annually outside the area of the HGMU and transported via (sub)surface run-off input processes (one or a combination of ditch, tributary, overland flow and (sub)surface drainage) to the HGMU. Quantification is difficult because various processes can alter the nutrient concentration of the (ground)water flow.	**P_i 1/x**	?
11c 12d	(4 5)|(4 5 6)	It is highly likely that the HGMU is receiving a nutrient input. It is unknown whether this input contains both N and P or only one of these nutrients. The source is inorganic fertiliser that is applied at least annually outside the area of the HGMU and transported via (sub)surface run-off input processes (one or a combination of ditch, tributary, overland flow and (sub)surface drainage) to the HGMU. It is unknown whether this input contains both N and P or only one of these nutrients. Quantification is difficult because various processes can alter the nutrient concentration of the (ground)water flow.	**Nut_i 1/x**	?
		DISCHARGE (ORGANIC FERTILISER)		
13c|14c	1 2 #3	It is highly likely that the HGMU is receiving a nutrient input. The source is organic fertiliser that is applied (either by spraying, spreading or injecting) at least annually outside the area of the HGMU and transported via subsurface water movement to the HGMU, where it reaches the surface by groundwater discharge. It is unlikely that the phosphate applied outside the HGMU reaches the HGMU via a groundwater flow. The high phosphorus adsorption capacity of the soil means that phosphorus will be bound to soil particles during its passage through the soil. Therefore the HGMU receives a nutrient input code for nitrogen only. As nitrate is much more mobile in the soil than ammonium (also ammonium binds to soil particles), it is likely that the nitrogen in the groundwater flow comprises mostly nitrate. Quantification is difficult because various processes can alter the nutrient concentration of the groundwater flow.	**N_i 1/x**	?
		RUN-OFF (ORGANIC FERTILISER)		
13c|14c	(4 5)|(4 5 6)	It is highly likely that the HGMU is receiving a nutrient input. The source is organic fertiliser that is applied (either by spraying, spreading or injecting) at least annually outside the area of the HGMU and transported via (sub)surface run-off input processes (one or a combination of ditch, tributary, overland flow and subsurface drainage) to the HGMU. Because nitrate is much more mobile in the soil than ammonium (ammonium binds to soil particles), it is likely that the nitrogen in the (ground)water flow comprises mostly nitrate.	**N_i 1/x** **P_i 1/x**	?

2.3.H.3 Look-up Table 1 cont'd

H.1.ACTION	H.2.ACTION	Rationale	Code	Quantification
		Quantification is difficult because various processes can alter the nutrient concentration of the groundwater flow.		
		DISCHARGE / RUN-OFF (ORGANIC WASTE)		
15c	1 2 #3	It is highly likely that the HGMU is receiving a nutrient input from organic industrial waste application. It is applied at least annually outside the area of the HGMU and transported via subsurface water movement to the HGMU, where it is exfiltrated by groundwater discharge. The nutrient composition of the organic industrial waste is not known. However, as it is unlikely that phosphorus applied outside the HGMU reaches the HGMU via a groundwater flow (phosphorus binds to soil particles), it can be excluded from the nutrient input code. As it is assumed that nitrogen is present in the waste, the HGMU receives a nutrient input code for nitrogen only. As nitrate is much more mobile in the soil than ammonium (also ammonium binds to soil particles), it is likely that the nitrogen in the groundwater flow comprises mostly nitrate. Quantification is difficult because the nutrient concentration depends highly on the type of organic industrial waste and, moreover, various processes can alter the nutrient concentration of the groundwater flow.	**N_i 1/x**	?
15c	(4 5)|(4 5 6)	It is highly likely that the HGMU is receiving nutrients from organic industrial waste application. It is applied at least annually outside the area of the HGMU and transported via (sub)surface run-off input processes (one or a combination of ditch, tributary, overland flow and subsurface drainage) to the HGMU. It is unknown whether this input contains both N and P or only one of these nutrients. Quantification is difficult because the nutrient concentration depends highly on the type of organic industrial waste and, moreover, various processes can alter the nutrient concentration of the groundwater flow.	**Nut_i 1/x**	?
		DISCHARGE (ANIMAL FAECES)		
16c	1 2 #3	It is highly likely that the HGMU is receiving nutrients from animal faeces. Livestock are kept outside the area of the HGMU and nutrients from their faeces are transported via subsurface water movement to the HGMU, where they are exfiltrated by groundwater discharge. It is unlikely that the phosphorus applied outside the HGMU reaches the HGMU via a groundwater flow. The high phosphorus adsorption capacity of the soil means that phosphorus will be bound to soil particles during its passage through the soil. Therefore the HGMU receives only a nutrient input code for nitrogen. As nitrate is much more mobile in the soil than ammonium (also ammonium binds to soil particles), it is likely that the nitrogen in the groundwater flow comprises mostly nitrate. Quantification is difficult because various processes can alter the nutrient concentration of the groundwater flow.	**N_i 1/x**	?

<table>
<tr><td></td><td></td><td>RUN-OFF (ANIMAL FAECES)</td><td></td><td></td></tr>
<tr><td>16c</td><td>(4 5)|(4 5 6)</td><td>It is highly likely that the HGMU is receiving nutrients from animal faeces. Livestock are kept outside the area of the HGMU and nutrients from their faeces are transported via (subsurface) run-off input processes (one or a combination of ditch, tributary, overland flow and subsurface drainage) to the HGMU.
Because nitrate is much more mobile in water than ammonium, it is likely that the nitrogen in the (ground)water flow comprises mostly nitrate.
Quantification is difficult because various processes can alter the nutrient concentration of the groundwater flow.</td><td>N_i 1/x
P_i 1/x</td><td>?</td></tr>
<tr><td></td><td></td><td>SURFACE WATER</td><td></td><td></td></tr>
<tr><td>17b|17d ((19b|20b| 21b) 22b)</td><td>N/A</td><td>It is highly likely that the HGMU is receiving nutrient input (both N and P) because of contact with nutrient-rich surface water.
Quantification is difficult because the nutrient input depends on many different aspects (discharge, HGMU/AA area, exact nutrient concentration of the water, etc.).</td><td>N_i 1/x
P_i 1/x</td><td>?</td></tr>
<tr><td>17b|17d (19b|20b| 21b)</td><td>N/A</td><td>It is highly likely that the HGMU is receiving nitrogen from nutrient-rich surface water supplied (artificially) to the HGMU.
Quantification is difficult because the nutrient input depends on many different aspects (discharge, HGMU/AA area, exact nutrient concentration of the water, etc.).</td><td>N_i 1/x</td><td>?</td></tr>
<tr><td>17b|17d 22b</td><td>N/A</td><td>It is highly likely that the HGMU is receiving phosphorus from nutrient-rich water supplied (artificially) to the HGMU.
Quantification is difficult because the nutrient input depends on many different aspects (discharge, HGMU/AA area, exact nutrient concentration of the water, etc.).</td><td>P_i 1/x</td><td>?</td></tr>
<tr><td>17c 18a ((19b|20b| 21b) 22b)</td><td>N/A</td><td>It is highly likely that the HGMU is receiving a nutrient input (both N and P) by periodic overbank flooding with nutrient-rich water from the river channel.
Quantification is difficult because the amount of nutrients received by the HGMU depends on many different factors (discharge, HGMU/AA area, exact nutrient concentration of the water, etc.).</td><td>N_i 1/x
P_i 1/x</td><td>?</td></tr>
<tr><td>17c 18a (19b|20b| 21b)</td><td>N/A</td><td>It is highly likely that the HGMU is receiving nitrogen from periodic overbank flooding with nutrient-rich water from the river channel.
Quantification is difficult because the amount of nutrients received by the HGMU depends on many different factors (discharge, HGMU/AA area, exact nutrient concentration of the water, etc.).</td><td>N_i 1/x</td><td>?</td></tr>
<tr><td>17c 18a 22b</td><td>N/A</td><td>It is highly likely that the HGMU is receiving phosphorus from periodic overbank flooding with nutrient-rich water from the river channel.
Quantification is difficult because the amount of nutrients received by the HGMU depends on many different factors (discharge, HGMU/AA area, exact nutrient concentration of the water, etc.).</td><td>P_i 1/x</td><td>?</td></tr>
</table>

2.3.H.3 Look-up Table 2

<table>
<tr><th>H.1.ACTION</th><th>H.2.ACTION</th><th>Rationale</th><th>Code</th></tr>
<tr><td>1a|5a|6a|7a|8a|#9|10a|11a|13a|14a|15a|16a|17a</td><td>N/A</td><td>There are no nutrients applied to or received by the HGMU, either directly or indirectly. (There is no application of inorganic fertiliser, organic fertiliser and organic industrial waste, no animal faeces and no nutrient-rich atmospheric deposition in the HGMU, or in the CA/AA. There is also no flooding or supply of surface water.)</td><td>Nut 2</td></tr>
<tr><td>#(1b 1c)(2b 3a)|(2c 4a))</td><td>N/A</td><td>There is a nutrient source applied directly to the HGMU. However, the amount of nutrients applied is too low to comprise a significant nutrient input.</td><td>Nut 2</td></tr>
<tr><td>(11b|11c) | (13b|13c) | (14b|14c) | (15b|15c) | (16b|16c)</td><td>(#1| #2| 3) (#4|#5|#6)</td><td>Although there is a nutrient source present in the neighbourhood of the HGMU, this nutrient source will not be received by the HGMU, because there is no groundwater discharge within or (sub)surface run-off into the HGMU.</td><td>Nut 2</td></tr>
<tr><td>11c 12c</td><td>1 2 #3 (#4|#5|#6)</td><td>Inorganic fertiliser (containing only P) is applied at least annually outside the area of the HGMU. It is unlikely that the P applied outside the HGMU reaches the HGMU via a groundwater flow. The high P adsorption capacity of the soil means that P will be bound to soil particles during its passage through the soil.</td><td>Nut 2</td></tr>
<tr><td>(17b|17c|17d|17e|18b|18c) (19a 20a 21a 22a)</td><td>N/A</td><td>Flooding or supply of surface water occurs in the HGMU, however, the nutrient concentration in this (flood) water is too low to comprise a significant nutrient input to the HGMU.</td><td>Nut 2</td></tr>
</table>

Note: The subscripts **d** or **i** refer respectively, to direct or indirect nutrient sources received by the HGMU in groundwater discharge, run-off, floodwater or managed water supplies (see above).

? refers to the lack of available quantitative information. See text on page 202.

not (significantly) receiving a nutrient input for the reasons described in the appropriate rationale statement(s), and a code **Nut 2** is assigned to the HGMU.

Note: This table works slightly differently from others in the procedures, so it is important to follow the instructions below very carefully:

Instructions for using Look-up Table 2
Start in the first column but instead of checking the answer combinations across each row of the look-up table in turn, first check all the rows of the first column: a combination of answers from *different* rows is possible and will lead to an assessment with outcome **Nut 2**. When this is the case you also have to combine the appropriate rationales. This is explained with the help of an example in which each row in the look-up table is set in a different type. The answers are given the same type as that of the row in the table where agreement was found. In consequence, the answer combination consists of different types indicating these rows and the subsequent rationale statements that apply.

Example answer combination:
H.1.ACTION

1a 5a 6a 7a 8a #9 10a **(11b|11c)** *13a 14a 15a 16a 17b* ***(19a 20a 21a 22a)***

H.2.ACTION

(#1|#2|3) (#4|#5|#6)

An HGMU with this answer combination will receive an assessment outcome **Nut 2**. The rationale will be a combination of those given in the look-up table as follows: The HGMU receives no direct nutrient inputs and **although inorganic fertiliser is applied somewhere in the surroundings of the HGMU, these nutrients do not reach the HGMU because there is no groundwater discharge or (subsurface) run-off to transport these nutrients**. ***Although there is contact with water from the main water body, nutrient concentrations in the water are too low to comprise a significant nutrient input***.

If you have assigned a code **Nut 2** to your HGMU you may wish to continue into the water quality functions (Nutrient Export or Nutrient Retention) assessment section in order to determine the *potential* of the HGMU to perform this function, though it is not actually being performed under current circumstances.

3. IT IS UNCERTAIN WHETHER A NUTRIENT INPUT IS BEING RECEIVED

If Look-up Tables 1 and 2 did not lead to an assessment, continue with Look-up Table 3. If any of these outputs are in agreement with your answers, then the HGMU may be receiving a nutrient input, but there is insufficient information to be certain.

You are advised not to continue to assess the HGMU for Nutrient Export or Nutrient Retention functions because the outcome will be that there is not enough information for assessing the process. This automatically leads to a nutrient input code **'2'** (a nutrient input is not (significantly) being received), although there is some chance that there may be a nutrient input.

If you are interested in the *potential* of the wetland environment to perform these functions and processes then proceed as instructed for the function of interest.

This completes the assessment of the nutrient inputs to the HGMU. This information may be useful directly to the wetland manager, but is used as important baseline information in the functional assessments in Section 3.

2.3.H.3 Look-up Table 3

<table>
<tr><th>H.1.ACTION</th><th>H.2.ACTION</th><th>Rationale</th><th>Code</th></tr>
<tr><td>1c 2a 3a 4d</td><td>N/A</td><td>The HGMU is in direct receipt of a nutrient supply from inorganic fertiliser (containing both N and P) at least annually.
However, the amount of N is too low to comprise a significant nitrogen input. The amount of P is unknown. It is not likely that this amount will be very high, when only a low amount of N is applied, but it is not certain.</td><td>Nut ?</td></tr>
<tr><td>1c 2a 3d 4a</td><td>N/A</td><td>The HGMU is in direct receipt of a nutrient supply from inorganic fertiliser (containing both N and P) at least annually. However, the amount of P is too low to comprise a significant phosphorus input. The amount of N is unknown. It is not likely that this amount will be very high, when only a low amount of P is applied, but it is not certain.</td><td>Nut ?</td></tr>
<tr><td>1b|5b|6b|7b|8b</td><td>N/A</td><td>The HGMU is not in direct receipt of a nutrient supply, but it has received a direct nutrient input within the last 20 years, or the HGMU is in receipt of direct nutrient inputs, but the nutrients are applied less than once a year. It is not certain whether this nutrient source (still) has an effect on the HGMU.</td><td>Nut ?</td></tr>
<tr><td>11b|13b|14b|15b|16b</td><td>(1 2 #3)|(4 5 6)</td><td>The HGMU does not receive an indirect nutrient supply in groundwater discharge or (sub)surface run-off, but it has received an indirect nutrient input within the last 20 years, or the HGMU is in receipt of an indirect nutrient input but the nutrients are applied less than once a year. It is not certain whether this nutrient source (still) has an effect on the HGMU.</td><td>Nut ?</td></tr>
<tr><td>(17e|18b|18c)
(19b|20b|21b|22b)</td><td>N/A</td><td>The HGMU is in receipt of nutrients because of contact with nutrient-rich surface water. However, contact takes place less than once a year. It is not certain whether this is frequent enough to be considered a significant nutrient input.</td><td>Nut ?</td></tr>
<tr><td>1d|5d|6d|7d|8d|10c|11d|13d|14d|15d|16d|17f|18e</td><td>N/A</td><td>It is unknown whether the HGMU is in receipt of nutrients, either directly or indirectly.</td><td>Nut ?</td></tr>
</table>

2.3.I Trace element input characterisation

This section is for the assessment of the Trace Element Export and Trace Element Retention functions (Sections 3.2.D and 3.2.E). Only answer the questions in this section if you wish to carry out an assessment for these functions.

The purpose of this section is to determine the trace element input into each HGMU. To enable the assessment of the water quality functions trace element storage and trace element export, the trace elements entering or likely to enter a HGMU must be characterised.

Trace element input is defined as an anthropogenically-derived source that results in trace elements being present above natural background concentration, and which is supplied to an HGMU by one or more of the following mechanisms:

- overbank flooding;
- direct and indirect application;
- groundwater discharge;
- run-off input.

For the purposes of functional assessment it is necessary to characterise both the *trace element source* and its ***mechanism(s) of supply*** (Fig. 2.3.20). This is achieved by identifying a potential trace element source, where it is being received within the HGMU, AA or CA (Action I.1) and the mechanism of supply for this potential source (Action I.2).

Trace element sources are identified in Action I.1. This section is divided into three parts:

- ***Part A:*** identifies trace element sources potentially received by the HGMU in floodwater.
- ***Part B:*** identifies trace element sources applied within the HGMU itself (direct trace element sources).
- ***Part C:*** identifies trace element sources applied outside of the HGMU, but within either the AA or the CA (indirect trace element sources) and potentially received by the HGMU.

When a trace element source originating from the area surrounding the HGMU is identified (i.e. positive answers in parts A and C above), you are also required to complete Action I.2 in order to identify the mechanism(s) of supply to the HGMU.

1. Answer consecutively the questions in Action I.1 for each HGMU. For *each* question answered *positively*, tick the appropriate box on **Recording Sheet 2d**. If a potential trace element source has been identified in Part A or C then go to Action I.2 and, unless otherwise instructed, answer all the questions there for each potential source identified.

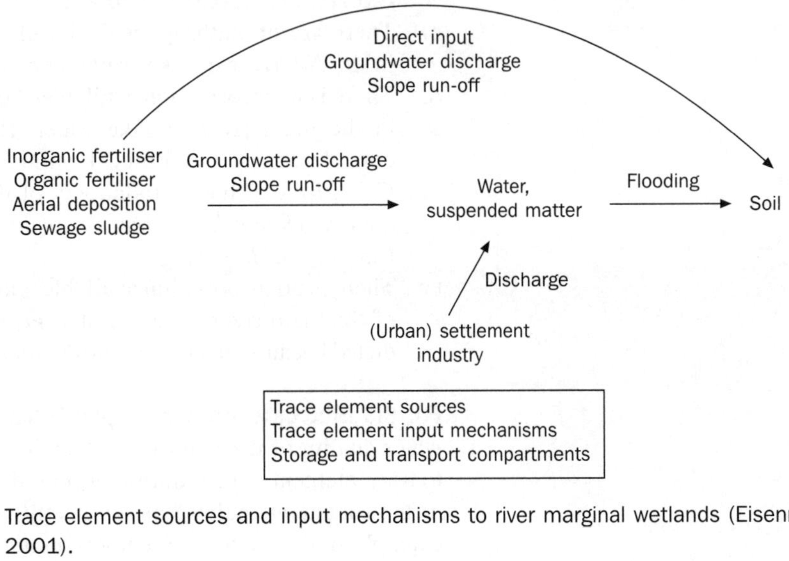

2.3.20 Trace element sources and input mechanisms to river marginal wetlands (Eisenmann, 2001).

2. Once all the questions in Action I.1 and Action I.2 have been answered for each HGMU being assessed, go to Action I.3 of the trace element input section and assign a trace element input code for each trace element input identified, according to instructions given there. Record this on **Recording Sheet 2e**.

Notes:

- The methodology for determining the trace element input code is very similar to that for functional assessment. It uses the same structure and syntax. Instructions and guidance on this and on using the look-up tables are given in the Introduction to Section 3. If a trace element input cannot be characterised because of insufficient or unreliable information, then the trace element storage and trace element export functions cannot be fully assessed. In these cases, only the potential for an HGMU to perform a function can be assessed and this is achieved by using the functional assessment section to evaluate the *potential* of the wetland to store or export trace elements.
- Cadmium (Cd) is used as an example of how to deal with other trace elements in the FAPs. Cadmium is commonly one of the most critical trace elements found in water and soils due to its high mobility and toxicity, but other elements such as arsenic and mercury are commonly found in wetlands. These should be assessed separately. See notes on trace element concentration on p. 104.

2.3.I.1 ACTION: Determine the presence of a potential trace element source

Record information: Recording Sheet 2d (*I.1. ACTION*)

When using maps to establish the presence of a potential trace element source, remember to overlay HGMU boundaries delineated on **AA Map 1** or **3** over the land use boundaries delineated in **AA Map 2**.

Part A: Trace element source in floodwater

Q1. What information is available about the water quality of the main water body (river or lake), especially with regard to trace element concentration or pollution that enables confirmation of any of the statements below?

a. There are anthropogenically elevated trace element concentrations in the AA river or lake water. *(Recording Sheet 1: C.3.20.14 (a))*
b. There is no information available about the trace element concentrations in the main river or lake water. Nevertheless the river or lake can be considered to be polluted by industrial inputs due to failure of EC Water Quality Directives. *(Recording Sheet 1: positive answers to C.3.20.1, C.3.20.2, C.3.20.3, C.3.20.4, C.3.20.5)*
c. There are no anthropogenically elevated trace element concentrations in the AA river or lake water. *(Recording Sheet 1: C.3.20.14 (b))*
d. There is no information available about the trace element concentrations in the main river or lake water. However, the river or lake can be considered as unpolluted by industrial inputs *(Recording Sheet 1: C.3.12b)* as it hasn't failed any of the EC Directives relating to water quality. *(Recording Sheet 1: negative answers to C.3.20.1, C.3.20.3, C.3.20.4, C.3.20.5)*
e. There is no information available about the trace element concentrations of the main river or lake water. Possible contamination cannot be predicted because of a lack of information. *(Recording Sheet 1: C.3.20.14 (c))*

For the Elbe case study example: What information is available about the water quality of the main water body (river or lake), especially with regard to trace element concentration or pollution that enables confirmation of any of the statements below? Note: quantities are given for the Elbe catchment example study site only. See notes on trace element concentration on p. 104.

a. There are anthropogenically elevated Cd concentrations in the AA river or lake water ($\geq$ 2.5 mg Cd kg^{-1} suspended matter). *(Recording Sheet 1: C.3.20.14 (a))*
b. There is no information available about the Cd concentrations in the main river or lake water. Nevertheless the river or lake can be considered to be polluted by industrial inputs due to failure of EC Water Quality Directives. *(Recording Sheet 1: positive answers to C.3.20.1, C.3.20.2, C.3.20.3, C.3.20.4, C.3.20.5)*
c. There are no anthropogenically elevated Cd concentrations in the AA river or lake water (< 2.5 mg Cd kg^{-1} suspended matter). *(Recording Sheet 1: C.3.20.14 (b))*
d. There is no information available about the Cd concentrations in the main river or lake water. However, the river or lake can be considered as unpolluted by industrial inputs *(Recording Sheet 1, C.3.12b)* as it hasn't failed any of the EC Directives relating to water quality. *(Recording Sheet 1: negative answers to C.3.20.1, C.3.20.3, C.3.20.4, C.3.20.5)*
e. There is no information available about the Cd concentrations of the main river or lake water. Possible contamination cannot be predicted because of a lack of information. *(Recording Sheet 1: C.3.20.14 (c))*

Note: these concentrations are specific to the Elbe catchment.

Part B: Direct trace element sources

Q2. How often is inorganic fertiliser (which can contain trace elements) applied within the HGMU*? (AA Map 2; Recording Sheet 1, C.3.2.1b)*
a. No history of application in the last 20 years
b. None currently applied but previously applied within the last 20 years
c. Currently applied once or more than once a year
d. Unknown

Q3. How often is organic industrial waste applied within the HGMU (including industrial organic waste; municipal waste; food processing waste)? *(AA Map 2; Recording Sheet 1, C.3.2.6b)*
a. No history of application in the last 20 years
b. None currently applied but previously applied within the last 20 years
c. Currently applied once or more than once a year
d. Unknown

Part C: Indirect trace element sources

Q4. How often is inorganic fertiliser (which can contain trace elements) applied within the AA*? (AA Map 2; Recording Sheet 1, C.3.2.1a)*
a. No history of application in the last 20 years
b. None currently applied but previously applied within the last 20 years
c. Currently applied once or more than once a year
d. Unknown

Q5. How often is organic industrial waste applied within the AA (including industrial organic waste; municipal waste; food processing waste)? *(AA Map 2; Recording Sheet 1, C.3.2.6a)*
a. No history of application in the last 20 years
b. None currently applied but previously applied within the last 20 years, or applied less than once a year
c. Currently applied once or more than once a year
d. Unknown

Q6. How often is inorganic fertiliser (which can contain trace elements) applied within the CA*? (CA Map 2; Recording Sheet 1, C.3.2.8)*
a. No history of application in the last 20 years
b. None currently applied but previously applied within the last 20 years
c. Currently applied once or more than once a year
d. Unknown

Q7. How often is organic industrial waste applied within the CA (including industrial organic waste; municipal waste; food processing waste)? *(CA Map 2; Recording Sheet 1, C.3.2.12)*
   a. No history of application in the last 20 years
   b. None currently applied but previously applied within the last 20 years, or applied less than once a year
   c. Currently applied once or more than once a year
   d. Unknown

2.3.I.2 *ACTION: Determine the mechanism of trace element supply for each trace element source identified outside the HGMU*

Record information: Recording Sheet 2d. *(I.2.ACTION)*

Part A: Overbank flooding

Only answer questions in Part A if you have identified an elevated trace element concentration in the main water body (river or lake) in Q1 in Action I.1 (positive response to a. b. or c.) and therefore the HGMU potentially could receive a trace element load originating from the main water body (river or lake). Otherwise go to Part B.

Q1. Is the hydrology of the HGMU characterised as having an overbank-flooding (**F4**) component (hydrological code: **H4**, **H7**, **H10**, **H12**, **H13**, **H14** or **H15**)? *(Recording Sheet 2a: F)*

Q.2. How often is the HGMU flooded with water from the main water body (river or lake)? (*Recording Sheet 1: C.3.19.8b*)
   a. Rarely (once every 10–50 years)
   b. Occasionally (once every 2–10 years)
   c. Frequently (once or more than once a year)
   d. Unknown

Only answer the questions in the following two sections (B and C) if you have identified trace element sources in Part C of Action I.1. Answer the questions for each source identified. Tick the correct answers to the questions in the boxes of **I.2.ACTION** on **Recording Sheet 2d**.

Part B: Groundwater discharge

To answer some of the following questions, it is necessary to determine whether any of the neighbouring, upslope (and therefore potentially influential) HGMUs has a trace element input. Before proceeding, please carry out the assessment to this point for all your upslope and adjacent HGMUs.

Q3. Is the hydrology of the HGMU characterised as having an **F1** groundwater discharge component (hydrological code **H1**, **H2**, **H3**, **H4**, **H11**, **H12**, **H14**, **H15**)? *(Recording Sheet 2a: F)*
[*If YES go to **Q4**, if NO go to Part C: Run-off or subsurface run-off input*]

Q4. Is the trace element source identified in Action I.1 applied to land within the AA or the CA upslope of the HGMU? *(AA Map 1; CA Map 1)*
[*If YES go to **Q5**, if NO go to Part C: Run-off or subsurface run-off input*]

Q5. If the trace element source is applied to an HGMU within the AA upslope of the HGMU being assessed, does this HGMU have a groundwater discharge component (hydrological code **H1**, **H2**, **H3**, **H4**, **H11**, **H12**, **H14**, **H15**)? *(Recording Sheet 2a: F; AA Map 1)*

[*Continue with Part C: Run-off or subsurface run-off input*]

Part C: Run-off or subsurface run-off input

Q6. Is the hydrology of the HGMU characterised as having a run-off input component (hydrological code **H2**, **H3**, **H4**, **H5**, **H6**, **H7**, **H13**, **H14**)? *(Recording Sheet 2a: F)*
[*If YES, go to **Q7**, if NO, return to start of Action I.2 and continue identification of mechanisms of supply of trace elements from other sources identified in Action I.1*]

Q7. If the trace element source identified is applied *either* elsewhere in the AA *or* in the CA, is one of the following true?

- It is applied to an area drained by either ditches, streams or subsurface artificial drains, which flow *uninterrupted* to the HGMU *(AA Map 2; CA Map 1)*
- The trace elements are applied to land within the AA or the CA that is upslope of the HGMU *(AA Map 1; CA Map 1)*

[*If YES, go to **Q8**, if NO, return to start of Action I.2 and continue identification of mechanisms of supply of trace elements from other sources identified in Action I.1*]

Q8. If the trace element source is applied to an HGMU within the AA upslope of the HGMU being assessed, does this HGMU have a run-off output component (hydrological code **H3**, **H4**, **H6**, **H8**, **H9**, **H11**, **H12**)? *(Recording Sheet 2a: F)*

[*Return to Action I.1 and continue identification of mechanisms of supply of trace elements from other sources identified in Action I.1*]

2.3.I.3 *ACTION: Characterisation of the trace element input*

Record information: Recording Sheet 2e (I.3.ACTION: trace element input code)

Compare the answers recorded for each HGMU on **Recording Sheet 2d**, Sections **I.1** and **I.2** with the possible outcomes below, and assign the appropriate trace element input code(s). Record the results obtained from the following assessment on the corresponding **Recording Sheet 2e**.

Look-up tables

1. IT IS HIGHLY LIKELY A TRACE ELEMENT INPUT IS BEING RECEIVED
If answers to the questions for this process are in agreement with those of any row in Look-up Table 1, then there is a high possibility that the HGMU is receiving a trace element input. The codes comprise two numbers based on the magnitude of a trace element input (first part of the code, e.g. '**1**/1') and the certainty with which it is occurring (second part of the code, e.g. 1/**1**). See Figure 2.3.21 for the different concentrations of trace element input sources based on the example of cadmium.

2. A TRACE ELEMENT INPUT IS NOT (significantly) BEING RECEIVED
If answers to the questions for this process are in agreement with those of any row in Look-up Table 2, then the HGMU may be in receipt of a trace element input, but to an insignificant degree.

3. THERE ARE CONSTRAINING FACTORS OR UNCERTAINTIES AS TO WHETHER A TRACE ELEMENT INPUT IS BEING RECEIVED
For any other combination of answers, apart from those in Look-up Table 3, it is not known if there is a trace element source to the HGMU. Assign code **Trae3** in these cases.

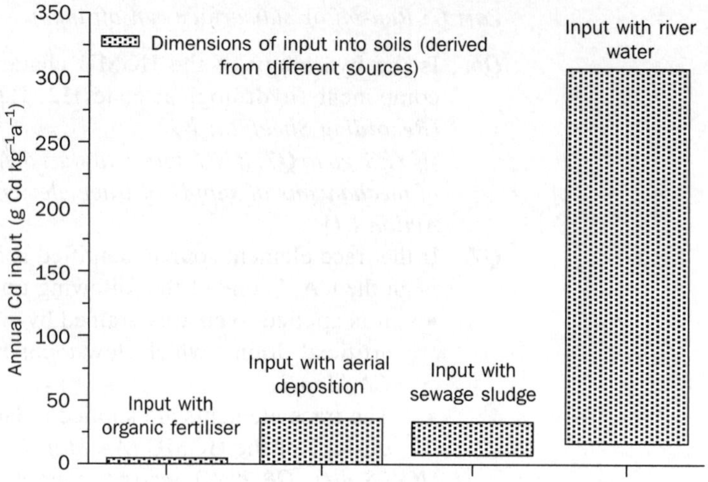

2.3.21 Differing concentrations of Cd input into wetland soils (according to Eisenmann, 2001).

4. A TRACE ELEMENT INPUT IS DEFINITELY NOT BEING RECEIVED

If answers to the questions for this process are in agreement with those of any row in Look-up Table 3, the HGMU is *definitely not* receiving a trace element input. You may wish to continue with the water quality functions assessment section in order to to determine the *potential* for an HGMU to perform this function, though it is not actually being performed under current circumstances.

2.3.I.3 Look-up Table 1

<table>
<tr><th>I.1.ACTION</th><th>I.2.ACTION</th><th>Rationale</th><th>Code</th></tr>
<tr><td>1a|1b</td><td>1 2c</td><td>The HGMU receives a trace element input from the main river or lake. The main river or lake has elevated concentrations of trace elements, and the HGMU is subject to regular flooding by the river or lake, i.e. once or more than once a year.</td><td>Trae1/1</td></tr>
<tr><td>2c|3c</td><td>N/A</td><td>The HGMU receives direct applications of trace elements in the form of inorganic fertilisers or organic industrial waste. Applications occur once or more than once a year.</td><td>Trae1/2</td></tr>
<tr><td>1a|1b</td><td>1 2b</td><td>The HGMU receives a trace element input from the main river or lake. The main river or lake has elevated concentrations of trace elements, and the HGMU is subject to occasional flooding by the river or lake, i.e. once every 2–10 years.</td><td>Trae1/3</td></tr>
<tr><td>4c|5c</td><td>(6 7)|(6 7 8)</td><td>Trace elements in the form of inorganic fertilisers or organic industrial waste are applied within the AA frequently, i.e. once or more than once a year, and it is likely that they are transported to the HGMU by surface run-off.</td><td>Trae1/4</td></tr>
<tr><td>4c|5c</td><td>3 4 #5</td><td>Trace elements in the form of inorganic fertilisers or organic industrial waste are applied within the AA frequently (i.e. once or more than once a year), and it is likely that they are transported to the HGMU by groundwater.</td><td>Trae1/5</td></tr>
<tr><td>6c|7c</td><td>(6 7)|(6 7 8)</td><td>Trace elements in the form of inorganic fertilisers or organic industrial waste are applied within the CA frequently (i.e. once or more than once a year), and it is likely that they are transported to the HGMU by surface run-off.</td><td>Trae1/6</td></tr>
<tr><td>6c|7c</td><td>3 4 #5</td><td>Trace elements in the form of inorganic fertilisers or organic industrial waste are applied within the CA frequently (i.e. once or more than once a year, and it is likely that they are transported to the HGMU by groundwater.</td><td>Trae1/7</td></tr>
</table>

2.3.I.3 Look-up Table 2

<table>
<tr><th>I.1.ACTION</th><th>I.2.ACTION</th><th>Rationale</th><th>Code</th></tr>
<tr><td>1a|1b</td><td>1 2a</td><td>The HGMU receives an insignificant trace element input from the main river or lake. The main river or lake has elevated concentrations of trace elements, but the HGMU is rarely subject to flooding by the river or lake, i.e. once every 11 or more years.</td><td>Trae2</td></tr>
<tr><td>2b|3b</td><td>N/A</td><td>The HGMU receives direct applications of trace elements in the form of inorganic fertilisers or organic industrial waste. However, the amounts received are likely to be small because applications are not frequent, but have occurred in the last 20 years.</td><td>Trae2</td></tr>
<tr><td>4b|5b|6b|7b</td><td>(6 7)|(6 7 8)</td><td>Trace elements in the form of inorganic fertilisers or organic industrial waste are applied within the AA or CA, and it is likely that they are transported to the HGMU by surface run-off. However, applications are infrequent, but have occurred within the past 20 years.</td><td>Trae2</td></tr>
<tr><td>4b|5b|6b|7b</td><td>3 4 #5</td><td>Trace elements in the form of inorganic fertilisers or organic industrial waste are applied within the AA or CA, and it is likely that they are transported to the HGMU by groundwater. However, applications are infrequent, but have occurred within the past 20 years.</td><td>Trae2</td></tr>
</table>

2.3.I.3 Look-up Table 3

<table>
<tr><th>I.1.ACTION</th><th>I.2.ACTION</th><th>Rationale</th><th>Code</th></tr>
<tr><td>1a | 1b</td><td>#1</td><td>The HGMU is not subject to surface water flooding and therefore does not receive any trace element inputs from the main river or lake.</td><td>Trae4</td></tr>
<tr><td>1c | 1d</td><td>*</td><td>There are no elevated concentrations of trace elements in the main river, so the HGMU cannot receive a trace element input from this source.</td><td>Trae4</td></tr>
<tr><td>2a 3a</td><td>N/A</td><td>There are no direct applications of trace elements to the HGMU, so the HGMU cannot receive an input from this source.</td><td>Trae4</td></tr>
<tr><td>4a 5a 6a 7a</td><td>N/A</td><td>There are no trace elements applied within the AA or CA, and so no trace elements can be transported to the HGMU from other areas of the AA or CA.</td><td>Trae4</td></tr>
<tr><td>4a | 5a | 6a | 7a</td><td>#6 | (6 #7) | (6 7 #8)</td><td>Trace elements in the form of inorganic fertilisers or organic industrial waste are applied within the AA or CA, but there is no mechanism of transport to the HGMU by surface run-off.</td><td>Trae4</td></tr>
<tr><td>4a | 5a | 6a | 7a</td><td>#(3 | 4) 5</td><td>Trace elements in the form of inorganic fertilisers or organic industrial waste are applied within the AA or CA, but there is no mechanism of transport to the HGMU by groundwater.</td><td>Trae4</td></tr>
</table>

Further reading and references

Böhm, E., Hillenbrand, T., Marscheider-Weidemann, F., Fuchs, S. and Scherer, U. (2000). *Input-Balancing of Prior Heavy Metals in Surface Waters (Bilanzierung des Eintrags prioritärer Schwermetalle in Gewässer)*, UBA Texte 29/01, Umweltbundesamt, Berlin, Germany.

Brandt, O., Geyer, S., Matheis, G. and Henle, K. (2000). Distribution and behaviour of trace elements in seepage water and groundwater of the River Mulde Marginal Wetland (Verteilung und Verhalten von Spurenelementen im Sicker- und Grundwasser der Muldeaue). In: Friese, K., Witter, B., Miehlich, G. and Rohe, M. (eds), *Stoffhaushalt von Auenökosystemen*, Springer Verlag, Berlin, Germany, 327–336.

Creaser C.S., Wood M.D., Alcock R., Copplestone D., Crook P and Barraclough D. (2007). *UKSHS Report No. 10: Environmental concentrations of polychlorinated dibenzo-p-dioxins and polychlorinated dibenzofurans (PCDDs/PCDFs) in UK soil and herbage*, Environment Agency, Bristol, available at http://publications.environment-agency.gov.uk/pdf/SCHO0607BMTD-e-e.pdf?lang=_e.

Eisenmann, V., Geyer, S., Henle, K., Rinklebe, J. and Scholz, M. (2001). *Binding Strength and Mobility of Trace Elements in River Marginal Wetland*, Unpublished report of EVALUWET project (see www.cordis.lu/en/home.html)

Ellenberg H., Weber H.E., Düll R., Wirth V., Werner W. and Paulissen D. (1992). Zeigerwerte von Pflanzen in Mittel-europa, *Scripta Geobotanica* **18**, 1–258.

Finck, A. (1992). *Fertiliser and Fertilisation (Dünger und Düngung)*, 2nd edn, VCH, Weinheim, Germany.

Fitter, R., Fitter, A. and Farrer, A. (1984). *Collins Guide to Grasses, Sedges, Rushes and Ferns*, Collins, London, UK.

Grime J.P. (2002). *Plant strategies, Vegetation processes and Ecosystem Properties.* Wiley, Chichester, UK.

Högbom, L., Nohrstedt, H.Ö. and Nordlund, S. (2001). Nitrogen Fertilization Effects on Stream Water Cadmium Concentration. *Journal of Environmental Quality* **30**, 189–193.

Hovmand, M.F. (1984). Cycling of Pb, Cd, Cu, Zn and Ni in Danish Agriculture. In: Berglund, S., Davis, D. and L'Hermite, P. (eds). *Utilization of Sewage Sludge on Land: Rates of Application and Long-Term Effects of Metals*, Proceedings of a Seminar held at Uppsala, June 7–9, 1983 (CEC), D. Reidel, Dordrecht, Boston, Lancaster, 166–185.

Hubbard, C.E. (1992). *Grasses*, Penguin Books, London, UK.

Ingwersen, J., Streck, T. Utermann J. and Richter, J. (2000). Ground water preservation by soil protection: determination of tolerable total Cd contents and Cd breakthrough times, *Journal of Plant Nutrition and Soil Science* **163**, 31–40.

Keddy, P.A. (2000). *Wetland Ecology. Principles and Conservation*, Cambridge Studies in Ecology, Cambridge University Press, Cambridge, UK.

Maltby, E., Hogan, D.V. and McInnes, R.J. (1996). *Functional Analysis of European Wetland Ecosystems – Phase 1 (FAEWE)*, European Commission, Ecosystems Research Report No. 18, Office for Official Publications of the European Communities, Luxembourg.

Matthess, G. (1984). Transport behaviour of pollutants in the soil (Transportverhalten von Schadstoffen im Boden), *Gewässerschutz, Wasser und Abwasser* **65**, 413–427.

Mitsch, W.J. and Gosselink, J.G. (2007). *Wetlands*, 4th edn, Wiley, New York.

Pyatt, D.G., Harrison, D. and Ford, A.S. (1969). Guide to the site types in forests of North- and Mid-Wales, *Forestry* **44**, 99–100.

Rose, F. and O'Reilly, C. (2006). *The Wild Flower Key. How to identify wild flowers, trees and shrubs in Britain and Ireland*, Frederick Warne, London, UK.

Sauerbeck, D. (1985). *Functions, Worth and Carrying Capacity of the Soil from Agrochemical View (Funktionen, Güte und Belastbarkeit des Bodens aus agrikulturchemischer Sicht)*, Materialien zur Umweltforschung, ed. by Council of experts on environmental questions authorized by the German Government (Rat von Sachverständigen für Umweltfragen), W. Kohlhammer, Stuttgart, Mainz, Germany.

Sauerbeck, D. and Rietz, E. (1981). On the cadmium contamination of mineral fertilisers in dependance of origin and production (Zur Cadmiumbelastung von Mineraldüngern in Abhängigkeit von Rohstoff und Herstellungsverfahren), *Landwirtsch. Forsch*, (Sonderheft 37), 685–696.

Sauerbeck, D. (1982). On the significance of cadmium in phosphate fertilisers (Zur Bedeutung des Cadmiums in Phosphatdüngemitteln), *Landbauforsch. Völkenrode* **32**, 192–197.

Schachtschabel, P., Blume, H.-P., Brümmer, G., Hartge, H.H. and Schwertmann, U. (1998). *Soil Science (Lehrbuch der Bodenkunde)*, 14th edn, F. Enke, Stuttgart, Germany.

SEPA (2000). *Environmental Quality Criteria for Lakes and Watercourses*, Report 4913, Swedish Environmental Protection Agency, Stockholm, Sweden.

Simpson, M. (2000). *A Functional Classification of European Wetland Eco-systems*, Unpublished PhD Thesis, Royal Holloway, University of London.

2.4 Recording sheets

The recording sheets should be used to record all information collected as part of the wetland database establishment as indicated in the field preparation, HGMU delineation and HGMU characterisation sections. The recording sheets, with the prepared maps, form the database that can be interrogated to determine the functional assessments.

The following recording sheets are included:

2.4.A Recording Sheet 1
2.4.B Recording Sheet 2a
Recording Sheet 2b
Recording Sheet 2c
Recording Sheet 2d
Recording Sheet 2e
2.4.C Recording Sheet 3
2.4.D Recording Sheet 4

2.4.A Recording Sheet 1

CLIMATIC INFORMATION													
C.3	**Date of field visit**												
C.3.1	**1i** Mean mon. prec.	J	F	M	A	M	J	J	A	S	O	N	D
	1ii Mean ann. prec.		**2** Prec. 4 Weeks Prior										
	3i Mean mon. evap.	J	F	M	A	M	J	J	A	S	O	N	D
	3ii Mean ann. evap.		**4** Mean wind speed	a	b	c							
	5 Mean soil temp	a	b	c	d	**6** Last spring frost mean date							
	7 First autumn frost		**8** Growing season										
	9 N conc. in atmos.	a	b	c	**10** Mean ann. temp. water	a	b	c					
	11 % of snow from precip.	a	b	c	**12** Arid or semi-arid region?	Y	N	U/NA					

LAND USE AND MANAGEMENT										
C.3.2	**1a** Inorg. fertiliser (AA)	a	b	c	d	**2a** Composition of fertiliser (AA)	a	b	c	d
	1b Inorg. fertiliser (HGMU)	a	b	c	d	**2b** Composition of fertiliser (HGMU)	a	b	c	d
	3ai Rate of N applied (AA)	a	b	c	d	**3aii** Rate of P applied (AA)	a	b	c	d
	3bi Rate of N applied (HGMU)	a	b	c	d	**3bii** Rate of P applied (HGMU)	a	b	c	d
	4a Org. fert. spray (AA)	a	b	c	d					
	4b Org. fert. spray (HGMU)	a	b	c	d					
	5a Org. inj. (AA)	a	b	c	d	**6a** Org. Ind. (AA)	a	b	c	
	5b Org. inj. (HGMU)	a	b	c	d	**6b** Org. Ind. HGMU	a	b	c	d
	7a Lime app. (AA)	a	b	c	d					
	7b Lime app. (HGMU)	a	b	c	d					
	8 Inorg. fertiliser (CA)	a	b	c	d	**9** Composition of fertiliser (CA) a	b	c	d	
	10 Org. spray (CA)	a	b	c	d	**11** Org. inj (CA)	a	b	c	d
	12 Org. ind (CA)	a	b	c	d	**13** Lime app (CA)	a	b	c	d
C.3.3	**1a** Burning frequency (AA)	a	b	c	d					
	1b Burning frequency (HGMU)	a	b	c	d					
C.3.4	**1a** Harvesting frequency (AA)	a	b	c	d					
	1b Harvesting frequency (HGMU)	a	b	c	d					
	2a Harvesting practices (AA)	a	b							
	2b Harvesting practices (HGMU)	a	b							
	3a Harvesting time and frequency (AA)	a	b	c	d	e				
	3b Harvesting time and frequency (HGMU)	a	b	c	d	e				
C.3.5	**1a** Ploughing frequency (AA)	a	b	c	d					
	1b Ploughing frequency (HGMU)	a	b	c	d					
	2 Ploughing frequency (CA)	a	b	c	d					
C.3.6	**1a** Forestry frequency (AA)	a	b	c	d					
	1b Forestry frequency (HGMU)	a	b	c	d					
C.3.7	**1a** Livestock density (AA)	a	b	c	d	**2a** Production (AA)	Y	N	U	
	1b Livestock density (HGMU)	a	b	c	d	**2b** Production (HGMU)	Y	N		
	3a Feed in rest/sleep area (AA)	Y	N	U	**5** Livestock density (CA)	a	b	c	d	
	3b Feed in rest/sleep area (HGMU)	Y	N	U/NA						
C.3.8	**1a** Large bird population (AA)	Y	N	U	**3a** Population density (AA)	a	b	c	d	e
	1b Large bird population (HGMU)	Y	N	U	**3b** Population density (HGMU)	a	b	c	d	e
C.3.9	**1** Surface drainage (AA)	Y	N	U	**3** Surface drainage (CA)	Y	N	U		
	5a Subsurface drainage (AA)	Y	N	U	**7** Subsurface drainage (CA)	Y	N	U		
	5b Subsurface drainage (HGMU)	Y	N	U						
C.3.10	**1a** Peat cutting and extraction (AA)	a	b	c	d					
	1b Peat cutting and extraction (HGMU)	a	b	c	d					
	2 Peat cutting and extraction (CA)	a	b	c	d					
C.3.11	**1a** Mining and extraction (AA)	a	b	c	d					
	1b Mining and extraction (HGMU)	a	b	c	d					
	2 Mining and extraction (CA)	a	b	c	d					
C.3.12	**1a** Frequency of inorg. waste app. (AA)	a	b	c	d					
	1b Frequency of inorg. waste app. (HGMU)	a	b	c	d					
	2 Frequency of inorg. waste app. (CA)	a	b	c	d					
C.3.13	**1a** Construction work (AA)	Y	N	U	**2** Construction work (CA)	Y	N	U		
	1b Construction work (HGMU)	Y	N	U						
C.3.14	**1** Extent of hunting, shooting, fishing	a	b	c	d					
C.3.15	**1** Extent of recreation (AA)	a	b	c	d					

CONSERVATION AND PROTECTED STATUS							
C.3.16	**1** Area of international conservation status	Y	N	U/NA			
	2 National conservation status	Y	N	U/NA			
	3 Local conservation status	Y	N	U/NA			
C.3.17	**1** Area surveyed for flora/fauna	Y	N	U/NA			
	2 Other flora/fauna records	Y	N	U/NA			
	3 Species of international importance	a	b	c	d	e	f
	4 Species of national importance	a	b	c	d	e	f
	5 Species of local importance	a	b	c	d	e	f
	6 Are the species no longer present ?	a	b	c	d	e	f
C.3.18	**1** Migratory invertebrates	Y	N	U/NA			
	2 Migratory fish	Y	N	U/NA			
	3 Migratory birds	Y	N	U/NA			
	4 Migratory mammals	Y	N	U/NA			

RIVER WATER INFORMATION							
C.3.19	**2a** Water level for frequent floods (< 2 yr)	m. relative to reference:	U/NA				
	2b Level for occasional floods (2–10 yr)	m. relative to reference:	U/NA				
	2c Level for rare floods (> 10 yr)	m. relative to reference:	U/NA				
	2d Water level for river bank full height	m. relative to reference:	U/NA				
	4a Discharge frequent floods (< 2 yr)	m^3 day^{-1}	U/NA				
	4b Discharge occasional floods (2–10 yr)	m^3 day^{-1}	U/NA				
	4c Discharge rare floods (> 10 yr)	m^3 day^{-1}	U/NA				
	5a Duration frequent floods (< 2 yr)	< 1 Week	1 W–1 M	> 1 Month	U/NA		
	5b Duration occasional floods (2–10 yr)	< 1 Week	1 W–1 M	> 1 Month	U/NA		
	5c Duration rare floods (> 10 yr)	< 1 Week	1 W–1 M	> 1 Month	U/NA		
	6 Months flooding occurs in 'normal' year						
	7a Main river or lake contact with AA	a	b	c	d	e	f
	7b Main river or lake contact with HGMU	a	b	c	d	e	f
	8a AA flooded with water from main river or lake	a	b	c	d		
	8b HGMU flooded with water from main river or lake	a	b	c	d		
C.3.20	**1** Fails EC freshwater fisheries directive	Y	N	U/NA			
	2 Fails EC dangerous substances directive	Y	N	U/NA			
	3 Fails EC urban waste water directive	Y	N	U/NA			
	4 Fails EC nitrate directive	Y	N	U/NA			
	5 Fails EC surface water abstraction directive	Y	N	U/NA			
	6 Mean nitrate–nitrogen con.	a	b	c			
	7 Mean ammonium–nitrogen	a	b	c			
	8 Mean phosphate–phosphorus	a	b	c			
	9 Mean total nitrogen	a	b	c			
	10 DOC conc. during base flow > 5 mg l^{-1}	Y	N	U/NA			
	11 DOC conc. during high flow > 5 mg l^{-1}	Y	N	U/NA			
	12i DOC conc. > 5 mg l^{-1} nearby watercourses	Y	N	U/NA			
	12ii DOC conc. < 5 mg l^{-1} nearby watercourses	Y	N	U/NA			
	13 POC conc. > 2 mg l^{-1}	a	b	c			
	14 CD concentration in river or lake water	a	b	c			

GROUNDWATER INFORMATION				
C.3.21	**6** Groundwater calcium level	Y	N	U/ NA
C.3.22	**1a** Mean groundwater level (AA)			U/NA
	1b Minimum groundwater level (AA)			U/NA
	1c Maximum groundwater level (AA)			U/NA
	2a Average surface water level (AA)			U/NA
	2b Minimum surface water level (AA)			U/NA
	2c Maximum surface water level (AA)			U/NA
	3 Top 20 cm soil water table fluctuations	Y	N	NA
	4a Maintained water level for conservation (AA)	Y	N	NA
	4b Maintained water level for conservation (HGMU)	Y	N	NA
C.3.23	**1** Fails EC Groundwater Directive	Y	N	NA

CATCHMENT LAND USE AND MANAGEMENT						
C.3.24	**1** Erosion features in upstream catchment	a	b	c	d	
	2 Eroding activities in upstream catchment	a	b	c	d	e
	3 Sediment reducing features upstream	a	b	c	d	
	4 Sediment transport enhancing features	a	b	c		
	5 Recent history of migrating channels	Y	N	NA		

TRACE ELEMENT INFORMATION						
C.3.25	**1a** Total soil trace element content (AA)	a	b	c	d	e
	1b Total soil trace element content (HGMU)	a	b	c	d	e
	2a Organic matter trace element content (AA)	a	b	c	d	e
	2b Organic matter content (HGMU)	a	b	c	d	e

2.4.B Recording Sheet 2a

C.2 Number	C.2 Drained	C.2 Landscape code	C.2 Gradient	C.2 Geomorphic code	F. Hydrol code	F1	F2	F3	F4

C.3 SOIL					
2. Topsoil moisture			6. Type of soil material		
3. Depth	4. Matrix colour	5. Mottle colour	7. Mineral particle size		8. Organic matter characteristics
0 cm			a	b	
			a	b	
			a	b	
			a	b	
			a	b	
25 cm			a	b	
			a	b	
			a	b	
			a	b	
			a	b	
50 cm			a	b	
General comments					
	4ii Munsell colour for mixed 30 cm of topsoil		7ii Average mineral particle size in Top 30 cm (for trace element functions)		

2.4.B Recording Sheet 2b
Nutrient input characterisation: Sections Action H.1 and Action H.2.

H.1.ACTION

PART A: DIRECT NUTRIENT SOURCES				
Q1	a	b	c	d
Q2	a	b	c	d
Q3	a	b	c	d
Q4	a	b	c	d
Q5	a	b	c	d
Q6	a	b	c	d
Q7	a	b	c	d
Q8	a	b	c	d
Q9	Y	N	U/NA	
Q10	a	b	c	

↓

PART B: INDIRECT NUTRIENT SOURCES				
Q11	a	b	c	d
Q12	a	b	c	d
Q13	a	b	c	d
Q14	a	b	c	d
Q15	a	b	c	d
Q16	a	b	c	d

↓

PART C: A NUTRIENT SOURCE REACHING THE HGMU VIA SURFACE WATER						
Q17	a	b	c	d	e	f
Q18	a	b	c	d		
Q19	a	b	c			
Q20	a	b	c			
Q21	a	b	c			
Q22	a	b	c			

H.2.ACTION

SECTION A: GROUNDWATER DISCHARGE			
Q1	Y	N	U/NA
Q2	Y	N	U/NA
Q3	Y	N	U/NA

↓

SECTION B: RUN-OFF AND SUBSURFACE RUN-OFF			
Q4	Y	N	U/NA
Q5	Y	N	U/NA
Q6	Y	N	U/NA

2.4.B Recording Sheet 2c

Nutrient input characterisation: Section H.3.ACTION → Assignment of a nutrient input code

Look-up Table 1: It is highly likely that a nutrient input is being received

	Direct input ($_d$ codes)					Indirect input ($_i$ codes)							
	Inorganic fertiliser	Organic fertiliser/ waste	Animal faeces	Bird faeces	Atmos. deposition	Discharge (inorg. fert.)	Run-off (inorg. fert.)	Discharge (org. fert.)	Run-off (org. fert.)	Discharge / Run-off (org. waste)	Discharge (faeces)	Run-off (faeces)	Surface water
Nitrogen codes: **(N)**													
Phosphorus codes: **(P)**													
N or P codes: **(Nut)**													

Look-up Table 2: A nutrient input is not significantly being received

Look-up Table 2: A nutrient input is not significantly being received

Look-up Table 3: It is uncertain whether a nutrient input is being received

Look-up Table 3: It is uncertain whether a nutrient input is being received

2.4.B Recording Sheet 2d
Trace element input characterisation: Sections I.1.ACTION and I.2.ACTION

I.1.ACTION

PART A: TRACE ELEMENT SOURCES IN FLOODWATER					
Q1	a	b	c	d	e

⇓

PART B: DIRECT TRACE ELEMENT SOURCES				
Q2	a	b	c	d
Q3	a	b	c	d

⇓

PART C: INDIRECT TRACE ELEMENT SOURCES				
Q4	a	b	c	d
Q5	a	b	c	d
Q6	a	b	c	d
Q7	a	b	c	d

I.2.ACTION

PART A: OVERBANK FLOODING				
Q1	Y	N	U/NA	
Q2	a	b	c	d

⇓

PART B: GROUNDWATER DISCHARGE			
Q3	Y	N	U/NA
Q4	Y	N	U/NA
Q5	Y	N	U/NA

2.4.B Recording Sheet 2e

Trace element input characterisation and mechanism of supply: Section I.1.ACTION → Characterisation of the trace element input, Section I.2.ACTION → Mechanism of trace element supply

Table 1: It is highly likely that a trace element input is being received

	Trace element source in floodwater	Direct trace element sources	Indirect trace element sources	
	Overbank flooding	Inorganic fertiliser and/or organic industrial waste	Groundwater discharge (inorg. fertil./org. ind. waste)	Run-off or subsurface run-off input (inorg. fertil./org. ind. waste)
codes				
codes				

Table 2: A trace element input is not significantly being received

	Trace element source in floodwater	Direct trace element sources	Indirect trace element sources	
	Overbank flooding	Inorganic fertiliser and/or organic industrial waste	Groundwater discharge (inorg. fertil./org. ind. waste)	Run-off or subsurface run-off input (inorg. fertil./org. ind. waste)
codes				

Table 3: It is uncertain whether a trace element input is being received

Table 3: It is uncertain whether a trace element input is being received

2.4.C Recording Sheet 3

A. Geomorphological indicators

A.1i Soil pH (10 cm)		**A.1ii** Soil pH (30 cm mixed)					
A.2 Surface Comp.	a	b	c	d	e	f	g
A.3 Microtopog.	1 negligible				2 significant		
A.4 Exposure	1	2	3	**A.5** Wind cropped trees	Y	N	

B. Hydrological indicators

B.1. Record HGMU Elevation				
B.1	**1** Ref. name:	**2** Ref. lev.:	**3** HGMU rel. lev.:	**4** HGMU abs. lev.:

B.2. Surface water flows						
B.2	**1** Flow into HGMU		**2** Flow from HGMU		**3** Watercourse from HGMU	
	4 Watercourse into HGMU		**5** Intermitt. from HGMU		**6** Perm. from HGMU	
	7 Drought flow		**8** Drought flow – river			

B.3 Springs and seepage				
B.3	**1** Active springs		**2** Waterlogged soil in drought	
	3 Standing water in frost		**4** Exceptionally snow-free	
	5 Concave profile		**6** Exceptionally lush vegetation	
	7 Mean surface water table depth		**8** Lowest surface water table depth (m)	m
	9 Highest surface water table depth (m)	m	**10** Mean groundwater depth (m)	m
	11 Lowest groundwater depth (m)	m	**12** Highest groundwater depth (m)	m
	13 Downslope of reservoir		**14** Brown or rust coloured standing water	
	15 Standing water covered by oily layer		**16** Soil flow rate to watercourse	a b c
	17 Soil flow rate traversing organic soils	a b c		

B.4. Inundation						
B.4.	**1** Inundation observed		**2** Never inundated in 5 years		**3** Permanently inund.	
	4 Overbank flooding		**5** Channel connections		**6** Flow velocity (m/s)	
	7 Water level after flood		**8** Local floatable matter		**9** Strand line	
	10 Old structures		**11** Flood marks		**12** Orchards	
	13 Water at footslope		**14** Embankments		**15** Other indicators	
	16 Mean river/lake level		**17** Mean freq. high level		**18** Mean occ. high level	
	19 Mean rare high level		**20** Rough vegetation		**21** Restricted outflow	
	22 Months waterlogging or flooding occur in year		**23** At the time of the site visit HGMU inundated?			

B.5. Erosion/sedimentation						
B.5.	**1** Erosion upsl. HGMU		**2** Erosion within HGMU		**3** Susp. sed. in river/lake	
	4 Recent dep. on surf.		**5** Recent dep. on veg.		**6** Susp. Sed. in runoff	
	7 Unvegetated		**8** Shifting course		**9** Bare or eroded land	

D. Vegetation / G. HGMU Area

Vegetation code(s) (with % cover)	Total Area m^2	Relative area %

C. Ecological indicators

C.1 Architectural diversity					
C.2 Litter Llayer	a	b	c	d	e
C.3 Reflects vegetation?	Y	N	U/NA		

C.4 Vegetation features	
a Canopy trees	
b Understorey trees	
c Shrubs or bushes	
d Tall herbaceous vegetation	
e Medium herbaceous vegetation	
f Low herbaceous vegetation	
g Dense ground vegetation	
h Sparse ground vegetation	
i Solitary trees or windbreaks	

C.5 Habitat features	
a Old, mature trees	
b Standing dead trees, little or no bark	
c Standing dead trees, with bark	
d Fallen dead trees, little or no bark	
e Fallen dead trees, with bark	
f Tree stumps	
g Tree wounds or sap runs	
h Rotten holes in trees	
i Hollow trees	
j Trees overhanging water	
k Gravel banks	
l Herbivore dung	
m Ant nest mounds	
n Grass tussocks	
o Wasp nests	
p Basking sites for reptiles (SE aspect)	
q Damp, shady areas	
r Abundant algae	
s River cliffs	
t Rocky outcrops	
u Derelict or uninhabited buildings	
v Caves, disused tunnels and mines	
w Old bridges, aqueducts or viaducts	
x Bat boxes	
y Dens, holts, lairs, domiciles	
z Animal tunnels or runways	
aa Beaver dams	
bb Boulders or rocks in the channel	
cc Rocks on lake shore	
ddi Deciduous vegetation that overhangs the watercourse	
ddii Deciduous vegetation that has the potential to overhang the watercourse	

2.4.D Recording Sheet 4 (copies of 2a)

C.2 Number	C.2 Drained	C.2 Landscape code	C.2 Gradient	C.2 Geomorphic code

F. Hydrol code	F1	F2	F3	F4

C.3 SOIL					
2. Topsoil moisture			6. Type of soil material		
3. Depth	4. Matrix colour	5. Mottle colour	7. Mineral particle size		8. Organic matter characteristics
0 cm			a	b	
			a	b	
			a	b	
			a	b	
			a	b	
25 cm			a	b	
			a	b	
			a	b	
			a	b	
			a	b	
50 cm			a	b	
General comments					
	4ii Munsell colour for mixed 30 cm of topsoil		7ii Average mineral particle size in Top 30 cm (for trace element functions)		

Section 3

Functional assessment

3.0 Introduction

Managers and decision makers can use functional assessment to determine what processes and functions (hydrological, biogeochemical and ecological) a wetland does or could perform and, where applicable, to what degree a function is being performed. The assessment can be carried out when all the information needed to examine a specific function or group of functions has been collected and recorded on Recording Sheets 1–4. The data collection methods are specified in Section 2.

The accompanying CD for this volume will generate a functional outcome automatically, provided that the 'function/process' entries required on the recording sheets and maps have been completed. If you do not wish to continue the assessment using the paper-based format, then follow the instructions on the CD. The computerised version requires an electronic version of the recording sheets and maps to be completed and will prompt you to do this.

Method

The functional assessment procedures (FAPs) provide a step-by-step technique for wetland assessment by requiring answers to a series of questions specified for each process or function. The answer to each question is recorded on an appraisal sheet, provided at the end of each process or function section. Once all the questions for a particular process or function have been completed, the answers are compared with options given in a series of look-up tables in which realistically possible combinations of answers have been set out, each giving an assessment outcome and an explanatory rationale statement. Examination of the rationale statements enables the user to question any apparent deficiencies in the assessment or assumptions made, and use the information to improve the accuracy of the procedures by feeding their insight back to the editors. When all the processes relevant to a particular function have been assessed, the outcomes are combined to give an overall assessment for that function. In detail, the following steps are required:

1. Read the notes at the top of the particular function or process you are assessing. These provide a description of the function or process, its significance and some guidance for conducting the assessment.
2. Answer the questions under each controlling variable (CV) *always* starting at the top (with ***CV1***, ***Q1***) and continuing in *strict numerical order* and without missing any, unless you are told otherwise.
3. The information needed to answer each question is to be found either on **Recording Sheets 1–4**, **AA Maps 1–3**, **CA Map 1** or in some cases from *assessments of other functions or processes* that may have been previously conducted or *assessments of near or neighbouring HGMUs* that may already have been assessed. References to data sources are always listed (in brackets) after the question.
4. For each question answered, circle the appropriate box on the corresponding appraisal sheet, which is found at the end of each individual process or function section.
5. Once all the questions have been answered for a particular process or function you can derive an assessment outcome using the look-up tables, which are found after the appraisal sheet. Combinations of answers that correspond to a particular assessment outcome are grouped in these tables together with an

assessment outcome code and rationale statement and, where appropriate, some indications of quantification. *The syntax for defining combinations of answers is described below.*

6. To find the correct assessment outcome, go to the first look-up table, which corresponds to a process or function *definitely being performed*. Go down the first column (***CV1***) of the chart looking for a combination of answers for this controlling variable that matches those of your completed appraisal sheet. Once you have found a match, go to the next column (***CV2***) and *continue from the point where you found the first **CV1** combination that matched*. Compare ***CV2*** combinations to your appraisal sheet until you find a match. Repeat this for all controlling variables in your process or function. Once you have found a match for all controlling variables, you will have determined the assessment outcome.
7. If you do not find a match in the first table (which corresponds to a process or function *definitely being performed*) go the next table (which corresponds to a process or function *not significantly being performed*) and repeat the procedure outlined above. If you still do not find a match, go the next table (which corresponds to a function or process *being performed but probably only to a small degree/there are constraining factors or uncertainties in quantification*). If you still do not find a match, your process or function is *definitely not being performed*, unless you are otherwise instructed.
8. Record the assessment outcome code in the appropriate box on your appraisal sheet.
9. Assessment of some functions requires combination of the assessments for a number of separate processes. Guidance on how to do this is given at the appropriate place.

The syntax for combinations of answers in look-up tables

The look-up table presents combinations of answers to the functional assessment questions that strongly indicate a certain outcome. These configurations are given using a logical syntax (derived from computer syntaxes) in order to avoid long and complicated sentences. The syntax is as follows:

space or (	= true if conditions before AND after space or 'opening bracket' apply
|	= true if conditions before OR after "|" apply
#	= true if the condition after "#" does NOT apply
()	= true if condition applies to GROUP OF ANSWERS between brackets
*	= true REGARDLESS; any condition in a category may apply
Any of	= true if ANY OF the following conditions applies
As above	= true for same condition stated in ABOVE BOX
–	= no conclusive information is available in this category

Example of the use of the syntax:

1 2|3 4	= true if the answers to questions Q1, Q2 and Q4 *or* Q1, Q3 and Q4 are 'Yes'
1 2|(3 4)	= true if the answers to questions Q1 and Q2 *or* Q1, Q3 and Q4 are 'Yes'
(1 2)|3|4	= true if the answers to questions Q1 and Q2 *and/or* Q3 *and/or* Q4 are 'Yes'
* # 3	= true for any combination of answers as long as the answer to Q3 is 'No'

An example of how to use the look-up tables

If we take the imaginary function of 'suitability of a water body for swimming' as an example, the questions to be answered for functional analysis may look as follows:

CV1 WATER CONDITIONS

Q1 Can the assessment area (AA) be described as a lake, pond or reservoir?
Q2 Has the AA depth more than 1.50 m?
Q3 Is the temperature of the water between 15 °C and 25 °C?

CV2 EVIDENCE OF SUITABILITY FOR SWIMMING

Q1 Is the water quality in the AA appropriate for swimming?
Q2 Is there an absence of dangerous wildlife (e.g. snakes, leeches)?
Q3 Is the substrate on the bottom of the AA sand?

CV3 EVIDENCE OF USE FOR SWIMMING

Q1 Has the AA ever been used for swimming?
Q2 Has the AA always been safe for swimmers?
Q3 Are there any facilities such as lifeguard post, changing rooms, toilets around the AA?

Look-up Table:
Example assessment chart:

CV1	CV2	CV3	Rationale	Code
1 2 3	1 2 3	1 2 3	All questions are answered with 'Yes' and it is certain that the AA provides suitable conditions for swimming.	**1/1**
1 2 3	1 2 3	*	All questions in **CV1** and **CV2** are answered with YES but the answers in **CV3** are unknown or uncertain. It is certain that the AA provides suitable conditions for swimming but there is no evidence of being used for swimming previously. It is therefore recommended that swimming occurs only with caution.	**1/2**
1 2 3	1(2|3)	*	All questions in **CV1** are answered with YES. The questions in **CV2** are answered as: "**Q1** is true and **Q2** or **Q3** is also true". It is therefore certain that the AA provides suitable conditions for swimming. However, there may be dangerous wildlife or an unsuitable substrate and so swimming only with extreme caution is recommended.	**1/3**
1 2 3	3	*	All questions in **CV1** are answered with YES; however, only question **Q3** in **CV2** is definitely true. It is therefore likely that the AA provides suitable conditions for swimming, as the morphology of the area is appropriate for swimming, but it might be dangerous owing to poor water quality or dangerous wildlife. Swimming only with extreme caution is recommended.	**1/x**
#1	*	*	The answer to **CV1**, **Q1** is negative. The AA is therefore not a water body appropriate for swimming. The answers to other questions are irrelevant in this context.	**4**
1 #(2|3)	*	*	The answer to **CV1**, **Q1** is positive but the answers to **Q2** and **Q3** are unknown or uncertain: the AA is a water body appropriate for swimming but is either not deep enough or does not have the correct temperature. The AA is therefore probably not suitable for swimming and the answers to the other questions are irrelevant.	**2**

Further reading and references

Adamus, P.R. (1983). *A Method for Wetland Functional Assessment, Volume II Critical Review and Evaluation Concepts*, FHWA Assessment Method, Rep. No. FHWA-IP-82-24, Federal Highway Administration, US Department of Transportation, Washington, DC.

Adamus, P.R. and Stockwell, L.T. (1983). *A Method for Wetland Functional Assessment: Volume I, Critical Review and Evaluation Concepts*, Rep. No. FHWA-IP-82-23, Federal Highway Administration, US Department of Transportation, Washington, DC.

Adamus, P.R., Stockwell, L.T., Clairain, E.J., Smith, R.D. and Young, R.E. (1987). *Wetland Evaluation Technique (WET), Volume II*, Operational Draft TRY-87, US Army Corps of Engineers, Waterways Experiment Station, Vicksburg, MS.

Adamus, P.R., Stockwell, L.T., Clairain, E.J., Morrow, M.E., Rozas, L.P. and Smith, R.D. (1991). *Wetland Evaluation Technique (WET), Volume I, Literature Review and Evaluation Rationale*, Technical Report WRP-DE-2, US Army Corps of Engineers, Waterways Experiment Station, Vicksburg, MS.

Boutin, C. and Keddy, P.A. (1993). A functional classification of wetland plants, *Journal of Vegetation Science* **4**, 591–600.

Brinson, M.M. (1993). *A Hydrogeomorphic Classification for Wetlands*, Wetlands Research Program Technical Report WRP-DE-4, US Army Corps of Engineers, Waterways Experiment Station, Vicksburg, MS.

Brinson, M.M., Hauer, F.R., Lee, L.C., Nutter, W.L., Rheinhardt, R.D., Smith, R.D. and Whigham, D. (1995). *A Guidebook for Application of Hygrogeomorphic Assessments to Riverine Wetlands*, US Army Corps of Engineers, Waterways Experiment Station, FHWA, Washington, DC.

Clapham, W.B. Jr (1973). *Natural Ecosystems*, Macmillan, New York.

Committee on Characterization of Wetlands (1995). *Wetlands, Characteristics and Boundaries*, National Academy Press, Washington, DC.

Cowardin, L.M. and Golet, F.C. (1995). US Fish and Wildlife Service wetland classification: A review, *Vegetatio* **118**, 139–152.

Cowardin, L.M., Carter, V., Golet, F.C. and LaRoe, E.T. (1979). *Classification of Wetlands and Deepwater Habitats of the United States*, US Fish & Wildlife Service Pub. FWS/OBS-79/31, Washington, DC.

Crabtree, A., Day, E., Gano, A. and Stevens, G. (1992). *Evaluation of Wetland Mitigation Measures*, US Office of the Natural Environment, Washington, DC.

Dansereau, P. (1959). Vascular aquatic plant communities of southern Quebec: A preliminary analysis, *Transactions of the Northeast Wildlife Conference* **10**, 25–54.

Doyle, A.F. (1986). The Charles River watershed: a dual approach to flood plain management. In: Kusler, J.A. and Riexinger, P. (eds) *Proc. National Wetland Assessment Symposium. Freshwater Wetlands, Ecological Processes and Management Potential*, Academic Press, New York, 38–45.

Dugan, P. (ed.) (1993). *Wetlands in Danger*, Oxford University Press, New York.

FHWA (1995). *The Federal Highway Administration as a Wetland Mitigation Bank User*, Federal Highway Administration, Washington, DC.

Fremlin, G. (ed.) (1974). *The National Atlas of Canada*, 4th edn, Macmillan, Toronto, Canada.

Goldsmith, F.B. and Harrison, C.M. (1976). Description and analysis of vegetation. In: Chapman, S.B. (ed.), *Methods in Plant Ecology*, Scientific Publishers, Oxford, 85–155.

Grime, J.P., Mason, G., Curtis, A., Rodman, J., Band, S.R., Mowforth, M.A.G., Neal, A.M. and Shaw, S. (1981). A comparative study of germination characteristics in a local flora. *Journal of Ecology* **69**, 1017–1059.

Hutchinson, G.E. (1975). *A Treatise on Limnology, Volume 3, Limnological Botany*, Wiley, New York.

Keddy, P.A. (2000). *Wetland Ecology: Principles & Conservation*, Cambridge University Press, Cambridge, UK.

Lane, P.A (1985). A food web approach to mutualism in lake communities. In: Boucher, D.H. (ed.), *The Biology of Mutualism: Ecology and Evolution*, Oxford University Press, New York, 344–374.

Larson, J.S., Adamus, P.R. and Clairain Jr, E.J. (1989). *Functional Assessment of Freshwater Wetlands: A Manual and Training Outline*, US Environmental Protection Agency, Corvallis, OR.

Lent, R.M., Weiskel, P.K., Lyford, F.P. and Armstrong, D.S. (1997). Hydrologic indices for non-tidal wetlands, *Wetlands* **17**, 19–30.

Loffler, H. and Malkhazova, S. (1990). Impacts of wetlands on man. In: Patten, B.C. (ed.) *Wetlands and Shallow Continental Water Bodies: Vol. 1 Nature and Human Relationships*, SOB Academic Publishing, The Hague, The Netherlands, 347–362.

Magee, D.W. (1998). *A Rapid Procedure for Assessing Wetland Functional Capacity*, Normandeau Associates, Bedford, NH.

Maltby, E., Hogan, D.V., Immirzi, C.P., Tellam, J.H. and Peijl, M.J. van der. (1994). Building a new approach to the investigation and assessment of wetland ecosystem functioning. In: Mitsch, W.J. (ed.), *Global Wetlands: Old World and New*, Elsevier, Amsterdam, The Netherlands, 637–658.

Maltby, E., Hogan, D.V. and McInnes, R.J. (1996). *Functional Analysis of European Wetlands Ecosystems – Phase 1 (FAEWE)*, European Commission, Ecosystems Research Report No. 18, European Commission, Brussels, Belgium.

Mitsch, W.J. and Gosselink, J.G. (2007). *Wetlands*, 4th edn, Wiley, New York.

Mitsch, W.J. and Wu, X. (1995). Wetlands and global change. In: Lal, R., Kimble, J., Levine, E. and Stewart, B.A. (eds.) *Advances in Soil Science, Soil Management and Greenhouse Effect*, Lewis Publishers, Boca Raton, FL, 205–230.

National Academy of Sciences National Research Council Wetlands Characterization Committee (1995). *Wetlands, Characteristics and Boundaries*, National Academy of Sciences Wetlands Task Force Report, Washington, DC.

Novitzki, R.P. (1982). *Hydrology of Wisconsin Wetlands*, Information Circular 40, Geological and Natural History Survey, University of Wisconsin, Madison, WI.

Pytlik, L.A., Quandt, L. and Grossman, R.B. (1995). *Some Simple Procedures for Field and Field-Laboratory Measurements with Emphasis on the Near Surface*, NRCS, NSSC, Lincoln, NE.

Raunkaier, C. (1937). *Plant life forms*, Translated by Gilbert-Carter, H., Clarendon Press, Oxford, UK.

Rigler, F.H. (1982). Recognition of the possible: an advantage of empiricism in ecology, *Canadian Journal of Fisheries and Aquatic Sciences* **39**, 1323–1331.

Smith, R.D., Amman, A., Bartoldus, C. and Brinson, M. (1995). *An Approach for Assessing Wetland Functions Using Hydrogeomorphic Classification, Reference Wetlands and Functional Indices*, US Army Corps of Engineers, Washington, DC.

Vitt, D.H., Yenhug, L. and Belland, R.J. (1995). Patterns of bryophyte diversity in peatlands of continental western Canada, *The Bryologist* **85**, 71–82.

Weinberg, G.M. (1975). *An Introduction to General Systems Thinking*, Wiley, New York.

Wellwe, M.W. (1981). *Freshwater Marshes, Ecology and Wildlife Management*, University of Minnesota Press, Minneapolis, MN.

Zhulidov, A.V., Headley, J.V., Roberts, R.D., Nikanorov, A.M. and Ischenko, A.A. (1997). *Atlas of Russian Wetlands*, Environment Canada, National Water Research Institute, Saskatoon, SK, Canada.

3.1

Hydrological functions

Introduction

Although their definition can be a cause for debate, wetlands are distinguished from other habitat types by their hydrology which, in turn, is a major factor determining their ecological and biogeochemical characteristics, such as the oxidation-reduction balance, nutrient availability and plant community composition. Indeed, Mitsch and Gosselink (2007) state that hydrology is the 'single most important determinant of the establishment and maintenance of specific types of wetlands and wetland processes'. Definitions of wetlands are discussed more fully elsewhere (Section 1.2) but, in general, they are characterised by the permanent or temporary presence of water above or close to the soil surface.

Wetland characteristics are maintained by periodic waterlogging, which is achieved when inputs of water exceed losses. This is known as a positive water balance. Wetlands can obtain their water from a number of sources:

- precipitation falling directly on the wetland;
- run-off from adjacent areas of land and inflow from streams and rivers;
- discharge of groundwater from an aquifer;
- flooding from adjacent surface waters;
- tidal inflow from the sea.

Wetlands can lose water by:

- evaporation;
- outflow to streams, rivers and the sea;
- drainage to adjacent surface waters;
- recharge of groundwater.

The relative importance of these water sources varies between wetland types, some examples of which are given in Fig. 3.1.1. The relatively slow flow through and storage of water in wetlands has a range of impacts on the hydrological cycle. Low velocities of flow mean that sediment particles and chemicals bound to them can settle in wetlands, and long residence times allow biogeochemical processes ample time to alter nutrient concentrations. In addition, water can infiltrate downwards, supplying underlying aquifers. Furthermore, the magnitude, duration and timing of flows in rivers receiving water from wetlands can be very different from those in areas where wetlands are absent. In many cases peak flows downstream are reduced and flows are sustained in dry periods. Wetlands are often, but not always, found in the transitional zones among terrestrial systems, groundwaters and surface waters. When located in such areas, they can act as buffers, altering the quantity and quality of water entering surface water or groundwater systems from adjacent land.

The role of wetlands in the hydrological cycle, and the importance of their hydrological functioning is reviewed by Bullock and Acreman (2003). Key results of their findings are given in Table 3.1.1. They conclude that many wetlands, particularly floodplains, can function to retain floodwaters, thereby reducing flood risk downstream. This flood attenuation property of floodplains is part of basic flood engineering knowledge (Natural Environment Research Council, 1975). In contrast, some wetlands, especially those in the headwater of catchments, may act as run-off source areas, and can generate floods. For example, the classic hydrological studies of Hewlett and Hibbert (1967) identified headwater wetlands

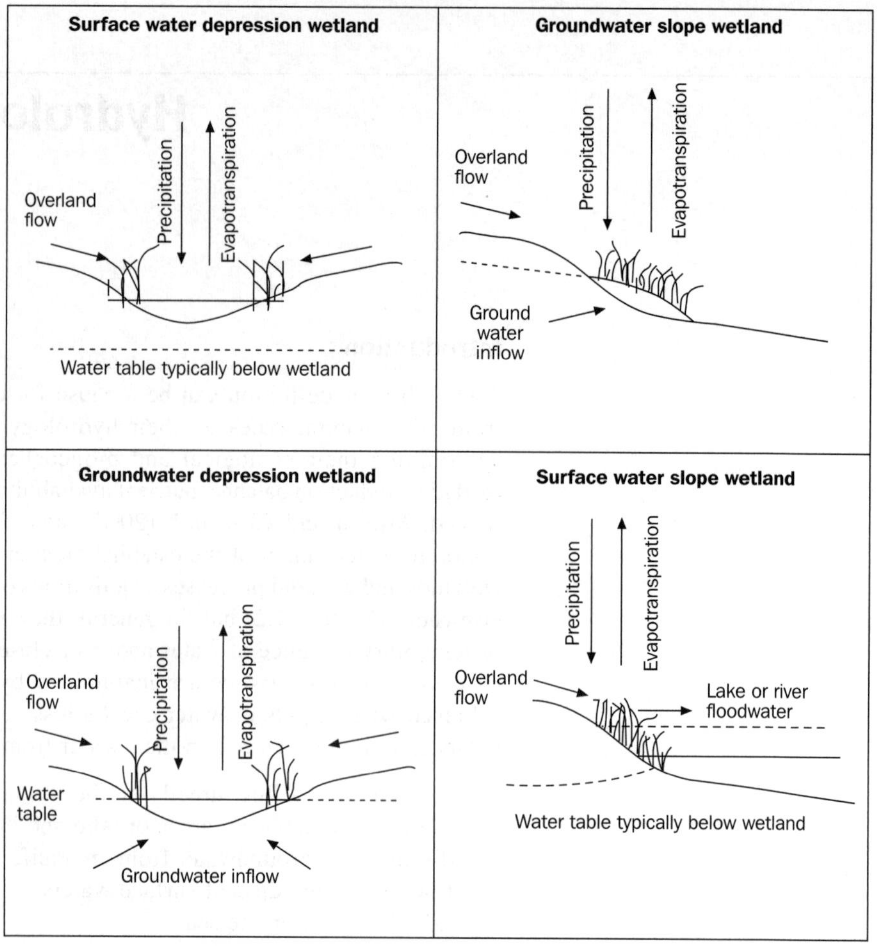

3.1.1 Water sources for various wetland types (modified from Novitzki, 1979).

Table 3.1.1 Key conclusions of Bullock and Acreman (2003)

Functioning	Role of wetlands in hydrological cycle
General impact on hydrology	81 % of statements conclude that wetlands significantly affect water cycle.
Floodwater detention	Most studies (82 %) show that floodplains reduce or delay floods. Certain headwater wetlands (45 %) reduce or delay flood peaks, though others (41 %) increase them.
Groundwater discharge	Many wetlands form where springs discharge groundwater.
Groundwater recharge	Certain wetlands (46 %) may contribute to recharge though, in other situations (26 %), recharge does not occur.

along river margins as flood generating areas. Burt (1995) concluded that '...most wetlands make very poor aquifers; ... accordingly, they yield little base flow, but in contrast, generate large quantities of flood run-off. Far from regulating river flow, wetlands usually provide a very flashy run-off regime'.

In addition, high evaporation rates from many wetlands can have impacts on water resources (Acreman *et al.*, 2003a). However, it is clear that wetlands have an important role in the hydrological cycle and need to be carefully considered in water management. The lack of uniformity underlies the importance of a procedural methodology to assess the hydrological functioning of active wetland components.

Floodwater detention

Some wetlands are subject to significant inputs of water from adjacent water bodies such as rivers and lakes. River marginal wetlands may be inundated during periods of high river discharge. Such flooding is a critical factor in the maintenance of floodplain ecosystems, as it facilitates the exchange of nutrients, carbon and species between the river and the floodplain. For example, floodwaters bring important supplies of phosphate to the floodplain, and inundation provides a spawning ground for some fish species. The importance of these processes is recognised in the flood pulse concept (Junk *et al.*, 1989). The frequency and duration of flood events depends on the hydrological regime of the river in question, and the geomorphological configuration of the valley bottom. In general, small low-order streams will have more numerous flood peaks of shorter duration than larger lowland rivers.

The long- or short-term storage of floodwaters in wetlands from run-off or overbank flooding can significantly delay and reduce peak flows (Fig. 3.1.2) and therefore reduce downstream flood damage. In Wisconsin, it has been suggested that catchments with 15 % wetlands and lakes have flood peaks 60–65 % lower than would be expected otherwise (Adamus and Stockwell, 1983). Acreman *et al.* (2003b) showed that removing floodplain storage on the River Cherwell, UK, by embanking the river, could increase flood peaks downstream by 50 %. The economic implications of this function are far-reaching and have been discussed above (Section 1.4). Summarising, floodplain wetlands reduce or delay floods, but headwater wetlands may increase peak flows (Bullock and Acreman, 2003).

Catchment hydrology can be altered by a number of human activities. The creation of dams to regulate flow, embankments or levees will disconnect the river from its floodplain. This will prevent wetlands there from retaining flood water and reducing peak flow, and will also have an impact on the biogeochemical and ecological functioning of wetlands, which is underpinned by periodic inundation.

Groundwater recharge and discharge

For a number of wetlands such as fens, upward or lateral movement of water from aquifers provides the chief water source, the mechanism for delivery depending on the type of aquifer. Indeed, in some wetlands, such as Langmere in the UK, wetland water levels correspond closely to groundwater levels in the underlying chalk aquifer (Acreman and José, 2000). Here, it is the presence of groundwater that generates waterlogged conditions and defines the wetland. The process of groundwater discharge to wetlands requires that the groundwater head is higher than the wetland water level and that any soil or rock layers between the two are permeable (Fig. 3.1.3).

Table 3.1.2 gives some examples of estimated groundwater interactions in selected wetlands. It should be noted that groundwaters may also be chemically

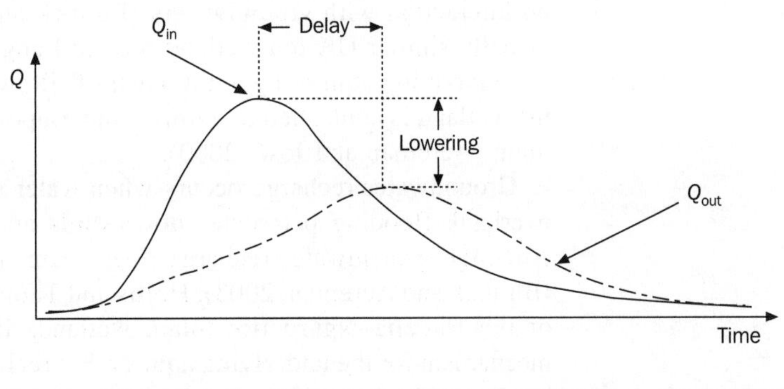

3.1.2 Delay and lowering of a flood peak while passing through a river/wetland system Q = discharge (i.e. volume of flow, e.g. m^3 per second); Q_{in} = inflow discharge; Q_{out} = outflow discharge (further explanation is given in Section 3.1.E.1).

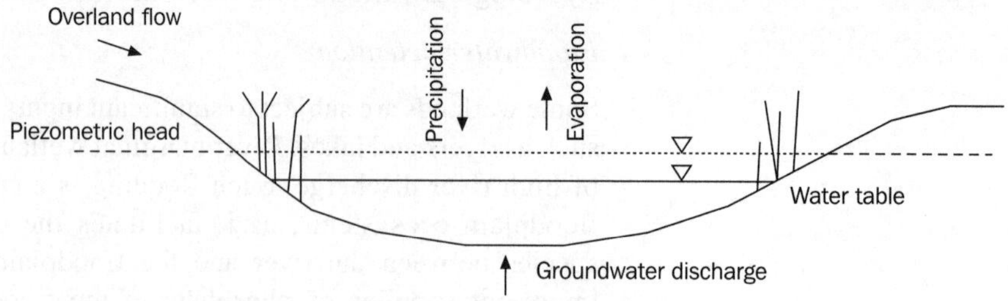

3.1.3 Groundwater discharge: when the groundwater piezometric head is above the wetland water level, groundwater is able to move upward into the wetland from the aquifer.

Table 3.1.2 Examples of groundwater/wetland interactions from a range of studies (from Gilvear and Bradley, in press)

Author	Location	Aquifer	Estimated groundwater contribution (%)	Estimated groundwater outflow (%)
Gehrels and Mulamootil (1990)	S. Ontario, USA	Glacial till	36	53
Koerselmann (1989)	Netherlands	Sand	45	4
Owen (1995)	Wisconsin, USA	Glacial till	1.5	4.3
Roulet (1990)	S. Ontario, USA	Moraine/till	95	0
Gilvear *et al.* (1993 and 1994)	East Anglia, England	Chalk	90	0
Gilvear *et al.* (1989)	East Anglia, England	Chalk Estuarine aquifers	80 12	< 1 < 1
Sacks *et al.* (1992)	S. Spain	Sands, silts and clay	30–40	5–10
Nuttle and Harvey (1995)	Virginia, USA	Estuarine clays	62	No information
Gerla and Matheney (1996)	N. Dakota, USA	Lacustrine sediments/till	19	No information

distinct from local surface waters, and that groundwater-fed wetlands may display distinctive biogeochemical and ecological characteristics that are different from surrounding areas (Wheeler and Shaw, 2001).

Although groundwater discharge is the primary source for many wetlands, it should be remembered that many wetlands form over impermeable rocks, permitting no interaction with groundwaters (Bullock and Acreman, 2003). At Fenmere, a visually similar UK fen wetland near to Langmere, the wetland is isolated from the underlying aquifer by a thin layer of silt, with the result that the hydrology of the wetland is controlled by rainfall and evaporation, with negligible groundwater input (Acreman and José, 2000).

Groundwater recharge occurs when water stored in wetlands from run-off or overbank flooding percolates downwards and replenishes underlying aquifers. Although groundwater recharge does occur, it does not appear to be the norm (Bullock and Acreman, 2003). Hollis and Thompson (1993) found that inundation of the Hadejia–Nguru floodplain wetlands in Nigeria was the main recharge mechanism for the underlying aquifer. For recharge to take place, the groundwater head must be lower than the wetland water level (Fig. 3.1.4). This can become particularly pronounced if there is abstraction of groundwater. For example, up until the 1970s, the Tables de Daimiel wetlands in central Spain were fed by

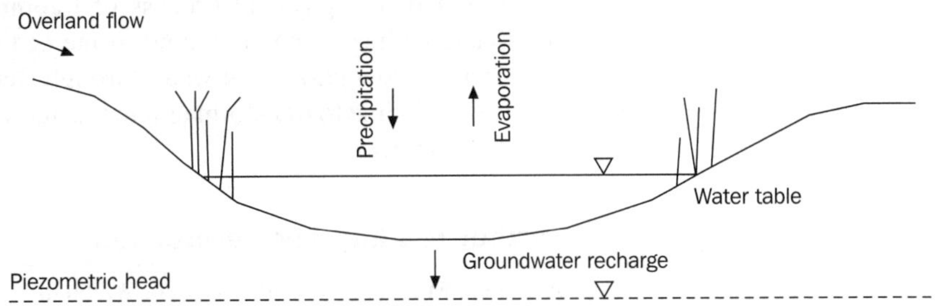

3.1.4 Groundwater recharge: when the groundwater piezometric head is below the wetland water level, groundwater is able to move downwards from the wetland into the aquifer.

groundwater discharging from the La Mancha aquifer. During the 1990s, when groundwater abstraction was high and rainfall low, the aquifer water table fell significantly and the wetland became a site of groundwater recharge (Bromley *et al.*, 2001) – see also Section 1.5, Fig. 1.5.1.

Changes to the wetland may also affect recharge rates. Lake Karla in Thessaly, Central Greece, was drained in 1963 and turned into cultivated land (Gerakis and Koutrakis, 1996). Although the depth of the lake varied considerably from year to year, it had contributed significantly to groundwater recharge. Following its drainage, there has been a serious reduction in groundwater levels and a number of other problems such as increased flooding. In the South of France, groundwater from the plain of La Crau feeds the coastal wetlands with nutrient-poor water of a constant temperature, which supports plant communities dominated by purple moor-grass (*Molinia caerulea*) and saw sedge (*Cladium mariscus*) (Skinner and Zalewski, 1995).

Sediment retention

Sediment dynamics are critical to wetland functioning. Not only can the movement of sediment be vital for the establishment and maintenance of some wetland systems, but sediments can also be important in the movement of nutrients and organic material. The retention of sediments within wetlands can have benefits to society, as excessive sediment in water can cause the blockage of navigation channels, and increase the treatment required for drinking water. In addition, it can cause riverbeds to change to a more silty condition, which may be detrimental to flora and fauna such as salmonid fish, which require clean gravels for spawning. This function is the net retention of sediments carried in suspension by water entering a wetland from overbank flooding or run-off.

The impact of wetlands on sediment loads can be dramatic. Novitzki (1979) demonstrated that, in Wisconsin river basins with 40 % wetland and lake area, sediment load was 90 % lower than basins with no lakes or wetlands. Furthermore, Kadlec and Hey (1994) showed that, for a range of wetlands, between 20 % and 90 % of inflowing sediments were retained.

Sediments can be derived from within a wetland (autochthonous sources) or from external (allochthonous) sources, and may be of mineral or organic origin. The relative importance of these sources varies from site to site, but is likely to reflect the hydrological regime. For example, floodplains are likely to receive a significant proportion of sediment from floodwaters, while other wetlands may receive most from run-off. Although sediment movements occur naturally within a catchment, modern agricultural and other practices have led to an increase in the amount of sediment mobilised.

The deposition of sediment in wetlands results from a reduction in the velocity of water movement. There are a number of characteristics of wetlands which make them suitable for deposition to occur. Some wetlands are typically broad and flat, and well suited to the retention of water and deposition of sediments. The presence of rough vegetation in a wetland is also an important factor in reducing flow rates and allowing settlement of particles.

As part of the physical process of trapping sediment, any material, such as phosphate or heavy metals bound to the sediment particles, is also captured. In addition the slow passage of water through wetlands allows many biogeochemical processes the time to transform or remove nutrients, thereby improving water quality (see Section 3.2).

Further reading and references

Acreman, M.C. and José, P.V. (2000). Wetlands. In: Acreman, M.C. (ed.). *The Hydrology of the UK*, Routledge, London, 204–224.

Acreman, M.C., Harding, R.J., Lloyd, C.R. and McNeil, D.D. (2003a). Evaporation characteristics of wetlands; experience from a wet grassland and a reedbed using eddy correlation measurements, *Hydrology and Earth System Science* **7** (1), 11–22.

Acreman, M.C., Booker, D.J. and Riddington, R. (2003b). Hydrological impacts of floodplain restoration: a case study of the river Cherwell, UK, *Hydrology and Earth System Sciences*. **7** (1), 75–86.

Adamus, P.R. and Stockwell, L.T. (1983). *A Method for Wetland Functional Assessment, Volume I, Critical Review and Evaluation Concepts*, Report No. FHWA-IP-82-23, Federal Highway Administration, US Department of Transportation, Washington, DC.

Bromley, J., Cruces, J., Acreman, M.C., Martinez, L. and Llamas, M.R. (2001). Problems of sustainable groundwater management in an area of over-exploitation: the Upper Guadiana catchment, central Spain, *International Journal of Water Resources Development* **17** (3), 379–396.

Bullock, A. and Acreman, M. (2003). The role of wetlands in the hydrological cycle, *Hydrology and Earth System Sciences* **7** (3), 358–389.

Burt, T. P. (1995). The role of wetlands in runoff generation from headwater catchments. In: Hughes, J.M.R. and Heathwaite, A.L. (eds). *Hydrology and Hydrochemistry of British Wetlands*, Wiley, Chichester, UK, 21–38.

Gehrels, J. and Mulamootil, M. (1990). Hydrological processes in a Southern Ontario wetland, *Hydrobiologia* **208**, 221–234.

Gerakis, P.A. and Koutrakis, E.T. (eds) (1996). *Greek Wetlands*, Commercial Bank of Greece, Athens.

Gerla, P.J. and Matheney, R.K. (1996). Seasonal variability and simulation of groundwater flow in a prairie wetland, *Hydrological Processes* **10**, 903–920.

Gilvear, D.J. and Bradley, C. (in press) Hydrological dynamics II: groundwater and hydrological connectivity. In: Maltby, E. and Barker, T. (eds), *The Wetlands Handbook*, Wiley-Blackwell, Oxford, UK.

Gilvear, D.J., Tellam, J.H., Lloyd, J.W. and Lerner, D.N. (1989). *The Hydrodynamics of East Anglian fen systems*, Final report to the Nature Conservancy Council, Anglian Water and Broads Authority, University of Birmingham Report.

Gilvear, D.J., Tellam, J.H., Lloyd, J.W. and Lerner, D.N. (1993). Quantification of the water balance and hydrological processes in the vicinity of a small groundwater-fed fen wetland, East Anglia, England, *Journal of Hydrology* **104**, 311–344.

Gilvear, D.J., Tellam, J.H., Lloyd, J.W. and Lerner, D.N. (1994). Wetland vulnerability in East Anglia: The range of validity of a generalised classification system, *Aquatic Conservation; Marine and Freshwater Ecosystems* **4**, 107–127.

Hewlett, J.D. and Hibbert A.R. (1967). Factors affecting the response of small watersheds to precipitation in humid areas. In: Sopper, W.E. and Lull, W.H. (eds), *International Symposium on Forest Hydrology*, Pergamon, New York, 275–290.

Hollis, G.E. and Thompson, J.R. (1993). Hydrological model of the floodplain. In: Hollis, G.E., Adams, W.M. and Aminu-Kanu, M. (eds) *The Hadejia–Nguru Wetlands: Environment, Economy and Sustainable Development of a Sahelian Floodplain Wetland*, IUCN, Switzerland and Cambridge, UK, 69–80.

Hughes, J.M.R. and Heathwaite, A.L. (1995). Introduction. In: Hughes, J.M.R. and Heathwaite, A.L. (eds) *Hydrology and Hydrochemistry of British Wetlands*, Wiley, Chichester, UK.

Junk, W.J., Bayley, P.B. and Sparks, R.E. (1989). The flood pulse concept in river-floodplain systems, *Canadian Journal of Fisheries and Aquatic Science* **106**, 110–127.

Kadlec, R.H. and Hey, D.L. (1994). Constructed wetlands for river water quality improvement, *Water Science and Technology* **29** (4), 159–168.

Koerselmann, W. (1989). Groundwater and surface water hydrology of a small groundwater-fed fen, *Wetland Ecology and Management* **1**, 31–43.

Mitsch, W.J. and Gosselink, J.G. (2007). *Wetlands*, 4th edn, Wiley, New York.

Natural Environment Research Council (1975). *Flood Studies Report. Volume 2 Flood Routing Studies*, Institute of Hydrology, Wallingford, UK.

Novitzki, R.P. (1979). The hydrologic characteristics of Wisconsin wetlands and their influence on floods, streamflow and sediment. In: Greeson, P.E., Clark, J.R. and Clark, J.E. (eds) *Wetland Functions and Values: the State of our Understanding*, American Water Resources Association, Minneapolis, MN, 377–388.

Nuttle, W.K. and Harvey, J.W. (1995). Fluxes of water and solute in a coastal wetland sediment. (I) the contribution of regional groundwater discharge, *Journal of Hydrology* **164**, 89–107.

Owen, C.R. (1995). Water budget and flow patterns in an urban wetland, *Journal of Hydrology* **169**, 171–187.

Roulet, N.T. (1990). Hydrology of a headwater basin wetland: groundwater discharge and wetland maintenance, *Hydrological Processes* **4**, 387–400.

Sacks, L.A., Herman, J.S., Konikow, L.F. and Vela, A.L. (1992). Seasonal dynamics of groundwater-lake interactions at Doñana National Park, Spain, *Journal of Hydrology* **136**, 123–154.

Salathé, T. (2000). The role of wetlands in maintaining 'good water status', *Implementing the EU Water Framework Directive*, Proceedings Seminar 2: The Role of Wetlands in River Basin Management. A seminar series on water, WWF, Brussels, Belgium.

Skinner, J. and Zalewski, S. (1995). *Functions and values of Mediterranean Wetlands*, Conservation of Mediterranean Wetlands Series, MedWet, Camarogue, France.

Wheeler, B.D. and Shaw, S.C. (2001). *A Wetland Framework for Impact Assessment at Statutory Sites in Eastern England*, Environment Agency R and D Report W6-068/TR1 and TR2, WRC, Medmenham, UK.

Hydrological functions assessed in these procedures

3.1.A Floodwater detention (function)

Further reading and references

Bendjoudi, H., Weng, P., Guerin, R. and Pastre, J.F. (2002) Riparian wetlands of the middle reach of the Seine river (France): historical development, investigation and present hydrologic functioning. A case study, *Journal of Hydrology* **263**, 131–155.

Bullock, A. and Acreman, M. (2003). The role of wetlands in the hydrological cycle, *Hydrology and Earth System Sciences* **7** (3), 358–389.

Carter, V. (1986). An overview of the hydrologic concerns related to wetlands in the United States, *Canadian Journal of Botany* **64**, 364–374.

Carter, V., Bedinger, M.S., Novitzki, R.P. and Wilen, W.O. (1979). Water resources and wetlands. In: Greeson, P.E., Clark, J.R. and Clark, J.E. (eds), *Wetland Functions and Values: the State of Our Understanding*, American Water Resources Association, Minneapolis, MN, 344–376.

Cole, A.C., Brooks, R.P. and Wardrop, D.H. (1997). Wetland hydrology as a function of hydrogeomorphic subclass, *Wetlands* **17**, 456–467.

Gilman, K. (1994). *Hydrology and Wetland Conservation*, Wiley, Chichester, UK.

Gosselink, J.G. and Turner, R.E. (1978). The role of hydrology in freshwater ecosystems. In: Good, R.E., Whigham, D.F. and Simpson, R.L. (eds), *Freshwater Wetlands, Ecological Processes and Management Potential*, Academic Press, New York, 249–287.

Hughes, J.M.R. and Heathwaite, A.L. (eds) (1996). *Hydrology and Hydrochemistry of British Wetlands*, Wiley, Chichester, UK.
Junk, W.J., Bayley, P.B. and Sparks, R.E. (1989). The flood pulse concept in river-floodplain systems, *Canadian Journal of Fisheries and Aquatic Science* **106**, 110–127.
Mitsch, W.J. and Gosselink, J.G. (2007). *Wetlands*, 4th edn, Wiley, New York.
Novitzki, R.P. (1979). The hydrologic characteristics of Wisconsin wetlands and their influence on floods, streamflow and sediment. In: Greeson, P.E., Clark, J.R. and Clark, J.E. (eds), *Wetland Functions and Values: the State of Our Understanding*, American Water Resources Association, Minneapolis, MN, 377–388.
Novitzki, R.P. (1985). The effects of lakes and wetlands on flood flows and base flows in selected northern and eastern states, *Proc. Conference Wetlands of the Chesapeake*, Easton, MD, Environmental Law Institute, Washington, DC, 143–154.

Introduction

Function description:

- Short-term (less than one week) storage of floodwater from overbank flooding or slope run-off.
- Long-term (more than one week) storage of floodwater.
- Detention of surface overland flow from surrounding slopes.

Significance – rationale for assessment: Marginal wetlands are often found on floodplains, which are usually capable of detaining water owing to their geomorphological configuration. They can thus decrease overland flow and mitigate against flood damage. Floodwater itself can be important for the provision of wildlife habitat and as a component in other functions such as sediment and nutrient retention.

- Storage of water may alleviate or prevent downstream flooding problems through the attenuation, and sometimes de-synchronisation, of flood peaks.
- Inundation provides a habitat for wildlife and may produce other ecologically favourable conditions.
- Flooding by river water can enable sediments, nutrients and pollutants, carried in the river, to be detained by the wetland.

Related functions, for which the outcome of this assessment may be required:

- Sediment retention
- Groundwater recharge
- Nutrient retention

For the Floodwater detention function to occur the following conditions must be met:

1. The level of the main river or lake must have significantly fluctuating water levels or, in some cases, adjoining higher areas of land must produce significant amounts of overland flow.
2. The HGMU must connect with the main river or lake during floods. Alternatively unimpeded overland flow from the HGMU to the channel must be possible.
3. Floods affecting the HGMU must have the potential to damage (economically or ecologically) the area downstream of the wetland, or provide hydrologically favourable conditions, or have high suspended-sediment loads.

If the HGMU does not meet any of the above requirements, or if the HGMU is of type **H1**, **H8**, **H11** or **H16** (*Recording Sheet 2a, F*), and therefore does not receive inputs from either run-off or overbank flooding, the function is unlikely to occur. In these cases, the functional assessment code for these HGMUs will be **4** and you need not carry out this assessment.

In all other cases, answer the following questions for each controlling variable. Complete all questions for which sufficient information is available. Circle the correct answer on the appraisal sheet. Use a new appraisal sheet for each HGMU. The location of the relevant information is given in brackets for each question.

Most wetlands, and certainly most marginal wetlands, meet these conditions and the Floodwater detention function is therefore commonly performed to some extent. The occurrence of hydrological inputs by overbank flooding or run-off has been determined and coded (*Recording Sheet 2a, F*) as follows:

- HGMUs likely to experience overbank flooding: **hydrological codes H4**, **H7**, **H9**, **H10**, **H12**, **H13**, **H14** or **H15**
- HGMUs that probably receive run-off inputs: **hydrological codes H2**, **H3**, **H4**, **H5**, **H6**, **H7**, **H13** or **H14**.

This assessment of flood water detention focuses on 'flood control' aspects since, though most wetlands will attenuate floods to a some degree, the amount of water stored in smaller areas is often negligible compared with the excess water generated during floods. The flood control function will therefore often be insignificant at the HGMU level. Although the function may be assessed, the outcome may not be very useful without qualification of its significance. To assess the significance of flood control, the following information must be available:

- Daily flood discharge in the river or lake *(Recording Sheet 1, C.3.19)*
- River or lake level near HGMU, relative to reference level *(Recording Sheet 3, B.4.17, B.4.18)*
- Average surface level in the HGMU *(Recording Sheet 3, B.1)*
- HGMU area *(Recording Sheet 3, G)*

The outcome of the assessment is a statement on the likelihood of occurrence of the function, and to some extent of its significance.

Controlling variables

CV1 Geomorphological evidence of flooding potential
CV2 Physical evidence of flooding
CV3 Historical evidence of flooding potential
CV4 Vegetation/habitat evidence of flooding
CV5 Other evidence for flooding

CV1 *Geomorphological evidence of flooding potential*

Q1. Is the HGMU classed as one of the following units with an 'E' prefix within the 'floodplain' component of the assessment area: a levee, terrace unit, or river or lake dune? *(Geomorphic codes **LV**, **TC or RD** – Recording Sheet 2a, C.2)*

Q2. Is the HGMU classed as one of the following units with an 'I' or 'D' prefix within the 'floodplain' component of the assessment area: flat area, backswamp, cut-off channel, ox-bow lake, former channel, point bar or tributary? *(Geomorphic codes **FA**, **BS**, **CO**, **OX**, **FC**, **PB** or **TB** – Recording Sheet 2a, C.2)*

CV2 *Physical evidence of flooding*

Go to ***CV3*** if no information is available.

Q1. Was the rainfall in the 4 weeks prior to the site visit approximately normal for the time of year (± 50 %)? *(Recording Sheet 1, C.3.1.1i; C.3.1.2)*

Q2. Have any of the following situations been observed in the HGMU?

- Both the HGMU and connections to the water body inundated. *(Recording Sheet 3, B.4.1, B.4.5)*
- River or lake water flowing into the HGMU as overbank flooding or through channels *(Recording Sheet 3, B.4.4 or B.4.5)*
- The HGMU inundated and the floodwater moving with a velocity greater than 0.1 m s^{-1} *(Recording Sheet 3, B.4.1, B.4.6)*

- Evidence of recently deposited mineral or organic material on the soil surface *(Recording Sheet 3, B.5.4)*
- Evidence of recently deposited mineral material on plant surfaces *(Recording Sheet 3, B.5.5)*
- A strand line observed in the HGMU, or the HGMU positioned between a strand line and the main river channel or lake *(Recording Sheet 3, B.4.9)*
- At the time of the site visit, was the HGMU inundated by a water source other than river or lake (with no flooded connections to the river channel or lake, with no inflow from the channel or lake, and no flow movement > 0.1 m s^{-1} within the HGMU)? *(Recording Sheet 3, B.4.1, B.4.4,* B.4.5, B.4.6, B.4.23*)*

CV3 *Historical evidence of flooding potential*

Go to ***CV4*** if no information is available. Perform this action only if data on flood discharges are available from Recording Sheet 1, C.3.19.

If possible, determine flood storage volume of the river or lake (*flood depth* × *HGMU area*, in m^3). The flood depth must be expressed in metres over the HGMU surface, and can be calculated from the flood level and the average surface level in the HGMU. Flood level and surface level must be relative to the same reference level, which can be m.O.D. or any local level. Flood levels for different flood frequencies are given in *Recording Sheet 1, C.3.19.2.* The HGMU surface level and the reference level are found on *Recording Sheet 3, B.1.*

Determine flood storage volume for:

- frequent (at least once every 2 years) flood events;
- occasional (once every 2–10 years) flood events;
- rare (once every 10–50 years) flood events;
- bankfull height – to the top of the levee.

Determine the flood storage volume as a fraction of the daily flood discharges as given in *Recording Sheet 1, C.3.19.4.*

Q1. Does the HGMU flood only rarely (less than once in 10 years)? *(Recording Sheet 1, C.3.19.2c; Recording Sheet 3, B.4.1)*
*[If No go to **Q3**]*

Q2. For rare flood events (less than once in 10 years), which of the following statements is true? *(Recording Sheet 1, C.3.19.2c; C.3.19.4c; Recording Sheet 3, B.4.1; ACTION (above).)*

a. The flood storage volume of the HGMU is less than 1 % of daily discharge
b. The flood storage volume of the HGMU is between 1 and 10 % of daily discharge
c. The flood storage volume of the HGMU is more than 10 % of daily discharge
d. Insufficient information is available to answer this question

Q3. Does the HGMU flood only occasionally (once every 2–10 years)? *(Recording Sheet 1, C.3.19.2b; Recording Sheet 3, B.4.1)*
*[If No go to **CV5**]*

Q4. For occasional (once every 2–10 years) flood events, which of the following statements is true? *(Recording Sheet 1, C.3.19.2b; C.3.19.4b; Recording Sheet 3, B.4.1; ACTION (above)).*

a. The flood storage volume of the HGMU is less than 1 % of daily discharge
b. The flood storage volume of the HGMU is between 1 and 10 % of daily discharge
c. The flood storage volume of the HGMU is more than 10 % of daily discharge
d. Insufficient information is available to answer this question

Q5. Does the HGMU flood frequently (at least once every 2 years)? *(Recording Sheet 1, C.3.19.2a; Recording Sheet 3, B.4.1)*

Q6. For frequent flood events, which of the following statements is true? *(Recording Sheet 1, C.3.19.2a; C.3.19.4a; Recording Sheet 3, B.4.1; ACTION (above))*

a. The flood storage volume of the HGMU is less than 1 % of daily discharge
b. The flood storage volume of the HGMU is between 1 and 10 % of daily discharge
c. The flood storage volume of the HGMU is more than 10 % of daily discharge
d. Insufficient information is available to answer this question

CV4 *Vegetation/habitat evidence of flooding*

Go to ***CV5*** if no information is available.

Q1. Does the HGMU vegetation have *vegetation code* ***A3***, ***A4***, ***A5***, ***A7***, ***G2***, ***G3***, ***G4***, ***M15***, ***S2***, ***S3***, ***W7*** or ***W9***? *(Recording Sheet 3, D)*

Q2. Does the HGMU vegetation have *vegetation code* ***A2–A10***, ***G1***, ***G5–11***, ***M8***, ***M9***, ***M13***, ***M14***, ***W4***, ***W5***, ***W6***, ***W8***, ***W10***, ***W11***, ***W12***? *(Recording Sheet 3, D)*

Q3. Does the HGMU vegetation have *vegetation code* ***G18–32***, ***S4–17***, ***W22–44***? *(Recording Sheet 3, D)*

CV5 *Other evidence of flooding*

Q1. Have the floodwaters, which inundate the HGMU, been observed to move at 0.1 m s^{-1} or more? *(Recording Sheet 3, B.4.6)*

Q2. Are any of the following true for the HGMU?

- The vegetation is dominated by scrub or dwarf shrubs *(Vegetation code S1–19, Recording Sheet 3, D)*
- The vegetation is dominated by woodland *(Vegetation code W1–44, Recording Sheet 3, D)*
- Hedgerows are found at intervals of less than 50 m *(AA Map 2)*
- Stock control or boundary structures such as fences or walls are present at intervals of less than 50 m *(AA Map 2)*
- Embankments or other solid flow-impeding structures are present within 100 m upstream or downstream of the HGMU? *(AA Map 1)*

Q3. Are any of the following statements true for the HGMU?

- The average water level in the HGMU remains at least 10 cm above the soil surface for at least a week after the flood recedes *(Recording Sheet 3, B.4.7)*
- The HGMU is permanently inundated *(Recording Sheet 3, B.4.3)*
- The average surface level of the HGMU is at least half a metre below the surface level of the surrounding HGMUs and the CA *(Recording Sheet 3, B.1; AA Map 1)*
- The HGMU is enclosed by unbreached levees, embankments or other physical structures that would prevent floodwater from returning directly to the main river or lake? *(AA Map 1)*

Q4. Are any of the following statements true for the HGMU?

- There are drains, ditches, streams or canals that may lead floodwater from the HGMU into adjacent HGMUs, or directly to a main stream channel or lake *(Recording Sheet 3, B.2.2; B.2.3; AA Map 1)*.
- The overall gradient of the ground surface is greater than 2° *(Recording Sheet 2a, C.2)*

Q5. Are all of the following statements true for the HGMU?

- There are no significant drains, streams or canals intercepting run-off which would otherwise have entered into the HGMU from most upslope units *(AA Map 1)*

- The combined surface area of upslope units adjacent to the HGMU is at least as large as the HGMU itself *(Recording Sheet 3, G; AA Map 1)*
- Upslope units adjacent to the HGMU have an overall gradient greater than 2° *(Gradient Code G1, Recording Sheet 2a, C.2; AA Map 1)*

Q6. Do the upslope areas have *hydrological codes* ***H3, H4, H6, H7, H8, H9, H11 or H12****? (Recording Sheet 2a, F)*

Assessment of the function

Compare the results recorded for each HGMU on the appraisal sheets with the possible outcomes below and assign an appropriate assessment code. Record this on the appraisal sheet. The syntax codes used are described in the introduction to Section 3.

Look-up tables

1. The function is definitely being performed
If answers to the questions for this function agree with any of the combinations listed in Look-up Table 1, the HGMU is definitely performing the function of detaining floodwater, and is probably reducing the severity of flooding downstream. The order in which the answers are given relates to both certainty of occurrence of the function (first part of the code, e.g. '**1**/1') and the degree to which it operates (second part of the code, e.g. '1/**1**'), based on the strength of evidence resulting from the combination of controlling variables: for code **1/1** the function is performed at the highest rate. Codes containing lower case letters (e.g. **1/4(a)**, **1/4(b)**, **1/4(c)**, **1/4(d)**) differ from each other in terms of the rationale statement but agree both in the certainty of occurrence and the degree of operation of the function.

2. The function is not (significantly) being performed
For any combination of answers that is not covered by either Look-up Table 1 or Look-up Table 2, the HGMU may be flooded by stream or lake water or inundated by run-off from upslope areas, but to an insignificant extent and not enough to affect functioning. Assign assessment code **2** to these HGMU

3. The function is probably being performed, but there are constraining factors or uncertainties
This statement is not applicable to the Floodwater Detention function.

4. The function is definitely not being performed
If answers to the questions for this function are in agreement with any of the combinations in Look-up Table 2, then the HGMU is never flooded by stream or lake water or inundated by run-off from upslope areas.

This completes the functional assessment for floodwater detention. The output codes can be recorded on the Functional Assessment Table in Section 4.0 and used to compare functions among HGMUs across the wetland. If you wish to assess all functions, move on to the next step of the FAPs: Groundwater recharge.

When all the HGMUs in the AA have been assessed, a general pattern of floodwater detention across the whole wetland will have been achieved. By manipulation of your answers to the controlling variable (CV) questions, you have the capacity to explore the probable outcomes of any management options under consideration, and relate them to specific functions. The electronic version of the FAPs can do this more quickly and easily.

3.1.A Flood Detention Appraisal Sheet

HGMU code: ______________________ **AA name:** ______________________

CV1 Geomorphological evidence of flooding potential			
Q1	Y	N	U/NA
Q2	Y	N	U/NA

CV4 Vegetation/Habitat evidence of flooding			
Q1	Y	N	U/NA
Q2	Y	N	U/NA
Q3	Y	N	U/NA

CV2 Physical evidence of flooding			
Q1	Y	N	U/NA
Q2	Y	N	U/NA
Q3	Y	N	U/NA

CV5 Other evidence of flooding			
Q1	Y	N	U/NA
Q2	Y	N	U/NA
Q3	Y	N	U/NA
Q4	Y	N	U/NA
Q5	Y	N	U/NA
Q6	Y	N	U/NA

CV3 Historical evidence of flooding potential				
Q1	Y	N	U/NA	
Q2	a	b	c	d
Q3	Y	N	U/NA	
Q4	a	b	c	d
Q5	Y	N	U/NA	
Q6	a	b	c	d

HGMU function assessment code: ______________________

3.1.A Look-up Table 1

<table>
<tr><th>CV1</th><th>CV2</th><th>CV3</th><th>CV4</th><th>CV5</th><th>Rationale</th><th>Code</th></tr>
<tr><td>1</td><td>2</td><td>*</td><td>#3</td><td>*</td><td>It is very likely that the HGMU floods with stream or lake water, but the frequency and the contribution to flood attenuation are unknown. Flow velocity and flood duration are unknown.</td><td>1/14</td></tr>
<tr><td>2</td><td>1 2</td><td>*</td><td>1|2</td><td>1</td><td>It is evident that the HGMU floods frequently with stream or lake water. It is not known what contribution is made to alleviation of downstream floods. Flow velocity is probably high and flood duration short.</td><td>1/4(a)</td></tr>
<tr><td>2</td><td>1 2</td><td>*</td><td>1|2</td><td>2</td><td>It is evident that the HGMU floods frequently with stream or lake water. It is not known what contribution is made to alleviation of downstream floods. Flow velocity is probably low.</td><td>1/4(b)</td></tr>
<tr><td>2</td><td>1 2</td><td>*</td><td>1|2</td><td>2 #4</td><td>It is evident that the HGMU floods frequently with stream or lake water. It is not known what contribution is made to alleviation of downstream floods. Flow velocity is probably low and flood duration long.</td><td>1/4(d)</td></tr>
<tr><td>2</td><td>1 2</td><td>*</td><td>1|2</td><td>2 4</td><td>It is evident that the HGMU floods frequently with stream or lake water. It is not known what contribution is made to alleviation of downstream floods. Flow velocity is probably low and flood duration short.</td><td>1/4(c)</td></tr>
<tr><td>2</td><td>1 2</td><td>*</td><td>1|2</td><td>*</td><td>It is evident that the HGMU floods frequently with stream or lake water. It is not known what contribution is made to alleviation of downstream floods. Flow velocity and flood duration are unknown.</td><td>1/4</td></tr>
<tr><td>2</td><td>2</td><td>*</td><td>1|2</td><td>*</td><td>It is evident that the HGMU floods with stream or lake water, but the frequency and the contribution to flood attenuation are unknown. Flow velocity and flood duration are unknown.</td><td>1/13</td></tr>
<tr><td>2</td><td>2</td><td>*</td><td>#3</td><td>*</td><td>It is very likely that the HGMU floods with stream or lake water, but the frequency and the contribution to flood attenuation are unknown. Flow velocity and flood duration are unknown.</td><td>1/14</td></tr>
<tr><td>2</td><td>*</td><td>1 2a</td><td>#3</td><td>*</td><td>It is evident that the HGMU floods with stream or lake water, but only rarely, during extreme events. It is evident that flooding provides a minor attenuation of extreme floods. Flow velocity and flood duration are unknown.</td><td>1/11</td></tr>
<tr><td>2</td><td>*</td><td>1 2b</td><td>#3</td><td>*</td><td>It is evident that the HGMU floods with stream or lake water, but only rarely, during extreme events. It is evident that flooding provides a moderate attenuation of floods. Flow velocity and flood duration are unknown.</td><td>1/10</td></tr>
<tr><td>2</td><td>*</td><td>1 2c</td><td>#3</td><td>*</td><td>It is evident that the HGMU floods with stream or lake water, but only rarely, during extreme events. It is evident that flooding provides a major attenuation of floods. Flow velocity and flood duration are unknown.</td><td>1/9</td></tr>
<tr><td>2</td><td>*</td><td>1 2d</td><td>#3</td><td>*</td><td>It is evident that the HGMU floods with stream or lake water, but only rarely, during extreme events. It is not evident what contribution is made to flood attenuation. Flow velocity and flood duration are unknown.</td><td>1/12</td></tr>
<tr><td>2</td><td>*</td><td>3 4a</td><td>#3</td><td>*</td><td>It is evident that the HGMU floods occasionally with stream or lake water. It is evident that flooding provides a minor attenuation of floods. Flow velocity and flood duration are unknown.</td><td>1/7</td></tr>
<tr><td>2</td><td>*</td><td>3 4b</td><td>#3</td><td>*</td><td>It is evident that the HGMU floods occasionally with stream or lake water. It is evident that flooding provides a moderate attenuation of floods. Flow velocity and flood duration are unknown.</td><td>1/6</td></tr>
<tr><td>2</td><td>*</td><td>3 4c</td><td>#3</td><td>*</td><td>It is evident that the HGMU floods occasionally with stream or lake water. It is evident that flooding provides a major attenuation of floods. Flow velocity and flood duration are unknown.</td><td>1/5</td></tr>
</table>

2	*	3 4d	#3	*	It is evident that the HGMU floods occasionally with stream or lake water. It is not evident what contribution is made to flood attenuation. Flow velocity and flood duration are unknown.	**1/8**
2	*	5 6a	#3	1	It is evident that the HGMU floods frequently with stream or lake water. It is evident that flooding provides a minor attenuation of frequent floods. Flow velocity is probably high and flood duration short.	**1/3(a)**
2	*	5 6a	#3	2	It is evident that the HGMU floods frequently with stream or lake water. It is evident that flooding provides a minor attenuation of frequent floods. Flow velocity is probably low.	**1/3(b)**
2	*	5 6a	#3	2 #4	It is evident that the HGMU floods frequently with stream or lake water. It is evident that flooding provides a minor attenuation of frequent floods. Flow velocity is probably low and flood duration long.	**1/3(d)**
2	*	5 6a	#3	2 4	It is evident that the HGMU floods frequently with stream or lake water. It is evident that flooding provides a minor attenuation of frequent floods. Flow velocity is probably low and flood duration short.	**1/3(c)**
2	*	5 6a	#3	*	It is evident that the HGMU floods frequently with stream or lake water. It is evident that flooding provides a minor attenuation of frequent floods. Flow velocity and flood duration are unknown.	**1/3**
2	*	5 6b	#3	1	It is evident that the HGMU floods frequently with stream or lake water. It is evident that flooding provides a moderate attenuation of frequent floods. Flow velocity is probably high and flood duration short.	**1/2(a)**
2	*	5 6b	#3	2	It is evident that the HGMU floods frequently with stream or lake water. It is evident that flooding provides a moderate attenuation of frequent floods. Flow velocity is probably low.	**1/2(b)**
2	*	5 6b	#3	2 #4	It is evident that the HGMU floods frequently with stream or lake water. It is evident that flooding provides a moderate attenuation of frequent floods. Flow velocity is probably low and flood duration long.	**1/2(d)**
2	*	5 6b	#3	2 4	It is evident that the HGMU floods frequently with stream or lake water. It is evident that flooding provides a moderate attenuation of frequent floods. Flow velocity is probably low and flood duration short.	**1/2(c)**
2	*	5 6b	#3	*	It is evident that the HGMU floods frequently with stream or lake water. It is evident that flooding provides a moderate attenuation of frequent floods. Flow velocity and flood duration are unknown.	**1/2**
2	*	5 6c	#3	1	It is evident that the HGMU floods frequently with stream or lake water. It is evident that flooding provides a major attenuation of frequent floods. Flow velocity is probably high and flood duration short.	**1/1(a)**
2	*	5 6c	#3	2	It is evident that the HGMU floods frequently with stream or lake water. It is evident that flooding provides a major attenuation of frequent floods. Flow velocity is probably low.	**1/1(b)**
2	*	5 6c	#3	2 #4	It is evident that the HGMU floods frequently with stream or lake water. It is evident that flooding provides a major attenuation of frequent floods. Flow velocity is probably low and flood duration long.	**1/1(d)**
2	*	5 6c	#3	2 4	It is evident that the HGMU floods frequently with stream or lake water. It is evident that flooding provides a major attenuation of frequent floods. Flow velocity is probably low and flood duration short.	**1/1(c)**
2	*	5 6c	#3	*	It is evident that the HGMU floods frequently with stream or lake water. It is evident that flooding provides a major attenuation of frequent floods. Flow velocity and flood duration are unknown.	**1/1**
2	*	5 6d	#3	1	It is evident that the HGMU floods frequently with stream or lake water. It is not known what contribution is made to alleviation of downstream floods. Flow velocity is probably high and flood duration short.	**1/4(a)**

3.1.A Look-up Table 1 cont'd

CV1	CV2	CV3	CV4	CV5	Rationale	Code
2	*	5 6d	#3	2	It is evident that the HGMU floods frequently with stream or lake water. It is not known what contribution is made to alleviation of downstream floods. Flow velocity is probably low.	**1/4(b)**
2	*	5 6d	#3	2 #4	It is evident that the HGMU floods frequently with stream or lake water. It is not known what contribution is made to alleviation of downstream floods. Flow velocity is probably low and flood duration long.	**1/4(d)**
2	*	5 6d	#3	2 4	It is evident that the HGMU floods frequently with stream or lake water. It is not known what contribution is made to alleviation of downstream floods. Flow velocity is probably low and flood duration short.	**1/4(c)**
2	*	5 6d	#3	*	It is evident that the HGMU floods frequently with stream or lake water. It is not known what contribution is made to alleviation of downstream floods. Flow velocity and flood duration are unknown.	**1/4**
2	*	*	1|2	*	It is very likely that the HGMU floods with stream or lake water but the frequency and the contribution to flood attenuation are unknown. Flow velocity and flood duration are unknown.	**1/14**
2	*	*	#3	5 6	It is likely that the HGMU is inundated regularly with run-off from upslope areas. Inundation with stream or lake water is likely, but its frequency is unknown. Flow velocity and flood duration are unknown.	**1/15**
*	1 2	1 2a	#3	*	It is evident that the HGMU floods with stream or lake water, but only rarely, during extreme events. It is evident that flooding provides a minor attenuation of extreme floods. Flow velocity and flood duration are unknown.	**1/11**
*	1 2	1 2b	#3	*	It is evident that the HGMU floods with stream or lake water, but only rarely, during extreme events. It is evident that flooding provides a moderate attenuation of floods. Flow velocity and flood duration are unknown.	**1/10**
*	1 2	1 2c	#3	*	It is evident that the HGMU floods with stream or lake water, but only rarely, during extreme events. It is evident that flooding provides a major attenuation of floods. Flow velocity and flood duration are unknown.	**1/9**
*	1 2	1 2d	#3	*	It is evident that the HGMU floods with stream or lake water, but only rarely, during extreme events. It is not evident what contribution is made to flood attenuation. Flow velocity and flood duration are unknown.	**1/12**
*	1 2	3 4a	#3	*	It is evident that the HGMU floods occasionally with stream or lake water. It is evident that flooding provides a minor attenuation of floods. Flow velocity and flood duration are unknown.	**1/7**
*	1 2	3 4b	#3	*	It is evident that the HGMU floods occasionally with stream or lake water. It is evident that flooding provides a moderate attenuation of floods. Flow velocity and flood duration are unknown.	**1/6**
*	1 2	3 4c	#3	*	It is evident that the HGMU floods occasionally with stream or lake water. It is evident that flooding provides a major attenuation of floods. Flow velocity and flood duration are unknown.	**1/5**
*	1 2	3 4d	#3	*	It is evident that the HGMU floods occasionally with stream or lake water. It is not evident what contribution is made to flood attenuation. Flow velocity and flood duration are unknown.	**1/8**
*	1 2	5 6a	#3	1	It is evident that the HGMU floods frequently with stream or lake water. It is evident that flooding provides a minor attenuation of frequent floods. Flow velocity is probably high and flood duration short.	**1/3(a)**

*	1 2	5 6a	#3	2	It is evident that the HGMU floods frequently with stream or lake water. It is evident that flooding provides a minor attenuation of frequent floods. Flow velocity is probably low.	**1/3(b)**
*	1 2	5 6a	#3	2 #4	It is evident that the HGMU floods frequently with stream or lake water. It is evident that flooding provides a minor attenuation of frequent floods. Flow velocity is probably low and flood duration long.	**1/3(d)**
*	1 2	5 6a	#3	2 4	It is evident that the HGMU floods frequently with stream or lake water. It is evident that flooding provides a minor attenuation of frequent floods. Flow velocity is probably low and flood duration short.	**1/3(c)**
*	1 2	5 6a	#3	*	It is evident that the HGMU floods frequently with stream or lake water. It is evident that flooding provides a minor attenuation of frequent floods. Flow velocity and flood duration are unknown.	**1/3**
*	1 2	5 6b	#3	1	It is evident that the HGMU floods frequently with stream or lake water. It is evident that flooding provides a moderate attenuation of frequent floods. Flow velocity is probably high and flood duration short.	**1/2(a)**
*	1 2	5 6b	#3	2	It is evident that the HGMU floods frequently with stream or lake water. It is evident that flooding provides a moderate attenuation of frequent floods. Flow velocity is probably low.	**1/2(b)**
*	1 2	5 6b	#3	2 #4	It is evident that the HGMU floods frequently with stream or lake water. It is evident that flooding provides a moderate attenuation of frequent floods. Flow velocity is probably low and flood duration long.	**1/2(d)**
*	1 2	5 6b	#3	2 4	It is evident that the HGMU floods frequently with stream or lake water. It is evident that flooding provides a moderate attenuation of frequent floods. Flow velocity is probably low and flood duration short.	**1/2(c)**
*	1 2	5 6b	#3	*	It is evident that the HGMU floods frequently with stream or lake water. It is evident that flooding provides a moderate attenuation of frequent floods. Flow velocity and flood duration are unknown.	**1/2**
*	1 2	5 6c	#3	1	It is evident that the HGMU floods frequently with stream or lake water. It is evident that flooding provides a major attenuation of frequent floods. Flow velocity is probably high and flood duration short.	**1/1(a)**
*	1 2	5 6c	#3	2	It is evident that the HGMU floods frequently with stream or lake water. It is evident that flooding provides a major attenuation of frequent floods. Flow velocity is probably low.	**1/1(b)**
*	1 2	5 6c	#3	2 #4	It is evident that the HGMU floods frequently with stream or lake water. It is evident that flooding provides a major attenuation of frequent floods. Flow velocity is probably low and flood duration long.	**1/1(d)**
*	1 2	5 6c	#3	2 4	It is evident that the HGMU floods frequently with stream or lake water. It is evident that flooding provides a major attenuation of frequent floods. Flow velocity is probably low and flood duration short.	**1/1(c)**
*	1 2	5 6c	#3	*	It is evident that the HGMU floods frequently with stream or lake water. It is evident that flooding provides a major attenuation of frequent floods. Flow velocity and flood duration are unknown.	**1/1**
*	1 2	5 6d	#3	1	It is evident that the HGMU floods frequently with stream or lake water. It is not known what contribution is made to alleviation of downstream floods. Flow velocity is probably high and flood duration short.	**1/4(a)**
*	1 2	5 6d	#3	2	It is evident that the HGMU floods frequently with stream or lake water. It is not known what contribution is made to alleviation of downstream floods. Flow velocity is probably low.	**1/4(b)**
*	1 2	5 6d	#3	2 #4	It is evident that the HGMU floods frequently with stream or lake water. It is not known what contribution is made to alleviation of downstream floods. Flow velocity is probably low and flood duration long.	**1/4(d)**

3.1.A Look-up Table 1 cont'd

<table>
<tr><th>CV1</th><th>CV2</th><th>CV3</th><th>CV4</th><th>CV5</th><th>Rationale</th><th>Code</th></tr>
<tr><td>*</td><td>1 2</td><td>5 6d</td><td>#3</td><td>2 4</td><td>It is evident that the HGMU floods frequently with stream or lake water. It is not known what contribution is made to alleviation of downstream floods. Flow velocity is probably low and flood duration short.</td><td>1/4(c)</td></tr>
<tr><td>*</td><td>1 2</td><td>5 6d</td><td>#3</td><td>*</td><td>It is evident that the HGMU floods frequently with stream or lake water. It is not known what contribution is made to alleviation of downstream floods. Flow velocity and flood duration are unknown.</td><td>1/4</td></tr>
<tr><td>*</td><td>2</td><td>*</td><td>1|2</td><td>*</td><td>It is very likely that the HGMU floods with stream or lake water, but the frequency and the contribution to flood attenuation are unknown. Flow velocity and flood duration are unknown.</td><td>1/14</td></tr>
<tr><td>*</td><td>*</td><td>1 2a</td><td>1|2</td><td>*</td><td>It is evident that the HGMU floods with stream or lake water, but only rarely, during extreme events. It is evident that flooding provides a minor attenuation of extreme floods. Flow velocity and flood duration are unknown.</td><td>1/11</td></tr>
<tr><td>*</td><td>*</td><td>1 2b</td><td>1|2</td><td>*</td><td>It is evident that the HGMU floods with stream or lake water, but only rarely, during extreme events. It is evident that flooding provides a moderate attenuation of floods. Flow velocity and flood duration are unknown.</td><td>1/10</td></tr>
<tr><td>*</td><td>*</td><td>1 2c</td><td>1|2</td><td>*</td><td>It is evident that the HGMU floods with stream or lake water, but only rarely, during extreme events. It is evident that flooding provides a major attenuation of floods. Flow velocity and flood duration are unknown.</td><td>1/9</td></tr>
<tr><td>*</td><td>*</td><td>1 2d</td><td>1|2</td><td>*</td><td>It is evident that the HGMU floods with stream or lake water, but only rarely, during extreme events. It is not evident what contribution is made to flood attenuation. Flow velocity and flood duration are unknown.</td><td>1/12</td></tr>
<tr><td>*</td><td>*</td><td>3 4a</td><td>1|2</td><td>*</td><td>It is evident that the HGMU floods occasionally with stream or lake water. It is evident that flooding provides a minor attenuation of floods. Flow velocity and flood duration are unknown.</td><td>1/7</td></tr>
<tr><td>*</td><td>*</td><td>3 4b</td><td>1|2</td><td>*</td><td>It is evident that the HGMU floods occasionally with stream or lake water. It is evident that flooding provides a moderate attenuation of floods. Flow velocity and flood duration are unknown.</td><td>1/6</td></tr>
<tr><td>*</td><td>*</td><td>3 4c</td><td>1|2</td><td>*</td><td>It is evident that the HGMU floods occasionally with stream or lake water. It is evident that flooding provides a major attenuation of floods. Flow velocity and flood duration are unknown.</td><td>1/5</td></tr>
<tr><td>*</td><td>*</td><td>3 4d</td><td>1|2</td><td>*</td><td>It is evident that the HGMU floods occasionally with stream or lake water. It is not evident what contribution is made to flood attenuation. Flow velocity and flood duration are unknown.</td><td>1/8</td></tr>
<tr><td>*</td><td>*</td><td>5 6a</td><td>1|2</td><td>1</td><td>It is evident that the HGMU floods frequently with stream or lake water. It is evident that flooding provides a minor attenuation of frequent floods. Flow velocity is probably high and flood duration short.</td><td>1/3(a)</td></tr>
<tr><td>*</td><td>*</td><td>5 6a</td><td>1|2</td><td>2</td><td>It is evident that the HGMU floods frequently with stream or lake water. It is evident that flooding provides a minor attenuation of frequent floods. Flow velocity is probably low.</td><td>1/3(b)</td></tr>
<tr><td>*</td><td>*</td><td>5 6a</td><td>1|2</td><td>2 #4</td><td>It is evident that the HGMU floods frequently with stream or lake water. It is evident that flooding provides a minor attenuation of frequent floods. Flow velocity is probably low and flood duration long.</td><td>1/3(d)</td></tr>
<tr><td>*</td><td>*</td><td>5 6a</td><td>1|2</td><td>2 4</td><td>It is evident that the HGMU floods frequently with stream or lake water. It is evident that flooding provides a minor attenuation of frequent floods. Flow velocity is probably low and flood duration short.</td><td>1/3(c)</td></tr>
</table>

<table>
<tr><td>*</td><td>*</td><td>5 6a</td><td>1|2</td><td>*</td><td>It is evident that the HGMU floods frequently with stream or lake water. It is evident that flooding provides a minor attenuation of frequent floods. Flow velocity and flood duration are unknown.</td><td>1/3</td></tr>
<tr><td>*</td><td>*</td><td>5 6b</td><td>1|2</td><td>1</td><td>It is evident that the HGMU floods frequently with stream or lake water. It is evident that flooding provides a moderate attenuation of frequent floods. Flow velocity is probably high and flood duration short.</td><td>1/2(a)</td></tr>
<tr><td>*</td><td>*</td><td>5 6b</td><td>1|2</td><td>2</td><td>It is evident that the HGMU floods frequently with stream or lake water. It is evident that flooding provides a moderate attenuation of frequent floods. Flow velocity is probably low.</td><td>1/2(b)</td></tr>
<tr><td>*</td><td>*</td><td>5 6b</td><td>1|2</td><td>2 #4</td><td>It is evident that the HGMU floods frequently with stream or lake water. It is evident that flooding provides a moderate attenuation of frequent floods. Flow velocity is probably low and flood duration long.</td><td>1/2(d)</td></tr>
<tr><td>*</td><td>*</td><td>5 6b</td><td>1|2</td><td>2 4</td><td>It is evident that the HGMU floods frequently with stream or lake water. It is evident that flooding provides a moderate attenuation of frequent floods. Flow velocity is probably low and flood duration short.</td><td>1/2(c)</td></tr>
<tr><td>*</td><td>*</td><td>5 6b</td><td>1|2</td><td>*</td><td>It is evident that the HGMU floods frequently with stream or lake water. It is evident that flooding provides a moderate attenuation of frequent floods. Flow velocity and flood duration are unknown.</td><td>1/2</td></tr>
<tr><td>*</td><td>*</td><td>5 6c</td><td>1|2</td><td>1</td><td>It is evident that the HGMU floods frequently with stream or lake water. It is evident that flooding provides a major attenuation of frequent floods. Flow velocity is probably high and flood duration short.</td><td>1/1(a)</td></tr>
<tr><td>*</td><td>*</td><td>5 6c</td><td>1|2</td><td>2</td><td>It is evident that the HGMU floods frequently with stream or lake water. It is evident that flooding provides a major attenuation of frequent floods. Flow velocity is probably low.</td><td>1/1(b)</td></tr>
<tr><td>*</td><td>*</td><td>5 6c</td><td>1|2</td><td>2 #4</td><td>It is evident that the HGMU floods frequently with stream or lake water. It is evident that flooding provides a major attenuation of frequent floods. Flow velocity is probably low and flood duration long.</td><td>1/1(d)</td></tr>
<tr><td>*</td><td>*</td><td>5 6c</td><td>1|2</td><td>2 4</td><td>It is evident that the HGMU floods frequently with stream or lake water. It is evident that flooding provides a major attenuation of frequent floods. Flow velocity is probably low and flood duration short.</td><td>1/1(c)</td></tr>
<tr><td>*</td><td>*</td><td>5 6c</td><td>1|2</td><td>*</td><td>It is evident that the HGMU floods frequently with stream or lake water. It is evident that flooding provides a major attenuation of frequent floods. Flow velocity and flood duration are unknown.</td><td>1/1</td></tr>
<tr><td>*</td><td>*</td><td>5 6d</td><td>1|2</td><td>1</td><td>It is evident that the HGMU floods frequently with stream or lake water. It is not known what contribution is made to alleviation of downstream or lake floods. Flow velocity is probably high and flood duration short.</td><td>1/4(a)</td></tr>
<tr><td>*</td><td>*</td><td>5 6d</td><td>1|2</td><td>2</td><td>It is evident that the HGMU floods frequently with stream or lake water. It is not known what contribution is made to alleviation of downstream or lake floods. Flow velocity is probably low.</td><td>1/4(b)</td></tr>
<tr><td>*</td><td>*</td><td>5 6d</td><td>1|2</td><td>2 #4</td><td>It is evident that the HGMU floods frequently with stream or lake water. It is not known what contribution is made to alleviation of downstream or lake floods. Flow velocity is probably low and flood duration long.</td><td>1/4(d)</td></tr>
<tr><td>*</td><td>*</td><td>5 6d</td><td>1|2</td><td>2 4</td><td>It is evident that the HGMU floods frequently with stream or lake water. It is not known what contribution is made to alleviation of downstream or lake floods. Flow velocity is probably low and flood duration short.</td><td>1/4(c)</td></tr>
<tr><td>*</td><td>*</td><td>5 6d</td><td>1|2</td><td>*</td><td>It is evident that the HGMU floods frequently with stream or lake water. It is not known what contribution is made to alleviation of downstream or lake floods. Flow velocity and flood duration are unknown.</td><td>1/4</td></tr>
<tr><td>*</td><td>*</td><td>*</td><td>#3*</td><td>5 6</td><td>It is likely that the HGMU is inundated regularly with run-off from upslope areas. It is unknown if flooding by stream or lake water occurs. Flow velocity and flood duration are unknown.</td><td>1/16</td></tr>
</table>

3.1.A Look-up Table 2

CV1	CV2	CV3	CV4	CV5	Rationale	Cod
The HGMU is of type H1, H8 , H11 or H16					There is evidence that the function is not performed.	**4**
*	*	*	3	*	There is evidence that the function is not performed.	**4**
#1#2	#(1 2)	#1	#1#2	*	There is no evidence that the function is performed.	**4**

Further reading and literature used or cited

Bullock, A. and Acreman, M.C. (2003). The role of wetlands in the hydrological cycle, *Hydrology and Earth System Sciences* **7**(3), 358–389.

Carter, V. (1986). An overview of the hydrologic concerns related to wetlands in the United States, *Canadian Journal of Botany* **64**, 364–374.

Carter, V., Bedinger, M.S., Novitzki, R.P. and Wilen, W.O. (1979). Water resources and wetlands. In: Greeson, P.E., Clark, J.R. and Clark, J.E. (eds), *Wetland Functions and Values: the State of our Understanding*, American Water Resources Association, Minneapolis, MN, 344–376.

Carter, V. and Novitzki, R.P. (1988). Some comments of the relation between ground water and wetlands. In: Hook, D.D., McKee Jr, W.H., Smith, H.K., Gregory, J., Burrell, V.G., DeVoe, M.R., Sojka, R.E., Gilbert, S., Banks, R., Stolzy, L.G., Brooks, C., Matthews, T.D., and Shear, T.H. (eds), *The Ecology and Management of Wetlands, Volume I*, Timber Press, Portland, OR, 68–86.

Gilman, K. (1994). *Hydrology and Wetland Conservation*, Wiley, Chichester, UK.

Hensel, B.R. and Miller, M.V. (1991). Effects of wetlands creation on groundwater flow, *Journal of Hydrology* **126**, 293–314.

Hughes, J.M.R. and Heathwaite, A.L. (eds) (1995). *Hydrology and Hydrochemistry of British Wetlands*, Wiley, Chichester, UK.

Mitsch, W.J. and Gosselink, J.G. (2007). *Wetlands*, 4th edn, Wiley, New York.

Siegel, D.I. (1988). A review of the recharge-discharge function of wetlands. In: Hook, D.D., McKee Jr, W.H., Smith, H.K., Gregory, J., Burrell, V.G., DeVoe, M.R., Sojka, R.E., Gilbert, S., Banks, R., Stolzy, L.G., Brooks, C., Matthews, T.D., and Shear, T.H. (eds), *The Ecology and Management of Wetlands, Volume I*, Timber Press, Portland, OR, 59–67.

Introduction

Function description: The recharge of groundwater by the detention and subsequent infiltration of floodwater or slope run-off into a wetland, and the percolation of that water into a significant aquifer.

Significance – rationale for assessment: The water stored in river and lake marginal wetlands from overbank flooding or slope run-off may percolate downwards into groundwater aquifers. This can replenish groundwater resources, maintain groundwater-dependent wetland ecosystems in discharge areas, or maintain base flow in rivers fed by discharge elsewhere.

Related functions, for which the outcome of this assessment may be required:

- Groundwater discharge

For the Groundwater recharge function to occur the following conditions must be met:

1. The HGMU must receive significant water inputs either from run-off or from overbank flooding. If the only inputs are from rainfall it is unlikely that recharge will occur at higher rates than in non-wetland areas.
2. Prior to water input, the water table must be sufficiently low to allow standing water to infiltrate the soil before it evaporates or runs off.
3. The HGMU must connect to an aquifer by means of a hydrological pathway. This usually means the HGMU must overlie the aquifer and not be separated from it by deposits of low permeability.
4. There can be no groundwater discharge.

These conditions are found mainly in wetlands where the regional water table is low, and which experience considerable dry periods. Examples are found in semi-arid or arid areas with permeable geology.

Functional assessment may not be necessary in the following cases, where the Groundwater recharge function is definitely not performed (code **4/n**) or not performed to a significant degree (code **2/n**):

- If the HGMU has geomorphic code **G2** or **G3** *(Recording Sheet 2a (Field), C.2)* it is probably too steep for recharge to occur. Assign code **2**.

- If the HGMU is of hydrological type **H1**, **H8**, **H11** or **H16** (*Recording Sheet 2a (Field), F)* it does not receive inputs from either run-off or overbank flooding. Recharge at a higher rate than in the adjoining non-wetland units is therefore unlikely. Assign code **4**.
- If the hydrological code is **H1**, **H2**, **H3**, **H4**, **H11**, **H12**, **H14** or **H15** *(Recording Sheet 2a (Field), F)* there is net upward movement of groundwater in the HGMU and recharge is impossible. Assign code **4**.

In all other cases, answer the following questions for each controlling variable. Complete all questions for which sufficient information is available. Circle the correct answer on the appraisal sheet. Use a new appraisal sheet for each HGMU. The location of the relevant information is given in brackets for each question.

Controlling variables

CV1 Flooding regime
CV2 Drift and solid geology
CV3 Groundwater dynamics and surface water inputs
CV4 Climate

CV1 *Flooding regime*

Q1. Is the HGMU frequently flooded with water from the main wetland (at least once in less than two years on average)? *(Recording Sheet 1, C.3.19.8b (answer a))*
Q2. Is the HGMU occasionally flooded with water from the main wetland (once every 2–10 years on average)? *(Recording Sheet 1, C.3.19.8b (answer b))*
Q3. Does the HGMU receive run-off inputs from an upslope area (in the AA or CA) at least as large as the HGMU itself? *(Hydrological codes* **H2**, **H3**, **H4**, **H5**, **H6**, **H7**, **H13** or **H14**; *Recording Sheet 2a, F; AA Map 1; CA Map 1).*
Q4. Does the HGMU have watercourses leading out of it to lower HGMUs or to the river or lake channel? *(Recording Sheet 3, B.2.3)*

CV2 *Drift and solid geology*

Q1. Are the following deposits present at or near the surface of the HGMU?
- Fine or medium particle size to 50 cm thickness of layer (*Recording Sheet 2a, C.3.7*)
- Peat to 50 cm depth *(Recording Sheet 2a, C.3.8)*

Q2. Are the following deposits present at or near the surface of the HGMU?
- Gravel particle size to 50 cm thickness of layer *(Recording Sheet 2a, C.3.7)*
- Coarse particle size to 50 cm thickness of layer *(Recording Sheet 2a, C.3.7)*

Q3. Is all or part of the HGMU underlain by any of the following?
- A recognised aquifer *(CA Map 2)*
- Solid rock likely to make a good aquifer *(CA Map 2)*
- Non-solid deposits likely to make a good aquifer *(CA Map 2)*

CV3 *Groundwater dynamics and surface water inputs*

Q1. Is at least one of the following statements true?
- Groundwater levels between inundation events drop to depths greater than 0.5 m below the ground surface *(Recording Sheet 3, B.3.11)*
- There are water courses (streams or lakes, channels and ditches) that lead water into and terminate within the HGMU *(Recording Sheet 3, B.2.4)*

CV4 *Climate*

Q1. The HGMU is situated in an arid or semi-arid region *(Recording Sheet 1, C.3.1.12)*

3.1.B Groundwater Recharge Appraisal Sheet

HGMU code: ______________________ | **AA name:** ______________________

CV1 Flooding regime			
Q1	Y	N	U/NA
Q2	Y	N	U/NA
Q3	Y	N	U/NA
Q4	Y	N	U/NA

CV3 Groundwater dynamics and surface water inputs			
Q1	Y	N	U/NA

CV2 Drift and solid geology			
Q1	Y	N	U/NA
Q2	Y	N	U/NA
Q3	Y	N	U/NA

CV4 Climate			
Q1	Y	N	U/NA

HGMU function assessment code: ____

3.1.B Look-up Table 1

<table>
<tr><th>CV1</th><th>CV2</th><th>CV3</th><th>CV4</th><th>Rationale</th><th>Code</th></tr>
<tr><td>1|(3 #4)</td><td>3</td><td>1</td><td>1</td><td>It is likely that recharge is extremely significant: a large recharge area in a (semi-) arid region is connected with a local aquifer and is inundated frequently. Field indicators confirm the recharge ability of the HGMU.</td><td>1/7</td></tr>
<tr><td>1|(3 #4)</td><td>3</td><td>1</td><td>#1</td><td>It is likely that recharge is extremely significant: a large recharge area is connected with a regional aquifer and is inundated frequently. Field indicators confirm the recharge ability of the HGMU.</td><td>1/6</td></tr>
<tr><td>1|(3 #4)</td><td>2|3</td><td>1</td><td>1</td><td>It is evident that recharge is extremely significant: a large recharge area in a (semi-) arid region is connected with a regional aquifer and is inundated frequently. Field indicators confirm the recharge ability of the HGMU.</td><td>1/1</td></tr>
<tr><td>1|(3 #4)</td><td>2|3</td><td>*</td><td>1</td><td>It is very likely that recharge is significant: a recharge area in a (semi-) arid region is connected with a regional aquifer and is inundated frequently.</td><td>1/5</td></tr>
<tr><td>1|(3 #4)</td><td>2|3</td><td>*</td><td>1</td><td>It is very likely that recharge is extremely significant: a large recharge area in a (semi-) arid region is connected with a regional aquifer and is inundated frequently.</td><td>1/3</td></tr>
<tr><td>2</td><td>2|3</td><td>1</td><td>1</td><td>It is evident that recharge is significant: a large recharge area in a (semi-) arid region is connected with a regional aquifer and is flooded occasionally for long durations. Field indicators confirm the recharge ability of the HGMU.</td><td>1/2</td></tr>
<tr><td>2</td><td>2|3</td><td>*</td><td>1</td><td>It is very likely that recharge is significant: a large recharge area in a (semi-) arid region is connected with a regional aquifer and is flooded occasionally for long durations.</td><td>1/4</td></tr>
</table>

3.1.B Look-up Table 2

<table>
<tr><th>CV1</th><th>CV2</th><th>CV3</th><th>CV4</th><th>Rationale</th><th>Code</th></tr>
<tr><td colspan="4">The HGMU is of type H1, H8, H11 or H16</td><td>There is evidence that the function is not performed.</td><td>4</td></tr>
<tr><td colspan="4">The HGMU is of type G2 G3</td><td>There is evidence that the function is not performed.</td><td>4</td></tr>
<tr><td>#1#2 #(3)</td><td>*</td><td>*</td><td>*</td><td>There is no evidence that the function is performed.</td><td>4</td></tr>
<tr><td>*</td><td>1</td><td>*</td><td>*</td><td>There is evidence that the function is not performed.</td><td>4</td></tr>
<tr><td>*</td><td>#2#3</td><td>*</td><td>*</td><td>There is no evidence that the function is performed.</td><td>4</td></tr>
</table>

Assessment of the function

Compare the results recorded for each HGMU on the appraisal sheets with the possible outcomes below, and assign an appropriate assessment code. Record this on the appraisal sheet. The syntax codes used are described in the introduction to Section 3.

Look-up tables

1. The function is definitely being performed
If the answers to the questions for this function agree with any of the combinations in Look-up Table 1, then the HGMU is definitely performing the function of recharge of groundwater to some extent. The order in which the answers are given relates to both the certainty of occurrence of the function (first part of the code, e.g. '**1**/1') and the degree to which it operates (second part of the code, e.g. '1/**1**') to sustain regional groundwater flow systems or water supply resources, based on the strength of evidence resulting from the combination of controlling variables: for code **1/1** the function is performed at the highest rate.

2. The function is not (significantly) being performed
For any combination of answers that is not covered by either Look-up Table 1 or Look-up Table 2, there may be recharge occurring in the HGMU, but only to a degree that is probably insignificant to functioning. Assign assessment code **2** to these HGMUs.

3. The function is probably being performed, but there are constraining factors or uncertainties
This statement is not applicable to this function.

4. The function is definitely not being performed
If answers to the questions for this function are in agreement with any of the combinations listed in Look-up Table 2, then recharge definitely does not occur in the HGMU.

This completes the functional assessment for groundwater recharge. The output codes can be recorded on the Functional Assessment Table in Section 4.0 and used to compare functions among HGMUs across the wetland. If you wish to assess all functions, move on to the next step of the FAPs: Groundwater discharge.

When all the HGMUs in the AA have been assessed, a general pattern of groundwater recharge across the whole wetland will have been achieved. By manipulation of your answers to the controlling variable (CV) questions, you have the capacity to explore the probable outcomes of any management options under consideration and relate them to specific functions. The electronic version of the FAPs can do this more quickly and easily.

3.1.C Groundwater discharge (function)

Further reading and references

Bullock, A. and Acreman, M.C. (2003). The role of wetlands in the hydrological cycle, *Hydrology and Earth System Sciences* **7**(3), 358–389.

Carter, V. (1986). An overview of the hydrologic concerns related to wetlands in the United States, *Canadian Journal of Botany* **64**, 364–374.

Carter, V., Bedinger, M.S., Novitzki, R.P. and Wilen, W.O. (1979). Water resources and wetlands. In: Greeson, P.E., Clark, J.R. and Clark, J.E. (eds) *Wetland Functions and Values: the State of Our Understanding*, American Water Resources Association, Minneapolis, MN, 344–376.

Carter, V. and Novitzki, R.P. (1988). Some comments of the relation between ground water and wetlands. In: Hook, D.D., McKee Jr, W.H., Smith, H.K., Gregory, J., Burrell, V.G., DeVoe, M.R., Sojka, R.E., Gilbert, S., Banks, R., Stolzy, L.G., Brooks, C., Matthews, T.D., and Shear, T.H. (eds), *The Ecology and Management of Wetlands, Volume I*, Timber Press, Portland, OR, 68–86.

Gilman, K. (1994) *Hydrology and Wetland Conservation*, Wiley, Chichester, UK.
Hughes, J.M.R. and Heathwaite, A.L. (eds) (1995). *Hydrology and Hydrochemistry of British Wetlands*, Wiley Chichester, UK.
Mitsch, W.J. and Gosselink, J.G. (2007). *Wetlands*, 4th edn, Wiley, New York.
Roulet, N.T. (1990). Hydrology of a headwater basin wetland: groundwater discharge and wetland maintenance, *Hydrological Processes* **4**, 387–400.
Siegel, D.I. (1988) A review of the recharge-discharge function of wetlands. In: Hook, D.D., McKee Jr, W.H., Smith, H.K., Gregory, J., Burrell, V.G., DeVoe, M.R., Sojka, R.E., Gilbert, S., Banks, R., Stolzy, L.G., Brooks, C., Matthews, T.D., and Shear, T.H. (eds), *The Ecology and Management of Wetlands, Volume I*, Timber Press, Portland, OR 59–67.

Introduction

Function description: The upward seepage of groundwater to the wetland surface.
Significance – rationale for assessment:

- Water may discharge from aquifers under specific hydrogeological conditions, emerging at the surface in springs or seepage zones. This may be important in the maintenance of river or lake base flow and lake surface water level, and for ecosystems dependent upon associated soil water regime or solutes.
- The discharge may maintain water levels or water quality in the wetland, thus supporting ecological and biogeochemical functions.
- Nutrients may be taken up from the discharged groundwater, and consequently removed from the hydrological system.

Related functions, for which outcome of this assessment may be required:

- Groundwater recharge

For the Groundwater discharge function to occur, the HGMU must meet the following conditions (Figs 3.1.5 and 3.1.6):

1. There must be a recharge area, on ground higher than the HGMU, where water can infiltrate.
2. There must be a source of water for infiltration in the recharge area. Usually this will be rainwater but in some cases it may be run-off input or even stream or lake water.
3. The recharge area must be connected to the HGMU through an aquifer.
4. The permeability of the deposits in the HGMU must be sufficient to allow significant upward seepage.

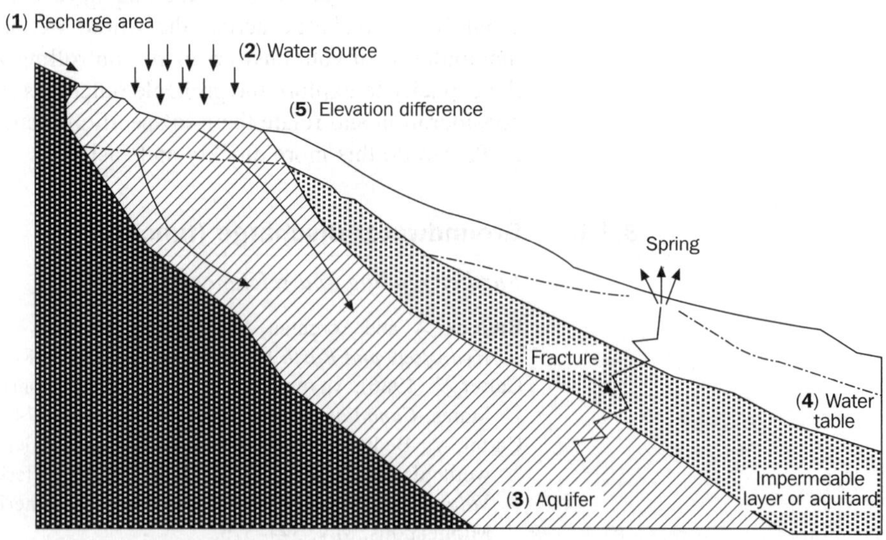

3.1.5 Prerequisites for groundwater discharge – confined aquifer. For the groundwater discharge function to occur the following conditions are necessary: (1) recharge area above the HGMU, (2) a water source, e.g. rainfall, run-off input, stream water, (3) recharge area connected with the HGMU through an aquifier, (4) high permeability of the HGMU deposits, (5) sufficient elevation difference between recharge and discharge areas.

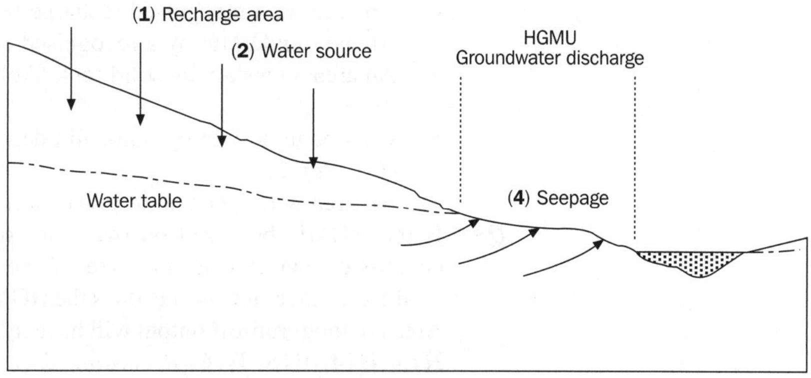

3.1.6 Prerequisites for groundwater discharge – unconfined aquifer.

5. Groundwater flow must not be intercepted between the recharge and discharge areas by factors such as natural discharge, artificial drainage, groundwater abstraction or vegetation uptake.
6. There must be a sufficient difference in elevation between the discharge and recharge areas to create a groundwater table gradient capable of producing significant discharge.

Functional assessment may not be necessary in the following cases, where either the groundwater discharge function is not performed (code **4/n**) or is not performed to a significant degree (code **2/n**):

- If the HGMU has a hydrological code of **H5**, **H6**, **H7**, **H8**, **H9**, **H10**, **H13**, **H16** *(Recording Sheet 2a, F)* the process, and therefore the function, of groundwater discharge is not performed.
- If the assessment code for the Groundwater recharge function (Section 3.1.B) is **1/n**, there is net downward movement of groundwater in the HGMU and discharge is impossible.

In all other cases, answer the following questions for each controlling variable. Complete all questions for which sufficient information is available. Circle the correct answer on the appraisal sheet. Use a new appraisal sheet for each HGMU. The location of the relevant information is given in brackets for each question.

The concept of 'significant discharge' poses a problem. The geology is often conducive to some groundwater flow, and some discharge will occur in most cases. However, it is often not a significant function, and may make only a minor contribution to the water budget of a wetland, which may be dominated by inputs from rainfall, run-off or flooding. For this reason a limited measure of its potential significance has been introduced.

Controlling variables

CV1 HGMU location and geometry
CV2 Drift and solid geological materials
CV3 Water table dynamics
CV4 Surface water flows

CV1 *HGMU location and geometry*

Q1. Is the elevation of the HGMU surface lower than that of the surrounding area (either in AA or CA), i.e. is the ground surface profile concave? *(Recording Sheet 3, B.3.5; AA Map 1; CA Map 1; Gradient code – Recording Sheet 2a, C.2)*

Q2. Is the HGMU located downslope of one of the following types of recharge area, which must be at least as large as the HGMU under study?

- An HGMU performing the 'groundwater recharge' function *(3.1.B, assessment codes **1/n**).*

- An area with recognised recharge potential *(CA Map 2)*
- An area underlain by a recognised aquifer *(CA Map 2)*
- An area underlain by solid rock likely to make a good aquifer *(CA Map 2)*
- An area underlain by non-solid deposits likely to make a good aquifer *(CA Map 2)*
- An area with a coarse or gravel soil *(Recording Sheet 2a, C.3.7)*

Q3. Is the HGMU bordered on two sides or more by higher ground (with a combined area at least the size of the HGMU) that does not appear to produce surface run-off, i.e. does the HGMU have a groundwater catchment? Areas without run-off output will have hydrological codes **H1**, **H2**, **H5**, **H10**, **H13**, **H14**, **H15**, **H16** *(Recording Sheet 2a, F)* or meet the criteria for a recharge area given in ***Q2.***

Q4. Does the HGMU have a surface water outlet but no surface water inlet? *(Recording Sheet 3, B.2.1, B.2.2; AA Map 1)*

Q5. Is the HGMU situated directly downslope from an artificial reservoir, pond or lake where water levels are kept higher than they would be under natural conditions? *(Recording Sheet 3, B.3.13; AA Map 1)*

CV2 *Drift and solid geological materials*

Q1. Is fine or medium material absent within 50 cm of the surface of the HGMU? *(Recording Sheet 2a, C.3.7)*

Q2. Is any part of the HGMU underlain by any of the following?
- A recognised aquifer *(CA Map 2)*
- Solid rock likely to make a good aquifer *(CA Map 2)*
- Non-solid deposits likely to make a good aquifer *(CA Map 2)*
- A soil which is indicative of underlying deposits with high permeability such as sand (coarse mineral material) or gravel *(Recording Sheet 2a, C.3.7)*

CV3 *Water table dynamics*

Groundwater flows underground until it reaches a discharge zone, an area where the water meets the land surface. Discharge may be clearly visible in the form of springs or less obvious as groundwater seepage into wetlands or contributing to stream or lake flows. If the water table is close to the land surface during the growing season, large amounts of groundwater can be removed through transpiration of the vegetation, while water at the surface will be lost through evaporation.

Q1. Is the water table at or above the HGMU surface most of the time? *(Recording Sheet 1, C.3.22.2a–c)*

Q2. Is the water table permanently less than 0.3 m below the HGMU surface? *(Recording Sheet 3, B.3.7–12)*

Q3. Are any of the following present within the HGMU?
- Springs: flowing water emerging from a very localised source area *(Recording Sheet 3, B.3.1)*
- Seepage zones as indicated by snow-free areas at times when snow covers surrounding areas *(Recording Sheet 3, B.3.3; B.3.4)*
- Permanent streams originating within the HGMU *(Recording Sheet 3, B.2.6; B.2.7)*
- Permanent standing water *(Recording Sheet 3, B.4.3)*

Q4. Is the vegetation in the HGMU predominantly of the following types, indicative of groundwater discharge conditions: *vegetation code **M6**, **M7**, **M10**, **M12**, **U14**?* *(Recording Sheet 3, D)*

Q5. Is the vegetation in the HGMU predominantly of the following types, indicative of a lack of groundwater discharge: *vegetation code **S2–19**, **G18–32**, **W22–44**?* (*Recording Sheet 3, D*)

3.1.C Groundwater Discharge Appraisal Sheet

HGMU code: ________________	**AA name:** ________________

CV1 HGMU location and geometry			
Q1	Y	N	U/NA
Q2	Y	N	U/NA
Q3	Y	N	U/NA
Q4	Y	N	U/NA
Q5	Y	N	U/NA

CV3 Water table dynamics			
Q1	Y	N	U/NA
Q2	Y	N	U/NA
Q3	Y	N	U/NA
Q4	Y	N	U/NA
Q5	Y	N	U/NA

CV2 Drift and solid geological materials			
Q1	Y	N	U/NA
Q2	Y	N	U/NA

CV4 Surface water flows			
Q1	Y	N	U/NA

HGMU function assessment code: ________________________________

3.1.C Look-up Table 1

<table>
<tr><th>CV1</th><th>CV2</th><th>CV3</th><th>CV4</th><th>Rationale</th><th>Code</th></tr>
<tr><td>1 2|3|4|5</td><td>1 2</td><td>1 2 3 4</td><td>1</td><td>It is evident that discharge occurs and is significant on a regional level. A large recharge area sustains permanent stream or lake flow. Field indicators confirm the groundwater origin of this water.</td><td>1/1</td></tr>
<tr><td>1 2|3|4|5</td><td>1 2</td><td>1 2 3 4</td><td>*</td><td>It is evident that discharge occurs and it is very likely that it is significant on a local or regional level. A large recharge area sustains flow from the HGMU. Field indicators confirm the groundwater origin of this water.</td><td>1/2</td></tr>
<tr><td>1 2|3|4|5</td><td>1 2</td><td>1 2</td><td>1</td><td>It is very likely that discharge occurs and is significant on a local or regional level. High water levels and sustained flow indicate groundwater origin.</td><td>1/3</td></tr>
<tr><td>1 2|3|4|5</td><td>1 2</td><td>1 2</td><td>*</td><td>It is evident that discharge occurs, but its local or regional significance is unknown. There are some field indicators for a groundwater origin.</td><td>1/5</td></tr>
<tr><td>1 2|3|4|5</td><td>1 2</td><td>3|4</td><td>1</td><td>It is very likely that discharge occurs and is significant on a local or regional level. There are some field indicators for a groundwater origin.</td><td>1/4</td></tr>
<tr><td>1 2|3|4|5</td><td>1 2</td><td>3|4</td><td>*</td><td>It is evident that discharge occurs, but its local or regional significance is unknown. There are some field indicators for a groundwater origin.</td><td>1/6</td></tr>
<tr><td>*</td><td>*</td><td>1 2 3|4</td><td>1</td><td>There are strong indications of groundwater discharge, which is locally or regionally significant, but its local or regional significance is unknown.</td><td>1/7</td></tr>
<tr><td>*</td><td>*</td><td>1 2 3|4</td><td>*</td><td>There are strong indications of groundwater discharge, but its origin and its local or regional significance is unknown.</td><td>1/8</td></tr>
</table>

3.1.C Look-up Table 2

<table>
<tr><th>CV1</th><th>CV2</th><th>CV3</th><th>CV4</th><th>Rationale</th><th>Code</th></tr>
<tr><td>#2#3</td><td>*</td><td>#1#2</td><td>*</td><td>There is no evidence that the function is performed.</td><td>4</td></tr>
<tr><td>*</td><td>#1</td><td>#1#2</td><td>*</td><td>There is no evidence that the function is performed.</td><td>4</td></tr>
<tr><td>*</td><td>*</td><td>5</td><td>*</td><td>There is evidence that the function is not performed.</td><td>4</td></tr>
</table>

CV4 *Surface water flows*

Q1. Does a stream or lake originate in the HGMU? *(Recording Sheet 3, B.2.3)*

Assessment of the function

Compare the results recorded for each HGMU on the appraisal sheets with the possible outcomes below, and assign an appropriate assessment code. Record this on the appraisal sheet. The syntax codes used are described in the introduction to Section 3.

Look-up tables

1. The function is definitely being performed
If the answers to the questions for this function agree with any of the combinations in Look-up Table 1, then the HGMU is definitely performing the function of groundwater discharge. The order in which the answers are given relates to both the certainty of occurrence of the function (first part of the code, e.g. '**1**/1') and the degree to which it operates (second part of the code, e.g. '1/**1**'), based on the strength of evidence resulting from the combination of controlling variables: for Code **1/1** the function is performed at the highest rate.

2. The function is not (significantly) being performed
For any combination of answers that is not covered by either Look-up Table 1 or Look-up Table 2, there may be recharge occurring in the HGMU, but only to a degree that is probably insignificant to functioning. Assign assessment code **2** to these HGMUs.

3. The function is probably being performed, but there are constraining factors or uncertainties
This statement is not applicable for the Groundwater Discharge function.

4. The function is definitely not being performed
If answers to the questions for this function are in agreement with any of the combinations listed in Look-up Table 2 then groundwater discharge definitely does not occur in the HGMU.

This completes the functional assessment for groundwater discharge. The output codes can be recorded on the Functional Assessment Table in Section 4.0 and used to compare functions among HGMUs across the wetland. If you wish to assess all functions, move on to the next step of the FAPs: Sediment retention.

When all the HGMUs in the AA have been assessed, a general pattern of groundwater discharge across the whole wetland will have been achieved. By manipulation of your answers to the controlling variable (CV) questions, you have the capacity to explore the probable outcomes of any management options under consideration and relate them to specific functions. The electronic version of the FAPs can do this more quickly and easily.

3.1.D Sediment retention (function)

Further reading and references

Bullock, A. and Acreman, M.C. (2003). The role of wetlands in the hydrological cycle, *Hydrology and Earth System Sciences* **7**(3), 358–389.

Carter, V. (1986). An overview of the hydrologic concerns related to wetlands in the United States, *Canadian Journal of Botany* **64**, 364–374.

Carter, V., Bedinger, M.S., Novitzki, R.P. and Wilen, W.O. (1979). Water resources and wetlands. In: Greeson, P.E., Clark, J.R. and Clark, J.E. (eds), *Wetland Functions and Values: the State of Our Understanding*, American Water Resources Association, Minneapolis, MN, 344–376.

Hughes, J.M.R. and Heathwaite, A.L. (eds) (1995). *Hydrology and Hydrochemistry of British Wetlands*, Wiley, Chichester, UK.

Kadlec, R.H. and Hey, D.L. (1994). Constructed wetlands for river water quality improvement, *Water Science and Technology* **29**(4), 159–168.
Mitsch, W.J. and Gosselink, J.G. (2007) *Wetlands*, 4th edn, Wiley, New York.
Novitzki, R.P. (1979). The hydrologic characteristics of Wisconsin wetlands and their influence on floods, streamflow and sediment. In: Greeson, P.E., Clark, J.R. and Clark, J.E. (eds), *Wetland Functions and Values: the State of Our Understanding*, American Water Resources Association, Minneapolis, MN, 377–388.

Introduction

Function description:

- Net retention of sediments carried in suspension by river or lake water during overbank flooding events, i.e. sedimentation of clay, silt and fine sand exceeds erosion – the HGMU acts as a sediment *sink* to the river or lake.
- Net retention of sediments carried in suspension by surface run-off from other wetland units or the CA – the HGMU acts as a sediment *buffer* to the river or lake.

Significance – rationale for assessment:

- River or lake water quality may be improved by reduction of suspended sediment load and turbidity.
- River or lake water quality may be improved by reduction of loads of pollutants and nutrients, such as heavy metals and phosphorus, which are part of or are adsorbed to the sediment.
- Sedimentation rates downstream of the wetland may be reduced.

Related functions, for which outcome of this assessment may be required:

- Nutrient retention

For the Sediment retention function to occur the following conditions must be met:

1. The HGMU must receive significant water inputs either from run-off or from overbank flooding.
2. The water inputs must have a significant suspended sediment load.
3. The flow velocity in the HGMU must be sufficiently low for the sediment to settle, but high enough to ensure continued inputs of sediments.
4. Erosion in the HGMU must be relatively small.

The amount of suspended sediment transported depends on the sediment particle size and the flow velocity of the water. A high flow velocity can transport a greater number of larger particles than can a slower current. Suspended sediment becomes deposited as flow velocity decreases.

Functional assessment may not be necessary in the following cases where either the sediment retention function is not performed (code **4/n**) or not performed to a significant degree (code **2/n**):

- If the HGMU has gradient code **G2** or **G3** *(Recording Sheet 2a (Field), C.2)* the chances of sedimentation of suspended sediment are small, as surface water flow velocities are likely to be high.
- If the HGMU has hydrological code **H1**, **H8**, **H11** or **H16** *(Recording Sheet 2a (Field), F)* it probably does not receive water inputs from either run-off or overbank flooding, and therefore no sediment is transported by this mechanism.

In any other case, answer the following questions for each controlling variable. Complete all questions for which sufficient information is available. Circle the correct answer on the appraisal sheet. Use a new appraisal sheet for each HGMU. The location of the relevant information is given in brackets for each question.

A complication with the Sediment Retention function is that it is highly variable in both time and space. Current sedimentation patterns and rates can be completely different from those in the recent past owing to changes in local flow or erosion conditions upstream or upslope. In many cases these changes cannot be easily detected. Also there can be difficulties interpreting the value of the function.

Controlling variables

CV1 Flooding regime
CV2 Suspended sediment load in floodwaters
CV3 Suspended sediment load in surface run-off
CV4 Sediment deposition
CV5 Erosion

CV1 *Flooding regime*

Q1. Is it evident that the HGMU is prone to frequent flooding (at least once in less than 2 years on average) by river or lake water? *(Recording Sheet 3, B.4.1; B.4.17; Recording Sheet 1, C.3.19.2a; FA procedure 3.1.A, Codes 1/1, 1/1a–d, 1/2, 1/2a–d, 1/3, 1/3a–d, 1/4, 1/4a–d)*

Q2. Is it evident that the HGMU is prone to occasional flooding (once every 2–10 years on average) by river or lake water? *(Recording Sheet 3, B.4.1; B.4.18; Recording Sheet 1, C.3.19.2b; FA procedure 3.1.A, Codes 1/5, 1/6, 1/7, 1/8)*

Q3. Is it evident that overbank flooding does not occur? (*Hydrological codes* **H1**, **H2**, **H3**, **H5**, **H6**, **H11**, **H16** – *Recording Sheet 2a, F; FA procedure 3.1.A, Codes 1/16, 2/n, 3/n)*

Q4. Is the flood storage volume known? *(See ACTION 1)*
[If YES, continue with **Q5**, if NO go to **Q8**]

Q5. Does the flood storage volume not exceed 1 % of daily peak discharge during occasional or frequent (at least once in 2–10 years) floods? *(See ACTION 1)*

Q6. Does the flood storage volume exceed 10 % of daily peak discharge during frequent (at least once in less than 2 years) floods? *(See ACTION 1)*

Q7. Does the flood storage volume exceed 10 % of daily peak discharge during occasional (at least once in 2-10 years) floods? *(See ACTION 1)*

Q8. Is the maximum flow velocity in the HGMU during floods less than 0.3 m s^{-1}? *(Recording Sheet 3, B.4.6)*

Q9. Is the HGMU mostly covered with 'rough' vegetation such as dense reed, bushes or forest, which will slow down floodwaters considerably? *(Recording Sheet 3, B.4.20)*

Q10. Is the HGMU more or less level? (*Gradient code* **G0** – *Recording Sheet 2a, C.2*)

Q11. If the HGMU has only one (potential) outlet for floodwater, which is clearly distinguishable in the field, is the width of the outlet less than 10 % of the average width of the flooded area throughout the HGMU? In other words, is the outflow from the HGMU restricted? *(Recording Sheet 3, B.4.21)*

ACTION 1 – Calculation of flood storage volume:
Perform this action only if data on flood discharges are available from *Recording Sheet 1 (C.3.19.4)* and if this action has not already been performed for the Flood Detention function.

If possible, determine flood storage volume of the river or lake (*flood depth* × *HGMU area*) in m^3. The flood depth must be expressed in metres over the HGMU surface, which can be calculated from the flood level and the average surface level in the HGMU. Flood level and surface level must be relative to the same reference level, which can be m.O.D. or any local level. Flood levels for different flood frequencies are given in *Recording Sheet 1, C.3.19.2*. The HGMU surface level and the reference level are given in *Recording Sheet 3, B.1*.

Determine flood storage volume for:

- frequent (at least once in less than 2 years) flood events;
- occasional (once every 2–10 years) flood events;
- rare (once every 10–50 years) flood events;
- bankfull height – to the top of the levee.

Determine the flood storage volume as a fraction of the daily flood discharges as given in *Recording Sheet 1, C.3.19.4.*

CV2 *Suspended sediment load in floodwaters*

Q1. Has the river or lake close to the HGMU (comprising less than 20 % of the total river length occurring upstream from the site) been observed to carry high concentrations of suspended sediments during peak discharges? *(Recording Sheet 3, B.5.3)*

Q2. Have any of the following activities occurred at a large scale in the upstream catchment (in more than 10 % of the catchment or bordering more than 10 % of the river channel), during the last 3 years *(Recording Sheet 1)*:
- Logging or afforestation *(C.3.24.2 (answer d))*
- Urban development or road building *(C.3.24.2 (answer b))*
- Mining *(C.3.24.2 (answer a))*
- Major river or lake channel modifications *(C.3.24.2 (answer c))*

Q3. Is a significant part (> 10 %) of the upstream catchment covered with one of the following? *(Recording Sheet 1)*:
- Mountainous areas *(C.3.24.1 (answer a))*
- Permanently or seasonally unvegetated areas *(C.3.24.1 (answer b))*
- Arid or semi-arid areas *(Topographic Maps; Recording Sheet 1, C.3.1.12)*
- Areas where annual potential evapotranspiration exceeds annual precipitation *(C.3.1.1; C.3.1.3)*
- Permafrost areas, where soils are frozen throughout the year ('pergelic' soils) *(C.3.1.5)*

Q4. Does the river directly upstream of the site (comprising at least 20 % of the river length) have a recent history of migrating channels (during the last 20 years)? *(Recording Sheet 1, C.3.24.5)*

CV3 *Suspended sediment load in surface run-off*

Q1. Are there indications of erosion in the area immediately upslope (in AA or CA) of the HGMU? *(Recording Sheet 3, B.5.1)*

Q2. Has surface run-off from the area immediately upslope (in AA or CA) been observed to carry high concentrations of suspended sediment? *(Recording Sheet 3, B.5.6)*

Q3. Have any of the following activities taken place at a large scale in the area immediately upslope (in AA or CA), during the last year? *(Recording Sheet 1)*:
- Logging or afforestation *(C.3.6.1a (answer c))*
- Urban development or road building *(C.3.13.1a; C.3.13.2)*
- Mining *(C.3.11.1a (answer c); C.3.11.2 (answer c))*
- Ploughing *(C.3.5.1a (answer c); C.3.5.2 (answer c))*

Q4. Is a significant part (> 50 %) of the area immediately upslope (in AA or CA) covered with one of the following *(Recording Sheet 1; AA Map 2)*:
- Mountainous areas *(AA Map 2)*
- Areas where bare rock commonly occurs *(AA Map 2)*
- Permanently or seasonally unvegetated areas (AA Map 2)
- Arid or semi-arid areas *(Recording Sheet 1, C.3.1.12)*
- Areas where annual potential evapotranspiration exceeds annual precipitation *(Recording Sheet 1, C.3.1.1; C.3.1.3)*
- Permafrost areas, where soils are frozen throughout the year *(Recording Sheet 1, C.3.1.5)*

CV4 *Sediment deposition*

Q1. Is there evidence of recently deposited mineral or organic material on the soil surface in the HGMU? *(Recording Sheet 3, B.5.4)*

Q2. Is there evidence of recently deposited mineral material on plant surfaces in the HGMU? *(Recording Sheet 3, B.5.5)*

Q3. Is the topsoil (0–10 cm) in the HGMU dominated by fine or medium mineral material? *(Recording Sheet 2a, C.3.7)*

Q4. Is the HGMU a 'levee', 'flat area' or 'backswamp', within the floodplain component? *(Recording Sheet 2a, C.2; Geomorphic codes:* ***LV***, ***FA***, ***BS****)*

CV5 *Erosion*

Q1. Is there evidence of erosion within the HGMU? This could include scouring of banks by action of the river, or loss of material in units not adjacent to the river/lake, by grazing animals or human activity. *(Recording Sheet 3, B.5.2)*

Q2. Do flow velocities in the HGMU exceed 0.3 m s^{-1}? *(Recording Sheet 3, B.4.6)*

Q3. Is the HGMU unvegetated or sparsely vegetated? *(Recording Sheet 3, B.5.7)*

Q4. Is the HGMU adjacent to a river that is known to shift its course locally, and is it situated on an outer bend of that river? *(Recording Sheet 3, B.5.8)*

Assessment of the function

Compare the results recorded for each HGMU on the appraisal sheets with the possible outcomes below, and assign an appropriate assessment code. Record this on the appraisal sheet. The syntax codes used are described in the introduction to Section 3.

Look-up tables

1. The function is definitely being performed
If the answers to the questions for this function agree with any of the combinations in Look-up Table 1 then the HGMU is definitely performing the function of sediment retention. The order in which the answers are given relates to both the certainty of occurrence of the function (first part of the code, e.g. '**1**/1') and the degree to which it operates (second part of the code, e.g. '1/**1**'), based on the strength of evidence resulting from the combination of controlling variables: for Code **1/1** the function is performed at the highest rate.

2. The function is not (significantly) being performed
For any combination of answers that is not covered by either Look-up Table 1 or Look-up Table 2 there may be sediment retention occurring in the HGMU, but probably only to a degree that is insignificant to functioning. Assign assessment code **2** to these HGMUs.

3. The function is probably being performed, but there are constraining factors or uncertainties
This statement is not applicable for this function.

4. The function is definitely not being performed
If answers to the questions for this function are in agreement with any of the combinations listed in Look-up Table 2 then sediment retention definitely does not occur in the HGMU.

This completes the functional assessment for sediment retention. The output codes can be recorded on the Functional Assessment Table in Section 4.0 and used to compare functions among HGMUs across the wetland. A section follows on the rapid quantification of hydrological functions. If you wish to assess all functions, move on to the next step of the FAPs: Biogeochemical functions.

When all the HGMUs in the AA have been assessed, a general pattern of sediment retention across the whole wetland will have been achieved. By manipulation of your answers to the controlling variable (CV) questions, you have the capacity to explore the probable outcomes of any management options under consideration and relate them to specific functions. The electronic version of the FAPs can do this more quickly and easily.

3.1.D Sediment Retention Appraisal Sheet

HGMU code: ____________	**AA name:** ____________

CV1 Flooding regime			
Q1	Y	N	U/NA
Q2	Y	N	U/NA
Q3	Y	N	U/NA
Q4	Y	N	U/NA
Q5	Y	N	U/NA
Q6	Y	N	U/NA
Q7	Y	N	U/NA
Q8	Y	N	U/NA
Q9	Y	N	U/NA
Q10	Y	N	U/NA
Q11	Y	N	U/NA

CV2 Suspended sediment load in flood waters			
Q1	Y	N	U/NA
Q2	Y	N	U/NA
Q3	Y	N	U/NA
Q4	Y	N	U/NA

CV3 Suspended sediment load in surface run-off			
Q1	Y	N	U/NA
Q2	Y	N	U/NA
Q3	Y	N	U/NA
Q4	Y	N	U/NA

CV4 Sediment deposition			
Q1	Y	N	U/NA
Q2	Y	N	U/NA
Q3	Y	N	U/NA
Q4	Y	N	U/NA

CV5 Erosion			
Q1	Y	N	U/NA
Q2	Y	N	U/NA
Q3	Y	N	U/NA
Q4	Y	N	U/NA

HGMU function assessment code: ____________________________

3.1.D Look-up Table 1

<table>
<tr><th>CV1</th><th>CV2</th><th>CV3</th><th>CV4</th><th>CV5</th><th>Rationale</th><th>Code</th></tr>
<tr><td>1(8|9|10 |11)</td><td>1|2|3|4</td><td>*</td><td>1|2|(3 4)</td><td>No more than 1.</td><td>The HGMU retains suspended sediments from stream or lake water during frequent floods, but it is unknown if this forms a significant proportion of the suspended sediment load in the stream or lake. A lack of erosion-indicators suggests that the retention will be long-term.</td><td>1/3</td></tr>
<tr><td>1(8|9|10 |11)</td><td>1|2|3|4</td><td>*</td><td>1|2|(3 4)</td><td>At least 1.</td><td>The HGMU retains suspended sediments from stream or lake water during frequent floods, but it is unknown if this forms a significant proportion of the suspended sediment load in the stream or lake. Some erosion-indicators suggest that the retention may be short-term.</td><td>1/13</td></tr>
<tr><td>1 6 (8|9 |10|11)</td><td>1|2|3|4</td><td>*</td><td>1|2|(3 4)</td><td>No more than 1.</td><td>The HGMU retains suspended sediments from stream or lake water during frequent floods, and this is likely to form a significant proportion of the suspended sediment load in the stream or lake. A lack of erosion-indicators suggests that the retention will be long-term.</td><td>1/1</td></tr>
<tr><td>1 6 (8|9 |10|11)</td><td>1|2|3|4</td><td>*</td><td>1|2|(3 4)</td><td>At least 1.</td><td>The HGMU retains suspended sediments from stream or lake water during frequent floods, and this is likely to form a significant proportion of the suspended sediment load in the stream or lake. Some erosion-indicators suggest that the retention may be short-term.</td><td>1/11</td></tr>
<tr><td>2(8|9|10 |11)</td><td>1|2|3|4</td><td>*</td><td>1|2|(3 4)</td><td>No more than 1.</td><td>The HGMU retains suspended sediments from stream or lake water during occasional floods, but it is unknown if this forms a significant proportion of the suspended sediment load in the stream or lake. A lack of erosion-indicators suggests that the retention will be long-term.</td><td>1/4</td></tr>
<tr><td>2(8|9|10 |11)</td><td>1|2|3|4</td><td>*</td><td>1|2|(3 4)</td><td>At least 1.</td><td>The HGMU retains suspended sediments from stream or lake water during occasional floods, but it is unknown if this forms a significant proportion of the suspended sediment load in the stream or lake. Some erosion-indicators suggest that the retention may be short-term.</td><td>1/14</td></tr>
<tr><td>2 7 (8|9 |10|11)</td><td>1|2|3|4</td><td>*</td><td>1|2|(3 4)</td><td>No more than 1.</td><td>The HGMU retains suspended sediments from stream or lake water during occasional floods, and this is likely to form a significant proportion of the suspended sediment load in the stream or lake. A lack of erosion-indicators suggests that the retention will be long-term.</td><td>1/2</td></tr>
<tr><td>2 7 (8|9 |10|11)</td><td>1|2|3|4</td><td>*</td><td>1|2|(3 4)</td><td>At least 1.</td><td>The HGMU retains suspended sediments from stream or lake water during occasional floods, and this is likely to form a significant proportion of the suspended sediment load in the stream or lake. Some erosion-indicators suggest that the retention may be short-term.</td><td>1/12</td></tr>
<tr><td>5</td><td>*</td><td>(1|2) (3|4)</td><td>1|2|3</td><td>No more than 1.</td><td>The HGMU retains suspended sediments but surface runoff is likely to be the only source. A lack of erosion-indicators suggests that the retention will be long-term.</td><td>1/8</td></tr>
<tr><td>5</td><td>*</td><td>(1|2) (3|4)</td><td>1|2|3</td><td>At least 1.</td><td>The HGMU retains suspended sediments, but surface run-off is likely to be the only source. Some erosion-indicators suggest that the retention may be short-term.</td><td>1/18</td></tr>
<tr><td>(8|9|10 |11)</td><td>1|2|3|4</td><td>*</td><td>1|2|3</td><td>No more than 1.</td><td>The HGMU is likely to retain suspended sediments from stream or lake water, but flooding is rare or the flooding frequency is unknown. A lack of erosion-indicators suggests that the retention will be long-term.</td><td>1/6</td></tr>
<tr><td>(8|9|10 |11)</td><td>1|2|3|4</td><td>*</td><td>1|2|3</td><td>At least 1.</td><td>The HGMU is likely to retain suspended sediments from stream or lake water, but flooding is rare or the flooding frequency is unknown. Some erosion-indicators suggest that the retention may be short-term.</td><td>1/16</td></tr>
</table>

3.1.D Look-up Table 1 cont'd

<table>
<tr><th>CV1</th><th>CV2</th><th>CV3</th><th>CV4</th><th>CV5</th><th>Rationale</th><th>Code</th></tr>
<tr><td>(8|9|10 |11)</td><td>1|2|3|4</td><td>*</td><td>1|2|(3 4)</td><td>No more than 1.</td><td>The HGMU retains suspended sediments from stream or lake water, but flooding is rare or the flooding frequency is unknown. A lack of erosion-indicators suggests that the retention will be long-term.</td><td>1/5</td></tr>
<tr><td>(8|9|10 |11)</td><td>1|2|3|4</td><td>*</td><td>1|2|(3 4)</td><td>At least 1.</td><td>The HGMU retains suspended sediments from stream or lake water, but flooding is rare or the flooding frequency is unknown. Some erosion-indicators suggest that the retention may be short-term.</td><td>1/15</td></tr>
<tr><td>*</td><td>At most 1.</td><td>(1|2) (3|4)</td><td>1|2|3</td><td>No more than 1.</td><td>The HGMU retains suspended sediments, but surface run-off is likely to be the only source. A lack of erosion-indicators suggests that the retention will be long-term.</td><td>1/9</td></tr>
<tr><td>*</td><td>At most 1.</td><td>(1|2) (3|4)</td><td>1|2|3</td><td>At least 1.</td><td>The HGMU retains suspended sediments, but surface run-off is likely to be the only source. Some erosion-indicators suggest that the retention may be short-term.</td><td>1/19</td></tr>
<tr><td>*</td><td>*</td><td>3 4</td><td>1|2|3</td><td>No more than 1.</td><td>The HGMU is likely to have surface run-off inputs with high suspended sediment loads, and is likely to retain at least some of the sediment. A lack of erosion-indicators suggests that the retention will be long-term.</td><td>1/10</td></tr>
<tr><td>*</td><td>*</td><td>3 4</td><td>1|2|3</td><td>At least 1.</td><td>The HGMU is likely to have surface run-off inputs with high suspended sediment loads, and is likely to retain at least some of the sediment. Some erosion-indicators suggest that the retention may be short-term.</td><td>1/20</td></tr>
<tr><td>*</td><td>*</td><td>*</td><td>1|2|3</td><td>No more than 1.</td><td>The HGMU is very likely to retain suspended sediments, but it is unknown if the sediment source is stream or lake water or surface run-off. A lack of erosion-indicators suggests that the retention will be long-term.</td><td>1/7</td></tr>
<tr><td>*</td><td>*</td><td>*</td><td>1|2|3</td><td>At least 1.</td><td>The HGMU retains suspended sediments, but it is unknown if the sediment source is stream or lake water or surface run-off. Some erosion-indicators suggest that the retention may be short-term.</td><td>1/17</td></tr>
</table>

3.1.D Look-up Table 2

<table>
<tr><th>CV1</th><th>CV2</th><th>CV3</th><th>CV4</th><th>CV5</th><th>Rationale</th><th>Code</th></tr>
<tr><td colspan="5">The HGMU is of type H1, H8, H11 or H16</td><td>There is evidence that the function is not performed.</td><td>4</td></tr>
<tr><td colspan="5">The HGMU is of type G2, G3</td><td>There is evidence that the function is not performed.</td><td>4</td></tr>
<tr><td>*</td><td>#(1|2 |3|4)</td><td>#(1|2 |3|4)</td><td>*</td><td>*</td><td>There is no evidence that the function can be performed.</td><td>4</td></tr>
<tr><td>*</td><td>*</td><td>*</td><td>#(1|2 |3|4)</td><td>1 2 3 4</td><td>There is evidence that the function is not performed.</td><td>4</td></tr>
</table>

3.1.E Rapid quantification of hydrological functions

Introduction

The most frequently asked questions in wetland hydrology relate to the magnitude of the functions performed, especially in floodwater detention and groundwater recharge. This is because of the obvious social and economic benefits that arise from flood control and replenishment of aquifers. However, there are also considerable benefits that result from the discharge of clean aquifer waters and the retention of sediment. Normally the quantification of these functions would demand considerable research time and expense, which may not be available to support decision making requirements. The following sections provide guidance on how these functions may be quantified and some of the techniques available for their modelling.

Structure of the rapid quantification section:

3.1.E.1 Rapid quantification of 3.1.A Floodwater detention
3.1.E.2 Rapid quantification of 3.1.B Groundwater recharge
3.1.E.3 Rapid quantification of 3.1.C Groundwater discharge
3.1.E.4 Rapid quantification of 3.1.D Sediment retention

3.1.E.1 *Floodwater detention (function)*

Introduction

Wetlands predominantly contribute towards mitigating floods through their ability to store floodwater. Hence, a quantitative evaluation of this function should answer the following questions:

- What is the water storage capacity of a wetland?
- How often do flood events occur during which water is stored in the wetland?
- What is the significance of the volume of floodwater retained by the wetland compared with the total volume of the flood?

The following proposes a simple assessment methodology for answering the above questions for both river marginal and lake marginal wetlands.

Quantification of Floodwater detention function in river marginal wetlands (RMW)

The ability of river marginal wetlands to perform the Floodwater detention function depends, as with lake marginal wetlands, on their ability to store excess floodwater that cannot be retained within the river channel. The performance of this function leads to both a delay in and a lowering of downstream flood peaks (Figs 3.1.7 and 3.1.8).

Step 1. Calculation of water storage capacity. The data needed at this stage are:

- A topographic map on which are drawn the boundaries of the river (R), river marginal wetland (RMW) and contributory area (CA).
- The geometry of the system and, consequently, the water storage capacity of its components. This may be obtained from either a series of cross-sections, each representing the geometry of a uniform segment of the system river channel, wetland and CA, or by using a Geographical Information System (GIS) for the construction of a digital terrain model (DTM).

The size and consequent characteristics of the river channel determine the approach to the collection of topographic data (Verhoeven *et al.*, 2003).

Step 2. Calculation of flood frequency. A simple procedure, the plotting position method, can be used to determine the frequency of flood events in a river marginal wetland. The plotting position method is a statistical analysis of maximum discharges, which gives the time needed for the recurrence of a flood event

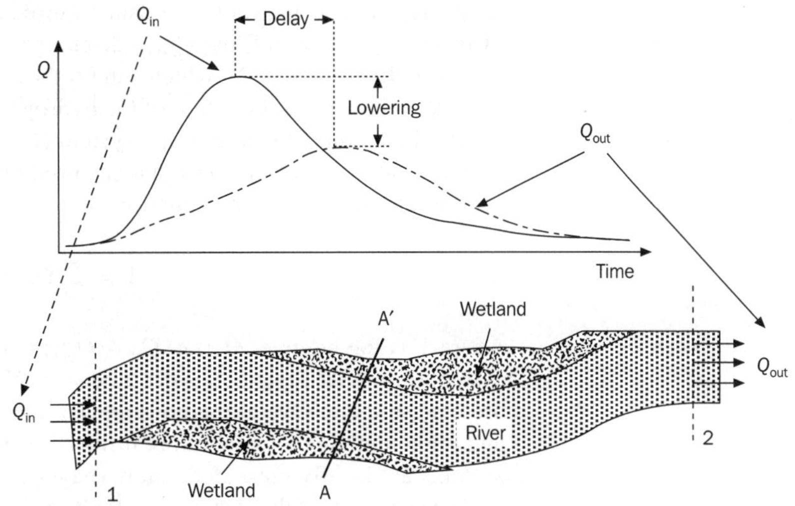

3.1.7 Delay and lowering of the flood peaks resulting from overbank flooding of RMWs. Q = discharge (i.e. volume of flow, e.g. m^3 per second); Q_{in} = inflow discharge; Q_{out} = outflow discharge.

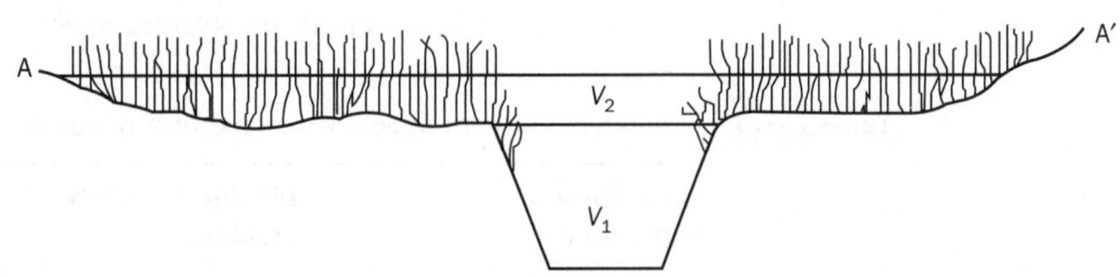

3.1.8 Cross-section of RMW (A–A′). V_1 is the maximum water volume that can be stored in the river channel. V_2 is the excess floodwater volume stored in the RMW during a flood event (this may be a part of, or the entire width of the floodplain depending on the size of the flood).

(discharge) greater than or equal to one specified. The time unit for this analysis most often used is a year. According to this method, the recurrence period (T), of a flood event in a time series is given by the following *Weibull formula* (Benson, 1962; Chow *et al.*, 1988):

$$T = \frac{n+1}{m}$$

where T is the time interval within which a specified magnitude of a flood event (discharge) will be equalled or exceeded, n is the total number of years for which measurements of flood discharges exist and, m is the 'position' of a given flood event within the time series. In order for the recurrence period T of flood events to be assessed, the time series of annual maximum flood discharges (Q_{max}) is used. The elements of the series are ranked in descending order and allocated a position number (m), the largest flood discharge having $m = 1$ (Table 3.1.3).

The graphical presentation of Table 3.1.3 is shown in Fig. 3.1.9. According to this graph, a flood event with peak discharge, for example, greater or equal to Q_3 can happen once in 10 years (arrowhead line). The relationship between discharge and return period can be defined mathematically, thus allowing extrapolation and accurate interpolation of values by assuming the pattern of return periods follows a standard statistical distribution. Many different types of probability distributions have been applied to the analysis of return periods, the most common being Gumbel, log-Normal, log Pearson Type III. The distribution which gives the best fit to the data should be chosen.

Step 3. Flood detention frequency assessment. This calculates the water volume that is stored in a river marginal wetland during a flood event, together with the frequency of such an event. The data required are measurements (or assessments) of inflow (Q_{in}) and outflow (Q_{out}) discharges during the annual maximum flood events (defined in Step 2), which can be shown graphically in a hydrograph (Figs 3.1.10a and b). The subtraction of the hydrograph ($Q_{out} - Q_{in}$) (Fig. 3.1.11a), gives the discharge that is stored in the system (Fig. 3.1.11b).

The total water volume (V) that is stored in the wetland system (Fig. 3.1.11c), can be estimated using the equation:

$$V = \sum_{t=1}^{t=n} Q_t \cdot \Delta_t$$

where V is the volume, Q_t (i.e. $Q_{in} - Q_{out}$) is the discharge stored in the wetland system (Fig. 3.1.11c) and Δ_t is the time step. Where inflow or outflow hydrographs are not available, several methods are available to derive estimates. A few simple and rapid methods for generating inflow–outflow hydrographs in natural wetland systems are briefly presented below (pages 289–293).

The geometry of the system (R–RMW–CA), as described by the cross-sections, together with the volume of storage calculated from discharge data, is used to derive the water volume that is stored by the wetland as well as by the other components of the system during a flood event of a given size (Table 3.1.4).

Figure 3.1.12 illustrates that once every T_x years a total water volume greater or equal to V_x is stored by the system R–RMW–CA. As the volumes of the three

Table 3.1.3 Annual maximum flood discharges for a period of n = 30 years

Position number	Maximum yearly discharge	Recurrence period (years)
m	Q_{max}	$T = (n+1)/m$
1	Q_1	30
2	Q_2	15
3	Q_3	10
n	Q_n	1

$Q_i \geq Q_{i+1}$

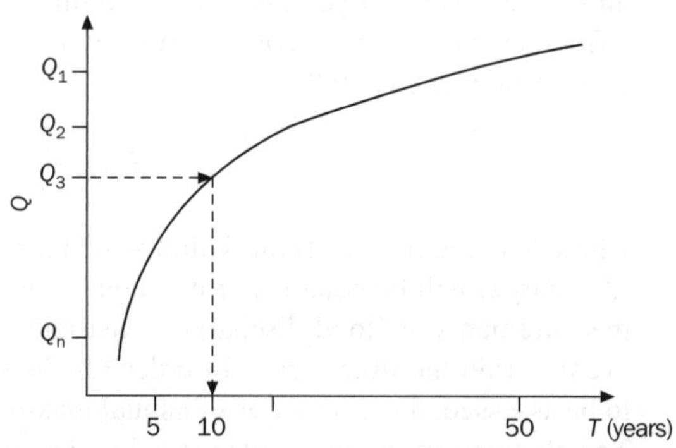

3.1.9 Graphical presentation of the recurrence period T for a time series composed of yearly maximum discharges. Q = discharge (i.e. volume of flow, e.g. m^3 per second).

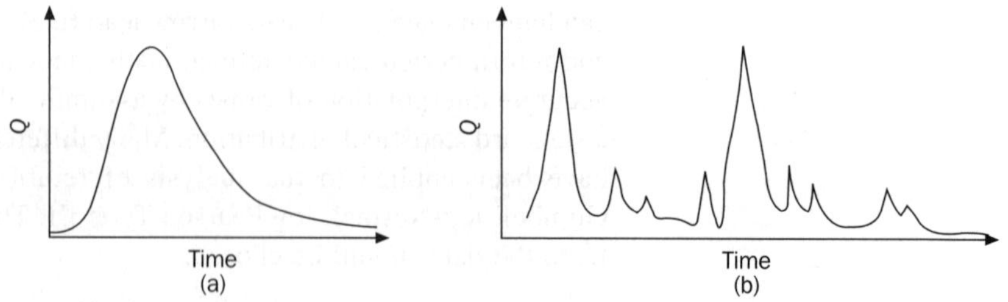

3.1.10 (a) Typical stream hydrograph of an individual precipitation event; (b) typical yearly hydrograph. Q = discharge (i.e. volume of flow, e.g. m^3 per second).

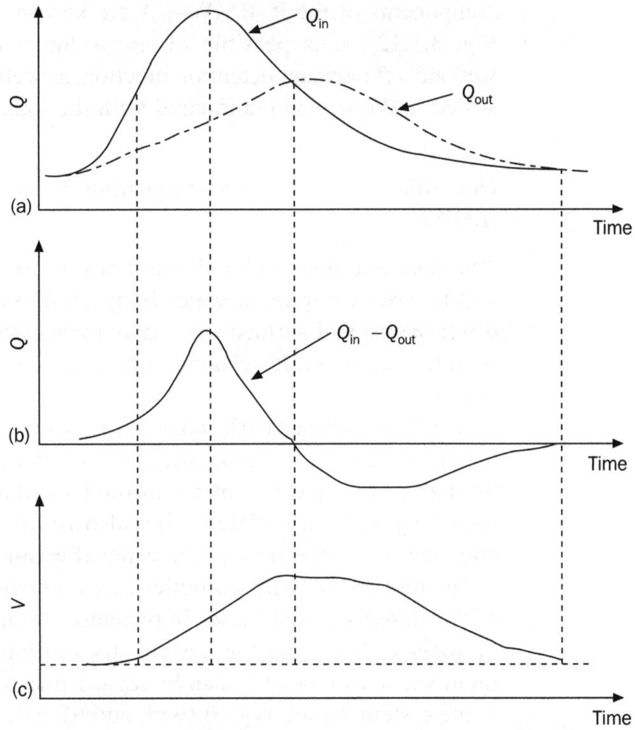

3.1.11 Temporal distribution of water stored in a wetland system during a flood event: (a) total inflow and outflow discharges; (b) difference between inflow – outflow discharges; (c) total water volume stored in the wetland system, during flood event. Q = discharge (i.e. volume of flow, e.g. m^3 per second); Q_{in} = inflow discharge; Q_{out} = outflow discharge.

Table 3.1.4 Annual maximum discharges, recurrence periods and stored water volumes for a period of n = 30 years

Annual maximum discharges	Corresponding stored water volumes	Recurrence period (years)
$\mathbf{Q_{max}}$	$\mathbf{V_s}$	$\mathbf{T = (n + 1)/m}$
Q_1	V_1	30
Q_2	V_2	15
Q_3	V_3	10
.	.	.
.	.	.
Q_n	V_n	1

$Q_i \geq Q_{i+1}$ and $V_i \geq V_{i+1}$

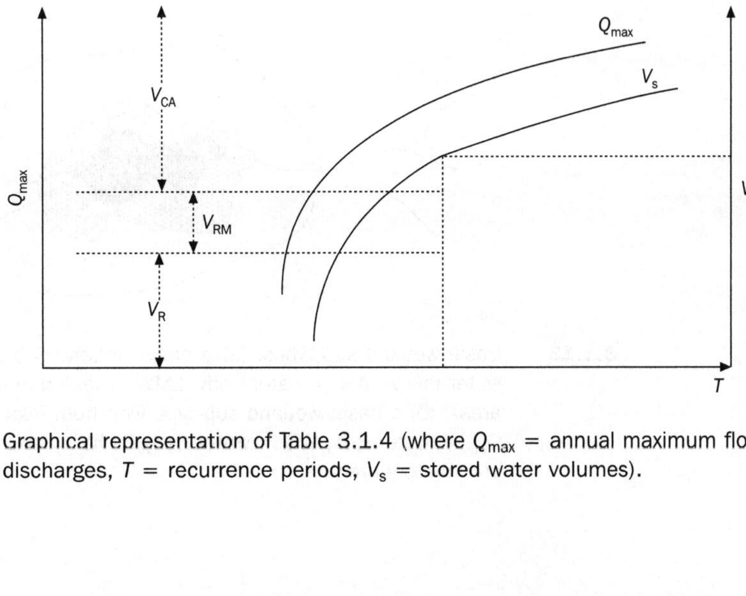

3.1.12 Graphical representation of Table 3.1.4 (where Q_{max} = annual maximum flood discharges, T = recurrence periods, V_s = stored water volumes).

components of the R–RMW–CA are known and constant (V_R, V_{RMW} and V_{CA} in Fig. 3.1.12), it is possible to assess the frequency of the performance of the wetland's floodwater detention function, as well as the significance of the floodwater stored in the wetland compared with the total volume of the flood event.

Quantification of floodwater detention function in lake marginal wetlands (LMW)

Two lake marginal wetland sub-types are recognised, as shown in Figs 3.1.13a and b. Both comprise a water body (WB) surrounded by a LMW and CA, but differ in terms of surface water dynamics in that Fig. 3.1.13a shows the case with an inlet but no outlet channel, and Fig. 3.1.13b that with both inlet and outlet channels.

The CA is defined as the area that drains directly into the area of interest, through either surface flow or groundwater flow. Water must be transported from the CA through local systems, but not through overland flooding from the main river, or regional groundwater systems. It is also assumed that the watershed for groundwater coincides with that for surface water (Section 2.1, Fieldwork preparation).

In sub-type A (with no outlet), as water flows (Q_{in}) into the system the level of the stored water (V) rises in response. Assuming that only a negligible amount of water is lost from the system (by infiltration, leakage and evaporation), the input water volume (V_{in}) can be regarded as being equal to the volume stored (V) in the system. In sub-type B (with outlet), however, the water volume that is stored (V) depends also on the water lost (Q_{out}) from the system through its outlet. Because of this difference in hydrological control, the quantification process has to be undertaken separately for each sub-type.

LMW sub-type A (with no outlet)

In this case, the ability of a wetland to mitigate floods depends on its ability to store part (or the whole amount) of the excess volume of water that inflows into the system (WB–LMW–CA) that cannot be stored by the water body (Fig. 3.1.13a).

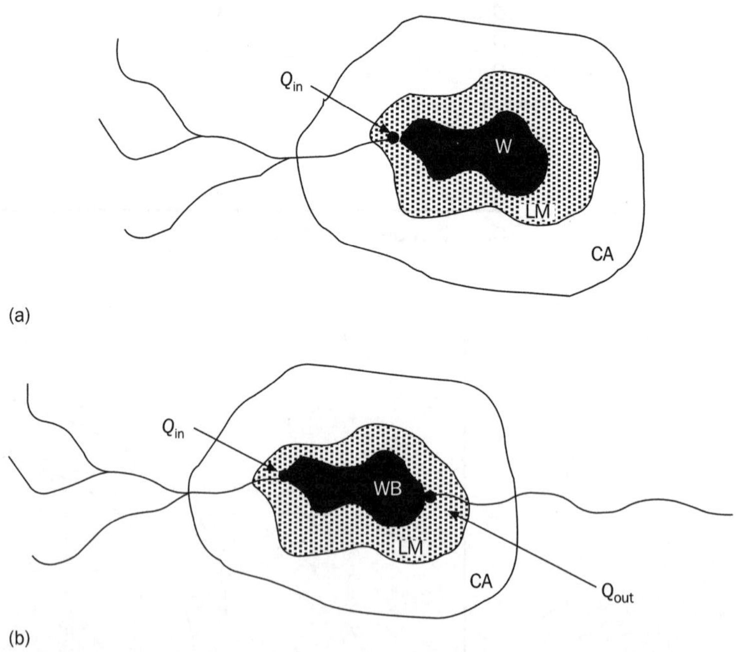

3.1.13 Basin wetland sub-types: (a) a basin wetland sub-type with inlet only (where Q_{in} = water inflow, WB = water body, LMW = lake marginal wetland and CA = contributory area); (b) a basin wetland sub-type with both inlet and outlet (where Q_{in} = water inflow, Q_{out} = water loss, WB = water body, LMW = lake marginal wetland and CA = contributory area).

Step 1. Calculation of water storage capacity. A topographic map on which are drawn the boundaries of the WB, LMW and CA is needed at this stage (Fig. 3.1.14, see colour Section).

The volume of water (*V*), that can be stored by the system WB–LMW–CA below any elevation (*H*), can be assessed either manually or using a GIS. In manual assessment, the mean of two areas bounded by successive contour isohytes multiplied by the contour height interval gives the interval volume, and the summation of interval volumes gives the total volume below any specified elevation. Using a GIS, the elevation contours together with the river network of the catchment area are digitised and processed using a GIS platform to enable a DTM of the catchment to be constructed. Applying a triangulation method (described below) on the data derived from the DTM, it is possible to estimate the volume of water that is stored below any elevation. It should be noted that the scale of the maps and the contour interval affect the accuracy of results in both methods. Maps using a scale from 1:25 000 to 1:5000 normally give satisfactory results.

Using the results of the above work, the relation $V = f(H)$ between the volume (*V*) that is stored in the system under any elevation (*H*) is quantified and transferred to a graph (Fig. 3.1.15). By determining the elevations of the boundaries of each component (WB, LMW and CA) on the topographic map and transferring them to the graph $V = f(H)$, one can determine both the wetland's water storage capacity and the potential for storage in each component of the system.

Valuable information can be derived from this graph:

- the volume threshold above which the wetland starts to store floodwater and thereby contributes to flood detention;
- The threshold of volume above which floodwater can cause damage during the storage of water in the CA.

Example of Step 1. Calculation of water storage capacity. Lake Cheimaditida is located in northern Greece (Prefecture of Florina). It is part of a wetland complex that includes another three lakes (Lake Zazari, Lake Petron and Lake Vegoritida). Its watershed covers an area of almost 34.7 km^2.

In the 1970s, a series of drainage works for agricultural purposes at the north-eastern part of Lake Cheimaditida was completed. This reduced the surface area of the lake to its present size. Among these works was the construction of a dyke along the north-east boundary of the lake in order to protect the adjacent agricultural land from flooding.

Lakes Zazari and Cheimaditida are connected hydrologically, by a channel constructed to allow water to flow from Lake Zazari (max. water elevation 599.2 m) to Lake Cheimaditida. The outlet of Lake Zazari is controlled manually by the operation of a sluice gate, while a spillway from Lake Cheimaditida (max elevation 591.3 m) channels water to Lake Petron.

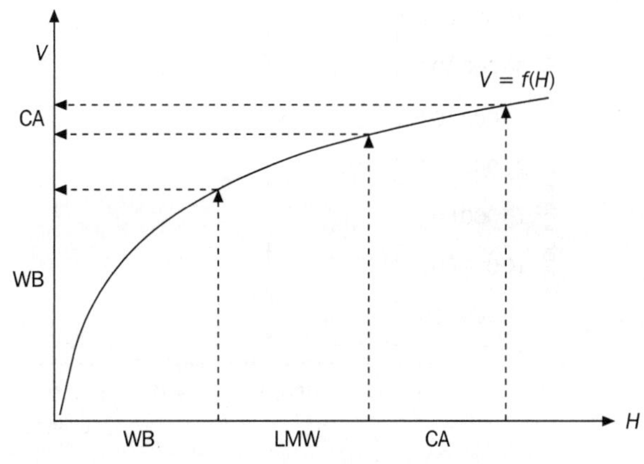

3.1.15 Graphical representation of the relation $V = f(H)$ between the stored water volume (*V*) in the system WB–LMW–CA and the corresponding elevation (*H*) of the water surface.

In order to assess the volume of water that could be stored by the system WB–LMW–CA of Lake Cheimaditida, a combination of topographic maps of different scale was used, 1:50 000 for the mountainous areas where slopes are steep and 1:5000 for the flat areas where the WB and LMW are located. The boundaries of the WB, LMW and CA were drawn on the map (Fig. 3.1.16, see colour Section).

The elevation contours together with the river network of the catchment were digitised and stored on the GIS, in order to construct the DTM of the area. Using the DTM, the relation $V = f(H)$ between the volume (V) stored in the system and any elevation (H) was quantified and transferred to a graph (Fig. 3.1.17). Using this graph, the following information was derived:

- the threshold of volume above which the wetland starts to store floodwater and hence contributes to flood mitigation is 2 031 133 m^3 ($H > 590.1$ m);
- the threshold of the volume above which flood water can cause damage (storage of water in the CA) is 10 262 826 m^3 ($H > 591.3$ m).

Step 2. Calculation of flood frequency. In order to calculate the frequency of flood events during which floodwater is stored in the wetland, it is necessary to obtain data on discharge measurements (or assessments) (Q) of inflows into the system. The inflow hydrograph can be used to estimate the total volume of water that is stored by the system as follows:

$$V = \sum_{t=1}^{t=n} Q_t \cdot \Delta_t$$

where V is the volume of inflow floodwater, Q_t is the inflow discharge as a function of time t and Δ_t ($= t_k - t_{k-1}$) is the time step of the inflow hydrograph (see Fig. 3.1.18). By transferring the total volume of water (V), that flows into the system to the $V = f(H)$ graph (Fig. 3.1.15), the volume of water stored in the wetland can be calculated, and its effectiveness in mitigating floods can be assessed.

In this way, by knowing the initial level of surface water in the system, before the flood event, one can evaluate the floodwater detention function of a wetland during an individual flood event. If the function needs to be assessed for a longer period of time, say one year, then an additional step should be included, to estimate the level of surfacewater in the system before any flood event. This makes it possible to monitor the water balance in the WB–LMW–CA system and consequently estimate its surface water elevation. This process can be expressed mathematically as follows:

$$V_i = V_{i-1} + Q_i - E_i - A_i$$

where V_i and V_{i-1} are stored volume of water by the system in days i and i–1, Q_i

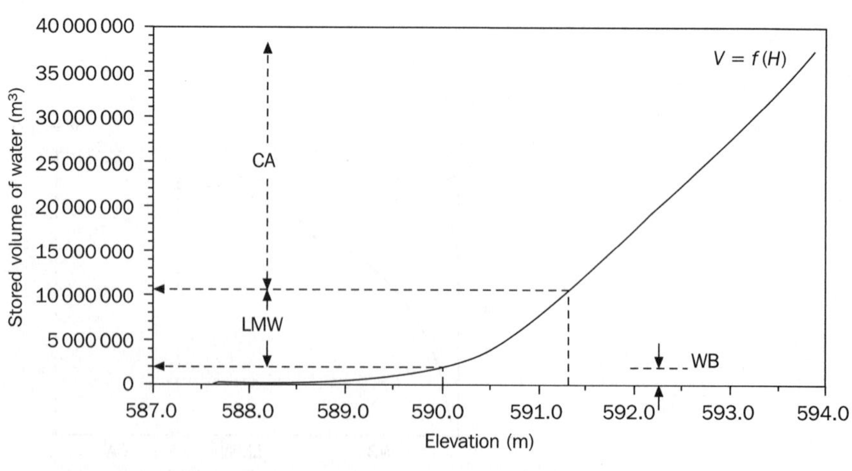

3.1.17 Graphical representation of the relation $V = f(H)$, between the stored water volume (V) in the system WB–LMW–CA and the corresponding elevation (H) of the water surface.

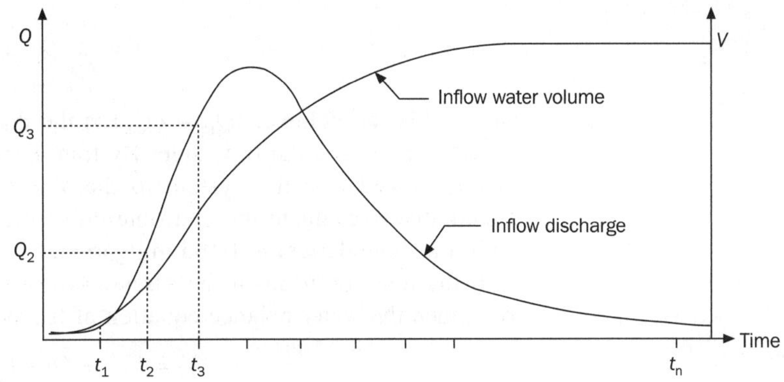

3.1.18 Discharge (*Q*) and corresponding cumulative volume (*V*) of inflow water.

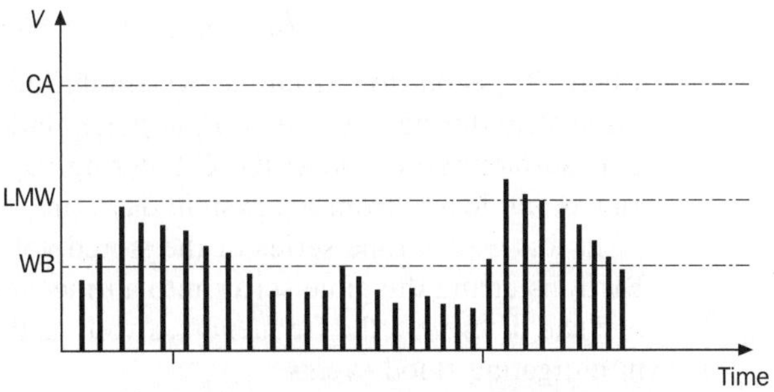

3.1.19 Daily water volume stored by the system WB–LMW–CA.

is inflow into the system during day i (m^3), E_i is net losses from the system during day i (m^3) and, A_i is abstraction from the system during day i (m^3).

The daily net losses from the system result from the following equation:

$$E_i = E_{o_i} + W_i - (R_i + q_i)$$

where E_{o_i} is evaporation from the system during day i (m^3), R_i is precipitation in the system during day i (m^3), q_i is surface run-off from the contribution area during day i (m^3) and, W_i is deep percolation and lateral losses from the system during day i (m^3).

In this way, a time series of the volume stored by the system can be obtained. By transferring this time series into a time–volume cartesian graph (Fig. 3.1.19), where the thresholds of water volume that can be stored in any of the three components (WB–LMW–CA) are represented as dashed lines, one can estimate the frequency as well as the size of contribution of a wetland in mitigating flood events.

LMW sub-type B (with outlet)

Here, the total water volume that can be stored by the system WB–LMW–CA depends on its outlet characteristics, which should be incorporated into the methodology described for sub-type A.

Step 1. Calculation of water storage capacity. The methodology is the same as that described for sub-type A (above).

Step 2. Calculation of flood frequency. The data needed in this step are measurements (or assessments) of both inflow (Q_{in}) and outflow (Q_{out}) discharges. The graphical representation of this property of water flow (Q) over time is the hydrograph (Fig. 3.1.10a). The subtraction of the hydrograph Q_{out} from Q_{in} (Fig. 3.1.11a), gives the discharge that is stored in the system (Fig. 3.1.11b). The total water volume (V) that is stored in the system (Fig. 3.1.11c), can be estimated using the equation:

$$V = \sum_{t=1}^{t=n} Q_t \cdot \Delta_t$$

where V is the volume, $(Q_{in} - Q_{out})$ is the discharge stored in the system (Fig. 3.1.11c), and Δ_t is the time step. By transferring the total volume (V), of water that is moved into the system to the $V = f(H)$ graph (Fig. 3.1.15), one can quantitatively evaluate the contribution of the wetland (volume of water stored within its boundaries) to flood mitigation.

If the function needs to be assessed for a longer period of time, such as one year, then the water balance equation of the system can be used:

$$V_i = V_{i-1} + Q_i - E_i - A_i$$

where V_i , V_{i-1}, Q_i , E_i , A_i are as described on p. 286 and the daily net losses (E_i), from the system result from the following equation:

$$E_i = Q_{out} + E_{o_i} + W_i - (R_i + q_i)$$

where Q_{out} is outflow from the system during day i (m^3), E_{o_i} is evaporation from the system during day i (m^3), R_i is precipitation in the system during day i (m^3), q_i is surface run-off from the CA during day i (m^3) and, W_i is deep percolation and lateral losses from the system day i (m^3).

In this way, a time series of the stored volume in the system can be obtained. By transferring this time series into a time–volume cartesian graph (Fig. 3.1.19), one can determine the frequency as well as the size of contribution of a wetland in mitigating flood events.

Example of Step 2. Calculation of flood frequency. The flood (detention) frequency has been assessed for the Cheimaditida wetland, for a 20-year period (January 1979–December 1998), by applying the water balance equation:

$$V_i = V_{i-1} + Q_i - E_i - A_i \qquad [1]$$

Where V_i and V_{i-1} are stored volume of water in the system in months i and i–1, Q_i is inflow into the system during month i (m^3), E_i is net losses from the system during month i (m^3) and, A_i is abstraction from the system during month i (m^3).

The net losses from the system were obtained from the following equation:

$$E_i = Q_{out} + E_{o_i} + W_i - (R_i + q_i)$$

where Q_{out} is outflow from the system during month i (m^3), E_{o_i} is evaporation from the system during month i (m^3), R_i is precipitation in the system during month i (m^3), q_i is surface run-off from the CA during month i (m^3) and, W_i is deep percolation and lateral losses from the system during month i (m^3). Water volume as inflows (overspilling from Lake Zazari) (Q_i) to and outflows (Q_{out}) from the system, are represented in Figs 3.1.20 and 3.1.21, respectively.

Precipitation inputs were calculated (a) as direct rainfall into the lake (R_i) and (b) as surface run-off (q_i) that flows into the lake (run-off coefficient: 0.2 for the period May-September and 0.4 for the remaining months). Water abstraction (A_i) for irrigation purposes takes place during summer. Abstracted water volumes were estimated.

Water losses through evaporation (E_{o_i}), were estimated using data gathered by an established adjacent meteorological station. Monthly deep percolation (W_i) along with lateral losses, was calculated as part of an existing hydrogeological study, and estimated to be 2500 m^3 $month^{-1}$. By applying equation (1) using the above-mentioned data, monthly water volumes stored in the system during the period January 79–December 98 were calculated.

The role of the LMW in mitigating floods can be assessed from Fig. 3.1.22, where the above calculations have been applied together with the thresholds of water volume that can be stored in any of the three components, WB, LMW and CA.

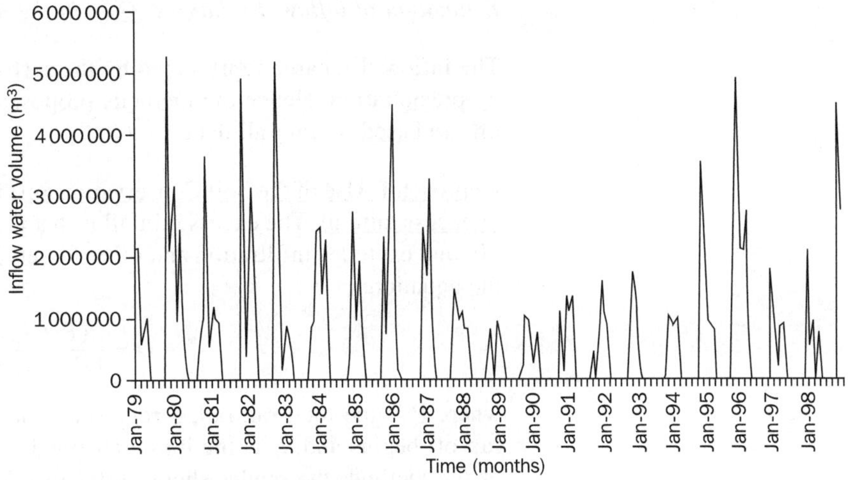

3.1.20 Monthly water volumes that flow into Lake Cheimaditida from Lake Zazari.

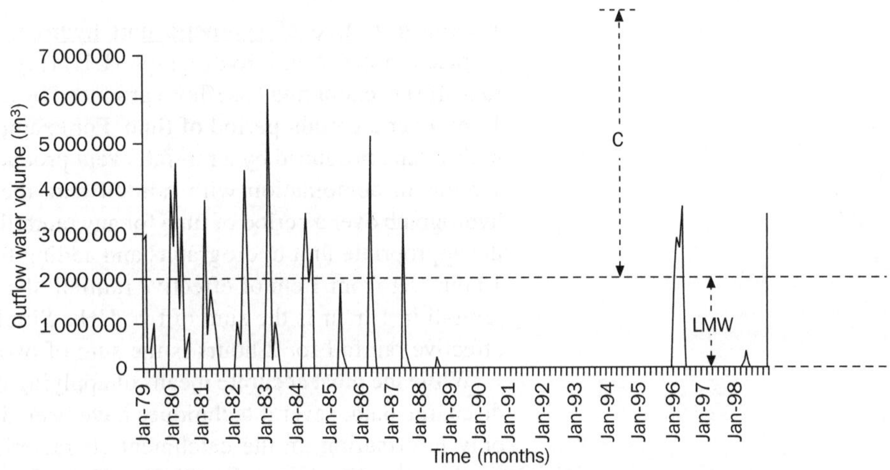

3.1.21 Monthly water volumes that flow from the Lake Cheimaditida area surrounding the system WB–LMW, evapotranspiration rate from each crop and rainfall data during the growing season.

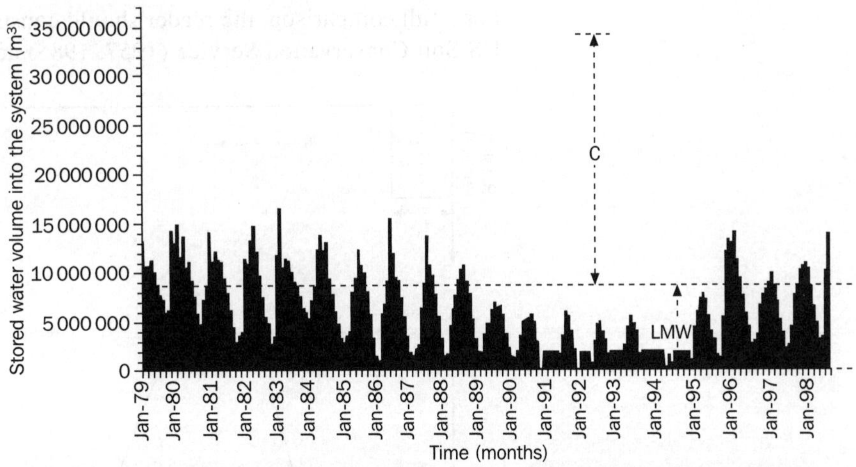

3.1.22 Graphical representation of the quantification of the Floodwater detention function.

Proposed methods for generating inflow–outflow hydrographs in natural wetland systems

For calculations shown in the previous sections, both inflow (Q_{in}) and outflow (Q_{out}) hydrographs were known. However, there are natural systems where no measurements of discharge exist. In these cases several methods are available for the assessment of both inflow and outflow hydrographs.

Estimation of inflow discharges (Q_{in}), in the absence of stream flow data

The inflow discharge represented by the surface run-off hydrograph is generated by precipitation. Hence the methods proposed for the estimation of surface run-off are based on rainfall data.

Approach 1. Use of the Soil Conservation Service (SCS) method for the estimation of excess rainfall. The excess rainfall is that proportion which becomes direct run-off, and excludes infiltration and other losses. Excess rainfall can be estimated by the equation:

$$Q = \frac{(P \pm I_a)^2}{(P - I_a) + S}$$

where P is the total rainfall, S represents the potential maximum retention after run-off begins and, I_a is the initial abstraction = 0.2 S. For a detailed description of the method, the reader should refer to US Soil Conservation Service (1972, 1981, 1985).

Approach 2. Use of synthetic unit hydrograph methods for the estimation of surface run-off. A unit hydrograph (UH) (Fig. 3.1.23) is the hydrograph of direct run-off (i.e. excluding baseflow) produced by a rainfall event of unit depth (usually 1 cm) over a certain period of time. For example, a 1 hour unit hydrograph is the hydrograph produced by a rainfall event producing one unit (1 cm) of rainfall over 1 hour. In combination with rainfall data the UH can be used to calculate the hydrograph over a period of time for any rainfall event of any duration by generating the appropriate unit hydrographs and adding them together. For example, given a 1 hour UH from 1 cm of effective rainfall, the hydrograph from 2 cm of effective rainfall in 1 hour is the sum of two UHs. Similarly, the hydrograph from 1 cm of effective rainfall for 2 hours is the sum of two UHs offset by 1 hour.

While the most accurate means of applying the UH is to calculate it from stream discharge data, several techniques have been developed to estimate the UH using only information on the catchment characteristics. These UH are referred to as synthetic unit hydrographs (SUHs). Two of these techniques are outlined here:

- synthetic unit hydrograph of Snyder (1938);
- synthetic unit hydrograph of SCS (1957).

For a full comparison, the reader should consult Snyder (1938), Sherman (1949), US Soil Conservation Service (1957, 1985) and Linsley *et al.* (1982).

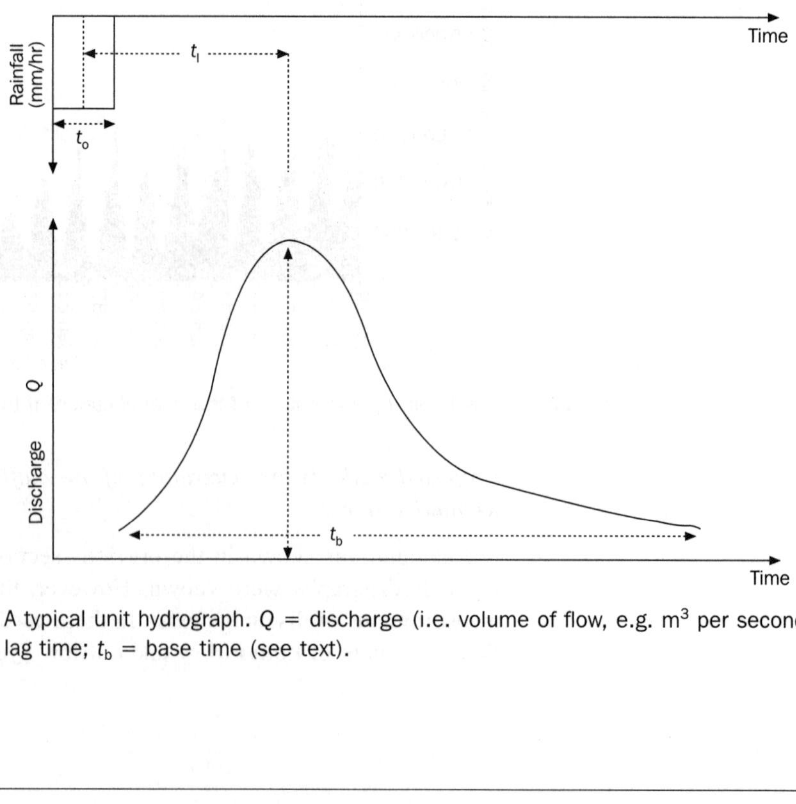

3.1.23 A typical unit hydrograph. Q = discharge (i.e. volume of flow, e.g. m^3 per second); t_l = lag time; t_b = base time (see text).

Snyder's synthetic unit hydrograph. Snyder (1938) was the first to investigate the relations between the features of the hydrograph and the geometry of catchments. To estimate the UH using Snyder's method the following steps are followed:

1. Determine the lag time (t_l) using the equation:

$$t_l = 0.75C_t(LL_c)^{0.3}$$

where C_t is a coefficient representing catchment characteristics (derived from similar neighbouring catchments, but estimates can be used where no data are available). Normally, C_t is 1.8–2.2 with lower numbers on steeper sloping catchments), L is the length of the main channel in km and L_c is the length from the downstream end of the catchment to the point on the channel nearest the centroid of the watershed in km.

2. The peak discharge (Q_p) of the UH is calculated from the following formula:

$$Q_p = \frac{2.75C_pA}{t_l}$$

where C_p is a coefficient reflecting the effects of flood wave retention and storage. This is calculated using data from neighbouring catchments or, if no data are available, use an estimate of 0.75. A is the area of the catchment in km^2.

3. The base time of the hydrograph (t_b) is calculated from:

$$t_b = 3 + \frac{t_l}{8}$$

where t_b is the base time in *days* and t_l is the lag time in *hours*.

Together, t_l, Q_p, and t_b allow the UH to be described for a *standard unit hydrograph*. This is a particular case where the duration of the UH (t_r) is defined by the following formula:

$$t_r = \frac{t_l}{5.5}$$

If this value of t_r is appropriate for the rainfall data to be used for the calculation of the hydrograph, then the calculated values of t_l, Q_p and t_b can be used to define the UH and $t_r = t_0$ in Fig. 3.1.24. If a different duration is required for the UH, the following procedure must be used to calculate the values of these parameters for the required duration:

4. The lag time of the required hydrograph (t_{lR}) is given by:

$$t_{lR} = t_l \pm \frac{t_r \pm t_0}{4}$$

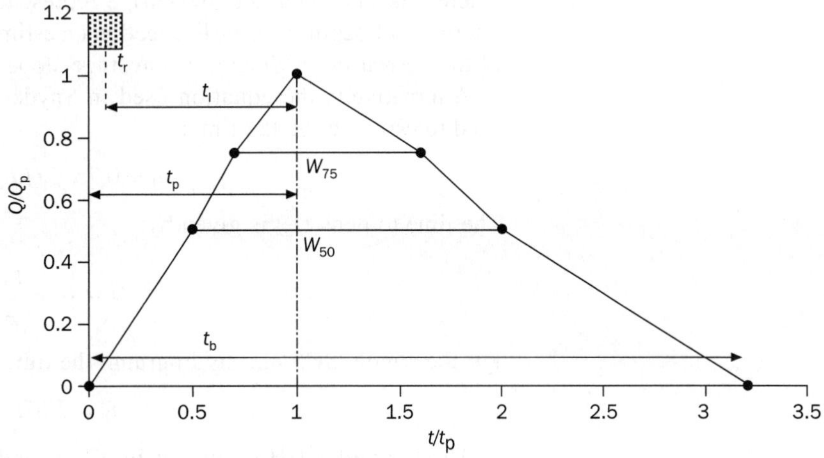

3.1.24 Snyder's synthetic unit hydrograph (see text).

where t_0 is the required UH duration and other terms are as defined previously.

5. Peak discharge of the required hydrograph (Q_{pR}) is given by:

$$Q_{pR} = \frac{Q_p t_l}{t_{lR}}$$

6. The base time of the hydrograph (t_{bR}) is given by:

$$t_{bR} = 3 + \frac{t_l}{8}$$

Although a simple triangle can be used to define the UH, improvements can be made by forcing the width of the hydrograph for certain values of the peak discharge (typically 50 % and 75 %). At 50 % of the peak discharge the time width (W) of the UH is given by:

$$W_{50} = 2.14\left[\frac{(Q_{pR})}{A}\right]^{\pm 1.08}$$

At 75 % the width is given by:

$$W_{75} = 1.22\left[\frac{(Q_{pR})}{A}\right]^{\pm 1.08}$$

This width is shifted so that a greater proportion of it is distributed after the peak. There are different distribution ratios suggested but, typically, it is distributed with a ratio of 1:2 before and after (see Fig. 3.1.24 for more clarification).

The Soil Conservation Service (SCS) synthetic unit hydrograph. The SCS synthetic unit hydrograph method can be applied using a simple triangular representation of the hydrograph or a non-linear UH using dimensionless ratios that describe the curve (Fig. 3.1.25). Both of these methods use the same calculations for peak discharge and time to peak.

The peak discharge (Q_p) is calculated by the equation:

$$Q_p = \frac{2.08\,A}{t_p}$$

where A is watershed area (km^2), t_p is time to peak in hours and, Q_p is peak discharge in m^3 s^{-1} (for a rainfall intensity of 1 cm hr^{-1}).

The lag time (t_l) can be estimated using several methods, such as the equation:

$$t_l = \frac{L^{0.8}(S+1)^{0.7}}{2.924\,y^{0.5}}$$

where L is catchment length (km), S represents the potential maximum retention after run-off begins (see earlier section on estimating excess rainfall for description of the derivation of S) and, y is average slope of the catchment (%).

Alternatively, the equation used in Snyder's method described above can be used to estimate the lag time:

$$t_l = 0.75\, C_t(LL_c)^{0.3}$$

The time to peak (t_p) is given by:

$$t_p = t_l + \frac{t_0}{2}$$

For the simple SCS unit hydrograph, the time base (t_b) is given by:

$$t_b = 2.67 t_l$$

The triangular UH described by Q_p, t_p and t_b can be applied without further modification but, for a more realistic description of a hydrograph, SCS proposed a dimensionless unit hydrograph that can be used to derive a more naturally shaped hydrograph. Figure 3.1.25 shows the shape of the SCS dimensionless UH.

To apply the dimensionless hydrograph, Q_p and t_p are calculated as described above. The shape of the hydrograph is then calculated from predefined ratios of Q/Q_p and t/t_p (shown in Table 3.1.5), derived from a large number of natural hydrographs that describe a dimensionless UH. For each point shown in Table 3.1.5, Q and t are calculated by multiplying the table entries by Q_p and t_p, respectively.

The shape of the UH derived using this method will, therefore, always be the same as that shown in Fig. 3.1.25, but stretched in the x and y directions depending on the values calculated for Q_p and t_p.

Outflow discharges (Q_{out}) from LMW, generated by individual precipitation events

The outflow (Q_{out}) discharge from a WB–LMW–CA system depends on the outlet characteristics of the system as well as on the relation $Q_{out} = f(H)$, where H is the water level in the system.

For a rapid assessment of outflow discharge in LMW, the Muskingum Routing Method (a hydrologic method) through a lake is proposed, where flood routing may be defined as the procedure whereby the time and magnitude of a flood wave, at a point on a stream, is determined from the known or assumed data at one or more points upstream. It was developed by G. T. McCarthy and others in connection

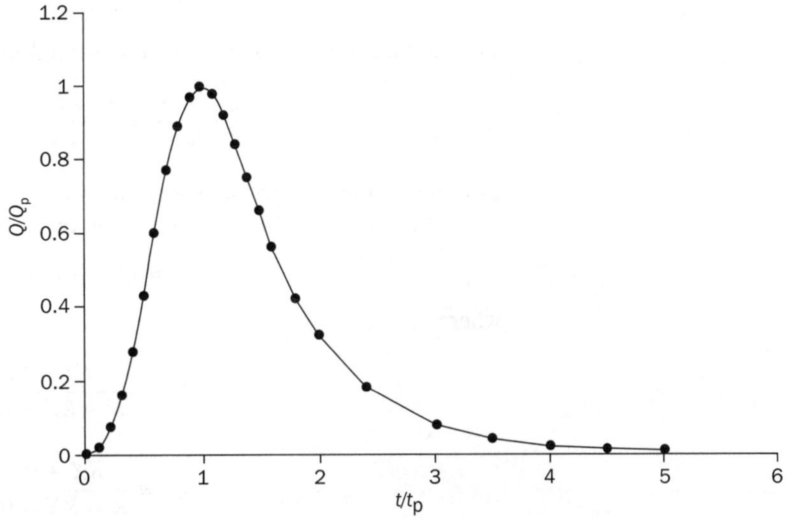

3.1.25 Shape of the SCS dimensionless synthetic unit hydrograph.

Table 3.1.5 Ratios of Q/Q_p and t/t_p for calculating SCS synthetic unit hydrograph

Q/Q_p	t/t_p	Q/Q_p	t/t_p
0	0	0.84	1.3
0.015	0.1	0.75	1.4
0.075	0.2	0.66	1.5
0.16	0.3	0.56	1.6
0.28	0.4	0.42	1.8
0.43	0.5	0.32	2
0.6	0.6	0.18	2.4
0.77	0.7	0.075	3
0.89	0.8	0.036	3.5
0.97	0.9	0.018	4
1	1	0.009	4.5
0.98	1.1	0.004	5
0.92	1.2		

with studies of the Muskingum Conservancy District Flood-Control Project in 1934–1935. Further information for this method can be found in Chow (1964). A brief description of the method is given below.

This method is based on the continuity equation, according to which:

$$\frac{dS}{dt} = I - O$$

where S is water storage, I is inflow discharge and, O is outflow discharge.

Storage within the system is separated into two parts: 'wedge' storage and 'prism' storage. When there is equilibrium between the inflow and outflow (steady flow) and a uniform water surface slope, the storage can be approximated by a prism as shown in Fig. 3.1.26. On the rising limb of a flood, however, there is an additional 'wedge' of storage due to the backwater effect, the magnitude of which is determined by the characteristics of the routing area. In the case of the falling limb of the flood, this wedge storage is negative. Assuming a linear relationship between storage and discharge and introducing a proportionality coefficient (X), which describes the split between the wedge and prism storage, the above equation takes the following form:

$$S_t = \underbrace{KO}_{\text{Prism storage}} + \underbrace{KX(I_t - O_t)}_{\text{Wedge storage}}$$

where K is the storage factor describing the travel time of the flood peak and has time units. Rearranging gives:

$$S_t = K(XI_t + (1 - X)O_t)$$

The continuity equation can be represented in different form as:

$$\frac{S_{t+1} - S_t}{\Delta t} = \frac{I_t + I_{t+1}}{2} - \frac{O_t + O_{t+1}}{2}$$

Combining the two previous equations for $t = 1$ and $t = 2$, the outflow discharge in the time step $t = 2$ (O_2), is given by the following equation:

$$O_2 = C_0 I_2 + C_1 I_1 + C_2 O_1$$

where:

$$C_0 = \frac{-KX + 0.5\,\Delta t}{K - KX + 0.5\,\Delta t}$$

$$C_1 = \frac{KX + 0.5\,\Delta t}{K - KX + 0.5\,\Delta t}$$

and

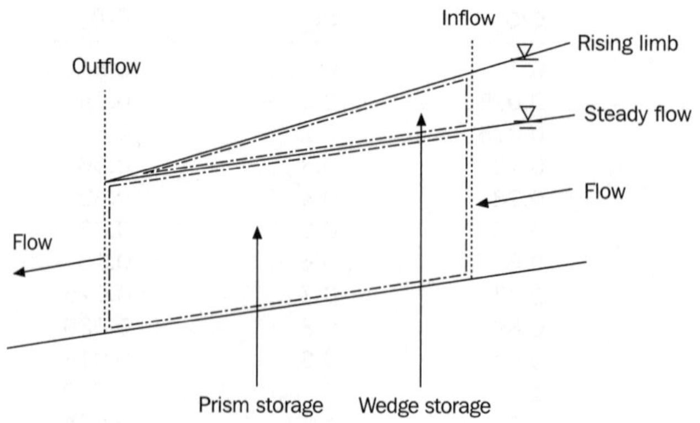

3.1.26 The Muskingum Routing Method.

$$C_2 = \frac{K - KX - 0.5\,\Delta t}{K - KX + 0.5\,\Delta t}$$

Therefore, for known values of I_t describing the inflow hydrograph and assuming $O_1 = I_1$ the outflow hydrograph can be calculated with estimates of K and X.

Where data are available, K is calculated by plotting S versus $(XI + (1 - X)O)$ for various arbitrarily selected values of X. As a result of the assumptions of the method, the value of X must be between 0 and 0.5. Typically, the plot of S versus $(XI + (1 - X)O)$ will show a hysteretic curve. The most appropriate values of K and X are those that generate the closest approximation to a straight line. The feasible region for the parameters is defined by the following relationship:

$$\frac{1}{2(1-X)} \le \frac{K}{\Delta t} \le \frac{1}{2X}$$

Where no outflow data are available for parameter estimation they may be estimated by comparison with similar catchments or by direct estimation taking into consideration the catchment characteristics. The value of X must be between 0 and 0.5. A value of 0 implies no wedge storage, i.e. a reservoir-type storage with no backwater. A low value for X would therefore be appropriate for many types of systems that include significant wetland areas. A value of 0.5 implies that the full length of the routing area is backwater. Approximations of K, the travel time of the flood wave, can be made from direct observation of flood peak timings.

Outflow discharges (Q_{out}) from RMW, generated by individual precipitation events

Flood routing methods can be used for the rapid assessment of outflow discharges from RMW. Flood routing methods can be categorised as either hydrologic or hydraulic. Hydrologic routing methods generally use a limited representation of the physical characteristics of the system and often use lumped attenuation and lag parameters. Mathematically, hydrologic routing techniques are based on the continuity equation, or the relation of flow being equal to the water velocity multiplied by cross-sectional area, and on the analytic or empirical relationship between the water storage in the system and the outlet discharge. On the other hand, hydraulic routing is performed using the fully dynamic wave equations, known as the Saint Venant equations, to quantify changing rates of channel flow.

Considering the dynamic characteristics of the flow into a channel, especially during flood events, the use of hydraulic flood routing methods is recommended here. Hydraulic routing methods may be used to predict the spatial and temporal patterns of the flow (usually a flood wave) as it moves down a given length of a channel. Several hydraulic routing methods available are:

- kinematic wave approximation
- diffusion wave approximation
- fully dynamic

A brief synopsis of these methods is given below; however, for a full comparison as well as a list of factors to consider in selecting the appropriate one, the reader should refer to Chow (1964) and US Army Corps of Engineers (1996).

Hydraulic routing methods for wave routing approximation are based on the combination of the momentum equation (1) and equation of continuity (2):

$$\frac{\partial y}{\partial x} + \frac{u}{g}\frac{\partial u}{\partial x} + \frac{1}{g}\frac{\partial u}{\partial t} = S_0 - S_f \qquad [1]$$

$$\frac{\partial A}{\partial t} + \frac{\partial Q}{\partial x} = q \qquad [2]$$

where x is horizontal distance alongside the channel, t is time, y is water depth, Q is water discharge, A is wetted cross-sectional area, q is lateral water inflow or outflow per unit length, u is water velocity (Q/A), g is gravity force, S_0 is channel slope and, S_f is friction slope or energy slope.

The above equations describe the one-dimensional non-steady water flow within an open channel, and they are commonly known as Saint Venant equations (Chow, 1959; Strelkoff, 1969, 1970; Ligget, 1975).

Kinematic wave approximation method. In the kinematic wave routing method, which is widely used for wave routing approximation, all the terms of the left side of equation (1) are assumed to be equal to zero. This implies that inertial and pressure forces are not important and there is no significant acceleration. Equation (1) therefore takes the following form:

$$S_f = S_0 \qquad [3]$$

The above assumption means that there is steady water flow within the channel (with frictional forces equal to gravitational forces).

Water discharge (Q) alongside the open channel may be calculated using Manning's equation:

$$Q = \frac{1}{n} AR^{2/3} S_0^{1/2} \qquad [4]$$

where Q is water discharge, R is wetted perimeter, A is wetted cross-sectional area and, n is Manning's coefficient.
Equation (4) may also be simplified and written as follows:

$$Q = \alpha A^m \qquad [5]$$

where α and m are functions of the wetted cross-section's shape and of the channel's wall friction.

Combining equations (2) and (5) gives:

$$\frac{\partial A}{\partial t} + amA^{(m-1)} \frac{\partial A}{\partial x} = q \qquad [6]$$

This equation can be solved using a finite difference approximation as implemented in the freely available HEC1 model (www.hec.usace.army.mil). For more details, the reader should consult DeVries and MacArthur (1979, 1985).

Where q is zero (i.e. there is no inflow along the length of the channel), there is no attenuation of the flood wave with the kinematic wave approximation method, and the calculation reduces to a simple translation of the initial wave downstream.

Diffusion wave approximation method. In the diffusion wave routing method, equation (1) is assumed to have the following simplified form:

$$\frac{\partial y}{\partial x} = S_0 - S_f \qquad [7]$$

With the combination and linearisation of equations (2) and (7), the diffusion wave equation, takes the following form (Miller and Cunge, 1975):

$$\frac{\partial Q}{\partial t} + c\frac{\partial Q}{\partial x} = m\frac{\partial Q^2}{\partial x^2} + cq \qquad [8]$$

where c is wave transfer speed in the x direction $c = \mathrm{d}Q/\mathrm{d}A$, μ is (hydraulic diffusivity) hydraulic diffusion coefficient $\mu = Q/2BS_0$ and, B is free water surface width. For details on the solution of the equations for this method see DeVries and MacArthur (1985).

Fully dynamic method. The dynamic routing wave approximation is the solution of the Saint Venant's equations (equations 1 and 2) above. The description and solution of these equations is beyond the scope of this document, but further information can be found in HEC (2002).

3.1.E.2 *Groundwater recharge (function)*

Introduction

According to the description of Groundwater recharge (FAPs, Section 3.1.B), the way that wetlands contribute towards recharging groundwater aquifers is mainly owing to their ability to store water, which subsequently infiltrates and percolates downwards into a significant aquifer. Hence, a quantitative evaluation of this function should answer the following questions:

- What is the water volume infiltrating and percolating from the wetland to the aquifer?
- What is the significance of this recharge volume, passing through the wetland compared with the total water volume recharged into the aquifer?

The following proposed methodology addresses the first question, aiming to provide a simple procedure for the rapid assessment of the water volume that recharges the aquifer. This method is applicable to both river marginal and lake marginal wetlands, since the rate of recharge is dependent mainly on the water volume that is stored in the wetland and not on its type.

In order for the second question to be answered, a detailed hydrogeological study of the aquifer is needed, in which its extent, depth, structural and hydraulic properties are estimated.

Rapid quantification of the Groundwater recharge function

In Figs 3.1.27 to 3.1.29, the hydrological relationship among the three parts of the system – water body (WB) – wetland (W) – aquifer (Aq) – is presented schematically.

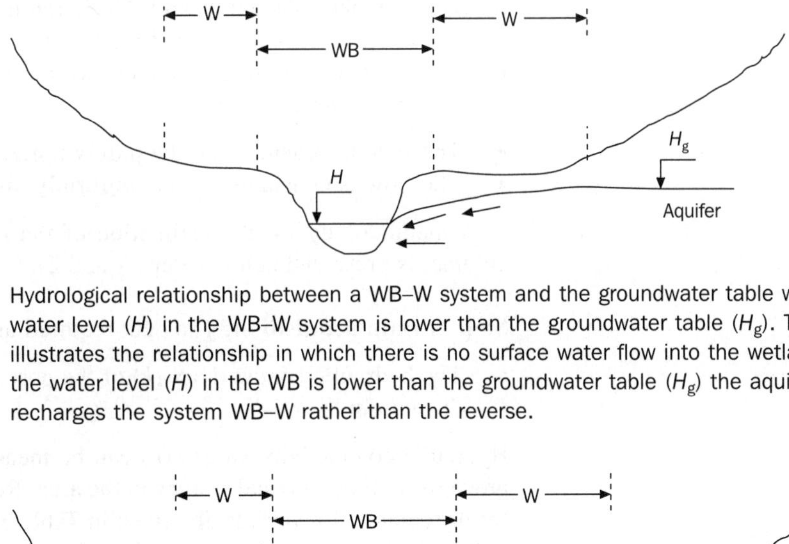

3.1.27 Hydrological relationship between a WB–W system and the groundwater table where the water level (H) in the WB–W system is lower than the groundwater table (H_g). This illustrates the relationship in which there is no surface water flow into the wetland. As the water level (H) in the WB is lower than the groundwater table (H_g) the aquifer recharges the system WB–W rather than the reverse.

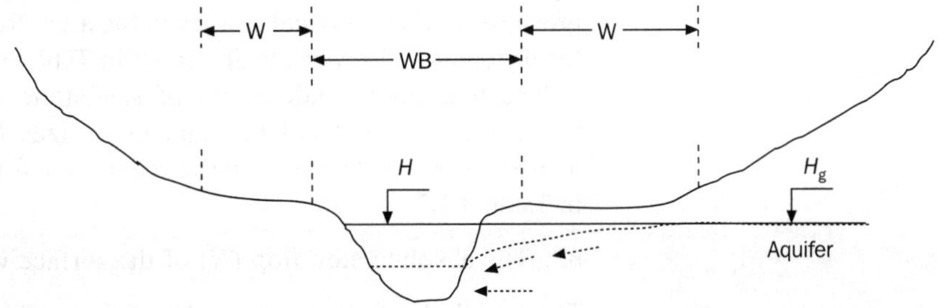

3.1.28 Hydrological relationship between a WB–W system and the groundwater table where the water level in the WB–W system (H) is in equilibrium with the groundwater table (H_g). This also illustrates the relationship where there is no surface water flow in the wetland but the surface of the main water body has risen to the level of the groundwater table. The flow from the aquifer to the system WB–W begins to decelerate and finally stops when equilibrium between the water level and the groundwater table has been established.

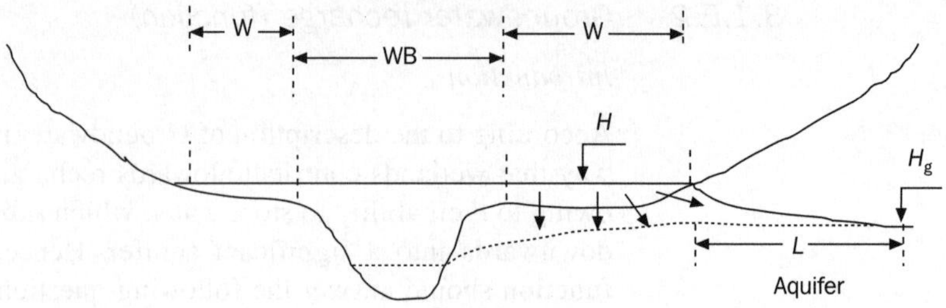

3.1.29 Hydrological relationship between a WB–W system and the groundwater table where the water level in the WB–W system (H) is above the groundwater table (H_g). There is surface water flow in the wetland, and water infiltrates and percolates to the aquifer, therefore the function of groundwater recharge is performed.

Taking into account the above scenarios, the following two conditions have to be met in order for the Groundwater recharge function to be performed:

- The hydraulic gradient $[(H–H_g)\ L^{-1}]$ between the water level (H) in the wetland and the groundwater table (H_g), has to be **positive**.
- The hydraulic conductivity of the material between the wetland and the aquifer has to be sufficient to allow water movement.

For the rapid estimation of the water discharge (Q) that enters the aquifer, Darcy's Law can be used. This is expressed by the equation:

$$Q = k\,A\,\frac{H - H_g}{L} \qquad [1]$$

where Q is discharge ($m^3\ s^{-1}$), k is hydraulic conductivity of the porous media (m s^{-1}), H is absolute elevation of the water level in the wetland (m), H_g is absolute elevation of the groundwater table (m), A is cross-sectional area through which the recharge takes place (m^2) and, L is horizontal distance between H and H_g (m). The above equation can be used making the following assumptions, commonly known as the Dupuit–Forchheimer assumptions (Bouwer, 1978; Terzidis and Karamouzis, 1985):

- The flow is considered to be purely horizontal.
- The flow is considered to be uniformly distributed with depth.

A methodology for the estimation of the recharged (into the aquifer) water volume, is presented below (Steps 1 and 2):

Step 1: Data acquisition. The data required are described below (a–e):

a. The hydraulic conductivity (k) of the porous media between the wetland and the aquifer

Hydraulic conductivity values (k) can be measured on site, or determined from previous hydrogeological studies in the area. Representative values of k (m day^{-1}) for different soil materials are given in Table 3.1.6.

The hydraulic conductivity of sandstone is considerably less than that of unconsolidated sand with the same grain sizes, because of cementation and higher density of the sandstone. General ranges of k for consolidated materials are shown in Table 3.1.7.

b. The absolute elevation (H) of the surface water in the wetland

The initial absolute elevation (H) of the surface water in the wetland should be identified for the period that the groundwater recharge function has to be estimated. This can be achieved either by a topographic survey or by measuring the water depth at a point of known absolute elevation (Fig. 3.1.30).

c. The relationship between the water volume $V = f(H)$ and the corresponding water surface area $A = f(H)$ that can be stored in the wetland below any elevation (H) (Figs 3.1.27 to 3.1.29)

Table 3.1.6 Representative values of hydraulic conductivity k (m day^{-1}) for different soil materials (Bouwer, 1978)

Soil material	k (m day^{-1})
Clay soils (surface)	0.01–0.2
Deep clay beds	10^{-8}–10^{-2}
Loam soils (surface)	0.1–1
Fine sand	1–5
Medium sand	5–20
Coarse sand	20–100
Gravel	100–1000
Sand and gravel mixes	5–100
Clay, sand and gravel mixes (e.g. glacial)	0.001–0.1

Table 3.1.7 General ranges of k for consolidated geological materials (Bouwer, 1978)

Consolidated material	k (m day^{-1})
Sandstone	0.001–1
Carbonate rock with secondary porosity	0.01–1
Shale	10^{-7}
Dense, solid rock	$< 10^{-5}$
Fractured or weathered rock (aquifers)	0.001–10
Fractured or weathered rock (core samples)	Almost 0–300
Volcanic rock	Almost 0–1000

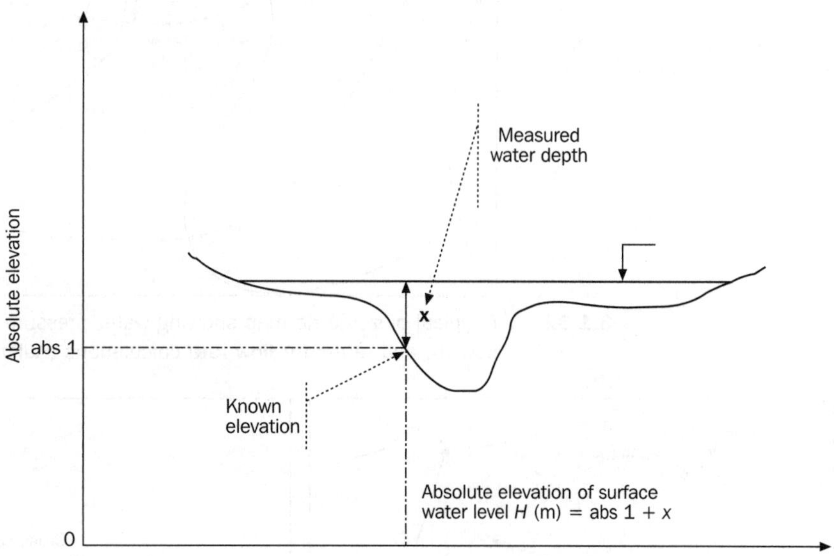

3.1.30 Estimation of the absolute elevation of surface water level.

The estimation of the relationships $V = f(H)$ and $A = f(H)$ can be obtained from an adequate topographic map either manually or using GIS (see p. 285 for assessment of water storage capacity.

d. The elevation (H_g) of the groundwater table in the area

The elevation of the groundwater table (H_g) can be assessed either by fieldwork or from any piezometric maps available (Fig. 3.1.32) of the surrounding area. These maps should represent at least the mean groundwater table for each season or for the dry and wet periods of a year. More detailed piezometric maps representing the groundwater table for each month require hydrogeological studies of the area.

e. The cross-sectional area through which the water recharges the aquifer

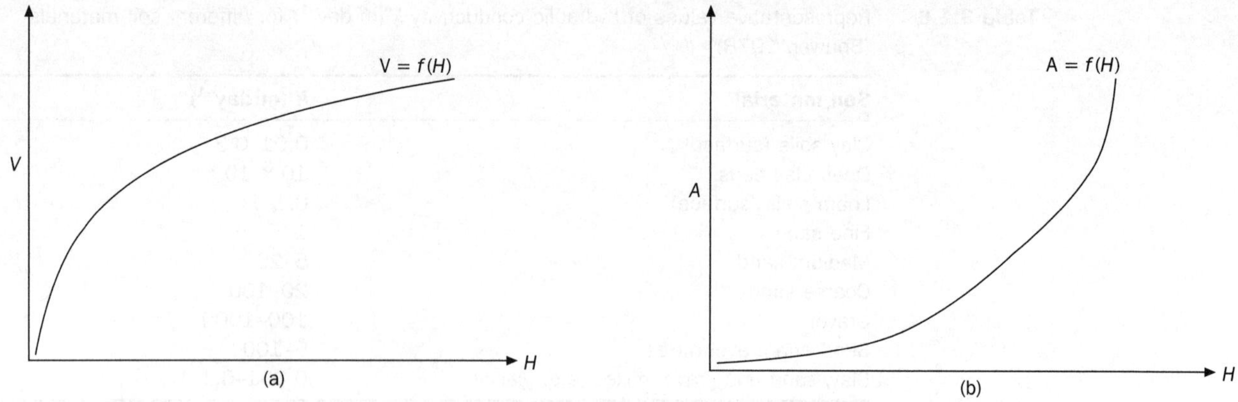

3.1.31 The relation between water volume/surface and corresponding elevation: (a) graphical representation of the relation $V = f(H)$ between the stored water volume (V) and the corresponding elevation (H); (b) graphical representation of the relation $A = f(H)$ between the water surface (A) and the corresponding elevation (H).

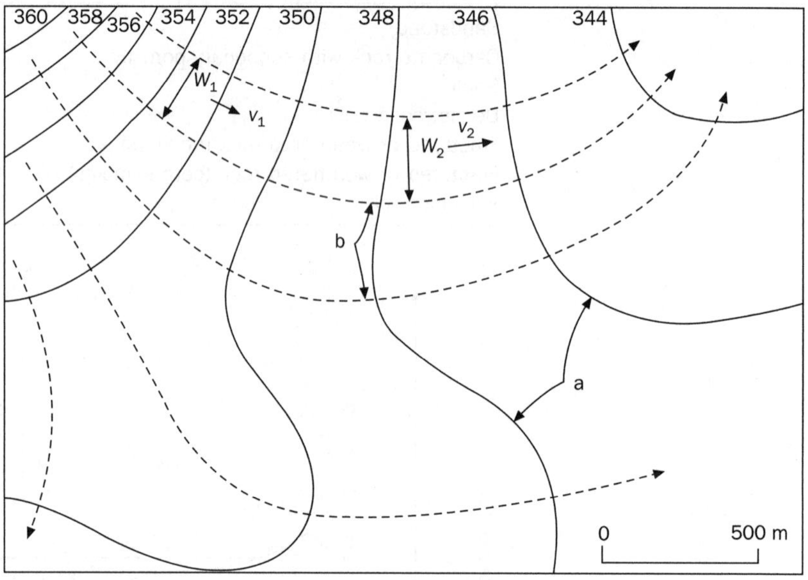

3.1.32 A typical piezometric map showing water pressure isobars (a), direction of groundwater flow (b) and terms for flow rate calculations (Kalergis, 1999).

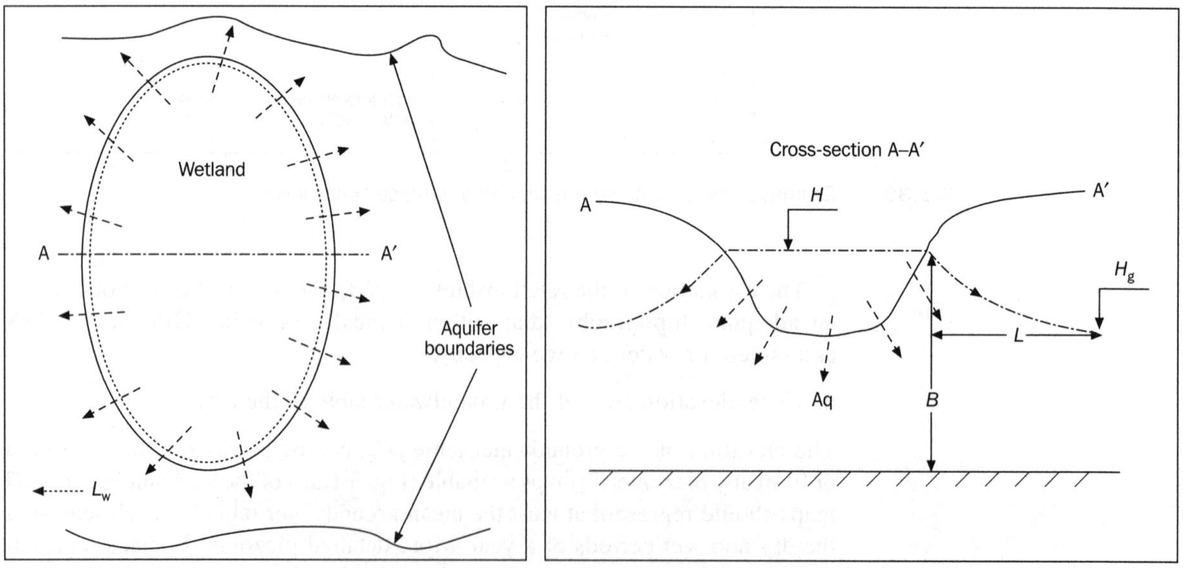

3.1.33 Ground view and cross-section of the system WB–W–Aq, where the recharge takes place through the whole of the wetland's wet perimeter. H = water level; Aq = aquifer; B = vertical dimension of aquifer; H_g = height of groundwater; L = lateral distance of recharge (wetland water height to groundwater height).

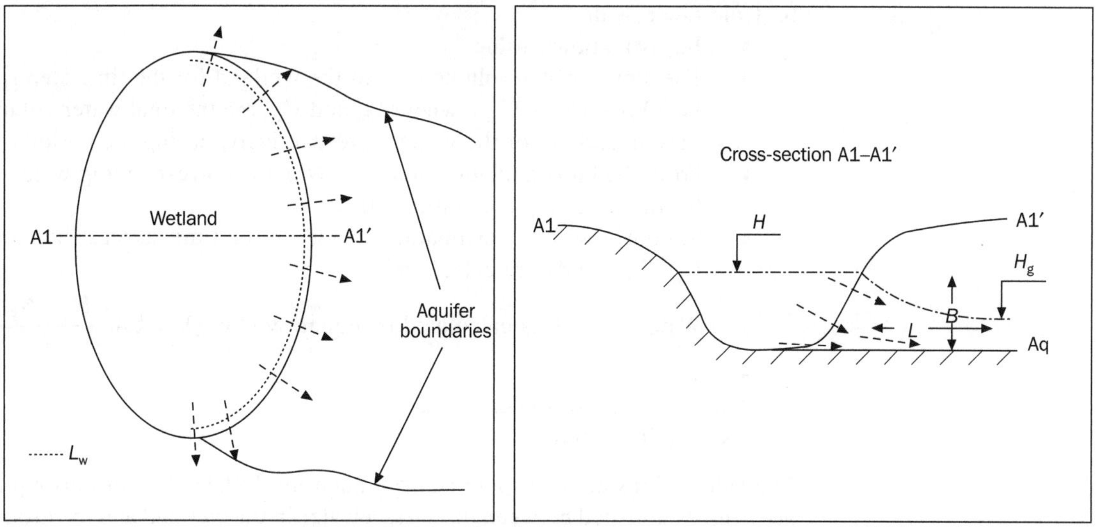

3.1.34 Ground view and cross-section of the system WB–W–Aq, where the recharge takes place through a section of the wetland's wet perimeter. H = water level; Aq = aquifer; B = vertical dimension of aquifer; H_g = height of groundwater; L = lateral distance of recharge (wetland water height to groundwater height).

The cross-sectional area can be estimated using the equation $A = L_w * B$ (see also Figs 3.1.33 and 3.1.34) where:

L_w is the length of the section of the wet perimeter of the wetland through which the recharge takes place (Figs 3.1.33 and 3.1.34). It has to be noted that once the groundwater recharge function begins, the water level in the wetland reduces and L_w is reduced accordingly. This has to be taken into account in the calculations.

B is the vertical distance between the bottom of the aquifer and the water surface in the wetland. The depth of the aquifer's bottom can be found either by fieldwork or by reference to previous hydrogeological studies in the area.

Step 2: Assessment of the Groundwater recharge function, using Darcy's Law. To estimate the Groundwater recharge function ($t_i = t_0 = 0$), water flow (Q_0) from the wetland to the aquifer is calculated by the Darcy's Law equation above (1), where H and H_g should be known values and $H > H_g$. Since the recharge function has determined the water volume, the water level in the wetland is being increased or decreased progressively, depending on the recharge value Q_i, together with the total water volume that flows into and out of the wetland. Assuming a time step (dt, days) upon which calculations are made, the estimation of the water level in the wetland (H_{t+1}) for the next time step (t + dt) is made by performing the following sequence of calculations. Note that for the completion of the following calculations, a series (a–c) of measured groundwater level values (H_g), are required.

a. Time $t_0 = 0$
- H_0 and H_{g0} should be measured (known values).
- From the known relationship $V = f(H)$, the stored wetland water volume (V_0) for the time step t_0 is estimated.
- From the known relationship $A = f(H)$, the length (L_{w0}) of the section of the wet perimeter through which recharge takes place for the time step t_0 is also estimated.
- The vertical distance (B_0) between the bottom of the aquifer and the water surface in the wetland for the time step t_0 is estimated (see above)
- $A_0 = L_{w0} * B_0$
- Water flow Q_0 is being calculated by equation (1): $Q_0 = k A_0 \dfrac{H_0 - H_{g0}}{L}$.
- The water volume that flows into the aquifer during time step t_0, is $V_{g0} = Q_0 * dt$.

b. Time $t_1 = t_0 + dt$

- H_{g1} is a known value.
- The stored water volume (V_1) in the wetland for the time step t_1 is $V_1 = V_0 - V_{g0} + (V_{in} - V_{out})$, where V_{in} and V_{out} are the total water volumes that flow in and out of the wetland, respectively, during the period $t_1 - t_0$.
- From the known relationship $V = f(H)$, the corresponding water level H_1 for the time step t_1 is estimated.
- Similarly with the estimations made on the time step t_0, the parameters L_{w1}, B_1 and A_1 are calculated.
- Water flow Q_1 is calculated by equation (1): $Q_1 = k A_1 \dfrac{H_1 - H_{g1}}{L}$.
- $V_{g1} = Q_1 * dt$.

c. Time $t_i = t_{i-1} + dt$ (where $i \geq 2$)
As described above.

The calculations above should be continued until $H_i = H_{gi}$ and thus equilibrium has been established between the surfacewater in the wetland and the groundwater table (aquifer). The total recharged water volume ($V_{g,\,total}$) into the aquifer can be estimated from the following sum:

$$V_{g,total} = \sum_{i=0}^{n} V_{gi} = \sum_{i=0}^{n} Q_i dt$$

An example of this procedure has been incorporated with the example of the Groundwater discharge function below.

3.1.E.3 *Groundwater discharge (function)*

Introduction

Groundwater discharge is described as the upward movement of groundwater to the wetland surface, which occurs under specific hydrogeological conditions. The performance of this function may be important in the maintenance of river baseflow or of ecosystems dependent upon associated soil water regime or solutes.

The following methodology provides a simple procedure for the rapid assessment of the water volume that discharges into a wetland through the upward movement of groundwater. This method is applicable to both river marginal and lake marginal wetlands, since the rate of discharge is mainly dependent on the water volume that is stored in the wetland, and not on its type.

Rapid quantification of the Groundwater discharge function

Figure 3.1.35 illustrates the case of the performance of the Groundwater discharge function. As the water level (H) in the WB is lower than the groundwater table (H_g) the aquifer recharges the system WB–W.

The following two conditions have to be met for the Groundwater discharge function to be performed:

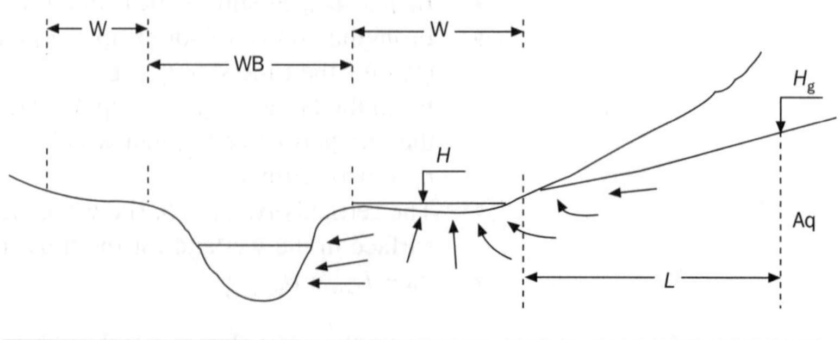

3.1.35 Representation of the Groundwater discharge function. W = wetland, H = water level, H_g = height of groundwater, Aq = aquifer and L = lateral distance of discharge.

- The hydraulic gradient $[(H–H_g)/L]$ between the water level in the wetland (H) and the groundwater table (H_g), has to be *negative*.
- The hydraulic conductivity of the material between the wetland and the aquifer has to be sufficient to allow water movement.

For the rapid estimation of the water that discharges from the aquifer to the wetland (Q), Darcy's Law can be used. This is expressed by the equation:

$$Q = k A \frac{H - H_g}{L} \qquad [1]$$

where
Q is discharge ($m^3 s^{-1}$), k is hydraulic conductivity ($m s^{-1}$), H is absolute elevation of the water level in the wetland (m), H_g is absolute elevation of the groundwater table (m), A is cross-sectional area through which the recharge takes place (m^2) and, L is horizontal distance between H and H_g (m).

The above equation can be used making the following assumptions, commonly known as the Dupuit–Forchheimer assumptions (Bouwer, 1978; Terzidis and Karamouzis, 1985):

- The flow is considered to be purely horizontal.
- The flow is considered to be uniformly distributed with depth.

The proposed methodology for the estimation of the discharged water volume, is presented below (Steps 1 and 2):

Step 1: Data acquisition. The data required are described below (a–e). Refer also to the details given in the equivalent section of the Groundwater recharge function (pages 298–301).

a. The hydraulic conductivity (k) of the porous media between the wetland and the aquifer (see above).
b. The absolute elevation (H) of the surface water in the wetland (see above).
c. The relationships between the water volume $V = f(H)$ and the corresponding water surface $A = f(H)$ that can be stored in the wetland below any elevation (H) (see above).
d. The elevation (H_g) of the groundwater table in the area (see above).
e. The cross-sectional area through which the water discharges the aquifer (see above).

The cross-sectional area can be estimated using the equation $A = L_w * B$, where:

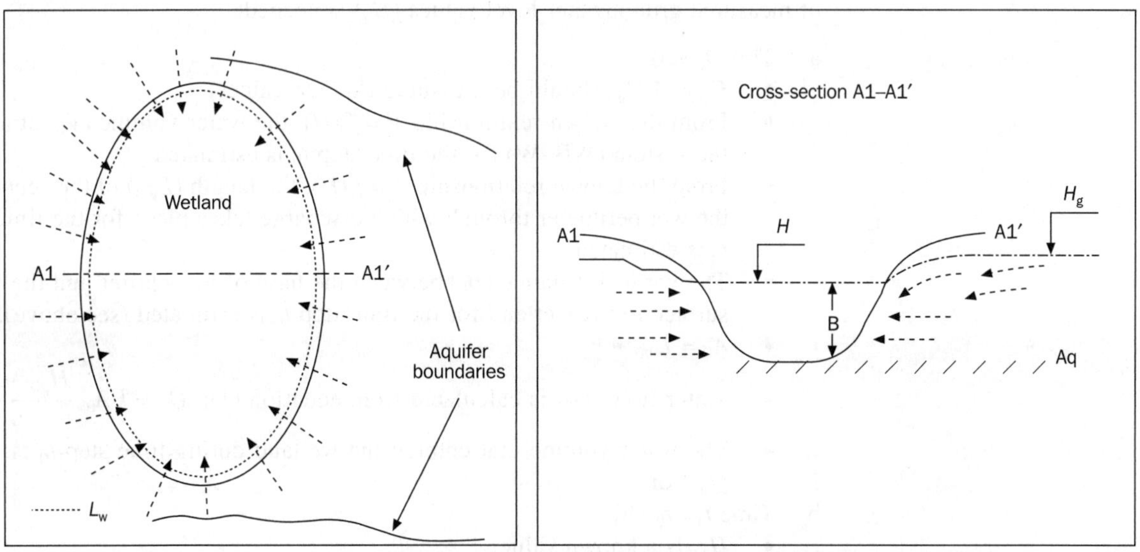

3.1.36 Ground view and cross-section of the system WB–W–Aq, where the discharge takes place through the whole of the wetland's wet perimeter. H = water level, H_g = height of aquifer, Aq = aquifer, B = distance between base of aquifer and H.

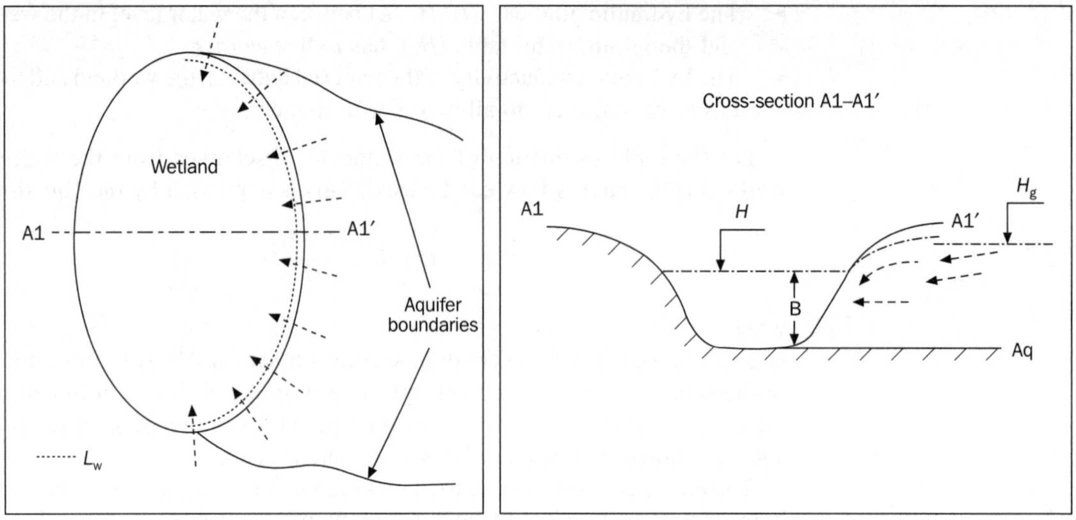

3.1.37 Ground view and cross-section of the system WB–W–Aq, where the discharge takes place through a section of the wetland's wet perimeter. H = water level, H_g = height of aquifer, Aq = aquifer, B = distance between base of aquifer and H.

L_w is the length of the section of the wet perimeter of the wetland through which the discharge takes place (Figs 3.1.36 and 3.1.37). It has to be noted that once the Groundwater discharge function begins, the water level in the wetland increases and L_w is increased accordingly. This has to be taken into account in the calculations.

B is the vertical distance between the base of the aquifer and the water surface in the wetland. The depth of the aquifer's base can be found either by fieldwork or by reference to previous hydrogeological studies.

Step 2: Assessment of the Groundwater discharge function, using Darcy's Law. To estimate the Groundwater discharge function ($t_i = t_0 = 0$), water flow (Q_0) from the aquifer to the wetland is calculated by the Darcy's Law equation (1), where H and H_g should be known values and $H < H_g$. The water volume and, accordingly, the water level in the wetland are being increased or decreased progressively, depending on the discharge value Q_i, together with the total water volume that flows into and out of the wetland. Assuming a time step (dt, e.g. days or months) upon which calculations take place, the estimation of the water level in the wetland (H_{t+1}) for the next time step (t + dt) is made by performing the following sequence of calculations. Note that for the completion of the following calculations, a series of measured groundwater level values (H_g) is needed.

a. Time $t_0 = 0$
   - H_0 and H_{g0} should be measured (known values).
   - From the known relationship $V = f(H)$, the water volume (V_0) stored in the system (WB–W) for the time step t_0 is estimated.
   - From the known relationship $A = f(H)$, the length (L_{w0}) of the section of the wet perimeter through which discharge takes place for the time step t_0 is estimated.
   - The vertical distance (B_0) between the base of the aquifer and the water surface in the wetland for the time step t_0 is estimated (see above).
   - $A_0 = L_{w0} * B_0$.
   - Water flow (Q_0) is calculated from equation (1): $Q_0 = k\,A_0 \dfrac{H_{g0} - H_0}{L}$.
   - The water volume that entered the wetland during time step t_0 is $V_{g0} = Q_0 * \mathrm{d}t$.

b. Time $t_1 = t_0 + \mathrm{d}t$
   - H_{g1} is a known value.
   - The stored water volume (V_1) in the wetland during the time step t_1 is $V_1 = V_0 + V_{g0} + (V_{in} - V_{out})$, where V_{in} and V_{out} are the total water volume that flows in and out of the wetland respectively, during the period t_1–t_0.

- From the known relationship $V = f(H)$, the corresponding water level H_1 for the time step t_1 is estimated.
- Similarly with the estimations made on the time step t_0, the parameters L_{w1}, B_1 and A_1 are calculated.
- Water flow (Q_1) is calculated from equation (1): $Q_1 = k A_1 \dfrac{H_{g1} - H_1}{L}$.
- $V_{g1} = Q_1 * dt$.

c. Time $t_i = t_{i-1} + dt$ (where $i \geq 2$)
As described above.

The calculations above are repeated for the period of interest. The function stops when $H_i = H_{gi}$ where the equilibrium between surface water (wetland) and groundwater table (aquifer) has been established.

The total discharged water volume ($V_{g,total}$) into the wetland can be estimated from the following sum:

$$V_{g,\,total} = \sum_{i=0}^{n} V_{gi} = \sum_{i=0}^{n} Q_i dt$$

The Groundwater recharge function is performed when the hydraulic gradient $[(H–H_g)/L] > 0$ and the Groundwater discharge function is performed when $[(H–H_g)/L] < 0$.

Groundwater discharge is calculated in parallel with groundwater recharge in order to assess the contribution of a wetland to the performance of these two functions for a given period. An example of the quantification of both functions is presented below in a worked example.

Worked example

The Groundwater recharge–discharge function has been assessed for the Cheimaditida wetland, for a six-year period (January 1993–December 1998). For more details regarding the hydrological system of Cheimaditida, see page 285.

According to the constructed DTM of the lake, a water volume less than 2 031 133 m^3 is stored in the WB, while any additional volume is stored in the wetland (W). In the agricultural area to the NE of Cheimaditida a dense drainage network exists that keeps groundwater level (H_g) to an absolute elevation of 589.5 m between May and September and 590 m for the rest of the year. The distance (L) between the drainage ditches and the boundaries of the wetland is approximately 25 m.

According to an existing hydrogeological survey in the area, the recharge–discharge function in Cheimaditida is being performed through a section of its northern and eastern shore with a length $L_w = 6500$ m (Fig. 3.1.38, see colour Section), which is assumed to remain constant for the whole period of calculations. Additionally, the absolute elevation of the base of the aquifer has been estimated to be 580.3 m.

The calculations for the estimation of the Groundwater recharge–discharge function for a selected time step of 1 month, are presented below:

a. Time $t_0 = 0$ (January 1993)
- At the beginning of January 1993, the absolute elevations of the water surfaces in the system WB–W were $H_0 = 591.2$ m and $H_{g0} = 590$ m.
- $H_0 > H_{g0}$ = > the Groundwater recharge function is being performed.
- From the known relationship $V = f(H)$, the water volume stored in the system WB–W equals $V_0 = f(H_0) = 9\,277\,884$ m^3.
- $L_{w0} = 6500$ m.
  Additionally, hydraulic conductivity (k) has been estimated at 0.0 000 124 m s^{-1}, while the horizontal distance $L = 25$ m.
- The vertical distance (B_0) is estimated as follows:

$$B_0 = H_0 - 580.3 = 591.2 - 580.3 = 10.9 \text{ m}$$

- $A_0 = L_{w0} * B_0 = 6.500 * 10.9 = 70\,850$ m^2.

- Water flow (Q_0) is calculated from equation (1):

$$Q_0 = k A_0 \frac{H_0 - H_{g0}}{L} = (0.0\,000124)(70\,850)\frac{591.2 - 590}{25}$$

$$= 0.0\,421\,699 \text{ m}^3/\text{s}$$

- The water volume that recharged the aquifer during January 1993 is equal to:

$$V_{g0} = Q_0 * dt = (0.0\ 421\ 699)\ (30 * 24 * 3600) = 109\ 304 \text{ m}^3$$

b. Time t_1 = 1 (February 1993)
- H_{g1} = 590 m
- The water volume (V_1) stored in the system WB–W at the beginning of the time step t_1 is

$$V_1 = V_0 - V_{g0} + (V_{in} - V_{out}) = V_0 - V_{g0} + (Q_{in} + R - E - Q_{out} - A)$$

where Q_{in} is surface run-off from the CA plus water inflow from Lake Zazari, during January 1993, R is precipitation into the system during January 1993, E is evaporation from the system during January 1993, Q_{out} is surface outflow from the system during January 1993 and, A is abstraction from the system during January 1993.

The parameters above are known from an existing hydrological survey.

Therefore, V_1 = 9 277 884 – 109 304 + (2 694 370 – 0) = 11 862 949 m^3, corresponding (by the known relationship $V = f(H)$) to an absolute elevation of H_1 = 591.5 m.
- $H_1 > H_{g1}$ => the Groundwater recharge function is being performed.
- As indicated above,
L_{w1} = 6500 m
k = 0.0 000 124 m s^{-1}
$B_1 = H_1 - 580.3 = 591.5 - 580.3 = 11.2$ m
- $A_1 = L_{w1} * B_1 = 72\ 800$ m^2
- $$Q_1 = k A_1 \frac{H_1 - H_{g1}}{L} = (0.0\,000124)(72\,800)\frac{591.5 - 590}{25}$$

$$= 0.0\ 541\ 632 \text{ m}^3/\text{s}$$

- The water volume that recharged the aquifer during February 1993 is equal to

$$V_{g1} = Q_1 * dt = (0.0\ 541\ 632)\ (30 * 24 * 3600) = 140\ 391 \text{ m}^3$$

Following the same procedure for the next period (March 1993–December 1998), the Groundwater recharge–discharge function is quantified. The results are presented in Table 3.1.8 and Figs 3.1.39 and 3.1.40, where negative values represent water movement from the aquifer into the system WB–W (performance of the Groundwater discharge function). During the calculations, recharged water volumes were subtracted from the total water budget (enabling the water volume of the next time step to be calculated), while discharged water volumes were added. From Figs 3.1.39 and 3.1.40, and taking into account Figs 3.1.3 and 3.1.4, the following may be concluded:

- The Groundwater recharge function is performed only by the wetland. During those periods that there is no water in the wetland, the function is not performed.
- The Groundwater discharge function is performed only by the WB, since there is no water in the W for the months that the aquifer discharges into the system WB–W.
- The role of the existing drainage network is significant for the performance of both functions, since it keeps the groundwater table at a constant level.

Table 3.1.8 Water volume stored in the wetland (V_i), recharged water volume (Q_{Ri}) and discharged water volume (Q_{Di}), for the period January 1993–December 1998

Month	V_{Total} (m^3)	V_i (m^3)	Recharged volume (Q_{Ri}, m^3)	Discharged volume (Q_{Di}, m^3)
Jan-93	9,277,884	7,246,751	109,304	
Feb-93	11,862,949	9,831,816	140,391	
Mar-93	13,708,214	11,677,081	161,951	
Apr-93	14,565,039	12,533,906	161,951	
May-93	13,333,368	11,302,235	198,302	
June-93	11,608,821	9,577,688	176,241	
July-93	8,251,325	6,220,192	144,402	
Aug-93	4,121,498	2,090,365	94,680	
Sep-93	883,130	0		–15,042
Oct-93	911,892	0		–52,647
Nov-93	988,385	0		–45,627
Dec-93	1,924,435	0		–
Jan-94	2,383,034	351,901	16,546	
Feb-94	4,110,577	2,079,444	51,644	
Mar-94	5,457,301	3,426,168	70,196	
Apr-94	6,013,039	3,981,906	70,196	
May-94	7,174,463	5,143,330	134,124	
June-94	5,441,620	3,410,487	104,290	
July-94	2,828,151	797,018	66,853	
Aug-94	244,559	0		–85,822
Sep-94	911,892	0		–15,042
Oct-94	911,892	0		–52,647
Nov-94	308,419	0		–113,650
Dec-94	488,834	0		–91,254
Jan-95	6,469,711	4,438,578	79,722	
Feb-95	9,625,607	7,594,474	109,304	
Mar-95	10,914,670	8,883,537	129,862	
Apr-95	11,720,982	9,689,849	140,391	
May-95	12,038,886	10,007,753	187,188	
June-95	10,375,677	8,344,544	165,461	
July-95	7,707,486	5,676,353	134,124	
Aug-95	5,002,791	2,971,658	104,290	
Sep-95	3,225,323	1,194,190	75,962	
Oct-95	1,726,255	0	0	0
Nov-95	1,478,120	0		–15,878
Dec-95	6,454,747	4,423,614	79,722	
Jan-96	11,984,448	9,953,315	140,391	
Feb-96	12,164,182	10,133,049	140,391	
Mar-96	12,487,017	10,455,884	140,391	
Apr-96	12,802,874	10,771,741	151,087	
May-96	12,658,085	10,626,952	198,302	
June-96	10,669,091	8,637,958	176,241	
July-96	7,852,232	5,821,099	134,124	
Aug-96	4,879,194	2,848,061	104,290	
Sep-96	2,582,963	551,830	66,853	
Oct-96	1,574,992	0		–8,022
Nov-96	1,717,678	0		–8,022
Dec-96	2,343,320	312,187	16,546	
Jan-97	5,191,300	3,160,167	60,836	
Feb-97	6,407,057	4,375,924	79,722	
Mar-97	6,626,622	4,595,489	79,722	
Apr-97	7,435,849	5,404,716	89,416	
May-97	8,332,168	6,301,035	144,402	
June-97	6,000,708	3,969,575	114,068	
July-97	3,429,622	1,398,489	75,962	
Aug-97	425,515	0		–62,424
Sep-97	911,892	0		–15,042
Oct-97	911,892	0		–52,647

Table 3.1.8 cont'd

Month	V_{Total} (m³)	V_i (m³)	Recharged volume (Q_{Ri}, m³)	Discharged volume (Q_{Di}, m³)
Nov-97	1,364,749	0		–15,878
Dec-97	2,139,258	108,125	8,189	
Jan-98	5,210,390	3,179,257	60,836	
Feb-98	5,928,500	3,897,367	70,196	
Mar-98	7,308,556	5,277,423	89,416	
Apr-98	7,056,663	5,025,530	89,416	
May-98	6,957,684	4,926,551	124,012	
June-98	5,981,733	3,950,600	114,068	
July-98	3,274,772	1,243,639	75,962	
Aug-98	114,318	0		–111,310
Sep-98	911,892	0		–15,042
Oct-98	911,892	0		–52,647
Nov-98	1,136,749	0		–31,087
Dec-98	8,692,092	6,660,959	99,277	
TOTAL			5,551,211	–859,728

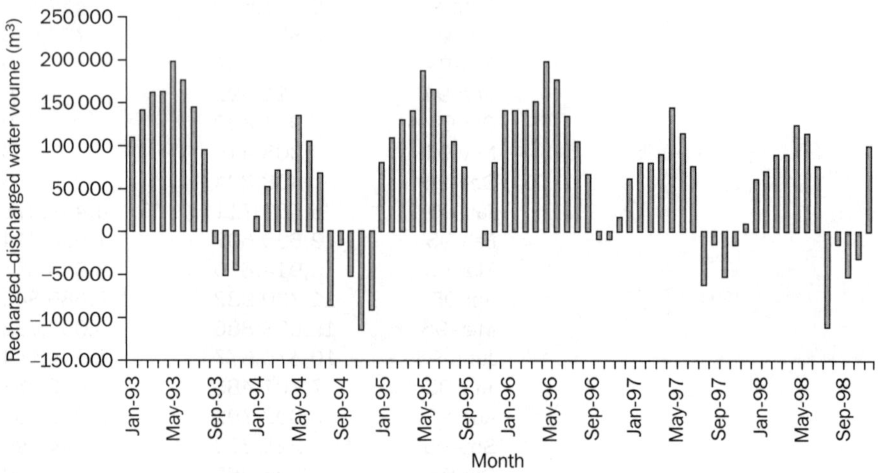

3.1.39 Graphical representation of the quantification of the Groundwater recharge–discharge functions.

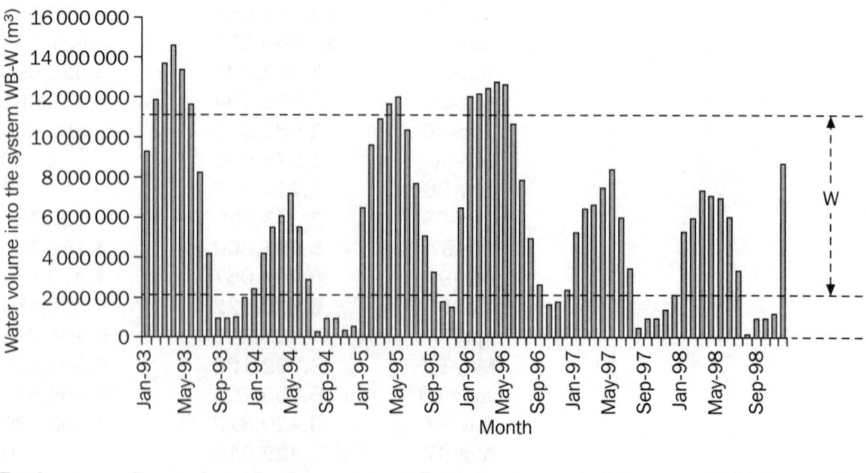

3.1.40 Total water volume stored in the system WB–W, for the period January 1993–December 1998.

3.1.E.4 Sediment retention (function)

Introduction

Water quality improvements such as the reduction of suspended sediments, the retention of pollutants and nutrients and the reduction of sedimentation rates downstream are values that wetlands can contribute in a catchment resulting from the Sediment retention function. A quantitative assessment of the sediment retention function must address the following questions:

- What is the retention capacity of a wetland?
- How often does retention occur?
- What is the significance of the reduced sediment load for the wetland compared to the whole sediment load of the flood?

This section clarifies the requirements for addressing these questions for river marginal wetlands.

Conditions required for sediment retention

Two hydrological systems are shown in Figs 3.1.41a and 3.1.41b supplying different sources of sediment. Figure 3.1.41a shows deposition of sediment transported in surface run-off from the CA. The river marginal wetland (RMW) acts as a sediment buffer to the river. Figure 3.1.41b shows retention of sediment carried as suspended load by river water during overbank flooding events. The RMW has, during these occasions, a temporary associated water body (WB) connected either to a (CA) or the river. The sediment load of the WB and the local hydraulic conditions of the RMW are such that the sedimentation rate exceeds the erosion rate. The RMW acts as a sediment sink to the river.

In Fig. 3.1.41a, the flow entering the RMW is surface run-off from a CA. The sediment load of the surface run-off and the hydraulic conditions of the RMW determine whether sedimentation takes place before water enters the river. In the second system (Fig. 3.1.41b), water flows over the riverbank into the RMW. The distribution of the water within the RMW, the flow velocity and the depth are factors that largely determine the conditions for sedimentation before water flows back to the river channel.

The ability of river marginal wetlands to contribute to sediment retention depends mainly on the local conditions within the RMW governing deposition and erosion. Erosion depends on the local hydrological conditions, whereas deposition or sedimentation depends on the concentration of the suspended sediment load as well as the local hydrological conditions. The wetland may act as a sediment sink depending on particle sizes, concentration of the sediment load and flow velocity. Salinity, temperature and a non-turbulent flow regime favour the sedimentation process.

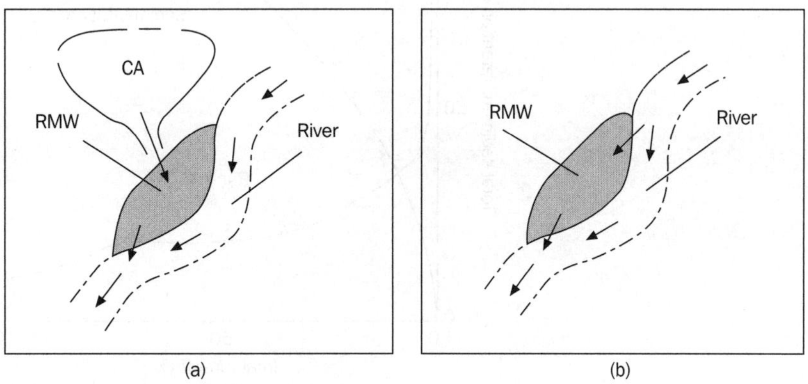

3.1.41 Two hydrological sources of sediment.

Dynamics in sedimentation

Sediment retention is a highly variable function in time and space. Figure 3.1.42 shows the idealised concentration of suspended sediments sampled at the inlet and the outlet of a wetland during a storm flood event. Run-off has also been measured during the flood event episode.

Soil erosion and concentration of suspended solids may increase with run-off. Streams carry more and coarser sediments as run-off increases. Flood events can often make up a large proportion of the sediment transport (Fig. 3.1.42). During a storm event, concentrations of sediments are often high initially and decrease later even if run-off is at the same rate. The system is, during these occasions, exposed to a so-called 'first flush'. Easily eroded sediments are transported downstream at an early stage of the flood event. Erosion then decreases, as the more easily available material diminishes. The proportion and location of arable land within the catchment, soil types and factors related to land-use all significantly influence erosion and the sediment load.

In Fig. 3.1.43, samples are sorted by sediment rate in decreasing order and then accumulated. Mean run-off flow index is calculated as accumulated run-off divided by the time for each new episode; sediment retention occurs mainly under storm run-off episodes.

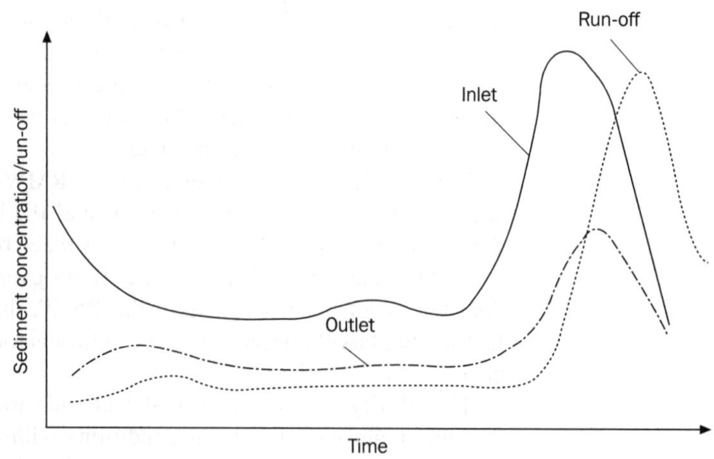

3.1.42 Sediment retention: idealised run-off and sediment concentration in inlet and outlet of a river marginal wetland.

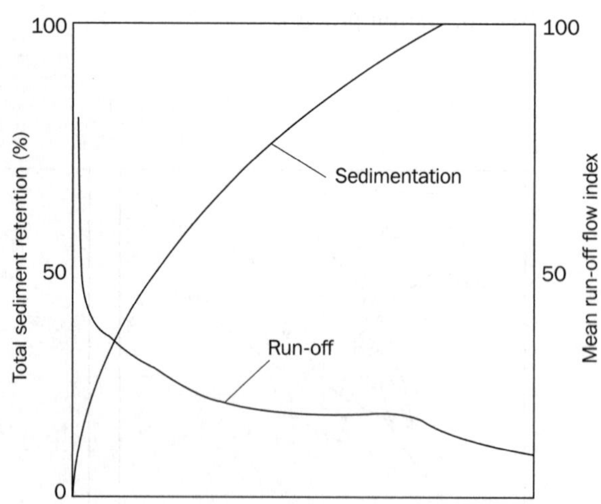

3.1.43 Relative values of sedimentation and run-off in a system similar to Fig. 3.1.41a.

Sampling methods and strategies

Measurements of loss, transport and deposition of sediments from the CA or WB require an intensive sampling strategy. Water flow proportional sampling systems in the inlet and the outlet ensure representative samples. In small watersheds, spot samples are recommended. Composite sampling is cheaper and is sufficient to calculate sediment mass-balance if sampling frequency is high and the sampling period long. However, the method requires laboratory work to isolate the sediment substrate. Recommended sampling intervals needed for some systems can be as frequent as four times a day or even hourly. It is, however, expensive to use intensive sampling, and not all natural or constructed systems are possible to assess in this way. Detailed information on sampling and sampling strategies can be found in Chen (1975), Hazen (1904) and Johnson (1992).

Quantitative assessment of sediment retention

With intensive sampling. Measurements of discharge inflow (Q) to the system are needed together with intensive sampling to determine the concentration of suspended sediments (C_S), at the inlet and outlet of the system. Sediment retention is calculated as the accumulated value of the product of mean discharge and the difference in suspended sediment concentration, $Q(C_{S,in} - C_{S,out})$.

Without intensive sampling. In smaller well-defined systems, where intensive sampling is not possible, a rapid quantification method with limited sampling is suggested. Run-off has a large influence on suspended sediment load and on sediment retention. In systems where intensive sampling is not possible, a measure of run-off can be used as a rapid quantification method to estimate the resulting sediment retention. Reference samples are taken to determine the suspended sediment concentration at the inlet and outlet of RMWs under conditions of low and high flow. Sediment retention is calculated using the intensive sampling method (above). An example of a reference graph is shown in Fig. 3.1.44, where reference samples are plotted to produce a reference curve, which is used for the rapid quantification of sediment retention. Run-off is plotted against the difference in sampled suspended sediment concentration (C_S), between the inlet and outlet. The difference in concentration of suspended load between the inlet and the outlet of the RMW is estimated from the reference curve by using the recorded values of run-off. Sediment retention is then calculated as above.

In other systems, detailed assessments of sediment retention and predicted retention capacity, frequency and the significance of reduced sediment load require

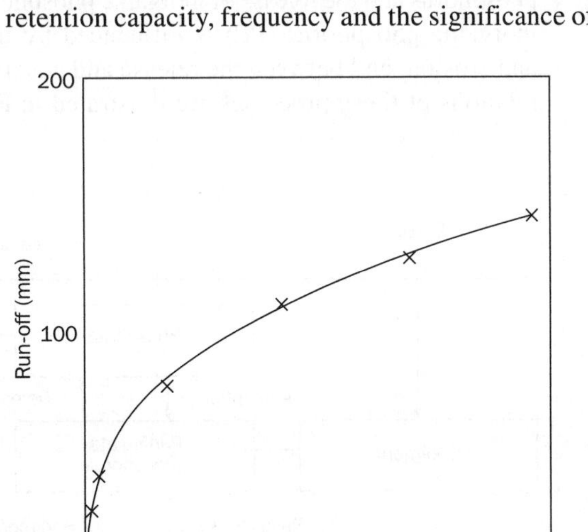

3.1.44 Plotted reference samples for establishment of a reference curve used in rapid quantification of sediment retention. Run-off is plotted against the difference in sampled suspended sediment concentration (C_S) between inlet and outlet.

the dynamic flows of the river to be considered. This is especially important for flooding events in systems similar to that in Fig. 3.1.41b when large amounts of sediments are often transported. A detailed quantitative assessment of sediment retention requires advanced hydrological methods in combination with sediment transport models and intensive sampling.

Sediment transport models

One-dimensional equations for the conversion of mass of dissolved or suspended material to total load (sediment transport rate and alluvial roughness) can be found in Ackers and White (1973) and Smart and Jaeggi (1983). For calculations of bed load and suspended load separately see Engelund and Fredsoe (1976) and Van Rijn (1984a,b). These sources of information contain detailed descriptions of methods and information on the selection of appropriate techniques.

Particulate phosphorus in relation to sedimentation

Phosphorus compounds are very reactive and easily form complexes with inorganic and organic compounds in soil, water and aquatic sediments. As a consequence, phosphorus is often the limiting nutrient in freshwater environments. Transport occurs to streams, lakes and wetlands as dissolved phosphate ($PO_4^{3\pm}$), dissolved organic compounds and as particulate-bound phosphorus. The mixtures of these forms vary between different environments and are, in general, determined by factors like land-use, soil type, soil aeration and vegetation cover. The risk of release of phosphorus is larger in soils and sediments that have a history of being over-supplied with phosphorus and is typical, for example, in areas receiving wastewater and on agricultural land with a long history of intensive animal and crop production. Topsoil usually contains the highest concentration of phosphorus, which can generate phosphorus-rich sediment transport in overland flow. Sedimentation occurs when flow reaches a wetland, lake or pond, where conditions favour particle settlement rather than continued suspension. The size, form and concentration of sediment particles, density of vegetation, depth of water and currents are the main factors controlling suitability for sedimentation.

The removal of nutrients by sedimentation is one way in which sediment retention can contribute to improved water quality. The concentration of dissolved phosphorus depends on a number of processes including uptake by vegetation, degradation of biological oxygen demand (BOD), adsorption of dissolved phosphorus and the release of inorganic phosphorus. The concentration of particulate inorganic phosphorus (PP) is influenced by the balance between sedimentation and erosion, and between the release and adsorption of dissolved phosphorus. The relations of these processes are illustrated in Fig. 3.1.45.

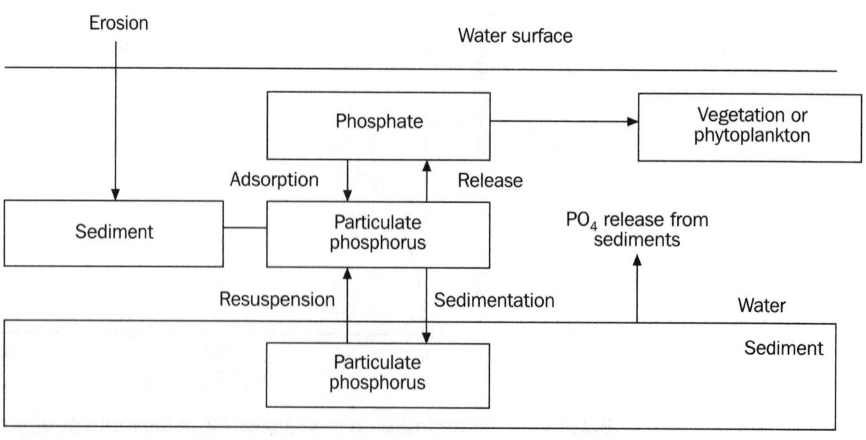

3.1.45 Phosphorus retention by sedimentation, resuspension by disturbance, and related phosphorus processes.

The formation of aggregates from finer particles, shallow depth and extensive vegetation cover increases rates of sediment retention. The shorter particle settling distance and vegetation which hinders the resuspension of sediment, while enabling uptake by plants, explains why wetlands are often more efficient than ponds in the removal of sediments and phosphorus. Factors outside the wetland environment itself, such as the input of phosphorus on arable land, also increase rates of phosphorus retention.

Phosphorus in two forms is sampled and measured in the suspended sediment load, total content of phosphorus (TOTP) and ortho-phosphate (PO). Particulate phosphorus (PP) is calculated as the difference between TOTP and PO.

$$PP = TOTP - PO$$

It is generally not necessary to investigate other forms of phosphorus except for detailed research purposes. Particulate phosphorus is quantified using data from sampling of TOTP and PO and measurements of run-off (Q).

$$PP = Q((TOTP - PO)_{in} - (TOTP - PO)_{out})$$

In addition to advanced hydrological methods, chemical and biological processes need to be considered. This calls for the use of sub-models.

In several models, calculations of phosphorus retention are integrated with sediment transport models. This is accomplished by a numerical integration routine that solves a system of differential equations describing physical, chemical and biological processes (Fig. 3.1.45).

For detailed information on phosphorus retention in wetlands and phosphorus modelling see Correll (1999), Kadlec and Knight (1996), Khalid *et al.* (1977), Ulén *et al.* (1998) and Richardson and Vaithiyanathan (in press).

Further reading and references

Ackers, P. and White, W.R. (1973). Sediment transport: new approach and analysis, *Proc. ASCE, JHD* **99**, HY11, 2041–2060.

Benson, M.A. (1962). Plotting Positions and Economics of Engineering Planning, *Journal of the Hydraulics Division ASCE* **88**, 57–71.

Bouwer, H. (1978). *Groundwater Hydrology*, McGraw-Hill, New York.

Bullock, A. and Acreman, M. (2003). The role of wetlands in the hydrological cycle, *Hydrology and Earth System Sciences* **7**(3), 358–389.

Chen, C. (1975). Design of sediment retention basins, *Proc. National Symposium on Urban Hydrology and Sediment Control*, University of Kentucky, Lexington, 28–31 July, Peasons Binckerhoff, Quade & Douglas, New York, 285–298.

Chow, V.T. (1959). *Open Channel Hydraulics*, McGraw-Hill, New York.

Chow, V.T. (1964). *Handbook of Applied Hydrology*, McGraw-Hill, New York.

Chow, V.T., Maidment, D.R. and Mays, L.W. (1988). *Applied Hydrology*, McGraw-Hill, New York.

Correll, D.L. (1999). Phosphorus: A rate limiting nutrient in surface waters, *Poultry Science* **78**, 674–682.

DeVries, J.J. and MacArthur, R.C. (1979). *Introduction and Application of Kinematic Wave Routing Techniques Using HEC-1*, Training Document, Hydrologic Engineering Center, McLean, VA.

DeVries, J.J. and MacArthur, R.C. (1985). *Flood Hydrograph Package, User Manual*, Training Document No. 10, Hydrologic Engineering Center, US Army Corps of Engineers, Davis, CA.

Engelund, F. and Fredsoe, J. (1976). A sediment transport model for straight alluvial channels, *Nordic Hydrology* **7**(5), 293–306.

Franti, T.G., Eisenhauer, D.E., McCullough, M.C., Stahr, L.M., Dosskey, M.G., Snow, D.D., Spalding, R.F. and Boldt, A.L. (2004). Watershed scale impacts of buffers and upland conservation practices on agrochemical delivery to streams, *Proceedings of the Self-sustaining Solutions for Streams, Wetlands, and Watersheds Conference*, St Paul, MN, 12–15 Sept, 323–332.

Gilvear, D.J. and Bradley, C. (in press). Hydrological dynamics II: groundwater and hydrological connectivity. In: Maltby, E. and Barker, T. (eds), *The Wetlands Handbook*, Wiley-Blackwell, Oxford.

Gilvear, D.J., Tellam, J.H., Lloyd, J.W. and Lerner, D.N. (1989). The hydrodynamics of

East Anglian fen systems. Final report to the Nature Conservancy Council, Anglian Water and Broads Authority. University of Birmingham Report.

Gilvear, D.J., Tellam, J.H., Lloyd, J.W. and Lerner, D.N. (1993). Quantification of the water balance and hydrological processes in the vicinity of a small groundwater-fed fen wetland, East Anglia, England, *Journal of Hydrology* **104**, 311–344.

Gilvear, D.J., Tellam, J.H., Lloyd, J.W. and Lerner, D.N. (1994). Wetland vulnerability in East Anglia: The range of validity of a generalised classification system, *Aquatic Conservation; Marine and Freshwater Ecosystems* **4**, 107–127.

Hazen, A. (1904). On sedimentation, *Transactions, American Society of Civil Engineers* **53**, 43–71.

HEC (2002). *HEC-RAS River Analysis System. Hydraulic Reference Manual*, US Army Corps of Engineers, Davis, CA.

ISCO (2003). Remote Monitoring of Runoff and Selection of Automatic Sampling to Ensure Sample Integrity, ISCO Inc., Lincoln, NE.

Johnson, R.C. (1992). Towards the design of a strategy for sampling suspended sediments in small headwater catchments. In: Borgen, J. Walling, D.E. and Day, T.J. (eds), *Erosion and Sediment Transport Monitoring Programmes in River Basins*, IAHS Red Book Publ. 210, IAHS Press, Centre for Ecology and Hydrology, Wallingford, UK, 225–232.

Kadlec, R.H. and Knight, R.L. (1996). *Treatment Wetlands*, Lewis, Boca Raton, FL.

Kalergis, G.A. (1999). *Applied Environmental Hydrogeology. Part A*, 2nd edn, Technical Chamber of Greece, Athens, Greece (in Greek).

Khalid, R.A., Patrick, W.H. and DeLaune, R.D. (1977). Phosphorus sorption characteristics of flooded soils, *Soil Science Society American Journal* **41**, 305–310.

Ligget, J.A. (1975). Basic equations of unsteady flow. In: *Unsteady Flow in Open Channels*, Water Resources Publication, Fort Collins, CO, 29–62.

Linsley, R., Kohler, M. and Paulhus, J. (1982). *Hydrology for Engineers*, 3rd edn, McGraw-Hill, New York.

Miller, W.A. and Cunge, J.A. (1975). Simplified equations of unsteady flow. In: Mahmood, K. and Yevjevich, V. (eds), *Unsteady Flow in Open Channels, Volume* 1, Water Resources Publication, Fort Collins, CO, 183–257.

Ramírez, J.A. (2000). Prediction and modelling of flood hydrology and hydraulics. In: Wohl, E. (ed.), *Inland Flood Hazards: Human, Riparian and Aquatic Communities*, Cambridge University Press, Cambridge, UK, 293–329.

Richardson, C.J. and Vaithiyanathan, P. (in press). Biogeochemical dynamics of phosphorus in wetlands. In: Maltby, E. and Barker, T. (eds), *The Wetlands Handbook*, Wiley-Blackwell, Oxford.

Rijn, L.C. van (1984a). Sediment transport, Part I: Bed load transport, *Journal of Hydraulic Engineering* **110** (10), 1431–1456.

Rijn, L.C. van (1984b). Sediment transport, Part II: Suspended load transport, *Journal of Hydraulic Engineering* **110** (11), 1613–1641.

Sherman, L.K. (1949). The unit hydrograph method. In: Meinzer, O.E. (ed.), *Hydrology*, Dover Publications, New York, 514–526.

Simpson, M. (2002). *A Proposed Functional Classification of European Wetlands: Development and Testing*, Volume II, PhD Thesis, Royal Holloway University of London.

Smart, G.M. and Jaeggi, M.N.R. (1983). *Sediment Transport on Steep Slopes*, Mitteilung nr 64, Laboratory of Hydraulics, Hydrology and Glaciology, Federal Technical University, Zürich, Switzerland.

Snyder, F.F. (1938). Synthetic unit graphs, *Transactions AGU* **19**, 447–454.

Strelkoff, T. (1969). The one-dimensional equations of open-channel flow, *Journal of the Hydraulics Division,* American Society of Civil Engineers, **95**(HY3), 861–876.

Strelkoff, T. (1970). Numerical solution of Saint-Venant equations, *Journal of the Hydraulics Division*, American Society of Civil Engineers **96**(1), 223–252.

Terzidis, G.A. and Karamouzis, D.N. (1985). *Groundwater Hydraulics*, Ziti Publications. Thessaloniki, Greece (in Greek).

Ulén, B., Shirmohammadi, A. and Bergström, L.F. (1998). Phosphorus transport through a clay soil, *Journal of Environmental Science and Health* **33**(1), 67–82.

US Army Corps of Engineers (1996). *Wetlands Hydrology and Hydraulics*, Study Draft as of 8/8/96, Hydrologic Engineering Center (HEC), Davis, CA.

US Soil Conservation Service (1957). *National Engineering Handbook*, Section 4, Hydrology, Supplement A, US Department of Agriculture, Washington, DC.

US Soil Conservation Service (1972). *National Engineering Handbook*, Section 4, Hydrology, US Department of Agriculture, Washington, DC.

US Soil Conservation Service (1981). *Urban Hydrology for Small Watersheds*, Tech. Release 55, US Department of Agriculture, Washington, DC.

US Soil Conservation Service (1985). *National Engineering Handbook*, Section. 4,

Hydrology, US Department of Agriculture, Washington, DC.
Verhoeven, R., Banasiak, R., Okruszko, T., Swiatek, D., Chormanski, J., Nowakowski, P., Kardel, I. and Stelmaszczyk, M. (2003). Hydraulic modelling of river flow-data collection and problem solving. In: Ignar, S., Nowakowski, P. and Okruszko, T. (eds), *WETHYDRO: Measurement Techniques and Data Assessment in Wetland Hydrology*, Warsaw Agricultural University Press, Warsaw, Poland, 9–23.

3.2
Biogeochemical functions

Introduction

The biogeochemical processes taking place in wetlands enable these ecosystems to play a key role in water quality, pollution control and the support of biodiversity. Biogeochemical transformations are not necessarily unique to wetlands, but the combinations and particular dynamics of biogeochemical cycles and processes operating within them impart special properties to wetland ecosystems. Biogeochemical functioning in wetlands is controlled fundamentally by hydrology, which affects not only the processes that take place within an individual wetland, but also those more remote interactions between wetlands and other systems, which influence the performance of some functions. Waterlogged (hydric) soils often have unique distributions of aerobic and anaerobic zones, which influence transformations of nutrients and metals that normally do not occur in other ecosystems. The combination of biological, chemical and physical processes that occur in wetlands result in biogeochemical changes that can mobilise, immobilise, otherwise transform and even remove completely from the wetland system a wide range of compounds and elements, including plant nutrients, organic carbon, heavy metals and other contaminants. Their location at the interface of terrestrial, groundwater and aquatic systems provides them with the opportunity to interact with and influence the quality of water in the wider environment. A generalised diagrammatic representation of wetland biogeochemical cycling is shown in Fig. 3.2.1.

Many of the transformation processes within wetlands are controlled by the oxidation/reduction (redox) potential in the soil. When flooding occurs, oxygen

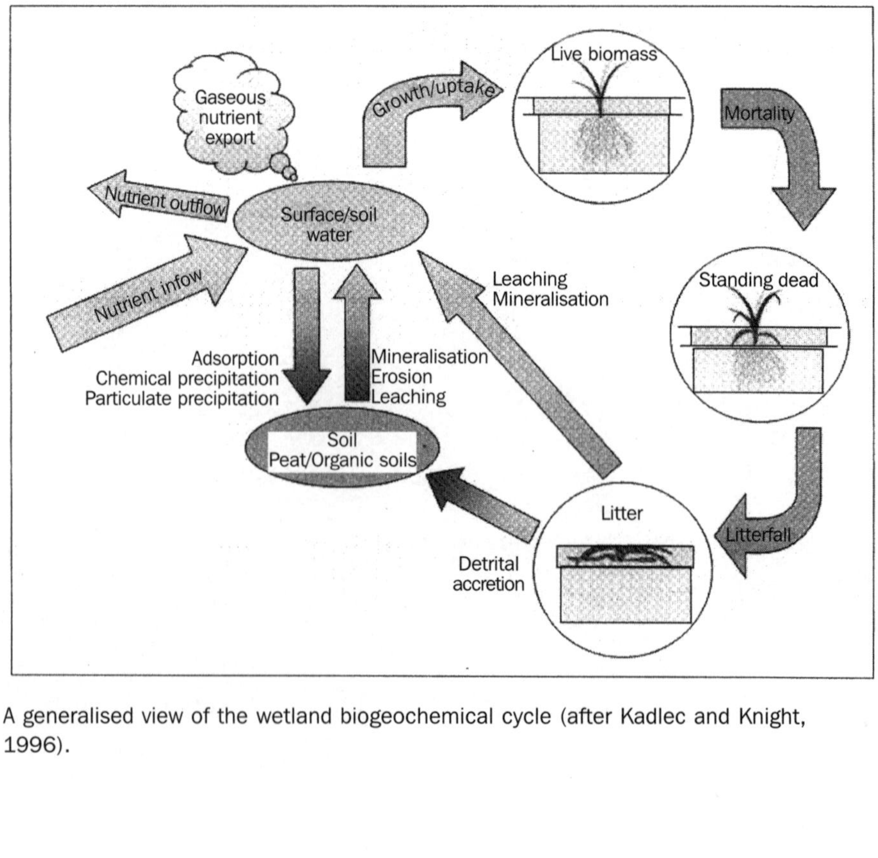

3.2.1 A generalised view of the wetland biogeochemical cycle (after Kadlec and Knight, 1996).

rapidly becomes depleted because, while oxygen consumption by micro-organisms continues, the rate of diffusion of oxygen into the soil is greatly reduced, being 10 000 times slower in a saturated soil than in a well drained soil (Greenwood, 1961). Redox potential is a measure of the electron availability in a solution. The term oxidation refers to either an uptake of oxygen by a compound or the loss of an electron or hydrogen from that compound, while reduction is the reverse process whereby a compound releases oxygen or takes up hydrogen or an electron. Following inundation of a wetland, significant oxygen depletion can occur over timescales from a few hours up to a few days, during which a well recognised sequence of transformation processes takes place, as illustrated in Fig. 3.2.2. The nature and rate of these transformations are fundamental to the rate and types of many of the biogeochemical functions performed by wetlands.

Another key controlling variable on the types of biogeochemical transformations that occur in wetlands is pH. The redox potential at which many compounds are stable is closely related to pH. When a soil floods, generally there is a move towards neutrality, but typically organic-rich and especially rain-fed wetlands have low pH, while those that have mineral soils and are groundwater fed are normally of higher pH.

A further factor important to the biogeochemical functioning of wetlands is the cation exchange capacity of the soil. This is a measure of the amount of exchangeable cations that a soil can hold, and is important in controlling adsorptive processes. Organic soils characteristic of many wetland systems typically have a high cation exchange capacity, although this declines in poorly humified peats. This relationship is illustrated in Fig. 3.2.3. Any positively charged molecule or particle has the potential to be adsorbed in soils with high cation exchange capacities, and this can be an important store for some compounds and elements in wetland soils.

High nutrient and heavy metal concentrations contribute to poor water quality and environmental degradation though, conversely, minimum concentrations of some chemicals are required to maintain surface water bodies in good ecological condition. Wetlands have the ability to regulate these concentrations in a variety of ways, and consequently can have considerable value in water quality maintenance and improvement. The important cycles with regard to water quality are those of nitrogen (N), phosphorus (P) and carbon (C). The ability of wetlands to influence these cycles and to regulate other potentially harmful elements and compounds underpins their functioning as buffer zones (see Section 1.4.2).

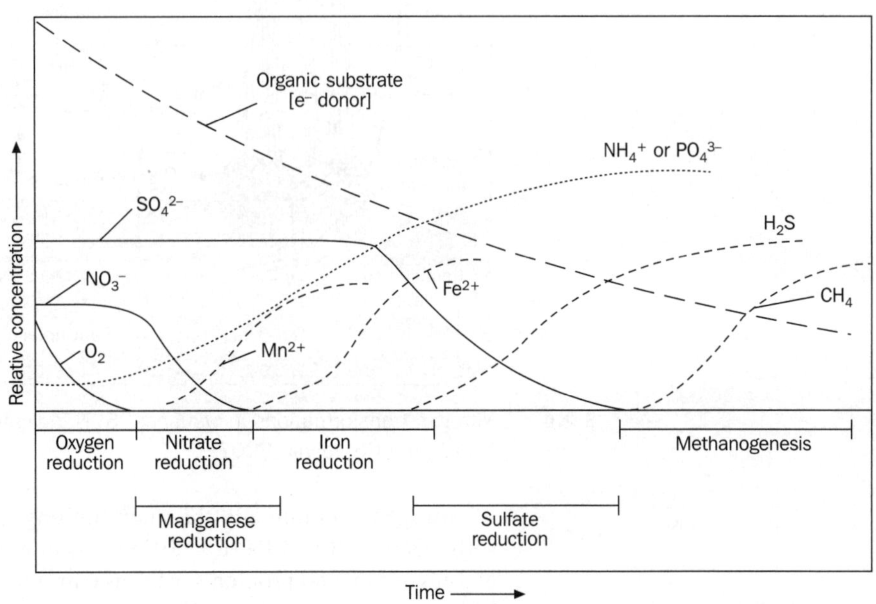

3.2.2 Sequence in time of transformations in soil following flooding (Reddy, K. R., in: Mitsch and Gosselink, 2007).

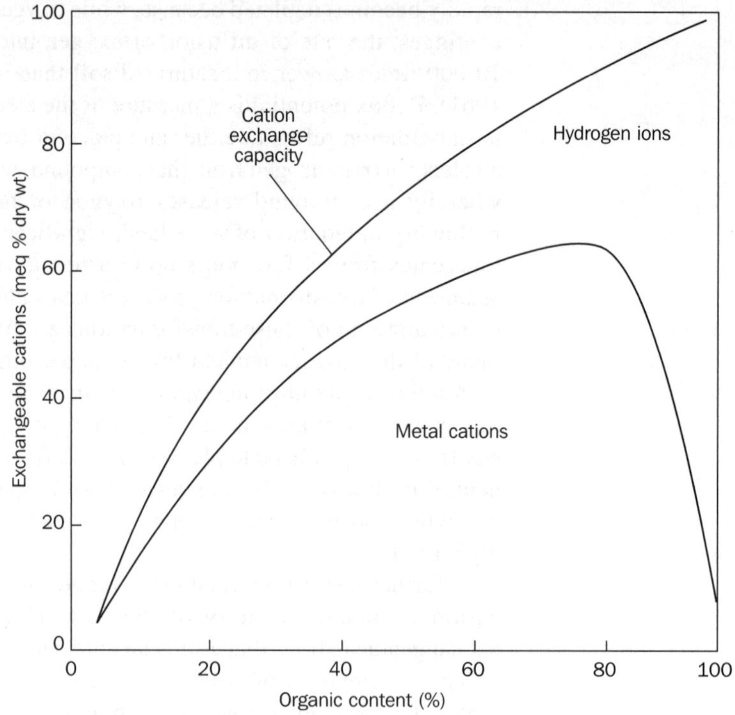

3.2.3 Relationship between cation exchange capacity and organic matter content in soils (Mitsch and Gosselink, 2007).

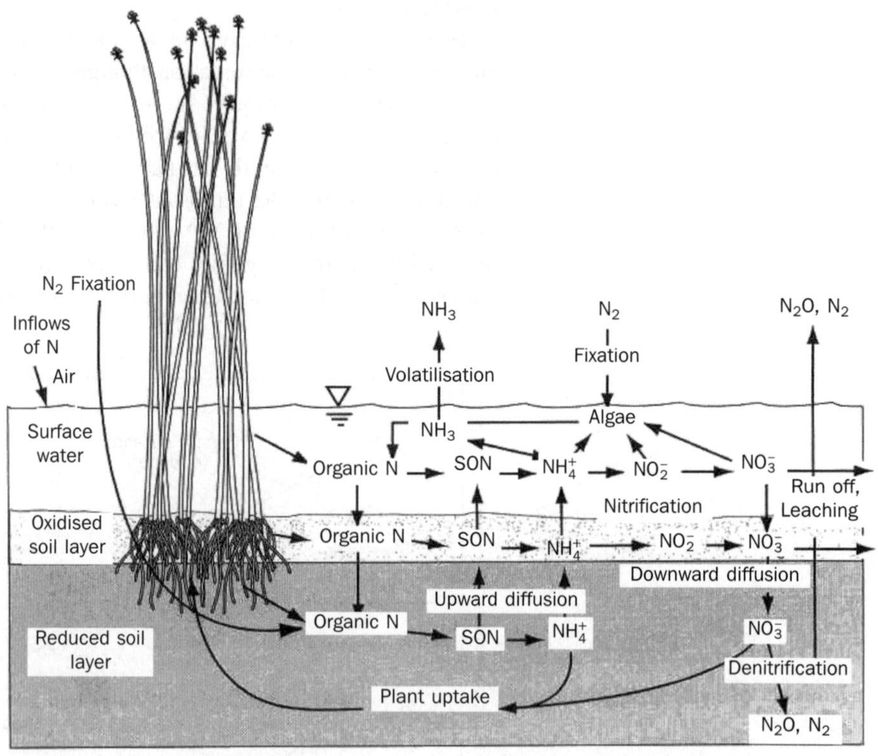

3.2.4 Nitrogen transformations in wetlands. SON indicates soluble organic nitrogen (source: Mitsch and Gosselink, 2007).

Nitrogen is commonly the main limiting nutrient for plant growth in wetland soils (Gambrell and Patrick, 1978) yet, in excess, it can cause severe ecological and environmental problems, such as eutrophication of surface- and groundwaters. There are numerous transformations that occur within wetlands involving N cycling, and these are summarised in Fig. 3.2.4. Typically, the primary form of mineralised N in flooded wetland soils is ammonium NH_4^+ although, in highly organic soils and peat, a large amount of N can be locked up in organic matter. The main transformation processes for N in wetlands are mineralisation, adsorption, plant uptake, nitrification, denitrification and fixation. Several of the pathways between these processes (which are often necessary for the performance of functions) rely upon the presence of an aerobic and oxidised soil layer above a lower anoxic layer (Fig. 3.2.4). The thickness of the oxidised layer can vary, and commonly is only a few millimetres or less. Important gaseous phases comprise part of the cycle, and are of particular significance in some wetlands with regard to their ability to export N as N_2 gas from the wetland system to the atmosphere.

Phosphorus is vital for plant growth and, like N, is often a limiting nutrient in wetlands. Also like N, it is subject to a wide variety of transformations in wetlands depending upon the particular conditions that prevail. The processes and transformations to which P is subject in wetlands are summarised in Fig. 3.2.5. Phosphorus occurs in organic and inorganic forms as both soluble and insoluble complexes. Unlike the nitrogen cycle, it is commonly stated that there is no gaseous phase in the phosphorus cycle, although evidence is growing that in some systems gaseous phosphine could be a significant pathway for P release to the atmosphere (Devai and DeLaune, 1995). The primary inorganic form of phosphorus is orthophosphate PO_4^{3-}, but its exact chemistry is largely pH dependent. Commonly, a large proportion of phosphorus in wetlands is locked up in organic matter or inorganic sediment. This results from precipitation under aerobic conditions with iron (Fe^{3+}), calcium or aluminium, adsorption onto soil particles, or inclusion in organic matter through plant uptake, all of which render it biologically unavailable. Unlike other elements, phosphorus is not directly affected by redox, but rather by its association with other elements, especially iron, the forms of which are redox-dependent. Consequently, anaerobic conditions can result in the mobilisation of previously precipitated and immobilised phosphorus. If released in excess to adjacent waters, phosphorus, like nitrogen, can result in severe eutrophication.

Carbon is also an important element and subject to transformation processes within wetlands. A summary of the types of transformations it can undergo in wetlands is shown in Fig. 3.2.6. Biodegradation is limited in wetlands by the

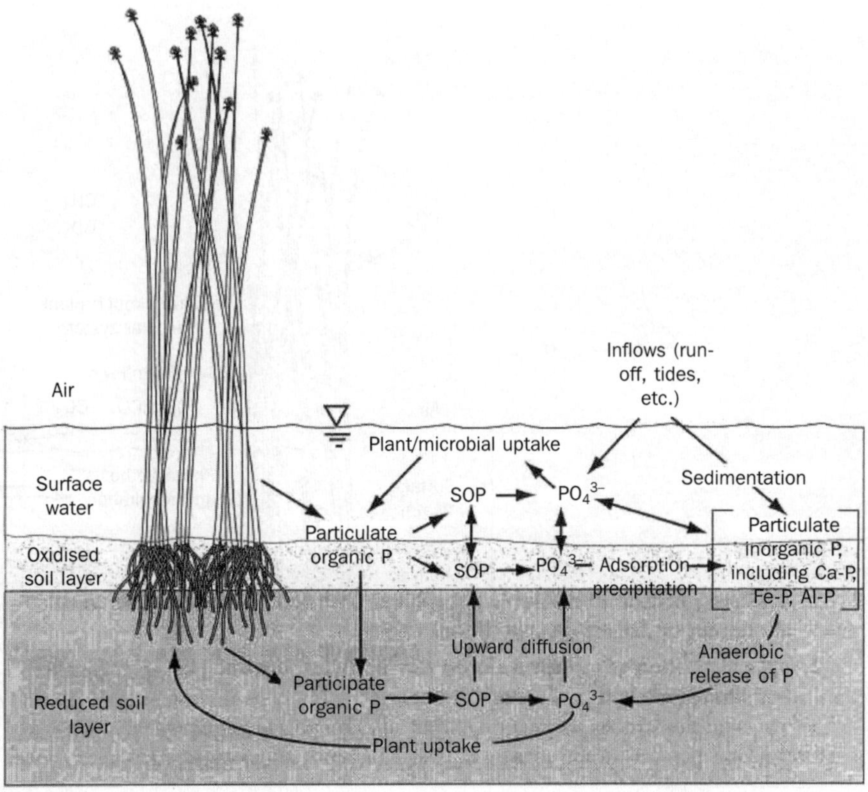

3.2.5 Phosphorus transformations in wetlands. SOP indicates soluble organic phosphorus (source: Mitsch and Gosselink, 2007).

characteristically reduced conditions that exist, leading to the accumulation of organic matter and the formation of peat. However, several anaerobic processes such as fermentation and methanogenesis can result in organic matter degradation, although the latter is strongly dependent upon sulphur concentrations. There are several reasons for this, including competition for substrates between sulphur and methane bacteria, inhibitory effects of sulphate and sulphide on methane bacteria, and the dependence of methane bacteria on the products of sulphur-reducing bacteria (Gambrell and Patrick, 1978). Carbon compounds are important for the biochemical transformation of many other elements in wetland systems, as they provide the essential substrate for the respiration of micro-organisms, and consequently are degraded in association with the transformation of other elements.

Many trace elements, in small quantities, are vital to the growth of flora and fauna, and consequently are commonly referred to as micronutrients. However, trace elements (including micronutrients) can reach toxic concentrations in soils and surfacewaters. The most toxic are mercury (Hg), cadmium (Cd) and lead (Pb), for which no biological function is known. Some elements are subject to biomagnification, a process by which concentrations increase in consumers among trophic levels of a food web, as they undergo biochemical transformations. For example, in many Scandinavian surfacewaters, Hg concentrations in fish have increased to levels that are now threatening their use for human consumption (Meili, 1997). Four major mechanisms operate in wetlands that enable interaction with trace elements. These are adsorption, precipitation, sediment retention and plant uptake, although some metals, especially Hg and selenium (Se), do have gaseous pathways. Problems associated with heavy metals in the environment are of particular concern in eastern Europe, where measures that can reduce trace metal mobility, such as the liming of agricultural soils, have been greatly reduced in recent years (Várallyay, 1993). Besides natural sources such as metalliferous rocks and degraded peatlands, trace elements in the environment also derive from artificial sources including fertilisers, manures, industrial waste and anthropogenic aerosols. The increase in mercury in a forest lake in Finland was attributed by Simola and Lodenius (1982) to peat drainage. Increased lead levels were found

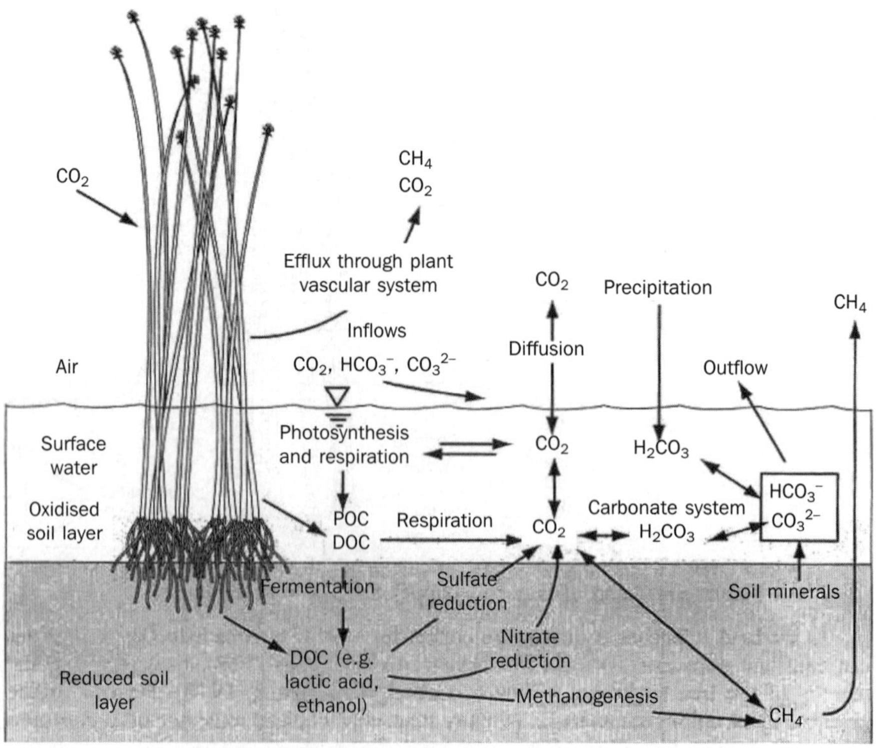

3.2.6 Carbon transformations in wetlands. POC indicates particulate organic carbon; DOC indicates dissolved organic carbon (source: Mitsch and Gosselink, 2007).

by Brännvall *et al.* (1997) in Swedish lakes next to drained organic soils and peat.

Biogeochemical functioning in wetlands

Nutrient retention

The retention of nutrients in wetlands can have significant effects on the quality of water moving through and draining from them, and is widely considered to be one of the most important functions they can perform. It enables wetlands to improve the quality of water, especially in those receiving either diffuse or point source inputs of polluted water. There are several processes by which nutrients can be retained in wetlands, including retention in plant material and soil organic matter. Some forms of N and P may be adsorbed to soil particles or chemically precipitated under certain soil environmental conditions or if, attached to or included in particulate matter, they may be retained through the hydrological function of sedimentation (see Section 3.1.D). Retention processes do not result in permanent removal of nutrients from a system, but provide a temporary store, over variable periods of time. For example, nutrients taken up by woody plants may be stored for tens or even hundreds of years, depending on the life cycle of the plant, before plant mortality leads to decay of the organic material and mineralisation, which releases the nutrients back into the ecosystem. Release of nutrients also can occur in the shorter term through leaf-fall and the dropping of small branches. However, following mortality or falling of leaves and branches, organic material may become incorporated into the wetland soil, and if conditions prevail that permit the accumulation of soil organic matter or even the development of peat, retention of nutrients taken up by either woody or herbaceous plants can persist for thousands of years.

Adsorption of nutrients in soils is dependent upon many variables, but the organic-rich soils typically found in wetlands are especially conducive to this process. This is because of the relationship between organic matter content and cation exchange capacity (Fig. 3.2.3). Nutrients must be in a form suitable for adsorption, e.g. nitrogen must usually be in the form of ammonium (NH_4^+) and phosphorus in the form of PO_4^{3+}. Phosphorus is also subject to precipitation under certain conditions, when it combines with any of several elements, but especially ferric iron, calcium or aluminium under aerobic conditions. However, if aerobic conditions do not persist, as is commonly the case in wetlands, then any previously precipitated phosphorus can be subject to remobilisation as a result of the chemical reduction of the combining element.

Some of the processes responsible for nutrient retention may be of a seasonal nature and, although the annual flux of nutrients into and out of a wetland potentially could be balanced in terms of mass, even temporary, short-term retention can be beneficial in some circumstances. For example, most eutrophication events occur in summer when temperatures are high and river discharges are low. Although the loads of nutrients in rivers at this time of year are often also low, high concentrations can easily result from relatively small inputs of nutrients from storm run-off or pollution events. If these nutrients can be temporarily stored, even if only until winter leaf-fall or herbaceous vegetation die-down, when nutrients taken up during the summer growing season are re-mineralised and released back to the aquatic system, then environmental benefits can be gained. In autumn and winter an increase in nutrient loading, especially of soluble nutrient species, can be found in wetlands. Much of this increase is due to wetland plants that shed litter or die back, and to higher rates of flow and erosion. The increase may not be particularly detrimental to aquatic ecology, provided that nutrients are flushed from the system, since nutrient uptake by the receiving aquatic plants and algae will be relatively small during autumn and winter. The flushing and dilution of these nutrient outputs are also likely to be relatively great at this time of year as river flows are high (Fisher and Acreman, 2004). This regulation of nutrient flow in water potentially can be significant in terms of avoiding eutrophication events in summer, without impacting annual nutrient fluxes. There may be significant changes in vegetation as a result of nutrient retention in the wetland, resulting often in monocultures of species that respond well to enrichment through 'luxury' growth. This can reduce overall biodiversity and alter the character of important wetland habitats.

Nutrient export

Nutrient export occurs through one of three principal processes: gaseous emission, harvesting of vegetation or erosion. The nutrient export function is particularly significant because, like nutrient retention, it provides mechanisms for maintaining existing water quality or improving the quality of polluted water but, unlike nutrient retention, it represents a permanent removal of nutrients from the system, with no risk of remobilisation should conditions in the wetland change. However, it should be emphasised that the nutrients are not being destroyed, but simply shifted from one ecosystem to another. Some of the processes that enable the export of nutrients can result in the release of harmful compounds, such as nitrous oxide, which is a greenhouse gas, or ammonia (NH_3), which can be deposited in nutrient-sensitive environments, resulting in degradation, as well as contributing to acid rain.

In-situ carbon retention

The carbon balance in wetland ecosystems is controlled primarily by environmental conditions, especially waterlogging, pH, nutrients and temperature, but anthropogenic impacts can also be important. Carbon contained in organic matter accumulates where the annual input of plant litter exceeds the annual breakdown. Accumulation is favoured by low temperatures, high acidity, oligotrophy and, generally most importantly, water saturation. The slow diffusion of oxygen in water exerts severe limitations on the rate of decomposition of organic matter in waterlogged environments, even if eutrophic or high temperature conditions exist, which otherwise encourage the microbial breakdown of organic material. Peat growth can be initiated under high rainfall conditions and by hydroseral development in open water or from colonisation of exposed mineral surfaces.

Many wetland soils are characterised by organic-rich horizons, which reflects the state of balance between the input of litter and the output of organic matter through decomposition and erosion. Where the organic soil material exceeds a certain minimum thickness or proportion, the profile qualifies as an organic or peat soil. Organic-rich wetland soils and peats can have considerable environmental, ecological, socio-economic and archaeological significance (Maltby, 1997).

Support of biodiversity

The variation in trophic status and pH across the range of peatland types results in significant variation in vegetation species lists and community assemblages. The distribution of invertebrates and higher animal species is frequently linked to this spatial variation of plants across hydrogeomorphic gradients, and commonly results in highly biodiverse faunal and floral populations. A number of plant species such as *Sphagnum* spp. (bog mosses), *Eriophorum* spp. (cotton grasses), *Carex* spp. (sedges) and characteristic grasses including *Molinia caerulea* and *Nardus stricta* actually promote the development of organic soil horizons and peat. Such plants not only have the ability to tolerate wet conditions but their dead remains are relatively difficult to break down and thus favour the accumulation of organic material. Wetlands can also provide special conditions conducive to the establishment of highly specialised carnivorous plants such as butterwort (*Pinguicula vulgaris*), sundew (*Drosera spp.)* and bladderworts (*Utricularia* spp.). Figure 3.2.7 shows mixed wetland vegetation including butterwort, sedge, bog bilberry (*Vaccinium uliginosum*), *Sphagnum* spp., liverworts and grasses.

Wetland and especially peat-forming ecosystems often uniquely preserve a sequential record of their own development together with evidence of features and

3.2.7 Common wetland species, including the insectivorous butterwort (*Pinguicula vulgaris*) (photo: T. Barker).

events in their contemporary environment. The combination of conditions favouring preservation (e.g. waterlogging, acidity, low microbial activity, relatively small temperature fluctuations), and the incremental patterns of accumulation of organic matter and sediment, make many wetlands exceptionally efficient at recording environmental changes at local and wider scales. Studies of pollen, macrofossils, human artefacts and cultural remains preserved in peats and wetland soils have contributed greatly to our knowledge of vegetation and landscape history, the nature and speed of change of past climates, and changes in human society. For more information see Section 1.4.4.

The value of wetlands as palaeoenvironmental and archaeological storehouses is dependent on the lack of disturbance to the stratigraphy and the progressive accumulation of organic matter and sediment. Both aspects are sensitive to human activities and climate change and can be severely impacted by the effects of drainage, reduction in sediment supply due to the isolation of floodplains from their river channels, or temperature increases sufficient to accelerate the decomposition of organic remains.

Early human communities often depended on wetlands for economic prosperity or even their survival. Some traditional uses still remain, such as cutting of reed for thatching, late-summer grazing of fen meadows, small-scale peat digging, hunting of wildfowl and fishing. Strong cultural traditions are often associated with such uses, giving rise to distinctive human communities and landscapes in different parts of Europe.

The particular benefit to archaeology of wetlands is that they preserve materials that would otherwise decompose through oxidation on dry land sites. Excellent examples are the prehistoric wooden trackways (Fig. 3.2.8), which were constructed across reedbeds in what is now the Glastonbury area of the Somerset Moors and Levels in the UK. These became abandoned with rising water levels and subsequently incorporated into deposits of fen peat (more information can be found in Coles and Coles, 1986; Brunning *et al.*, 2000; Brunning, 2001).

Trace element storage

Trace elements in the environment can be highly toxic to a wide range of organisms, but wetlands provide a potential sink for them. The sustainability of this function depends on many factors. For example, if the retention process predominantly is through contaminated sediment retention, the sediment retention capacity of any

3.2.8 Excavation of the Neolithic Sweet Track, Somerset, UK, discovered in 1970. This is one of the world's oldest wooded trackways, initial construction dated by dendrochronology to 3807/3806 BCE (photo: Somerset Levels Project).

wetland is finite, and eventually will be exceeded. Performance of this function also incurs potential hazards, whereby accumulation of heavy metals in a wetland system can eventually result in degradation of the system itself, and potentially result in biomagnification (see above). As with nutrient retention, trace element retention represents only a temporary retention of potential pollutants and, if conditions within a wetland change, remobilisation and possibly toxic flushes are always a possibility.

Trace element export

The permanent removal of trace elements from surface waters is highly dependent upon their form within the wetland. Trace elements incorporated into vegetation can be removed by management processes such as harvesting, and this enables controlled removal. However, if stores of trace elements have accumulated in wetlands, it is possible that they can become remobilised and exported. This may take the form of physical erosion by either wind or water, or the biogeochemical remobilisation and transport (usually via water) of trace elements which may be bound to soil particles or precipitated in the soil. Changes in environmental conditions, such as in acidity or redox potential, can cause releases of trace elements, and result in wetlands becoming sources, where they previously acted as sinks.

Organic carbon concentration control

The amount of dissolved organic carbon (DOC) in surface waters is an important source of energy for an aquatic system, but can also impact the quality of water in several ways. It can affect the turbidity of surface waters, have a strong influence on pH and act as a strong complexing agent, affecting the transport of many chemicals, especially trace metals. Excessive concentrations of DOC can cause problems in the treatment of water for human consumption. Peatlands are highly concentrated stores of carbon, and their degradation can have dramatic impacts on the quality of water draining from them, especially with regard to DOC concentration.

Consequences of degradation

Disturbance of the carbon dynamics in wetlands can have adverse effects for both environment and human society, with consequences at local, regional and global scales. Organic deposits are natural stores not only for carbon but also for plant nutrients, accumulated and partially recycled through decomposition, together with a range of pollutant inputs including acidity and sulphur compounds originating from the burning of fossil fuels, nitrogen oxides from vehicle exhausts, NH_3 from intensive agriculture and metals from industrial sources (Proctor, 1995). Such contaminants are stable whilst the organic layers are intact, but rapid mineralisation or excessive drainage disturbance can have far-reaching effects (Proctor and Maltby, 1998).

Further reading and references

Brännvall, M.-L., Bindler, R., Emteryd, O., Nilsson, M. and Renberg, I. (1997). Stable isotope and concentration records of atmospheric lead pollution in peat and lake sediments in Sweden, *Water, Air, and Soil Pollution* **100** (3/4), 243–252.

Brunning, R. (2001). *Archaeology and Peat Wastage on the Somerset Moors*, Unpublished report for the Environment Agency, Somerset County Council, Taunton, UK.

Brunning, R., Hogan, D.V., Jones, A.M, Jones, J., Maltby, E., Robinson, M. and Straker, V. (2000). Saving the Sweet Track. The *in-situ* preservation of a Neolithic wooden trackway, Somerset, UK, *Conservation and Management of Archaeological Sites* **4**. 3–20.

Coles, B.J. and Coles, J.M. (1986). *Sweet Track to Glastonbury*, Thames and Hudson, London, UK.

Devai, I. and DeLaune, R.D. (1995). Evidence for phosphine production and emission from Louisiana and Florida marsh soils, *Organic Geochemistry* **23**(3), 277–279.

Fisher, J. and Acreman, M.C. (2004). Wetland nutrient removal: a review of the evidence, *Hydrology and Earth System Sciences* **8**(4), 673–685.

Gambrell, R.P. and Patrick, W.H. Jr (1978). Chemical and microbiological properties of anaerobic soils and sediments. In: Hooke, D.D. and Crawford, R.M.M. (eds), *Plant Life in Anaerobic Environments*, Ann Arbor Science, Michigan, 375–423.

Greenwood, D.J. (1961). The effect of oxygen concentration on the decomposition of organic materials in soil, *Plant and Soil* **14**, 360–376.

Kadlec, R.H. and Knight, R.L. (1996). *Treatment Wetlands*, Lewis, Boca Raton, FL.

Maltby, E. (1997). Peatlands: The science case for conservation and sound management. In: Parkyn, L., Stoneman, R.E. and Ingram, H.A.P. (eds), *Conserving Wetlands*, CABI, Wallingford, UK, 121–131.

Meili, M. (1997). Mercury in lakes and rivers. In: Sigel, A. and Sigel, H. (eds), *Mercury and Its Effects on Environment and Biology*, Metal Ions in Biological Systems, Volume 34, Marcel Dekker, New York, 21–51.

Mitsch, W.J. and Gosselink, J.G. (2007). *Wetlands*, 4th edn, Wiley, New York.

Proctor, M.C.F. (1995). The ombrogenous bog environment. In: Wheeler, B.D., Shaw, S.C., Fojt, W.J. and Robertson, R.A. (eds), *Restoration of Temperate Wetlands*, Wiley, Chichester, UK, 287–303.

Proctor, M.C.F. and Maltby, E. (1998). Relations between acid atmospheric deposition and the surface pH of some ombrotrophic bogs in Britain, *Journal of Ecology* **86**(2), 329–340.

Simola, H. and Lodenius, M. (1982). Recent increase in mercury sedimentation in a forest lake attributable to peatland drainage, *Bulletin of Environmental Contamination and Toxicology* **29**, 298–305.

Várallyay, G. (1993). *Soil Data Bases, Soil Mapping Soil Information – and Soil Monitoring Systems in Hungary*, National Report for the FAO/ECE International Workshop of Soil Conservation Monitoring Systems, Budapest, 14–17 Sept., FAO.

Biogeochemical functions and processes assessed in these procedures

3.2.A Nutrient retention (water quality function)

3.2.A.1 Long-term retention of nutrients through plant uptake (process)

Further reading and references

Bischoff, J.M., Bukaveckas, P., Mitchell, M. J. and Hurd, T. (2001). N storage and cycling in vegetation of a forested wetland: implications for watershed N processing, *Water, Air, and Soil Pollution* **128**, 97–114.

Denny, P. (1987). Mineral cycling by wetland plants – a review, *Archiv für Hydrobiologie–Beiheft Ergebnisse der Limnologie* **27**, 1–25.

Fennessy, M.S. and Cronk, J.K. (2001). The effectiveness and restoration potential of riparian ecotones for the management of non-point source pollution, particularly nitrate, *Critical Reviews in Environmental Science and Technology* **27**(4), 285–317.

Fisher, J. and Acreman, M.C. (2004). Wetland nutrient removal: a review of the evidence, *Hydrology and Earth System Sciences* **8**(4), 673–685.

Hefting, M., Clément, J.C., Bienkowski, P., Dowrick, D., Guénat, C., Butturini, A., Topa, S., Pinay, G. and Verhoeven, J. (2005). The role of vegetation and litter in the nitrogen dynamics of riparian buffer zones in Europe, *Ecological Engineering* **24**(5), 465–482.

Johnston, C.A. (1991). Sediment and nutrient retention by freshwater wetlands: effects on surface water quality, *Critical Reviews in Environmental Science and Technology* **5,6** (21), 491–565.

Klopatek, J.M. (1978). Nutrient dynamics of freshwater riverine marshes and the role of emergent macrophytes. In: Good, R.E., Whigham, D.F. and Simpson, R.L. (eds),

Freshwater Wetlands; Ecological Processes and Management Potential, Academic Press, New York, 194–216.

Oorschot, M.M.P. van (1990). De Rol van Vegetatie bij het Verwijderen van Nutriënten uit Water. In: Oorschot, M.M.P. van and Verhoeven, J.T.A. (eds), *Moerassen voor de Zuivering van Water*, PAO-cursus Moerassen voor de zuivering van water, PAO-cursus, The Utrecht Plant Ecology News Report 11, 64–85 (in Dutch).

Orleans, A.B.M., Mugge, F.L.T., van der Meij, T., ter Keurs, W.J. (1994). *Minder nutriënten in het oppervlaktewater door bufferstroken: een literatuuranalyse*, Milieubiologie Rijksunversiteit, Leiden, The Netherlands (in Dutch).

Pinay, G., Ruffinoni, C. and Fabre, A. (1995). Nitrogen cycling in two riparian forest soils under different geomorphic conditions, *Biogeochemistry* **30**, 9–29.

Richardson, C.J. and Vaithiyanathan, P. (in press). Biogeochemical dynamics of phosphorus in wetlands. In, Maltby, E. and Barker, T. (eds), *The Wetlands Handbook*, Wiley-Blackwell, Oxford, UK.

Vitousek, P.M. and Reiners, W.A. (1975). Ecosystem succession and nutrient retention: A hypothesis, *BioScience* **25**(6), 376–381.

White, J.R. and Reddy, K.R. (in press). Nitrogen cycling in wetlands. In: Maltby, E. and Barker, T. (eds), *The Wetlands Handbook*, Wiley-Blackwell, Oxford, UK.

Introduction

Process description:

- Many wetland plants can retain excess nutrients (nitrogen or phosphorus) by assimilation of nutrients into their biomass (see Fig. 3.2.9 for a general description of nutrient dynamics in successional systems). However, at the end of the growing season, the death and decomposition of plant material will result in the release of these nutrients back into soil and water (this is not as significant in coniferous or evergreen vegetation as in deciduous or herbaceous vegetation). Nutrients incorporated into woody tissue will be retained for a longer period, as the tissue will not die after each growing season. Nutrients assimilated in herbaceous vegetation may be exported from the system through land-use management (this process is assessed separately in Section 3.2.B.2).
- One requirement for the process to occur is that the hydrogeomorphic unit (HGMU) is in receipt of nutrients (from direct application of nutrient-rich material or fertilisers or interaction with nutrient-rich waters). Vegetation indicative of a high rate of nutrient uptake can use these excess nutrients for its own growth.
- Areas of low disturbance are likely to optimise the uptake of excess nutrients.

Significance – rationale for assessment: The retention of excess nutrients by assimilation into woody biomass will result in an improvement of the quality of water bodies receiving flows from the wetland. The uptake of excess nutrients by herbaceous plant material will result only in a seasonal improvement of water quality because at the end of the growing season the nutrients will be re-released. The assessment deals predominantly with the long-term retention of nutrients in woody tissue.

Controlling variables

CV1 Nutrient input
CV2 Vegetation type
CV3 Degree of disturbance

Answer all the questions below for *each* HGMU; unless otherwise instructed, continue from one question to the next. On the appraisal sheet circle the correct answer. The location of the relevant information is given in brackets after each question. Use a new appraisal sheet for each HGMU.

If there are no significant nutrient inputs identified for the HGMU, or if the nutrient input is uncertain (nutrient input codes **Nut 2** or **Nut ?**), you may wish to assess the potential of the HGMU to perform this process. To do this, conduct the assessment as normal but answer questions only in ***CV2*** and ***CV3***. Establish

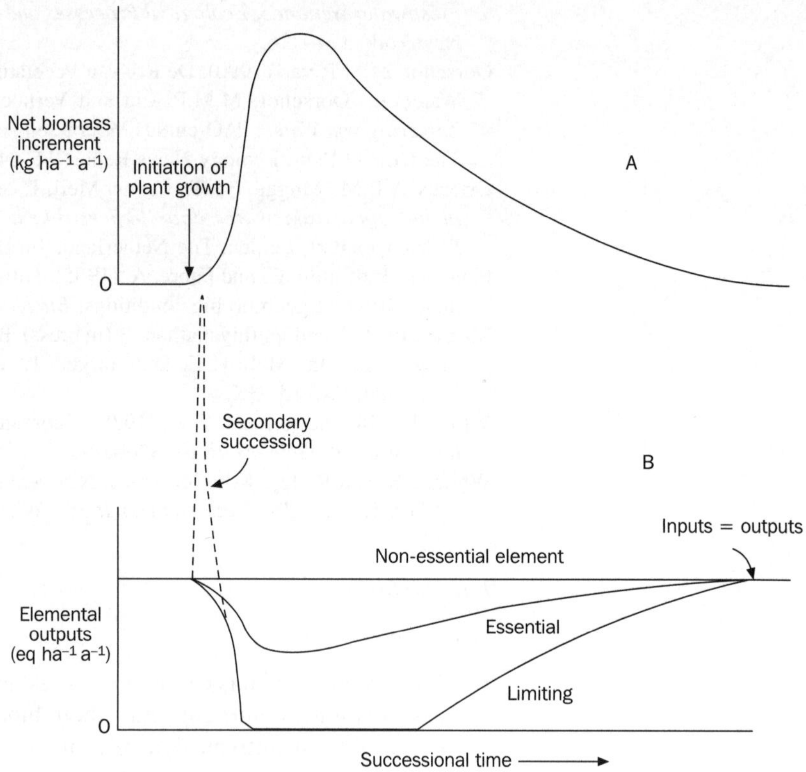

3.2.9 Simplified representation of biomass increment and outputs of non-essential, essential and limiting elements in a primary successional sequence. Y-axis values are given in equivalent measures. A: Variation in net biomass increment with time in a primary successional sequence. Biomass accumulates rapidly at first, and then the rate of biomass accretion declines gradually to zero.
B: Variation in nutrient losses with time in a primary successional sequence. As storage pools of nutrients grow in biomass and soil, nutrients are retained within an ecosystem, but when storage pools reach steady state, nutrient outputs must equal inputs. Elements limiting to plant growth (often nitrogen or phosphorus) are retained most strongly; losses of these elements may decline to near zero. Elements essential but not limiting to plant growth will be significantly retained within ecosystems, but outputs will not approach zero. Outputs of elements that are non-essential to plants and which are not accumulated within ecosystems will vary little over the course of succession. The dotted line represents very high rates of nutrient loss immediately after disturbance in a secondary successional sequence (e.g. forestry). Total biomass in the ecosystem would decrease during this period of high losses. Timescales of succession in forests are variable. The total sequence indicated in the graphs ranges from ± 50 years (pine stands in the subtropics) to ± 300 years in the temperate deciduous forests. (Vitousek and Reiners, 1975).

an assessment outcome as normal but ignore the need for positive answers for ***CV1***. The rationale statement can be used as normal but ignore the reference to receipt of nutrient inputs.

CV1 *Nutrient input*

Q1. Has the HGMU been assigned a nutrient input code indicating a direct input of nitrogen, i.e. nutrient input codes $\mathbf{N_d}$ **1/a**, $\mathbf{N_d}$ **1/b** or $\mathbf{N_d}$ **1/x**? *(Recording Sheet, 2c)*

Q2. Has the HGMU been assigned a nutrient input code indicating a direct input of phosphorus, *i.e.* nutrient input codes $\mathbf{P_d}$ **1/a**, $\mathbf{P_d}$ **1/b** or $\mathbf{P_d}$ **1/x**? *(Recording Sheet 2c)*

Q3. Has the HGMU been assigned a nutrient input code indicating a direct input of nutrients, not knowing whether this input consists of both nitrogen and phosphorus or only one of these nutrients, i.e. nutrient input code $\mathbf{Nut_d}$ **1/x**? *(Recording Sheet 2c)*

Q4. Has the HGMU been assigned a nutrient input code indicating an indirect input of nitrogen, i.e. nutrient input code $\mathbf{N_i}$ **1/x**? *(Recording Sheet 2c)*

Q5. Has the HGMU been assigned a nutrient input code indicating an indirect input of phosphorus, i.e. nutrient input code $\mathbf{P_i}$ **1/x**? *(Recording Sheet 2c)*

Q6. Has the HGMU been assigned a nutrient input code indicating an indirect input of nutrients, not knowing whether this input consists of both nitrogen and phosphorus or only one of these nutrients, i.e. nutrient input code $\mathbf{Nut_i}$ **1/x**? *(Recording Sheet 2c)*

Q7. Has the HGMU been assigned a nutrient input code indication that it is unknown or uncertain if nutrients are received, i.e. nutrient input code **Nut ?** *(Recording Sheet 2c)*

CV2 *Vegetation type*

Q1. Has the HGMU been assigned an unvegetated or sparsely vegetated code (**Ux** codes)? *(Recording Sheet 3, D)*
[*If YES do not continue with the following questions, but go directly to the Look-up table*]

Q2. What area of the HGMU is covered by trees or shrubs? *(AA Map 2, Recording Sheet 3, D)*

a. There are (almost) no trees or shrubs present in the HGMU (< 5 %) [*Continue with Q5*]
b. Trees or shrubs cover 5–< 30 % of the HGMU [*Continue with Q3*]
c. Trees or shrubs cover 30–< 60 % of the HGMU [*Continue with Q3*]
d. Trees or shrubs cover 60 % or more of the HGMU [*Continue with Q3*]

Q3. Has the vegetation been assigned one of the following shrub (**Sx**) codes? *(Recording Sheet 3, D)*

a. **S3**
b. **S18**
c. Other **S** code
d. No **S** code has been assigned to the HGMU

Q4. Has the vegetation been assigned one of the following woodland (**Wx**) codes? *(Recording Sheet 3, D)*

a. **W5–W12, W14, W15, W17–W19**
b. Other **W** code
c. No **W** code has been assigned to the HGMU

Q5. Has the vegetation been assigned one of the following grassland (**Gx**) codes? *(Recording Sheet 3, D)*

a. **G8–G10, G14, G30–G32**
b. Other **G** code
c. No **G** code has been assigned to the HGMU

Q6. Has the vegetation been assigned one of the following mire (**Mx**) codes? *(Recording Sheet 3, D)*

a. **M8, M9, M12-M15**
b. Other **M** code
c. No **M** code has been assigned to the HGMU

Q7. Has the vegetation been assigned one of the following aquatic (**Ax**) codes? *(Recording Sheet 3, D)*

a. **A4, A6, A7, A8**
b. Other **A** code
c. No **A** code has been assigned to the HGMU

CV3 *Degree of disturbance*

Q1. How often does the AA experience drought (defined as less than 25 mm rainfall for the month prior to, and during each month of, the growing season)?
[*Determine the growing season (Recording Sheet 1, C.3.1.6, C.3.1.7, C.3.1.8) and compare this to the mean monthly rainfall record (Recording Sheet 1, C.3.1.1i)*]

a. Not in the last 20 years
b. Every year in the last 20 years
c. Other or unknown

Q2. How often does the AA experience prolonged waterlogging or flooding throughout the growing season (stress occurs if soils are flooded for more than two successive months during the growing season)?
[***For flooding:*** *determine the growing season (Recording Sheet 1, C.3.1.6, C.3.1.7, C.3.1.8) and compare to months when flooding typically occurs (Recording Sheet 1, C.3.19.6).* ***For waterlogging:*** *Recording Sheet 3, B.3.2*]
a. Never in the last 20 years
b. Every year in the last 20 years
c. Other or unknown

Q3. How often is the HGMU subjected to disturbance by ploughing? *(Recording Sheet 1, C.3.5.1b; AA Map 2)*
a. No history of ploughing in the last 20 years / not applicable
b. Not currently ploughed but ploughed within the last 20 years, or ploughed but less than once a year
c. Currently ploughed once or more than once a year
d. Unknown

Q4. Are there any current mining or extraction activities occurring that disturb the surface of the HGMU? *(Recording Sheet 1, C.3.11.1b; AA Map 2)*
a. No indication of mining or extraction / not applicable
b. Historical areas of mining or extraction, currently inactive
c. Currently active areas of mining or extraction
d. Unknown

Q5. Is there any current construction work occurring that is disturbing the surface of the HGMU? *(Recording Sheet 1, C.3.13.1b; AA Map 2)*

Q6. How often is the vegetation within the HGMU burnt as part of land management practices? *(Recording Sheet 1, C.3.3.1b; AA Map 2)*
a. No history of burning in the last 20 years/not applicable
b. Not currently burnt but previously burnt within the last 20 years, or burnt but less than once in 5 years
c. Burnt once or more than once in the last 5 years
d. Unknown

Q7. How often does felling and removal of timber occur in woodland areas within the HGMU? *(Recording Sheet 1, C.3.6.1b; AA Map 2)*
a. Never/not applicable
b. No current felling and removal but previously practised within the last 20 years, or felling and removal occurs, but less than once in 5 years
c. Felling and removal has taken place once or more than once in the last 5 years
d. Unknown

Assessment of the process

Compare the results recorded for each HGMU on the appraisal sheets with the possible outcomes below and assign an appropriate assessment code. Record this on the appraisal sheet. The syntax codes used are described in the introduction to Section 3.

Look-up tables

1. The process is definitely being performed
If answers to the questions for this process agree with any of the combinations in Look-up Table 1, then the HGMU is definitely performing the process of long-term retention of excess nutrients (N and P) through plant uptake, and an estimation of the amount of nutrients 'removed' by this process is given.

In Look-up Table 1 the *total* uptake and subsequent storage of nutrients in the bole/branches per year is given, *not* the *extra* storage of nutrients resulting from input of *excess* nutrients. These figures are based on data collected during empirical

3.2.A.1 *Long-term Retention of Nutrients (N & P) Through Plant Uptake Appraisal Sheet*

HGMU code: ______________	**AA name:** ______________

CV1 Nutrient input			
Q1	Y	N	U/NA
Q2	Y	N	U/NA
Q3	Y	N	U/NA
Q4	Y	N	U/NA
Q5	Y	N	U/NA
Q6	Y	N	U/NA
Q7	Y	N	U/NA

CV3 Degree of disturbance				
Q1	a	b	c	
Q2	a	b	c	
Q3	a	b	c	d
Q4	a	b	c	d
Q5	Y	N	U/NA	
Q6	a	b	c	d
Q7	a	b	c	d

CV2 Vegetation type				
Q1	Y	N	U/NA	
Q2	a	b	c	d
Q3	a	b	c	d
Q4	a	b	c	
Q5	a	b	c	
Q6	a	b	c	
Q7	a	b	c	

HGMU process assessment code: ______________

3.2.A.1 Look-up Table 1

CV1	CV2	CV3	Rationale	Code	Quantification
1|2|3 |4|5|6	2b 3a	1a 2a (3a|3b) (4a|4b) #5 (6a|6b) (7a|7b)	The HGMU receives either a direct or indirect nutrient input. Shrubs indicative of a high rate of nutrient uptake are present on less than 30 % of the HGMU. The vegetation is not adapted to prolonged waterlogging, but this never occurs in the HGMU. The HGMU experiences a low degree of disturbance from drought, ploughing, mining or extraction, construction work, burning of vegetation and felling and removal of trees, so the ability of plants to take up nutrients is not affected. NB: The amount of nutrients retained in wood per hectare is dependent upon the age of trees and shrubs: young trees or shrubs take up relatively more nutrients than older trees or shrubs. In mature woodlands tree and shrub mortality (a process by which nutrients stored in wood return to the system) can be as or even more important than uptake and storage (see Fig. 3.2.9).	**1/c**	> 5 but < 25 kg N ha^{-1} a^{-1} > 0.5 but < 1.5 kg P ha^{-1} a^{-1}
1|2|3|4 |5|6	2c 3a	1a 2a (3a|3b) (4a|4b) #5 (6a|6b) (7a|7b)	The HGMU receives either a direct or indirect nutrient input. Shrubs indicative of a high rate of nutrient uptake are present on 30–< 60 % of the HGMU. The vegetation is not adapted to prolonged waterlogging, but this never occurs in the HGMU. The HGMU experiences a low degree of disturbance from drought, ploughing, mining or extraction, construction work, burning of vegetation and felling and removal of trees, so the ability of plants to take up nutrients is not affected. NB: The amount of nutrients retained in wood per hectare is dependent upon the age of trees and shrubs: young trees or shrubs take up relatively more nutrients than older trees or shrubs. In mature woodlands tree and shrub mortality (a process by which nutrients stored in wood return to the system) can be as or even more important than uptake and storage (see Fig. 3.2.9).	**1/b**	> 10 but < 50 kg N ha^{-1} a^{-1} > 1 but < 3 kg P ha^{-1} a^{-1}
1|2|3|4 |5|6	2d 3a	1a 2a (3a|3b) (4a|4b) #5 (6a|6b) (7a|7b)	The HGMU receives either a direct or indirect nutrient input. Shrubs indicative of a high rate of nutrient uptake are present on 60 % or more of the HGMU. The vegetation is not adapted to prolonged waterlogging, but this never occurs in the HGMU. The HGMU experiences a low degree of disturbance from drought, ploughing, mining or extraction, construction work, burning of vegetation and felling and removal of trees, so the ability of plants to take up nutrients is not affected. NB: The amount of nutrients retained in wood per hectare is dependent upon the age of trees and shrubs: young trees or shrubs take up relatively more nutrients than older trees or shrubs. In mature woodlands tree and shrub mortality (a process by which nutrients stored in wood return to the system) can be as or even more important than uptake and storage (see Fig. 3.2.9).	**1/a**	> 10 but < 80 kg N ha^{-1} a^{-1} > 1 but < 5 kg P ha^{-1} a^{-1}
1|2|3|4 |5|6	2b (3b|4a)	1a (3a|3b) (4a|4b) #5 (6a|6b) (7a|7b)	The HGMU receives either a direct or indirect nutrient input. Shrubs or trees indicative of a high rate of nutrient uptake are present on less than 30 % of the HGMU. The HGMU experiences a low degree of disturbance from drought, ploughing, mining or extraction, construction work, burning of vegetation and felling and removal of trees, so the ability of plants to take up nutrients is not affected. The vegetation present in the HGMU is adapted to waterlogged soils, therefore waterlogging is not a stress factor for the plants. NB: The amount of nutrients retained in wood per hectare is dependent upon the age of trees and shrubs: young trees or shrubs take up relatively more nutrients than older trees or shrubs. In mature woodlands tree and shrub mortality (a process by which nutrients stored in wood return to the system) can be as or even more important than uptake and storage (see Fig. 3.2.9).	**1/c**	> 5 but < 20 kg N ha^{-1} a^{-1} > 0.5 but < 1.5 kg P ha^{-1} a^{-1}

<table>
<tr><td>1|2|3|4
|5|6</td><td>2c
(3b|4a)</td><td>1a
(3a|3b)
(4a|4b)
#5
(6a|6b)
(7a|7b)</td><td>The HGMU receives a direct nutrient input. Shrubs or trees indicative of a high rate of nutrient uptake are present on 30–< 60 % of the HGMU. The HGMU experiences a low degree of disturbance from drought, ploughing, mining or extraction and construction work. The vegetation present in the HGMU is adapted to waterlogged soils, therefore waterlogging is not a stress factor for the plants. The vegetation is not burnt. Felling or removal of trees does not take place.
NB: The amount of nutrients retained in wood per hectare is dependent upon the age of trees and shrubs: young trees or shrubs take up relatively more nutrients than older trees or shrubs. In mature woodlands tree and shrub mortality (a process by which nutrients stored in wood return to the system) can be as or even more important than uptake and storage (see Fig. 3.2.9).</td><td>1/b</td><td>$>$ 10 but $<$ 60 kg N ha^{-1} a^{-1}
$>$ 1 but $<$ 3 kg P ha^{-1} a^{-1}</td></tr>
<tr><td>1|2|3|4
|5|6</td><td>2d
(3b|4a)</td><td>1a
(3a|3b)
(4a|4b)
#5
(6a|6b)
(7a|7b)</td><td>The HGMU receives a direct nutrient input. Shrubs or trees indicative of a high rate of nutrient uptake are present on 60 % or more of the HGMU. The HGMU experiences a low degree of disturbance from drought, ploughing, mining or extraction and construction work. The vegetation present in the HGMU is adapted to waterlogged soils, therefore waterlogging is not a stress factor for the plants. The vegetation is not burnt. Felling or removal of trees does not take place.
NB: The amount of nutrients retained in wood per hectare is dependent upon the age of trees and shrubs: young trees or shrubs take up relatively more nutrients than older trees or shrubs. In mature woodlands tree and shrub mortality (a process by which nutrients stored in wood return to the system) can be as or even more important than uptake and storage (see Fig. 3.2.9).</td><td>1/a</td><td>$>$ 10 but $<$ 80 kg N ha^{-1} a^{-1}
$>$ 1 but $<$ 5 kg P ha^{-1} a^{-1}</td></tr>
</table>

3.2.A.1 Look-up Table 2

CV1	CV2	CV3	Rationale	Code
# (1 2 3 4 5 6)|7	*	*	The HGMU does not receive any nutrient additions or it is uncertain whether the HGMU receives a nutrient input. The process of uptake of nutrients can occur, but only as part of the natural nutrient cycle.	**2**
*	1	*	There is no or little vegetation present in the HGMU.	**2**
*	3c|4b	*	Trees or shrubs indicative of a low rate of nutrient uptake are present.	**2**
*	5b|6b |7b	*	Vegetation indicative of a low rate of nutrient uptake is present. NB: An increase in nutrient additions in the HGMU can cause a shift in species composition. Mowing or grazing can help to overcome this problem.	**2**

3.2.A.1 Look-up Table 3

<table>
<tr><th>CV1</th><th>CV2</th><th>CV3</th><th>Rationale</th><th>Code</th></tr>
<tr><td>1|2|3|4
|5|6</td><td>(2b|2c|
2d)
3d|4c</td><td>1a 2a
(3a|3b)
(4a|4b)
#5
(6a|6b)</td><td>The HGMU receives either a direct or indirect nutrient input and experiences a low degree of disturbance from drought, waterlogging, ploughing, mining or extraction, construction work or burning of vegetation. Although shrubs or trees are present in (part of) the HGMU, the vegetation has not been assigned a shrub or woodland code. The presence of woody plants results in long-term retention of nutrients in the HGMU. Because of lack of information about the shrubs or trees present, quantification is not possible.</td><td>1/x</td></tr>
<tr><td>1|2|3|4
|5|6</td><td>5a</td><td>1a 2a
(3a|3b)
(4a|4b)
#5
(6a|6b)</td><td>The HGMU receives either a direct or indirect nutrient input. Grassland vegetation indicative of a high rate of nutrient uptake is present. The HGMU experiences a low degree of disturbance from drought, waterlogging, ploughing, mining or extraction, construction work or burning of vegetation, so the ability of plants to take up nutrients is not affected.
Seasonal retention of nutrients will occur. However, at the end of the growing season the nutrients taken up by the plants will return to the system. Seasonal retention can nevertheless have beneficial effects on the water system, e.g. plants retain nutrients in the growing season that will not become available for algal growth in a water body that receives water from the wetland.</td><td>1/x</td></tr>
<tr><td>1|2|3|
4|5|6</td><td>6a|7a</td><td>1a
(3a|3b)
(4a|4b)
#5
(6a|6b)</td><td>The HGMU receives direct nutrient additions. Mire or aquatic vegetation indicative of a high rate of nutrient uptake is present. The HGMU experiences a low degree of disturbance from drought, ploughing, mining or extraction or burning, so the ability of plants to take up nutrients is not affected.
Seasonal retention of nutrients will occur. However, at the end of the growing season the nutrients taken up by the plants will return to the system. Seasonal retention can nevertheless have beneficial effects on the water system, e.g. plants retain nutrients in the growing season that will not become available for algal growth in a water body that receives water from the wetland.</td><td>1/x</td></tr>
<tr><td>*</td><td>*</td><td>1b|2b|
3c|4c|5</td><td>Other factors may be suitable, but disturbance or stress because of drought, waterlogging, ploughing, mining or extraction or construction work reduce uptake of nutrients by plants.</td><td>1/x</td></tr>
<tr><td>*</td><td>*</td><td>6c</td><td>Although some of the nutrients in plant biomass are exported to the atmosphere through burning, nutrient-rich ashes also fertilise the soil. Rapid regrowth of plants after the burning event will result in uptake of these nutrients, but outflow or leaching from the system is also possible. Therefore it is very difficult to give quantification.</td><td>1/x</td></tr>
<tr><td>*</td><td>*</td><td>7c</td><td>Nutrients stored in biomass are exported. However, the disturbance caused by these management practices also result in an extra input of nutrients into the system (leaves and branches with a high concentration of nutrients fall on the ground and decompose rapidly, and nutrients are released back into the wetland system, see Fig. 3.2.9)</td><td>1/x</td></tr>
<tr><td>*</td><td>*</td><td>1c|2c|
3d|4d|
6d|7d</td><td>Other factors may be suitable, but it is not known whether disturbance or stress (drought, waterlogging, ploughing, mining or extraction, construction work, burning of vegetation or forestry) takes place, which reduces uptake of nutrients by trees or shrubs.</td><td>1/x</td></tr>
</table>

studies carried out in the EU projects leading to the production of these FAPs (FAEWE, FAEWE II and PROTOWET), and published research related to this topic. In the situations described in Look-up Table 1 *extra* uptake definitely takes place, but it is very hard to distinguish between uptake of excess nutrients and uptake of nutrients naturally present in the wetland system.

Moreover, it is not possible to distinguish between the reactions of plants upon direct or indirect nutrient additions. Therefore in the quantification, no distinction has been made between nutrients from these different sources.

2. The process is not (significantly) being performed
If answers to the questions for this process agree with any of the combinations in Look-up Table 2, then the HGMU is not significantly performing the process of plant uptake of excess nutrients.

3. The process is being performed, but there are constraining factors or uncertainties
If Look-up Tables 1 and 2 did not lead to an assessment, the process of retention of excess nutrients by plant uptake is performed, but quantification is not possible because there are one or more constraining factors or uncertainties. In Look-up Table 3 you can find out how many constraining factors influence the process, or which factors cause the uncertainty.

4. The process is definitely not being performed
This statement is not applicable to this process.

This completes the functional assessment for the process of excess nutrient retention through plant uptake. The output codes and quantification values can be recorded on the Functional Assessment Table at the end of Section 3.2.A.6.

3.2.A.2 *Storage of nutrients (N and P) in soil organic matter (process)*

Further reading and references

Armentano, T.V. and Verhoeven, J.T.A. (1990). Biogeochemical cycles: global. In: Patten, B.C. (ed.), *Wetlands and Shallow Continental Water Bodies, Volume 1*, SPB Academic, The Hague, The Netherlands, 281–311.

Bischoff, J.M., Bukaveckas, P., Mitchell, M.J. and Hurd, T. (2001). N storage and cycling in vegetation of a forested wetland: Implications for watershed N processing, *Water, Air, and Soil Pollution* **128**, 97–114.

Richardson, C.J. (1990). Biogeochemical cycles: regional. In: Patten, B.C. (ed.), *Wetlands and shallow Continental Water Bodies, Volume 1*, SPB Academic Publishing, The Hague, The Netherlands, 259–279.

Richardson, C.J. and Vaithiyanathan, P. (in press). Biogochemical dynamics of phosphorus in wetlands. In: Maltby, E. and Barker, T. (eds), *The Wetlands Handbook*, Wiley-Blackwell, Oxford, UK.

Swift, M.J.S., Heal, O.W. and Anderson, J.M. (1979). Decomposition processes in terrestrial ecosystems. In: Swift, M.J.S., Heal, O.W. and Anderson, J.M. (eds), *Decomposition in Terrestrial Ecosystems*, Blackwell, Oxford, UK, 1–46.

Introduction

Process description:

- The retention of excess nutrients (nitrogen or phosphorus) in a wetland by assimilation into plant biomass and their subsequent storage in the soil in a partially decomposed or undecomposed form.
- One requirement for the process to occur is that the HGMU is in receipt of nutrients (from direct application of nutrient-rich material or fertilisers or interaction with nutrient-rich waters), which are taken up by plants and subsequently stored in dead biomass.
- Soil conditions must be conducive to accumulation of organic matter, i.e. waterlogged or anaerobic soil conditions must prevail.

- The process is optimal in HGMUs in which there is a high degree of interaction between the soil and the nutrient inputs and in areas of low disturbance, so that the ability of plants to take up nutrients is not affected.

Significance – rationale for assessment: Because nutrients are stored in soil organic matter, water quality is improved in water bodies receiving surface or subsurface water from the wetland.

Controlling variables

CV1 Nutrient input
CV2 Organic matter status of soil
CV3 Soil water regime
CV4 Vegetation type
CV5 Landform

Answer all the questions below for *each* HGMU; unless otherwise instructed continue from one question to the next. On the appraisal sheet circle the correct answer. The location of the relevant information is given in brackets after each question. Use a new appraisal sheet for each HGMU.

If there are no significant nutrient inputs identified for the HGMU, or if the nutrient input is uncertain (nutrient input codes **Nut 2** or **Nut ?**), you may wish to assess the potential of the HGMU to perform this process. To do this, conduct the assessment as normal but answer questions only in ***CV2***, ***CV3***, ***CV4*** and ***CV5***. Establish an assessment outcome as normal but ignore the need for positive answers for ***CV1***. The rationale statement can be used as normal but ignore the reference to receipt of nutrient inputs.

CV1 *Nutrient input*

Q1. Has the HGMU been assigned a nutrient input code indicating a direct input of nitrogen, i.e. nutrient input codes $\mathbf{N_d}$ **1/a**, $\mathbf{N_d}$ **1/b** or $\mathbf{N_d}$ **1/x**? *(Recording Sheet 2c)*

Q2. Has the HGMU been assigned a nutrient input code indicating a direct input of phosphorus, i.e. nutrient input codes $\mathbf{P_d}$ **1/a**, $\mathbf{P_d}$ **1/b** or $\mathbf{P_d}$ **1/x**? *(Recording Sheet 2c)*

Q3. Has the HGMU been assigned a nutrient input code indicating a direct input of nutrients, but it is not known if this input comprises both nitrogen and phosphorus or only one of these nutrients, i.e. nutrient input code $\mathbf{Nut_d}$ **1/x**? *(Recording Sheet 2c)*

Q4. Has the HGMU been assigned a nutrient input code indicating an indirect input of nitrogen, i.e. nutrient input code $\mathbf{N_i}$ **1/x**? *(Recording Sheet 2c)*

Q5. Has the HGMU been assigned a nutrient input code indicating an indirect input of phosphorus, i.e. nutrient input code $\mathbf{P_i}$ **1/x**? *(Recording Sheet 2c)*

Q6. Has the HGMU been assigned a nutrient input code indicating an indirect input of nutrients, but it is not known if this input comprises both nitrogen and phosphorus or only one of these nutrients, i.e. nutrient input code $\mathbf{Nut_i}$ **1/x**? *(Recording Sheet 2c)*

Q7. Has the HGMU been assigned a nutrient input code indicating that it is unknown or uncertain whether there is a nutrient input, i.e. nutrient input code **Nut ?** ? *(Recording Sheet 2c)*

CV2 *Organic matter status of soil*

Q1. Is the soil described as organic? *(Recording Sheet 2a, C.3.6)*

Q2. What is the predominant description of the top 30 cm of the soil? *(Recording Sheet 2a, C.3.8)*

a. Fibrous peat
b. Amorphous peat
c. Semi-fibrous peat

d. Humus
e. Humose
f. Other

Q3. Is the top 20 cm of the soil predominantly described as black or very dark? *(Recording Sheet 2a, C.3.4)*

Q4. How often is land within the HGMU ploughed? *(Recording Sheet 1, C.3.5.1b; AA Map 2)*

a. No history of ploughing in the last 20 years/not applicable
b. Not currently ploughed but previously ploughed within the last 20 years, or ploughed but less than once a year
c. Currently ploughed once or more than once a year
d. Unknown

Q5. How often is peat cut and removed from the HGMU? *(Recording Sheet 1, C.3.10.1b; AA Map 2)*

a. No peat cutting occurs
b. Hand cutting occurs at a local or traditional scale
c. Mechanised cutting of peat occurs at an industrial or commercial scale
d. Unknown

CV3 *Soil water regime*

Q1. Is the moisture status of the top 20 cm of the soil described as either flooded or wet? *(Recording Sheet 2a, C.3.2)*

Q2. Has the HGMU been assigned a hydrological code indicating groundwater discharge (**F1**), i.e. **H1**, **H2**, **H3**, **H4**, **H11**, **H12**, **H14** or **H15**? *(Recording Sheet 2a, F1)*

Q3. Has the HGMU been assigned a hydrological code indicating (sub)surface run-off input (**F2**), i.e. **H2**, **H3**, **H4**, **H5**, **H6**, **H7**, **H13** or **H14**? *(Recording Sheet 2a, F2)*

Q4. Has the HGMU been assigned a hydrological code indicating overbank flooding (**F4**), i.e. **H4**, **H7**, **H9**, **H10**, **H12**, **H13**, **H14** or **H15**? *(Recording Sheet 2a, F4)*

Q5. Is the top 20 cm of the soil prone to water table fluctuations (alternate periods of waterlogged and non-waterlogged soil conditions, which may be due to natural, seasonal fluctuations, or water table management), which result in non-waterlogged soil conditions for at least half of the year? *(Recording Sheet 1, C.3.22.3)*

CV4 *Vegetation type*

Q1. Does the vegetation classification indicate the presence of a mire, i.e. code **Mx**? *(Recording Sheet 3, D)*

Q2. Does the vegetation classification indicate the dominance of bog moss (*Sphagnum* spp.), i.e. codes **M1–M7**, **M10–M12**, **A1** or **W4**? *(Recording Sheet 3, D)*

Q3. Does the vegetation classification indicate the presence of humid grassland, i.e. codes **G2–G9**, **G11**, **G14**? *(Recording Sheet 3, D)*

Q4. Does the vegetation classification indicate the presence of woodland either on perennially waterlogged soil or prone to flooding, i.e. codes **W4–W8**, **W10–W12**? *(Recording Sheet 3, D)*

Q5. Does the vegetation classification indicate the presence of vegetation growing under dry conditions, i.e. codes **G18–G32**, **S5–S18**, **U2**, **U4–U13**, **W22–W44**? *(Recording Sheet 3, D)*

Q6. Has the HGMU been classed as unvegetated or sparsely vegetated, i.e. **U** codes? *(Recording Sheet 3, D)*

CV5 *Landform*

Q1. Is the HGMU classed as a peat dome with the geomorphic code **PD**? *(Recording Sheet 2a, C.2)*

3.2.A.2 *Storage of Nutrients (N and P) in Soil Organic Matter Appraisal Sheet*

HGMU code: ______________________	**AA name:** ______________________

CV1 Nutrient input			
Q1	Y	N	U/NA
Q2	Y	N	U/NA
Q3	Y	N	U/NA
Q4	Y	N	U/NA
Q5	Y	N	U/NA
Q6	Y	N	U/NA
Q7	Y	N	U/NA

CV4 Vegetation type			
Q1	Y	N	U/NA
Q2	Y	N	U/NA
Q3	Y	N	U/NA
Q4	Y	N	U/NA
Q5	Y	N	U/NA
Q6	Y	N	U/NA

CV2 Organic matter status of soil						
Q1	Y	N	U/NA			
Q2	a	b	c	d	e	f
Q3	Y	N	U/NA			
Q4	a	b	c	d		
Q5	a	b	c	d		

CV5 Landform			
Q1	Y	N	U/NA
Q2	Y	N	U/NA

CV3 Soil water regime			
Q1	Y	N	U/NA
Q2	Y	N	U/NA
Q3	Y	N	U/NA
Q4	Y	N	U/NA
Q5	Y	N	U/NA

HGMU process assessment code: ____

3.2.A.2 Look-up Table 1

<table>
<tr><th>CV1</th><th>CV2</th><th>CV3</th><th>CV4</th><th>CV5</th><th>Rationale</th><th>Code</th><th>Quantification</th></tr>
<tr><td>1|2|3|4|5|6</td><td>(1|2a|2b|2c|2d|2e|3) 4a (5a|5b)</td><td>(1|2|3|4) #5</td><td>1|2</td><td>1|2</td><td>The HGMU receives either a direct or indirect nutrient input. The soil has been identified as a peat or organic soil, and presently is not cultivated or extensively mined. Waterlogging or inundation predominate, producing anaerobic soil conditions which favour accumulation of organic matter. The presence of mire vegetation is indicative of high organic matter accumulation. The landform is indicative of organic matter accumulation, or the depressional nature of the landform favours waterlogging and anaerobic conditions.
When organic matter accumulates, nutrients bound within it become stored.</td><td>1/a</td><td>> 10 but < 50 kg N ha^{-1} a^{-1}
> 2 but < 10 kg P ha^{-1} a^{-1}</td></tr>
<tr><td>1|2|3|4|5|6</td><td>(1|2a|2b|2c|2d|2e|3) 4a (5a|5b)</td><td>(1|2|3|4) #5</td><td>3|4</td><td>1|2</td><td>The HGMU receives either direct or indirect nutrient input. The soil has been identified as a peat or organic soil, and presently is not cultivated or extensively mined. Waterlogging or inundation predominate, producing conditions which favour accumulation of organic matter. The vegetation is anaerobic soil indicative of organic matter accumulation. The landform is indicative of organic matter accumulation, or the depressional nature of the landform favours waterlogging and anaerobic conditions.
When organic matter accumulates, nutrients bound within it become stored.</td><td>1/b</td><td>> 5 but < 25 kg N ha^{-1} a^{-1}
> 1 but < 5 kg P ha^{-1} a^{-1}</td></tr>
<tr><td>1|2|3|4|5|6</td><td>(1|2a|2b|2c|2d|2e|3) 4a (5a|5b)</td><td>(1|2|3|4) #5</td><td>1|2</td><td>*</td><td>The HGMU receives either direct or indirect nutrient input. The soil has been identified as a peat or organic soil, and presently is not cultivated or extensively mined. Waterlogging or inundation predominate, producing anaerobic soil conditions which favour accumulation of organic matter. The presence of mire vegetation is indicative of high organic matter accumulation.
When organic matter accumulates, nutrients bound within it become stored.</td><td>1/a</td><td>> 10 but < 50 kg N ha^{-1} a^{-1}
> 2 but < 10 kg P ha^{-1} a^{-1}</td></tr>
<tr><td>1|2|3|4|5|6</td><td>(1|2a|2b|2c|2d|2e|3) 4a (5a|5b)</td><td>(1|2|3|4) #5</td><td>3|4</td><td>*</td><td>The HGMU receives a direct nutrient input. The soil has been identified as a peat or organic soil, and presently is not cultivated or extensively mined. Waterlogging or inundation predominate, producing anaerobic soil conditions which favour accumulation of organic matter. The vegetation is indicative of organic matter accumulation.
When organic matter accumulates, nutrients bound within it become stored.</td><td>1/b</td><td>> 5 but < 25 kg N ha^{-1} a^{-1}
> 1 but < 5 kg P ha^{-1} a^{-1}</td></tr>
</table>

3.2.A.2 Look-up Table 2

<table>
<tr><th>CV1</th><th>CV2</th><th>CV3</th><th>CV4</th><th>CV5</th><th>Rationale</th><th>Code</th></tr>
<tr><td>#(1 2 3 4 5 6)|7</td><td>*</td><td>*</td><td>*</td><td>*</td><td>The HGMU does not receive a nutrient input or it is uncertain whether the HGMU receives a nutrient input. Storage of nutrients in soil organic matter can occur, but only as part of the natural internal nutrient cycle of the HGMU.</td><td>2</td></tr>
<tr><td>*</td><td>#(1 2a 2b 2c 2d 2e 3)</td><td>*</td><td>*</td><td>*</td><td>The soil has not been identified as a peat or organic soil.</td><td>2</td></tr>
<tr><td>*</td><td>4c|5c</td><td>*</td><td>*</td><td>*</td><td>The soil is used for cultivation or is extensively mined for peat. Conditions conducive to oxidation of organic matter occur. This process releases nutrients.</td><td>2</td></tr>
<tr><td>*</td><td>*</td><td>#(1 2 3 4)</td><td>*</td><td>*</td><td>Waterlogging or inundation do not predominate, indicating aerobic soil conditions, which favour degradation of organic matter. This process releases nutrients.</td><td>2</td></tr>
<tr><td>*</td><td>*</td><td>*</td><td>5|6</td><td>*</td><td>The vegetation is indicative of conditions unfavourable for organic matter accumulation, or there is no vegetation (and therefore no organic matter) present.</td><td>2</td></tr>
</table>

3.2.A.2 Look-up Table 3

<table>
<tr><th>CV1</th><th>CV2</th><th>CV3</th><th>CV4</th><th>CV5</th><th>Rationale</th><th>Code</th></tr>
<tr><td>*</td><td>4b|4d|5d</td><td>*</td><td>*</td><td>*</td><td>It is unknown if the HGMU is subject to ploughing or mining, or if the influence of previous or infrequent ploughing is still important. Because ploughing or mining cause unfavourable conditions for organic matter accumulation, it is not possible to quantify the process properly without more detailed information.</td><td>1/x</td></tr>
<tr><td>*</td><td>*</td><td>5</td><td>*</td><td>*</td><td>Fluctuating groundwater levels cause an alternation between waterlogged (anaerobic) and non-waterlogged (aerobic) soil conditions. The former is favourable for organic matter accumulation, while the latter is conducive to oxidation of organic matter. Therefore it is uncertain if the net result will be accumulation and storage of nutrients in soil organic matter.</td><td>1/x</td></tr>
<tr><td>*</td><td>*</td><td>*</td><td>#(1 2 3 4 5 6)</td><td>*</td><td>Vegetation other than that mentioned in the questions in CV4 is present. It is uncertain if this vegetation is indicative of organic matter accumulation and storage of nutrients in soil organic matter.</td><td>1/x</td></tr>
</table>

Q2. Does the HGMU have the geomorphic prefix code **D** indicating that it is a depressional unit, i.e. codes **BS**, **CO**, **OX**, **FC**, **LK**, **AD**, **UD**? *(Recording Sheet 2a, C.2)*

Assessment of the process

Compare the results recorded for each HGMU on the appraisal sheets with the possible outcomes below, and assign an appropriate assessment code. Record this on the appraisal sheet. The syntax codes used are described in the introduction to Section 3.

Look-up tables

1. The process is definitely being performed
If answers to the questions for this process agree with any combination in Look-up Table 1, then the HGMU is definitely performing the process of storage of excess nutrients in soil organic matter, and an estimation of the amount of nutrients stored by this process is given.

In Look-up Table 1 the *total* storage of nutrients in soil organic matter per year is given, *not* the *extra* storage of nutrients because of input of *extra* nutrients. These figures are based on data collected during empirical studies carried out in the EU projects leading to the production of these FAPs (FAEWE, FAEWE II and PROTOWET), and published research related to this topic. In the situations described in Look-up Table 1 *extra* storage definitely takes place, but it is very hard to distinguish between *extra* storage and *normal* storage. Moreover, it is not possible to distinguish between the storage of nutrients supplied from either direct or indirect nutrient sources. Therefore, no distinction is made in the quantification of storage of nutrients supplied directly or indirectly.

2. The process is not (significantly) being performed
If answers to the questions for this process agree with any of the combinations in Look-up Table 2, then the HGMU is not significantly performing the process of storage of excess nutrients in soil organic matter.

3. The process is being performed, but there are constraining factors or uncertainties
If Look-up Tables 1 and 2 did not lead to an assessment, the process of storage of excess nutrients in soil organic matter is performed, but quantification is difficult because there are constraining factors or uncertainties. In Look-up Table 3 you can find out how many constraining factors influence the process, or which factors cause the uncertainty.

4 The process is definitely not being performed
This statement is not applicable to this process.

This completes the functional assessment for the process of storage of excess nutrients in soil organic matter. The output codes and quantification values can be recorded on the Functional Assessment Table at the end of Section 3.2.A.6.

3.2.A.3 Adsorption of nitrogen as ammonium (process)

Further reading and references

Bak, A., Oorschot, M.M.P. van and Verhoeven, J.T.A. (1996). *Helofytenfilters voor Waterkwaliteitsverbetering ten behoeve van de Vinex-Lokatie Leidsche Rijn*, Universiteit Utrecht, The Netherlands (in Dutch).

Kadlec, R.H. (1995). Nitrogen processing in treatment wetlands. In: Vymazal, J. (ed.), *Nutrient Cycling and Retention in Wetlands and their Use for Waste Water Treatment*, Workshop, Ecology and Use of Wetlands, Prague and Botanical Institute, 6–9 September, Třeboň, Czech Republic, 101–123.

Meuleman, A.F.M. (1987). *Waterkwaliteitsverbetering door Helofytenfilters; Een Studie naar de Toepassingsmogelijkheden van Helofytenfilters in de Krimpenerwaard*, Studie- en Informatiecentrum TNO voor Milieuonderzoek, Delft, Universiteit Utrecht, The Netherlands (in Dutch).

Ridder, R.P. de (1996). *Helofytenfilters, Integratie van Oppervlaktewaterzuivering, Natuur en andere Functies in Moerassen*, LBL-mededeling 206, Utrecht, The Netherlands (in Dutch).

Schlesinger, W.H. (1997). *Biogeochemistry, an Analysis of Global Change*, 2nd edn, Academic Press, San Diego, CA.

Verhoeven, J.T.A., Beltman, B. and Vermeer, J.G.H. (1995). Species composition of small fens in relation to the nutrient status of the peat soil and the groundwater, *Colloques Phytosociologiques* **13**, 815–824

Wang, F.L. and Alva, A.K. (2000). Ammonium adsorption and desorption in sandy soils, *Soil Science Society of America Journal* **64**, 1669–1674.

Introduction

Process description:

- The permanent and semi-permanent adsorption of nitrogen as ammonium to soil mineral and/or organic material.
- One requirement for the process to occur is that the HGMU is in direct or indirect receipt of ammonium, or other nitrogenous materials if there is a capability to transform the latter to ammonium.
- The process is optimised in soils with a pH of 4–7, soils with a high clay content or soils with a high, well decomposed organic matter content. Clay minerals and highly decomposed organic matter in the soil provide adsorption sites for ammonium.
- The process of ammonium adsorption has a threshold after which further addition of ammonium will not lead to more adsorption. Also, it is possible for ammonium to be gradually released from the soil colloids over time if soil conditions change. Adsorption of nitrogen as ammonium will therefore result in only temporary retention.
- The importance of ammonium adsorption as a retention mechanism is poorly understood because little information about this process has been published. Therefore only a qualitative assessment of ammonium adsorption is possible.

Significance – rationale for assessment: Improvement of water quality in water bodies receiving surface or subsurface water from the wetland by retention of nitrogen in the form of ammonium within the soil.

Controlling variables

CV1 Nutrient input
CV2 Potential of soil for interaction with ammonium
CV3 Soil composition
CV4 Soil water regime
CV5 Soil pH

Answer all the questions below for *each* HGMU; unless otherwise instructed continue from one question to the next. On the appraisal sheet circle the correct answer. The location of the relevant information is given in brackets after each question. Use a new appraisal sheet for each HGMU.

If there are no significant nutrient inputs identified for the HGMU, or if the nutrient input is uncertain (nutrient input codes **Nut 2** or **Nut ?**), you may wish to assess the potential of the HGMU to perform this process. To do this, conduct the assessment as normal but answer questions only in ***CV2***, ***CV3***, ***CV4*** and ***CV5***. Establish an assessment outcome as normal but ignore the need for positive answers for ***CV1***. The rationale statement can be used as normal but ignore reference to receipt of nutrient inputs.

CV1 *Nutrient input*

Q1. Is the HGMU in receipt of a nitrogen nutrient input? Circle all the codes that are applicable:

a. **N_d 1/a, N_d 1/b** or **N_d 1/x**
b. **N_i 1/x**
c. **Nut_d 1/b** or **Nut_d 1/x**
d. **Nut_i 1/x**
e. **Nut ?**

(Recording Sheet 2c)

[*If YES to **Q1 a, b, c** or **d** continue with **Q2**. If NO to all answers above or if YES to **Q1e**, then continuing with this assessment will give information only about an HGMU's **potential** to perform this process. To do this continue with **CV2***]

Q2. What is the source of the nutrient input received by the HGMU? Circle all those applicable.

a. Inorganic fertiliser applied directly to the HGMU *(Recording Sheet 2b, H1 Action, Q1)*
b. Organic fertiliser applied directly on the surface of the HGMU *(Recording Sheet 2b, H1 Action, Q5)*
c. Organic fertiliser injected into the surface of the HGMU *(Recording Sheet 2b, H1 Action, Q6)*
d. Organic industrial waste applied within the HGMU *(Recording Sheet 2b, H1 Action, Q7)*
e. Manure from livestock kept within the HGMU *(Recording Sheet 2b, H1 Action, Q8)*
f. Excrement from large populations of birds *(Recording Sheet 2b, H1 Action, Q9)*
g. Atmospheric deposition *(Recording Sheet 2b, H1 Action, Q10)*
h. Contact with surface water containing more than 1 mg l^{-1} ammonium-nitrogen *(Recording Sheet 2b, H1 Action, Q20)*
i. Contact with surface water containing more than 1 mg l^{-1} nitrate-nitrogen *(Recording Sheet 2b, H1 Action, Q19)*
j. Contact with surface water with a total nitrogen concentration of more than 2.2 mg l^{-1} *(Recording Sheet 2b, H1 Action, Q21)*
k. Groundwater discharge *(Recording Sheet 2b, H2 Action, Q1–3)*
l. Run-off or subsurface run-off *(Recording Sheet 2b, H2 Action, Q4–6)*

[*If NO to **Q2 h, i, j, k** and **l**, and YES to either **Q2 a, b, c, d, e** and/or **f**, then continue with **CV3**. In all other cases continue with **CV2***]

CV2 *Potential of soil for interaction with ammonium*

Q1. Has the HGMU been assigned a geomorphic code (**D**) indicating it is a depressional landform, i.e. **BS, CO, OX, FC, PB**? *(Recording Sheet 2a, C.2; AA Map 3)*

Q2. Is the HGMU adjacent to any HGMUs or areas of land not included in the AA that are topographically higher? *(Recording Sheet 2a, C.2; AA Map 3)*

Q3. Is the gradient of the HGMU coded as **G0** or **G1**? *(Recording Sheet 2a, C.2; AA Map 3)*

Q4. Is the microtopography of the HGMU described as significant? *(Recording Sheet 3, A.3)*

Q5. Is the groundwater level artificially maintained at high levels in the HGMU at any time (e.g. for nature conservation purposes)? *(Recording Sheet 1, C.3.22.4b; AA Map 2)*

Q6. Does the HGMU have a soil profile described as having at least two of the following characteristics:

- Coarse particle size throughout its depth?
- Gravel particle size throughout its depth?
- Mineral soil material?

(Recording Sheet 2a, C.3.7)

Q7. Does the HGMU have artificial subsurface drainage? *(Recording Sheet 1, C.3.9.5b; AA Map 2 or 3)*

Only answer ***Q8*** if you have answered YES to ***Q2* h, i** or **j** in ***CV1***.

Q8. Does enriched surface water have contact with the HGMU for at least 3 successive days? *(Recording Sheet 1, C.3.19.5a → answer should be (**1 week – 1 month**) or (**> 1 month**). If answered (**< 1 week**), only answer **Q8** positively when more details are available.)*

CV3 *Soil composition*

Q1. Is the upper 20 cm of the soil profile described as predominantly a mineral soil with a fine or medium particle size? *(Recording Sheet 2a, C.3.7)*

Q2. Is the upper 20 cm of the soil profile described as predominantly dark, humose and organic? *(Recording Sheet 2a, C.3.8)*

Q3. Is the upper 20 cm of the soil profile composed of predominantly organic material described as:

a. Amorphous peat or humus?

b. Semi-fibrous or fibrous peat?

(Recording Sheet 2a, C.3.8)

CV4 *Soil water regime*

Q1. Is the upper 20 cm of the soil profile described as having greyish, greenish or black colour? *(Recording Sheet 2a, C.3.4)*

Q2. Are there mottles in the upper 20 cm of the soil profile described as reddish, ochreous or rusty? *(Recording Sheet 2a, C.3.5)*

Q3. Is the soil profile composed of two different coloured layers, the lower being greyish and the upper being brownish, reddish, ochreous or rusty? *(Recording Sheet 2a, C.3.4)*

Q4. Is the upper 20 cm of the soil prone to water table fluctuations (alternate periods of waterlogged and non-waterlogged soil conditions, which may be due to natural seasonal fluctuations, or water table management), which result in non-waterlogged soil conditions for at least half of the year? *(Recording Sheet 1, C.3.22.3; AA Map 2)*

CV5 *Soil pH*

Q1. Is the soil pH between 4 and 7? *(Recording Sheet 3, A.1i)*

Q2. Is the vegetation within the HGMU coded as anything other than **S13**, **U12**, **U19** and **W40**? *(Recording Sheet 3, D)*

Assessment of the process

Compare the results recorded for each HGMU on the appraisal sheets with the possible outcomes below, and assign an appropriate assessment code. Record this on the appraisal sheet. The syntax codes used are described in the introduction to Section 3.

Look-up tables

1. The process is definitely being performed

If the answers to the questions for this process agree with any of the combinations in Look-up Table 1, then the HGMU is definitely performing the process of adsorption of excess ammonium. The overall importance of ammonium adsorption as a retention mechanism in wetlands is still poorly documented. Because there is little quantitative information about ammonium adsorption in literature, this process cannot be assessed quantitatively, so only a qualitative assessment is given.

2. The process is not (significantly) being performed

If the answers to the questions for this process agree with any of the combinations in Look-up Table 2, then the HGMU is not significantly performing the process of adsorption of excess ammonium.

3.2.A.3 *Adsorption of Nitrogen as Ammonium Appraisal Sheet*

HGMU code: ____________________	**AA name:** ____________________

CV1 Nutrient input						
Q1	a	b	c	d	e	
Q2	a	b	c	d	e	f
	g	h	i	j	k	l

CV4 Soil water regime			
Q1	Y	N	U/NA
Q2	Y	N	U/NA
Q3	Y	N	U/NA
Q4	Y	N	U/NA

CV2 Potential of soil for interaction with ammonium			
Q1	Y	N	U/NA
Q2	Y	N	U/NA
Q3	Y	N	U/NA
Q4	Y	N	U/NA
Q5	Y	N	U/NA
Q6	Y	N	U/NA
Q7	Y	N	U/NA
Q8	Y	N	U/NA

CV5 Soil pH			
Q1	Y	N	U/NA
Q2	Y	N	U/NA

CV3 Soil composition			
Q1	Y	N	U/NA
Q2	Y	N	U/NA
Q3a	Y	N	U/NA
Q3b	Y	N	U/NA

HGMU process assessment code: ________________________________

3.2.A.3 Look-up Table 1

<table>
<tr><th>CV1</th><th>CV2</th><th>CV3</th><th>CV4</th><th>CV5</th><th>Rationale</th><th>Code</th></tr>
<tr><td>1a 2a</td><td>N/A</td><td>1</td><td>1 #(2 3 4)</td><td>1|2</td><td>The HGMU receives an input of nitrogen (probably consisting of or including ammonium) from inorganic fertiliser applied directly to the surface of the HGMU. Clay minerals are present in the soil profile to provide adsorption sites. Wet soil decreases rates of conversion of ammonium to nitrate, enhancing the duration of retention. Moreover, a wet soil is favourable for dissimilatory nitrate reduction, a process in which nitrate is reduced to ammonium. Soil pH is conducive to adsorption.</td><td>1</td></tr>
<tr><td>1a (2b|c |e|f)</td><td>N/A</td><td>1</td><td>1 #(2 3 4)</td><td>1|2</td><td>The HGMU receives an input of ammonium either from organic fertiliser applied directly to the surface of the HGMU or from livestock or bird excrement. Clay minerals are present in the soil profile to provide adsorption sites. Wet soil decreases rates of conversion from ammonium to nitrate, enhancing the duration of retention. Soil pH is conducive to adsorption.</td><td>1</td></tr>
<tr><td>1b (2h|j)</td><td>8</td><td>1</td><td>1 #(2 3 4)</td><td>1|2</td><td>The HGMU receives an input of nitrogen (probably consisting of or including ammonium) via contact with water with high nitrogen or ammonium concentration. The water is present long enough in the HGMU to sufficiently interact with the HGMU soil. Clay minerals are present in the soil profile to provide adsorption sites. Wet soil decreases rates of conversion of ammonium to nitrate, enhancing the duration of retention. Soil pH is conducive to adsorption.</td><td>1</td></tr>
<tr><td>1a 2a</td><td>N/A</td><td>(2|3a)</td><td>1 #(2 3 4)</td><td>1|2</td><td>The HGMU receives an input of nitrogen (probably consisting of or including ammonium) from inorganic fertiliser applied directly to the surface of the HGMU. Highly decomposed organic matter is present in the soil profile to provide adsorption sites. Wet soil decreases rates of conversion of ammonium to nitrate, enhancing the duration of retention. Moreover, a wet soil is favourable for dissimilatory nitrate reduction, a process in which nitrate is reduced to ammonium. Soil pH is conducive to adsorption.</td><td>1</td></tr>
<tr><td>1a (2b|c |e|f)</td><td>N/A</td><td>(2|3a)</td><td>1 #(2 3 4)</td><td>1|2</td><td>The HGMU receives an input of ammonium from organic fertiliser applied directly to the surface of the HGMU or from livestock or bird excrement. Highly decomposed organic matter is present in the soil profile to provide adsorption sites. Wet soil decreases rates of conversion of ammonium to nitrate, enhancing the duration of retention. Soil pH is conducive to adsorption.</td><td>1</td></tr>
<tr><td>1b (2h|j)</td><td>8</td><td>(2|3a)</td><td>1 #(2 3 4)</td><td>1|2</td><td>The HGMU receives an input of nitrogen (probably consisting of or including ammonium) from floodwater or supplied water with high nitrogen or ammonium concentration. The water is present long enough in the HGMU to sufficiently interact with the HGMU soil. Highly decomposed organic matter is present in the soil profile to provide adsorption sites. Wet soil decreases rates of conversion of ammonium to nitrate, enhancing the duration of retention. Soil pH is conducive to adsorption.</td><td>1</td></tr>
<tr><td>1a 2a</td><td>N/A</td><td>1 (2|3a)</td><td>1 #(2 3 4)</td><td>1|2</td><td>The HGMU receives an input of nitrogen (probably consisting of or including ammonium) from inorganic fertiliser applied directly to the surface of the HGMU. Clay minerals and highly decomposed organic matter are present in the soil profile to provide adsorption sites. Wet soil decreases rates of conversion of ammonium to nitrate, enhancing the duration of retention. Moreover, a wet soil is favourable for dissimilatory nitrate reduction, a process in which nitrate is reduced to ammonium. Soil pH is conducive to adsorption.</td><td>1</td></tr>
</table>

3.2.A.3 Look-up Table 1 cont'd

<table>
<tr><th>CV1</th><th>CV2</th><th>CV3</th><th>CV4</th><th>CV5</th><th>Rationale</th><th>Code</th></tr>
<tr><td>1a (2b|c |e|f)</td><td>N/A</td><td>1 (2|3a)</td><td>1 #(2 3 4)</td><td>1|2</td><td>The HGMU receives an input of ammonium from organic fertiliser applied directly to the surface of the HGMU, or from livestock or bird excrement. Clay minerals and highly decomposed organic matter are present in the soil profile to provide adsorption sites. Wet soil decreases rates of conversion of ammonium to nitrate, enhancing the duration of retention. Soil pH is conducive to adsorption.</td><td>1</td></tr>
<tr><td>1b (2h|j)</td><td>8</td><td>1 (2|3a)</td><td>1 #(2 3 4)</td><td>1|2</td><td>The HGMU receives an input of nitrogen (probably consisting of or including ammonium) from floodwater or supplied water with high nitrogen or ammonium concentration. The water is present long enough in the HGMU to sufficiently interact with the HGMU soil. Clay minerals and highly decomposed organic matter are present in the soil profile to provide adsorption sites. Wet soil decreases rates of conversion of ammonium to nitrate, enhancing the duration of retention. Soil pH is conducive to adsorption.</td><td>1</td></tr>
</table>

3.2.A.3 Look-up Table 2

CV1	CV2	CV3	CV4	CV5	Rationale	Code
#(1a-d)|1e|#(2a-l)	*	*	*	*	There is no nutrient source entering the HGMU containing a suitable ammonium source (or the nutrient source is uncertain). Adsorption of ammonium will be occurring in the soil as part of natural nitrogen cycling. The HGMU might have the potential to store nitrogen through adsorption of ammonium if an appropriate nutrient source is received.	**2**
*	*	#(1 2 3a 3b)	*	*	There is no soil component, either mineral or organic, that is suitable for the adsorption of ammonium in the soil profile.	**2**

3.2.A.3 Look-up Table 3

<table>
<tr><th>CV1</th><th>CV2</th><th>CV3</th><th>CV4</th><th>CV5</th><th>Rationale</th><th>Code</th></tr>
<tr><td>1a 2g</td><td>*</td><td>*</td><td>*</td><td>*</td><td>The HGMU is in direct receipt of ammonium through ammonium-rich atmospheric deposition. If atmospheric deposition is the only input of ammonium into the system, adsorption of ammonium can take place, but only to a small degree.</td><td>1/x</td></tr>
<tr><td>1c|1d</td><td>*</td><td>*</td><td>*</td><td>*</td><td>The HGMU is in receipt of a nutrient supply. However, it is uncertain if this source contains ammonium, or other inorganic nitrogen compounds.</td><td>1/x</td></tr>
<tr><td>2i</td><td>8</td><td>*</td><td>*</td><td>*</td><td>The HGMU receives a nitrate supply from floodwater. For ammonium adsorption to occur, nitrate must be reduced to ammonium. Reduction of nitrate to ammonium can occur in wetlands (dissimilatory nitrate reduction), but there is very little known about this process.</td><td>1/x</td></tr>
<tr><td>2h|2i|2j</td><td>#8</td><td>*</td><td>*</td><td>*</td><td>The water is present in the HGMU only for a short time. It is uncertain if this is long enough for sufficient interaction of the nutrient input with the HGMU soil for significant adsorption to occur.</td><td>1/x</td></tr>
<tr><td>2k|2l</td><td>(1|2|3|4|5) (#6 #7)</td><td>*</td><td>*</td><td>*</td><td>The HGMU receives nutrients in groundwater discharge or in (sub)surface run-off. Wetland morphology ensures interaction of the nutrient input with the HGMU soil. Because nitrate is more mobile in water than ammonium, it is likely that the nitrogen in the (ground)water flow consists mostly of nitrate. For ammonium adsorption to occur, nitrate must be reduced to ammonium. Reduction of nitrate to ammonium can occur in wetlands (dissimilatory nitrate reduction), but there is very little known about this process.</td><td>1/x</td></tr>
<tr><td>*</td><td>#(1 2 3 4 5)</td><td>*</td><td>*</td><td>*</td><td>Wetland morphology is not optimal for interaction of the nutrient input with the HGMU soil.</td><td>1/x</td></tr>
<tr><td>*</td><td>6|7</td><td>*</td><td>*</td><td>*</td><td>Well-drained soil or enhanced drainage may reduce nutrient interaction with the HGMU soil and constrain process rates. Also nitrification of ammonium occurs because of aerobic soil conditions.</td><td>1/x</td></tr>
<tr><td>*</td><td>*</td><td>3b</td><td>*</td><td>*</td><td>Semi-fibrous or fibrous organic material provides fewer adsorption sites for ammonium than does amorphous peat or humus.</td><td>1/x</td></tr>
<tr><td>*</td><td>*</td><td>*</td><td>2|3|4</td><td>*</td><td>Soil saturation might be only seasonal and this will reduce the duration of ammonium retention due to conversion to nitrate during dry periods.</td><td>1/x</td></tr>
<tr><td>*</td><td>*</td><td>*</td><td>*</td><td>#(1 2)</td><td>pH is not optimal for the process to occur.</td><td>1/x</td></tr>
</table>

3. The process is being performed, but there are constraining factors or uncertainties

If Look-up Tables 1 and 2 above did not lead to an assessment, the process of ammonium adsorption is performed, but there are constraining factors or uncertainties. In Look-up Table 3 you can find out how many constraining factors influence the process, or which factors cause the uncertainty.

4. The process is definitely not being performed

This statement is not applicable to this process.

This completes the functional assessment for the process of adsorption of N as ammonium. The output codes can be recorded on the Functional Assessment Table at the end of Section 3.2.A.6.

3.2.A.4 Adsorption and precipitation of P in the soil (process)

Further reading and references

Bak, A. and Waardenburg, H.W. (1999). *Onderzoek toekomstige Waterkwaliteit en Waterzuivering in Leidsche Rijn*, Rapport nr. 99.33, Bureau Waardenburg bv., Culemborg, The Netherlands (in Dutch).

Bak, A., Oorschot M.M.P. van and Verhoeven, J.T.A. (1996). *Helofytenfilters voor Waterkwaliteitsverbetering ten behoeve van de Vinex-Lokatie Leidsche Rijn*, Universiteit Utrecht, The Netherlands (in Dutch).

Borggaard, O.K. (1983). The influence of iron oxides on phosphate adsorption by soil, *Journal of Soil Science* **34**, 333–341.

Bridgham, S.D, Johnston, C.A, Schubauer-Berigan, J.P. *et al.* (2001). Phosphorus sorption dynamics in soils and coupling with surface and pore water in riverine wetlands, *Soil Science Society of America Journal* **65**(2), 577–588.

Edwards, A.C. and Withers, P.J.A. (1998). Soil phosphorus management and water quality: a UK perspective, *Soil Use and Management* **14**, 124–130.

Farahbakhshazad, N. and Morrison, G.M. (2003). Phosphorus removal in a vertical upflow constructed wetland system, *Water Science and Technology* **48**(5), 43–50.

Fisher, J. and Acreman, M.C. (2004). Wetland nutrient removal: a review of the evidence, *Hydrology and Earth System Sciences* **8**(4), 673–685.

House, W.A., Jickells, T.D., Edwards, A.C., Praska, K.E. and Denison, F.H. (1998). Reactions of phosphorous with sediments in fresh and marine waters, *Soil Use and Management* **14**, 139–146.

Kadlec, R.H. (1997). An abiotic wetland phosphorus model, *Ecological Engineering* **8**, 145–172.

Kadlec, R.H. (1999). The limits of phosphorus removal in wetlands, *Wetlands Ecology and Management* **7**, 165–175.

Koopmans, G.F., Chardon, W.J., de Willigen, P. and van Riemsdijk, W.H. (2004). Phosphorus desorption dynamics in soil and the link to a dynamic concept of bioavailability, *Journal of Environmental Quality* **33**(4), 1393–1402.

Lijklema, L. (1990). Mass balances and cycling of nitrogen and phosphorus in flooded soils. In: Oorschot, M.M.P. van and Verhoeven, J.T.A. (eds), *Moerassen voor de Zuivering van Water*, PAO-cursus Moerassen voor de zuivering van water, PAO-cursus, The Utrecht Plant Ecology News Report 11, 44–51.

Morel, C., Tunney, P., Plenet, D. and Pellerin, S. (2000). Transfer of phosphate ions between soil and solution: perspectives in soil testing, *Journal of Environmental Quality* **29**(1), 50–59.

Olde Venterink, H., Wiegman, F., Lee, G. E. M. van der and Vermaat, J. E. (2003). Role of active floodplains for nutrient retention in the River Rhine, *Journal of Environmental Quality* **32**, 1430–1435

Ponnamperuma, F.N. (1972). The chemistry of submerged soils, *Advances in Agronomy* **24**, 29–96.

Richardson, C.J., Qian, S.S. and Craft, C.B. (1995). In: Vymazal, J. (ed.), *Nutrient Cycling and Retention in Wetlands and their Use for Waste Water Treatment, Ecology and Use of Wetlands*, Prague and Botanical Institute, Třeboň, 101–123.

Richardson, C.J. and Qian, S.S. (1999). Comments: Limits of phosphorus removal in wetlands (Kadlec, previous issue 165–175). *Wetlands Ecology and Management* **7**, 235–238.

Richardson, C.J. and Vaithiyanathan, P. (in press). Biogochemical dynamics of phosphorus in wetlands. In: Maltby, E. and Barker, T. (eds), *The Wetlands Handbook*, Wiley-Blackwell, Oxford, UK.

Valk, A.G. van der, Davis, C.B., Baker, J.L. and Beer, C.E. (1978). Natural freshwater wetlands as nitrogen and phosphorus traps for land runoff. In: Greeson, P.E., Clark, J.R. and Clark, J.E. (eds), *Wetland Values and Functions: the State of our Understanding*, American Water Resources Association, Minneapolis, MN, 457–467.

Introduction

Process description:

- The adsorption of excess phosphate–P to soil particles by physico-chemical processes resulting in temporary retention in the wetland.
- One requirement for the process to occur is that the HGMU is in receipt of phosphate from direct or indirect sources.
- The process is optimised when aluminium, calcium or iron are present; P can bind with these elements. Acidic conditions (pH 4–6.5) indicate the presence of iron or aluminium, while alkaline conditions (pH 6.5–8) indicate the presence of calcium.
- In clay, limestone and organic soils phosphate adsorption capacity is greater than in sandy or gravelly soils.
- The process of adsorption of P in a soil is finite. When all adsorption sites are saturated an extra input of P will not lead to more adsorption. Even the reverse of the process is possible: P can be released gradually from the soil over time. This happens when P concentrations in the overlying water column are lower than in the soil pore water. P can also be released from adsorption sites when other elements are present that bind P more strongly than soil particles do. For example sulphide (produced in anaerobic soils when sulphate is reduced) reacts with iron, iron hydroxides and iron phosphate complexes, thereby reducing the availability of iron for phosphate-binding and releasing bound phosphate.
- It is difficult to quantify the process of adsorption of phosphorus because of its high temporal and spatial variability and relationship to site history. Therefore only a qualitative assessment of adsorption of phosphorus in the soil is given.
- As well as being adsorbed, phosphate may also be precipitated. These two processes can occur simultaneously in the soil. Adsorption binds phosphate (in an easily reversible manner) to charged adsorption sites in soil particles, while precipitation is a chemical reaction in which phosphate ions form insoluble salts with calcium, iron or aluminium ions. It is hard to distinguish between the two processes, and consequently the texts in this section refer only to adsorption, but in fact both processes are likely to be occurring.
- Precipitation can also take place in the water column when calcium, aluminium or iron are present. Under these conditions, phosphate will form insoluble phosphate salts, which precipitate. In this section precipitation in the water column is not considered.

Significance – rationale for assessment: Improvement of water quality in water bodies receiving surface or subsurface water from the wetland by retention of phosphorus within the soil.

Controlling variables

CV1 Nutrient input
CV2 Soil type
CV3 Soil water regime
CV4 Landform

Answer all the questions below for *each* HGMU; unless otherwise instructed continue from one question to the next. On the appraisal sheet circle the correct

answer. The location of the relevant information is given in brackets after each question. Use a new appraisal sheet for each HGMU.

If there are no significant nutrient inputs identified for the HGMU, or if the nutrient input is uncertain (nutrient input codes **Nut 2** or **Nut ?**), you may wish to assess the potential of the HGMU to perform this process. To do this, conduct the assessment as normal but answer questions only in ***CV2***, ***CV3*** and ***CV4.*** Establish an assessment outcome as normal but ignore the need for positive answers for ***CV1***. The rationale statement can be used as normal but ignore the reference to receipt of nutrient inputs.

CV1 *Nutrient input*

Q1. Has the HGMU been attributed one of the following nutrient input codes, indicating an input of phosphorus? Circle all those applicable:

a. $\mathbf{P_d}$ **1/a**, $\mathbf{P_d}$ **1/b**, $\mathbf{P_d}$ **1/x**
b. $\mathbf{P_i}$ **1/x**
c. $\mathbf{Nut_d}$ **1/b**, $\mathbf{Nut_d}$ **1/x**, $\mathbf{Nut_i}$ **1/x**
d. **Nut ?**

(Recording Sheet 2c)

[If YES to ***Q1 a, b*** or ***c*** continue with ***Q2***. If NO to all answers above or YES to ***Q1d***, then continuing with this assessment will give information only about an HGMU's **potential** to perform this process; to do this continue with ***CV2***.]

Q2. What is the source of the nutrient input received by the HGMU? Circle all those applicable:

a. Inorganic fertiliser applied directly to the HGMU *(Recording Sheet 2b, H1 Action, Q1)*
b. Organic fertiliser applied directly on the surface of the HGMU *(Recording Sheet 2b, H1 Action, Q5)*
c. Organic fertiliser injected into the surface of the HGMU *(Recording Sheet 2b, H1 Action, Q6)*
d. Organic industrial waste applied within the HGMU *(Recording Sheet 2b, H1 Action, Q7)*
e. Manure from livestock kept within the HGMU *(Recording Sheet 2b, H1 Action, Q8)*
f. Excrement from large populations of birds *(Recording Sheet 2b, H1 Action, Q9)*
g. Contact with water containing > 0.1 mg l^{-1} phosphate–phosphorus *(Recording Sheet 2b, H1 Action, Q22)*
h. Groundwater discharge *(Recording Sheet 2b, H2 Action, Q1–3)*
i. Run-off or subsurface run-off *(Recording Sheet 2b, H2 Action, Q4–6)*

CV2 *Soil type*

Q1. Is the soil described as a mineral soil? *(Recording Sheet 2a, C.3.6)*
[*If YES continue with* ***Q2***. *If NO continue with* ***Q7***.]

Q2. Is the soil particle size predominantly fine (= clayey)? *(Recording Sheet 2a, C.3.7)*

Q3. Is the soil particle size described as predominantly coarse or gravelly (= sandy or containing small stones)? *(Recording Sheet 2a, C.3.7)*

Q4. Is the soil colour described as predominantly greyish, or greenish? *(Recording Sheet 2a, C.3.4)*

Q5. Is the soil colour described as predominantly rusty? *(Recording Sheet 2a, C.3.4)*

Q6. Are the mottles in the soil described as reddish, ochreous or rusty? *(Recording Sheet 2a, C.3.5)*

Q7. Is the soil described as organic? *(Recording Sheet 2a, C.3.6)*
[*If YES continue with* ***Q8***. *If NO continue with* ***Q11***.]

Q8. What description is given to the organic material? *(Recording Sheet 2a, C.3.8)*:

a. Amorphous peat, humus, humose
b. Semi-fibrous, fibrous

Q9. If present, is standing water brown or rust-coloured? *(Recording Sheet 3, B.3.14)*

Q10. If present, is standing water covered with a thin oily layer (= iron bacteria)? NB The layer breaks apart when touched. *(Recording Sheet 3, B.3.15)*

Q11. How often is limestone added to the soil of the HGMU*? (Recording Sheet 1, C.3.2.7b, AA Map 2)*

a. No history of application of limestone in the last 10 years/not applicable
b. None currently, but previously applied within the last 10 years, or applied but less than once a year
c. Currently applied once or more than once a year
d. Unknown

Q12. Is it likely that the groundwater is rich in calcium (e.g. because groundwater flow originates in or passes through areas of marine sand or limestone)? *(Recording Sheet 1, C.3.21.6)*

Q13. Are limestone fragments prominent within the soil profile? *(Recording Sheet 2a, General comments)*

Q14. Do any reddish, ochreous or rusty hard indurate layers occur within the soil profile? *(Recording Sheet 2a, General comments)*

Q15. What is the pH of the soil? *(Recording Sheet 3, A.1i)*

a. < 4
b. 4–6.5
c. > 6.5

CV3 *Soil water regime*

Q1. Is the HGMU permanently flooded with standing water or is the groundwater level at or near the soil surface (i.e. within 20 cm) for most of the year? *(Recording Sheet 3, B.4.3 and B.3.10 and/or AA Map 2)*

Q2. Is the upper 20 cm of the soil prone to water table fluctuations (alternate periods of waterlogged and non-waterlogged soil conditions, which may be due to natural, seasonal fluctuations, or water table management), which result in non-waterlogged soil conditions for at least half of the year? *(Recording Sheet 1, C.3.22.3)*

CV4 *Landform*

Q1. Has the HGMU been assigned a geomorphic prefix code **D**, indicating that it is one of the depressional features listed below? *(Recording Sheet 2a, C.2)*

a. **BS, CO, FC, AD, UD**
b. **OX, LK, DT, CN**

Q2. Has the HGMU been assigned a gradient code **G0** indicating that it is flat or gentle (0°–2°)? *(Recording Sheet 2a, C.2)*

Q3. Is the microtopography in the HGMU described as significant? *(Recording Sheet 3, A.3)*

Assessment of the process

Compare the results recorded for each HGMU on the appraisal sheets with the possible outcomes below, and assign an appropriate assessment code. Record this on the appraisal sheet. The syntax codes used are described in the introduction to Section 3.

Look-up tables

1. The process is definitely being performed

If the answers to the questions for this process agree with any of the combinations in Look-up Table 1, then the HGMU is definitely performing the processes of

3.2.A.4 Adsorption and Precipitation of P in the Soil Appraisal Sheet

HGMU code: ____________	**AA name:** ____________

CV1 Nutrient input					
Q1	a	b	c	d	
Q2	a	b	c	d	e
	f	g	h	i	

CV3 Soil water regime			
Q1	Y	N	U/NA
Q2	Y	N	U/NA

CV2 Soil type				
Q1	Y	N	U/NA	
Q2	Y	N	U/NA	
Q3	Y	N	U/NA	
Q4	Y	N	U/NA	
Q5	Y	N	U/NA	
Q6	Y	N	U/NA	
Q7	Y	N	U/NA	
Q8	a	b		
Q9	Y	N	U/NA	
Q10	Y	N	U/NA	
Q11	a	b	c	d
Q12	Y	N	U/NA	
Q13	Y	N	U/NA	
Q14	Y	N	U/NA	
Q15	a	b	c	

CV4 Landform			
Q1a	Y	N	U/NA
Q1b	Y	N	U/NA
Q2	Y	N	U/NA
Q3	Y	N	U/NA

HGMU process assessment code: ____________

3.2.A.4 Look-up Table 1

<table>
<tr><th>CV1</th><th>CV2</th><th>CV3</th><th>CV4</th><th>Rationale</th><th>Code</th></tr>
<tr><td>1a (2a|2b|2c|2d|2e|2f)</td><td>(1 2) (5|6|14) (15b|15c)</td><td>#1 | 2</td><td>1a|2|3</td><td>The HGMU receives a direct input of phosphorus. Rusty mottles or the rusty colour of the soil or presence of a reddish, ochreous or rusty hard indurate layer indicate the presence of iron. Phosphate can bind to iron. Also clay particles can bind phosphate. The soil is not permanently anaerobic, which would otherwise reduce the retention capacity of the soil. Wetland morphology ensures interaction of the nutrient input with the HGMU.</td><td>1</td></tr>
<tr><td>1a (2a|2b|2c|2d|2e|2f)</td><td>(1 2) (5|6|14) (15b|15c)</td><td>#1 | 2</td><td>*</td><td>The HGMU receives a direct input of phosphorus. Rusty mottles or the rusty colour of the soil or presence of a reddish, ochreous or rusty hard indurate layer indicate the presence of iron. Phosphate can bind to iron. Also clay particles can bind phosphate. The soil is not permanently anaerobic, which would otherwise reduce the retention capacity of the soil.</td><td>1</td></tr>
<tr><td>1a (2a|2b|2c|2d|2e|2f)</td><td>(1 2) (11b|11c|12|13) 15c</td><td>*</td><td>1a|2|3</td><td>The HGMU receives a direct input of phosphorus. There is evidence of the presence of calcium. Phosphate can bind to calcium. Also clay particles can bind phosphate. Wetland morphology ensures interaction of the nutrient input with the HGMU.</td><td>1</td></tr>
<tr><td>1a (2a|2b|2c|2d|2e|2f)</td><td>(1 2) (11b|11c|12|13) 15c</td><td>*</td><td>*</td><td>The HGMU receives a direct input of phosphorus. There is evidence of the presence of calcium. Phosphate can bind to calcium. Also clay particles can bind phosphate.</td><td>1</td></tr>
<tr><td>1a (2a|2b|2c|2d|2e|2f)</td><td>(7 8a) (9|10) (15b|15c)</td><td>#1 | 2</td><td>1a|2|3</td><td>The HGMU receives a direct input of phosphorus. Brown or rusty coloured water or water with an oily layer on it indicates the presence of iron. Phosphate can bind to iron. Also organic particles can bind phosphorus. The soil is not permanently anaerobic, which would otherwise reduce the retention capacity of the soil. Wetland morphology ensures interaction of the nutrient input with the HGMU.</td><td>1</td></tr>
<tr><td>1a (2a|2b|2c|2d|2e|2f)</td><td>(7 8a) (9|10) (15b|15c)</td><td>#1 | 2</td><td>*</td><td>The HGMU receives a direct input of phosphorus. The presence of brown or rusty coloured water or water with an oily layer on it indicates the presence of iron and/or there is evidence of the presence of calcium. Phosphate can bind to iron. Also organic particles can bind phosphorus. The soil is not permanently anaerobic, which would otherwise reduce the retention capacity of the soil.</td><td>1</td></tr>
<tr><td>1a (2a|2b|2c|2d|2e|2f)</td><td>(7 8a) (11b|11c|12|13) 15c</td><td>*</td><td>1a|2|3</td><td>The HGMU receives a direct input of phosphorus. The soil contains limestone or there is evidence of the presence of calcium to which phosphate can bind. Also organic particles can bind phosphorus. Wetland morphology ensures interaction of the nutrient input with the HGMU.</td><td>1</td></tr>
<tr><td>1a (2a|2b|2c|2d|2e|2f)</td><td>(7 8a) (11b|11c|12|13) 15c</td><td>*</td><td>*</td><td>The HGMU receives a direct input of phosphorus. The soil contains limestone or there is evidence of the presence of calcium to which phosphate can bind. Also organic particles can bind phosphorus.</td><td>1</td></tr>
<tr><td>1a (2a|2b|2c|2d|2e|2f)</td><td>(5|6|14) (15b|15c)</td><td>#1 | 2</td><td>1a|2|3</td><td>The HGMU receives a direct input of phosphorus. Rusty mottles or the rusty colour of the soil or presence of a reddish, ochreous or rusty hard indurate layer indicate the presence of iron. Phosphate can bind to iron. The soil is not permanently anaerobic, which would otherwise reduce the retention capacity of the soil. Wetland morphology ensures interaction of the nutrient input with the HGMU.</td><td>1</td></tr>
</table>

<table>
<tr><td>1a
(2a|2b|
2c|2d|
2e|2f)</td><td>(5|6|14)
(15b|15c)</td><td>#1 | 2</td><td>*</td><td>The HGMU receives a direct input of phosphorus. Rusty mottles or the rusty colour of the soil or presence of a reddish, ochreous or rusty hard indurate layer indicate the presence of iron. Phosphate can bind to iron. The soil is not permanently anaerobic, which would otherwise reduce the retention capacity of the soil.</td><td>1</td></tr>
<tr><td>1a
(2a|2b|
2c|2d|
2e|2f)</td><td>(11b|11c|
12|13)
15c</td><td>*</td><td>1a|2|3</td><td>The HGMU receives a direct input of phosphorus. The soil contains limestone or there is evidence for the presence of calcium, which provides adsorption sites for phosphate. Wetland morphology ensures interaction of the nutrient input with the HGMU.</td><td>1</td></tr>
<tr><td>1a
(2a|2b|
2c|2d|
2e|2f)</td><td>(11b|11c|
12|13)
15c</td><td>*</td><td>*</td><td>The HGMU receives a direct input of phosphorus. The soil contains limestone or there is evidence for the presence of calcium, which provides adsorption sites for phosphate.</td><td>1</td></tr>
<tr><td>1b 2g</td><td>(1 2)
(5|6|14)
(15b|15c)</td><td>#1 | 2</td><td>1a|2|3</td><td>The HGMU receives phosphorus because it has contact with phosphate-rich water. Rusty mottles or the rusty colour of the soil or presence of a reddish, ochreous or rusty hard indurate layer indicate the presence of iron. Phosphate can bind to iron. Also clay particles can bind phosphate. The soil is not permanently anaerobic, which would otherwise reduce the retention capacity of the soil. Wetland morphology ensures interaction of the nutrient input with the HGMU.
NB: When sulphate does not occur naturally in the HGMU or when the HGMU does not regularly receive sulphate-rich water, an occasional inflow of sulphate-rich water (from either river water or surface run-off) can cause mobilisation of phosphorus, because the phosphorus bound to iron in soil and sediments will be replaced by sulphide (produced by sulphate reduction in anaerobic soils) (= internal eutrophication/ internal loading).</td><td>1</td></tr>
<tr><td>1b 2g</td><td>(1 2)
(5|6|14)
(15b|15c)</td><td>#1 | 2</td><td>*</td><td>The HGMU receives phosphorus because it has contact with phosphate-rich water. Rusty mottles or the rusty colour of the soil or presence of a reddish, ochreous or rusty hard indurate layer indicate the presence of iron. Phosphate can bind to iron. Also clay particles can bind phosphate. The soil is not permanently anaerobic, which would otherwise reduce the retention capacity of the soil.
NB: When sulphate does not occur naturally in the HGMU or when the HGMU does not regularly receive sulphate-rich water, an occasional inflow of sulphate-rich water (from either river water or surface run-off) can cause mobilisation of phosphorus, because the phosphorus bound to iron in soil and sediments will be replaced by sulphide (produced by sulphate reduction in anaerobic soils) (= internal eutrophication/ internal loading).</td><td>1</td></tr>
<tr><td>1b 2g</td><td>(1 2)
(11b|11c|
12|13)15c</td><td>*</td><td>1a|2|3</td><td>The HGMU receives phosphorus because it has contact with phosphate-rich water. There is evidence of the presence of calcium, to which phosphate can bind. Also clay particles can bind phosphate. Wetland morphology ensures interaction of the nutrient input with the HGMU.</td><td>1</td></tr>
<tr><td>1b 2g</td><td>(1 2)
(11b|11c|
12|13)15c</td><td>*</td><td>*</td><td>The HGMU receives phosphorus because it has contact with phosphate-rich water. There is evidence of the presence of calcium, to which phosphate can bind. Also clay particles can bind phosphate.</td><td>1</td></tr>
</table>

3.2.A.4 Look-up Table 1 cont'd

<table>
<tr><th>CV1</th><th>CV2</th><th>CV3</th><th>CV4</th><th>Rationale</th><th>Code</th></tr>
<tr><td>1b 2g</td><td>(7 8a)
(9|10)
(15b|15c)</td><td>#1|2</td><td>1a|2|3</td><td>The HGMU receives phosphorus because it has contact with phosphate-rich water. The brown or rusty coloured water or water with an oily layer on it indicates presence of iron. Phosphate binds to iron. Also organic particles can bind phosphorus. The soil is not permanently anaerobic, which would otherwise reduce the retention capacity of the soil. Wetland morphology ensures interaction of the nutrient input with the HGMU.
NB: When sulphate does not occur naturally in the HGMU or when the HGMU does not regularly receive sulphate-rich water, an occasional inflow of sulphate-rich water (from either river water or surface run-off) can cause mobilisation of phosphorus, because the phosphorus bound to iron in soil and sediments will be replaced by sulphide (produced by sulphate reduction in anaerobic soils) (= internal eutrophication/internal loading).</td><td>1</td></tr>
<tr><td>1b 2g</td><td>(7 8a)
(9|10)
(15b|15c)</td><td>#1|2</td><td>*</td><td>The HGMU receives phosphorus because it has contact with phosphate-rich water. The brown or rusty coloured water or water with an oily layer on it indicates the presence of iron. Phosphate binds to iron. Also organic particles can bind phosphorus. The soil is not permanently anaerobic, which would otherwise reduce the retention capacity of the soil.
NB: When sulphate does not occur naturally in the HGMU or when the HGMU does not regularly receive sulphate-rich water, an occasional inflow of sulphate-rich water (from either river water or surface run-off) can cause mobilisation of phosphorus, because the phosphorus bound to iron in soil and sediments will be replaced by sulphide (produced by sulphate reduction in anaerobic soils) (= internal eutrophication/internal loading).</td><td>1</td></tr>
<tr><td>1b 2g</td><td>(7 8a)
(11b|11c|
12|13) 15c</td><td>*</td><td>1a|2|3</td><td>The HGMU receives phosphorus because it has contact with phosphate-rich water. There is evidence of the presence of calcium, to which phosphate can bind. Also organic particles can bind phosphorus. Wetland morphology ensures interaction of the nutrient input with the HGMU.</td><td>1</td></tr>
<tr><td>1b 2g</td><td>(7 8a)
(11b|11c|
12|13) 15c</td><td>*</td><td>*</td><td>The HGMU receives phosphorus because it has contact with phosphate-rich water. There is evidence of the presence of calcium, to which phosphate can bind. Also organic particles can bind phosphorus.</td><td>1</td></tr>
<tr><td>1b 2g</td><td>(5|6|14)
(15b|15c)</td><td>#1|2</td><td>1a|2|3</td><td>The HGMU receives phosphorus because it has contact with phosphate-rich water. Rusty mottles or the rusty colour of the soil or presence of a reddish, ochreous or rusty hard indurate layer indicate the presence of iron. Phosphate can bind to iron. The soil is not permanently anaerobic, which would otherwise reduce the retention capacity of the soil. Wetland morphology ensures interaction of the nutrient input with the HGMU.
NB: When sulphate does not occur naturally in the HGMU or when the HGMU does not regularly receive sulphate-rich water, an occasional inflow of sulphate-rich water (from either river water or surface run-off) can cause mobilisation of phosphorus, because the phosphorus bound to iron in soil and sediments will be replaced by sulphide (produced by sulphate reduction in anaerobic soils) (= internal eutrophication/internal loading).</td><td>1</td></tr>
<tr><td>1b 2g</td><td>(5|6|14)
(15b|15c)</td><td>#1|2</td><td>*</td><td>The HGMU receives phosphorus because it has contact with phosphate-rich water. Rusty mottles or the rusty colour of the soil or presence of a reddish, ochreous or rusty hard indurate layer indicate the presence of iron. Phosphate can bind to iron. The soil is not permanently anaerobic, which would otherwise reduce the retention capacity of the soil.</td><td>1</td></tr>
</table>

<table>
<tr><td></td><td></td><td></td><td></td><td>NB: When sulphate does not occur naturally in the HGMU or when the HGMU does not regularly receive sulphate-rich water, an occasional inflow of sulphate-rich water (from either river water or surface run-off) can cause mobilisation of phosphorus, because the phosphorus bound to iron in soil and sediments will be replaced by sulphide (produced by sulphate reduction in anaerobic soils) (=internal eutrophication/internal loading).</td><td></td></tr>
<tr><td>1b 2g</td><td>(11b|11c|
12|13)
15c</td><td>*</td><td>1a|2|3</td><td>The HGMU receives phosphorus because it has contact with phosphate-rich water. The soil contains limestone, which provides adsorption sites for phosphate. Wetland morphology ensures interaction of the nutrient input with the HGMU.</td><td>1</td></tr>
<tr><td>1b 2g</td><td>(11b|11c|
12|13)
15c</td><td>*</td><td>*</td><td>The HGMU receives phosphorus because it has contact with phosphate-rich water. The soil contains limestone, which provides adsorption sites for phosphate.</td><td>1</td></tr>
<tr><td>1b 2h|2i</td><td>(1 2)
(5|6|14)
(15b|15c)</td><td>#1 | 2</td><td>1a|2|3</td><td>The HGMU receives phosphorus via groundwater discharge or (sub)surface run-off. Rusty mottles or the rusty colour of the soil or presence of a reddish, ochreous or rusty hard indurate layer indicate the presence of iron. Phosphate can bind to iron. Also clay particles can bind phosphate. The soil is not permanently anaerobic, which would otherwise reduce the retention capacity of the soil. Wetland morphology ensures interaction of the nutrient input with the HGMU.</td><td>1</td></tr>
<tr><td>1b 2h|2i</td><td>(1 2)
(5|6|14)
(15b|15c)</td><td>#1 | 2</td><td>*</td><td>The HGMU receives phosphorus via groundwater discharge or (sub)surface run-off. Rusty mottles or the rusty colour of the soil or presence of a reddish, ochreous or rusty hard indurate layer indicate the presence of iron. Phosphate can bind to iron. Also clay particles can bind phosphate. The soil is not permanently anaerobic, which would otherwise reduce the retention capacity of the soil.</td><td>1</td></tr>
<tr><td>1b 2h|2i</td><td>(1 2)
(11b|11c|
12|13) 15c</td><td>*</td><td>1a|2|3</td><td>The HGMU receives phosphorus via groundwater discharge or (sub)surface run-off. There is evidence of the presence of calcium, to which phosphate can bind. Also clay particles can bind phosphate. Wetland morphology ensures interaction of the nutrient input with the HGMU.</td><td>1</td></tr>
<tr><td>1b 2h|2i</td><td>(1 2)
(11b|11c|
12|13) 15c</td><td>*</td><td>*</td><td>The HGMU receives phosphorus via groundwater discharge or (sub)surface run-off. There is evidence of the presence of calcium, to which phosphate can bind. Also clay particles can bind phosphate.</td><td>1</td></tr>
<tr><td>1b 2h|2i</td><td>(7 8a)
(9|10)
(15b|15c)</td><td>#1|2</td><td>1a|2|3</td><td>The HGMU receives phosphorus via groundwater discharge or (sub)surface run-off. The brown or rusty coloured water or water with an oily layer on it indicates presence of iron. Phosphate binds to iron. Also organic particles can bind phosphorus. The soil is not permanently anaerobic, which would otherwise reduce the retention capacity of the soil. Wetland morphology ensures interaction of the nutrient input with the HGMU.</td><td>1</td></tr>
<tr><td>1b 2h|2i</td><td>(7 8a)
(9|10)
(15b|15c)</td><td>#1|2</td><td>*</td><td>The HGMU receives phosphorus via groundwater discharge or (sub)surface run-off. The brown or rusty coloured water or water with an oily layer on it indicates the presence of iron. Phosphate binds to iron. Also organic particles can bind phosphorus. The soil is not permanently anaerobic, which would otherwise reduce the retention capacity of the soil.</td><td>1</td></tr>
<tr><td>1b 2h|2i</td><td>(7 8a)
(11b|11c|
12|13) 15c</td><td>*</td><td>1a|2|3</td><td>The HGMU receives phosphorus via groundwater discharge or (sub)surface run-off. There is evidence of the presence of calcium, to which phosphate can bind. Also organic particles can bind phosphorus. Wetland morphology ensures interaction of the nutrient input with the HGMU.</td><td>1</td></tr>
</table>

3.2.A.4 Look-up Table 1 cont'd

<table>
<tr><th>CV1</th><th>CV2</th><th>CV3</th><th>CV4</th><th>Rationale</th><th>Code</th></tr>
<tr><td>1b 2h|2i</td><td>(5|6|14) (15b|15c)</td><td>#1|2</td><td>1a|2|3</td><td>The HGMU receives phosphorus via groundwater discharge or (sub)surface run-off. Rusty mottles or the rusty colour of the soil or presence of a reddish, ochreous or rusty hard indurate layer indicate the presence of iron. Phosphate can bind to iron. The soil is not permanently anaerobic, which would otherwise reduce the retention capacity of the soil. Wetland morphology ensures interaction of the nutrient input with the HGMU.</td><td>1</td></tr>
<tr><td>1b 2h|2i</td><td>(5|6|14) (15b|15c)</td><td>#1|2</td><td>*</td><td>The HGMU receives phosphorus via groundwater discharge or (sub)surface run-off. Rusty mottles or the rusty colour of the soil or presence of a reddish, ochreous or rusty hard indurate layer indicate the presence of iron. Phosphate can bind to iron. The soil is not permanently anaerobic, which would otherwise reduce the retention capacity of the soil.</td><td>1</td></tr>
<tr><td>1b 2h|2i</td><td>(11b|11c| 12|13) 15c</td><td>*</td><td>1a|2|3</td><td>The HGMU receives phosphorus via groundwater discharge or (sub)surface run-off. The soil contains limestone, which provides adsorption sites for phosphate. Wetland morphology ensures interaction of the nutrient input with the HGMU.</td><td>1</td></tr>
<tr><td>1b 2h|2i</td><td>(11b|11c| 12|13) 15c</td><td>*</td><td>*</td><td>The HGMU receives phosphorus via groundwater discharge or (sub)surface run-off. The soil contains limestone, which provides adsorption sites for phosphate.</td><td>1</td></tr>
</table>

3.2.A.4 Look-up Table 2

CV1	CV2	CV3	CV4	Rationale	Code
#(1a 1b 1c) 1d	*	*	*	The HGMU does not receive any phosphorus additions, or input of phosphorus is uncertain.	**2**
*	3	*	*	The soil is unlikely to retain phosphorus. The soil is likely to be very permeable, hence reducing contact time.	**2**
*	1 2 4 #6	*	*	The grey/green colour of the soil and absence of mottles indicates a reduced environment. Such conditions are unfavourable for phosphorus retention.	**2**
*	15a	*	*	The low pH indicates unfavourable conditions for adsorption of phosphate to soil particles.	**2**
*	*	1	*	The soil is permanently inundated with water and likely to be anoxic. Such conditions are unfavourable for phosphorus retention.	**2**
*	*	*	1b	The soil is permanently inundated with water and likely to be anoxic. Such conditions are unfavourable for phosphorus retention.	**2**

3.2.A.4 Look-up Table 3

CV1	CV2	CV3	CV4	Rationale	Code
1c	*	*	*	The HGMU receives a nutrient input, but it is uncertain whether this input contains phosphorus.	**1/x**
*	(1 2)|(7 8a) #(3 4 5 6 9 10 11b 11c 12 13)	*	*	There is no visible evidence of presence of iron or calcium to which phosphorus can bind. However, phosphorus can also be adsorbed to clay and organic material, as well as elements that are not identifiable in the field (e.g. aluminium).	**1/x**
*	8b	*	*	Less humified organic matter provides less adsorption sites than humified organic matter.	**1/x**

adsorption and precipitation of phosphorus in soil. The overall importance of phosphorus adsorption and precipitation is poorly documented. Because there is only a small amount of quantitative information about phosphorus adsorption and precipitation available, this process cannot be assessed quantitatively, so only a qualitative assessment is given.

2. The process is not (significantly) being performed

If the answers to the questions for this process agree with any of the combinations in Look-up Table 2, then the HGMU is not significantly performing the processes of adsorption and precipitation of phosphorus in soil.

3. The process is probably being performed, but there are constraining factors or uncertainties

If Look-up Tables 1 and 2 did not lead to an assessment, the process of phosphorus adsorption and precipitation is performed, but there are constraining factors or uncertainties. In Look-up Table 3 you can find out how many constraining factors influence the process, or which factors cause the uncertainty.

4. The process is definitely not being performed

This statement is not applicable to this process.

This completes the functional assessment for the process of adsorption and precipitation of P in soil. The output codes can be recorded on the Functional Assessment Table at the end of Section 3.2.A.6.

3.2.A.5 Retention of particulate nutrients (N and P) (process)

Further reading and references

Brenner, M., Keenan, L.W., Miller, S.J. and Schelske, C.L. (1999). Spatial and temporal patterns of sediment and nutrient accumulation in shallow lakes of the Upper St. Johns River Basin, Florida, *Wetlands Ecology and Management* **6**, 221–240.

Fisher, J. and Acreman, M.C. (2004). Wetland nutrient removal: a review of the evidence, *Hydrology and Earth System Sciences* **8**(4), 673–685.

Johnston, C.A. (1991). Sediment and nutrient retention by freshwater wetlands: effects on surface water quality, *Critical Reviews in Environmental Science and Technology* **5**,**6**(21), 491–565.

Maltby, E., Hogan, D.V. and McInnes, R.J. (1996). *Functional Analysis of European Wetland Ecosystems; Phase 1 (FAEWE)*, European Commission, Brussels.

Olde Venterink, H. (2000). *Nitrogen, Phosphorus and Potassium Flows Controlling Plant Productivity and Species Richness*, PhD Dissertation, Utrecht University, The Netherlands.

Olde Venterink, H., Vermaat, J.E., Pronk, M., Wiegman, F., Lee, G.E.M. van der, Hoorn, M. van den, Higler, L.W.G. and Verhoeven, J.T.A. (2006). Importance of sediment deposition and denitrification for nutrient retention in floodplain wetlands, *Applied Vegetation Science* **9**, 163–174.

Olde Venterink, H., Wiegman, F., Lee, G. E. M. van der and Vermaat, J. E. (2003). Role of active floodplains for nutrient retention in the River Rhine, *Journal of Environmental Quality* **32**, 1430–1435

Wienk, L.D., Verhoeven, J.T.A., Coops, H. and Portielje, R. (2000). Peilbeheer en Nutriënten; Literatuurstudie naar de Effecten van Peildynamiek op de Nutriëntenhuishouding van Watersystemen, RIZA rapport 2000.012, RIZA, Lelystad, The Netherlands (in Dutch).

Introduction

Process description:

- The process is considered to occur in an HGMU when nutrients (nitrogen and phosphorus), which are physico-chemically bound to particulate matter transported by in-flowing water, are deposited and retained within the wetland.
- It is difficult to quantify the process of retention of particulate nutrients, because the process is highly variable over time and space. Therefore only a qualitative assessment of retention of particulate nutrients is made.

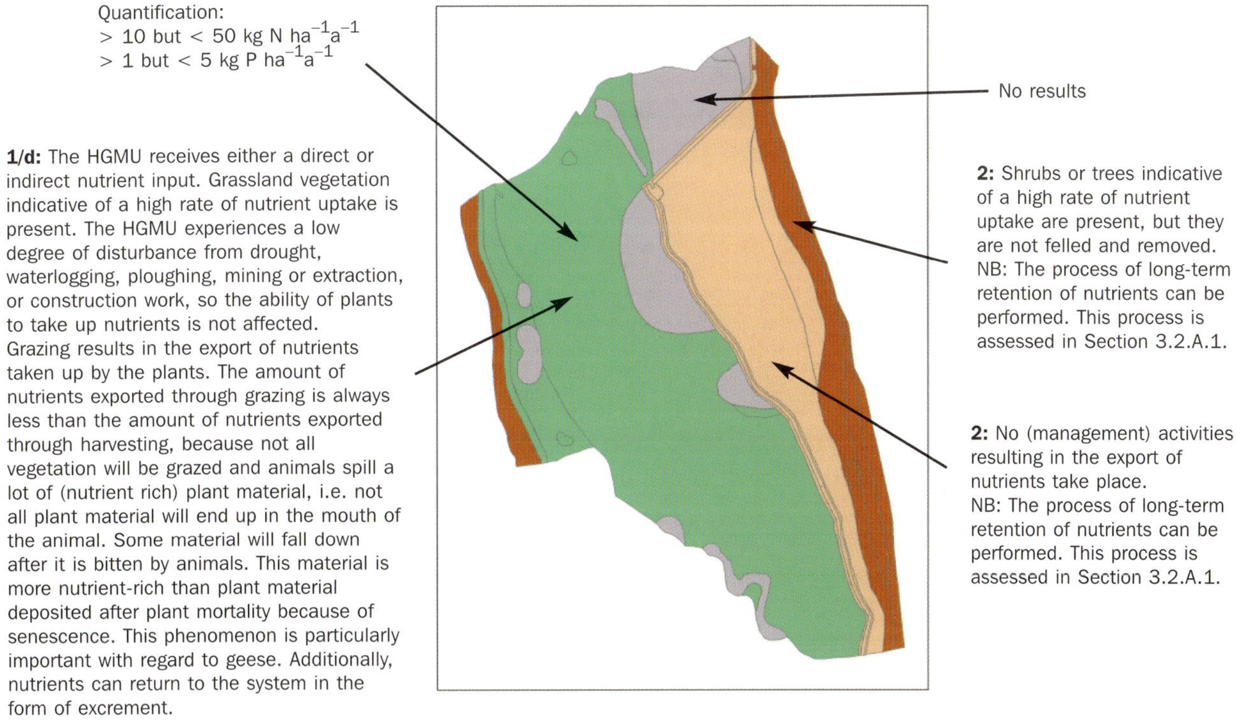

1.8.2 Example of functional assessment outcomes for the function of nutrient removal at a study site in the River Tamar catchment, UK. This output has been imported into the WEDSS (see Preface), which displays the outputs spatially. Colours are used to indicate different HGMUs. '1/d' and '2' refer to the output codes in the look-up tables.

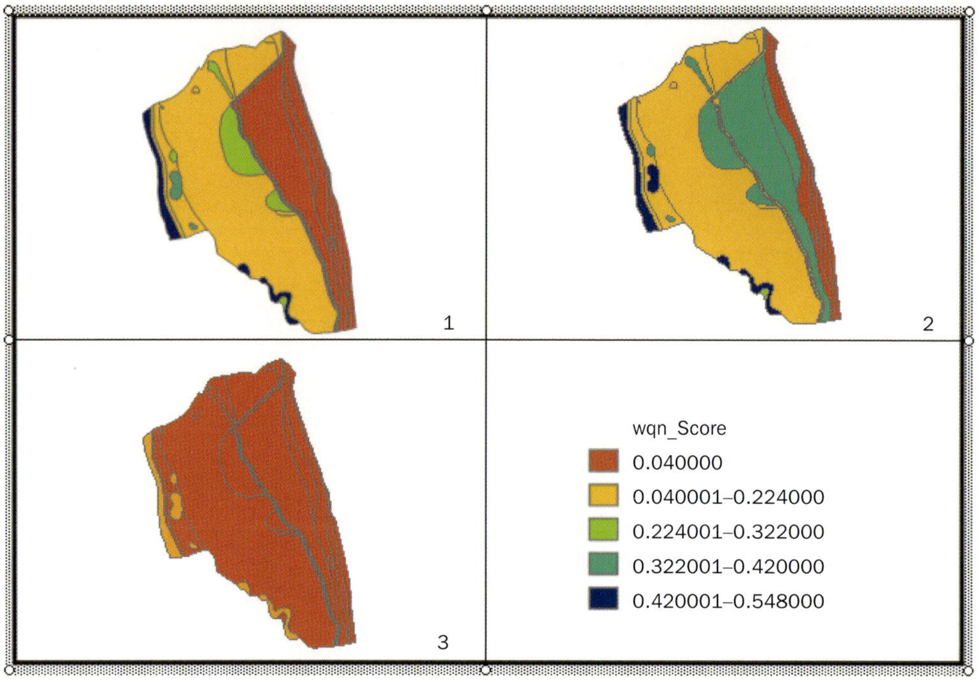

1.10.1 Comparison of water quality enhancement after biogeochemical removal of nitrogen in a wetland in the Tamar catchment, Devon, UK. Three possible management scenarios are illustrated. No. 1 (top left) refers to 'Business as Usual', where management continues as in the past; No. 2 is the 'Degraded' scenario and 3 is a 'Green' scenario, where management is optimised for environmental quality. The scoring (bottom right) is part of the outputs of the CD-ROM version of the FAPs that were imported into the WEDSS, from which this figure is taken.

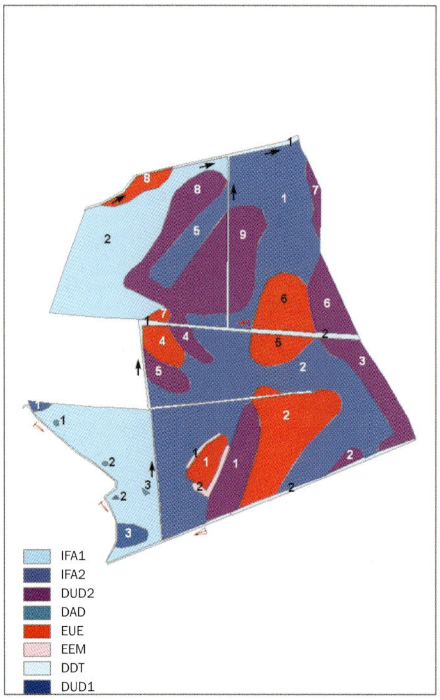

1.10.2 Map of the site at Sutton Common (Humberhead Levels), which was mapped using FAPs. Colours refer to landforms, and the legend codes begin with either D (depression), E (elevation) or I (intermediate), followed by features such as FA (flat area), UD (unidentified depression), DT (ditch), etc. as described in the text. The numbers on the map indicate where the landscape is separated by a boundary, e.g. a ditch, and are sequential for each colour-coded landscape feature.

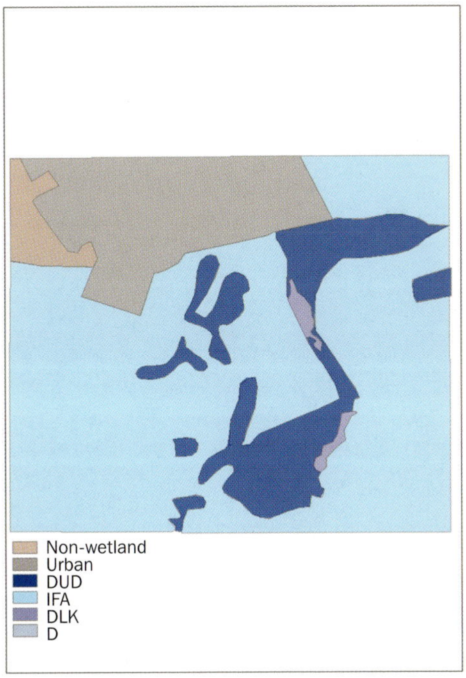

1.10.3 Scaling up of wetland functional landscape units in the Sutton Common area of the Humberhead Levels. Information was sourced from Ordnance Survey, soil maps and SSSI descriptions, as well as by applying the landscape patterns established at the study sites and site observations.

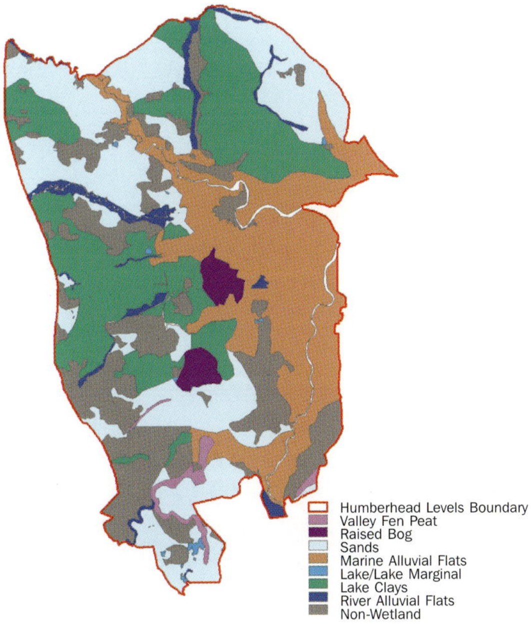

1.10.4 Extrapolation of functional assessments from pilot sites and their surrounding areas to the entire Humberhead Levels, using soil and geology maps to infer landscape types and GIS for mapping.

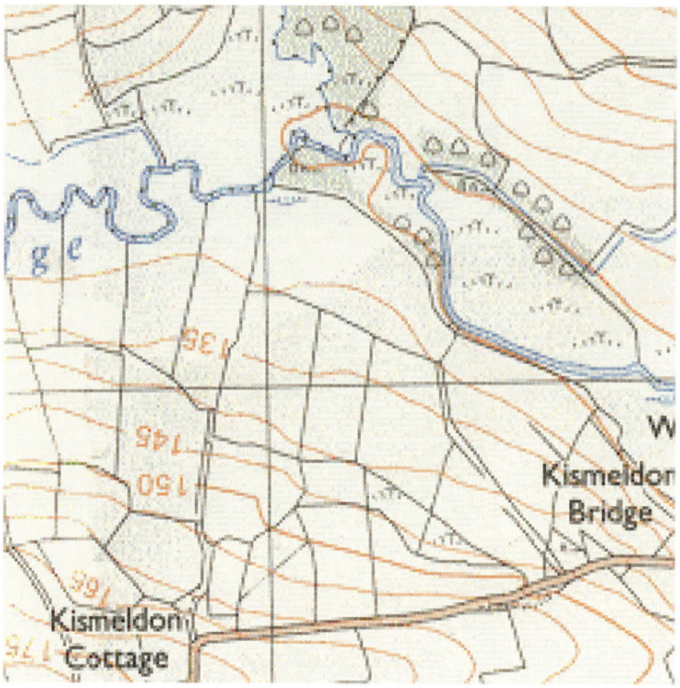

2.1.1 Ordnance Survey (OS) 1:25 000 Map (Pathfinder series 1273, dated 1990) of assessment area (AA). This 1:25,000 OS map covers the same area as the aerial photograph (Fig. 2.1.2). In combination, the two can be used to gain an impression of the topography, land use and spatial relationships among features of the wetland. Reproduced by permission of Ordnance Survey on behalf of HMSO.

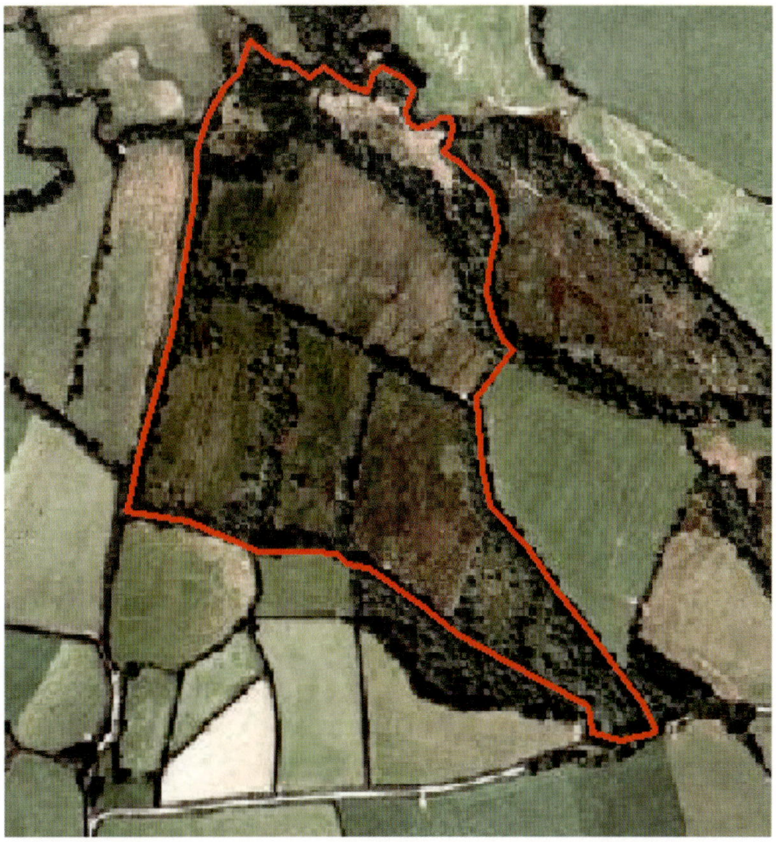

2.1.2 Aerial photograph of the assessment area (photo courtesy of English Nature).

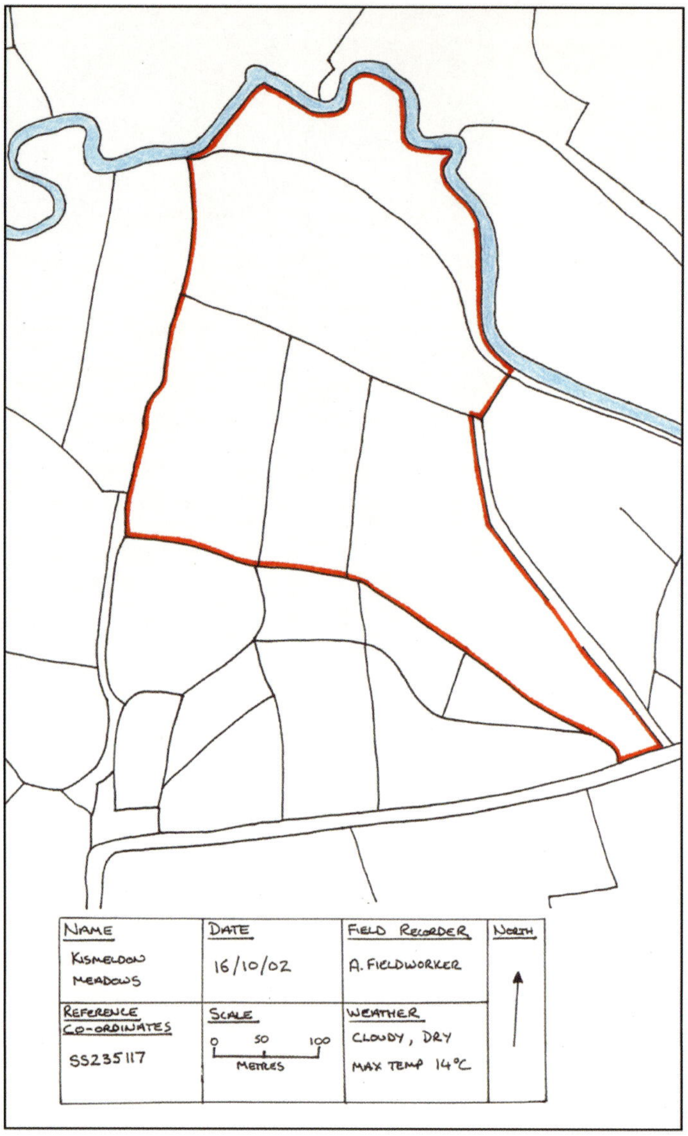

For clarity, the main features that are distinguishable from the aerial photograph and the OS map (e.g. field boundaries, roads, rivers, etc.) have been copied to create this base-map.
At this stage the following features have been added:

- boundary of AA (red)
- main rivers (blue)
- archive information box

No other features (e.g. embankments, ditches, streams, drains or springs) have been identified using the information available at this stage.

2.1.3 Map of the assessment area. Reproduced by permission of Ordnance Survey on behalf of HMSO.

NAME	DATE	FIELD RECORDER	NORTH
KISMELDON MEADOWS	16/10/02	A. FIELDWORKER	↑
REFERENCE CO-ORDINATES	SCALE	WEATHER	
SS 235117	0 100m 200m	CLOUDY, DRY MAX. TEMP. 14°C	

The CA map is at a smaller scale than the AA maps, and has been produced using a digital scanning of an OS 1:25,000 map. Both the AA and CA have been delineated (in red).
The archive information box on the AA maps has been copied to this map.
At this stage no details about slope have been added to the map because the data from the map is not of high enough resolution to do this accurately. The upslope limit of the CA is delineated by a road, which at this stage is assumed to be bounded by ditches, which intercept run-off from upslope and restrict the size of the CA. This information is to be checked in the field.

2.1.4 Map of the contributory area (CA). Reproduced by permission of Ordnance Survey on behalf of HMSO. © Crown copyright 2008. All rights reserved. Ordnance Survey Licence number 100047942.

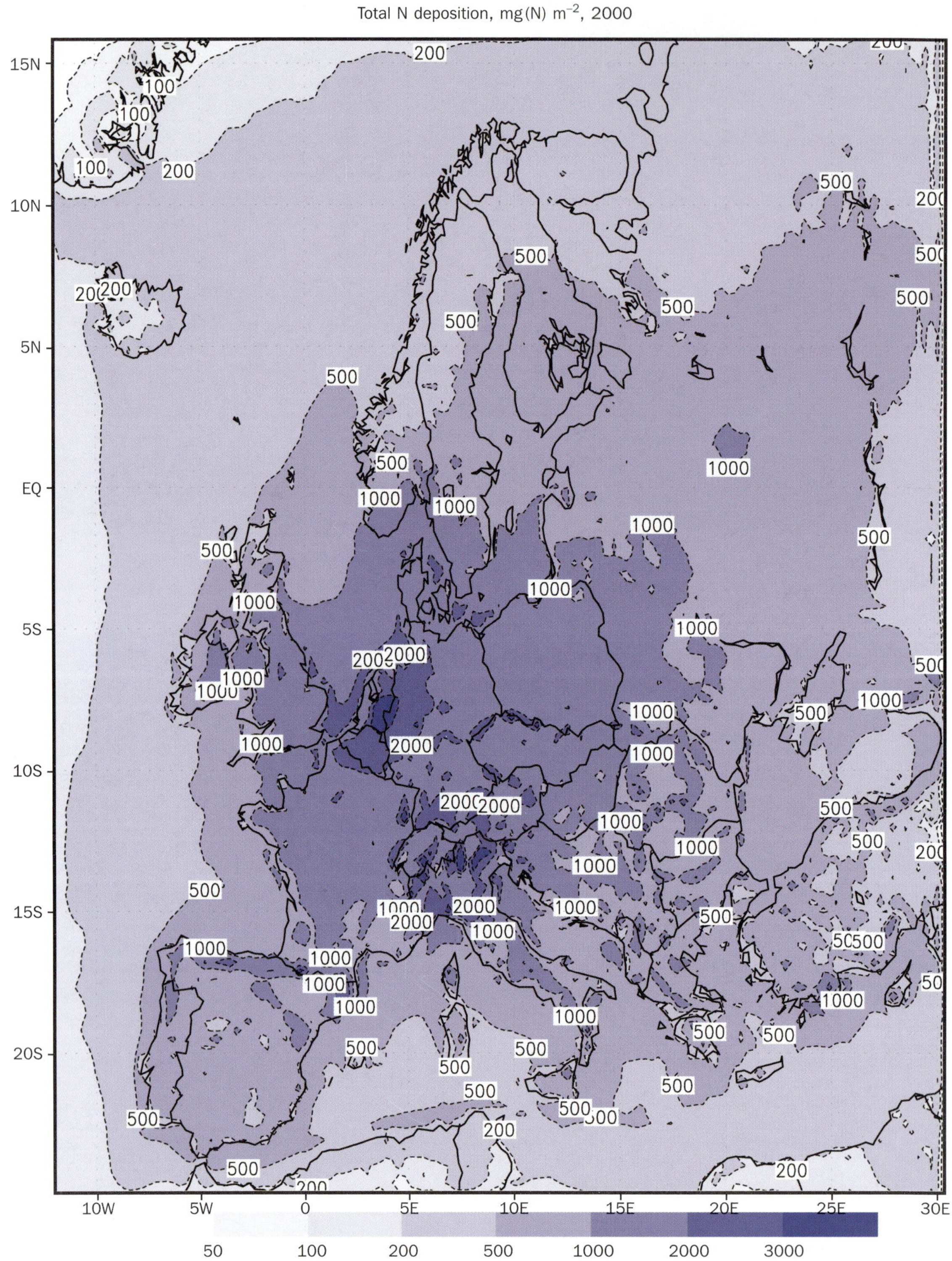

2.1.5 Total nitrogen deposition in Europe for the year 2000. Data are in mg m^{-2} total nitrogen (map courtesy of M. Hongisto, unpublished data).

Digital map of European ecological regions (DMEER)

- Aegean & West Turkey sclerophyllous and mixed forest
- Alps conifer and mixed forests
- Anatolian conifer and deciduous mixed forests
- Appenine deciduous montane forests
- Arabian desert and East Sahero-Arabian xeric shrub
- Arctic desert
- Azerbaijan shrub desert and steppe
- Balkan mixed forests
- Baltic mixed forests
- Caledon coniferous forests
- Cantabrian mixed forests
- Carpathian montane coniferous forests
- Caspian Hyrcanian mixed forests
- Caspian Lowland desert
- Caucasus mixed forests
- Celtic broadleaf forests
- Central Anatolian deciduous forests
- Central Anatolian steppe
- Central European mixed forests
- Corsican montane broadleaf and mixed forests
- Crete Mediterranean forests
- Crimean submediterranean forest complex
- Cyprus Mediterranean forests
- Dinaric Mountains mixed forests
- East European forest steppe
- Eastern Anatolian deciduous forests
- Eastern Anatolian montane steppe
- Eastern Mediterranean coniferous/ sclerophyllous/broadleaf forests
- Elburz Range forest steppe
- English Lowlands beech forests
- Euxine-Colchic deciduous forest
- Faroe Islands boreal grasslands
- Iberian conifer forests
- Iberian sclerophyllous and semi-deciduous forests
- Iceland boreal birch forest and alpine tundra
- Illyrian deciduous forests
- Italian sclerophyllous and semi-deciduous forests
- Kazakh semi-desert
- Kazakh steppe
- Kola Peninsula tundra
- Mesopotamian shrub desert
- Middle East steppe
- North Atlantic moist mixed forests
- Northeastern Spain & Southern France Mediterranean
- Northern Temperate Atlantic
- Northern Anatolian conifer and deciduous forests
- Northwest Iberian montane forests
- Northwest Russian/Novaya Zemlya tundra
- Pannonian mixed forests
- Pindus Mountains mixed forests
- Po Basin mixed forests
- Pontic steppe
- Pyrenees conifer and mixed forests
- Red Sea Nubo-Sindian tropical desert and semi-desert
- Rodope montane mixed forests
- Sarmatic mixed forests
- Scandinavian and Russian taiga
- Scandinavian coastal coniferous forests
- Scandinavian montane birch forest and grasslands
- Sea
- South Appenine mixed montane forests
- Southeastern Iberian shrubs and woodlands
- Southern Anatolian montane conifer and deciduous forests
- Southern Temperate Atlantic
- Southwest Iberian Mediterranean sclerophyllous and mixed forests
- Tyrrhenian-Adriatic sclerophyllous and mixed forests
- Urals montane tundra and taiga
- Western European broadleaf forests
- Yamalagydanskaja tundra
- Outside data coverage

2.1.6 Map of European Ecological Regions, courtesy of the European Environment Agency (EEA). Copyright EEA, Copenhagen, 2007. http://www.eea.europa.eu

The major landscape components, in this case the slope and floodplain, of the AA have been delineated (red).

2.2.2 AA Map 1. Reproduced by permission of Ordnance Survey on behalf of HMSO. © Crown copyright 2008. All rights reserved. Ordnance Survey Licence number 100047942.

Transect lines (yellow) have been added to the map.

2.2.3 AA Map 2. Reproduced by permission of Ordnance Survey on behalf of HMSO. © Crown copyright 2008. All rights reserved. Ordnance Survey Licence number 100047942.

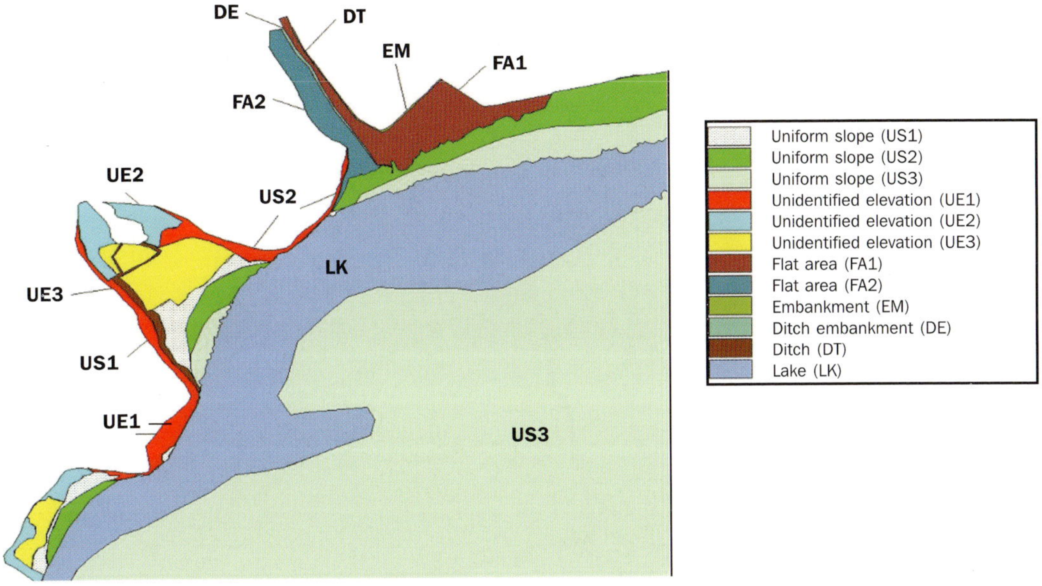

2.2.7 Example map of lake marginal wetland (LMW) HGMUs using a colour-coded key.

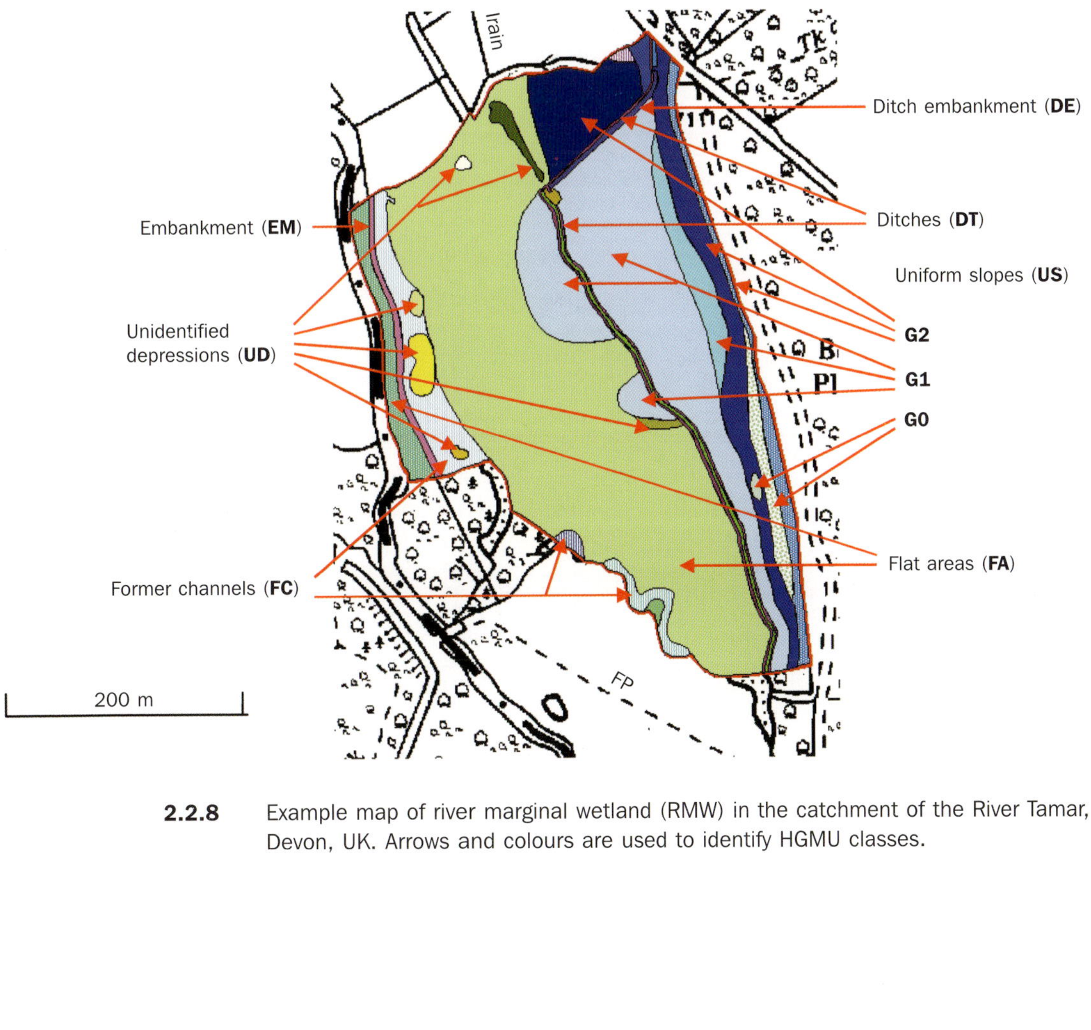

2.2.8 Example map of river marginal wetland (RMW) in the catchment of the River Tamar, Devon, UK. Arrows and colours are used to identify HGMU classes.

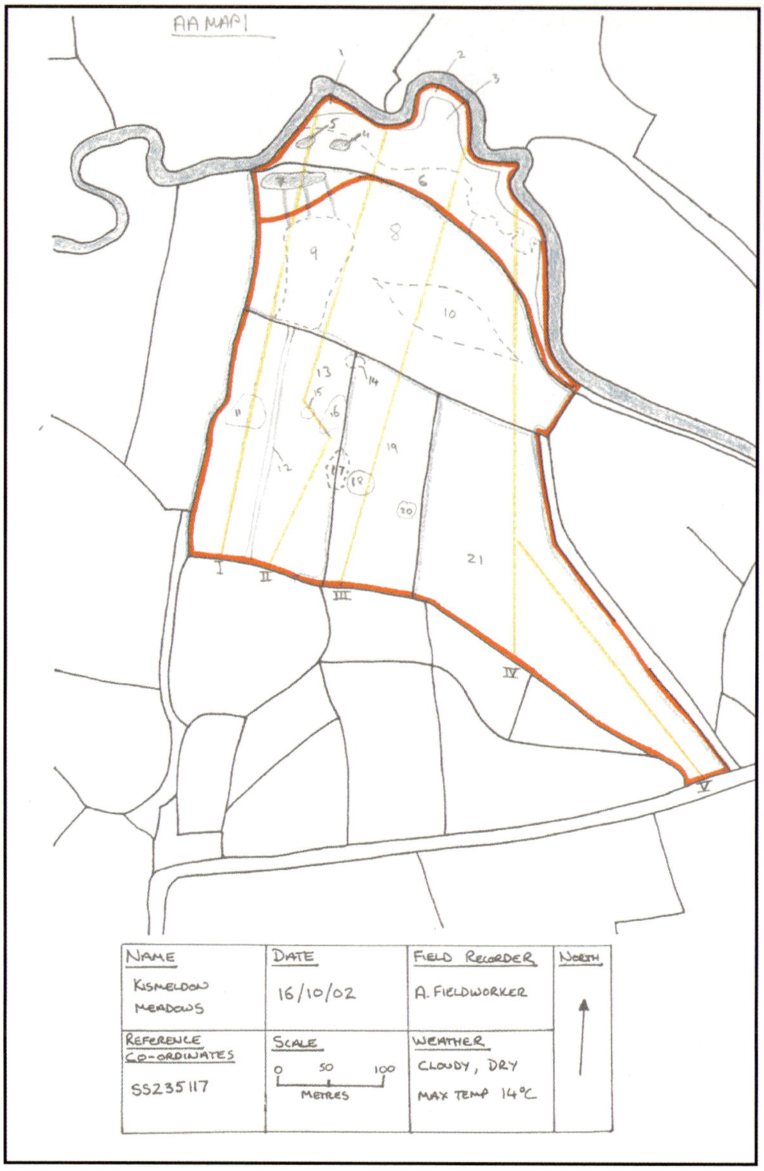

HGMU No.	Landscape component	Gradient code	Geomorphic prefix code	Geomorphic code
1	F	**G0**	**E**	**LV**
2	F	**G0**	**E**	**LV**
3	F	**G0**	**I**	**FA**
4	F	**G1**	**D**	**LK**
5	F	**G1**	**D**	**LK**
6	F	**G2**	**I**	**FA**
7	F	**G1**	**D**	**LK**
8	S	**G1**	**I**	**US**
9	S	**G1**	**I**	**US**
10	S	**G1**	**I**	**US**
11	S	**G1**	**E**	**PD**
12	S	**G1**	**D**	**DT**
13	S	**G1**	**I**	**US**
14	S	**G1**	**I**	**US**
15	S	**G1**	**E**	**PD**
16	S	**G1**	**E**	**PD**
17	S	**G1**	**I**	**US**
18	S	**G1**	**E**	**PD**
19	S	**G1**	**I**	**US**
20	S	**G1**	**E**	**PD**
21	S	**G1**	**I**	**US**

2.2.9 Example AA map showing transect lines (yellow). Slope and floodplain areas are delineated in red. HGMUs are drawn on the map and numbered as described in the table. Reproduced by permission of Ordnance Survey on behalf of HMSO.

(a)

(b)

2.2.10 (a) Peaty soil profile showing greyish pallid horizon above ochreous subsoil with grey and reddish mottles, applying to the grey and reddened layers beneath; (b) soil profile horizons and Munsell Colour Chart.

All HGMUs have been identified and delineated, and their boundaries confirmed through the inspection of soil profiles, the locations of which have been marked by a cross on the map. In this case the soil information did not indicate any need for further subdivisions of any of the HGMUs.

2.2.11 Final AA Map. Reproduced by permission of Ordnance Survey on behalf of HMSO.

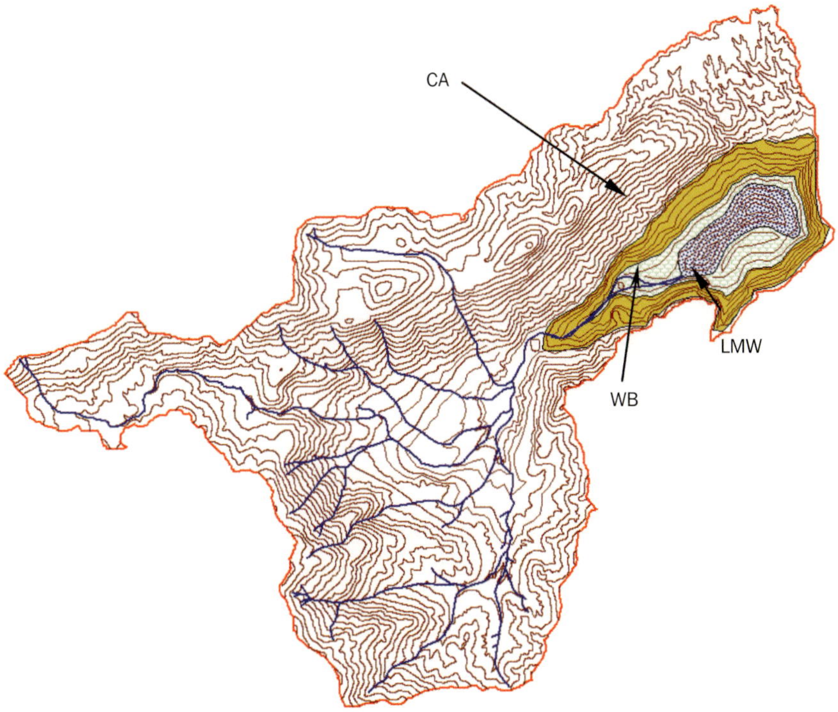

3.1.14 Topographic map with elevation contours, showing the boundaries of the WB, LMW and CA.

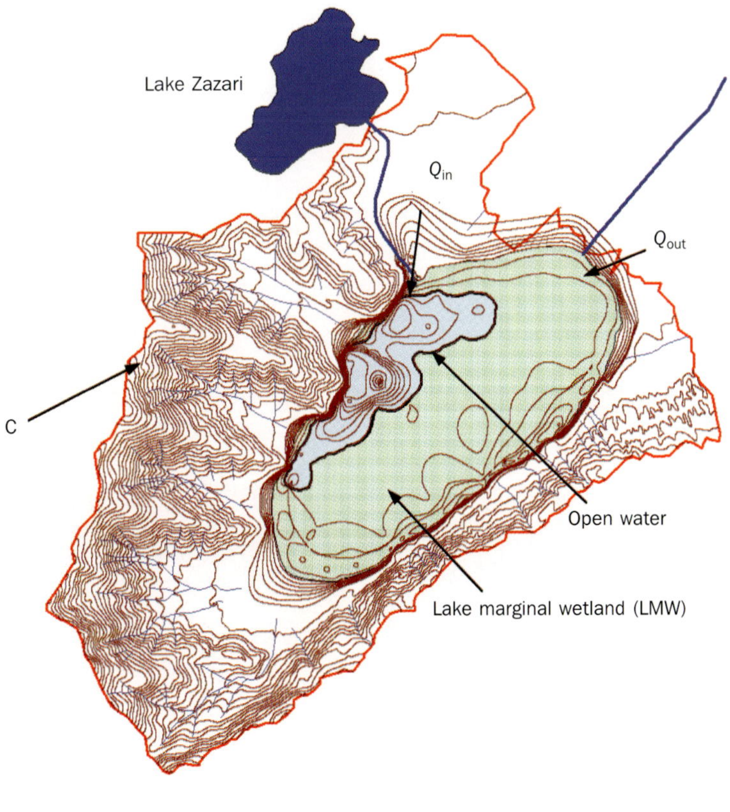

3.1.16 Topographic map which shows the boundaries of the WB and LMW. Q = inflow and outflow.

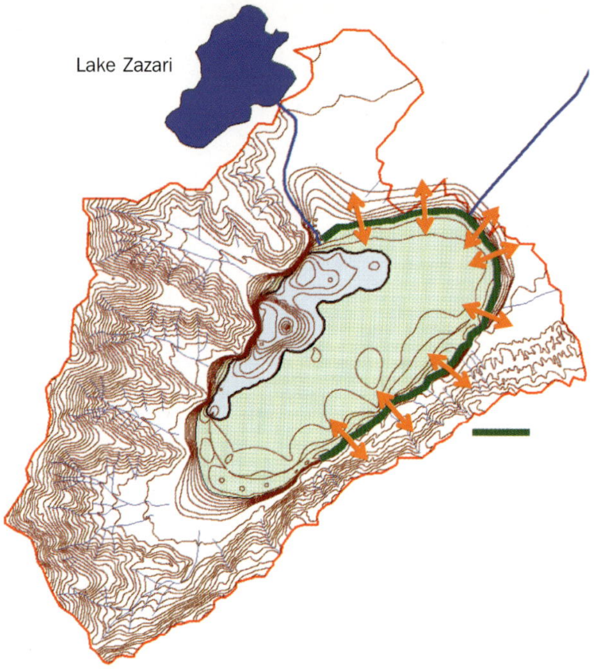

3.1.38 The section of the wet perimeter through which the recharge takes place.

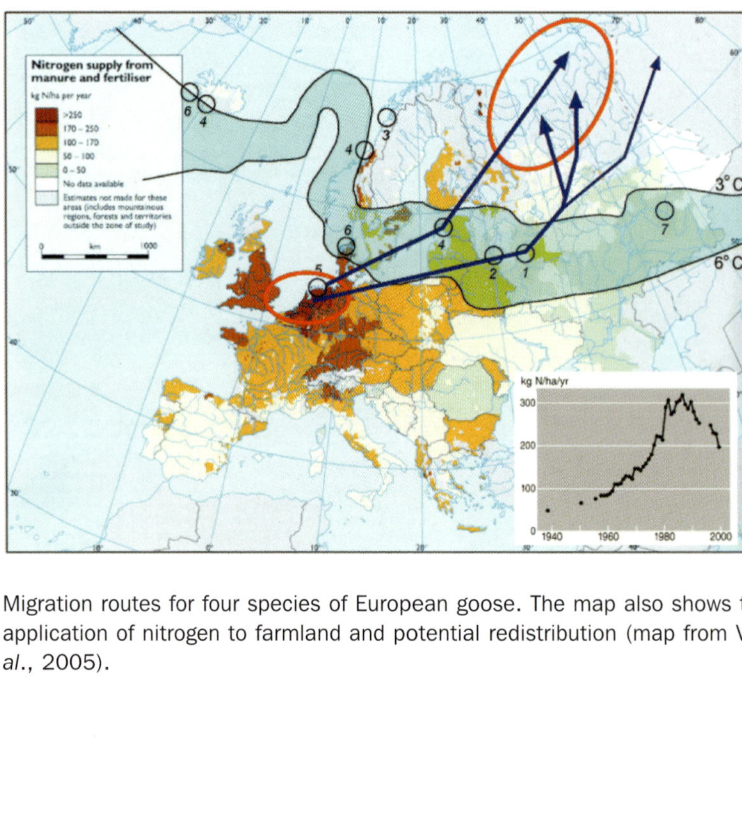

3.3.4 Migration routes for four species of European goose. The map also shows the application of nitrogen to farmland and potential redistribution (map from Van Eerden *et al.*, 2005).

Significance – rationale for assessment: Improvement of water quality in water bodies receiving surface or subsurface water from the wetland by retention of nutrients bound to sediment retained within the wetland.

Controlling variables

CV1 Nutrient input
CV2 Flooding regime
CV3 Particulate nutrient load
CV4 Soil/sediment deposition
CV5 Soil/sediment erosion

Answer all the questions below for *each* HGMU; unless otherwise instructed continue from one question to the next. On the appraisal sheet circle the correct answer. The location of the relevant information is given in brackets after each question. Use a new appraisal sheet for each HGMU.

If there are no significant nutrient inputs identified for the HGMU, or if the nutrient input is uncertain (nutrient input codes **Nut 2** or **Nut ?**), you may wish to assess the potential of the HGMU to perform this process. To do this conduct the assessment as normal but answer questions only in ***CV2***, ***CV3***, ***CV4*** and ***CV5***. Establish an assessment outcome as normal, but ignore the need for positive answers for ***CV1***. The rationale statement can be used as normal but ignore reference to receipt of nutrient inputs.

CV1 *Nutrient input*

Q1. Is the HGMU in receipt of an indirect nutrient input, i.e. codes **N_i 1/x**, **P_i 1/x** or **Nut_i 1/x**? *(Recording Sheet 2c)*
*[If YES continue with **Q2.** If NO, then continuing with this assessment will give information only about an HGMU's **potential** to perform this process, not the actual performance of this process. To do this go to **CV2**]*

Q2. What is the source of the nutrient input received by the HGMU? Circle all those applicable:

a. Contact with water containing > 1 mg l^{-1} ammonium–nitrogen *(Recording Sheet 2b, H1 Action, Q20)*
b. Contact with water containing > 1 mg l^{-1} nitrate–nitrogen *(Recording Sheet 2b, H1 Action, Q19)*
c. Contact with water with a total nitrogen concentration of > 2.2 mg l^{-1} *(Recording Sheet 2b, H1 Action, Q21)*
d. Contact with water with a phosphate–phosphorus concentration of > 0.1 mg l^{-1} *(Recording Sheet 2b, H1 Action, Q22)*
e. Groundwater discharge *(Recording Sheet 2b, H2 Action, Q1–3)*
f. Surface run-off or subsurface run-off *(Recording Sheet 2b, H2 Action, Q4–6)*

CV2 *Flooding regime*

Q1. Has the HGMU been assigned a hydrological code indicating that overbank flooding occurs (**F4**), i.e. **H4**, **H7**, **H9**, **H10**, **H12**, **H13**, **H14** or **H15**? *(Recording Sheet 2a, F)*

Q2. Has the HGMU been assigned a hydrological code indicating that there is a (sub)surface run-off input, but *no* (sub)surface run-off output, i.e. hydrological codes **H2**, **H5**, **H13**, **H14**? *(Recording Sheet 2a, F)*

Only answer ***Q3*** and ***Q4*** if you have answered YES to ***Q1*** in ***CV2***.

Q3. Are flow velocities in the HGMU during floods below 0.3 ms^{-1}? (*Recording Sheet 3, B.4.6*)

Q4. If the HGMU has only one (potential) outlet for flood water, is the width of the outlet less than 10 % of the average width of the area usually flooded in the HGMU i.e. is the outflow from the HGMU restricted? (*Recording Sheet 3, B.4.21, AA Map 3*)

CV3 *Particulate nutrient load*

Only answer ***Q1–Q4*** if you have answered YES to ***Q1*** in ***CV2***. Otherwise continue with ***Q5*** (read the note before ***Q5*** first).

Q1. Has the river that floods the HGMU (within a distance upstream of the site of less than 20 % of the total upstream river length) been observed to carry high concentrations of suspended sediment during peak discharges? *(Recording Sheet 3, B.5.3)*

Q2. Have any of the activities listed below taken place in more than 10 % of the upstream catchment directly along the river channel during the last 3 years? *(Recording Sheet 1, C.3.24.2 (answers a–d))*

- Logging or afforestation
- Urban development or road building
- Mining
- Major river channel improvements

Q3. Is over 10 % of the upstream catchment occupied by one or more of the types of area listed below? *(Recording Sheet 1)*:

- Mountainous areas (*C.3.24.1 (answer a)*)
- Sparsely vegetated areas (*C.3.24.1 (answer b)*)
- Areas where mean annual potential evapotranspiration exceeds mean total annual precipitation (*C.3.1.1, C.3.1.3i*) [*Compare the average monthly rainfall to the monthly average evapotranspiration records*]

Q4. Does the river or area directly upstream of the site (comprising less than 20 % of the total river length or catchment area) have a recent history of migrating channels (during the last 20 years)? (*Recording Sheet 1, C.3.24.5*)

Only answer ***Q5–Q9*** if you have answered YES to ***Q2*** in ***CV2***. Otherwise continue with ***CV4***.

Q5. Are there indications of erosion in the area immediately upslope (in AA and CA) of the HGMU? (*Recording Sheet 3, B.5.1*)

Q6. Has surface run-off from the area immediately upslope of the HGMU (in AA and CA) been observed to carry high concentrations of suspended sediment? (*Recording Sheet 3, B.5.6*)

Q7. Has the area upslope from the HGMU been assigned a gradient code **G2** or **G3** indicating that it is moderate or steep (> 5°)? *(Recording Sheet 2a, C.2)*

Q8. Have any of the activities listed below taken place across more than 50 % of the area immediately upslope (in AA and CA), during the last three years? (*Recording Sheet 1, C.3.24.2 (answers a–d)*)

- Logging or afforestation
- Urban development or road building
- Mining
- Major river channel improvements

Q9. Is over 50 % of the area immediately upslope (in AA and CA) occupied by one or more of the types of land listed below? *(Recording Sheet 1)*

- Mountainous areas (*C.3.24.1 (answer a)*)
- Sparsely vegetated areas (*C.3.24.1 (answer b)*)
- Areas where annual potential evapotranspiration exceeds annual precipitation (*C.3.1.1, C.3.1.3i*) [*Compare the average monthly rainfall to the monthly average evapotranspiration records*]
- Permafrost areas, where soils are frozen throughout the year ('pergelic' soils) (*C.3.1.5 (answer a)*)

CV4 *Soil/sediment deposition*

Q1. Is there evidence of recently deposited mineral or organic material at the soil surface in the HGMU? (*Recording Sheet 3, B.5.4*)

Q2. Is there evidence of recently deposited mineral material on plant surfaces in the HGMU? (*Recording Sheet 3, B.5.5*)

Q3. Is the soil in the upper 10 cm of the soil profile in the HGMU described as fine? *(Recording Sheet 2a, C.3.7)*

Q4. Is the HGMU mostly covered with 'rough' vegetation such as dense reed, bushes, shrubs or forest, which will slow-down (flood) waters considerably? *(Recording Sheet 3, B.4.20)*

Q5. Has the HGMU been assigned a geomorphic prefix code **D**, indicating that it is a depressional feature, i.e. **BS**, **CO**, **OX**, **FC**, **LK**, **DT**, **CN**, **AD** or **UD**? *(Recording Sheet 2a, C.2)*

Q6. Is the microtopography in the HGMU described as significant? *(Recording Sheet 3, A.3)*

CV5 *Soil/sediment erosion*

Q1. Is the vegetation classed as unvegetated or sparsely vegetated, i.e. vegetation code **Ux**? (*Recording Sheet 3, D*)

Q2. Are there any indicators of erosion within the HGMU? This could be scouring in units not adjacent to the river, or steep exposed banks in units adjacent to the river (e.g. levees). (*Recording Sheet 3, B.5.2*)

Only answer ***Q3*** if you have answered YES to ***Q1*** in ***CV2***.

Q3. Is the HGMU adjacent to a river that is known to shift its course locally, and is situated on an outer bend? (*Recording Sheet 3, B.5.8*)

Assessment of the process

Compare the results recorded for each HGMU on the appraisal sheets with the possible outcomes below, and assign an appropriate assessment code. Record this on the appraisal sheet. The syntax codes used are described in the introduction to Section 3.

Look-up tables

1. The process is definitely being performed

If the answers to the questions for this process agree with any of the combinations in Look-up Table 1, then the HGMU is definitely performing the process of retention of particulate nutrients. It is difficult to quantify the process of retention of particulate nutrients, because the process is temporally and spatially highly variable. Therefore only a qualitative assessment of retention of particulate nutrients is given.

2. The process is not (significantly) being performed

If the answers to the questions for this process agree with any of the combinations in Look-up Table 2, then the HGMU is not significantly performing the process of retention of particulate nutrients.

3. The process is being performed, but there are constraining factors or uncertainties

If Look-up Tables 1 and 2 did not lead to an assessment, the process of retention of particulate nutrients is performed, but there are constraining factors or uncertainties. In Look-up Table 3 you can find out which constraining factors affect the process, or which factors cause the uncertainty.

4. The process is definitely not being performed

This statement is not applicable to this process.

This completes the functional assessment for the process of retention of particulate N and P. The output codes can be recorded on the Functional Assessment Table at the end of Section 3.2.A.6.

3.2.A.5 *Retention of Particulate Nutrients (N and P) Appraisal Sheet*

HGMU code: ____________ **AA name:** ____________

CV1 Nutrient input						
Q1	a		b		c	
Q2	a	b	c	d	e	f

CV2 Flooding regime			
Q1	Y	N	U/NA
Q2	Y	N	U/NA
Q3	Y	N	U/NA
Q4	Y	N	U/NA

CV3 Particulate nutrient load			
Q1	Y	N	U/NA
Q2	Y	N	U/NA
Q3	Y	N	U/NA
Q4	Y	N	U/NA
Q5	Y	N	U/NA
Q6	Y	N	U/NA
Q7	Y	N	U/NA
Q8	Y	N	U/NA
Q9	Y	N	U/NA

CV4 Soil/sediment deposition			
Q1	Y	N	U/NA
Q2	Y	N	U/NA
Q3	Y	N	U/NA
Q4	Y	N	U/NA
Q5	Y	N	U/NA
Q6	Y	N	U/NA

CV5 **Soil/sediment erosion**			
Q1	Y	N	U/NA
Q2	Y	N	U/NA
Q3	Y	N	U/NA

HGMU process assessment code: ____________

3.2.A.5 Look-up Table 1

<table>
<tr><th>CV1</th><th>CV2</th><th>CV3</th><th>CV4</th><th>CV5</th><th>Rationale</th><th>Code</th></tr>
<tr><td>1 (2a|2b|2c|2d)</td><td>1 (3|4)</td><td>1|2|3|4</td><td>1|2|3|4|5|6</td><td>#(1 2 3)</td><td>The HGMU receives a nutrient input via overbank flooding. River water is likely to be retained at low flow velocities and it is likely that the in-flowing water contains a high particulate load. Because of the high concentration of nutrients in the river water it is expected that nutrients will be adsorbed to the particulates. There is evidence that the particulates are retained and there is no evidence of erosion and loss of particulate-bound nutrients, indicating retention may be long-term.
NB: The amount of nutrients retained depends upon the flood water residence time and the depth of the floodwater.</td><td>1</td></tr>
<tr><td>1 2f</td><td>2</td><td>5|6|7|8|9</td><td>1|2|3|4|5|6</td><td>#(1 2)</td><td>The HGMU receives a nutrient input via run-off. It is likely that run-off water contains a high particulate load. Because of the high concentration of nutrients in the water it is expected that nutrients will be adsorbed to the particulates. There is evidence that the particulates are retained and there is no evidence of erosion and loss of particulate bound nutrients.
NB: The amount of nutrients retained depends upon the run-off residence time and the depth of the floodwater.</td><td>1</td></tr>
<tr><td>1 (2a|2b|2c|2d) 2f</td><td>1 2 (3|4)</td><td>(1|2|3|4) (5|6|7|8|9)</td><td>1|2|3|4|5|6</td><td>#(1 2 3)</td><td>The HGMU receives a nutrient input via both floodwater and run-off water. River floodwater and run-off water are likely to be retained at low flow velocities and it is likely that the in-flowing water contains a high particulate load. Because of the high concentration of nutrients in the water it is expected that nutrients will be adsorbed to the particulates. There is evidence that the particulates are retained and there is no evidence of erosion and loss of particulate-bound nutrients, indicating retention may be long-term.
NB: The amount of nutrients retained depends upon the hydraulic residence time and the depth of the floodwater.</td><td>1</td></tr>
</table>

3.2.A.5 Look-up Table 2

CV1	CV2	CV3	CV4	CV5	Rationale	Code
# (1 2a 2b 2c 2d 2f) | 1b	*	*	*	*	The HGMU does not receive a nutrient input via overbank flooding from the river or via run-off from upslope areas, or it is uncertain whether the HGMU receives a nutrient input.	**2**
2e	*	*	*	*	The HGMU receives an indirect nutrient input only via groundwater discharge. Retention of particulate nutrients does not take place with this type of nutrient input.	**2**
*	#(1 2)	*	*	*	Overbank flooding or (sub)surface run-off from upslope areas does not occur. Therefore there is no transportation mechanism for particulate material to reach the HGMU.	**2**
*	1	#(1 2 3 4)	*	*	The HGMU receives a nutrient input via overbank flooding from the river, but in-flowing waters are not likely to contain a high particulate load.	**2**
*	2	#(5 6 7 8 9)	*	*	The HGMU receives a nutrient input via run-off from upslope areas. In-flowing waters are not likely to contain a high particulate load.	**2**
*	*	*	#(1 2 3 4 5 6)	*	There is no evidence that particulates are retained.	**2**

3.2.A.5 Look-up Table 3

CV1	CV2	CV3	CV4	CV5	Rationale	Code
*	*	*	*	1|2|3	There is some evidence of erosion and loss of particulate-bound nutrients, indicating retention may be short-term.	**1/x**

3.2.A.6 Functional assessment: Nutrient retention

The purpose of this section is to determine whether the Nutrient Retention function is occurring and if so, to what degree. Follow the method below for *each* HGMU. On the Functional Assessment Table at the end of this section, record the functional outcome for each HGMU within your AA.

Nutrient retention is definitely being performed

The function 'nutrient retention' is definitely being performed by the HGMU if one or more of the component processes of this function are occurring, i.e. if:

Any of the process assessment codes = **1/n** (results from Look-up Table 1 (definitely performed) and/or **1/x** (performed but with uncertainties or constraints)

The statement *nutrient retention is definitely being performed* is a broad statement. The statement does not include any quantitative or semi-quantitative information. Not all processes contribute equally to the function. Therefore if only one process with a low maximum retention rate is contributing to the function, a smaller amount of nutrient will be retained than if several processes or one process with a high maximum retention rate are contributing to the function. To estimate the degree of performance of the function, the contribution of the processes to the function has to be 'calculated'. Guidance for this, which uses the quantitative information, is given below in *summing of processes to determine a function.*

Nutrient retention is not significantly being performed

The function 'nutrient retention' is not significantly being performed by an HGMU if *none* of the component processes of this function are occurring, i.e. if:

All process assessment codes = **2**

Summing of processes to determine a function

In order to be able to take account of the outcomes of all the different processes, both quantified *and* non-quantified processes are assigned an *estimated* quantitative value. A separate functional assessment is performed for nitrogen (N) retention and for phosphorus (P) retention as N and P are involved in different processes and they react differently under varied circumstances.

The estimated value of a quantified process consists of the (arithmetic) mean of two end-points of the quantitative range given for the assessment of that process. For instance, where the assessment outcome code is **1/a**, the average value for 'retention by storage of nutrients in soil organic matter' is 30 kg ha^{-1} a^{-1} for nitrogen ((10 + 50)/2) and 6 kg ha^{-1} a^{-1} for phosphorus ((2 + 10)/2).

The values for the non-quantified processes or process outcomes (outcomes assessed with code **1/x** or code **1**) have all been estimated. The values for processes with assessment outcome code **1** are based on expert judgement: the importance of these processes is scaled relative to the rates of the quantified processes. Values of processes with code **1/x** have been estimated as 50 % of the lowest quantitative value given to any of the codes **1/c**, **1/b** or **1**. This may underestimate the retention rate in a number of cases, but it is intended as a conservative quantitative estimate of the process.

In the nitrogen and phosphorus tables in the following Section (Tables 3.2.1–3.2.3), the average (estimated) values are given in the light grey rows. In the case of quantified processes the minimum and maximum rates are also given. All estimated values (i.e. all values that could not be found in the assessment tables of the process) are indicated with * in the table.

Functional Assessment Tables: 3.2.A Nutrient retention

Tables 3.2.1 and 3.2.2 on pages 371 and 372 enable calculation of the total value for the function 'nutrient retention' for nitrogen and phosphorus, respectively. The tables can be used for one HGMU only. Therefore, it is necessary to make copies of these tables for each HMGU before you start the assessment.

How to use the tables is explained step by step below.

Step 1

On the appraisal sheets, look up which codes have been assigned to the following processes:

- 3.2.A.1 Long-term retention of nutrients through plant uptake (N and P)
- 3.2.A.2 Storage of nutrients in soil organic matter (N and P)
- 3.2.A.3 Adsorption of nitrogen as ammonium (N)
- 3.2.A.4 Adsorption and precipitation of phosphorus in the soil (P)
- 3.2.A.5 Retention of particulate nutrients (N and P)

Step 2

For each process, look up the estimated average values for nitrogen and phosphorus accompanying your codes in the tables (light grey).
Place a circle around these values.

Step 3

In the last row of the nitrogen and the phosphorus tables (estimated values NITROGEN / estimated values PHOSPHORUS) write the circled values. For instance if you have assigned a code **1/a** to the process 'storage of nutrients in soil organic matter', encircle **30 kg N ha^{-1} a^{-1}** and **6 kg P ha^{-1} a^{-1}** in the third row of the column 'storage in soil organic matter' in the nitrogen table and phosphorus table, respectively. Next, write down **30** and **6** on the dotted lines in the last rows of these columns. Only those processes relevant to retention of either nitrogen or phosphorus are shown.

Step 4

After you have written down four values (one value for each process) in the last row, add up these values and record the total in the appropriate light grey box below the table. The values in the light grey boxes are the final outcome for the function 'nutrient retention' in the HGMU under consideration.

Step 5

Repeat this action for all HGMUs.

Notes

1. The summed value in the light grey boxes is derived from a summation of estimated values and should therefore be seen as a very rough approximation. It can be used as an approximate indication of performance of the function. No calculations should be performed with the values in contexts other than this assessment.
2. The information present in the other rows in the tables (min and max) is needed if you want to (re)calculate the values yourself.
   - 'Min' = the lowest rate of the range accompanying the code
   - 'Max' = the highest rate of the range accompanying the code
   - 'Average' = the average of the lowest and highest rates of the range

   Values in the table marked with * are not derived from the assessment charts, but are estimated from expert knowledge. For more information, see text above.

Step 6

To obtain an indication of the relative degree of performance of the function in the HGMU under consideration, compare the total values for nitrogen retention and phosphorus retention in your HGMU with the categories Table 3.2.3 below.

Table 3.2.1 Nitrogen retention

NITROGEN		Plant uptake (N)	Storage in soil organic matter (N)	Adsorption of N as ammonium	Retention of particulate nutrients (N)
Code **1/a**	Min	10* kg N ha^{-1} a^{-1}	10 kg N ha^{-1} a^{-1}		
	Max	80 kg N ha^{-1} a^{-1}	50 kg N ha^{-1} a^{-1}		
	Average	45 kg N ha^{-1} a^{-1}	30 kg N ha^{-1} a^{-1}		
Code **1/b**	Min	10* kg N ha^{-1} a^{-1}	5 kg N ha^{-1} a^{-1}		
	Max	50 kg N ha^{-1} a^{-1}	25 kg N ha^{-1} a^{-1}		
	Average	30 kg N ha^{-1} a^{-1}	15 kg N ha^{-1} a^{-1}		
Code **1/c**	Min	5* kg N ha^{-1} a^{-1}			
	Max	25 kg N ha^{-1} a^{-1}			
	Average	15 kg N ha^{-1} a^{-1}			
Code **1/d**	Min				
	Max				
	Average				
Code **1/x**	Estimated average	7.5* kg N ha^{-1} a^{-1}	7.5* kg N ha^{-1} a^{-1}	15* kg N ha^{-1} a^{-1}	25* kg N ha^{-1} a^{-1}
Code **1**	Estimated average			30 kg N ha^{-1} a^{-1}	50 kg N ha^{-1} a^{-1}
Code **2**	Average	0 kg N ha^{-1} a^{-1}	0 kg N ha^{-1} a^{-1}	0 kg N ha^{-1} a^{-1}	0 kg N ha^{-1} a^{-1}
Estimated values NITROGEN		________kg N ha^{-1} a^{-1}	________kg N ha^{-1} a^{-1}	________kg N ha^{-1} a^{-1}	________kg N ha^{-1} a^{-1}

Sum of values NITROGEN

⇓

____________ kg N ha^{-1} a^{-1}

Table 3.2.2 Phosphorus retention

PHOSPHORUS		Plant uptake (P)	Storage in soil organic matter (P)	Adsorption of P	Retention of particulate nutrients (P)
Code **1/a**	Min	1* kg P ha^{-1} a^{-1}	2 kg P ha^{-1} a^{-1}		
	Max	5 kg P ha^{-1} a^{-1}	10 kg P ha^{-1} a^{-1}		
	Average	3 kg P ha^{-1} a^{-1}	6 kg P ha^{-1} a^{-1}		
Code **1/b**	Min	1* kg P ha^{-1} a^{-1}	1 kg P ha^{-1} a^{-1}		
	Max	3 kg P ha^{-1} a^{-1}	5 kg P ha^{-1} a^{-1}		
	Average	2 kg P ha^{-1} a^{-1}	3 kg P ha^{-1} a^{-1}		
Code **1/c**	Min	0.5* kg P ha^{-1} a^{-1}			
	Max	1.5 kg P ha^{-1} a^{-1}			
	Average	1 kg P ha^{-1} a^{-1}			
Code **1/d**	Min				
	Max				
	Average				
Code **1/x**	Estimated average	0.5* kg P ha^{-1} a^{-1}	1.5* kg P ha^{-1} a^{-1}	2.5* kg P ha^{-1} a^{-1}	5* kg P ha^{-1} a^{-1}
Code **1**	Estimated average			5* kg P ha^{-1} a^{-1}	10* kg P ha^{-1} a^{-1}
Code **2**	Average	0 kg P ha^{-1} a^{-1}	0 kg P ha^{-1} a^{-1}	0 kg P ha^{-1} a^{-1}	0 kg P ha^{-1} a^{-1}
Estimated values PHOSPHORUS		________ kg P ha^{-1} a^{-1}	________ kg P ha^{-1} a^{-1}	________ kg P ha^{-1} a^{-1}	________ kg P ha^{-1} a^{-1}

Sum of values PHOSPHORUS

⇓

______________ kg P ha^{-1} a^{-1}

Table 3.2.3 Nitrogen/phosphorus retention comparison

Degree of performance	Nitrogen	Phosphorus
Not performed	0 kg ha^{-1} a^{-1}	0 kg ha^{-1} a^{-1}
Low performance	> 0–<15 kg ha^{-1} a^{-1}	> 0–<2 kg ha^{-1} a^{-1}
Intermediate performance	15–<100 kg ha^{-1} a^{-1}	2–<10 kg ha^{-1} a^{-1}
High performance	100–155 kg ha^{-1} a^{-1}	10–24 kg ha^{-1} a^{-1}

When all contributing processes are assigned a code **2**, the function is *not performed* and no nitrogen or phosphorus is retained in the HGMU. Retention of less than 15 kg N ha^{-1} a^{-1} or less than 2 kg P ha^{-1} a^{-1}, indicates a *low performance* of the function. The function is *performed to an intermediate level* when values are between 15 and 100 kg N ha^{-1} a^{-1} or between 2 and 10 kg P ha^{-1} a^{-1}. The function is considered as having a *high performance* when values of retention exceed 100 kg N ha^{-1} a^{-1} or 10 kg P ha^{-1} a^{-1}. Theoretically the Nutrient Retention function is performed to the maximum level when all quantified processes have been assigned a code **1/a** and all non-quantified processes a code **1**. Adding up the accompanying values for these codes leads to a maximum retention value of 155 kg ha^{-1} a^{-1} for nitrogen and 24 kg ha^{-1} a^{-1} for phosphorus.

Step 7

Make a note of the degree of performance (not performed, low, intermediate or high) of the function 'nutrient retention' in your HGMU: for transfer to the Functional Analysis Summary table in Section 4.0.

The importance of the performance of the function 'nutrient retention' should be considered in combination with the performance of the function 'nutrient export'. Both functions contribute to the same 'goal', purification of water from nutrients. Whether the retention and export of nutrients in the HGMU contributes to improved water quality in the wetland or more widely downstream in the catchment depends mainly on hydrological factors, such as the rate of water flow and the residence time in the HGMU. Spatial interaction, such as the extent and duration of exchange of water between the HGMU and the wetland or water body also play a role. Site-specific information is needed to make exact predictions. However, by comparing function outcomes of different HGMUs, the user can get insight into which HGMU types contribute most to good water quality.

The outcomes above should be interpreted with care. A high performance rate does not automatically lead to best results for water quality. One should also take into account the input of nutrients. Retention in HGMUs with a direct nutrient input (i.e. HGMUs in agricultural use) may not result in better water quality, as the input rate might exceed the retention rate. Retention of nutrients in HGMUs with an indirect input of nutrients normally does contribute to a better water quality, as these HGMUs capture nutrients that have been released from other areas (such as diffuse agricultural pollution or sewage facilities) and would otherwise lead to eutrophication.

This completes the functional assessment for nutrient retention. The output codes can be recorded on the Functional Assessment Table in Section 4.0 and used to compare functions among HGMUs across the wetland. If you wish to assess all functions, move on to the next step of the FAPs: Nutrient Export.

When all the HGMUs in the assessment area have been assessed, a general pattern of nutrient retention across the whole wetland will be evident. By manipulation of your answers to the controlling variable (CV) questions, you have the capacity to explore the probable outcomes of any management options under consideration and relate them to specific functions. The electronic version of the FAPs can do this more quickly and easily.

Functional Assessment Table: 3.2.A Nutrient retention

HGMU code	Nutrient retention

AA name: ______________________

3.2.B Nutrient export (water quality function)

3.2.B.1.1 *Gaseous export of N (denitrification) (process)*

Further reading and references

Armentano, T.V. and Verhoeven, J.T.A. (1990). Biogeochemical cycles: global. In: Patten, B.C. (ed.), *Wetlands and Shallow Continental Water Bodies, Volume 1*, SPB Academic, The Hague, The Netherlands, 281–311.

Baker, C.J. and Maltby, E. (1995). Nitrate removal by river marginal wetlands: factors affecting the provision of a suitable denitrification environment. In: Hughes, J.M.R. and Heathwaite, A.L. (eds) *Hydrology and Hydrochemistry of British Wetlands*, Wiley, Chichester, UK, 291–313.

Blackwell, M.S.A. and Maltby, E. (1998). Identification and field testing of zones of enhanced denitrification (ZEDs). In: McComb, A.J. and Davis, J.A. (eds), *Wetlands for the Future*, contributions from INTECOL's V International Wetlands Conference, Gleneagles Publishing, Adelaide, Australia, 443–456.

Cosandy, A.C., Maitre, V. and Guenat, C. (2003). Temporal denitrification patterns in different horizons of two riparian soils, *European Journal of Soil Science* **54**, 5–37.

Fennessy, M.S. and Cronk, J.K. (2001). The effectiveness and restoration potential of riparian ecotones for the management of non-point source pollution, particularly nitrate, *Critical Reviews in Environmental Science and Technology*, **27**(4), 285–317.

Flynn, N., Maltby, E. and Gardner, P.J. (1998). Nitrogen interactions between floodwater and floodplain soils. In: McComb, A.J. and Davis, J.A. (eds), *Wetlands for the Future*, contributions from INTECOL's V International Wetlands Conference, Gleneagles Publishing, Adelaide, Australia, 433–442.

Flynn, N., Gardner, P.J. and Maltby, E. (1999). The measurement and analysis of denitrification rates obtained using soil columns from river marginal wetlands, *Soil Use and Management* **15**, 150–156.

Gardner, P.J., Flynn, N. and Maltby, E. (2000). A temperature correction procedure for field-based soil denitrification experiments, *Soil Use and Management* **16**, 260–263.

Haag, D. and Kaupenjohann, M. (2001). Landscape fate of nitrate fluxes and emissions in Central Europe; a critical review of concepts, data, and models for transport and retention, *Agriculture, Ecosystems and Environment* **86**, 1–21.

Hefting, M.M. and Klein, J.J.M. de (1998). Nitrogen removal in buffer strips along a lowland stream in the Netherlands: a pilot study, *Environmental Pollution* **102**(S1), 521–526.

Hefting, M.M., Verhoeven, J.T.A., Bienkowksi, P., Clement, J.C., Dowrick, D., Guenat, C., Nin, E. and Sorana, T. (2005). The role of vegetation and litter in the nitrogen dynamics of riparian buffer zones in Europe, *Ecological Engineering* **24**, 465–482.

IPCC (2001). *Climate Change 2001: Impacts, Adaptation, and Vulnerability*, Cambridge University Press, Cambridge, UK.

Johnston, C.A. (1991). Sediment and nutrient retention by freshwater wetlands: effects on surface water quality, *Critical Reviews in Environmental Science and Technology* **5,6**(21), 491–565.

Klein, C.A.M. de and Logtestijn, R. van (1994). Denitrification in the topsoil of managed grasslands in The Netherlands in relation to soil type and fertilizer level, *Plant and Soil* **163**, 33–44.

Knowles, R. (1981). Denitrification, *Ecology Bulletin* **33**, 315–329.

Knowles, R. (1982). Denitrification, *Microbiological Reviews* **46**(1), 43–70.

Koerselman, W. (1989). *Hydrology and Nutrient Budgets of Fens in an Agricultural Landscape*, PhD Dissertation, Utrecht University, The Netherlands.

Luijn, F. (1997). *Nitrogen Removal by Denitrification in the Sediments of a Shallow Lake*, PhD Dissertation, Landbouw Universiteit Wageningen, The Netherlands.

Meuleman, A.F.M. (1993). *Performance of Treatment Wetlands*, PhD Dissertation, Utrecht University, The Netherlands.

Mitsch, W.J. (1994). *Global Wetlands: Old World and New*, Elsevier Amsterdam, The Netherlands.

Mitsch, W.J. and Gosselink, J.G. (2007). *Wetlands*, 4th edn, Wiley, New York.

Morris, J.T. (1991). Effects of nitrogen loading on wetland ecosystems with particular reference to atmospheric deposition, *Annual Review of Ecological Systems* **22**, 57–279.

Oorschot, M.M.P. van (1996). *Effects of the Vegetation on Carbon, Nitrogen and Phosphorus Dynamics in English and French Riverine Grasslands*, PhD Dissertation, Utrecht University, The Netherlands.

Olde Venterink, H. (2000). *Nitrogen, Phosphorus and Potassium Flows Controlling Plant Productivity and Species Richness*, PhD Dissertation, Utrecht University, The Netherlands.

Orleans, A.B.M., Mugge, F.L.T., van der Meij, T., ter Keurs, W.J. (1994). *Minder nutriënten in het oppervlaktewater door bufferstroken: een literatuuranalyse*, Milieubiologie Rijksunversiteit, Leiden, The Netherlands (in Dutch).

Pinay, G., Roques, L. and Fabre, A. (1993). Spatial and temporal patterns of denitrification in a riparian forest, *Journal of Applied Ecology* **30**, 581–591.

Pinay, G., Ruffinoni, C. and Fabre, A. (1995). Nitrogen cycling in two riparian forest soils under different geomorphic conditions, *Biogeochemistry* **30**, 9–29.

Richardson, C.J. (1990). Biogeochemical cycles: regional. In: Patten, B.C. (ed.) *Wetlands and Shallow Continental Water Bodies, Volume 1*, SPB Academic, The Hague, The Netherlands, 259–279.

Schipper, L.A., Cooper, A.B., Harfoot, C.G. and Dyck, W.J. (1993). Regulators of denitrification in an organic riparian soil, *Soil Biology Biochemistry* **25**(7), 925–933.

Schreijer, M., Kampf, R., Verhoeven, J.T.A. and Toet, S. (2000). *Nabehandeling van RWZI-Effluent tot bruikbaar Oppervlaktewater in een Moerassysteem*, Hoogheemraadschap der Uitwaterende Sluizen in Hollands Noorderkwartier, Utrecht University, The Netherlands (in Dutch).

Velthof, G.L. (1997). *Nitrous Oxide Emission From Intensively Managed Grasslands*, PhD Dissertation, Landbouw Universiteit Wageningen, The Netherlands.

Verhoeven, J.T.A., Whigham, D.F., Kerkhoven, M. van, O'Neill, J. and Maltby, E. (1994). Comparative study of nutrient-related processes in geographically separated wetlands: towards a science base for functional assessment procedures. In: W.J. Mitsch (ed.) *Global Wetlands: Old World and New*, Elsevier, Amsterdam, 91–106.

Verhoeven, J.T.A., Keuter, A., Kerkhoven, M.B. van and Wassen, M. (1996). Control of local nutrient dynamics in mires by regional and climatic factors: a comparison of Dutch and Polish sites, *Journal of Ecology* **84**, 647–656.

Verhoeven, J.T.A., Bogaards, H. and Spink, A. (2001). Initial estimates of nutrient-related process rates in floodplains along modified rivers in the Netherlands. In: Nienhuis, P.H., Leuven, R.S.E.W. and Ragas, A.M.J. (eds), *New Concepts for Sustainable Management of River Basins*, Backhuys, Leiden, The Netherlands, 229–240.

Verhoeven, J.T.A., Whigham, D.F., Logtestijn, R. van and O'Neill, J. (2001). A comparative study of nitrogen and phosphorus cycling in tidal and non-tidal riverine wetlands, *Wetlands* **21**(2), 210–222.

Introduction

Process description:

- Denitrification is the bacterial reduction of nitrate ($NO_3^{\pm}$) to nitrous oxide (N_2O) and dinitrogen (atmospheric nitrogen, N_2) and their subsequent export from the HGMU into the atmosphere.
- One requirement for the process to occur is that the HGMU is in receipt of nitrate from directly applied nitrate-rich fertiliser, directly applied nitrogenous materials or fertilisers, which are transformed to produce nitrate, or interaction with nitrate- or nitrogenous material-bearing waters.
- Optimal soil conditions for the process include the presence of a carbon source for bacterial respiration, an anaerobic soil atmosphere, soil interstitial water with pH above 6 and soil temperature greater than 8 °C. Figure 3.2.10 shows the approximate relationship between water-filled pore space and N_2O emissions.
- When the nitrogen input does not consist of nitrate, but of ammonium (NH_4) or organic nitrogen, these nitrogenous materials must be converted into nitrate before denitrification can occur. Organic nitrogen will first be mineralised into ammonium. Under aerobic conditions ammonium can be oxidised to form nitrate. This process of oxidation of ammonium into nitrate is called *nitrification*. The nitrate produced in the process of nitrification will diffuse into the anaerobic soil layers where it will be reduced into nitrous oxide and finally dinitrogen gas (*denitrification*). When the nutrient input consists of ammonium or organic nitrogen, aerobic as well as anaerobic soil layers have to be present (either in space or time) in order to provide conditions suitable for both nitrification and denitrification.

In this section, the total gaseous export of nitrogen will be assessed, i.e. both

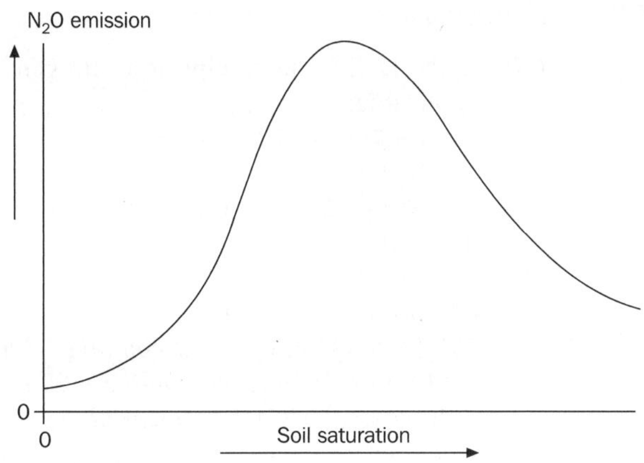

3.2.10 N_2O emission in relation to water-filled pore space. Hypothetical relationship between mean water-filled pore space and N_2O emission from the soil surface. The shape of the curves may vary somewhat with soil type and conditions. The figure describes the active zone of the soil (< 1 m depth) (adapted from Velthof, 1997).

dinitrogen (N_2) and nitrous oxide (N_2O) emissions. Because of insufficient or ambiguous research results it is not possible to predict the proportion in which both gases are released from wetland soils although, at low pH, low temperature, high nitrate concentrations and when alternating aerobic and anaerobic soil conditions occur, the ratio of $N_2O{:}N_2$ increases. High emissions of nitrous oxide contribute to (1) global warming, because N_2O has 296 times the global warming potential of CO_2 (IPCC, 2001), and (2) the destruction of stratospheric ozone, which increases the amount of biologically harmful ultraviolet radiation received at the surface of the earth. Therefore, conditions that are expected to be favourable for relatively high emissions of nitrous oxide are indicated.

Significance – rationale for assessment: Improvement of water quality in water bodies receiving surface or subsurface water from the wetland by removal of nitrogen by denitrification.

Controlling variables

CV1 Nutrient input
CV2 Potential for interaction with nitrate
CV3 Soil temperature
CV4 Soil carbon
CV5 Soil oxygen status
CV6 Soil pH

Answer all the questions below for *each* HGMU; unless otherwise instructed, continue from one question to the next. Circle the correct answer on the appraisal sheet. The location of the relevant information is given in brackets after each question. Use a new appraisal sheet for each HGMU.

If there are no nutrient inputs identified for the HGMU, or if the nutrient input is uncertain (nutrient input codes **Nut 2** or **Nut ?**), the process of denitrification can still occur. However, in this case the denitrification rates will be far lower than with an external nutrient input. Besides assessing the actual situation with no nutrient input, you may wish to assess the potential of the HGMU to perform this process. To do this conduct the assessment as normal but only answer questions in ***CV3***, ***CV4***, ***CV5*** and ***CV6***. Establish an assessment outcome as normal, but ignore the need for positive answers for ***CV1*** and ***CV2*** in the assessment chart. The rationale statement can be used as normal but ignore reference to receipt of nutrient inputs.

CV1 *Nutrient input*

Q1. Is the HGMU in receipt of a nitrogen nutrient input? Circle all the codes applicable:

a. $\mathbf{N_d}$ **1/a**, $\mathbf{N_d}$ **1/b**, $\mathbf{N_d}$ **1/x**
b. $\mathbf{N_i}$ **1/x**
c. $\mathbf{Nut_d}$ **1/b**, $\mathbf{Nut_d}$ **1/x**
d. $\mathbf{Nut_i}$ **1/x**
e. **Nut ?**

(Recording Sheet 2c)

[*If YES to **Q1 a, b, c** or **d** continue with **Q2**. If NO to all answers above or if YES to **Q1e**, then continuing with this assessment will give information only about an HGMU's **potential** to perform this process. To do this continue with **CV2***]

Q2. What is the source of the nutrient input received by the HGMU? Circle all those applicable:

a. Inorganic fertiliser applied directly to the HGMU *(Recording Sheet 2b, H1 Action, Q1)*
b. Organic fertiliser applied directly on the surface of the HGMU *(Recording Sheet 2b, H1 Action, Q5)*
c. Organic fertiliser injected into the surface of the HGMU *(Recording Sheet 2b, H1 Action, Q6)*
d. Organic industrial waste applied within the HGMU *(Recording Sheet 2b, H1 Action, Q7)*
e. Manure of livestock kept within the HGMU *(Recording Sheet 2b, H1 Action, Q8)*
f. Excrement from large populations of birds *(Recording Sheet 2b, H1 Action, Q9)*
g. Atmospheric deposition *(Recording Sheet 2b, H1 Action, Q10)*
h. Contact with water containing > 1 mg l^{-1} ammonium–nitrogen *(Recording Sheet 2b, H1 Action, Q20)*
i. Contact with water containing > 1 mg l^{-1} nitrate–nitrogen *(Recording Sheet 2b, H1 Action, Q19)*
j. Contact with water with a total nitrogen concentration of > 2.2 mg N l^{-1} *(Recording Sheet 2b, H1 Action, Q21)*
k. Groundwater discharge *(Recording Sheet 2b, H2 Action, Q1–3)*
l. Run-off or subsurface run-off *(Recording Sheet 2b, H2 Action, Q4–6)*

[*If NO to **Q2 h, i, j, k** and **l,** and YES to any of **Q2 a, b, c, d, e** or **f**, then continue with **CV3**. In all other cases continue with **CV2***]

CV2 *Potential for interaction with nitrate*

Only answer the questions in this section if you have identified an indirect nutrient input (answers **b** or **d** to **Q1** in **CV1**). If you have identified a direct nutrient input (answers **a** or **c** to **Q1** in **CV1**), ignore **CV2** and continue with **CV3**.

Q1. Has the HGMU been assigned a geomorphic prefix code **D**, indicating it is a depressional landform, i.e. **BS, CO, OX, FC, PB**? *(Recording Sheet 2a, C.2; AA Map 3)*

Q2. Is the HGMU adjacent to any HGMUs or areas of land not included in the AA that are topographically higher? *(Recording Sheet 2a, C.2; AA Map 3)*

Q3. Is the gradient of the HGMU coded as **G0** or **G1**? *(Recording Sheet 2a, C.2; AA Map 3)*

Q4. Is the microtopography of the HGMU described as significant? *(Recording Sheet 3, A.3)*

Q5. Is the groundwater level artificially maintained at high levels in (part of) the AA at any time (e.g. for nature conservation purposes)? *(Recording Sheet 1, C.3.22.4a)*

Q6. Does the HGMU have a soil profile described as having at least two of the following characteristics:

- coarse particle size throughout its depth
- gravel particle size throughout its depth
- mineral soil material?

(Recording Sheet 2a, C.3.7)

Q7. Is the HGMU under-drained by subsurface pipes? *(Recording Sheet 1, C.3.9.5b; AA Map 2 or 3)*

Only answer ***Q8*** if you have answered YES to ***Q2*** **h**, **i** or **j** in ***CV1***.

Q8. Does enriched surface water have contact with the HGMU for at least 3 successive days? *(Recording Sheet 1, C.3.19.5a → answer should be (**1 week–1 month**) or (**> 1 month**). If answered (**< 1 week**), only answer* **Q8** *positively when more details are available)*

CV3 *Soil temperature*

Q1. What is the mean annual soil temperature status? *(Recording Sheet 1, C.3.1.5)*

a. ″ 0 °C (pergelic)

b. > 0 to 8 °C (cryic)

c. > 8 °C (mesic or thermic)

CV4 *Soil carbon*

Q1. Is the source of the nutrient input organic fertiliser spraying, spreading or injection, organic industrial waste or animal faeces arising from livestock? *(Recording Sheet 2b, H1 ACTION, Q5–8)*

Q2. Is part or all of the top 20 cm of the soil characterised by the presence of significant amounts of organic matter described as humus or amorphous, fibrous or semi-fibrous peat? *(Recording Sheet 2a, C.3.8)*

Q3. Is the HGMU characterised as unvegetated or with limited soil development, i.e. does it have one of the following vegetation codes: **U2**, **U4–U13**, **U15**, **U16**, **U19**, **S2**? *(Recording Sheet 3, D)*

CV5 *Soil oxygen status*

Q1. Does the nutrient input received by the HGMU have a groundwater discharge (**F1**) component, i.e. Hydrological Code **H1**, **H2**, **H3**, **H4**, **H11**, **H12**, **H14**, **H15**?

(Recording Sheet 2a, F1)

Q2. Does one of the statements listed below apply to the soil profile?

- the colour of the soil matrix is predominantly blue/grey or greenish
- mottles are present in at least the top 20 cm of the soil profile
- the soil material is described as being predominantly organic
- soil organic material in at least the top 20 cm is described as fibrous peat, amorphous peat, semi-fibrous peat or humose

(Recording Sheet 2a, C.3.4, C.3.5, C3.8, C3.6)

Q3. Is the vegetation within the HGMU classed as one of the following:
G2–G4; G6–G9; G11; G14; M1–M15; W4–W8; W11; W12; U14?
(Recording Sheet 3, D)
Note: aquatic systems are excluded because they are beyond the scope of the assessment methodology developed to date, not because denitrification is not performed by them.

Q4. Is the upper 20 cm of the soil prone to water table fluctuations (alternate periods of waterlogged and non-waterlogged soil conditions, which may be due to natural, seasonal fluctuations, or water table management), which result in non-waterlogged soil conditions for at least half of the year? *(Recording Sheet 1, C.3.22.3)*

Q5. Is the average depth of the groundwater table at or just below the soil surface (i.e. within 20 cm) for most of the year? *(Recording Sheet 3, B.3.10 and/or AA Map 2)*

3.2.B.1.1 Gaseous Export of Nitrogen (N): Denitrification Appraisal Sheet

HGMU code: ______________	**AA name:** ______________

CV1 Nutrient input						
Q1	a	b	c	d	e	
Q2	a	b	c	d	e	f
	g	h	i	j	k	l

CV2 Potential for interaction with nitrate			
Q1	Y	N	U/NA
Q2	Y	N	U/NA
Q3	Y	N	U/NA
Q4	Y	N	U/NA
Q5	Y	N	U/NA
Q6	Y	N	U/NA
Q7	Y	N	U/NA
Q8	Y	N	U/NA

CV3 Soil temperature			
Q1	a	b	c

CV4 Soil carbon			
Q1	Y	N	U/NA
Q2	Y	N	U/NA
Q3	Y	N	U/NA

CV5 Soil oxygen status			
Q1	Y	N	U/NA
Q2	Y	N	U/NA
Q3	Y	N	U/NA
Q4	Y	N	U/NA
Q5	Y	N	U/NA

CV6 Soil pH			
Q1	a	b	c
Q2	Y	N	U/NA

HGMU process assessment code: ______________________________

3.2.B.1.1 Look-up Table 1

<table>
<tr><th>CV1</th><th>CV2</th><th>CV3</th><th>CV4</th><th>CV5</th><th>CV6</th><th>Rationale</th><th>Code</th><th>Quantification</th></tr>
<tr><td>1a 2a</td><td>N/A</td><td>1c</td><td>1|2|#3</td><td>(1|2|3)
(4|5)</td><td>1c 2</td><td>A direct input of nitrogen (probably containing nitrate) is applied to the surface of the HGMU at least annually.
Carbon, soil oxygen, pH and soil temperature are conducive to denitrification. Anaerobic soil conditions or alternation between aerobic and anaerobic soil conditions prevail, which are favourable for denitrification.
Significant N_2 and N_2O emissions are likely.</td><td>1/b</td><td>> 10 but < 80 kg N ha^{-1} a^{-1}</td></tr>
<tr><td>1a 2a</td><td>N/A</td><td>1c</td><td>1|2|#3</td><td>5</td><td>1c 2</td><td>A direct input of nitrogen (probably containing nitrate) is applied to the surface of the HGMU at least annually.
Carbon, soil oxygen, pH and soil temperature are conducive to denitrification. Anaerobic soil conditions prevail, which are favourable for denitrification.
Because of the predominantly anaerobic soil conditions N_2O production is low.</td><td>1/b</td><td>> 10 but < 80 kg N ha^{-1} a^{-1}</td></tr>
<tr><td>1a
(2b|2c|
2e|2f)</td><td>N/A</td><td>1c</td><td>1|2|#3</td><td>1|2|3|4</td><td>1c 2</td><td>Organic nitrogen is applied directly to the surface of the HGMU at least annually. Carbon, soil oxygen, pH and soil temperature are conducive to denitrification. Soil conditions alternate between aerobic and anaerobic, enabling alternating periods of nitrification (transforming ammonium into nitrate) and denitrification (conversion of nitrate into gaseous nitrogen components).
Significant N_2 and N_2O emissions are likely.</td><td>1/b</td><td>> 10 but < 80 kg N ha^{-1} a^{-1}</td></tr>
<tr><td>(1a 2g)
#(2a|2b|
2c|2d|2e|
2f|2h|2i|
2j|2k|2l)</td><td>N/A</td><td>1c</td><td>1|2|#3</td><td>(1|2|3)
(4|5)</td><td>1c 2</td><td>Nitrogen is received through atmospheric deposition. Carbon, soil oxygen, pH and soil temperature are conducive to denitrification. Anaerobic soil conditions or alternation between aerobic and anaerobic soil conditions prevail, which are favourable for denitrification.</td><td>1/c</td><td>< 15 kg N ha^{-1} a^{-1}</td></tr>
<tr><td>1b (2k|2l)</td><td>(1|2|3|
4|5)
(#6 #7)</td><td>1c</td><td>1|2|#3</td><td>(1|2|3)
(4|5)</td><td>1c 2</td><td>It is highly likely that a nitrogen input is received from outside the HGMU via groundwater discharge or through (sub)surface run-off. Residence time is long enough for significant interaction between the nitrogen input and the soil. Carbon, soil oxygen, pH and soil temperature are conducive to denitrification. Anaerobic soil conditions or alternation between aerobic and anaerobic soil conditions prevail, which are favourable for denitrification.
Significant N_2 and N_2O emissions are likely.
Many different situations fall into this category.</td><td>1/a</td><td>> 10 but < 300 kg N ha^{-1} a^{-1}</td></tr>
</table>

3.2.B.1.1 Look-up Table 1 cont'd

<table>
<tr><th>CV1</th><th>CV2</th><th>CV3</th><th>CV4</th><th>CV5</th><th>CV6</th><th>Rationale</th><th>Code</th><th>Quantification</th></tr>
<tr><td></td><td></td><td></td><td></td><td></td><td></td><td>Therefore the quantification class is very broad. In general, the higher the nitrogen input, the higher the denitrification rate. Only with extremely high nitrogen inputs is it possible that other factors may become limiting.</td><td></td><td></td></tr>
<tr><td>1b
(2h|2i|2j)</td><td>(1|2|3|
4|5)
(#6 #7)
8</td><td>1c</td><td>1|2|#3</td><td>(1|2|3)
(4|5)</td><td>1c 2</td><td>It is highly likely that a nitrogen input is received from outside the HGMU via overbank flooding or (artificial) supply of nutrient-rich water. Residence time is long enough for significant interaction between the nitrogen input and the soil. Carbon, soil oxygen, pH and soil temperature are conducive to denitrification. Anaerobic soil conditions or alternation between aerobic and anaerobic soil conditions prevail, which are favourable for denitrification.
Significant N_2 and N_2O emissions are likely.
Many different situations fall into this category.
Therefore the quantification class is very broad. In general, the higher the nitrogen input, the higher the denitrification rate. Only with extremely high nitrogen inputs is it possible that other factors may become limiting.</td><td>1/a</td><td>> 10 but < 300 kg N ha^{-1} a^{-1}</td></tr>
</table>

3.2.B.1.1 Look-up Table 2

CV1	CV2	CV3	CV4	CV5	CV6	Rationale	Code
#(1a–d)| 1e	*	*	*	*	*	The HGMU does not receive any nutrient input, or it is uncertain whether the HGMU receives a nutrient input. If other controlling variables are conducive, the process can still occur, but only as part of natural nitrogen cycling. Maximum denitrification rates will be around 2 kg N ha^{-1} a^{-1}.	**2**
*	*	1a	*	*	*	Other factors may be suitable, but the temperature of the soil will prevent denitrification from occurring at all.	**2**
*	*	*	#1 #2 3	*	*	Other factors may be suitable, but the HGMU soil has little or no organic matter and therefore no carbon is available to denitrifying bacteria and the process cannot be performed.	**2**
*	*	*	*	#(1 2 3 4 5)	*	Other factors may be suitable, but there is too much oxygen present in the soil. The aerobic conditions prevent denitrification from occurring.	**2**
*	*	*	*	*	1a	Other factors may be suitable, but the low pH will prevent denitrification occurring at all.	**2**

3.2.B.1.1 Look-up Table 3

<table>
<tr><th>CV1</th><th>CV2</th><th>CV3</th><th>CV4</th><th>CV5</th><th>CV6</th><th>Rationale</th><th>Code</th></tr>
<tr><td>1c (2a|2d)</td><td>*</td><td>*</td><td>*</td><td>*</td><td>*</td><td>A nutrient input derived from artificial fertiliser or organic industrial waste is applied directly to the surface of the HGMU, but it is uncertain whether this input contains nitrogen.</td><td>1/x</td></tr>
<tr><td>1d (2k|2l)</td><td>*</td><td>*</td><td>*</td><td>*</td><td>*</td><td>The HGMU receives an indirect nutrient input derived from artificial fertiliser or organic industrial waste, but it is uncertain whether this input contains nitrogen.</td><td>1/x</td></tr>
<tr><td>2b|2c|
2d|2e|2f</td><td>*</td><td>*</td><td>*</td><td>5</td><td>*</td><td>Because of a constant high groundwater level, coupled (alternating) nitrification and denitrification cannot occur (or can take place only in micro-sites). This means denitrification is likely to become limited by nitrate supply. Because of the predominantly anaerobic soil conditions N_2O production is low.</td><td>1/x</td></tr>
<tr><td>2h|2i|2j
|2k|2l</td><td>#(1 2 3
4 5)</td><td>*</td><td>*</td><td>*</td><td>*</td><td>Wetland morphology is not suitable for interaction of the nutrient input with the HGMU.</td><td>1/x</td></tr>
<tr><td>2h|2i|2j|
2k|2l</td><td>(6|7)</td><td>*</td><td>*</td><td>*</td><td>*</td><td>Well-drained soil or enhanced drainage may reduce nutrient input contact time and constrain the process rate.</td><td>1/x</td></tr>
<tr><td>2h|2i|2j</td><td>#8</td><td>*</td><td>*</td><td>*</td><td>*</td><td>Flood duration is not long enough for significant contact of the nutrient input with the HGMU.</td><td>1/x</td></tr>
<tr><td>*</td><td>*</td><td>1b</td><td>*</td><td>*</td><td>*</td><td>Soil temperature is not optimal for denitrification. This will constrain the process rate.
At low temperatures the proportion of N_2O produced compared with N_2 increases.</td><td>1/x</td></tr>
<tr><td>*</td><td>*</td><td>*</td><td>*</td><td>*</td><td>1b | #2</td><td>pH is not optimal for denitrification. This will constrain the process rate.
At low pH the proportion of N_2O produced increases compared with N_2.</td><td>1/x</td></tr>
</table>

CV6 *Soil pH*

Q1. What is the pH of the soil?

a. < 4
b. 4–6
c. > 6

(Recording Sheet 3, A.1i)

Q2. Is the vegetation within the HGMU coded as anything other than **A1**, **G5**, **G6**, **G13**, **U5**, **M1–M6**, **M8**, **M11**, **S1**, **W14**, **W16**, **W21**, **W26**, **W27**? *(Recording Sheet 3, D)*

Assessment of the process

Compare the results recorded for each HGMU on the appraisal sheets with the possible outcomes below, and assign an appropriate assessment code. Record this on the appraisal sheet. The syntax codes used are described in the introduction to Section 3.

Look-up tables

1. The process is definitely being performed

If the answers to the questions for this process agree with any of the combinations in Look-up Table 1, then the HGMU is definitely performing the process of denitrification and an estimation of the amount of nitrogen removed by this process is given.

N_2O emission is affected by the nitrate supply: a high nitrate availability results in a high N_2O:N_2 ratio. Moreover, relatively more N_2O is produced at low pH, at low temperature and when aerobic and anaerobic soil conditions alternate. The quantification figures are based on data collected during empirical studies carried out in the EU projects leading to the production of these FAPs (FAEWE, FAEWE II and PROTOWET), and published research related to this topic.

2. The process is not (significantly) being performed

If the answers to the questions for this process agree with any of the combinations in Look-up Table 2, then the HGMU is not significantly performing the process of denitrification.

3. The process is being performed, but there are constraining factors or uncertainties

If Look-up Tables 1 and 2 did not lead to an assessment, the process of denitrification is performed, but there are one or more constraining factors. In Look-up Table 3 you can find out how many reducing factors play a role in the process, or which factors cause the uncertainty. If answers agree with any of the following options, then the HGMU is performing the process, but quantification is not possible.

4. The process is definitely not being performed

This statement is not applicable to this process.

This completes the functional assessment for the process of gaseous export of N (denitrification). The output codes and quantification values can be recorded on the Functional Assessment Table at the end of this Section 3.2.B.4.

3.2.B.1.2 Gaseous export of N (ammonia volatilisation) (process)

Further reading and references

King, D.L. (1985). Nutrient cycling by wetlands and possible effects of water levels. In: Prince, H.H. and D'Itri, F.M. (eds), *Coastal Wetlands*, Lewis, Chelsea, MI, 69–86.

Mikkelsen, D.S., De Datta, S.K. and Obcemea, W.N. (1978). Ammonia volatilization losses from flooded rice soils, *Soil Science Society of America Journal* **42**, 725–730.

Reddy, K.R. and Patrick, W.H., Jr (1984). Nitrogen transformations and loss in flooded soils and sediments, *CRC Critical Reviews in Environmental Control* **13**(4), 273–309.

Richardson, C.J. (1990). Biogeochemical cycles: regional. In: Patten, B.C. (ed.), *Wetlands and Shallow Continental Water Bodies, Volume 1*, SPB Academic, The Hague, The Netherlands, 259–279.

Smil, V. (1999). *Elementaire Kringlopen*, Natuur & Techniek, Amsterdam, The Netherlands.

Tian, G., Cai, Z., Cao, J. and Li, X. (2001). Factors affecting ammonia volatilisation from a rice-wheat rotation system, *Chemosphere* **42**, 123–129.

Vlek, P.L.G. and Stumpe, J.M. (1978). Effects of solution chemistry and environmental conditions on ammonia volatilization losses from aqueous systems, *Soil Science Society of America Journal* **42**, 416–421.

White, J.R. and Reddy, K.R. (In press). Nitrogen cycling in wetlands. In: Maltby, E. and Barker, T. (eds), *The Wetlands Handbook*, Wiley-Blackwell, Oxford, UK.

Wienk, L.D., Verhoeven, J.T.A., Coops, H. and Portielje, R. (2000). Peilbeheer en Nutriënten; Literatuurstudie naar de Effecten van Peildynamiek op de Nutriëntenhuishouding van Watersystemen, RIZA rapport 2000.012, RIZA Lelystad, The Netherlands (in Dutch).

Introduction

Process description:

- The physico-chemical release of excess ammonium or ammonia nitrogen as gaseous ammonia and export from the HGMU into the atmosphere.
- One requirement for the process to occur is that an HGMU is in receipt of ammonium or ammonia by direct input to the soil surface (especially the urea excreted by farm livestock and other mammals or of the uric acid excreted by birds). It may also occur from shallow water bodies that are in receipt of ammonium- or ammonia-rich water, material or fertilisers, when conditions are optimal.
- Conditions are optimal when pH in soil water or in the water column is above 7, soil or water temperature is above 8 °C, the soil is composed predominantly of sandy or gravelly mineral material and the soil or water body is exposed so that the diffusion gradient from the soil or water body to atmosphere is enhanced.
- The overall importance of ammonia volatilisation as a loss mechanism in wetlands is poorly documented. Because of the small amount of information on rates of ammonia volatilisation from wetlands this process cannot be assessed quantitatively, so only a qualitative assessment is provided.

Large emissions of ammonia lead to an increase in N-deposition, which will acidify and eutrophicate many ecosystems, both aquatic and terrestrial.

Significance – rationale for assessment: Improvement of water quality in water bodies receiving surface or subsurface water from the wetland by export of nitrogen by loss of ammonia into the atmosphere.

Controlling variables
CV1 Nutrient input
CV2 Water column/soil environment
CV3 Exposure

Answer all the questions below for *each* HGMU; unless otherwise instructed continue from one question to the next. Circle the correct answer on the appraisal sheet. The location of the relevant information is given in brackets after each question. Use a new appraisal sheet for each HGMU.

If there are no nutrient inputs identified for the HGMU, or if the nutrient input is uncertain (nutrient input codes **Nut 2** or **Nut ?**), you may wish to assess the potential of the HGMU to perform this process. To do this conduct the assessment as normal but answer questions only in ***CV2*** and ***CV3***. Establish an assessment outcome as normal but ignore the need for positive answers for ***CV1***. The rationale statement can be used as normal but ignore reference to receipt of nutrient inputs.

CV1 *Nutrient input*

Q1. Has the HGMU been attributed one of the nutrient input codes listed below, indicating an input of nitrogen? Circle all those applicable:

a. **N_d 1/a, N_d 1/b, N_d 1/x**
b. **N_i 1/x**
c. **Nut_d 1/b, Nut_d 1/x**
d. **Nut_i 1/x**
e. **Nut ?**
(Recording Sheet 2c)
[*If YES to **Q1a**, **b**, **c** or **d**, continue with **Q2**. If NO to all answers above or if YES to **Q1e**, then continuing with this assessment will give information only about an HGMU's **potential** to perform this process. To do this continue with **CV2***]

Q2. What is the source of the nutrient input received by the HGMU? Circle all those applicable:
a. Inorganic fertiliser applied directly to the HGMU *(Recording Sheet 2b, H1 Action, Q1)*
b. Organic fertiliser applied directly on the surface of the HGMU *(Recording Sheet 2b, H1 Action, Q5)*
c. Organic fertiliser injected into the surface of the HGMU *(Recording Sheet 2b, H1 Action, Q6)*
d. Organic industrial waste applied within the HGMU *(Recording Sheet 2b, H1 Action, Q7)*
e. Manure from livestock kept within the HGMU *(Recording Sheet 2b, H1 Action, Q8)*
f. Excrements from large populations of birds *(Recording Sheet 2b, H1 Action, Q9)*
g. Atmospheric deposition *(Recording Sheet 2b, H1 Action, Q10)*
h. Contact with water containing > 1 mg l^{-1} ammonium–nitrogen *(Recording Sheet 2b, H1 Action, Q20)*
i. Contact with water containing > 1 mg l^{-1} nitrate–nitrogen *(Recording Sheet 2b, H1 Action, Q19)*
j. Contact with water with a total nitrogen concentration of > 2.2 mg l^{-1} *(Recording Sheet 2b, H1 Action, Q21)*
k. Groundwater discharge *(Recording Sheet 2b, H2 Action, Q1–3)*
l. Run-off or subsurface run-off *(Recording Sheet 2b, H2 Action, Q4–6)*
[*If NO to **Q2 h**, **i**, **j**, **k** and **l**, and YES to any of **Q2 a**, **b**, **c**, **d**, **e** or **f**, then continue with **CV3**. In all other cases continue with **CV2***]

CV2 *Water column/soil environment (see Fig. 3.2.11)*

Q1. Is more than 75 % of the HGMU covered with surface water or is the HGMU frequently flooded? *(AA map 2)*
a. Yes [*Continue with **Q2***]
b. No [*Continue with **Q4***]

Q2. Is the pH in the water column > 7? *(Recording Sheet 3, A.1ii)*

Q3. What is the mean annual temperature of the water column? *(Recording Sheet 1, C.3.1.10)*
a. < 8 °C *(answer 10a)*
b. 8 °C *(answer 10b)*
If this information is not available you can use the mean annual soil temperature as a surrogate (Recording Sheet 1, C.3.1.5)]

Q4. Is the pH of the soil > 7? *(Recording Sheet 3, A.1i)*

Q5. Is the vegetation of a type that would characterise pH conditions > pH 7 within the HGMU, e.g. vegetation code **G18**, **G24**, **G27**, **G28**, **W40**, **U12**? *(Recording Sheet 3, D)*

Q6. What is the mean annual temperature of the soil? *(Recording Sheet 1, C.3.1.5)*
a. ″ 0 °C (pergelic)
b. > 0 and ″ 8 °C (cryic)
c. > 8 °C (mesic or thermic)

Q7. Is the top 20 cm of the soil in the HGMU described as having a coarse or gravelly texture? *(Recording Sheet 2a, C.3.7)*

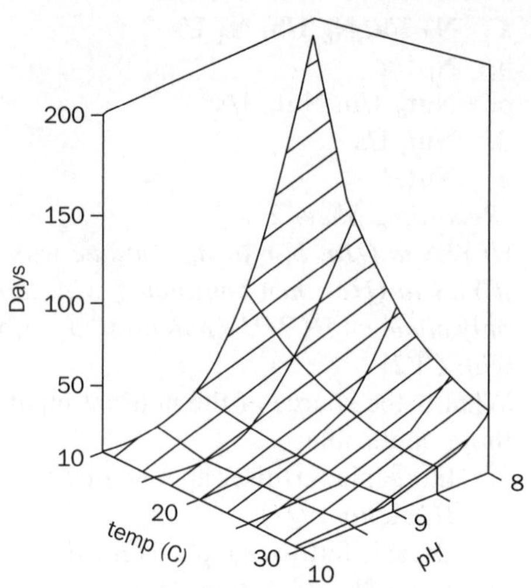

3.2.11 Ammonia volatilisation in relation to temperature and pH (days required to volatilise one-half of the ammonia in excess of atmospheric equilibrium from the water to the atmosphere as a function of water pH and temperature from a two-metre-deep system subjected to a surface wind of 2 m s^{-1}) (From: King, 1985).

CV3 *Exposure*

Q1. How is the exposure of the HGMU characterised?

a. Sheltered

b. Moderate or exposed

(Recording Sheet 3, A.4)

Q2. Are wind-cropped trees present in the HGMU? *(Recording Sheet 3, A.5)*

Assessment of the process

Compare results recorded for each HGMU on the appraisal sheets with the possible outcomes in the tables below, and assign an appropriate assessment code. Record this on the appraisal sheet. The syntax codes used are described in the introduction to Section 3.

Look-up tables

1. The process is definitely being performed

If the answers to the questions for this process agree with any of the combinations in Look-up Table 1, then the HGMU is definitely performing the process of ammonia volatilisation. The overall importance of ammonia volatilisation as a loss mechanism in wetlands is poorly documented. Because of the small amount of information available on rates of ammonia volatilisation, this process cannot be assessed quantitatively, so only a qualitative assessment is provided.

2. The process is not (significantly) being performed

If the answers to the questions for this process agree with any of the combinations in Look-up Table 2, then the HGMU is not significantly performing the process of ammonia volatilisation.

3. The process is being performed, but there are constraining factors or uncertainties

If Look-up Tables 1 and 2 did not lead to an assessment, the process of ammonia volatilisation is performed, but there are constraining factors or uncertainties. In Look-up Table 3 you can find out how many constraining factors influence the process, or which factors cause the uncertainty.

4. The process is definitely not being performed

This statement is not applicable to this process.

3.2.B.1.2 Gaseous Export of N (Ammonia Volatilisation) Appraisal Sheet

HGMU code: ______________	**AA name:** ______________

CV1 Nutrient input						
Q1	a	b	c	d	e	
Q2	a	b	c	d	e	f
	g	h	i	j	k	l

CV3 Exposure			
Q1	a	b	
Q2	Y	N	U/NA

CV2 Water column/soil environment			
Q1	a	b	
Q2	Y	N	U/NA
Q3	a	b	
Q4	Y	N	U/NA
Q5	Y	N	U/NA
Q6	a	b	c
Q7	Y	N	U/NA

HGMU process assessment code: ______________________________

3.2.B.1.2 Look-up Table 1

CV1	CV2	CV3	Rationale	Code
1a (2b|2e|2f)	*	*	Excrement from cattle or birds or application of organic material or fertilisers result in availability of highly volatile ammonium or ammonia compounds in the topsoil of the HGMU. Ammonia volatilisation is a rapid process. Ammonia losses are thus expected to be high during the first few days after N-application or input. NB: High emissions of ammonia lead to an increased N-deposition, which may acidify and eutrophicate adjacent (including terrestrial) ecosystems.	**1**
1b 2h	1a 2 3b	1b|2	The HGMU receives an input of ammonium via contact with water with a high ammonium concentration. The pH of the water is likely to be alkaline and the average temperature is above 8 °C, both of which are optimal for ammonia volatilisation. The HGMU is exposed, which optimises air movement and maintains a high diffusion gradient from the water surface to the atmosphere. Ammonia volatilisation is a rapid process. Ammonia losses are thus expected to be high during the first few days after N-application or N-input. NB: High emissions of ammonia lead to increased N-deposition, which may acidify and eutrophicate adjacent (including terrestrial) ecosystems.	**1**
1b 2h	1b (4|5) 6c 7	1b|2	The HGMU receives an input of ammonium via contact with water with a high ammonium concentration. Soil pH is likely to be alkaline, the average temperature is above 8 °C, and the soil is composed of coarse material minimising chemical adsorption of ammonia compounds to soil particles, all of which are optimal for ammonia volatilisation. The HGMU is exposed, which optimises air movement and maintains a high diffusion gradient from the soil surface to the atmosphere. Ammonia volatilisation is a rapid process. Ammonia losses are thus expected to be high during the first few days after N-application or N-input. NB: High emissions of ammonia lead to an increased N-deposition, which may acidify and eutrophicate adjacent (including terrestrial) ecosystems.	**1**
1b 2j	1a 2 3b	1b|2	The HGMU receives an input of nitrogen (probably consisting of or including ammonium) via contact with water with a high nitrogen concentration. The pH of the water is likely to be alkaline and the average temperature is above 8 °C, both of which are optimal for ammonia volatilisation. The HGMU is exposed, which optimises air movement and maintains a high diffusion gradient from the water surface to the atmosphere. Ammonia volatilisation is a rapid process. Ammonia losses are thus expected to be high during the first few days after N-application or N-input. NB: High emissions of ammonia lead to an increased N-deposition, which may acidify and eutrophicate adjacent (including terrestrial) ecosystems.	**1**
1b 2j	1b (4|5) 6c 7	1b|2	The HGMU receives an input of nitrogen (probably consisting of or including ammonium) via contact with water with a high concentration. Soil pH is likely to be alkaline, the average temperature is above 8 ºC, and the soil is composed of coarse material minimising chemical adsorption of ammonia compounds to soil particles, all of which are optimal for ammonia volatilisation. The HGMU is exposed which optimises air movement and therefore maintains a high diffusion gradient from the soil surface to the atmosphere. Ammonia volatilisation is a rapid process. Ammonia losses are thus expected to be high during the first few days after N-application or N-input. NB: High emissions of ammonia lead to an increased N-deposition, which may acidify and eutrophicate adjacent (including terrestrial) ecosystems.	**1**

3.2.B.1.2 Look-up Table 2

<table>
<tr><th>CV1</th><th>CV2</th><th>CV3</th><th>Rationale</th><th>Code</th></tr>
<tr><td>#(1a–d)|1e</td><td>*</td><td>*</td><td>There is no nutrient source received that is suitable for the process to occur, or it is uncertain whether a nutrient source is present. The HGMU may have the potential to remove nitrogen through ammonia volatilisation, if an appropriate nutrient source were received.</td><td>2</td></tr>
<tr><td>1d 2d</td><td>*</td><td>*</td><td>The HGMU receives an indirect nutrient input derived from organic industrial waste. It is uncertain if this source contains nitrogen and ammonium in particular.</td><td>2</td></tr>
<tr><td>2a</td><td>*</td><td>*</td><td>A nutrient supply is derived from artificial fertiliser. Ammonia volatilisation from artificial fertiliser is low (only in the case of fertilisation with urea is ammonia volatilisation likely to occur).</td><td>2</td></tr>
<tr><td>2g</td><td>*</td><td>*</td><td>The HGMU is in receipt of ammonium through ammonium-rich atmospheric deposition. If atmospheric deposition is the only input of ammonium into the system, ammonia volatilisation is not significant.</td><td>2</td></tr>
<tr><td>2i</td><td>*</td><td>*</td><td>The HGMU receives a nitrate input in nitrate-rich surface water. For ammonia volatilisation to occur, nitrate should be reduced to ammonium. Reduction of nitrate to ammonium can occur in wetlands (dissimilatory nitrate reduction), but there is little known about this process.</td><td>2</td></tr>
<tr><td>2k|2l</td><td>*</td><td>*</td><td>The HGMU receives nutrients in groundwater discharge or in (sub)surface run-off. Because nitrate is more mobile in water than ammonium, it is likely that the nitrogen in the (ground)water flow consists mostly of nitrate. For ammonia volatilisation to occur, nitrate should be reduced to ammonium. Reduction of nitrate to ammonium can occur in wetlands (dissimilatory nitrate reduction), but there is little known about this process.</td><td>2</td></tr>
<tr><td>#(1c|1d)#2c</td><td>#2 |(#4 #5)</td><td>*</td><td>Soil or water pH is too low for the process to occur.</td><td>2</td></tr>
<tr><td>*</td><td>6a</td><td>*</td><td>Soil temperature is too low to allow the process to occur.</td><td>2</td></tr>
</table>

3.2.B.1.2 Look-up Table 3

<table>
<tr><th>CV1</th><th>CV2</th><th>CV3</th><th>Rationale</th><th>Code</th></tr>
<tr><td>1c 2d</td><td>*</td><td>*</td><td>A nutrient supply derived from organic industrial waste is being applied directly to the surface of the HGMU. It is uncertain if this source contains ammonium, or even nitrogen in general.</td><td>1/x</td></tr>
<tr><td>2c</td><td>*</td><td>*</td><td>A nutrient supply with (probably) a high concentration of ammonia is being injected below the surface of the HGMU. Because of injection into the soil, ammonia will not easily volatilise.</td><td>1/x</td></tr>
<tr><td>*</td><td>3a|6b</td><td>*</td><td>Temperature is not optimal for the process to occur.</td><td>1/x</td></tr>
<tr><td>*</td><td>#7</td><td>*</td><td>Silt, clay or organic material provide adsorption sites for ammonium and therefore reduce or slow the process of ammonia volatilisation.</td><td>1/x</td></tr>
<tr><td>*</td><td>*</td><td>1a</td><td>The HGMU is classed as sheltered, which means that there is little air movement and therefore a small diffusion gradient from the HGMU surface to the atmosphere.</td><td>1/x</td></tr>
</table>

This completes the functional assessment for the process of gaseous export of N (ammonia volatilisation). The output codes can be recorded on the Functional Assessment Table at the end of Section 3.2.B.4.

3.2.B.2 Export of nutrients through vegetation management (process)

Further reading and references

Duel, H. and Saris, F.J.A. (1986). Waterzuivering door Macro-Helofytenfilters, *Landschap* **3**(4), 295–305.

Hefting, M., Clément, J.C., Bienkowski, P., Dowrick, D., Guénat, C., Butturini, A., Topa, S., Pinay, G. and Verhoeven, J. (2005). The role of vegetation and litter in the nitrogen dynamics of riparian buffer zones in Europe, *Ecological Engineering* **24**(5), 465–482.

Johnston, C.A. (1991). Sediment and nutrient retention by freshwater wetlands: effects on surface water quality, *Critical Reviews in Environmental Science and Technology* **5,6** (21), 491–565.

Koerselman, W. (1989). *Hydrology and Nutrient Budgets of Fens in an Agricultural Landscape*, PhD Dissertation, Utrecht University, The Netherlands.

Meuleman, A.F.M. (1999). *Performance of Treatment Wetlands.* PhD Dissertation, Utrecht University, The Netherlands.

Olde Venterink, H. (2000). *Nitrogen, Phosphorus and Potassium Flows Controlling Plant Productivity and Species Richness*, PhD Dissertation, Utrecht University, The Netherlands.

Olde Venterink, H., Vermaat, J.E., Pronk, M., Wiegman, F., Lee, G.E.M. van der, Hoorn, M. van de, Higler, L.W.G. and Verhoeven, J.T.A. (2006). Importance of sediment deposition and denitrification for nutrient retention in floodplain wetlands, *Applied Vegetation Science* **9**, 163–174.

Oomes, M.J.M., Kuikman, P.J. and Jacobs, F.H.H. (1997). Nitrogen availability and uptake by grassland in mesocosms at two water levels and two water qualities, *Plant and Soil* **192**, 249–259.

Oorschot, M.M.P. van (1990). De rol van vegetatie bij het verwijderen van nutriënten uit water. In: Oorschot, M.M.P. van and Verhoeven, J.T.A. (eds), *Moerassen voor de Zuivering van Water*, PAO-cursus Moerassen voor de zuivering van water, PAO-cursus, The Utrecht Plant Ecology News Report **11**, 44–51.

Oorschot, M.M.P. van (1994). Plant production, nutrient uptake and mineralization in river marginal wetlands: The impact of nutrient additions due to former land-use. In: Mitsch, W. J. (ed.), *Global Wetlands: Old World and New*, Elsevier Science B.V., Amsterdam, The Netherlands, 133–150.

Oorschot, M.M.P. van (1996). *Effects of the Vegetation on Carbon, Nitrogen and Phosphorus Dynamics in English and French Riverine Grasslands*, PhD Dissertation, Utrecht University, The Netherlands.

Oorschot, M.M.P. van, Hayes, C. and Strien, I. van (1998). The influence of soil desiccation on plant production, nutrient uptake and plant nutrient availability in two French floodplain grasslands, *Regulated Rivers: Reseach and Management* **14**, 313–327.

Orleans, A.B.M., Mugge, F.L.T., van der Meij, T., ter Keurs, W.J. (1994). *Minder nutriënten in het oppervlaktewater door bufferstroken: een literatuuranalyse*, Milieubiologie Rijksunversiteit, Leiden, The Netherlands (in Dutch).

Richardson, C.J. and Vaithiyanathan, P. (in press). Biogeochemical dynamics of phosphorus in wetlands. In: Maltby, E. and Barker, T. (eds), *The Wetlands Handbook*, Wiley-Blackwell, Oxford, UK.

Sather, J.H., Smith, R.D. and Larson, J.S. (1990). Natural values of wetlands. In: Patten, B.C. (ed.), *Wetlands and Shallow Continental Water Bodies*, SPB Academic, The Hague, The Netherlands, 373–388.

Verhoeven, J.T.A. (1990). De Nutriëntenhuishouding van Zoetwatermoerassen, speciaal met betrekking tot Verrijking. In: Oorschot, M.M.P. van and Verhoeven, J.T.A. (eds), *Moerassen voor de Zuivering van Water*, PAO-cursus Moerassen voor de zuivering van water, PAO-cursus, The Utrecht Plant Ecology News Report 11, 5–21.

Verhoeven, J.T.A., Keuter, A., Kerkhoven, M.B. van and Wassen, M. (1996). Control of local nutrient dynamics in mires by regional and climatic factors: a comparison of Dutch and Polish sites, *Journal of Ecology* **84**, 647–656.

Introduction

Process description:

- Plants retain excess nutrients (including nitrogen and phosphorus) by assimilation into their biomass. However, this is a short-term retention of nutrients, since at the end of the growing season, the death of the plants and their subsequent decomposition results in a return of the nutrients to the wetland system. Through management, the nutrients retained in vegetation can be exported from the wetland system before they are released naturally.

- Long-term retention of excess nutrients in plants occurs only in woody vegetation and is assessed in Section 3.2.A.1. Additionally, long-term retention of nutrients can occur if plant biomass is stored in soil organic matter (see Section 3.2.A.2).
- One requirement for the process to occur is that the HGMU is in receipt of excess nutrients from direct applications of nutrient-rich material or fertiliser or interaction with nutrient-bearing waters. These excess nutrients must be available to plants capable of a high rate of nutrient uptake. Subsequent management of the vegetation can result in the removal from the wetland system of plant biomass and the nutrients stored therein.
- The process of nutrient uptake is optimised in HGMUs experiencing low levels of disturbance so that the ability of plants to take up nutrients is not affected.
- The main vegetation management practices considered here are harvesting of biomass and grazing by livestock.
- Burning of vegetation is a management practice that can result in the export of nutrients via gaseous emissions. However, the amount of nutrients lost by this process is relatively small compared with the other management techniques, and therefore is not separately assessed.

Significance – rationale for assessment: The assimilation of excess nutrients into plant tissue and subsequent removal from the wetland system may improve water quality in water bodies receiving surface or subsurface run-off from the wetland.

Controlling variables

CV1 Nutrient input
CV2 Vegetation type
CV3 Vegetation management
CV4 Degree of disturbance

Answer all the questions below for *each* HGMU; unless otherwise instructed continue from one question to the next. Circle the correct answer on the appraisal sheet. The location of the relevant information is given in brackets after each question. Use a new appraisal sheet for each HGMU.

If there are no nutrient inputs identified for the HGMU, or if the nutrient input is uncertain (nutrient input codes **Nut 2** or **Nut ?**) you may wish to assess the potential of the HGMU to perform this process. To do this conduct the assessment as normal but answer questions only in ***CV2***, ***CV3*** and ***CV4***. Establish an assessment outcome as normal but ignore the need for positive answers for ***CV1*** in the assessment chart. The rationale statement can be used as normal but ignore the reference to receipt of nutrient inputs.

If the HGMU is not subject to vegetation management practices the potential export may still be assessed. Conduct the assessment as normal but answer questions only in ***CV1***, ***CV2*** and ***CV4***. Establish an assessment outcome as normal but ignore the answers for ***CV3*** and ignore the reference to vegetation management practices.

CV1 *Nutrient input*

Q1. Has the HGMU been assigned a nutrient input code indicating a direct input of nitrogen, i.e. nutrient input codes $\mathbf{N_d}$ **1/a**, $\mathbf{N_d}$ **1/b** or $\mathbf{N_d}$ **1/x**? *(Recording Sheet 2c)*

Q2. Has the HGMU been assigned a nutrient input code indicating a direct input of phosphorus, i.e. nutrient input codes $\mathbf{P_d}$ **1/a**, $\mathbf{P_d}$ **1/b** or $\mathbf{P_d}$ **1/x**? *(Recording sheet 2c)*

Q3. Has the HGMU been assigned a nutrient input code indicating a direct input of nutrients, not knowing whether this input consists of both nitrogen and phosphorus or only one of these nutrients, i.e. nutrient input codes **Nut_d 1/x**? *(Recording sheet 2c)*

Q4. Has the HGMU been assigned a nutrient input code indicating an indirect input of nitrogen, i.e. nutrient input codes **N_i 1/x**? *(Recording sheet 2c)*

Q5. Has the HGMU been assigned a nutrient input code indicating an indirect input of only phosphorus, i.e. nutrient input codes **P_i 1/x**? *(Recording sheet 2c)*

Q6. Has the HGMU been assigned a nutrient input code indicating an indirect input of nutrients, not knowing whether this input consists of both nitrogen and phosphorus or only one of these nutrients, i.e. nutrient input codes **Nut_i 1/x**? *(Recording sheet 2c)*

Q7. Has the HGMU been assigned a nutrient input code **Nut ?**? *(Recording sheet 2c)*

CV2 *Vegetation type*

Q1. Has the vegetation been classed as unvegetated or sparsely vegetated, i.e. coded as **U**?
(Recording Sheet 3, D) [*If YES do not finish this questionnaire, but go to Look-up Table 2 on p. 400 'The process in not (significantly) being performed'*]

Q2. Has the vegetation been classed as one of the following grassland codes? *(Recording Sheet 3, D)*
a. **G8–G11**, **G14**, **G30–G32**
b. Other **G** code

Q3. Has the vegetation been classed as one of the following mire codes? *(Recording Sheet 3, D)*
a. **M8**, **M9**, **M12–M15**
b. Other **M** code

Q4. Has the vegetation been classed as one of the following aquatic codes? *(Recording Sheet 3, D)*
a. **A4**, **A6**, **A7**, **A8**
b. Other **A** code

Q5. Has the vegetation been classed as one of the following shrub codes? *(Recording Sheet 3, D)*
a. **S3**
b. **S18**
c. Other **S** code

Q6. Has the vegetation been classed as one of the following woodland codes? *(Recording Sheet 3, D)*
a. **W5–W12**, **W14**, **W15**, **W17–W19**
b. Other **W** code

CV3 *Vegetation management*

Q1. Is the vegetation in the HGMU subject to harvesting, burning or grazing by livestock? *(Recording Sheet 1, C.3.3.1b; C.3.4.1b; C.3.7.1b)*
[*If YES continue with **Q2**. If NO go to Look-up Table 2 on p. 400 'The process is not (significantly) being performed', or assess the **potential** of the HGMU for export through vegetation management practices by ignoring answers for **CV3** in the Look-up table*]

Q2. How often is herbaceous vegetation within the HGMU harvested? *(Recording Sheet 1, C.3.4.1b)*
a. No history of harvesting of herbaceous vegetation in the last 20 years/not applicable [*If YES continue with **Q4***]
b. Not currently harvested but previously harvested within the last 20 years, or harvested but less than once a year [*If YES continue with **Q4***]

c. Currently harvested once or more in the last two years [*If YES continue with **Q3***]
d. Unknown [*If YES continue with **Q4***]

Q3. How often and when does harvesting take place each year? *(Recording Sheet 1, C.3.4.3b)*
a. Once, only in spring (± April–June)
b. Once, only in summer (± July–August)
c. Once, only in autumn (± September–November)
d. Once, only in winter (± December–March)
e. Twice or more a year

Q4. How often does felling and removal of timber occur in woodland within the HGMU? *(Recording Sheet 1, C.3.6.1b)*
a. Never/not applicable
b. No current felling and removal but previously practised within the last 20 years
c. Felling and removal has occurred once or more in the last 5 years
d. Unknown

Q5. Are any of the following forms of livestock kept within the HGMU? *(AA Map 2; Recording Sheet 1, C.3.7.1b)*

- Cattle or horses or donkeys — 1 ha^{-1}
- Pigs — 3 ha^{-1}
- Sheep or goats — 10 ha^{-1}
- Turkeys or geese — 30 ha^{-1}
- Chickens or ducks — 100 ha^{-1}

a. No history of these livestock types in the last 20 years/not applicable [*If YES skip **Q6** and **Q7***]
b. None currently kept but previously kept within the last 20 years, or kept but at a lower density than the figure indicated above [*If YES skip **Q6** and **Q7***]
c. Currently kept with a density equal to or higher than the figure indicated above [*If YES continue with **Q6***]
d. Unknown [*If YES skip **Q6** and **Q7***]

Q6. Is most of the manure produced by the livestock deposited outside the HGMU (e.g. in a stable or in an area outside the HGMU where animals ruminate)? *(Negative answer to Recording Sheet 1, C.3.7.3b)*

Q7. Are any of the livestock listed above in ***Q3*** kept for the purpose of meat, milk or wool production? *(Recording Sheet 1, C.3.7.2b)*

Q8. How often is vegetation within the HGMU burnt as part of land management practices? *(Recording Sheet 1, C.3.3.1b)*
a. No history of burning in the last 20 years/not applicable
b. Not currently burnt but previously burnt within the last 20 years
c. Currently burnt once or more than once in the last 5 years
d. Unknown

CV4 *Degree of disturbance*

Q1. How often does the HGMU suffer from stress because of drought (defined as less than 25 mm rainfall for the month prior to, and during each month of, the growing season)?
[*Determine the growing season (Recording Sheet 1, C.3.1.6, C.3.1.7, C.3.1.8) and compare this to the rainfall record (Recording Sheet 1, C.3.1.1i)*]
a. Never or rarely
b. Every year or most years
c. Occasionally or unknown

Q2. How often does the HGMU suffer stress from prolonged waterlogging and flooding throughout the growing season (stress occurs if soils are flooded or waterlogged for more than two successive months during the growing season)?

HGMU code: ______________________	**AA name:** ______________________

CV1 Nutrient input			
Q1	Y	N	U/NA
Q2	Y	N	U/NA
Q3	Y	N	U/NA
Q4	Y	N	U/NA
Q5	Y	N	U/NA
Q6	Y	N	U/NA
Q7	Y	N	U/NA

CV3 Vegetation management					
Q1	Y	N	U/NA		
Q2	a	b	c	d	
Q3	a	b	c	d	e
Q4	a	b	c	d	
Q5	a	b	c	d	
Q6	Y	N	U/NA		
Q7	Y	N	U/NA		
Q8	a	b	c	d	

CV2 Vegetation type			
Q1	Y	N	U/NA
Q2a	Y	N	U/NA
Q2b	Y	N	U/NA
Q3a	Y	N	U/NA
Q3b	Y	N	U/NA
Q4a	Y	N	U/NA
Q4b	Y	N	U/NA
Q5a	Y	N	U/NA
Q5b	Y	N	U/NA
Q5c	Y	N	U/NA
Q6a	Y	N	U/NA
Q6b	Y	N	U/NA

CV4 Degree of disturbance				
Q1	a	b	c	
Q2	a	b	c	
Q3	a	b	c	d
Q4	Y	N	U/NA	
Q5	Y	N	U/NA	

HGMU process assessment code: ______________________________

[**For flooding:** *determine the growing season (Recording Sheet 1, C.3.1.6, C.3.1.7, C.3.1.8) and compare with months when flooding typically occurs (Recording Sheet 1, C.3.19.6).* **For waterlogging:** *(Recording Sheet 3, B.3.2)*]

a. Never or rarely
b. Every year or most years
c. Occasionally or unknown

Q3. How frequently is the HGMU subjected to disturbance by ploughing? *(Recording Sheet 1, C.3.5.1b; AA Map 2)*

a. No history of ploughing in the last 20 years/not applicable
b. Not currently ploughed but ploughed within the last 20 years, or ploughed but less than once a year
c. Currently ploughed once or more a year
d. Unknown

Q4. Are there any current mining or extraction activities occurring that disturb the surface of the HGMU? *(Recording Sheet 1, C.3.11.1b, AA Map 2)*

Q5. Is there any current construction work occurring that is disturbing the surface of the HGMU? *(Recording Sheet 1, C.3.13.1b; AA Map 2)*

Assessment of the process

Compare the results recorded for each HGMU on the appraisal sheets with the possible outcomes below, and assign an appropriate assessment code. Record this on the appraisal sheet. The syntax codes used are described in the introduction to Section 3.

Look-up tables

1. The process is definitely being performed
If the answers to the questions for this process agree with any of the combinations in Look-up Table 1, then the HGMU is definitely performing the process of export of nutrients (N and P) through vegetation management, and an estimation of the amount of nutrients exported by this process is given.

In Look-up Table 1, the *total* export of nutrients per year is given, not only the export of *excess* nutrients. In the situations described, export of *excess* nutrients definitely takes place, but it is very hard to distinguish between removal of excess nutrients and removal of nutrients already present in the wetland system. Moreover, it is not possible to take into account the difference between export of vegetation storing nutrients received from direct sources as opposed to indirect nutrient sources. Therefore in Look-up Table 1, no distinction is made between nutrients from these different sources. The quantification figures are based on data collected during empirical studies carried out in the EU projects leading to the production of these FAPs (FAEWE, FAEWE II and PROTOWET), and published research related to this topic.

2. The process is not (significantly) being performed
If the answers to the questions for this process agree with any of the combinations in Look-up Table 2, then the HGMU is not significantly performing the process of export of *excess* nutrients (N and P) through vegetation management.

3. The process is being performed, but there are constraining factors or uncertainties
If Look-up Tables 1 and 2 did not lead to an assessment, the process of export of excess nutrients by vegetation management is being performed, but quantification is not possible, because there are one or more constraining factors, or uncertainties. In Look-up Table 3 you can find out how many constraining factors influence the process, or which factors cause the uncertainty.

4. The process is definitely not being performed
This statement is not applicable to this process.

3.2.B.2 Look-up Table 1

<table>
<tr><th>CV1</th><th>CV2</th><th>CV3</th><th>CV4</th><th>Rationale</th><th>Code</th><th>Quantification</th></tr>
<tr><td>1|2|3|4|5|6</td><td>2a</td><td>1 2c (3a|3b|3c)</td><td>1a 2a (3a|3b) #4 #5</td><td>The HGMU receives either a direct or indirect nutrient input. Grassland vegetation indicative of a high rate of nutrient uptake is present. The HGMU experiences a low degree of disturbance from drought, waterlogging, ploughing, mining or extraction or construction work, so the ability of plants to take up nutrients is not affected. The vegetation is harvested annually in spring, summer or autumn, resulting in export of nutrients retained in above-ground plant biomass.</td><td>1/b</td><td>> 30 but < 200 kg N ha^{-1} a^{-1}
> 3 but < 20 kg P ha^{-1} a^{-1}</td></tr>
<tr><td>1|2|3|4|5|6</td><td>2a</td><td>1 2c 3d</td><td>1a 2a (3a|3b) #4 #5</td><td>The HGMU receives either a direct or indirect nutrient input. Grassland vegetation indicative of a high rate of nutrient uptake is present. The HGMU experiences a low degree of disturbance from drought, waterlogging, ploughing, mining or extraction or construction work, so the ability of plants to take up nutrients is not affected. The vegetation is harvested annually in winter, resulting in export of nutrients retained in above-ground plant biomass.</td><td>1/d</td><td>> 10 but < 50 kg N ha^{-1} a^{-1}
> 1 but < 5 kg P ha^{-1} a^{-1}</td></tr>
<tr><td>1|2|3|4|5|6</td><td>2a</td><td>1 2c 3e</td><td>1a 2a (3a|3b) #4 #5</td><td>The HGMU receives either a direct or indirect nutrient input. Grassland vegetation indicative of a high rate of nutrient uptake is present. The HGMU experiences a low degree of disturbance from drought, waterlogging, ploughing, mining or extraction or construction work, so the ability of plants to take up nutrients is not affected. The vegetation is harvested more than once a year, resulting in large export of nutrients retained in above-ground plant biomass.</td><td>1/a</td><td>> 50 but < 400 kg N ha^{-1} a^{-1}
> 5 but < 40 kg P ha^{-1} a^{-1}</td></tr>
<tr><td>1|2|3|4|5|6</td><td>3a|4a</td><td>1 2c 3a</td><td>1a (3a|3b) #4 #5</td><td>The HGMU receives either a direct or indirect nutrient input. Mire or aquatic vegetation indicative of a high rate of nutrient uptake is present. The HGMU experiences a low degree of disturbance from drought, ploughing, mining or extraction or construction work, so the ability of plants to take up nutrients is not affected. The vegetation is harvested annually in spring, resulting in export of nutrients retained in above-ground plant biomass. Removal is less than when harvested at peak standing crop (summer/start of autumn).</td><td>1/c</td><td>> 20 but < 100 kg N ha^{-1} a^{-1}
> 2 but < 10 kg P ha^{-1} a^{-1}</td></tr>
<tr><td>1|2|3|4|5|6</td><td>3a|4a</td><td>1 2c (3b|3c)</td><td>1a (3a|3b) #4 #5</td><td>The HGMU receives either a direct or indirect nutrient input. Mire or aquatic vegetation indicative of a high rate of nutrient uptake is present. The HGMU experiences a low degree of disturbance from drought, ploughing, mining or extraction or construction work, so the ability of plants to take up nutrients is not affected. The vegetation is harvested annually in summer or autumn, resulting in export of nutrients retained in above-ground plant biomass.
NB: When reed (Phragmites australis) is harvested in summer, the long-term vitality of the reed stand can be affected. To assure its long-term vitality, it is better to harvest in autumn.</td><td>1/a</td><td>> 50 but < 400 kg N ha^{-1} a^{-1}
> 5 but < 40 kg P ha^{-1} a^{-1}</td></tr>
<tr><td>1|2|3|4|5|6</td><td>3a|4a</td><td>1 2c 3d</td><td>1a (3a|3b) #4 #5</td><td>The HGMU receives either a direct or indirect nutrient input. Mire or aquatic vegetation, indicative of a high rate of nutrient uptake, is present. The HGMU experiences a low degree of disturbance from drought, ploughing, mining or</td><td>1/c</td><td>> 20 but < 100 kg N ha^{-1} a^{-1}
> 2 but < 10 kg P ha^{-1} a^{-1}</td></tr>
</table>

<table>
<tr><td></td><td></td><td></td><td></td><td>extraction or construction work, so the ability of plants to take up nutrients is not affected. The vegetation is harvested annually in winter, resulting in export of nutrients retained in above-ground plant biomass. In winter, many plants translocate nutrients from above-ground plant tissue to below-ground plant tissue as a reserve for the next growing season. Harvesting in winter therefore results in less nutrient removal than in summer or autumn, before the nutrients are translocated.</td><td></td><td></td></tr>
<tr><td>1|2|3|4|5|6</td><td>3a|4a</td><td>1 2c 3e</td><td>1a (3a|3b) #4 #5</td><td>The HGMU receives either a direct or indirect nutrient input. Mire or aquatic vegetation, indicative of a high rate of nutrient uptake, is present. The HGMU experiences a low degree of disturbance from drought, ploughing, mining or extraction or construction work, so the ability of plants to take up nutrients is not affected. The vegetation is harvested more than once a year, resulting in large export of nutrients retained in above-ground plant biomass.</td><td>1/a</td><td>> 50 but < 400 kg N ha^{-1} a^{-1}
> 5 but < 40 kg P ha^{-1} a^{-1}</td></tr>
<tr><td>1|2|3|4|5|6</td><td>2a</td><td>5c (6|7)</td><td>1a 2a (3a|3b) #4 #5</td><td>The HGMU receives either a direct or indirect nutrient input. Grassland vegetation, indicative of a high rate of nutrient uptake, is present. The HGMU experiences a low degree of disturbance from drought, waterlogging, ploughing, mining or extraction or construction work, so the ability of plants to take up nutrients is not affected. Grazing results in the export of nutrients taken-up by the plants. The amount of nutrients exported through grazing is always less than the amount of nutrients exported through harvesting, because not all vegetation will be grazed and some animals can spill a considerable amount of (nutrient-rich) plant material. This material is more nutrient-rich than plant material deposited after plant senescence. This phenomenon is particularly important with regard to geese. Additionally, nutrients are returned to the system in the form of excrement.</td><td>1/d</td><td>> 10 but < 50 kg N ha^{-1} a^{-1}
> 1 but < 5 kg P ha^{-1} a^{-1}</td></tr>
</table>

3.2.B.2 Look-up Table 2

<table>
<tr><th>CV1</th><th>CV2</th><th>CV3</th><th>CV4</th><th>Rationale</th><th>Code</th></tr>
<tr><td>#(1 2 3 4 5 6|7)</td><td>*</td><td>*</td><td>*</td><td>The HGMU does not receive any nutrient additions, or it is uncertain whether the HGMU receives a nutrient input, i.e. there are no excess nutrients, thus the process of export of excess nutrients through vegetation management is not performed.
When harvesting or grazing takes place, nutrients taken up by plants (as part of the normal nutrient cycle) nevertheless will be exported from the wetland.</td><td>2</td></tr>
<tr><td>*</td><td>1</td><td>*</td><td>*</td><td>There is little or no vegetation available for removal from the HGMU.</td><td>2</td></tr>
<tr><td>*</td><td>2b|3b|4b|5c|6b</td><td>*</td><td>*</td><td>Vegetation indicative of a low rate of nutrient uptake is present. Therefore excess nutrients will not be taken up.
When harvesting or grazing takes place, nutrients takenup by the plants (as part of the normal nutrient cycle) nevertheless will be exported from the wetland.
Harvesting or grazing may be necessary when the system does receive excess nutrients. Without management activities, an increase in nutrient additions may cause an unwanted shift in species composition.</td><td></td></tr>
<tr><td>*</td><td>5a|5b|6a</td><td>#4c</td><td>*</td><td>Shrubs or trees indicative of a high rate of nutrient uptake are present, but they are not felled and removed.
NB: The process of long-term retention of nutrients can be performed. This process is assessed in Section 3.2.A.1.</td><td>2</td></tr>
<tr><td>*</td><td>*</td><td>#1</td><td>*</td><td>No (management) activities resulting in the export of nutrients take place.
NB: The process of long-term retention of nutrients can be performed. This process is assessed in Section 3.2.A.1.</td><td>2</td></tr>
<tr><td>*</td><td>*</td><td>5c #6 #7</td><td>*</td><td>The vegetation is grazed by livestock, but there is no export of nutrients because the animals are not kept for the purpose of meat, milk or wool and/or the nutrient-rich manure of the animals is deposited in the HGMU so most nutrients are re-mobilised within the HGMU.</td><td>2</td></tr>
</table>

3.2.B.2 Look-up Table 3

<table>
<tr><th>CV1</th><th>CV2</th><th>CV3</th><th>CV4</th><th>Rationale</th><th>Code</th></tr>
<tr><td>*</td><td>3a|4a</td><td>5c 6</td><td>*</td><td>Mire or aquatic vegetation indicative of a high rate of nutrient uptake is present. This type of vegetation is normally not grazed by commercial livestock, but rather by ducks or geese (see CV3, Q5). Although these birds eat (nutrient-rich) plant material, they can also spill a considerable amount. This material is more nutrient-rich than plant material deposited after plant senescence. This phenomenon is particularly important with regard to geese. Additionally, nutrients are returned to the system in the form of excrement. It is difficult to state that there is real export and, if there is, it is very difficult to quantify.</td><td>1/x</td></tr>
<tr><td>*</td><td>5a|5b|6a</td><td>4c</td><td>*</td><td>Shrubs or trees indicative of a high rate of nutrient uptake are present. Felling or removal of trees results in the export of nutrients. A large amount of nutrients are exported in this way, but most of these nutrients had been taken up over a long period prior to harvesting and therefore its significance is difficult to quantify. Moreover, during felling or removal, a large amount of nutrient-rich tree tissue (leaves and branches) can fall and become mineralised, resulting in some return of nutrients to soil and water. All these factors make it difficult to quantify this process.</td><td>1/x</td></tr>
<tr><td>*</td><td>*</td><td>8c</td><td>*</td><td>Some nutrients in plant biomass are exported to the atmosphere through burning but nutrient-rich ashes also fertilise the soil. Fast re-growth of plants after the burning event will result in uptake of these nutrients, but outflow or leaching from the system is also possible. Therefore it is very difficult to quantify this process.</td><td>1/x</td></tr>
<tr><td>*</td><td>*</td><td>*</td><td>1b|2b|
3c|4|5</td><td>Other factors may be suitable, but disturbance or stress because of drought, waterlogging, ploughing, mining, extraction or construction work may reduce the uptake of nutrients by plants.</td><td>1/x</td></tr>
<tr><td>*</td><td>*</td><td>*</td><td>1c|2c|
3d|6d</td><td>Other factors may be suitable, but it is not known whether disturbance or stress (drought, waterlogging, ploughing, mining, extraction, construction work, or burning of vegetation) takes place. These disturbances might reduce the uptake of nutrients by plants.</td><td>1/x</td></tr>
</table>

This completes the functional assessment for the process of export of nutrients through vegetation management. The output codes and quantification values can be recorded on the Functional Assessment Table at the end of Section 3.2.B.4.

3.2.B.3 Export of nutrients via water- and wind-mediated processes (process)

Further reading and references

Campbell, D.R., Lavoie, C. and Rochefort, L. (2002). Wind erosion and surface stability in abandoned milled peatlands, *Canadian Journal of Soil Science* **82**(1), 85–95.

Heathwaite, A.L. and Dils, R.M. (2000). Characterising phosphorus loss in surface and subsurface hydrological pathways, *Science of the Total Environment* **251/252**, 523–538.

Heathwaite, A.L., Sharpley, A.N. and Gburek, W.J. (2000). A conceptual approach for integrating phosphorus and nitrogen management at watershed scales, *Journal of Environmental Quality* **29**, 158–166.

Heckrath, G., Brookes, P.C., Poulton, P.R. and Goulding, K.W.T. (1995). Phosphorus leaching from soils containing different phosphorus concentrations in the Broadbalk experiment, *Journal of Environmental Quality* **24**, 904–910.

Lyles, L. (1975). Possible effects of wind erosion on soil productivity, *Journal of Soil and Water Conservation* **30**, 279–283.

Palis, R.G., Okwach, G. and Rose, C.W. (1990). Soil-erosion processes and nutrient loss, 1. The interpretation of enrichment ratio and nitrogen loss in runoff sediment, *Australian Journal of Soil Research* **28**(4), 623–639.

Selkirk, J.M. and Saffigna, L.J. (1999). Wind and water erosion of a peat and sand area on sub-antarctic Macquarie Island, *Arctic Antarctic Alpine Research* **31**(4), 412–420.

Sharpley, A.N. (1985). Depth of surface soil runoff interaction as affected by rainfall, soil slope and management, *Soil Science Society of America Journal* **49**, 1010–1015.

Toy, T.J., Foster, G.R. and Renard, K.G. (2002). *Soil Erosion: Processes, Prediction, Measurement and Control*, Wiley, New York.

Warburton, J. (2003). Wind-splash erosion of bare peat on UK upland moorlands, *Catena* **52**(3–4), 191–207.

Zhang, G.H., Liu, B.Y. and Liu, G.B. (2003). Detachment of undisturbed soil by shallow flow. *Soil Science Society America Journal* **67**(3), 713–719.

Introduction

Process description:

- The transport of excess nutrients (nitrogen or phosphorus) either dissolved in water or chemically bound to particulate matter, by water- or wind-mediated processes and subsequent export from the HGMU.
- The process is considered to occur when an HGMU has an input of, or has retained nutrients from, either direct application of fertiliser or interaction with nutrient-rich waters, or when nutrients have previously been retained by other wetland processes, and these are removed by water or wind transport processes.
- Desorption of nutrients from soil particles (i.e. the mobilisation of previously adsorbed nutrients) resulting in nutrients becoming available in the water column is not considered in this section.
- It is difficult to quantify the process of export of nutrients via water- and wind-mediated processes because the process is both temporally and spatially highly variable. Therefore only a qualitative assessment of export of nutrients via water- and wind-mediated processes is given.

The wetland may be acting as a source of nutrients, which can contaminate water bodies.

Significance – rationale for assessment: The wetland may be supporting systems elsewhere by providing a nutrient supply.

Note: In most cases the amount of nutrients exported by wind-mediated processes is likely to be small in comparison with the amount exported by water-mediated processes.

Controlling variables

CV1 Nutrient input
CV2 Hydrological regime
CV3 Landform
CV4 Soil/sediment erosion

Answer all the questions below for *each* HGMU; unless otherwise instructed continue from one question to the next. On the appraisal sheet circle the correct answer. The location of the relevant information is given in brackets after each question. Use a new appraisal sheet for each HGMU.

If there are no significant nutrient inputs identified for the HGMU, or if the nutrient input is uncertain (nutrient input codes **Nut 2** or **Nut ?**), you may wish to assess the potential of the HGMU to perform this process. To do this, conduct the assessment as normal but answer questions only in ***CV2***, ***CV3*** and ***CV4***. Establish an assessment outcome as normal but ignore the need for positive answers for ***CV1***. The rationale statement can be used as normal but ignore the reference to receipt of nutrient inputs.

CV1 *Nutrient input*

Q1. Which of the following nutrient input codes applies to the HGMU?
a. $\mathbf{N_d}$ **1/a**, $\mathbf{N_d}$ **1/b** or $\mathbf{N_d}$ **1/x**
b. $\mathbf{P_d}$ **1/a**, $\mathbf{P_d}$ **1/b** or $\mathbf{P_d}$ **1/x**
c. $\mathbf{Nut_d}$ **1/b** or $\mathbf{Nut_d}$ **1/x**
d. $\mathbf{N_i}$ **1/x**
e. $\mathbf{P_i}$ **1/x**
f. $\mathbf{Nut_i}$ **1/x**
g. **Nut ?**
h. No input (**Nut 2**)
(Recording Sheet 2c)

Only answer ***Q2–Q5*** if you have already conducted a functional assessment of function 3.2.A, Nutrient retention. Otherwise go to ***CV2***.

Q2. Has the HGMU been classed as **1/a**, **1/b** or **1/x** for the assessment of process 3.2.A.2, storage of nutrients in soil organic matter? *(Functional Assessment 3.2.A.2)*

Q3. Has the HGMU been classed as **1** or **1/x** for the assessment of process 3.2.A.3, adsorption of N as ammonium? *(Functional Assessment 3.2.A.3)*

Q4. Has the HGMU been classed as **1** or **1/x** for the assessment of process 3.2.A.4, adsorption and precipitation of P in the soil? *(Functional Assessment 3.2.A.4)*

Q5. Has the HGMU been classed as **1** (and not as **1/x**) for the assessment of process 3.2.A.5, retention of particulate nutrients (N and P)? *(Functional Assessment 3.2.A.5)*

CV2 *Hydrological regime*

Q1. Has the HGMU been assigned a hydrological code indicating that run-off output (**F3**) occurs, i.e. **H3**, **H4**, **H6**, **H7**, **H8**, **H9**, **H11** or **H12**? *(Recording Sheet 2a, F3)*

Q2. Has the HGMU been assigned a hydrological code indicating that overbank flooding (**F4**) occurs, i.e. **H4**, **H7**, **H9**, **H10**, **H12**, **H13**, **H14** or **H15**? *(Recording Sheet 2a, F4)*
[*If YES continue with **Q3**. If NO continue with **Q4***]

Q3. Do flow velocities in the HGMU exceed 0.3 m s^{-1}? (*Recording Sheet 3, B.4.6.*)

Q4. Which term is used to describe the topsoil moisture content status? *(Recording Sheet 2a, C.3.2)*
a. Dry
b. Moist
c. Wet
d. Flooded

Q5. Is the average groundwater table ≥ 1 m below the surface? *(Recording Sheet 3, B.3.10, and/or AA Map 2)*

Q6. Is the total annual precipitation < 500 mm? *(Recording Sheet 1, C.3.1.1i)*

Only answer ***Q7*** if you have already conducted a functional assessment of function 3.1.B, Groundwater recharge. Otherwise go to ***CV3***.

Q7. Has the HGMU been assigned a score of **1/n** for function of 3.1.B, Groundwater recharge? *(Functional Assessment 3.1.B)*

CV3 *Landform*

Q1. Does the HGMU share a boundary with the water body or with an HGMU that has been assigned one of the following geomorphic codes: **TB**, **FC**, **DT**, **LK**, **OX**, **BS**, **CL** or **CN**? *(AA Map 2; Recording Sheet 2a, C.2)*

Q2. Is there a clear flow path without any obstacles or barriers that could prevent water flowing or draining into neighbouring HGMUs (if present) and hence to a nearby (river) channel or water body? *(Recording Sheet 3, B.2.3; AA Map 2)*

Q3. Has the HGMU been assigned a gradient code **G2** or **G3,** indicating that it is moderate or steep (> 5°)? *(Recording Sheet 2a, C.2)*

Q4. How is the topsoil particle size described? *(Recording Sheet 2a, C.3.7)*

a. Fine
b. Medium
c. Coarse or Gravel

Q5. Within the soil profile, is there a distinct horizon of indurated material (such as a hard pan) that could prevent surface water draining through the soil? *(Recording Sheet 2a, C.3 General comments)*

Q6. What is the topographic exposure of the HGMU? *(Recording Sheet 3, A.4)*

a. 3 (Sheltered)
b. 2 (Moderate)
c. 1 (Exposed)

Q7. What is the average wind speed in the AA? *(Recording Sheet 1, C.3.1.4)*

a. ≤ 4.8 m s^{-1}
b. > 4.8 and ″ 6.6 m s^{-1}
c. > 6.6 m s^{-1}

CV4 *Soil/sediment erosion*

Q1. Are there any indicators of erosion within the HGMU, e.g. scouring in units not adjacent to a river, or in riparian units, undercutting of the bank or the presence of steep exposed banks? *(Recording Sheet 3, B.5.2)*

Q2. Are there any substantial patches of bare or eroded land within the HGMU? *(Recording Sheet 3, B.5.9)*

Q3. Is the habitat unvegetated or sparsely vegetated, i.e. Coded **U**? (*Recording Sheet 3, D*)

Q4. If any part of the HGMU currently experiences a high degree of disturbance, what is its cause? *(Recording Sheet 1)*

a. Harvesting of herbaceous vegetation *(C.3.4.1b (answer c))*
b. Ploughing of land *(C.3.5.1b (answer c))*
c. Woodland management and forestry *(C.3.6.1b (answer c))*
d. Trampling by livestock *(C.3.7.1b (answer c))*
e. Recreational activities (question of the AA) *(C.3.15.1 (answer c))*
f. None

Q5. If there are trees in the HGMU are they wind-cropped? *(Recording Sheet 3, A.5)*

Assessment of the process

Compare the results recorded for each HGMU on the appraisal sheet with the possible outcomes below, and assign an appropriate assessment code. Record this

3.2.B.3 *Export of Nutrients (N and P) via Wind- and Water-mediated Processes Appraisal Sheet*

HGMU code: ______________________	**AA name:** ______________________

CV1 Nutrient input				
Q1	a	b	c	d
	e	f	g	h
Q2	Y	N	U/NA	
Q3	Y	N	U/NA	
Q4	Y	N	U/NA	
Q5	Y	N	U/NA	

CV2 Hydrological regime				
Q1	Y	N	U/NA	
Q2	Y	N	U/NA	
Q3	Y	N	U/NA	
Q4	a	b	c	d
Q5	Y	N	U/NA	
Q6	Y	N	U/NA	
Q7	Y	N	U/NA	

CV3 Landform			
Q1	Y	N	U/NA
Q2	Y	N	U/NA
Q3	Y	N	U/NA
Q4	a	b	c
Q5	Y	N	U/NA
Q6	a	b	c
Q7	a	b	c

CV4 Soil/sediment erosion						
Q1	Y	N	U/NA			
Q2	Y	N	U/NA			
Q3	Y	N	U/NA			
Q4	a	b	c	d	e	f
Q5	Y	N	U/NA			

HGMU process assessment code: ______________________________________

3.2.B.3 Look-up Table 1

<table>
<tr><th>CV1</th><th>CV2</th><th>CV3</th><th>CV4</th><th>Rationale</th><th>Code</th></tr>
<tr><td>1a|1b|1c</td><td>1</td><td>(1|2|3)
(4a|4b|5)</td><td>(1|2)|
((1|2)
(3|4a-e))</td><td>The HGMU receives a direct input of nutrients. Nutrients are likely to be exported to nearby rivers or water bodies in run-off as particulate matter. There is evidence that erosion occurs within the HGMU. When nitrate or nitrogenous compounds that are easily transformed into nitrate are the main nutrient input, it is likely that this nitrate is exported as dissolved matter.
NB: The wetland acts as a source of nutrients and therefore supplies water bodies receiving water from the wetland.</td><td>1</td></tr>
<tr><td>1a</td><td>1</td><td>(1|2|3)
(4a|5)</td><td>*</td><td>The HGMU receives a direct input of nitrogen. When nitrate or nitrogenous compounds that are easily transformed into nitrate are the main nutrient input, it is likely that this nitrate is exported as dissolved matter.
NB: The wetland acts as a source of nitrate and therefore supplies water bodies receiving water from the wetland.</td><td>1</td></tr>
<tr><td>1a|1b|1c</td><td>2 3</td><td>(1|2|3)
(4a|4b|5)</td><td>(1|2)|
((1|2)
(3|4a|4b|
4c|4d|4e))</td><td>The HGMU receives a direct input of nutrients. Nutrients are likely to be exported into the river or water body in retreating floodwater as particulate matter. There is evidence that erosion occurs within the HGMU, or activities take place likely to generate erosion and sediment movement. When nitrate or nitrogenous compounds that are easily transformed into nitrate are the main nutrient input or activities are likely to generate erosion or sediment movement, it is likely that this nitrate is exported as dissolved matter.
NB: The wetland acts as a source of nutrients and therefore supplies them to water bodies that receive water from the wetland.</td><td>1</td></tr>
<tr><td>1a</td><td>2 3</td><td>(1|2|3)
(4a|5)</td><td>*</td><td>The HGMU receives a direct input of nitrogen. When nitrate or nitrogenous compounds that are easily transformed into nitrate are the main nutrient input, it is likely that this nitrate is exported into the river or water body in retreating floodwater as dissolved matter.
NB: The wetland acts as a source of nitrate and therefore supplies it to water bodies that receive water from the wetland.</td><td>1</td></tr>
<tr><td>1a</td><td>(4a 5)|7</td><td>(4c) #5</td><td>*</td><td>The HGMU receives a direct input of nitrogen. There is evidence that groundwater recharge occurs within the HGMU. When nitrate or nitrogenous compounds that are easily transformed into nitrate are the main nutrient input, it is likely that this nitrate is exported from the HGMU into the groundwater.
NB: High concentrations of nitrate in the groundwater may cause problems for drinking water supply.</td><td>1</td></tr>
<tr><td>1a|1b|1c</td><td>4a 6</td><td>6c 7c</td><td>2|3|4a|4b
|4c|4d|
4e|5</td><td>The HGMU receives a direct input of nutrients. Conditions conducive to wind erosion are present. The HGMU is exposed and mean wind velocity is high. There is evidence that erosion occurs within the HGMU. Particulate nutrients are likely to be exported from the HGMU via wind erosion.</td><td>1</td></tr>
<tr><td>(1d|1e|
1f)|(2|3|
4 #5)</td><td>4a 6</td><td>6c 7c</td><td>2|3|4a|
4b|4c|
4d|4e|5</td><td>The HGMU receives an indirect input of nutrients, or excess nutrients are retained in the soil of the HGMU. Conditions conducive to wind erosion are present. The HGMU is exposed and mean wind velocity is high. There is evidence that erosion occurs within the HGMU. Particulate nutrients are likely to be exported from the HGMU via wind erosion.</td><td>1</td></tr>
</table>

<table>
<tr><td>1d</td><td>(4a 5)|7</td><td>(4c) #5</td><td>*</td><td>The HGMU receives an indirect input of nitrogen (probably nitrate). There is evidence that groundwater recharge occurs within the HGMU. Dissolved nitrate is likely to be exported from the HGMU into the groundwater.
High concentrations of nitrate in the groundwater may cause problems for drinking water supply.</td><td>1</td></tr>
<tr><td>(2|3|4)
#5</td><td>1</td><td>(1|2|3)
(4a|4b|5)</td><td>(1|2)|
((1|2)
(3|4a|4b|
4c|4d|4e))</td><td>Excess nutrients are retained in the soil of the HGMU. Nutrients are likely to be exported into the river or water body in run-off water as particulate material. There is evidence that erosion occurs within the HGMU.
NB: The wetland acts as a source of nutrients and therefore supplies water bodies receiving water from the wetland.</td><td>1</td></tr>
<tr><td>(2|3|4)
#5</td><td>2 3</td><td>(1|2|3)
(4a|4b|5)</td><td>(1|2)|
((1|2)
(3|4a|4b|
4c|4d|4e))</td><td>Excess nutrients are retained in the soil of the HGMU. Nutrients are likely to be exported into the river or water body in retreating floodwater as particulate matter. There is evidence that erosion occurs within the HGMU.
NB: The wetland acts as a source of nutrients and therefore supplies them to water bodies that receive water from the wetland.</td><td>1</td></tr>
</table>

3.2.B.3 Look-up Table 2

<table>
<tr><th>CV1</th><th>CV2</th><th>CV3</th><th>CV4</th><th>Rationale</th><th>Code</th></tr>
<tr><td>1b
(# 1a
1c–h 2
3 4)</td><td>1| (2 3)</td><td>(1|2|3)
(4a|4b|5)</td><td>#(1 2 3
4a–e)</td><td>The HGMU receives a direct input of phosphorus. It is unlikely that phosphorus is exported to nearby rivers or water bodies in run-off or retreating floodwater as dissolved matter, as phosphorus will be adsorbed to soil particles or will precipitate during transportation.</td><td>2</td></tr>
<tr><td>1h
#(2 3 4)</td><td>*</td><td>*</td><td>*</td><td>The HGMU does not receive an input of nutrients or retain dissolved nutrients.</td><td>2</td></tr>
<tr><td>5</td><td>*</td><td>*</td><td>*</td><td>Particulate nutrients are retained and erosion does not take place.</td><td>2</td></tr>
<tr><td>*</td><td>#(1 2 3
4a 5 6 7)</td><td>*</td><td>*</td><td>There is no apparent pathway for nutrients to be exported.</td><td>2</td></tr>
<tr><td>*</td><td>#(1 2 3 7)
4a 6</td><td>6a|7a</td><td>*</td><td>There is no floodwater or run-off water that can transport eroded nutrients. Low wind velocities or sheltering of the HGMU mean that wind erosion is not an important factor.</td><td>2</td></tr>
<tr><td>#1c</td><td>*</td><td>*</td><td>#(1 2 3
4a–e 5)
4f</td><td>There is no evidence of erosion.</td><td>2</td></tr>
</table>

3.2.B.3 Look-up Table 3

<table>
<tr><th>CV1</th><th>CV2</th><th>CV3</th><th>CV4</th><th>Rationale</th><th>Code</th></tr>
<tr><td>1c
(# 1a–b
1d–h
2 3 4)</td><td>1|(2 3)</td><td>(1|2|3)
(4a|4b|5)</td><td>#(1 2 3
4a|4b|4c|
4d|4e)</td><td>The HGMU receives a direct input of nutrients. It is uncertain if the input contains nitrogen. When nitrate or nitrogenous compounds that are easily transformed into nitrate are the main nutrient input, it is likely that this nitrate is exported into the river or water body in run-off or retreating floodwater as dissolved matter. However, when the input is phosphorus only, dissolved export in run-off or retreating floodwater is unlikely, as phosphorus will be adsorbed to soil particles or will precipitate during transportation.</td><td>1/x</td></tr>
<tr><td>1g</td><td>*</td><td>*</td><td>*</td><td>It is not certain whether the HGMU receives nutrients. If the wetland received nutrients in the past, these nutrients may be stored in the wetland. The process of erosion will export these nutrients. It is possible that nutrients (especially P) applied to or received by the HGMU in the past are adsorbed to soil particles. Changes in hydrological regime that result in, for example, a normally aerobic soil becoming anaerobic, or a supply of sulphate-rich surface water to a normally sulphate-poor environment, can result in mobilisation and possibly export, of adsorbed nutrients.</td><td>1/x</td></tr>
<tr><td>1d|1e|1f</td><td>1</td><td>(1|2|3)
(4a|4b|5)</td><td>(1|2)|
((1|2)
(3|4a|4b|
4c|4d|
4e))</td><td>The HGMU receives an indirect input of nutrients. Nutrients are likely to be exported into the river or water body in run-off water as particulate material. There is evidence that erosion occurs within the HGMU. When nitrate or nitrogenous compounds that are easily transformed into nitrate are the main nutrient input, it is likely that this nitrate is exported as dissolved matter.
As there is (indirect) input of nutrients as well as export, it is uncertain if the result is a net export of nutrients.
NB: The wetland acts as a source of nutrients and therefore supplies them to water bodies that receive water from the wetland.</td><td>1/x</td></tr>
<tr><td>1d|1e|1f</td><td>2 3</td><td>(1|2|3)
(4a|4b|5)</td><td>(1|2)|
((1|2)
(3|4a|4b|
4c|4d|
4e))</td><td>The HGMU receives an indirect input of nutrients or excess nutrients that are retained in the soil of the HGMU. Nutrients are likely to be exported into the river or water body in retreating floodwater as particulate matter.
There is evidence that erosion occurs within the HGMU. When nitrate or nitrogenous compounds that are easily transformed into nitrate are the main nutrient input, it is likely that this nitrate is exported as dissolved matter.
As there is (indirect) input of nutrients as well as export, it is uncertain if the result is a net export of nutrients.
NB: The wetland acts as a source of nutrients and therefore supplies water bodies receiving water from the wetland.</td><td>1/x</td></tr>
<tr><td>*</td><td>1|(2 3)</td><td>#(1 2 3)|
#(4a 4b 5)</td><td>*</td><td>Evidence for erosion is not very strong.</td><td>1/x</td></tr>
<tr><td>*</td><td>1|(2 3)</td><td>*</td><td>#(1 2)</td><td>Evidence for erosion is not very strong.</td><td>1/x</td></tr>
<tr><td>*</td><td>2 #3</td><td>*</td><td>*</td><td>Flow velocity is not fast enough to cause erosion.</td><td>1/x</td></tr>
<tr><td>*</td><td>4a 6</td><td>6b|7b</td><td>*</td><td>Conditions for wind erosion are not optimal.</td><td>1/x</td></tr>
<tr><td>*</td><td>(4a|5)|7</td><td>4a|4b|5</td><td>*</td><td>Conditions for exportation of nutrients via groundwater recharge are not optimal.</td><td>1/x</td></tr>
</table>

on the appraisal sheet. The syntax codes used are described in the introduction to Section 3.

Look-up tables

1. The process is definitely being performed

If the answers to the questions for this process agree with any of the combinations in Look-up Table 1, then the HGMU is definitely performing the process of export of nutrients via water- and wind-mediated processes. It is difficult to quantify this, because it is both temporally and spatially highly variable. Therefore only a qualitative assessment is given (the code consists only of one number, the quantitative information being excluded).

2. The process is not (significantly) being performed

If the answers to the questions for this process agree with any of the combinations in Look-up Table 2, then the HGMU is not significantly performing the process of nutrient export via water- and wind-mediated processes.

3. The process is being performed, but there are constraining factors or uncertainties

If Look-up Tables 1 and 2 did not lead to an assessment, the process of nutrient export by water- and wind-mediated processes is performed, but there are constraining factors or uncertainties. In Look-up Table 3 you can find out how many constraining factors influence the process, or which factors cause the uncertainty.

4. The process is definitely not being performed

This statement is not applicable to this process.

This completes the functional assessment for the process of export of nutrients through water- and wind-mediated processes. The output codes and quantification values can be recorded on the Functional Assessment Table at the end of Section 3.2.B.4.

3.2.B.4 Functional assessment: Nutrient export

The purpose of this section is to determine whether the Nutrient Export function is occurring and, if so, to what degree. Follow the method below for *each* HGMU. On the Functional Assessment Table at the end of this section, record the functional outcome for each HGMU within your AA.

Nutrient export is definitely being performed

The function 'nutrient export' is definitely being performed by the HGMU if one or more of the component processes of this function are occurring, i.e. if:

Any of the process assessment codes = **1/n** (results from Look-up Table 1 (definitely performed) and/or **1/x** (performed but with uncertainties or constraints)

The statement *nutrient export is definitely being performed* is only a generalised statement. It does not indicate any quantitative or semi-quantitative information. Not all processes contribute equally to the function. Therefore, if only one process with a low maximum export rate is contributing to the function, a smaller amount of nutrient will be exported than if several processes, or one process with a high maximum export rate, are contributing to the function. To estimate the degree of performance of the function, the contribution of the processes to the function has to be calculated. Guidance for this calculation, which uses the quantitative information, is given below.

Nutrient export is not significantly being performed

The function 'nutrient export' is not significantly being performed by an HGMU if *none* of the component processes of this function is occurring, i.e. if:

All process assessment codes = **2**

Summing of processes to determine a function

In order to be able to take account of the outcomes of all the different processes, both quantified *and* non-quantified processes are assigned an *estimated* quantitative value. A separate functional assessment is performed for nitrogen (N) export and for phosphorus (P) export as N and P are involved in different processes and they react differently in varied circumstances.

The estimated value of a quantified process consists of the mean of two end-points of the quantitative range given for the assessment of that process, e.g. when the assessment outcome code is **1/a**, the average value for 'export through vegetation management' is 225 kg ha^{-1} a^{-1} for nitrogen ((50 + 400)/2) and 22.5 kg ha^{-1} a^{-1} for phosphorus ((5 + 40)/2).

The values for the non-quantified processes or process outcomes (outcomes assessed with code **1** or code **1/x**) have all been estimated. The values for processes with assessment outcome code **1** are based on expert judgement: the importance of these processes is scaled relative to the rates of the quantified processes. Values of processes with code **1/x** have been estimated as 50 % of the lowest quantitative value given to any of the codes **1/c**, **1/b** or **1**. This may underestimate the export rate in a number of cases, but it is intended as a conservative quantitative estimation of the process.

In the nitrogen and phosphorus tables in the following Section (Tables 3.2.4–3.2.6), the average (estimated) values are given in the light grey rows. In case of quantified processes, the minimum and maximum rates are also given. All estimated values (i.e. all values that could not be found in the assessment tables of the process) are indicated with * in the table.

Functional assessment tables: 3.2.B Nutrient export

Tables 3.2.4, 3.2.5 and 3.2.6 on pages 412 to 414 enable calculation of the total value for the function 'nutrient export' for nitrogen and for phosphorus, respectively. The tables can be used for one HGMU only. Therefore, it is necessary to make copies of the tables for each HMGU, before you start the assessment. Use of the tables is explained stepwise.

Step 1

On the appraisal sheets, look up which codes have been assigned to the following processes:

- 3.2.B.1.1 Denitrification (N)
- 3.2.B.1.2 Ammonia volatilisation (N)
- 3.2.B.2 Export of nutrients through vegetation management (N and P)

The process 'export of nutrients via wind- and water-mediated processes' is not taken into account in this functional assessment for the following reasons:

- The process is highly variable in space and time and therefore it is not possible to make even a rough estimation.
- All other nutrient export and nutrient retention processes result in water *purification*. However, export of nutrients via wind- and water-mediated processes results either in water purification (wind erosion) *or* in water *contamination* (erosion by water-mediated processes).

Step 2

For each process, look up the estimated average values for nitrogen and phosphorus accompanying your codes in the tables (light grey rows). Circle these values.

Table 3.2.4 Nitrogen export

NITROGEN		**Denitrification (N)**	**Ammonia volatilisation**	**Vegetation management (N)**
Code **1/a**	Min	10* kg N ha^{-1} a^{-1}		50 kg N ha^{-1} a^{-1}
	Max	300 kg N ha^{-1} a^{-1}		400 kg N ha^{-1} a^{-1}
	Average	155 kg N ha^{-1} a^{-1}		225 kg N ha^{-1} a^{-1}
Code **1/b**	Min	10* kg N ha^{-1} a^{-1}		30 kg N ha^{-1} a^{-1}
	Max	80 kg N ha^{-1} a^{-1}		200 kg N ha^{-1} a^{-1}
	Average	45 kg N ha^{-1} a^{-1}		115 kg N ha^{-1} a^{-1}
Code **1/c**	Min	5* kg N ha^{-1} a^{-1}		20 kg N ha^{-1} a^{-1}
	Max	15 kg N ha^{-1} a^{-1}		100 kg N ha^{-1} a^{-1}
	Average	10 kg N ha^{-1} a^{-1}		60 kg N ha^{-1} a^{-1}
Code **1/d**	Min			10 kg N ha^{-1} a^{-1}
	Max			50 kg N ha^{-1} a^{-1}
	Average			30 kg N ha^{-1} a^{-1}
Code **1/x**	Estimated average	5* kg N ha^{-1} a^{-1}	20* kg N ha^{-1} a^{-1}	15* kg N ha^{-1} a^{-1}
Code **1**	Estimated average		40* kg N ha^{-1} a^{-1}	
Code **2**	Average	0 kg N ha^{-1} a^{-1}	0 kg N ha^{-1} a^{-1}	0 kg N ha^{-1} a^{-1}
Estimated values NITROGEN		______________ kg N ha^{-1} a^{-1}	______________ kg N ha^{-1} a^{-1}	______________ kg N ha^{-1} a^{-1}

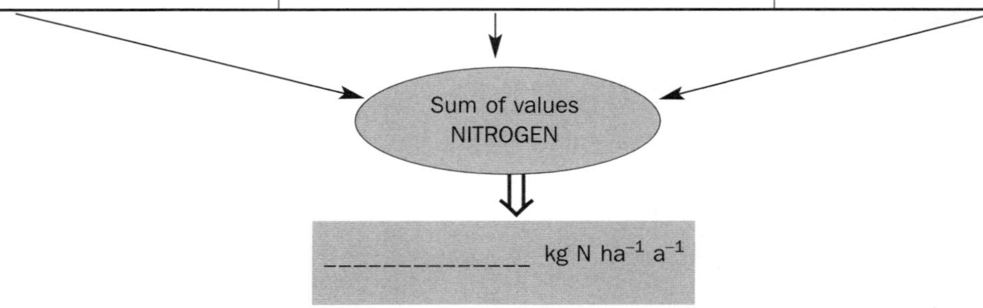

Table 3.2.5 Phosphorus export

PHOSPHORUS		**Vegetation management (P)**
Code **1/a**	Min	5 kg P ha^{-1} a^{-1}
	Max	40 kg P ha^{-1} a^{-1}
	Average	22.5 kg P ha^{-1} a^{-1}
Code **1/b**	Min	3 kg P ha^{-1} a^{-1}
	Max	20 kg P ha^{-1} a^{-1}
	Average	11.5 kg P ha^{-1} a^{-1}
Code **1/c**	Min	2 kg P ha^{-1} a^{-1}
	Max	10 kg P ha^{-1} a^{-1}
	Average	6 kg P ha^{-1} a^{-1}
Code **1/d**	Min	1 kg P ha^{-1} a^{-1}
	Max	5 kg P ha^{-1} a^{-1}
	Average	3 kg P ha^{-1} a^{-1}
Code **1/x**	Estimated average	1.5* kg P ha^{-1} a^{-1}
Code **1**	Estimated average	
Code **2**	Average	0 kg P ha^{-1} a^{-1}
Estimated values PHOSPHORUS		________________ kg P ha^{-1} a^{-1}

Sum of values PHOSPHORUS

________________ kg P ha^{-1} a^{-1}

Step 3

In the last row of the nitrogen and the phosphorus tables (estimated values NITROGEN / estimated values PHOSPHORUS) write the circled values. For instance if you have assigned a code **1/a** to the process 'export of nutrients through vegetation management', circle **250 kg N ha^{-1} a^{-1}** and **25 kg P ha^{-1} a^{-1}** in the third row of the column 'vegetation management' in the nitrogen table and phosphorus table, respectively. Next write down **250** and **25** on the dotted lines in the last rows of these columns. Only those processes relevant to export of nitrogen or phosphorus are shown.

Step 4

After you have written down three values (one value for each process) in the last row in the nitrogen table (Table 3.2.4), add up these values and record the total in the appropriate light grey box below the table. In the phosphorus table (Table 3.2.5), copy the value in the last row to the light grey box below the table. The values in the light grey boxes are the final outcome for the function 'nutrient export' in the HGMU under consideration.

Step 5

Repeat this action for all HGMUs.

Notes:

1. The summed value in the light grey boxes is derived from a summation of estimated values and should be considered only as a rough approximation. It can be used as an indication of performance of the function. No calculations should be performed with the values in contexts other than this assessment.
2. The information present in the other rows in the tables (min and max) is needed if you want to (re)calculate the values yourself.
   - 'Min' = the lowest rate of the range accompanying the code.
   - 'Max' = the highest rate of the range accompanying the code.
   - 'Average' = the average between the lowest and highest rate of the range.

Values in the table marked with * are not derived from the look-up tables, but are estimated by experts. For more information see text above.

Step 6

To obtain an indication of the relative degree of performance of the function in the HGMU under consideration, compare the summed values for nitrogen export and phosphorus export in your HGMU with the categories in Table 3.2.6.

When all contributing processes are assigned a code **2**, the function is *not performed* and no nitrogen or phosphorus is exported from the HGMU. Export of less than 15 kg N ha^{-1} a^{-1} or less than 2 kg P ha^{-1} a^{-1}, indicates a *low performance* of the function. The function is *performed intermediately* when values are between 15 and 100 kg N ha^{-1} a^{-1} or in between 2 and 10 kg P ha^{-1} a^{-1}. The function is qualified with *high performance* when values of export exceed 100 kg N ha^{-1} a^{-1} or 10 kg P ha^{-1} a^{-1}. Theoretically the export is performed maximally when all quantified processes have been assigned a code **1/a** and all non-quantified processes a code **1**. Summing the accompanying values for these codes leads to a maximum retention value of 420 kg ha^{-1} a^{-1} and 22.5 kg ha^{-1} a^{-1} for nitrogen and phosphorus, respectively.

Table 3.2.6 Nitrogen/phosphorus export comparison

Degree of performance	Nitrogen	Phosphorus
Not performed	0 kg ha^{-1} a^{-1}	0 kg ha^{-1} a^{-1}
Low performance	> 0 < 15 kg ha^{-1} a^{-1}	> 0 < 2 kg ha^{-1} a^{-1}
Intermediate performance	≥ 15 < 100 kg ha^{-1} a^{-1}	≥ 2 < 10 kg ha^{-1} a^{-1}
High performance	≥ 100 ″ 420 kg ha^{-1} a^{-1}	≥ 10 ″ 22.5 kg ha^{-1} a^{-1}

Functional Assessment Table: 3.2.B Nutrient export

HGMU code	Nutrient export

AA name: ________________

Make a note of the degree of performance (not performed, low, intermediate or high) of the function 'nutrient export' in your HGMU for transfer to the Functional Analysis Summary Table in Section 4.0.

The importance of the performance of the function 'nutrient export' should be considered in combination with the performance of the function 'nutrient retention'. Both functions contribute to the same 'goal': purification of water from nutrients. Whether the export and retention of nutrients in the HGMU contributes to improved quality in the wetland or more widely downstream in the catchment depends mainly on hydrological factors, such as the rate of water flow and the residence time of water in the HGMU. Spatial interactions, such as the extent and duration of exchange of water between the HGMU and the wetland or water body also play a role. More information is needed to make exact predictions. However, by comparing function outcomes of different HGMUs, the user can gain insight into which HGMU types contribute most to good water quality.

The outcomes above should be interpreted with care. A high performance rate does not automatically lead to best results for water quality. One should also take into account the input of nutrients. Export in HGMUs with a direct nutrient input (i.e. HGMUs in agricultural use) may not result in better water quality, as the input rate might exceed the export rate. Export of nutrients in HGMUs with an indirect input of nutrients normally does contribute to a better water quality, as these HGMUs export nutrients, which have been released from other areas (i.e. diffuse agricultural pollution or sewage facilities) and would otherwise lead to eutrophication.

This completes the functional analysis for nutrient export. The output codes and quantification values can be recorded on the Functional Analysis Summary Table in Section 4.0, and used to compare functions among HGMUs across the wetland. If you wish to assess all functions, move on to the next step of the FAPs: *In-situ* carbon retention.

When all the HGMUs in the assessment area have been assessed, a general pattern of nutrient export across the whole wetland will be evident. By manipulation of your answers to the controlling variable (CV) questions, you have the capacity to explore the probable outcomes of any management options under consideration and relate them to specific functions. The electronic version of the FAPs can do this more quickly and easily.

3.2.C *In-situ* carbon retention (function)

3.2.C.1 Organic matter accumulation (process)

Further reading and references

Armentano, T.V. and Verhoeven, J.T.A. (1990). Biogeochemical cycles: global. In: Patten, B.C. (ed.), *Wetlands and Shallow Continental Water Bodies, Volume 1*, SPB Academic, The Hague, The Netherlands, 281–312.

Bouwman, A.F. (1990). *Soils and the Greenhouse Effect*, Wiley, Chichester, UK.

Burgerhart, N. (2001). *Mogelijkheden voor Koolstofopslag in Nederlandse Ecosystemen*, Leerstoelgroep Natuurbeheer en Plantenecologie, Wageningen Universiteit, The Netherlands.

Dise, N.B. (in press). The critical role of carbon in wetlands. In: Maltby, E. and Barker, T. (eds), *The Wetlands Handbook*, Wiley-Blackwell, Oxford, UK.

Goosen, H.E., Ruijgrok, C.M., Mager, S., Rozema, J., Hoosbeek, M.R., Breemen, N. Van, Aiking, H. and Vellinga, P. (1996). *Natuurontwikkeling en de Mogelijkheden voor Koolstofopslag: Samenvatting en Beleidsaanbeveling*, R-96/07, Instituut voor Milieuvraagstukken, Amsterdam, The Netherlands.

Nykänen, N., Alm, J., Lang, K., Silvola, J. and Martikainen, P. (1995). Emissions of CH_4, N_2O and CO_2 from a virgin fen and a fen drained for grassland in Finland, *Journal of Biogeography* **22**, 351–357.

Pol, A. van den, Beusichem, M.L. and Oenema, O. (1997). Effects of grassland management on the emission of methane from intensively managed grasslands on peat soil, *Plant and Soil* **189**, 1–9.

Pol, A. van den, Beusichem, M.L. and Oenema, O. (1999). Methane emissions from wet grasslands on peat soil in a nature preserve, *Biogeochemistry* **44**, 205–220.

Pol, A. van den, Beusichem, M.L. and Oenema, O. (1999). Determinants of spatial variability of methane emissions from wet grasslands on peat soil, *Biogeochemistry* **44**, 221–237.

Richardson, C.J. (1990). Biogeochemical cycles: regional. In: Patten, B.C. (ed.), *Wetlands and Shallow Continental Water Bodies, Volume 1*, SPB Academic, The Hague, The Netherlands, 259–279.

Wienk, L.D., Verhoeven, J.T.A., Coops, H. and Portielje, R. (2000). *Peilbeheer en Nutriënten; Literatuurstudie naar de Effecten van Peildynamiek op de Nutriëntenhuishouding van Watersystemen*, RIZA rapport 2000.012, RIZA, Lelystad, The Netherlands.

Introduction

The function of *In-situ* carbon retention consists of just one process, that of the accumulation of organic matter.

Process description:

- The accumulation of partially decomposed biomass in the soil.
- The process is performed when hydrological and geomorphological conditions in the HGMU result in wet soils, anaerobic conditions and relatively low rates of decomposition. These conditions favour the accumulation of organic matter. The accumulation of organic matter occurs commonly in the form of peat deposits.

Significance – rationale for assessment:

- Carbon in the form of carbon dioxide is a greenhouse gas and therefore partially responsible for the processes that lead to global warming. Carbon can be stored for long periods of time in peat and therefore removed from the global carbon cycle, preventing acceleration of the greenhouse effect.
- Peat deposits can be important as records of historical changes in environmental conditions through the preservation of pollen. Additionally, they sometimes contain organic archaeological remains such as ancient trackways, preserved in the anaerobic environment.

Conditions favourable for peat accumulation may also be favourable for emissions of methane (CH_4). Methane is naturally present in the atmosphere, but increasing rates of methane emission enhance the greenhouse effect. Each molecule of CH_4 in the atmosphere contributes 32 times more to the earth's greenhouse effect relative to each molecule of CO_2 (Bouwman, 1990). Methane is produced in highly reduced (anaerobic) environments during the degradation of organic matter, when nitrate, manganese, iron or sulphate are absent from the soil. For degradation of organic matter, 'electron acceptors' are needed. Under completely anaerobic conditions CO_2 is used as an electron acceptor by which methane is produced that can be released to the atmosphere. However, nitrate, manganese, iron and sulphate are more efficient as electron acceptors and, as long as one of these elements is present in the soil in sufficient concentrations, methane production will not take place.

The process of methane production is very complex, e.g. when CH_4 passes through a soil layer where oxygen is present, it will be oxidised to CO_2 before it is released to the atmosphere. This assessment does not deal with methane emissions from wetlands because precise understanding of methanogenesis is still limited.

Controlling variables

CV1 Organic matter status of soil
CV2 Soil water regime
CV3 Vegetation type
CV4 Landform
CV5 Climatic regime

Answer all the questions below for *each* HGMU; unless otherwise instructed continue from one question to the next. On the appraisal sheet circle the correct

answer. The location of the relevant information is given in brackets after each question. Use a new appraisal sheet for each HGMU.

Note: The potential performance of this function *cannot* be assessed.

CV1 *Organic matter status of soil*

Q1. Is the soil described as organic? *(Recording Sheet 2a, C.3.6)*

Q2. What is the predominant description of the top 30 cm of the soil? *(Recording Sheet 2a, C.3.8)*

a. Fibrous peat
b. Amorphous peat
c. Semi-fibrous peat
d. Humus
e. Humose
f. Other

Q3. Is the top 20 cm of the soil predominantly described as black or very dark? *(Recording Sheet 2a, C.3.4)*

Q4. How often is the HGMU ploughed? *(Recording Sheet 1, C.3.5.1b; AA Map 2)*

a. No history of ploughing
b. Not currently ploughed but previously ploughed within the last 20 years, or ploughed but less than once a year
c. Currently ploughed once or more a year
d. Unknown

Q5. How often is peat cut and removed from the AA? *(Recording Sheet 1, C.3.10.1a; AA Map 2)*

a. No peat cutting occurs
b. Hand cutting occurs at a local or traditional scale
c. Mechanised cutting of peat occurs at an industrial or commercial scale
d. Unknown

CV2 *Soil water regime*

Q1. Is the moisture of the top 20 cm of the soil described as either flooded or wet? *(Recording Sheet 2a, C.3.2)*

Q2. Has the HGMU been assigned a hydrological code indicating groundwater discharge (**F1**), i.e. **H1**, **H2**, **H3**, **H4**, **H11**, **H12**, **H14** or **H15**? *(Recording Sheet 2a, F1)*

Q3. Has the HGMU been assigned a hydrological code indicating (sub)surface run-off input (**F2**), i.e. **H2**, **H3**, **H4**, **H5**, **H6**, **H7**, **H13** or **H14**? *(Recording Sheet 2a, F2)*

Q4. Has the HGMU been assigned a hydrological code, indicating overbank flooding (**F4**) i.e. **H4**, **H7**, **H9**, **H10**, **H12**, **H13**, **H14** or **H15** or artificial supply of surface water? *(Recording Sheet 2a, F4, or Recording Sheet 2b, H.1.ACTION, Q17d)*

Q5. Is the top 20 cm of the soil prone to water table fluctuations (alternate periods of waterlogged and non-waterlogged soil conditions, which may be due to natural or seasonal fluctuations, or water table management), which result in non-waterlogged conditions for at least half of the year? (*Recording Sheet 1, C.3.22.3)*

CV3 *Vegetation type*

Q1. Does the vegetation classification indicate the presence of a mire, i.e. code **Mx**? *(Recording Sheet 3, D)*

Q2. Does the vegetation classification indicate the presence of bog moss (*Sphagnum* spp.), i.e. codes **M1–M7**, **M10–M12**, **A1** or **W4**? *(Recording Sheet 3, D)*

Q3. Does the vegetation classification indicate the presence of humid grassland, i.e. codes **G2–G4**, **G6–G9**, **G11**, **G14**? *(Recording Sheet 3, D)*

Q4. Does the vegetation classification indicate the presence of woodland either on perennially waterlogged soil or a site prone to flooding, i.e. codes **W4–W8**, **W10–W12**? *(Recording Sheet 3, D)*

Q5. Does the vegetation classification indicate the presence of vegetation growing under dry conditions, i.e. codes **G18–G32**, **S5–S18**, **U2**, **U4–U13**, **W22–W44**? *(Recording Sheet 3, D)*

Q6. Has the HGMU been classified as unvegetated or sparsely vegetated, i.e. code **Ux**? *(Recording Sheet 3, D)*

CV4 *Landform*

Q1. Is the HGMU classed as a peat dome with the geomorphic code **PD**? *(Recording Sheet 2a, C.2)*

Q2. Does the HGMU have a geomorphic code indicating that it is a depressional unit, i.e. **BS**, **CO**, **OX**, **FC**, **LK**, **AD**, **UD**? *(Recording Sheet 2a, C.2)*

CV5 *Climatic regime*

Q1. Is the average monthly precipitation at the nearest meteorological recording site greater than 75 mm per month for 9 or more months each year? *(Recording Sheet 1, C.3.1.1)*

Assessment of the process

Compare the results recorded for each HGMU on the appraisal sheet with the possible outcomes below, and assign an appropriate assessment code. Record this on the appraisal sheet. The syntax codes used are described in the introduction to Section 3.

Look-up tables

1. The process is definitely being performed
If the answers to the questions for this process agree with any of the combinations in Look-up Table 1, then the HGMU is definitely performing the process of carbon retention via accumulation of organic matter, and estimation is given of the amount of carbon retained. These figures are based on data collected during empirical studies carried out in the EU projects leading to the production of these FAPs (FAEWE, FAEWE II and PROTOWET), and published research related to this topic.

2. The process is not (significantly) being performed
If the answers to the questions for this process agree with any of the combinations in Look-up Table 2, then the HGMU is not significantly performing the process of accumulation of organic matter.

3. The process is being performed, but there are constraining factors or uncertainties
If Look-up Tables 1 and 2 did not lead to an assessment, the process of organic matter accumulation is performed, but quantification is difficult because there are one or more constraining factors or uncertainties. In Look-up Table 3 you can find out how many constraining factors influence the process, or which factors cause the uncertainty.

4. The process is definitely not being performed
This statement is not applicable to this process.

This completes the functional analysis for *in-situ* carbon retention. The output codes and quantification values can be recorded on the Functional Analysis

3.2.C.1 *Organic Matter Accumulation (In-Situ Carbon Retention) Appraisal Sheet*

HGMU code: ______________	AA name: ______________

CV1 Organic matter status of soil						
Q1	Y	N		U/NA		
Q2	a	b	c	d	e	f
Q3	Y	N		U/NA		
Q4	a	b	c	d		
Q5	a	b	c	d		

CV4 Landform			
Q1	Y	N	U/NA
Q2	Y	N	U/NA

CV2 Soil water regime			
Q1	Y	N	U/NA
Q2	Y	N	U/NA
Q3	Y	N	U/NA
Q4	Y	N	U/NA
Q5	Y	N	U/NA

CV5 Climatic regime			
Q1	Y	N	U/NA

CV3 Vegetation type			
Q1	Y	N	U/NA
Q2	Y	N	U/NA
Q3	Y	N	U/NA
Q4	Y	N	U/NA
Q5	Y	N	U/NA
Q6	Y	N	U/NA

HGMU process assessment code: ______________________________

3.2.C.1 Look-up Table 1

<table>
<tr><th>CV1</th><th>CV2</th><th>CV3</th><th>CV4</th><th>CV5</th><th>Rationale</th><th>Code</th><th>Quantification</th></tr>
<tr><td>(1|2a|2b|2c|2d|3) 4a (5a|5b)</td><td>(1|2|3|4) #5</td><td>1|2</td><td>1|2</td><td>1</td><td>The soil has been identified as a peat or organic soil, and presently is not cultivated or extensively mined. Waterlogging or inundation predominate, producing anaerobic soil conditions that favour accumulation of organic matter. The vegetation is indicative of a high rate of organic matter accumulation. The landform is indicative of organic matter accumulation, or the depressional nature of the landform favours waterlogging and anaerobic soil conditions. Finally climatic conditions favour accumulation of organic matter.</td><td>1/a</td><td>200–1000 kg C ha^{-1} a^{-1}</td></tr>
<tr><td>(1|2a|2b|2c|2d|3) 4a (5a|5b)</td><td>(1|2|3|4) #5</td><td>3|4</td><td>1|2</td><td>1</td><td>The soil has been identified as a peat or organic soil, and presently is not cultivated or extensively mined. Waterlogging or inundation predominate, producing anaerobic soil conditions that favour accumulation of organic matter. The vegetation is indicative of organic matter accumulation. The landform is indicative of organic matter accumulation, or the depressional nature of the landform favours waterlogging and anaerobic soil conditions. Finally climatic conditions favour accumulation of organic matter.</td><td>1/b</td><td>100–500 kg C ha^{-1} a^{-1}</td></tr>
<tr><td>(1|2a|2b|2c|2d|3) 5a (5a|5b)</td><td>(1|2|3|4) #5</td><td>1|2</td><td>1|2</td><td>*</td><td>The soil has been identified as a peat or organic soil, and presently is not cultivated or extensively mined. Waterlogging or inundation predominate, producing anaerobic soil conditions that favour accumulation of organic matter. The vegetation is indicative of a high rate of organic matter accumulation. The landform is indicative of organic matter accumulation, or the depressional nature of the landform favours waterlogging and anaerobic soil conditions.</td><td>1/a</td><td>200–1000 kg C ha^{-1} a^{-1}</td></tr>
<tr><td>(1|2a|2b|2c|2d|3) 4a (5a|5b)</td><td>(1|2|3|4) #5</td><td>3|4</td><td>1|2</td><td>*</td><td>The soil has been identified as a peat or organic soil, and presently is not cultivated or extensively mined. Waterlogging or inundation predominate, producing anaerobic soil conditions that favour accumulation of organic matter. The vegetation is indicative of organic matter accumulation. The landform is indicative of organic matter accumulation, or the depressional nature of the landform favours waterlogging and anaerobic soil conditions.</td><td>1/b</td><td>100–500 kg C ha^{-1} a^{-1}</td></tr>
<tr><td>(1|2a|2b|2c|2d|3) 4a (5a|5b)</td><td>(1|2|3|4) #5</td><td>1|2</td><td>*</td><td>*</td><td>The soil has been identified as a peat or organic soil, and presently is not cultivated or extensively mined. Waterlogging or inundation predominate, producing anaerobic soil conditions that favour accumulation of organic matter. The vegetation is indicative of a high rate of organic matter accumulation.</td><td>1/a</td><td>200–1000 kg C ha^{-1} a^{-1}</td></tr>
<tr><td>(1|2a|2b|2c|2d|3) 4a (5a|5b)</td><td>(1|2|3|4) #5</td><td>3|4</td><td>*</td><td>*</td><td>The soil has been identified as a peat or organic soil, and presently is not cultivated or extensively mined. Waterlogging or inundation predominate, producing anaerobic soil conditions that favour accumulation of organic matter. The vegetation is indicative of organic matter accumulation.</td><td>1/b</td><td>100–500 kg C ha^{-1} a^{-1}</td></tr>
</table>

3.2.C.1 Look-up Table 2

CV1	CV2	CV3	CV4	CV5	Rationale	Code
#(1 2a–d 3)	*	*	*	*	The soil has not been identified as a peat or organic soil.	**2**
4c|5c	*	*	*	*	The soil is used for cultivation or is extensively mined for peat. Conditions conducive to oxidation of organic matter occur. NB: Degradation of organic matter results in emission of CO_2 from the wetland. As CO_2 is a greenhouse gas, the wetland contributes to the acceleration of the greenhouse effect.	**2**
*	#(1 2 3 4)	*	*	*	Waterlogging or inundation do not predominate, resulting in aerobic soil conditions that favour degradation of organic matter. NB: Degradation of organic matter results in emission of CO_2 from the wetland. As CO_2 is a greenhouse gas, the wetland contributes to the acceleration of the greenhouse effect.	**2**
*	*	5|6	*	*	The vegetation is indicative of conditions that are unfavourable for organic matter accumulation, or there is no vegetation (and therefore no organic matter) present.	**2**

3.2.C.1 Look-up Table 3

CV1	CV2	CV3	CV4	CV5	Rationale	Code
4b|4d|5d	*	*	*	*	It is unknown whether the HGMU is subjected to ploughing or mining, or whether the influence of previous or infrequent ploughing is still noticeable. Because ploughing or mining cause unfavourable conditions for organic matter accumulation, it is not possible to quantify the process properly without information on these impacts.	**1/x**
*	5	*	*	*	Fluctuating groundwater levels cause an alternation between waterlogged (anaerobic) and non-waterlogged (aerobic) soil conditions. The former is favourable for organic matter accumulation, while the latter is conducive to oxidation of organic matter. Therefore it is uncertain if the net result will be accumulation.	**1/x**
*	*	#(1 2 3 4 5 6)	*	*	Vegetation other than that mentioned in the questions in ***CV3*** is present. It is uncertain if this vegetation is indicative of organic matter accumulation.	**1/x**

Summary Table in Section 4.0, and used to compare functions among HGMUs across the wetland. If you wish to assess all functions, move on to the next step of the FAPs: Trace element storage.

When all the HGMUs in the AA have been assessed, a general pattern of *in situ* carbon retention across the whole wetland will be evident. By manipulation of your answers to the controlling variable (CV) questions, you have the capacity to explore the probable outcomes of any management options under consideration and relate them to specific functions. The electronic version of the FAPs can do this more quickly and easily.

3.2.D Trace element storage (function)

3.2.D.1 Physical retention of trace elements (process)

Further reading and references

Banus, M., Valiela, I.J. and Teal, M. (1974). Export of lead from salt marshes, *Marine Pollution Bulletin* **5**(1), 6–9.

Böhm, E., Hillenbrand, T., Marscheider-Weidemann, F., Fuchs, S. and Scherer, U. (2000). Bilanzierung des Eintrags prioritärer Schwermetalle in Gewässer, UBA Texte 29/01, Umweltbundesamt, Berlin, Germany (in German).

Eisenmann, V., Geyer, S., Henle, K., Rinklebe, J. and Scholz, M. (2001). Binding strength and mobility of trace elements in river marginal wetland, unpublished, Helmholtz Centre for Environmental Research, UFZ, Leipzig, Germany.

Petersen, W., Willer, E. and Willamowski, C. (1997). Remobilization of trace elements from polluted anoxic sediments after resuspension in oxic water, *Water, Air and Soil Pollution* **99**, 515–522.

Ross, S.M., Wood, M.D., Copplestone, D., Warriner, M. and Crook, P. (2007) *Environmental Concentrations of Heavy Metals in UK Soil and Herbage*, UK Soil and Herbage Pollutant Survey Report No. 7, Environment Agency, Bristol, UK.

Wallschläger, D., Desai, M.V.M. and Wilken, R.-D. (1996). The role of humic substances in the aqueous mobilization of mercury from contaminated floodplain soils, *Water, Air, Soil Pollution* **90**, 507–520.

Winde, F. (2000). Der hochwassergebundene Schwermetalltransport als Ursache der Bodenkontamination in der Saaleaue bei Halle. In: Friese, K., Witter, B., Miehlich, G. and Rode, M. (eds), *Stoffhaushalt von Auenökosystemen. Böden und Hydrologie, Schadstoffe, Bewertungen*, Springer, Berlin, 237–246.

Introduction

Process description:

- The process is considered to occur in an HGMU when anthropogenically-derived trace elements, which are physico-chemically bound to particulate matter transported in water from surface water bodies or other HGMUs within the AA or CA, are deposited and retained within the HGMU.
- It is difficult to quantify the process of retention of trace elements because the process is highly temporally and spatially variable. Therefore only a qualitative assessment of retention of trace elements is given.

Significance – rationale for assessment: Improvement of water quality in water bodies receiving surface or subsurface water from the wetland by retention of trace elements bound to retained sediment.

Controlling variables

CV1 Trace element input
CV2 Flooding regime
CV3 Particulate trace element load
CV4 Soil/sediment deposition
CV5 Soil/sediment erosion

Answer all the questions below for each HGMU; unless otherwise instructed continue from one question to the next. On the appraisal sheet, circle the correct

answer. The location of the relevant information is given in brackets after each question. Use a new appraisal sheet for each HGMU.

If there are no significant anthropogenic trace element inputs identified for the HGMU, or if the trace element input is uncertain (code **3n** and **4n**, *Recording Sheet 2e)*, you may still assess the potential of the HGMU to perform this process. To do this, conduct the assessment as normal but answer questions only in ***CV2***, ***CV3***, ***CV4*** and ***CV5***. Establish an assessment outcome as normal, but ignore the need for positive answers for ***CV1***. The rationale statement in the assessment section can be used as normal, but ignore reference to receipt of trace element inputs and their interaction with the HGMU.

CV1 *Trace element input*

Q1. What trace element input code has been assigned to the HGMU? (*Recording Sheet, 2e, 1*)

a. **Trae 1/1** or **Trae 1/3**
b. **Trae 1/4, Trae 1/5, Trae 1/6** or **Trae 1/7**
c. Other

CV2 *Flooding regime*

Q1. Has the HGMU been assigned a hydrological code indicating that overbank flooding occurs (**F4**), i.e. **H4**, **H7**, **H9**, **H10**, **H12**, **H13**, **H14** or **H15**? *(Recording Sheet 2a, F)*

Q2. Has the HGMU been assigned a hydrological code indicating that there is a surface run-off input, but *no* surface run-off output, i.e. hydrological codes **H2**, **H5**, **H13**, **H14**? *(Recording Sheet 2a, F)*

Only answer ***Q3*** and ***Q4*** if you have answered YES to ***Q1*** in ***CV2***.

Q3. Are flow velocities less than 0.3 m s^{-1} in the HGMU during flood events? (*Recording Sheet 3, B.4.6*)

Q4. If the HGMU has only one (potential) outlet for flood water, is the width of the outlet less than 10 % of the average width of the area usually flooded in the HGMU, i.e. is the outflow from the HGMU restricted? (*Recording Sheet 3, B.4.21; AA Map 3*)

CV3 *Particulate trace element load*

Only answer ***Q1–Q4*** if you have answered YES to ***Q1*** in section ***CV2***. Otherwise continue with ***Q5*** (read the note before ***Q5*** first).

Q1. Has the river that floods the HGMU been observed to carry high concentrations of suspended sediment during peak discharges? Observations of the river should be within 20 % of the total upstream length lying closest to the HGMU. *(Recording Sheet 3, B.5.3)*

Q2. Have any of the activities listed below taken place along more than 10 % of the upstream river channel during the last 3 years? *(Recording Sheet 1, C.3.24.2 (answers a–d))*

- Logging or afforestation
- Urban development or road building
- Mining
- Major river channel improvements

Q3. Is more than 10 % of the upstream catchment occupied by one or more of the land types listed below? *(Recording Sheet 1)*:

- Mountainous areas (*C.3.24.1 (answer a)*)
- Sparsely vegetated areas (*C.3.24.1 (answer b)*)
- Areas where mean annual potential evapotranspiration exceeds mean total annual precipitation (*C.3.1.1, C.3.1.3i*) [*Compare the average monthly rainfall with the average monthly evapotranspiration records*]

Q4. Does the river directly upstream of the site (within the 20 % of the total upstream river length lying closest to the HGMU) indicate migrating channels during the last 20 years? (*Recording Sheet 1, C.3.24.5*)

Only answer ***Q5–Q9*** if you have answered YES to ***Q2*** in ***CV2***. Otherwise continue with ***CV4***.

Q5. Are there indications of erosion in the area immediately upslope (in AA or CA) of the HGMU? (*Recording Sheet 3, B.5.1*)

Q6. Has surface run-off from the area immediately upslope of the HGMU (in AA or CA) been observed to carry high concentrations of suspended sediment? (*Recording Sheet 3, B.5.6*)

Q7. Has the area upslope from the HGMU been assigned a gradient code **G2** or **G3** indicating that it is moderate or steep (> 5°)? *(Recording Sheet 2a, C.2)*

Q8. Have any of the activities listed below taken place at a large scale (> 50 %) in the area immediately upslope (in AA or CA), during the last 3 years? (*Recording Sheet 1, C.3.24.2 (answers a–d)*)

- Logging or afforestation
- Urban development or road building
- Mining
- Major river channel improvements

Q9. Is over 50 % of the area immediately upslope (in AA or CA) occupied by one or more of the land types listed below? *(Recording Sheet 1)*

- Mountainous areas (*C.3.24.1 (answer a)*)
- Sparsely vegetated areas (*C.3.24.1 (answer b)*)
- Areas where annual potential evapotranspiration exceeds annual precipitation (*C.3.1.1, C.3.1.3i*) [*Compare the average monthly rainfall to the average monthly evapotranspiration records*]
- Permafrost areas, where soils are frozen throughout the year ('pergelic' soils) (*C.3.1.5 (answer a)*)

CV4 *Soil/sediment deposition*

Q1. Is there evidence of recently deposited mineral or organic material at the soil surface in the HGMU? (*Recording Sheet 3, B.5.4*)

Q2. Is there evidence of recently deposited mineral material on plant surfaces in the HGMU? (*Recording Sheet 3, B.5.5*)

Q3. Is the soil in the upper 10 cm of the soil profile in the HGMU described as fine or medium? *(Recording Sheet 2a, C.3.7)*

Q4. Is the HGMU mostly covered with 'rough' vegetation such as dense reed, bushes, shrubs or forest, which will slow down the passage of (flood) waters considerably? *(Recording Sheet 3, B.4.20)*

Q5. Has the HGMU been assigned a geomorphic code indicating that it is a depressional feature, i.e. **BS**, **CO**, **OX**, **FC**, **LK**, **DT**, **CN**, **AD** or **UD**? *(Recording Sheet 2a, C.2)*

Q6. Is the microtopography in the HGMU described as significant? *(Recording Sheet 3, A.3)*

CV5 *Soil/sediment erosion*

Q1. Is the vegetation classed as unvegetated or sparsely vegetated, i.e. vegetation code **Ux**? (*Recording Sheet 3, D*)

Q2. Are there any indicators of erosion within the HGMU? This could be scouring in units not adjacent to the river, or steep exposed banks in units adjacent to the river (e.g. levees). (*Recording Sheet 3, B.5.2*)

Only answer ***Q3*** if you have answered YES to ***Q1*** in section ***CV2***.

Q3. Is the HGMU adjacent to a river that is known to shift its course locally, and situated on an outer bend? (*Recording Sheet 3, B.5.8*)

3.2.D.1 Physical Retention of Trace Elements Appraisal Sheet

HGMU code: ______________________	**AA name:** ______________________

CV1 Trace element input			
Q1	a	b	c

CV2 Flooding regime			
Q1	Y	N	U/NA
Q2	Y	N	U/NA
Q3	Y	N	U/NA
Q4	Y	N	U/NA

CV3 Particulate trace element load			
Q1	Y	N	U/NA
Q2	Y	N	U/NA
Q3	Y	N	U/NA
Q4	Y	N	U/NA
Q5	Y	N	U/NA
Q6	Y	N	U/NA
Q7	Y	N	U/NA
Q8	Y	N	U/NA
Q9	Y	N	U/NA

CV4 Soil/sediment deposition			
Q1	Y	N	U/NA
Q2	Y	N	U/NA
Q3	Y	N	U/NA
Q4	Y	N	U/NA
Q5	Y	N	U/NA
Q6	Y	N	U/NA

CV5 Soil/sediment erosion			
Q1	Y	N	U/NA
Q2	Y	N	U/NA
Q3	Y	N	U/NA

HGMU process assessment code: ____

3.2.D.1 Look-up Table 1

<table>
<tr><th>CV1</th><th>CV2</th><th>CV3</th><th>CV4</th><th>CV5</th><th>Rationale</th><th>Code</th></tr>
<tr><td>1a</td><td>1 (3|4)</td><td>1|2|3|4</td><td>1|2|3|4|5|6</td><td>#(1 2 3)</td><td>The HGMU receives a trace element input via overbank flooding. River water is likely to be retained at low flow velocities and it is likely that the in-flowing water contains a high particulate load. Because of the high concentration of trace elements in the river water, it is expected that trace elements will be adsorbed to the particulates. There is evidence that the particulates are retained, and there is no evidence of erosion or loss of particulate-bound trace elements, indicating that retention may be long-term.
NB: The amount of trace elements retained depends upon the hydraulic residence time and the depth of the floodwater.</td><td>1</td></tr>
<tr><td>1b</td><td>2</td><td>5|6|7|8|9</td><td>1|2|3|4|5|6</td><td>#(1 2)</td><td>The HGMU receives a nutrient input via run-off. It is likely that run-off water contains a high particulate load. Because of the high concentration of trace elements in the water, it is expected that trace elements will be adsorbed to the particulates. There is evidence that the particulates are retained, and there is no evidence of erosion or loss of particulate-bound trace elements, indicating that retention may be long-term.
NB: The amount of trace elements retained depends upon the hydraulic residence time and the depth of the floodwater.</td><td>1</td></tr>
<tr><td>1a|1b</td><td>1 2 (3|4)</td><td>(1|2|3|4) (5|6|7|8|9)</td><td>1|2|3|4|5|6</td><td>#(1 2 3)</td><td>The HGMU receives a trace element input via both floodwater and run-off water. River floodwater and run-off water are likely to be retained at low flow velocities and it is likely that the in-flowing water contains a high particulate load. Because of the high concentration of trace elements in the water, it is expected that trace elements will be adsorbed to the particulates. There is evidence that the particulates are retained, and there is no evidence of erosion or loss of particulate-bound trace elements, indicating that retention may be long-term.
NB: The amount of trace elements retained depends upon the hydraulic residence time and the depth of the floodwater.</td><td>1</td></tr>
</table>

3.2.D.1 Look-up Table 2

CV1	CV2	CV3	CV4	CV5	Rationale	Code
*	1	#(1 2 3 4)	*	*	The HGMU could receive a trace element input via overbank flooding from the river, but in-flowing waters are not likely to contain a high particulate load.	**2**
*	2	#(5 6 7 8 9)	*	*	The HGMU could receive a trace element input via run-off from upslope areas. In-flowing waters are not likely to contain a high particulate load.	**2**
*	*	*	#(1 2 3 4 5 6)	*	There is no evidence that particulates are retained.	**2**

3.2.D.1 Look-up Table 3

CV1	CV2	CV3	CV4	CV5	Rationale	Code
*	*	*	*	1|2|3	There is some evidence of erosion and loss of particulate-bound trace elements, indicating that retention may be short-term.	**1/x**

3.2.D.1 Look-up Table 4

CV1	CV2	CV3	CV4	CV5	Rationale	Code
1c	*	*	*	*	The HGMU does not receive a trace element input via overbank flooding from the river or via run-off from upslope areas, or it is uncertain whether the HGMU receives a trace element input.	**4**
*	#(1 2)	*	*	*	Overbank flooding or (sub)surface run-off from upslope areas does not occur. Therefore there is no transportation mechanism for particulate material to reach the HGMU.	**4**

Assessment of the process

Compare the results recorded for each HGMU on the appraisal sheet with the possible outcomes below, and assign an appropriate assessment code. Record this on the appraisal sheet. The syntax codes used are described in the introduction to Section 3.

Look-up tables

1. The process is definitely being performed
If the answers to the questions for this process agree with any of the combinations in Look-up Table 1, then the HGMU is definitely performing the process of physical retention of trace elements. It is difficult to quantify the process because it is temporally and spatially highly variable. Therefore only a qualitative assessment of retention of trace elements is given.

2. The process is not (significantly) being performed
If the answers to the questions for this process agree with any of the combinations in Look-up Table 2, then the HGMU is not significantly performing the process of physical retention of trace elements.

3. The process is being performed, but there are constraining factors or uncertainties
If Look-up Tables 1 and 2 did not lead to an assessment, the process of physical retention of trace elements is performed, but there are constraining factors or uncertainties. In Look-up Table 3 you can find out how many constraining factors affect the process, or which factors cause the uncertainty.

4. The process is definitely not being performed
If the answers to the questions for this process agree with any of the combinations in Look-up Table 4, then the HGMU is definitely not performing the process of physical retention of trace elements.

This completes the functional assessment for the process of physical retention of trace elements. The output codes can be recorded on the Functional Assessment Table at the end of Section 3.2.D.4.

3.2.D.2 Biogeochemical retention of trace elements (process)

Further reading and references

Finnern, H., Grottenthaler, W. and Kühn, D. (2005). Bodenkundliche Kartieranleitung. Ad-hoc-Arbeitsgruppe Boden der Geologischen Landesämter und der Bundeanstalt für Geowissenschaften und Rohstoffe der Bundesrepublik Deutschland, 5. Aufl., E. Schweizerbartsche Verlagsbuchhandlung, Stuttgart, Switzerland (in German).

Ahnstrom, Z.S. and Parker, D.R. (1999). Development and assessment of a sequential extraction procedure for the fractionation of soil cadmium, *Soil Science Society of America Journal* **63**, 1650–1658.

Blume, H.-P. (ed.) (2005). *Handbook of Soil Conservation (Handbuch des Bodenschutzes)*, 3rd edn, Ecomed Publishing Landsberg/Lech, Germany.

Blume, H.-P. and Brümmer, G. (1991). Prediction of heavy-metal behaviour in soil by means of simple field tests, *Ecotoxicology and Environmental safety*, **22**, 164–174.

Fränzle, O. (1993). *Contaminants in Terrestrial Environments*, Springer Series in Physical Environment 13, Springer, Berlin, Heidelberg, New York, London, Paris, Tokyo.

Gibbs, R.J. (1973). Mechanisms of trace metal transport in rivers, *Science* **180**, 71–73.

Harres, H.-P. (1990). Schwermetalle in Böden. In: Darmstädter Kleingartenanlangen – Entwicklung, Nutzung und Belastung aus soziologischer und geoökologischer Sicht. In: *Geoöko-Test* **1**, 67–79, Bensheim, o.J. (in German).

Herms, U. (1982). *Untersuchungen zur Schwermetallöslichkeit in kontaminierten Böden und kompostierten Siedlungsabfällen in Abhängigkeit von Bodenreaktion, Redoxbedingungen und Stoffbestand*, Dissertation, University of Kiel, Germany (in German).

Hodgson, J.M. (ed.) (1976). *Soil Survey field handbook*, Soil Survey Technical Monograph, No. 5, Rothamsted Experimental Station, Harpenden, UK.

Hornburg, V. and Brümmer, G.W. (1993). Verhalten von Schwermetallen in Böden. 1. Untersuchungen zur Schwermetallmobilität, *Zeitschrift fur Pflanzenphysiologie und Bodenk* **156**, 467–477 (in German).

Jenkins, D.A. and Wyn Jones, R.G. (1980). Trace elements in rocks, soils, plants and animals. In: Davies, B.E. (ed.), *Applied Soil Trace Elements*, Wiley, Chichester, New York, Brisbane, Toronto, 1–20.

Kabata-Pendias, A. (2000). *Trace Elements in Soils and Plants*, CRC Press, Boca Raton, FL.

McGeehan, S.L., Fendorf, S.E. and Naylor, D.V. (1998). Alteration of arsenic sorption in flooded-dried soils. *Soil Science Society of America Journal* **62**, 828–833.

Miehlich, G. and Gröngröft, A. (1989). Heavy metal contents in the soil solution (Schwermetallgehalte in der Bodenlösung). In: Behrens, D. and Wiesner, J. (eds), *Beurteilung von Schwermetallkontaminationen im Boden (Assessment of Heavy Metal Contamination in Soils)*, Dechema Fachgespräche Umweltschutz, Dechema, Frankfurt/Main, 199–218 (in German).

Naidu, R. and Harter, R.D. (1998). Effect of different organic ligands on cadmium sorption by and extractability from soils, *Soil Science Society of America Journal* **62**, 644–650.

Pickering, W.F. (1981). Selective chemical extraction of soil components and bound metal species, *Critical Reviews Analytical Chemistry* **12**, 233–245.

Pluquet, E., Filipinski, M. and Grupe, M. (1990). Zur Cadmiumaufnahme von Kulturpflanzen aus geogen und anthropogen angereicherten Böden, *Z. f. Kulturtechnik und Landentwicklung* **31**, 105–111 (in German).

Ross, S.M., Wood, M.D., Copplestone, D., Warriner, M. and Crook, P. (2007). *Environmental Concentrations of Heavy Metals in UK Soil and Herbage*, UK Soil and Herbage Pollutant Survey Report No. 7, Environment Agency, Bristol, UK.

Schachtschabel, P., Blume, H.-P., Brümmer, G., Hartge, H.H. and Schwertmann, U. (1998). *Lehrbuch der Bodenkunde*, 14. Auflage, F. Enke, Stuttgart, Germany (in German).

Schoeneberger, P.J., Wysocki, D.A., Benham, E.C. and Broderson, W.D. (1998). *Field Book for Describing and Sampling Soils*, Natural Resources Conservation Service, United States Department of Agriculture, National Soil Survey Center, Lincoln, NE.

Schwarz, A., Wilcke, W., Styk, J. and Zech, W. (1999). Heavy metal release from soils in batch pH (stat) experiments, *Soil Science Society of America Journal* **63**, 290–296.

Sims, J.L. and Patrick Jr, W.H. (1978). Micronutrient cations in soil under varying redox potential and pH, *Soil Science Society of America Journal* **42**, 258–262.

Shuman, L.M. and Hargrove, W.L. (1985). Effect of tillage on the distribution of manganese, copper, iron and zinc in soil fractions, *Soil Science Society of America Journal* **49**, 1117–1121.

Sposito, G., Lund, L.J. and. Chang, A.C (1982). Trace metal chemistry in arid-zone field soils amended with sewage-sludge: I. Fractionation of Ni, Cu, Zn, Cd and Pb in solid phases, *Soil Science Society of America Journal* **46**, 5260–264.

Tessier, A., Campbell, P.G. and Bisson, M. (1979). Sequential extraction procedure for the speciation of particulate trace metals, *Analytical Chemistry* **51**, 844–851.

Winde, F. (2000). Der hochwassergebundene Schwermetalltransport als Ursache der Bodenkontamination in der Saaleaue bei Halle. In: Friese, K., Witter, B., Miehlich, G. and Rode, M. (eds), *Stoffhaushalt von Auenökosystemen. Böden und Hydrologie, Schadstoffe, Bewertungen*, Springer, Berlin, 237–246 (in German).

Zeien, H. (1995). *Chemische Extraktion zur Bestimmung der Bindungsformen von Schwermetallen in Böden*, Bonner Bodenk. Abh. 17, Bonn, Germany (in German).

Zeien, H. and Brümmer, G. (1989). Chemische Extraktionen zur Bestimmung von Schwermetallbindungsformen in Böden, *Mitteilungen der Deutschen Bodenkundlichen Gesellschaft* **59**, 505–510 (in German).

Introduction

Process description:

- This process describes the fixation or the biogeochemical retention of anthropogenically-derived trace elements (heavy metals and metalloids such as arsenic) in the solid phase of river or lake marginal wetland soils, mostly by sorption processes. The process refers solely to the maintenance of this fixation of trace elements following their introduction to the site.
- Biogeochemical retention is controlled by soil pH, redox potential, organic

matter content, particle size and hydrology. Because most retention mechanisms are reversible, changes in soil conditions (especially pH or redox potential) can result in the mobilisation of trace elements, especially those that are adsorbed (see Process 3.2.E.3, biogeochemical remobilisation of trace elements).

Significance – rationale for assessment: The biogeochemical retention of trace elements reduces their mobility and availability to plants, and the quality of surface and groundwater bodies is protected.

Controlling variables

CV1 Trace element input or soil concentration
CV2 Soil texture and organic matter content
CV3 Soil environmental conditions

Answer all the questions below for each HGMU; unless otherwise instructed continue from one question to the next. On the appraisal sheet, circle the correct answer. The location of the relevant information is given in brackets after each question. Use a new appraisal sheet for each HGMU.

Cadmium (Cd) is used as an example trace element because it is toxic, often abundant and easily mobilised. Cadmium is a relevant trace contaminant for many European alluvial wetlands, and this module can be expected to have wide applicability as a result. The trace element functions have been developed following extensive work in a case study site in the catchment of the Elbe in Germany, which has elevated concentrations of Cd. The concentration ranges used for classifying concentrations relative to background levels are therefore based on the Elbe data. The interpretation of the interactions between physico-chemical processes and their influence on the behaviour of Cd is based on empirical data from these studies. There are other trace elements that may need to be considered in wetlands and these include, but are not limited to, arsenic, mercury and selenium. Similar processes of assessment may be used for these, but it should be noted that background concentrations of trace elements are likely to be location-specific due to regional differences in mineralogy. In addition, the assessor will need to consider the modifying influences of physico-chemical processes within the wetland on the speciation and resultant behaviour of the trace element under consideration. As a result, prescriptive guidance based on concentration ranges may not be applicable across all regions of Europe. For this reason, the user should identify appropriate regional concentration data for the specific trace elements in their HGMU, then undertake a comparative assessment of measured wetland concentrations accordingly. A literature search will give guidance on processes and trace elements of relevance in European wetlands. Suitable starting references are given in Sections 3.2.D and 3.2.E.

CV1 *Trace element input or soil concentration*

Q1. Does the soil trace element concentration indicate that anthropogenically-derived contamination of the HGMU has taken place? *(Recording Sheet 1, C.3.25.1b)*

a. Yes, the soil trace element concentration is much higher than typical background concentrations.
b. Yes, the soil trace element concentration is moderately higher than typical background concentrations.
c. Yes, the soil trace element concentration is slightly higher than typical background concentration.
d. No, the soil trace element concentration is not elevated.
e. Unknown, no soil analytical data are available.

e.g. for the Elbe catchment example site (you will have to determine regional mean concentrations for the trace element you are concerned with; see note above):

Q1. Does the soil Cd concentration indicate that anthropogenically-derived contamination of the HGMU has taken place? *(Recording Sheet 1, C.3.25.1b)*

a. Yes, the soil Cd concentration is much higher than typical background concentrations (> 20 mg kg^{-1} soil).
b. Yes, the soil Cd concentration is moderately higher than typical background concentrations (10–20 mg kg^{-1} soil).
c. Yes, the soil Cd concentration is slightly higher than typical background concentration (> 5 but < 10 mg kg^{-1} soil).
d. No, the soil Cd concentration is not elevated (≤ 5 mg kg^{-1} soil).
e. Unknown, no soil analytical data are available.

Note: This is specific to the Elbe catchment. In the UK, for example, typical background Cd concentrations are around 1–2 mg kg^{-1} soil.

Q2. Has the HGMU been assigned a trace element input code indicating an input of trace elements? *(Recording Sheet, 2e, 1)*

a. Yes, trace element input code **Trae 1/1**, **1/2**, **1/3**, **1/4**, **1/5**, **1/6** or **1/7**.
b. Yes, trace element input code **Trae 2**.
c. Uncertain, trace element input code **Trae 3**.
d. No, trace element input code **Trae 4**.

CV2 *Soil texture and organic matter content*

Q1. What is the mineral particle size of the topsoil (an average of the top 30 cm)?

a. Coarse or gravel
b. Medium or fine

(Recording Sheet 2a, C.3.7ii)

Q2. Are there any laboratory analyses of organic matter content in the soils of the HGMU? *(Recording Sheet 1, C.3.25.2b (answers a–d))*

*[If Yes, go to **Q4**. If No, go to **Q3**]*

Note: To answer this question requires the use of Munsell Soil Color Charts (Munsell Color Company Inc., Baltimore, Maryland 21218, USA). In all cases, the moist soil colour is described unless otherwise stated (Hodgson, 1976).

Q3. Establish the organic matter content of the topsoil by determining which of the following statements is most applicable to a soil sample homogenised over the top 30 cm of the profile. *(Recording Sheet 2a)*

a. If the mineral particle size of the topsoil is coarse *and* the Munsell Color Chroma is ≥ 3.5 *or* the mineral particle size of the topsoil is medium or fine *and* the Munsell Color Chroma is ≥ 4, it can be assumed that the soil organic matter content of the topsoil is < 2%. *(Recording Sheet 2a, C.3.7 and C.3.4ii)*
b. If the mineral particle size of the topsoil is coarse *and* Munsell Color Chroma is < 3.5 *or* the mineral particle size is medium or fine *and* Munsell Color Chroma is < 3 it can be assumed that the organic matter content of the topsoil is 2–8 %. *(Recording Sheet 2a, C.3.7 and C.3.4ii)*
c. If the topsoil is a mineral soil with a greyish, very dark or black colour (but not manganiferous) *(Recording Sheet 2a, C.3.4) and* one or more of the following conditions are true, it can be assumed that the soil organic matter content is > 8 but < 15 %:
    - The HGMU is poorly drained; therefore the soils are flooded, wet or moist. *(Recording Sheet 2a, C.3.2)*
    - The HGMU comprises grassland with intensive root growth (vegetation codes with a **G** prefix). *(Recording Sheet 3, D)*
    - The HGMU is dominated by trees, and the litter decomposition rate is low (either it is a coniferous forest or there is a forest debris litter layer). *(Recording Sheet 3, C.2. (answer c))*
d. If the topsoil is organic (fibrous, amorphous or semi-fibrous peat), it can be assumed that the organic matter content of the topsoil is ≥ 15 %. *(Recording Sheet 2a, C.3.8)*

*[Go to **CV3**]*

Q4. What is the organic matter content of the topsoil, obtained by chemical analysis (mean of the upper 30 cm or the plough layer)? (Recording Sheet 1, C.3.25.2b)

a. < 2 %
b. 2–8 %
c. > 8 but < 15 %
d. ≥ 15 %

CV3 *Soil environmental conditions (pH and redox potential)* (see Figs 3.2.12 and 3.2.13)

Q1. What is the predominant pH of the upper 30 cm of the soil? *(Recording Sheet 3, A.1ii)*

a. 7 or higher
b. 6–< 7
c. 5–< 6
d. 4–< 5
e. 3–< 4

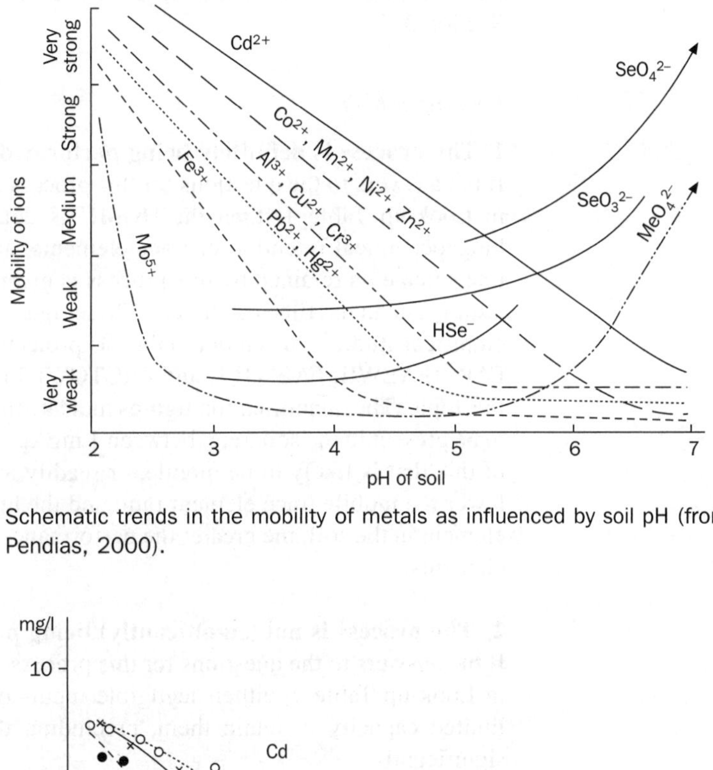

3.2.12 Schematic trends in the mobility of metals as influenced by soil pH (from Kabata-Pendias, 2000).

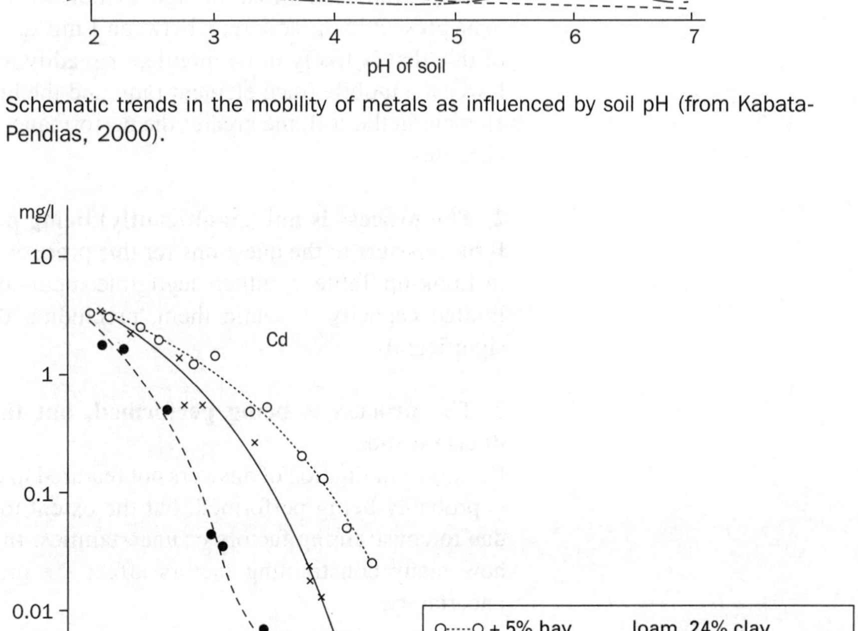

3.2.13 The effect on pH of trace element immobilisation by adding organic substances to the soil (From Harres 1990).

f. < 3
g. No analytical data available

Q2. Does the Munsell colour code of a mixed sample of the upper 30 cm of soil indicate the presence of an elevated iron oxide concentration? (*Recording Sheet 2a, C.3.4ii*)

a. Yes, Munsell colour hue is ≤ 7.5 YR and the chroma:value ratio is 1 to 1.5.
b. Yes, Munsell colour hue is ≤ 7.5 YR and the chroma:value ratio is > 1.5.
c. No, Munsell colour hue is > 7.5 YR.

Q3. Do the soils exhibit redoximorphic features indicative of either predominantly or at least temporarily wet conditions, i.e. 5 % or more of the soil profile is characterised by areas that are greyer or lighter coloured than the adjacent matrix, while at the same time containing some iron oxide mottles, as described in ***Q2***? *(Recording Sheet 2a, C.3.4ii)*

Assessment of the process

Compare the results recorded for each HGMU on the appraisal sheet with the possible outcomes below, and assign an appropriate assessment code. Record this on the appraisal sheet. The syntax codes used are described in the introduction to Section 3.

Look-up tables

1. The process is definitely being performed
If the answers to the questions for this process agree with any of the combinations in Look-up Table 1, then the HGMU is definitely performing the process of biogeochemical retention of trace elements, and an estimation of the amount of trace elements retained by this process is given. An example is given using Cd in a study site in the Elbe catchment. These figures are based on data collected during empirical studies carried out in the EU projects leading to the production of these FAPs (FAEWE, FAEWE II and PROTOWET), and published research related to this topic. The quantification figures indicate the concentrations of cadmium likely to be present in the soil, (e.g. between 1 mg kg^{-1} and 6 mg kg^{-1}), and the proportion of this that is likely to be mobile, or readily available (e.g. 10 %). Therefore the lower the mobile trace element ratio and the higher the concentration of the trace element in the soil, the greater the performance of the process of retention of trace elements.

2. The process is not (significantly) being performed
If the answers to the questions for this process agree with any of the combinations in Look-up Table 2, either negligible inputs of trace elements occur or there is limited capacity to retain them, precluding the process from being performed significantly.

3. The process is being performed, but there are constraining factors or uncertainties
For any combination of answers not featured in Look-up Tables 1 and 2, the process is probably being performed, but the extent to which it is occurring is unknown due to constraining factors or uncertainties. In Look-up Table 3 you can find out how many constraining factors affect the process, or which factors cause the uncertainty.

4. The process is definitely not being performed
If the answers to the questions for this process agree with any of the combinations in Look-up Table 4, then the process is definitely not being performed.

This completes the functional assessment for the process of biogeochemical

3.2.D.2 Biogeochemical Retention of Trace Elements Appraisal Sheet

HGMU code: ______________	**AA name:** ______________

CV1 Trace element input or soil concentration					
Q1	a	b	c	d	e
Q2	a	b	c	d	

CV3 Soil environmental conditions							
Q1	a	b	c	d	e	f	g
Q2	a	b	c				
Q3	Y	N	U/NA				

CV2 Soil texture and organic matter content				
Q1	a	b		
Q2	Y	N	U/NA	
Q3	a	b	c	d
Q4	a	b	c	d

HGMU process assessment code: ______________

3.2.D.2 Look-up Table 1

CV1	CV2	CV3	Rationale	Code	Quantification
1a	*	*	High concentrations of trace elements have been measured in the soil, indicating that the HGMU must be performing the process of biogeochemical retention of trace elements to a high degree. It is likely that all controlling variables are optimal.	**1/1**	e.g. for cadmium *in the Elbe* Mobile Cd ratio 0–10 %, maximum mobile Cd amount > 2 mg kg^{-1}
1b 2a	1b (3c|3d|4c|4d)	(1a|1b) (2a|2b) #3	Moderately high concentrations of trace elements have been measured in the soil, indicating that the HGMU must be performing the process of biogeochemical retention of trace elements to a moderately high degree. A high trace element input has been identified, and neither soil pH, redox potential, organic matter content or particle size are likely to restrict the retention process.	**1/2**	e.g. for cadmium *in the Elbe* Mobile Cd ratio 10–30 %, minimum mobile Cd amount 1 mg kg^{-1}, maximum mobile Cd amount 6 mg kg^{-1}
1b 2a	1a (3c|3d|4c|4d)	(1a|1b) (2a|2b) #3	Moderately high concentrations of trace elements have been measured in the soil, indicating that the HGMU must be performing the process of biogeochemical retention of trace elements to a moderately high degree. A high trace element input has been identified, but the retention capacity is restricted by the limited number of fixation sites within the soil because of the coarse or gravelly texture of the soil, but not by the organic matter content, which is relatively high. Soil pH and redox potential do not restrict the retention of trace elements.	**1/3**	e.g. for cadmium *in the Elbe* Mobile Cd ratio 10–30 %, minimum mobile Cd amount 1 mg kg^{-1}, maximum mobile Cd amount 6 mg kg^{-1}
1b 2a	1b (3c|3d|4c|4d)	(1c|1d) (2a|2b) #3	Moderately high concentrations of trace elements have been measured in the soil, indicating that the HGMU must be performing the process of biogeochemical retention of trace elements to a moderately high degree. A high trace element input has been identified, and the retention capacity is not restricted by the limited number of fixation sites within the soil because of the high organic matter content, and the medium or fine texture of the soil. Low soil pH restricts the retention capacity. Soil redox potential does not restrict the retention of trace elements.	**1/4**	e.g. for cadmium *in the Elbe* Mobile Cd ratio 10–30 %, minimum mobile Cd amount 1 mg kg^{-1}, maximum mobile Cd amount 6 mg kg^{-1}
1b 2a	1a (3c|3d|4c|4d	(1c|1d) (2a|2b) #3	Moderately high concentrations of trace elements have been measured in the soil, indicating that the HGMU must be performing the process of biogeochemical retention of trace elements to a moderately high degree. A high trace element input has been identified, but the retention capacity is restricted by the limited number of fixation sites within the soil because of the coarse or gravelly texture of the soil, but not by the organic matter content, which is relatively high. Low soil pH also restricts the retention capacity. Soil redox potential does not restrict the retention of trace elements.	**1/5**	e.g. for cadmium *in the Elbe* Mobile Cd ratio 10–30 %, minimum mobile Cd amount 1 mg kg^{-1}, maximum mobile Cd amount 6 mg kg^{-1}
1b 2a	1b (3b|4b)	(1a|1b) (2a|2b) #3	Moderately high concentrations of trace elements have been measured in the soil, indicating that the HGMU must be performing the process of biogeochemical retention of trace elements to a moderately high degree. A high trace element input has been identified, but the retention capacity is restricted by the limited number of fixation sites within the soil because of the moderately low organic matter content, but the medium or fine texture of the soil promotes retention. Soil pH and redox potential do not restrict the retention of trace elements.	**1/6**	e.g. for cadmium *in the Elbe* Mobile Cd ratio 10–30 %, minimum mobile Cd amount 1 mg kg^{-1}, maximum mobile Cd amount 6 mg kg^{-1}
1b 2a	1b (3a|4a)	(1a|1b) (2a|2b)	Moderately high concentrations of trace elements have been measured in the soil, indicating that the HGMU must be performing the process of biogeochemical retention of trace elements	**1/7**	e.g. for cadmium *in the Elbe* Mobile Cd ratio 10–30 %,

		#3	to a moderately high degree. A high trace element input has been identified, but the retention capacity is restricted by the limited number of fixation sites within the soil because of the low organic matter content, but the medium or fine texture of the soil promotes retention. Soil pH and redox potential do not restrict the retention of trace elements.		minimum mobile Cd amount 1 mg kg^{-1}, maximum mobile Cd amount 6 mg kg^{-1}
1b 2a	1b (3a|4a)	(1c|1d) (2a|2b) #3	Moderately high concentrations of trace elements have been measured in the soil, indicating that the HGMU must be performing the process of biogeochemical retention of trace elements to a moderately high degree. A high trace element input has been identified, but the retention capacity is restricted by the limited number of fixation sites within the soil because of the low organic matter content, but the medium or fine texture of the soil promotes retention. Low soil pH also restricts the retention capacity. Soil redox potential does not restrict the retention of trace elements.	**1/8**	e.g. for cadmium *in the Elbe* Mobile Cd ratio 10–30 %, minimum mobile Cd amount 1 mg kg^{-1}, maximum mobile Cd amount 6 mg kg^{-1}
1b 2a	1b (3b|4b)	(1c|1d) (2a|2b) #3	Moderately high concentrations of trace elements have been measured in the soil, indicating that the HGMU must be performing the process of biogeochemical retention of trace elements to a moderately high degree. A high trace element input has been identified, but the retention capacity is restricted by the limited number of fixation sites within the soil because of the moderately low organic matter content, but the medium or fine texture of the soil promotes retention. Low soil pH also restricts the retention capacity. Soil redox potential does not restrict the retention of trace elements.	**1/9**	e.g. for cadmium *in the Elbe* Mobile Cd ratio 10–30 %, minimum mobile Cd amount 1 mg kg^{-1}, maximum mobile Cd amount 6 mg kg^{-1}
1b 2a	1b (3c|3d|4c|4d)	1g (2a|2b) #3	Moderately high concentrations of trace elements have been measured in the soil, indicating that the HGMU must be performing the process of biogeochemical retention of trace elements to a moderately high degree. A high trace element input has been identified, and the retention capacity is not restricted by the limited number of fixation sites within the soil because of the high organic matter content, and the medium or fine texture of the soil. Soil redox potential does not restrict the retention of trace elements, and it is unknown if pH affects retention capacity.	**1/10**	e.g. for cadmium *in the Elbe* Mobile Cd ratio 10–30 %, minimum mobile Cd amount 1 mg kg^{-1}, maximum mobile Cd amount 6 mg kg^{-1}
1b 2a	1b (3b|4b)	1g (2a|2b) #3	Moderately high concentrations of trace elements have been measured in the soil, indicating that the HGMU must be performing the process of biogeochemical retention of trace elements to a moderately high degree. A high trace element input has been identified, but the retention capacity is restricted by the limited number of fixation sites within the soil because of the moderately low organic matter content, but the medium or fine texture of the soil promotes retention. Soil redox potential does not restrict the retention of trace elements, and it is unknown if pH affects retention capacity.	**1/11**	e.g. for cadmium *in the Elbe* Mobile Cd ratio 10–30 %, minimum mobile Cd amount 1 mg kg^{-1}, maximum mobile Cd amount 6 mg kg^{-1}
1b 2a	1a (3c|3d|4c|4d)	1g (2a|2b) #3	Moderately high concentrations of trace elements have been measured in the soil, indicating that the HGMU must be performing the process of biogeochemical retention of trace elements to a moderately high degree. A high trace element input has been identified, but the retention capacity is restricted by the limited number of fixation sites within the soil because of the coarse or gravelly texture of the soil, but not the organic matter content, which is relatively high. Soil redox potential does not restrict the retention of trace elements, and it is unknown if pH affects retention capacity.	**1/12**	e.g. for cadmium *in the Elbe* Mobile Cd ratio 10–30 %, minimum mobile Cd amount 1 mg kg^{-1}, maximum mobile Cd amount 6 mg kg^{-1}

3.2.D.2 Look-up Table 1 cont'd

<table>
<tr><th>CV1</th><th>CV2</th><th>CV3</th><th>Rationale</th><th>Code</th><th>Quantification</th></tr>
<tr><td>1b 2a</td><td>1b
(3a|4a)</td><td>1g
(2a|2b)
#3</td><td>Moderately high concentrations of trace elements have been measured in the soil, indicating that the HGMU must be performing the process of biogeochemical retention of trace elements to a moderately high degree. A high trace element input has been identified, but the retention capacity is restricted by the limited number of fixation sites within the soil because of the low organic matter content, but the medium or fine texture of the soil promotes retention. Soil redox potential does not restrict the retention of trace elements, and it is unknown if pH affects retention capacity.</td><td>1/13</td><td>e.g. for cadmium in the Elbe Mobile Cd ratio 10–30 %, minimum mobile Cd amount 1 mg kg^{-1}, maximum mobile Cd amount 6 mg kg^{-1}</td></tr>
<tr><td>1b 2a</td><td>1a
(3b|4b)</td><td>1g
(2a|2b)
#3</td><td>Moderately high concentrations of trace elements have been measured in the soil, indicating that the HGMU must be performing the process of biogeochemical retention of trace elements to a moderately high degree. A high trace element input has been identified, but the retention capacity is restricted by the limited number of fixation sites within the soil because of the coarse or gravelly texture of the soil, or the moderately low organic matter content. Soil redox potential does not restrict the retention of trace elements, and it is unknown if pH affects retention capacity.</td><td>1/14</td><td>e.g. for cadmium in the Elbe Mobile Cd ratio 10–30 %, minimum mobile Cd amount 1 mg kg^{-1}, maximum mobile Cd amount 6 mg kg^{-1}</td></tr>
<tr><td>1b 2a</td><td>1a
(3b|4b)</td><td>(1a|1b)
(2a|2b)
#3</td><td>Moderately high concentrations of trace elements have been measured in the soil, indicating that the HGMU must be performing the process of biogeochemical retention of trace elements to a moderately high degree. A high trace element input has been identified, but the retention capacity is restricted by the limited number of fixation sites within the soil because of the coarse or gravelly texture of the soil, or the moderately low organic matter content. Soil pH and redox potential do not restrict the retention of trace elements.</td><td>1/15</td><td>e.g. for cadmium in the Elbe Mobile Cd ratio 10–30 %, minimum mobile Cd amount 1 mg kg^{-1}, maximum mobile Cd amount 6 mg kg^{-1}</td></tr>
<tr><td>1b 2a</td><td>1a
(3b|4b)</td><td>(1c|1d)
(2a|2b)
#3</td><td>Moderately high concentrations of trace elements have been measured in the soil, indicating that the HGMU must be performing the process of biogeochemical retention of trace elements to a moderately high degree. A high trace element input has been identified, but the retention capacity is restricted by the limited number of fixation sites within the soil because of the coarse or gravelly texture of the soil, or the moderately low organic matter content. Low soil pH also restricts the retention capacity. Soil redox potential does not restrict the retention of trace elements.</td><td>1/16</td><td>e.g. for cadmium in the Elbe Mobile Cd ratio 10–30 %, minimum mobile Cd amount 1 mg kg^{-1}, maximum mobile Cd amount 6 mg kg^{-1}</td></tr>
<tr><td>1b 2b</td><td>*</td><td>*</td><td>Moderately high concentrations of trace elements have been measured in the soil, indicating that the HGMU must be performing the process of biogeochemical retention of trace elements to a moderately high degree. However, inputs of trace elements are negligible or rare, suggesting accumulation over a long period of time, with consequent long-term biogeochemical retention.</td><td>1/17</td><td>e.g. for cadmium in the Elbe Mobile Cd ratio 10–30 %, minimum mobile Cd amount 1 mg kg^{-1}, maximum mobile Cd amount 6 mg kg^{-1}</td></tr>
<tr><td>1b 2c</td><td>*</td><td>*</td><td>Moderately high concentrations of trace elements have been measured in the soil, indicating that the HGMU must be performing the process of biogeochemical retention of trace elements to a moderately high degree. The nature of the trace element input is unknown, but conditions generally must be conducive to retention of trace elements within the soil.</td><td>1/18</td><td>e.g. for cadmium in the Elbe Mobile Cd ratio 10–30 %, minimum mobile Cd amount 1 mg kg^{-1}, maximum mobile Cd amount 6 mg kg^{-1}</td></tr>
</table>

<table>
<tr><td>1c 2a</td><td>1b
(3c|3d|
4c|4d)</td><td>(1a|1b)
(2a|2b)
#3</td><td>Slightly elevated concentrations of trace elements have been measured in the soil, indicating that the HGMU must be performing the process of biogeochemical retention of trace elements to a small degree. A high trace element input has been identified, and neither soil pH, redox potential, organic matter content nor particle size are likely to restrict the retention process, therefore the reason for only a small retention capacity is unknown.</td><td>1/19</td><td>e.g. for cadmium in the Elbe
Mobile Cd ratio 30–50 %,
mobile Cd amount > 1.5 mg kg^{-1} but < 5 mg kg^{-1}</td></tr>
<tr><td>1c 2a</td><td>1a
(3c|3d|
4c|4d)</td><td>(1a|1b)
(2a|2b)
#3</td><td>Slightly elevated concentrations of trace elements have been measured in the soil, indicating that the HGMU must be performing the process of biogeochemical retention of trace elements to a small degree. A high trace element input has been identified, but the retention capacity is restricted by the restricted number of fixation sites within the soil because of the coarse or gravelly texture of the soil, but not the organic matter content, which is relatively high. Soil pH and redox potential do not restrict the retention of trace elements.</td><td>1/20</td><td>e.g. for cadmium in the Elbe
Mobile Cd ratio 30–50 %,
mobile Cd amount > 1.5 mg kg^{-1} but < 5 mg kg^{-1}</td></tr>
<tr><td>1c 2a</td><td>1b
(3c|3d|
4c|4d)</td><td>(1c|1d)
(2a|2b)
#3</td><td>Slightly elevated concentrations of trace elements have been measured in the soil, indicating that the HGMU must be performing the process of biogeochemical retention of trace elements to a small degree. A high trace element input has been identified, and the retention capacity is not restricted by the restricted number of fixation sites within the soil because of the high organic matter content, and the medium or fine texture of the soil. Low soil pH restricts the retention capacity. Soil redox potential does not restrict the retention of trace elements.</td><td>1/21</td><td>e.g. for cadmium in the Elbe
Mobile Cd ratio 30–50 %,
mobile Cd amount > 1.5 mg kg^{-1} but < 5 mg kg^{-1}</td></tr>
<tr><td>1c 2a</td><td>1a
(3c|3d|
4c|4d)</td><td>(1c|1d)
(2a|2b)
#3</td><td>Slightly elevated concentrations of trace elements have been measured in the soil, indicating that the HGMU must be performing the process of biogeochemical retention of trace elements to a small degree. A high trace element input has been identified, but the retention capacity is restricted by the limited number of fixation sites within the soil because of the coarse or gravelly texture of the soil, but not the organic matter content, which is relatively high. Low soil pH also restricts the retention capacity. Soil redox potential does not restrict the retention of trace elements.</td><td>1/22</td><td>e.g. for cadmium in the Elbe
Mobile Cd ratio 30–50 %,
mobile Cd amount > 1.5 mg kg^{-1} but < 5 mg kg^{-1}</td></tr>
<tr><td>1c 2a</td><td>1b
(3b|4b)</td><td>(1a|1b)
(2a|2b)
#3</td><td>Slightly elevated concentrations of trace elements have been measured in the soil, indicating that the HGMU must be performing the process of biogeochemical retention of trace elements to a small degree. A high trace element input has been identified, but the retention capacity is restricted by the limited number of fixation sites within the soil because of the moderately low organic matter content, but the medium or fine texture of the soil promotes retention. Soil pH and redox potential do not restrict the retention of trace elements.</td><td>1/23</td><td>e.g. for cadmium in the Elbe
Mobile Cd ratio 30–50 %,
mobile Cd amount > 1.5 mg kg^{-1} but < 5 mg kg^{-1}</td></tr>
<tr><td>1c 2a</td><td>1b
(3a|4a)</td><td>(1a|1b)
(2a|2b)
#3</td><td>Slightly elevated concentrations of trace elements have been measured in the soil, indicating that the HGMU must be performing the process of biogeochemical retention of trace elements to a small degree. A high trace element input has been identified, but the retention capacity is restricted by the limited number of fixation sites within the soil because of the low organic matter content, but the medium or fine texture of the soil promotes retention. Soil pH and redox potential do not restrict the retention of trace elements.</td><td>1/24</td><td>e.g. for cadmium in the Elbe
Mobile Cd ratio 30–50 %,
mobile Cd amount > 1.5 mg kg^{-1} but < 5 mg kg^{-1}</td></tr>
<tr><td>1c 2a</td><td>1b
(3a|4a)</td><td>(1c|1d)
(2a|2b)
#3</td><td>Slightly elevated concentrations of trace elements have been measured in the soil, indicating that the HGMU must be performing the process of biogeochemical retention of trace elements to a small degree. A high trace element input has been identified, but the retention capacity</td><td>1/25</td><td>e.g. for cadmium in the Elbe
Mobile Cd ratio 30–50 %,
mobile Cd amount > 1.5 mg</td></tr>
</table>

3.2.D.2 Look-up Table 1 cont'd

CV1	CV2	CV3	Rationale	Code	Quantification
			is restricted by the limited number of fixation sites within the soil because of the low organic matter content, but the medium or fine texture of the soil promotes retention. Low soil pH also restricts the retention capacity. Soil redox potential does not restrict the retention of trace elements.		kg^{-1} but < 5 mg kg^{-1}
1c 2a	1b (3b|4b)	(1c|1d) (2a|2b) #3	Slightly elevated concentrations of trace elements have been measured in the soil, indicating that the HGMU must be performing the process of biogeochemical retention of trace elements to a small degree. A high trace element input has been identified, but the retention capacity is restricted by the limited number of fixation sites within the soil because of the moderately low organic matter content, but the medium or fine texture of the soil promotes retention. Low soil pH also restricts the retention capacity. Soil redox potential does not restrict the retention of trace elements.	**1/26**	e.g. for cadmium *in the Elbe* Mobile Cd ratio 30–50 %, mobile Cd amount > 1.5 mg kg^{-1} but < 5 mg kg^{-1}
1c 2a	1b (3c|3d| 4c|4d)	1g (2a|2b) #3	Slightly elevated concentrations of trace elements have been measured in the soil, indicating that the HGMU must be performing the process of biogeochemical retention of trace elements to a small degree. A high trace element input has been identified, and the retention capacity is not restricted by the limited number of fixation sites within the soil, because of the high organic matter content, and the medium or fine texture of the soil. Soil redox potential does not restrict the retention of trace elements, and it is unknown if pH affects retention capacity.	**1/27**	e.g. for cadmium *in the Elbe* Mobile Cd ratio 30–50 %, mobile Cd amount > 1.5 mg kg^{-1} but < 5 mg kg^{-1}
1c 2a	1b (3b|4b)	1g (2a|2b) #3	Slightly elevated concentrations of trace elements have been measured in the soil, indicating that the HGMU must be performing the process of biogeochemical retention of trace elements to a small degree. A high trace element input has been identified, but the retention capacity is restricted by the limited number of fixation sites within the soil, because of the moderately low organic matter content, but the medium or fine texture of the soil promotes retention. Soil redox potential does not restrict the retention of trace elements, and it is unknown if pH affects retention capacity.	**1/28**	e.g. for cadmium *in the Elbe* Mobile Cd ratio 30–50 %, mobile Cd amount > 1.5 mg kg^{-1} but < 5 mg kg^{-1}
1c 2a	1a (3c|3d |4c|4d)	1g (2a|2b) #3	Slightly elevated concentrations of trace elements have been measured in the soil, indicating that the HGMU must be performing the process of biogeochemical retention of trace elements to a small degree. A high trace element input has been identified, but the retention capacity is restricted by the limited number of fixation sites within the soil, because of the coarse or gravelly texture of the soil, but not the organic matter content, which is relatively high. Soil redox potential does not restrict the retention of trace elements, and it is unknown if pH affects retention capacity.	**1/29**	e.g. for cadmium *in the Elbe* Mobile Cd ratio 30–50 %, mobile Cd amount > 1.5 mg kg^{-1} but < 5 mg kg^{-1}
1c 2a	1b (3a|4a)	1g (2a|2b) #3	Slightly elevated concentrations of trace elements have been measured in the soil, indicating that the HGMU must be performing the process of biogeochemical retention of trace elements to a small degree. A high trace element input has been identified, but the retention capacity is restricted by the limited number of fixation sites within the soil, because of the low organic matter content, but the medium or fine texture of the soil promotes retention. Soil	**1/30**	e.g. for cadmium *in the Elbe* Mobile Cd ratio 30–50 %, mobile Cd amount > 1.5 mg kg^{-1} but < 5 mg kg^{-1}

			redox potential does not restrict the retention of trace elements, and it is unknown if pH affects retention capacity.		
1c 2a	1a (3b|4b)	1g (2a|2b) #3	Slightly elevated concentrations of trace elements have been measured in the soil, indicating that the HGMU must be performing the process of biogeochemical retention of trace elements to a small degree. A high trace element input has been identified, but the retention capacity is restricted by the limited number of fixation sites within the soil, because of the coarse or gravelly texture of the soil, or the moderately low organic matter content. Soil redox potential does not restrict the retention of trace elements, and it is unknown if pH affects retention capacity.	**1/31**	e.g. for cadmium *in the Elbe* Mobile Cd ratio 30–50 %, mobile Cd amount > 1.5 mg kg^{-1} but < 5 mg kg^{-1}
1c 2a	1a (3b|4b)	(1a|1b) (2a|2b) #3	Slightly elevated concentrations of trace elements have been measured in the soil, indicating that the HGMU must be performing the process of biogeochemical retention of trace elements to a small degree. A high trace element input has been identified, but the retention capacity is restricted by the limited number of fixation sites within the soil, because of the coarse or gravelly texture of the soil, or the moderately low organic matter content. Soil pH and redox potential do not restrict the retention of trace elements.	**1/32**	e.g. for cadmium *in the Elbe* Mobile Cd ratio 30–50 %, mobile Cd amount > 1.5 mg kg^{-1} but < 5 mg kg^{-1}
1c 2a	1a (3b|4b)	(1c|1d) (2a|2b) #3	Slightly elevated concentrations of trace elements have been measured in the soil, indicating that the HGMU must be performing the process of biogeochemical retention of trace elements to a small degree. A high trace element input has been identified, but the retention capacity is restricted by the limited number of fixation sites within the soil, because of the coarse or gravelly texture of the soil, or the moderately low organic matter content. Low soil pH also restricts the retention capacity. Soil redox potential does not restrict the retention of trace elements.	**1/33**	e.g. for cadmium *in the Elbe* Mobile Cd ratio 30–50 %, mobile Cd amount > 1.5 mg kg^{-1} but < 5 mg kg^{-1}
1c 2b	*	*	Slightly elevated concentrations of trace elements have been measured in the soil, indicating that the HGMU must be performing the process of biogeochemical retention of trace elements to a small degree. However, inputs of trace elements are negligible or rare, suggesting that trace elements in the soil must have accumulated over a long period, and therefore biogeochemical retention of trace elements must be sustained over a long period.	**1/34**	e.g. for cadmium *in the Elbe* Mobile Cd ratio 30–50 %, mobile Cd amount > 1.5 mg kg^{-1} but < 5 mg kg^{-1}
1c 2c	*	*	Slightly elevated concentrations of trace elements have been measured in the soil, indicating that the HGMU must be performing the process of biogeochemical retention of trace elements to a small degree. The nature of the trace element input is unknown, and it could be a limited supply or other factors that are restricting the retention of trace elements within the soil.	**1/35**	e.g. for cadmium *in the Elbe* Mobile Cd ratio 30–50 %, mobile Cd amount > 1.5 mg kg^{-1} but < 5 mg kg^{-1}
1e 2a	1b (3c|3d|4c|4d)	(1a|1b) (2a|2b) #3	The concentration of trace elements in the soil is not known. However, it is likely that the HGMU is performing the process of biogeochemical retention of trace elements because a high trace element input has been identified, and neither soil pH, redox potential, organic matter content nor particle size are likely to restrict the retention process.	**1/36**	e.g. for cadmium *in the Elbe* Mobile Cd ratio > 50 %
1e 2a	1a (3c|3d|4c|4d)	(1a|1b) (2a|2b) #3	The concentration of trace elements in the soil is not known. However, it is likely that the HGMU is performing the process of biogeochemical retention of trace elements because a high trace element input has been identified, but the retention capacity is restricted by the	**1/37**	e.g. for cadmium *in the Elbe* Mobile Cd ratio > 50 %

3.2.D.2 Look-up Table 1 cont'd

<table>
<tr><th>CV1</th><th>CV2</th><th>CV3</th><th>Rationale</th><th>Code</th><th>Quantification</th></tr>
<tr><td></td><td></td><td></td><td>limited number of fixation sites within the soil, because of the coarse or gravelly texture of the soil, but not the organic matter content, which is relatively high. Soil pH and redox potential do not restrict the retention of trace elements.</td><td></td><td></td></tr>
<tr><td>1e 2a</td><td>1b
(3c|3d|
4c|4d)</td><td>(1c|1d)
(2a|2b)
#3</td><td>The concentration of trace elements in the soil is not known. However, it is likely that the HGMU is performing the process of biogeochemical retention of trace elements because a high trace element input has been identified, and the retention capacity is not restricted by the limited number of fixation sites within the soil, because of the high organic matter content, and the medium or fine texture of the soil. Low soil pH restricts the retention capacity. Soil redox potential does not restrict the retention of trace elements.</td><td>1/38</td><td>e.g. for cadmium in the Elbe
Mobile Cd ratio > 50 %</td></tr>
<tr><td>1e 2a</td><td>1a
(3c|3d|
4c|4d</td><td>(1c|1d)
(2a|2b)
#3</td><td>The concentration of trace elements in the soil is not known. However, it is likely that the HGMU is performing the process of biogeochemical retention of trace elements because a high trace element input has been identified, but the retention capacity is restricted by the limited number of fixation sites within the soil, because of the coarse or gravelly texture of the soil, but not the organic matter content, which is relatively high. Low soil pH also restricts the retention capacity. Soil redox potential does not restrict the retention of trace elements.</td><td>1/39</td><td>e.g. for cadmium in the Elbe
Mobile Cd ratio > 50 %</td></tr>
<tr><td>1e 2a</td><td>1b
(3b|4b)</td><td>(1a|1b)
(2a|2b)
#3</td><td>The concentration of trace elements in the soil is not known. However, it is likely that the HGMU is performing the process of biogeochemical retention of trace elements because a high trace element input has been identified, but the retention capacity is restricted by the limited number of fixation sites within the soil because of the moderately low organic matter content, but the medium or fine texture of the soil promotes retention. Soil pH and redox potential do not restrict the retention of trace elements.</td><td>1/40</td><td>e.g. for cadmium in the Elbe
Mobile Cd ratio > 50 %</td></tr>
<tr><td>1e 2a</td><td>1b
(3a|4a)</td><td>(1a|1b)
(2a|2b)
#3</td><td>The concentration of trace elements in the soil is not known. However, it is likely that the HGMU is performing the process of biogeochemical retention of trace elements because a high trace element input has been identified, but the retention capacity is restricted by the limited number of fixation sites within the soil because of the low organic matter content, but the medium or fine texture of the soil promotes retention. Soil pH and redox potential do not restrict the retention of trace elements.</td><td>1/41</td><td>e.g. for cadmium in the Elbe
Mobile Cd ratio > 50 %</td></tr>
<tr><td>1e 2a</td><td>1b
(3a|4a)</td><td>(1c|1d)
(2a|2b)
#3</td><td>The concentration of trace elements in the soil is not known. However, it is likely that the HGMU is performing the process of biogeochemical retention of trace elements because a high trace element input has been identified, but the retention capacity is restricted by the limited number of fixation sites within the soil because of the low organic matter content, but the medium or fine texture of the soil promotes retention. Low soil pH also restricts the retention capacity. Soil redox potential does not restrict the retention of trace elements.</td><td>1/42</td><td>e.g. for cadmium in the Elbe
Mobile Cd ratio > 50 %</td></tr>
<tr><td>1e 2a</td><td>1b
(3b|4b)</td><td>(1c|1d)
(2a|2b)
#3</td><td>The concentration of trace elements in the soil is not known. However, it is likely that the HGMU is performing the process of biogeochemical retention of trace elements because a high trace element input has been identified, but the retention capacity is restricted by the</td><td>1/43</td><td>e.g. for cadmium in the Elbe
Mobile Cd ratio > 50 % %</td></tr>
</table>

<table>
<tr><td></td><td></td><td></td><td>limited number of fixation sites within the soil, because of the moderately low organic matter content, but the medium or fine texture of the soil promotes retention. Low soil pH also the retention capacity. Soil redox potential does not restrict the retention of trace elements.</td><td></td><td></td></tr>
<tr><td>1e 2a</td><td>1a
(3b|4b)</td><td>(1a|1b)
(2a|2b)
#3</td><td>The concentration of trace elements in the soil is not known. However, it is likely that the HGMU is performing the process of biogeochemical retention of trace elements because a high trace element input has been identified, but the retention capacity is likely to be restricted by the limited number of fixation sites within the soil because of the coarse or gravelly texture of the soil, or the moderately low organic matter content. Soil pH and redox potential do not restrict the retention of trace elements.</td><td>1/44</td><td>e.g. for cadmium in the Elbe
Mobile Cd ratio > 50 %</td></tr>
<tr><td>1e 2a</td><td>1a
(3b|4b)</td><td>(1c|1d)
(2a|2b)
#3</td><td>The concentration of trace elements in the soil is not known. However it is likely that the HGMU is performing the process of biogeochemical retention of trace elements because a high trace element input has been identified, but the retention capacity is restricted by the limited number of fixation sites within the soil because of the coarse or gravelly texture of the soil, or the moderately low organic matter content. Low soil pH also restricts the retention capacity. Soil redox potential does not restrict the retention of trace elements.</td><td>1/45</td><td>e.g. for cadmium in the Elbe
Mobile Cd ratio > 50 %</td></tr>
</table>

3.2.D.2 Look-up Table 2

CV1	CV2	CV3	Rationale	Code
1e 2b	*	*	The concentration of trace elements in the soil is not known. Inputs of trace elements are negligible or rare, and therefore the process might be performed, but only to a very small extent, and only if other controlling variables are favourable.	**2**
*	*	2c|3	The soil is predominantly anaerobic and has a low redox potential, which severely limits its capacity to retain trace elements.	**2**
*	*	1e|1f	The pH of the soil is very low and this severely limits its capacity to retain trace elements.	**2**

3.2.D.2 Look-up Table 3

<table>
<tr><th>CV1</th><th>CV2</th><th>CV3</th><th>Rationale</th><th>Code</th></tr>
<tr><td>1e 2a</td><td>1b (3c|3d|4c|4d)</td><td>1g (2a|2b) #3</td><td>The concentration of trace elements in the soil is not known. However, it is likely that the HGMU is performing the process of biogeochemical retention of trace elements because a high trace element input has been identified, and the retention capacity is not restricted by the limited number of fixation sites within the soil, due to the high organic matter content, and the medium or fine texture of the soil. Soil redox potential does not restrict the retention of trace elements, but it is unknown if pH affects retention capacity.</td><td>1/x</td></tr>
<tr><td>1e 2a</td><td>1b (3b|4b)</td><td>1g (2a|2b) #3</td><td>The concentration of trace elements in the soil is not known. However, it is likely that the HGMU is performing the process of biogeochemical retention of trace elements because a high trace element input has been identified, but the retention capacity is restricted by the limited number of fixation sites within the soil, due to the moderately low organic matter content, but the medium or fine texture of the soil promotes retention. Soil redox potential does not restrict the retention of trace elements, but it is unknown if pH affects retention capacity.</td><td>1/x</td></tr>
<tr><td>1e 2a</td><td>1b (3a|4a)</td><td>1g (2a|2b)</td><td>The concentration of trace elements in the soil is not known. However, it is likely that the HGMU is performing the process of biogeochemical retention of trace elements because a high trace element input has been identified, but the retention capacity is restricted by the limited number of fixation sites within the soil, due to the low organic matter content, but the medium or fine texture of the soil promotes retention. Soil redox potential does not restrict the retention of trace elements, but it is unknown if pH affects retention capacity.</td><td>1/x</td></tr>
<tr><td>1e 2a</td><td>1a (3c|3d|4c|4d)</td><td>1g (2a|2b) #3</td><td>The concentration of trace elements in the soil is not known. However, it is likely that the HGMU is performing the process of biogeochemical retention of trace elements because a high trace element input has been identified, but the retention capacity is restricted by the limited number of fixation sites within the soil, due to the coarse or gravelly texture of the soil, but not by the organic matter content, which is relatively high. Soil redox potential does not restrict the retention of trace elements, but it is unknown if pH effects retention capacity.</td><td>1/x</td></tr>
<tr><td>1e 2a</td><td>1a (3b|4b)</td><td>1g (2a|2b) #3</td><td>The concentration of trace elements in the soil is not known. However, it is likely that the HGMU is performing the process of biogeochemical retention of trace elements because a high trace element input has been identified, but the retention capacity is restricted by the limited number of fixation sites within the soil, due to the coarse or gravelly texture of the soil, or the moderately low organic matter content. Soil redox potential does not restrict the retention of trace elements, and it is unknown if pH affects retention capacity.</td><td>1/x</td></tr>
<tr><td>(1a|1b|1c) 2b|2d</td><td>*</td><td>*</td><td>Elevated concentrations of trace elements have been measured in the soil, but currently there is no, or negligible, input of trace elements. The elevated concentrations must be a historic feature, and conditions within the soil must currently be conducive to the retention of trace elements.</td><td>1/x</td></tr>
<tr><td>1e 2c</td><td>*</td><td>*</td><td>The concentration of trace elements in the soil is not known. The nature of the trace element input is unknown, and therefore it is not possible to say whether the process is being performed.</td><td>1/x</td></tr>
</table>

3.2.D.2 Look-up Table 4

CV1	CV2	CV3	Rationale	Code
1d	*	*	Soil trace element concentrations are not elevated above typical background concentrations, and therefore the process of trace element retention is not being performed.	**4**
*	1a (3a|4a)	*	The soil particle size is coarse or gravelly, and there is negligible organic matter in the soil. Consequently the ability of the soil to bind trace elements is severely restricted, and therefore the process of trace element retention cannot be performed.	**4**

retention of trace elements. The output codes can be recorded on the Functional Assessment Table at the end of Section 3.2.D.4.

3.2.D.3 Long-term retention of trace elements through plant uptake (process)

Further reading and references

Brümmer, G.W. (1986). Heavy metal species, mobility and availability in soils. In: Bernhard, M., Brinckmann, F.E., Sadler, P.J. (eds), *The Importance of Chemical "Speciation" in Environmental Processes*, Report on the Dahlem Workshop Berlin 1984, Springer, Berlin, Heidelberg, New York, London, Paris, Tokyo, 169–192.

Dunbabin, J.S. and Bowmer, K.H. (1992). Potential use of constructed wetlands for the treatment of industrial wastewaters containing metals, *Science of the Total Environment* **111** (2–3), 151–168.

Kabata-Pendias, A.H. (2000). *Trace Elements in Soils and Plants*, CRC, Boca Raton, FL.

Kadlec, R.H. (1995). Overview: Surface flow constructed wetlands, *Water Science and Technology* **32**(3), 1–12.

Kadlec, R.H. and Knight, R.L. (1996). *Treatment Wetlands*, CRC Boca Raton, FL.

Karpiscak, M.M., Whiteaker, L.R. and Artiola, J.F. (2001). Nutrient and heavy metal uptake and storage in constructed wetland systems in Arizona, *Water Science and Technology* **44**(11–12), 455–462.

King, R.F, Royle, A, Putwain, P.D. and Dickinson, N.M. (2006). Changing contaminant mobility in a dredged canal sediment during a three-year phytoremediation trial, *Environmental Pollution* **143**, 318–326.

Kloke, A., Sauerbeck, D. and Vetter, H. (1984). The contamination of plants and soils with heavy metals, and the transport of metals in terrestrial food chains. In: Nriagu, J.O. (ed.), *Changing Metal Cycles and Human Health*, Dahlem Konferenzen 1984, Springer, Berlin, Heidelberg, New York, Tokyo, 113–141.

Kuntze, H., Fleige, H., Gruppe, M., Koch, D., Pluquet, E. and Stumpfe, A. (1995). Kennzeichnung der Empfindlichkeit der Böden gegenüber Schwermetallen unter Berücksichtigung von lithogenem Grundgehalt, pedogener An- und Abreicherung sowie anthropogener Zusatzbelastung. In: *Mobilität von Schwermetallen in Böden mit erhöhtem geogenen Grundgehalt und anthropogener Zusatzbelastung*, UBA Texte 55/96, Umweltbundesamt, Berlin, Germany.

Mitsch, WJ. and Wise, K.M. (1998). Water quality, fate of metals and predictive model validation of a constructed wetland treating acid mine drainage, *Water Resources* **32**(6), 1888–1900.

Ross, S.M., Wood, M.D., Copplestone, D., Warriner, M. and Crook, P. (2007) *Environmental Concentrations of Heavy Metals in UK Soil and Herbage*, UK Soil and Herbage Pollutant Survey Report No. 7, Environment Agency, Bristol, UK.

Sauerbeck, D. and Styperek, P. (1988). *Schadstoffe im Boden, insbesondere Schwermetalle und organische Schadstoffe aus langjähriger Anwendung von Siedlungsabfällen*, Teilbericht Schwermetalle, UBA Texte 16/88, Umweltbundesamt, Berlin, Germany.

Introduction

Process description:

- Many wetland plants can retain trace elements by assimilation into their biomass. However, at the end of their growing season, the death and decomposition of plant material will result in the release of trace elements that have been taken up by plants back into the soil and soil water (this is not as significant with coniferous and evergreen vegetation as it is with deciduous or herbaceous vegetation). Trace elements incorporated into woody tissue will be retained for a longer period, as the tissue does not die after each growing season. Trace elements assimilated in herbaceous vegetation may be exported from the system through land-use management (this process is assessed separately in Section 3.2.E.1).
- A requirement for the process to occur is that the HGMU either is in receipt of trace elements in a form that is available for uptake, or that the soil contains a store of trace elements that are available for uptake. Consequently, the speciation of trace metals in the soil solution is an important controlling variable, along with the degree of biogeochemical retention.

- Plant uptake of trace elements also depends on the density of the root network within contaminated soil horizons or layers, and on the water consumption of plants. Therefore the type of vegetation will also influence the amount of trace element uptake from soils.
- Areas of low disturbance are likely to optimise the uptake of trace metals.

Note: The uptake of trace elements by plants poses a potential risk to humans and animals if the plant is ingested, and also can result in the poisoning of the plant.

Potential values – rationale for assessment:

- The long-term assimilation of anthropogenically-derived trace elements into plant tissue may improve water quality in water bodies receiving surface or subsurface run-off from the wetland.
- Plant uptake represents the export of trace elements from soil, resulting in an improvement in soil quality and buffering capacity.

In this process, the density and type of vegetation is used to give a simple indication of the wetland's capacity for long-term retention of trace elements through plant uptake. However, the process is highly complex and, if more detailed statements are required, further analyses should be pursued. In further analyses, the more dynamic relationships between plant species and trace elements (Fig. 3.2.14) should be considered, as should the age of the plant, the stage of development, the depth of the roots, the density of the root network and the water consumption of plants.

Occurrence of trace elements frequently is locally specific, but guidance for these is not applicable across Europe because of regional variations. Concentrations quoted below are specific to the Elbe case study. The user should determine regional concentration means for the specific trace elements in their HGMU (sourced from the national environmental protection body or scientific literature), then assess comparative concentrations accordingly. See note on page 432.

Note: Before this process can be assessed, it is necessary first to carry out an assessment of the biogeochemical retention of trace elements (Process 3.2.D.2).

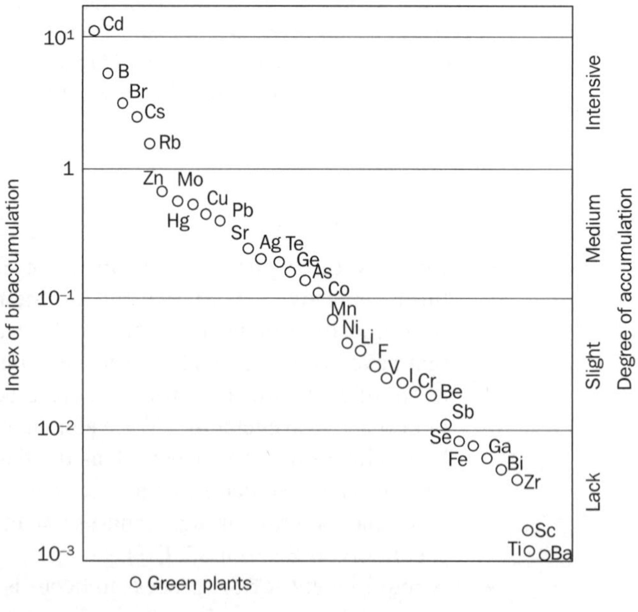

3.2.14 Bioaccumulation of trace elements by plants from soils. Index of bioaccumulation (BAC, IBA) was calculated as the ratio of trace elements in plants to their concentration in soils. The calculation is based on data from a range of plants and soils (from Kabata-Pendias, 2000).

Controlling variables

CV1 Trace element input/concentration in soil
CV2 Vegetation type
CV3 Degree of disturbance

Answer all the questions below for *each* HGMU; unless otherwise instructed continue from one question to the next. On the appraisal sheet, circle the correct answer. The location of the relevant information is given in brackets after each question. Use a new appraisal sheet for each HGMU.

CV1 *Trace element input/concentration in soil*

Q1. Does the soil trace element concentration indicate that anthropogenically-derived contamination of the HGMU has taken place? *(Recording Sheet 1, C.3.25.1b)*

a. Yes, the soil trace element concentration is much higher than typical background concentrations.
b. Yes, the soil trace element concentration is moderately higher than typical background concentrations.
c. Yes, the soil trace element concentration is slightly higher than typical background concentration.
d. No, the soil trace element concentration is not elevated.
e. Unknown, no soil analytical data are available.

e.g. for the Elbe catchment example site (you will have to determine regional mean concentrations for the trace element you are concerned with; see note on page 432):

Q1. Does the soil Cd concentration indicate that anthropogenically-derived contamination of the HGMU has taken place? *(Recording Sheet 1, C.3.25.1b)*

a. Yes, the soil Cd concentration is much higher than typical background concentrations (> 20 mg kg^{-1} soil).
b. Yes, the soil Cd concentration is moderately higher than typical background concentrations (10–20 mg kg^{-1} soil).
c. Yes, the soil Cd concentration is slightly higher than typical background concentration (> 5 but < 10 mg kg^{-1} soil).
d. No, the soil Cd concentration is not elevated (≤ 5 mg kg^{-1} soil).
e. Unknown, no soil analytical data are available.

Note: This is specific to the Elbe catchment. In the UK, for example, typical background Cd concentrations are around 1–2 mg kg^{-1} soil.

Q2. Has the HGMU been assigned a trace element input code indicating an input of trace elements? *(Recording Sheet, 2e, 1)*

a. Yes, trace element input code **Trae 1/1**, **1/2** or **1/3**.
b. Yes, trace element input code **Trae 2**.
c. Uncertain, trace element input code **Trae 3**.
d. No, trace element input code **Trae 4**.

Q3. Which code for biogeochemical retention of trace elements has been assigned to the HGMU? (*Functional Assessment 3.2.D.2*)

a. **1/1–1/18**, **1/36–1/38**, **1/40** or **1/41**
b. **1/34**, **1/35**, **1/39**, **1/42–1/46**, **2**, **1/x** or **4**
c. **1/19–1/33**

CV2 *Vegetation type*

Q1. Has the HGMU been classed as unvegetated or sparsely vegetated, i.e. coded **U**? *(Recording Sheet 3, D)*
[If YES do not continue with the following questions, but go directly to Look-up Table 2 on p. 456, 'The process is not (significantly) being performed']

Q2. What area of the HGMU is covered by trees or shrubs? *(AA Map 2; Recording Sheet 3.D)*

a. There are (almost) no trees or shrubs present in the HGMU (< 5 %). [*Continue with* ***Q5***]
b. Trees or shrubs cover 5–< 30 % of the HGMU [*Continue with* ***Q3***]
c. Trees or shrubs cover 30–< 60 % of the HGMU [*Continue with* ***Q3***]
d. Trees or shrubs cover 60 % or more of the HGMU. [*Continue with* ***Q3***]

Q3. Has the vegetation been assigned one of the following shrub (**S**) codes? *(Recording Sheet 3, D)*
a. **S3**
b. **S18**
c. Other **S** code
d. No **S** code has been assigned to the HGMU

Q4. Has the vegetation been assigned one of the following woodland (**W**) codes? *(Recording Sheet 3, D)*
a. **W5–W12, W14, W15, W17–W19**
b. Other **W** code
c. No **W** code has been assigned to the HGMU

Q5. Has the vegetation been assigned one of the following grassland (**G**) codes? *(Recording Sheet 3, D)*
a. **G8–G10, G14, G30–G32**
b. Other **G** code
c. No **G** code has been assigned to the HGMU

Q6. Has the vegetation been assigned one of the following mire (**M**) codes? *(Recording Sheet 3, D)*
a. **M8, M9, M12–M15**
b. Other **M** code
c. No **M** code has been assigned to the HGMU

Q7. Has the vegetation been assigned one of the following aquatic (**A**) codes? *(Recording Sheet 3, D)*
a. **A4, A6, A7, A8**
b. Other **A** code
c. No **A** code has been assigned to the HGMU

CV3 *Degree of disturbance*

Q1. How often does the AA experience drought (defined as less than 25 mm rainfall for the month prior to, and during each month of, the growing season)?
[*Determine the growing season (Recording Sheet 1, C.3.1.6a, C.3.1.7a, C.3.1.8a) and compare this to the mean monthly rainfall record (Recording Sheet 1, C.3.1.1i)*]
a. Never in the last 20 years
b. Every year in the last 20 years
c. Other/unknown

Q2. How often does the AA experience prolonged waterlogging or flooding throughout the growing season (stress occurs if soils are flooded for more than two successive months during the growing season)?
[***For flooding:*** *determine the growing season (Recording Sheet 1, C.3.1.6a, C.3.1.7a, C.3.1.8a) and compare to months when flooding typically occurs (Recording Sheet 1, C.3.19.6).* ***For waterlogging:*** *Recording Sheet 3, B.3.2*]
a. Never in the last 20 years
b. Every year in the last 20 years
c. Other/unknown

Q3. How often is the HGMU subjected to disturbance by ploughing? *(Recording Sheet 1, C.3.5.1b; AA Map 2)*
a. No history of ploughing in the last 20 years/not applicable
b. Not currently ploughed but ploughed within the last 20 years, or ploughed but less than once a year
c. Currently ploughed once or more than once a year
d. Unknown

Q4. Are there any current mining or extraction activities occurring that disturb the surface of the HGMU? *(Recording Sheet 1, C.3.11.1b; AA Map 2)*

a. No indication of mining or extraction/not applicable
b. Historical areas of mining or extraction, currently inactive
c. Currently active areas of mining or extraction
d. Unknown

Q5. Is there any current construction work occurring that is disturbing the surface of the HGMU? *(Recording Sheet 1, C.3.13.1b; AA Map 2)*

Q6. How often is the vegetation within the HGMU burnt as part of land management practices? *(Recording Sheet 1, C.3.3.1b; AA Map 2)*

a. No history of burning in the last 20 years/not applicable
b. Not currently burnt but previously burnt within the last 20 years, or burnt but less than once in 5 years
c. Burnt once or more than once in the last 5 years
d. Unknown

Q7. How often does felling and removal of timber occur in woodland areas within the HGMU? *(Recording Sheet 1, C.3.6.1b; AA Map 2)*

a. Never/not applicable
b. No current felling and removal but previously practised within the last 20 years, or felling and removal occurs, but less than once in 5 years
c. Felling and removal has taken place once or more than once in the last 5 years
d. Unknown

Assessment of the process

Compare the results recorded for each HGMU on the appraisal sheet with the possible outcomes below, and assign an appropriate assessment code. Record this on the appraisal sheet. The syntax codes used are described in the introduction to Section 3.

Woody plants have a longer lifespan than herbaceous plants and, as a result, can usually store trace elements for a longer time period. Although herbaceous plants generally are faster growing than woody vegetation (and therefore can take up greater proportions of trace elements), they have a relatively short lifespan. In fast-growing species, long-term retention is not necessarily achieved and the trace elements may be returned to the soil with plant senescence. However, if the area is harvested or grazed, trace elements can be removed by export rather than retention on site. See Process 3.2.E.1 (export of trace elements through vegetation management).

Look-up tables

1. The process is definitely being performed
If answers to the questions for this process agree with any of the combinations in Look-up Table 1, then the HGMU is definitely performing the process of plant uptake of trace elements. The outputs are in order of the CV answers, not the output codes themselves.

2. The process is not (significantly) being performed
If answers to the questions for this process agree with any of the combinations in Look-up Table 2, then the HGMU is not significantly performing the process of plant uptake of trace elements.

3. The process is being performed, but there are constraining factors or uncertainties
For any combination of answers not present in Look-up Tables 1 and 2, then the HGMU is performing the process but probably only to a very small degree. There are constraining factors or uncertainties in assessing the degree to which it is being performed, which can be determined from Look-up Table 3.

HGMU code: ______________	**AA name:** ______________

	CV1 Trace element input/availability				
Q1	a	b	c	d	e
Q2	a	b	c	d	
Q3	a	b	c		

	CV3 Degree of disturbance			
Q1	a	b	c	
Q2	a	b	c	
Q3	a	b	c	d
Q4	a	b	c	d
Q5	Y	N	U/NA	
Q6	a	b	c	d
Q7	a	b	c	d

	CV2 Vegetation type			
Q1	Y	N	U/NA	
Q2	a	b	c	d
Q3	a	b	c	d
Q4	a	b	c	
Q5	a	b	c	
Q6	a	b	c	
Q7	a	b	c	

HGMU process assessment code: ______________

3.2.D.3 Look-up Table 1

<table>
<tr><th>CV1</th><th>CV2</th><th>CV3</th><th>Rationale</th><th>Code</th></tr>
<tr><td>1a|2a 3c</td><td>2b 3a</td><td>1a 2a
(3a|3b)
(4a|4b)
#5
(6a|6b)
(7a|7b)</td><td>The HGMU has either a high trace element input or a high concentration of trace elements in the soil. The biogeochemical retention capacity of the soil is low, so trace elements will be readily available for plant uptake. Shrubs indicative of a high rate of trace element uptake are present on less than 30 % of the HGMU. The vegetation is not adapted to prolonged waterlogging, but this never occurs in the HGMU. The HGMU experiences a low degree of disturbance from drought, ploughing, mining or extraction, construction work, burning of vegetation and felling and removal of trees, so the ability of plants to take up trace elements is not affected.</td><td>1/3</td></tr>
<tr><td>1a|2a 3c</td><td>2c 3a</td><td>1a 2a
(3a|3b)
(4a|4b)
#5
(6a|6b)
(7a|7b)</td><td>The HGMU has either a high trace element input or a high concentration of trace elements in the soil. The biogeochemical retention capacity of the soil is low, so trace elements will be readily available for plant uptake. Shrubs indicative of a high rate of trace element uptake are present on 30–60 % of the HGMU. The vegetation is not adapted to prolonged waterlogging, but this never occurs in the HGMU. The HGMU experiences a low degree of disturbance from drought, ploughing, mining or extraction, construction work, burning of vegetation and felling and removal of trees, so the ability of plants to take up trace elements is not affected.</td><td>1/2</td></tr>
<tr><td>1a|2a 3c</td><td>2d 3a</td><td>1a 2a
(3a|3b)
(4a|4b)
#5
(6a|6b)
(7a|7b)</td><td>The HGMU has either a high trace element input or a high concentration of trace elements in the soil. The biogeochemical retention capacity of the soil is low, so trace elements will be readily available for plant uptake. Shrubs indicative of a high rate of trace element uptake are present on more than 60 % of the HGMU. The vegetation is not adapted to prolonged waterlogging, but this never occurs in the HGMU. Moreover the HGMU experiences a low degree of disturbance from drought, ploughing, mining or extraction, construction work, burning of vegetation and felling and removal of trees, so the ability of plants to take up trace elements is not affected.</td><td>1/1</td></tr>
<tr><td>(1b|1c| 2b) 3c</td><td>2b 3a</td><td>1a 2a
(3a|3b)
(4a|4b)
#5
(6a|6b)
(7a|7b)</td><td>The HGMU has either a low trace element input, or a moderate or low concentration of trace elements in the soil. The biogeochemical retention capacity of the soil is low, so trace elements will be readily available for plant uptake. Shrubs indicative of a high rate of trace element uptake are present on less than 30 % of the HGMU. The vegetation is not adapted to prolonged waterlogging, but this never occurs in the HGMU. The HGMU experiences a low degree of disturbance from drought, ploughing, mining or extraction, construction work, burning of vegetation and felling and removal of trees, so the ability of plants to take up trace elements is not affected.</td><td>1/9</td></tr>
<tr><td>(1b|1c| 2b) 3c)</td><td>2c 3a</td><td>1a 2a
(3a|3b)
(4a|4b)
#5
(6a|6b)
(7a|7b)</td><td>The HGMU has either a low trace element input, or a moderate or low concentration of trace elements in the soil. The biogeochemical retention capacity of the soil is low, so trace elements will be readily available for plant uptake. Shrubs indicative of a high rate of trace element uptake are present on 30–60 % of the HGMU. The vegetation is not adapted to prolonged waterlogging, but this never occurs in the HGMU. The HGMU experiences a low degree of disturbance from drought, ploughing, mining or extraction, construction work, burning of vegetation and felling and removal of trees, so the ability of plants to take up trace elements is not affected.</td><td>1/8</td></tr>
<tr><td>(1b|1c| 2b) 3c)</td><td>2d 3a</td><td>1a 2a
(3a|3b)
(4a|4b)
#5
(6a|6b)
(7a|7b)</td><td>The HGMU has either a low trace element input, or a moderate or low concentration of trace elements in the soil. The biogeochemical retention capacity of the soil is low, so trace elements will be readily available for plant uptake. Shrubs indicative of a high rate of trace element uptake are present on more than 60 % of the HGMU. The vegetation is not adapted to prolonged waterlogging, but this never occurs in the HGMU. Moreover the HGMU experiences a low degree of disturbance from drought, ploughing, mining or extraction, construction work, burning of vegetation, and felling and removal of trees, so the ability of plants to take up trace elements is not affected.</td><td>1/7</td></tr>
</table>

3.2.D.3 Look-up Table 1 cont'd

CV1	CV2	CV3	Rationale	Code
1a|2a 3c	2b (3b|4a)	1a (3a|3b) (4a|4b) #5 (6a|6b) (7a|7b)	The HGMU has either a high trace element input or a high concentration of trace elements in the soil. The biogeochemical retention capacity of the soil is low, so trace elements will be readily available for plant uptake. Shrubs or trees indicative of a high rate of trace element uptake are present on less than 30 % of the HGMU. The HGMU experiences a low degree of disturbance from drought, ploughing, mining or extraction, construction work, burning of vegetation and felling and removal of trees, so the ability of plants to take up trace elements is not affected. The vegetation present in the HGMU is adapted to waterlogged soils, therefore waterlogging is not a stress factor for the plants. NB: The amount of trace elements retained in wood per hectare is dependent upon the ages of trees and shrubs: young trees and shrubs take up more trace elements than do older trees and shrubs. In mature woodlands, tree or shrub mortality (a process by which trace elements stored in wood return to the system) can be as important, or even more so, than uptake and storage.	**1/6**
1a|2a 3c	2c (3b|4a)	1a (3a|3b) (4a|4b) #5 (6a|6b) (7a|7b)	The HGMU has either a high trace element input or a high concentration of trace elements in the soil. The biogeochemical retention capacity of the soil is low, so trace elements will be readily available for plant uptake. Shrubs or trees indicative of a high rate of trace element uptake are present on 30–60 % of the HGMU. The HGMU experiences a low degree of disturbance from drought, ploughing, mining or extraction and construction work. The vegetation present in the HGMU is adapted to waterlogged soils, therefore waterlogging is not a stress factor for the plants. The vegetation is not burnt. Felling or removal of trees does not take place. NB: The amount of trace elements retained in wood per hectare is dependent upon the ages of trees and shrubs: young trees and shrubs take up more trace elements than do older trees and shrubs. In mature woodlands, tree or shrub mortality (a process by which trace elements stored in wood return to the system) can be as important, or even more so, than uptake and storage.	**1/5**
1a|2a 3c	2d (3b|4a)	1a (3a|3b) (4a|4b) #5 (6a|6b) (7a|7b)	The HGMU has either a high trace element input or high concentration of trace elements in the soil. The biogeochemical retention capacity of the soil is low, so trace elements will be readily available for plant uptake. Shrubs or trees indicative of a high rate of trace element uptake are present on more than 60 % of the HGMU. The HGMU experiences a low degree of disturbance from drought, ploughing, mining or extraction and construction work. The vegetation present in the HGMU is adapted to waterlogged soils, therefore waterlogging is not a stress factor for the plants. The vegetation is not burnt. Felling or removal of trees does not take place. NB: The amount of trace elements retained in wood per hectare is dependent upon the ages of trees and shrubs: young trees and shrubs take up more trace elements than do older trees and shrubs. In mature woodlands, tree or shrub mortality (a process by which trace elements stored in wood return to the system) can be as important, or even more so, than uptake and storage.	**1/4**
(1b|1c| 2b) 3c)	2d (3b|4a)	1a 2a (3a|3b) (4a|4b) #5 (6a|6b) (7a|7b)	The HGMU has either a low trace element input or a moderate or low concentration of trace elements in the soil. The biogeochemical retention capacity of the soil is low, so trace elements will be readily available for plant uptake. Shrubs or trees indicative of a high rate of trace element uptake are present on less than 30 % of the HGMU. The vegetation is not adapted to prolonged waterlogging, but this never occurs in the HGMU. The HGMU experiences a low degree of disturbance from drought, ploughing, mining or extraction, construction work, burning of vegetation, and felling and removal of trees, so the ability of plants to take up trace elements is not affected. NB: The amount of trace elements retained in wood per hectare is dependent upon the ages of trees and shrubs: young trees and shrubs take up more trace elements than do older trees and shrubs. In mature woodlands tree or shrub	**1/12**

			mortality (a process by which trace elements stored in wood return to the system) can be as important, or even more so, than uptake and storage.	
(1b|1c|2b) 3c	2d (3b|4a)	1a 2a (3a|3b) (4a|4b) #5 (6a|6b) (7a|7b)	The HGMU has either a low trace element input or a moderate or low concentration of trace elements in the soil. The biogeochemical retention capacity of the soil is low, so trace elements will be readily available for plant uptake. Shrubs or trees indicative of a high rate of trace element uptake are present on 30–60 % of the HGMU. The vegetation is not adapted to prolonged waterlogging, but this never occurs in the HGMU. The HGMU experiences a low degree of disturbance from drought, ploughing, mining or extraction, construction work, burning of vegetation, and felling and removal of trees, so the ability of plants to take up trace elements is not affected. NB: The amount of trace elements retained in wood per hectare is dependent upon the ages of trees and shrubs: young trees and shrubs take up more trace elements than do older trees and shrubs. In mature woodlands tree or shrub mortality (a process by which trace elements stored in wood return to the system) can be as important, or even more so, than uptake and storage.	**1/11**
(1b|1c|2b) 3c	2d (3b|4a)	1a 2a (3a|3b) (4a|4b) #5 (6a|6b) (7a|7b)	The HGMU has either a low trace element input or a moderate or low concentration of trace elements in the soil. The biogeochemical retention capacity of the soil is low, so trace elements will be readily available for plant uptake. Shrubs or trees indicative of a high rate of trace element uptake are present on more than 60 % of the HGMU. The vegetation is not adapted to prolonged waterlogging, but this never occurs in the HGMU. Moreover, the HGMU experiences a low degree of disturbance from drought, ploughing, mining or extraction, construction work, burning of vegetation, and felling and removal of trees, so the ability of plants to take up trace elements is not affected. NB: The amount of trace elements retained in wood per hectare is dependent upon the ages of trees and shrubs: young trees and shrubs take up more trace elements than do older trees and shrubs. In mature woodlands tree or shrub mortality (a process by which trace elements stored in wood return to the system) can be as important, or even more so, than uptake and storage.	**1/10**

3.2.D.3 Look-up Table 2

<table>
<tr><th>CV1</th><th>CV2</th><th>CV3</th><th>Rationale</th><th>Code</th></tr>
<tr><td>*</td><td>1</td><td>*</td><td>There is no or little vegetation present in the HGMU; therefore there is no significant uptake.</td><td>2</td></tr>
<tr><td>*</td><td>3c|4b</td><td>*</td><td>Trees or shrubs indicative of a low rate of productivity are present, and therefore the uptake of trace elements is likely to be small.</td><td>2</td></tr>
<tr><td>*</td><td>5b|6b|7b</td><td>*</td><td>Vegetation indicative of a low rate of productivity is present, and therefore the uptake of trace elements is likely to be small.</td><td>2</td></tr>
</table>

3.2.D.3 Look-up Table 3

<table>
<tr><th>CV1</th><th>CV2</th><th>CV3</th><th>Rationale</th><th>Code</th></tr>
<tr><td>*</td><td>2a</td><td>*</td><td>There are almost no trees or shrubs present in the HGMU, and therefore opportunities for plant uptake of trace elements are highly restricted.</td><td>1/x</td></tr>
<tr><td>1e|2c</td><td>*</td><td>*</td><td>It is unknown whether the HGMU has elevated concentrations of trace elements in the soil, or if a trace element input is received by the HGMU.</td><td>1/x</td></tr>
<tr><td>3b</td><td>*</td><td>*</td><td>Biogeochemical retention of trace elements in the soil is likely to affect any plant uptake to some extent, or is unknown.</td><td>1/x</td></tr>
<tr><td>1a|2a
3c</td><td>5a</td><td>1a 2a
(3a|3b)
(4a|4b)
#5
(6a|6b)</td><td>The HGMU has either a high trace element input or high concentration of trace elements in the soil. The biogeochemical retention capacity of the soil is low, so trace elements will be readily available for plant uptake. Grassland vegetation indicative of a high rate of productivity is present. The HGMU experiences a low degree of disturbance from drought, waterlogging, ploughing, mining or extraction, construction work, or burning of vegetation, so the ability of plants to takeup trace elements is not affected.
Seasonal retention of trace elements will occur. However, at the end of the growing season the trace elements taken up by the plants will return to the system.</td><td>1/x</td></tr>
<tr><td>(1b|1c|
2b) 3c</td><td>5a</td><td>1a 2a
(3a|3b)
(4a|4b)
#5
(6a|6b)</td><td>The HGMU has either a low trace element input or a moderate or low concentration of trace elements in the soil. The biogeochemical retention capacity of the soil is low, so trace elements will be readily available for plant uptake. Grassland vegetation indicative of a high rate of productivity is present. The HGMU experiences a low degree of disturbance from drought, waterlogging, ploughing, mining or extraction, construction work, or burning of vegetation, so the ability of plants to takeup trace elements is not affected.
Seasonal retention of trace elements will occur. However, at the end of the growing season the trace elements taken up by the plants will return to the system.</td><td>1/x</td></tr>
<tr><td>1a|2a
3c</td><td>6a|7a</td><td>1a
(3a|3b)
(4a|4b)
#5
(6a|6b)</td><td>The HGMU has either a high trace element input or high concentration of trace elements in the soil. The biogeochemical retention capacity of the soil is low, so trace elements will be readily available for plant uptake. Mire or aquatic vegetation indicative of a high rate of productivity is present. The HGMU experiences a low degree of disturbance from drought, ploughing, mining or extraction or burning, so the ability of plants to take up trace elements is not affected.
Seasonal retention of trace elements will occur. However, at the end of the growing season the trace elements taken up by the plants will return to the system.</td><td>1/x</td></tr>
<tr><td>(1b|1c|
2b) 3c</td><td>6a|7a</td><td>1a
(3a|3b)
(4a|4b)
#5
(6a|6b)</td><td>The HGMU has either a low trace element input or a moderate or low concentration of trace elements in the soil. The biogeochemical retention capacity of the soil is low, so trace elements will be readily available for plant uptake. Mire or aquatic vegetation indicative of a high rate of productivity is present. The HGMU experiences a low degree of disturbance from drought, ploughing, mining or extraction or burning, so the ability of plants to take up trace elements is not affected.
Seasonal retention of trace elements will occur. However, at the end of the growing season the trace elements taken up by the plants will return to the system.</td><td>1/x</td></tr>
<tr><td>*</td><td>*</td><td>1b|2b|3c|
4c|5</td><td>Other factors may be suitable, but disturbance or stress because of drought, waterlogging, ploughing, mining or extraction, or construction work reduce uptake of trace elements by plants.</td><td>1/x</td></tr>
</table>

3.2.D.3 Look-up Table 3 cont'd

CV1	CV2	CV3	Rationale	Code
*	*	6c	Although some of the trace elements in plant biomass are exported to the atmosphere through burning, trace element-rich ashes return trace elements to the soil. Rapid re-growth of plants after the burning event will result in uptake of these trace elements, but the extent to which this occurs is unknown.	**1/x**
*	*	7c	Trace elements stored in biomass are exported. However, the disturbance caused by these management practices also results in recycling of some trace elements into the system (leaves and branches with a high concentration of trace elements fall on the ground and decompose rapidly, and trace elements are released back into the wetland system).	**1/x**
*	*	1c|2c|3d|4d|6d|7d	Other factors may be suitable, but it is not known whether disturbance or stress (drought, waterlogging, ploughing, mining or extraction, construction work, burning of vegetation, or forestry) takes place, which reduces uptake of trace elements by trees or shrubs.	**1/x**

3.2.D.3 Look-up Table 4

CV1	CV2	CV3	Rationale	Code
# (1a 1b 1c 2a 2b)	*	*	The HGMU is not in receipt of any trace elements and trace element concentrations in the soil are not elevated.	**4**
3a	*	*	Biogeochemical retention of trace elements in the soil is high, meaning that any trace elements present in the soil will be strongly bound and therefore not available for plant uptake.	**4**

4. The process is definitely not being performed

If answers to the questions for this process agree with any of the combinations in Look-up Table 4, then the HGMU is definitely not performing the process of plant uptake of trace elements.

This completes the functional assessment for the process of long-term retention of trace elements through plant uptake. The output codes can be recorded on the Functional Assessment Table at the end of Section 3.2.D.4.

3.2.D.4 Functional assessment: Trace element storage

To determine the overall performance of the Trace Element Storage function, the assessment must consider the operation of the component processes:

- physical retention of trace elements (D1)
- biogeochemical retention of trace elements (D2)
- long-term retention of trace elements through plant uptake (D3)

Trace element storage is definitely being performed (high and intermediate performance)

The function 'trace element storage' is definitely being performed by the HGMU if one or more of the component processes (D1–D3) of this function are occurring, i.e. if:

Any of the process assessment codes = **1/n** or **1/x** (results from Look-up Tables 1 and/or 3 in 3.2.D)

The statement 'trace element storage is definitely being performed' is a broad statement. The statement does not include any quantitative or semi-quantitative information. Not all processes contribute equally to the function. Therefore if only one process with a low maximum retention rate is contributing to the function, less of the trace metals will be retained than if several of these low retention processes, or if one process with a high maximum retention rate is contributing to the function. To estimate the performance of the function, all the processes contributing to the function have to be 'calculated'. This is done by comparing the combination of assessment outcome codes determined for each process with the list in Table 3.2.7.

Trace element storage is not (significantly) being performed (low performance)

The function 'trace element storage' is not being significantly performed by an HGMU if none of the component processes of this function are occurring, that is if:

All process assessment codes = **2**

Trace element storage is definitely not being performed (not performed)

The function 'trace element storage' is definitely not being performed by an HGMU if none of the component processes of this function are occurring, that is if:

all process assessment codes = **4**

This completes the functional analysis for trace element storage. The output codes can be recorded on the Functional Analysis Summary Table in Section 4.0, and used to compare functions among HGMUs across the wetland.

The same process can be used for other trace elements known to be present in the wetland assessment area provided sufficient data can be collected from the

Table 3.2.7 Functional assessment of trace element storage. Summing of trace element storage processes to determine a function through semi-quantification. Compare the combinations of process output codes (e.g. **1**, **1/x**, **2**, **4**) and read off the semi-quantitative score to gain an overall assessment. **1/n** = 1/any number

Trace element storage – process outcome codes				**Trace element storage**	
Process	**Process**	**Process**	**Score**	**Performance**	**Degree of performance**
1/n	**1/n**	**1/n**	1	Definitely performed	High performance
1/n	**1/n**	**1/x**	0.9		
1/n	**1/n**	**2**	0.9		
1/n	**1/n**	**4**	0.9		
1/n	**1/x**	**1/x**	0.8		
1/n	**1/x**	**2**	0.8		
1/n	**1/x**	**4**	0.8		
1/n	**2**	**2**	0.8		
1/n	**2**	**4**	0.8		
1/n	**4**	**4**	0.8		
1/x	**1/x**	**1/x**	0.5	Definitely performed but with constraining factors or uncertainties	Intermediate performance
1/x	**1/x**	**2**	0.4		
1/x	**1/x**	**4**	0.4		
1/x	**2**	**2**	0.3		
1/x	**2**	**4**	0.3		
1/x	**4**	**4**	0.2		
2	**2**	**2**	0.1	Not significantly performed	Low performance
4	**4**	**4**	0	Definitely not performed	Not performed

catchment. If you wish to assess all functions, move on to the next step of the FAPs: Trace element export.

When all the HGMUs in the assessment area have been assessed, a general pattern of trace element storage for this metal across the whole wetland will be evident. By manipulation of your answers to the controlling variable (CV) questions, you have the capacity to explore the probable outcomes of any management options under consideration and relate them to specific functions. The electronic version of the FAPs can do this more quickly and easily.

3.2.E Trace element export (function)

3.2.E.1 Export of trace elements through vegetation management (process)

Further reading and references

Kabata-Pendias, A. (2000). *Trace Elements in Soils and Plants*, CRC, Boca Raton, FL.

Kadlec, R.H. (1995). Overview: Surface flow constructed wetlands, *Water Science and Technology* **32**(3), 1–12.

Kadlec, R.H. and Knight, R.L. (1996). *Treatment Wetlands*, CRC, Boca Raton, FL.

Karpiscak, M.M., Whiteaker, L.R. and Artiola, J.F. (2001). Nutrient and heavy metal uptake and storage in constructed wetland systems in Arizona, *Water Science and Technology* **44**(11–12), 455–462.
King, R.F., Royle, A., Putwain, P.D. and Dickinson, N.M. (2006). Changing contaminant mobility in a dredged canal sediment during a three-year phytoremediation trial, *Environmental Pollution* **143**, 318–326.
Kloke, A., Sauerbeck, D. and Vetter, H. (1984). The contamination of plants and soils with heavy metals, and the transport of metals in terrestrial food chains. In: Nriagu, J.O. (ed.), *Changing Metal Cycles and Human Health*, Dahlem Konferenzen 1984, Springer, Berlin, Heidelberg, New York, Tokyo, 113–141.
Mitsch, WJ. and Wise, K.M. (1998). Water quality, fate of metals and predictive model validation of a constructed wetland treating acid mine drainage, *Water Resources* **32**(6), 1888–1900.

Introduction

Process description:

- Many plants can retain excess trace elements by assimilation into their biomass. However, often this may be only short-term unless the trace elements become stored in woody material. At the end of the growing season, the death of the plants, or the shedding of leaves and their subsequent decomposition, results in a return of the trace elements to the wetland system. However, through management, the trace elements retained in vegetation can be exported from the wetland system before they are released.
- A requirement for the process to occur is that the HGMU either is in receipt of trace elements in a form that is available for uptake, or that the soil contains a store of trace elements that are available for uptake. Consequently the speciation of trace metals in the soil solution is an important controlling variable, along with the degree of biogeochemical retention.
- Subsequent management of the vegetation can result in the removal from the wetland system of plant biomass and the trace elements stored therein.
- The process of trace element uptake is optimised when HGMUs experience low levels of disturbance (ploughing, mining, prolonged waterlogging or drought conditions) so that the ability of plants to take up trace elements is not affected.
- The main vegetation management practices considered here are harvesting of biomass and grazing by livestock.
- Burning of vegetation is a management practice that can result in the export of trace elements via gaseous emissions. However, the amount of trace elements lost by this process is relatively small compared with the other management techniques, and therefore is not assessed in this section.

Note: The uptake of trace elements by plants poses a potential risk to humans and animals if the plant is ingested, and may also result in poisoning of the plant.

Significance – rationale for assessment:

- The assimilation of excess trace elements into plant tissue and subsequent removal from the wetland system may improve water quality in water bodies receiving surface or subsurface run-off from the wetland.
- Plant uptake of trace elements and the subsequent removal of biomass represents the export of trace elements from soil, resulting in an improvement in soil quality.

In this process, the density and type of vegetation is used to give a simple indication of the wetland's capacity for long-term retention of trace elements through plant uptake. However, the process is highly complex and if more detailed statements are required, further research should be pursued. In particular, the more dynamic relationships between plant species and trace elements should be considered, as should the age of the plant, the stage of development, the depth of the roots, the density of the root network and the water consumption of plants.

Note: Before this process can be assessed, it is necessary first to carry out an assessment of the biogeochemical retention of trace elements (Process 3.2.D.2).

Controlling variables

CV1 Trace element input/concentration in soil
CV2 Vegetation type
CV3 Vegetation management
CV4 Degree of disturbance

Answer all the questions below for *each* HGMU; unless otherwise instructed continue from one question to the next. On the appraisal sheet circle the correct answer. The location of the relevant information is given in brackets after each question. Use a new appraisal sheet for each HGMU.

If there are no trace element inputs identified for the HGMU, or if the trace element input is insignificant or uncertain (trace element input codes **Trae 2/x**, **3** or **4**) you may wish to assess the potential of the HGMU to perform this process. To do this, conduct the assessment as normal but answer questions only in ***CV2***, ***CV3*** and ***CV4***. Establish an assessment outcome but ignore the need for positive answers for ***CV1*** in the look-up table. The rationale statement can be used as normal but ignore the reference to receipt of trace element inputs.

If the HGMU is not subject to vegetation management practices, the potential export may still be assessed. Conduct the assessment as normal but answer questions only in ***CV1***, ***CV2*** and ***CV4***. Establish an assessment outcome but ignore the answers for ***CV3*** and ignore the reference to vegetation management practices.

CV1 *Trace element input/concentration in soil*

Q1. Does the soil trace element concentration indicate that anthropogenically-derived contamination of the HGMU has taken place? *(Recording Sheet 1, C.3.25.1b)*

a. Yes, the soil trace element concentration is much higher than typical background concentrations.
b. Yes, the soil trace element concentration is moderately higher than typical background concentrations.
c. Yes, the soil trace element concentration is slightly higher than typical background concentration.
d. No, the soil trace element concentration is not elevated.
e. Unknown, no soil analytical data are available.

e.g. for the Elbe catchment example site (you will have to determine regional mean concentrations for the trace element you are concerned with; see note on p. 432):

Q1. Does the soil Cd concentration indicate that anthropogenically-derived contamination of the HGMU has taken place? *(Recording Sheet 1, C.3.25.1b)*

a. Yes, the soil Cd concentration is much higher than typical background concentrations (> 20 mg kg^{-1} soil).
b. Yes, the soil Cd concentration is moderately higher than typical background concentrations (10–20 mg kg^{-1} soil).
c. Yes, the soil Cd concentration is slightly higher than typical background concentration (> 5 but < 10 mg kg^{-1} soil).
d. No, the soil Cd concentration is not elevated (″5 mg kg^{-1} soil).
e. Unknown, no soil analytical data are available.

Note: This is specific to the Elbe catchment. In the UK, for example, typical background Cd concentrations are around 1–2 mg kg^{-1} soil.

Q2. Has the HGMU been assigned a code indicating an input of trace elements? *(Recording Sheet, 2e, 1)*

a. Yes, trace element input code **Trae 1/1**, **1/2**, **1/3**, **1/4**, **1/5**, **1/6** or **1/7**.

b. Yes, trace element input code **Trae 2**.
c. Uncertain, trace element input code **Trae 3**.
d. No, trace element input code **Trae 4**.

Q3. Which code for biogeochemical retention of trace elements has been assigned to the HGMU? (*Functional Assessment 3.2.D.2*)
a. **1/1–1/18**, **1/36–1/38**, **1/40** or **1/41**
b. **1/19–1/33**
c. **1/34**, **1/35**, **1/39**, **1/42–1/46**, **2**, **1/x**, or **4**

CV2 *Vegetation type*

Q1. Has the vegetation been classed as unvegetated or sparsely vegetated, i.e. coded **U**? *(Recording Sheet 3, D)*
[*If YES do not finish this questionnaire, but go to Look-up Table 2 on page 471, 'The function is not (significantly being performed'*]

Q2. Has the vegetation been classed as one of the following grassland codes? *(Recording Sheet 3, D)*
a. **G8–G10**, **G14**, **G30–G32**
b. Other **G** code

Q3. Has the vegetation been classed as one of the following mire codes? *(Recording Sheet 3, D)*
a. **M8**, **M9**, **M12–M15**
b. Other **M** code

Q4. Has the vegetation been classed as one of the following aquatic codes? *(Recording Sheet 3, D)*
a. **A4**, **A6**, **A7**, **A8**
b. Other **A** code

Q5. Has the vegetation been classed as one of the following shrub codes? *(Recording Sheet 3, D)*
a. **S3**
b. **S18**
c. Other **S** code

Q6. Has the vegetation been classed as one of the following woodland codes? *(Recording Sheet 3, D)*
a. **W5–W12**, **W14**, **W15**, **W17–W19**
b. Other **W** code

CV3 *Vegetation management*

Q1. Is the vegetation in the HGMU subject to harvesting or grazing by livestock? (*Recording Sheet 1, C.3.4b; C.3.7.b)*
[*If YES continue with **Q2**. If NO go to Look-up Table 2, on page 471, 'the function is not (significantly) being performed', or assess the **potential** of the HGMU for export through vegetation management practices by ignoring answers for **CV3** in the Look-up table*]

Q2. How often is herbaceous vegetation within the HGMU harvested? *(Recording Sheet 1, C.3.4.1b)*
a. No history of harvesting of herbaceous vegetation in the last 20 years [*If YES continue with **Q4***]
b. Not currently harvested but previously harvested within the last 20 years, or harvested but less than once a year [*If YES continue with **Q4***]
c. Currently harvested once or more than once in the last two years [*If YES continue with **Q3***]
d. Unknown [*If YES continue with **Q4***]

Q3. How often and when does harvesting take place each year? *(Recording Sheet 1, C.3.4.3b)*
a. Once, only in spring (April–June)
b. Once, only in summer (July–August)
c. Once, only in autumn (September–November)

d. Once, only in winter (December–March)
e. Twice or more than twice a year

Q4. How often does felling and removal of timber occur in woodland within the HGMU? *(Recording Sheet 1, C.3.6.1b)*

a. Not in the last 20 years
b. No current felling and removal but previously practised within the last 20 years
c. Felling and removal has occurred once or more than once in the last 5 years
d. Unknown

Q5. Are any of the following forms of livestock kept within part or all of the HGMU? *(AA Map 2; Recording Sheet 1, C.3.7.1b)*

- Cattle or horses or donkeys — 1 ha^{-1}
- Pigs — 3 ha^{-1}
- Sheep or goats — 10 ha^{-1}
- Turkeys or geese — 30 ha^{-1}
- Chickens or ducks — 100 ha^{-1}

a. No history of livestock kept in the last 20 years [*If YES skip **Q6** and **Q7***]
b. None currently kept but previously kept within the last 20 years, or kept but at a lower density than the figure indicated above [*If YES skip **Q6** and **Q7***]
c. Currently kept with a density equal to or higher than the figure indicated above [*If YES continue with **Q6***]
d. Unknown [*If YES skip **Q6** and **Q7***]

Q6 If the answer to ***Q5c*** is YES, do the livestock feed/graze in the same area as where they ruminate, rest or sleep? (Recording Sheet 1, C.3.7.3b)

Q7. Are any of the livestock listed above in ***Q5*** kept for the purpose of meat, milk or wool production? *(Recording Sheet 1, C.3.7.2b)*

CV4 *Degree of disturbance*

Q1. How often does the HGMU suffer from stress because of drought (defined as less than 25 mm rainfall for the month prior to, and during each month of, the growing season)?
[*Determine the growing season (Recording Sheet 1, C.3.1.6, C.3.1.7, C.3.1.8) and compare this to the rainfall record (Recording Sheet 1, C.3.1.1i)*]

a. Never in the last 20 years
b. Every year in the last 20
c. Other/unknown

Q2. How often does the HGMU suffer stress from prolonged waterlogging and flooding throughout the growing season (stress occurs if soils are flooded or waterlogged for more than two successive months during the growing season)?
*[**For flooding:** determine the growing season (Recording Sheet 1, C.3.1.6, C.3.1.7, C.3.1.8) and compare to months when flooding typically occurs (Recording Sheet 1, C.3.19.6). **For waterlogging:** (Recording Sheet 3, B.3.2)]*

a. Never in the last 20 years
b. Every year in the last 20
c. Other/unknown

Q3. How frequently is the HGMU subjected to disturbance by ploughing? *(Recording Sheet 1, C.3.5.1b; AA Map* 2)

a. No history of ploughing in the last 20 years/not applicable
b. Not currently ploughed but ploughed within the last 20 years, or ploughed but less than once a year
c. Currently ploughed once or more than once a year
d. Unknown

Q4. Are there any current mining or extraction activities occurring that disturb the surface of the HGMU? *(Recording Sheet 1, C.3.11.1b, AA Map 2)*

HGMU code: ______________	**AA name:** ______________

CV1 Trace element input/availability					
Q1	a	b	c	d	e
Q2	a	b	c	d	
Q3	a	b	c		

CV3 Vegetation management					
Q1	Y	N	U/NA		
Q2	a	b	c	d	
Q3	a	b	c	d	e
Q4	a	b	c	d	
Q5	a	b	c	d	
Q6	Y	N	U/NA		
Q7	Y	N	U/NA		

CV2 Vegetation type			
Q1	Y	N	U/NA
Q2a	Y	N	U/NA
Q2b	Y	N	U/NA
Q3a	Y	N	U/NA
Q3b	Y	N	U/NA
Q4a	Y	N	U/NA
Q4b	Y	N	U/NA
Q5a	Y	N	U/NA
Q5b	Y	N	U/NA
Q5c	Y	N	U/NA
Q6a	Y	N	U/NA
Q6b	Y	N	U/NA

CV4 Degree of disturbance				
Q1	a	b	c	
Q2	a	b	c	
Q3	a	b	c	d
Q4	Y	N	U/NA	
Q5	Y	N	U/NA	

HGMU process assessment code: ______________________________

3.2.E.1 Look-up Table 1

CV1	CV2	CV3	CV4	Rationale	Code
1a|2a 3c	2a	1 2c (3a|3b|3c)	1a 2a (3a|3b) #4 #5	The HGMU has either a high trace element input or a high concentration of trace elements in the soil. The biogeochemical retention capacity of the soil is low, so trace elements will be readily available for plant uptake. Grassland vegetation indicative of a high rate of trace element uptake is present. The HGMU experiences a low degree of disturbance from drought, waterlogging, ploughing, mining or extraction, or construction work, so the ability of plants to take up trace elements is not affected. The vegetation is harvested annually in spring, summer or autumn, resulting in export of trace elements retained in above-ground plant biomass.	**1/2**
1a|2a 3c	2a	1 2c 3d	1a 2a (3a|3b) #4 #5	The HGMU has either a high trace element input or a high concentration of trace elements in the soil. The biogeochemical retention capacity of the soil is low, so trace elements will be readily available for plant uptake. Grassland vegetation indicative of a high rate of trace element uptake is present. The HGMU experiences a low degree of disturbance from drought, waterlogging, ploughing, mining or extraction, or construction work, so the ability of plants to take up trace elements is not affected. The vegetation is harvested annually in winter, resulting in export of trace elements retained in above-ground plant biomass.	**1/4**
1a|2a 3c	2a	1 2c 3e	1a 2a (3a|3b) #4 #5	The HGMU has either a high trace element input or a high concentration of trace elements in the soil. The biogeochemical retention capacity of the soil is low, so trace elements will be readily available for plant uptake. Grassland vegetation indicative of a high rate of trace element uptake is present. The HGMU experiences a low degree of disturbance from drought, waterlogging, ploughing, mining or extraction, or construction work, so the ability of plants to take up trace elements is not affected. The vegetation is harvested more than once a year, resulting in large export of trace elements retained in above-ground plant biomass.	**1/1**
1a|2a 3c	3a|4a	1 2c 3a	1a (3a|3b) #4 #5	The HGMU has either a high trace element input or a high concentration of trace elements in the soil. The biogeochemical retention capacity of the soil is low, so trace elements will be readily available for plant uptake. Mire or aquatic vegetation indicative of a high rate of trace element uptake is present. The HGMU experiences a low degree of disturbance from drought, ploughing, mining or extraction, or construction work, so the ability of plants to take up trace elements is not affected. The vegetation is harvested annually in spring, resulting in export of trace elements retained in above-ground plant biomass. Removal is less than when harvested at peak standing crop (summer / beginning of autumn).	**1/3**
1a|2a 3c	3a|4a	1 2c (3b|3c)	1a (3a|3b) #4 #5	The HGMU has either a high trace element input or a high concentration of trace elements in the soil. The biogeochemical retention capacity of the soil is low, so trace elements will be readily available for plant uptake. Mire or aquatic vegetation indicative of a high rate of trace element uptake is present. The HGMU experiences a low degree of disturbance from drought, ploughing, mining or extraction, or construction work, so the ability of plants to take up trace elements is not affected. The vegetation is harvested annually in summer or autumn, resulting in export of trace elements retained in above-ground plant biomass. NB: When reed (*Phragmites australis*) is harvested in summer, the long-term vitality of the reed stand can be adversely affected. To assure its long-term vitality it is better to harvest in autumn.	**1/1**

1a|2a 3c	3a|4a	1 2c 3d	1a (3a|3b) #4 #5	The HGMU has either a high trace element input or high concentration of trace elements in the soil. The biogeochemical retention capacity of the soil is low, so trace elements will be readily available for plant uptake. Mire or aquatic vegetation indicative of a high rate of trace element uptake is present. The HGMU experiences a low degree of disturbance from drought, ploughing, mining or extraction, or construction work, so the ability of plants to take up trace elements is not affected. The vegetation is harvested annually in winter resulting in export of trace elements retained in above-ground plant biomass. In winter many plants translocate trace elements from above- to below-ground plant tissue as a reserve for the next growing season. Harvesting in winter therefore results in less removal of trace elements than it would in summer or autumn, before the trace elements are translocated.	**1/3**
1a|2a 3c	3a|4a	1 2c 3e	1a (3a|3b) #4 #5	The HGMU has either a high trace element input or a high concentration of trace elements in the soil. The biogeochemical retention capacity of the soil is low, so trace elements will be readily available for plant uptake. Mire or aquatic vegetation indicative of a high rate of trace element uptake is present. The HGMU experiences a low degree of disturbance from drought, ploughing, mining or extraction, or construction work, so the ability of plants to take up trace elements is not affected. The vegetation is harvested more than once a year, resulting in large export of trace elements retained in above-ground plant biomass.	**1/1**
1a|2a 3c	2a	5c (6|7)	1a 2a (3a|3b) #4 #5	The HGMU has either a high trace element input or a high concentration of trace elements in the soil. The biogeochemical retention capacity of the soil is low, so trace elements will be readily available for plant uptake. Grassland vegetation indicative of a high rate of trace element uptake is present. The HGMU experiences a low degree of disturbance from drought, waterlogging, ploughing, mining or extraction, or construction work, so the ability of plants to take up trace elements is not affected. Grazing results in the export of trace elements taken up by the plants. The amount of trace elements exported through grazing is always less than that exported through harvesting, because not all vegetation will be grazed, and animals may spill and trample a proportion of (trace element-rich) plant material. Some material will fall to the ground after it is bitten by animals. This is more trace element-rich than plant material deposited due to senescence. This phenomenon is particularly important with regard to grazing by geese. Additionally, trace elements can return to the system in the form of excrement.	**1/4**
1b|1c| 2b 3c	2a	1 2c (3a|3b|3c)	1a 2a (3a|3b) #4 #5	The HGMU has either a low trace element input or a moderate or low concentration of trace elements in the soil. The biogeochemical retention capacity of the soil is low, so trace elements will be readily available for plant uptake. Grassland vegetation indicative of a high rate of trace element uptake is present. The HGMU experiences a low degree of disturbance from drought, waterlogging, ploughing, mining or extraction, or construction work, so the ability of plants to take up trace elements is not affected. The vegetation is harvested annually in spring, summer or autumn, resulting in export of trace elements retained in above-ground plant biomass.	**1/6**
1b|1c| 2b 3c	2a	1 2c 3d	1a 2a (3a|3b) #4 #5	The HGMU has either a low trace element input or a moderate or low concentration of trace elements in the soil. The biogeochemical retention capacity of the soil is low, so trace elements will be readily available for plant uptake. Grassland vegetation indicative of a high rate of trace element uptake is present. The HGMU experiences a low degree of disturbance from drought, waterlogging, ploughing, mining or extraction, or construction work, so the ability of plants to take up trace elements is not affected. The vegetation is harvested annually in winter, resulting in export of trace elements retained in above-ground plant biomass.	**1/8**

3.2.E.1 Look-up Table 1 cont'd

CV1	CV2	CV3	CV4	Rationale	Code
1b|1c| 2b 3c	2a	1 2c 3e	1a 2a (3a|3b) #4 #5	The HGMU has either a low trace element input or a moderate or low concentration of trace elements in the soil. The biogeochemical retention capacity of the soil is low, so trace elements will be readily available for plant uptake. Grassland vegetation indicative of a high rate of trace element uptake is present. The HGMU experiences a low degree of disturbance from drought, waterlogging, ploughing, mining or extraction, or construction work, so the ability of plants to take up trace elements is not affected. The vegetation is harvested more than once a year, resulting in large export of trace elements retained in above-ground plant biomass.	**1/5**
1b|1c| 2b 3c	3a|4a	1 2c 3a	1a (3a|3b) #4 #5	The HGMU has either a low trace element input or a moderate or low concentration of trace elements in the soil. The biogeochemical retention capacity of the soil is low, so trace elements will be readily available for plant uptake. Mire or aquatic vegetation indicative of a high rate of trace element uptake is present. The HGMU experiences a low degree of disturbance from drought, ploughing, mining or extraction, or construction work, so the ability of plants to take up trace elements is not affected. The vegetation is harvested annually in spring, resulting in export of trace elements retained in above-ground plant biomass. Removal is less than when harvested at peak standing crop (summer / beginning of autumn).	**1/7**
1b|1c| 2b 3c	3a|4a	1 2c (3b|3c)	1a (3a|3b) #4 #5	The HGMU has either a low trace element input or a moderate or low concentration of trace elements in the soil. The biogeochemical retention capacity of the soil is low, so trace elements will be readily available for plant uptake. Mire or aquatic vegetation indicative of a high rate of trace element uptake is present. The HGMU experiences a low degree of disturbance from drought, ploughing, mining or extraction, or construction work, so the ability of plants to take up trace elements is not affected. The vegetation is harvested annually in summer or autumn resulting in export of trace elements retained in above-ground plant biomass. NB: When reed (*Phragmites australis*) is harvested in summer, the long-term vitality of the reed stand can be adversely affected. To assure its long-term vitality it is better to harvest in autumn.	**1/5**
1b|1c| 2b 3c	3a|4a	1 2c 3d	1a (3a|3b) #4 #5	The HGMU has either a low trace element input or a moderate/low concentration of trace elements in the soil. The biogeochemical retention capacity of the soil is low, so trace elements will be readily available for plant uptake. Mire or aquatic vegetation indicative of a high rate of trace element uptake is present. The HGMU experiences a low degree of disturbance from drought, ploughing, mining or extraction, or construction work, so the ability of plants to take up trace elements is not affected. The vegetation is harvested annually in winter resulting in export of trace elements retained in above-ground plant biomass. In winter many plants translocate trace elements from above- to below-ground plant tissue as a reserve for the next growing season. Harvesting in winter therefore results in less removal of trace elements than it would in summer or autumn, before trace elements are translocated.	**1/7**
1b|1c| 2b 3c	3a|4a	1 2c 3e	1a (3a|3b) #4 #5	The HGMU has either a low trace element input or a moderate or low concentration of trace elements in the soil. The biogeochemical retention capacity of the soil is low, so trace elements will be readily available for plant uptake. Mire or aquatic vegetation indicative of a high rate of trace element uptake is present. The HGMU experiences a low degree of disturbance from drought, ploughing, mining or extraction, or construction work, so the ability of plants to take up trace elements is not affected. The vegetation is harvested more than once a year, resulting in large export of trace elements retained in above-ground plant biomass.	**1/5**

<table>
<tr><td>1b|1c|
2b 3c</td><td>2a</td><td>5c (6|7)</td><td>1a 2a
(3a|3b)
#4 #5</td><td>The HGMU has either a low trace element input or a moderate or low concentration of trace elements in the soil. The biogeochemical retention capacity of the soil is low, so trace elements will be readily available for plant uptake. Grassland vegetation indicative of a high rate of trace element uptake is present. The HGMU experiences a low degree of disturbance from drought, waterlogging, ploughing, mining or extraction, or construction work, so the ability of plants to take up trace elements is not affected. Grazing results in the export of trace elements taken up by the plants. The amount of trace elements exported through grazing is always less than that exported through harvesting, because not all vegetation will be grazed and animals may spill and trample a proportion of (trace element-rich) plant material. Some material will fall to the ground after it is bitten by animals. This is more trace element-rich than plant material deposited owing to senescence. This phenomenon is particularly important with regard to grazing by geese. Additionally, trace elements can return to the system in the form of excrement.</td><td>1/8</td></tr>
</table>

3.2.E.1 Look-up Table 2

<table>
<tr><th>CV1</th><th>CV2</th><th>CV3</th><th>CV4</th><th>Rationale</th><th>Code</th></tr>
<tr><td>1d 2d</td><td>*</td><td>*</td><td>*</td><td>The HGMU does not receive any trace element additions, or the soil trace element concentration is not elevated, i.e. there are no excess trace elements, therefore the process of export of excess trace elements through vegetation management is not performed.
When harvesting or grazing takes place, trace elements taken up by plants as part of the normal trace element cycle nevertheless will be exported from the wetland system.</td><td>2</td></tr>
<tr><td>3a</td><td>*</td><td>*</td><td>*</td><td>The biogeochemical retention capacity of the soil is high, and therefore any trace elements present in the HGMU are likely to be tightly bound within the soil, and not available for plant uptake.</td><td>2</td></tr>
<tr><td>*</td><td>1</td><td>*</td><td>*</td><td>There is little or no vegetation present in the HGMU.</td><td>2</td></tr>
<tr><td>*</td><td>2b|3b|
4b|5c|6b</td><td>*</td><td>*</td><td>Vegetation indicative of a low rate of trace element uptake is present. Therefore excess trace elements will not be taken up.
When harvesting or grazing takes place, trace elements taken up by the plants (as part of the normal trace element cycle) nevertheless will be exported. This results in the export of trace elements from the wetland system.</td><td>2</td></tr>
<tr><td>*</td><td>5a|5b|6a</td><td>#4c</td><td>*</td><td>Shrubs or trees indicative of a high rate of trace element uptake are present, but they are not felled and removed.
NB: The process of long-term retention of trace elements can be performed. This process is assessed in Section 3.2.D.2.</td><td>2</td></tr>
<tr><td>*</td><td>*</td><td>#1</td><td>*</td><td>No (management) activities resulting in the export of trace elements take place.
NB: The process of long-term retention of trace elements can be performed. This process is assessed in Section 3.2.D.2.</td><td>2</td></tr>
<tr><td>*</td><td>*</td><td>5c #6 #7</td><td>*</td><td>The vegetation is grazed by livestock, but there is no export of trace elements because the animals are not kept for the purposes of meat, milk or wool, or the trace element-rich manure of the animals is deposited in the HGMU, so most trace elements are re-mobilised within the HGMU.</td><td>2</td></tr>
</table>

3.2.E.1 Look-up Table 3

CV1	CV2	CV3	CV4	Rationale	Code
1e 2c	*	*	*	It is unknown whether there is a source of trace elements within the HGMU.	**1/x**
*	*	2d 4d 5d	*	It is unknown if there are any management activities that could result in the export of trace elements.	**1/x**
*	3a|4a	5c #6	*	Mire or aquatic vegetation indicative of a high rate of trace element uptake is present. This type of vegetation is normally not grazed by livestock. The only grazing of this type of vegetation is by ducks or geese (see ***CV3***, ***Q5***). Although ducks and geese eat plant material, they also spill a significant amount. This material is more trace element-rich than plant material deposited owing to senescence. This phenomenon is particularly important with regard to grazing by geese. Additionally, trace elements can return to the system in the form of excrement.	**1/x**
*	5a|5b|6a	4c	*	Shrubs or trees indicative of a high rate of trace element uptake are present. Felling or removal of trees results in the export of trace elements. A large amount of trace elements can be exported in this way, but most of these will have been taken up over a long period prior to harvesting, and therefore its significance is difficult to quantify. Moreover, during felling and removal, a large amount of trace element-rich tree tissue (leaves and branches) can fall and will become mineralised, resulting in a return of trace elements to soil and water.	**1/x**
*	*	*	1b|2b|3c|4|5	Other factors may be suitable, but disturbance or stress because of drought, waterlogging, ploughing, mining or extraction, or construction work may reduce the uptake of trace elements by plants.	**1/x**
*	*	*	1c|2c|3d|4|5	Other factors may be suitable, but it is not known whether disturbance or stress (drought, waterlogging, ploughing, mining or extraction, construction work, or burning of vegetation) takes place. These disturbances might reduce the uptake of trace elements by plants.	**1/x**

Q5. Is there any current construction work occurring that is disturbing the surface of the HGMU? *(Recording Sheet 1, C.3.13.1b, AA Map 2)*

Assessment of the process

Compare the results recorded for each HGMU on the appraisal sheet with the possible outcomes below, and assign an appropriate assessment code. Record this on the appraisal sheet. The syntax codes used are described in the introduction to Section 3.

Look-up tables

1. The process is definitely being performed
If answers to the questions for this process agree with any of the combinations in Look-up Table 1, then the HGMU is definitely performing the process of export of excess trace elements through vegetation management. The outputs are in order of the CV answers, not the output codes themselves.

2. The process is not (significantly) being performed
If answers to the questions for this process agree with any of the combinations in Look-up Table 2, then the HGMU is not significantly performing the process of export of *excess* trace elements through land-use management.

3. The process is being performed, but there are constraining factors or uncertainties
If Look-up Tables 1 and 2 did not lead to an assessment, the process of export of excess trace elements by vegetation management is being performed, but there are one or more constraining factors, or uncertainties. In Look-up Table 3 you can find out how many constraining factors influence the process, or which factors cause the uncertainty. If this table does not lead to an assessment, there are constraining factors or uncertainties but it is not clear what these are. In these cases, also assign code **'1x'**.

4. The process is definitely not being performed
This outcome is not applicable for this process.

This completes the functional assessment for the process of export of excess trace elements through vegetation managements. The output codes should be compared with those in Table 3.2.8 on page 485 and the appropriate information entered in the Functional Assessment Table in Section 4.0.

3.2.E.2 *Physical remobilisation of trace elements (process)*

Further reading and references

Davies, B.E. (1980). Trace element pollution. In: Davies, B.E. (ed.), *Applied Soil Trace Elements*, Wiley, Chichester, New York, Brisbane, Toronto, 287–352.

Dunbabin, J.S. and Bowmer, K.H. (1992). Potential use of constructed wetlands for the treatment of industrial wastewaters containing metals, *Science of the Total Environment* **111**(2–3), 151–168.

Kabata-Pendias, A. (2000). *Trace Elements in Soils and Plants*, CRC, Boca Raton, FL.

Kadlec, R.H. and Knight, R.L. (1996). *Treatment Wetlands*, CRC, Boca Raton, FL.

King, R.F., Royle, A., Putwain, P.D. and Dickinson, N.M. (2006). Changing contaminant mobility in a dredged canal sediment during a three-year phytoremediation trial, *Environmental Pollution* **143**, 318–326.

Mitsch, W.J. and Wise, K.M. (1998). Water quality, fate of metals and predictive model validation of a constructed wetland treating acid mine drainage, *Water Resources* **32**(6), 1888–1900.

Rupp, H., Krüger, F., Büttner, O., Kunert, M., Meißner, R., Muhs, K. and Witter B. (2000). Wirkung von Hochwasserereignissen auf die Schadstoffbelastung von Auen und kulturwirtschaftlich genutzten Böden im Überschwemmungsbereich von Oka und Elbe.

– Bericht für das Bundesministerium f. Bildung und Forschung (BMBF) (FKZ 02 WT 9617/0).

Winde, F. (2000). Der hochwassergebundene Schwermetalltransport als Ursache der Bodenkontamination in der Saaleaue bei Halle. In: Friese *et al.* (Hrsg.), *Stoffhaushalt von Auenökosystemen*, Böden und Hydrologie, Schadstoffe, Bewertungen, Springer, Berlin, 237–246.

Introduction

Process description:

- The process involves the transport of anthropogenically-derived, excess trace elements chemically bound to particulate matter in the HGMU, by water- or wind-mediated processes, and subsequent export from the HGMU.
- The process is considered to occur when an HGMU has an input of, or has retained, trace elements from either direct application or indirect application, and these are removed by water- or wind-transport processes.
- The remobilisation of bound trace elements by dissolution into the water column is not included in this section, but is included in Section 3.2.E.3 Biogeochemical remobilisation of trace elements.
- It is difficult to quantify the process of physical remobilisation of trace elements because it is highly variable both temporally and spatially. Therefore only a qualitative assessment of the process is given.

Significance – rationale for assessment: The wetland may be acting as a source of trace elements, which can contaminate water bodies.

Note: In most cases the amount of trace elements exported by wind-mediated processes is likely to be small in comparison with the amount exported by water-mediated processes.

Controlling variables

CV1 Trace element input/concentration in soil
CV2 Hydrological regime
CV3 Landform
CV4 Soil/sediment erosion

Answer all the questions below for *each* HGMU; unless otherwise instructed continue from one question to the next. On the appraisal sheet circle the correct answer. The location of the relevant information is given in brackets after each question. Use a new appraisal sheet for each HGMU.

If there are no significant trace element inputs or concentrations in the soil identified for the HGMU, or if the trace element input is uncertain, you may wish to assess the potential of the HGMU to perform this process. To do this, conduct the assessment as normal but answer questions only in ***CV2***, ***CV3*** and ***CV4***. Establish an assessment outcome as normal but ignore the need for positive answers for ***CV1***. The rationale statement can be used as normal but ignore the reference to receipt or concentration of trace elements.

Cadmium (Cd) is used as an example trace element because it is toxic, often abundant and easily mobilised. Cadmium is a relevant trace contaminant for many European alluvial wetlands, and this module can be expected to have wide applicability as a result. The trace element functions have been developed following extensive work in a case study site in the catchment of the Elbe in Germany, which has elevated concentrations of Cd. The concentration ranges used for classifying concentrations relative to background levels are therefore based on the Elbe data. The interpretation of the interactions between physico-chemical processes and their influence on the behaviour of Cd is based on empirical data from these studies. There are other trace elements that may need to be considered in wetlands and these include, but are not limited to, arsenic, mercury, and selenium. Similar processes of assessment may be used for these but it should be noted that background concentrations of trace elements are likely to be location-specific due to regional differences in mineralogy. In addition, the assessor will need to consider the modifying influences of physico-chemical processes within the wetland on the speciation and resultant behaviour of the trace element under consideration. As a result, prescriptive guidance based on concentration ranges may not be applicable across all regions of Europe. For this reason, the user should identify appropriate regional concentration data for the specific trace elements in their HGMU, then undertake a comparative assessment of measured wetland concentrations accordingly. A literature search will give guidance on processes and trace elements of relevance in European wetlands. Suitable starting references are given in Sections 3.2.D and 3.2.E.

CV1 *Trace element input/concentration in soil*

Q1. Does the soil trace element concentration indicate that anthropogenically-derived contamination of the HGMU has taken place? *(Recording Sheet 1, C.3.25.1b)*

a. Yes, the soil trace element concentration is much higher than typical background concentrations.
b. Yes, the soil trace element concentration is moderately higher than typical background concentrations.
c. Yes, the soil trace element concentration is slightly higher than typical background concentration.
d. No, the soil trace element concentration is not elevated.
e. Unknown, no soil analytical data are available.

e.g. for the Elbe catchment example site (you will have to determine regional mean concentrations for the trace element you are concerned with; see note above):

Q1. Does the soil Cd concentration indicate that anthropogenically-derived contamination of the HGMU has taken place? *(Recording Sheet 1, C.3.25.1b)*

a. Yes, the soil Cd concentration is much higher than typical background concentrations (> 20 mg kg^{-1} soil).
b. Yes, the soil Cd concentration is moderately higher than typical background concentrations (10–20 mg kg^{-1} soil).
c. Yes, the soil Cd concentration is slightly higher than typical background concentration (> 5 but < 10 mg kg^{-1} soil).
d. No, the soil Cd concentration is not elevated (≤ 5 mg kg^{-1} soil).
e. Unknown, no soil analytical data are available.

Note: This is specific to the Elbe catchment. In the UK, for example, typical background Cd concentrations are around 1–2 mg kg^{-1} soil.

Q2. Has the HGMU been assigned a code indicating an input of trace elements? *(Recording Sheet, 2e, 1)*

a. Yes, trace element input code **Trae 1/1**, **1/2**, **1/3**, **1/4**, **1/5**, **1/6** or **1/7**.
b. Yes, trace element input code **Trae 2**.
c. Uncertain, trace element input code **Trae 3**.
d. No, trace element input code **Trae 4**.

CV2 *Hydrological regime*

Q1. Has the HGMU been assigned a hydrological code indicating that run-off output (**F3**) occurs, i.e. **H3**, **H4**, **H6**, **H7**, **H8**, **H9**, **H11** or **H12**? *(Recording Sheet 2a, F3)*

Q2. Has the HGMU been assigned a hydrological code indicating that overbank flooding (**F4**) occurs, i.e. **H4**, **H7**, **H9**, **H10**, **H12**, **H13**, **H14** or **H15**? *(Recording Sheet 2a, F4)*
[*If YES continue with **Q3**. If NO continue with **Q4***]

Q3. Do flow velocities in the HGMU exceed 0.3 m s^{-1}? *(Recording Sheet 3, B.4.6)*

Q4. Which term is used to describe the topsoil moisture content status? *(Recording Sheet 2a, C.3.2)*

a. Dry
b. Moist
c. Wet
d. Flooded

Q5. Is the average groundwater table $\geq$ 1 m below the surface? *(Recording Sheet 3, B.3.10, and/or AA Map 2)*

Q6. Is the total annual precipitation < 500 mm? *(Recording Sheet 1, C.3.1.1i)*

CV3 *Landform*

Q1. Does the HGMU share a boundary with the water body or with an HGMU, which has been assigned one of the following geomorphic codes: **TB**, **FC**, **DT**, **LK**, **OX**, **BS**, **CO** or **CN**? *(AA Map 2; Recording Sheet 2a, C.2)*

Q2. Is there a clear flow path without any obstacles or barriers that could prevent water flowing or draining into neighbouring HGMUs (if present) and hence to a nearby (river) channel or water body? *(Recording Sheet 3, B.2.3; AA Map 2)*

Q3. Has the HGMU been assigned a gradient code **G2** or **G3,** indicating that it is moderate or steep (> 5°)? *(Recording Sheet 2a, C.2)*

Q4. How is the topsoil particle size described? *(Recording Sheet 2a, C.3.7)*

a. Fine
b. Medium
c. Coarse or gravel

Q5. Within the soil profile, is there a distinct horizon of indurated material (such as a hard pan) that could prevent surface water draining through the soil? *(Recording Sheet 2a, C.3 General comments)*

Q6. What is the topographic exposure of the HGMU? *(Recording Sheet 3, A.4)*

a. 3 (Sheltered)
b. 2 (Moderate)
c. 1 (Exposed)

Q7. What is the average wind speed in the AA? *(Recording Sheet 1, C.3.1.4)*

a. ≤ 4.8 m s^{-1}
b. > 4.8 and ≤ 6.6 m s^{-1}
c. > 6.6 m s^{-1}

CV4 *Soil/sediment erosion*

Q1. Are there any indicators of erosion within the HGMU, e.g. scouring in units not adjacent to a river, or in riparian units, undercutting of the bank or the presence of steep exposed banks? *(Recording Sheet 3, B.5.2)*

Q2. Are there any substantial patches of bare or eroded land within the HGMU? *(Recording Sheet 3, B.5.9)*

Q3. Is the habitat classed as unvegetated or sparsely vegetated, i.e. vegetation code **Ux**? (*Recording Sheet 3, D.*)

Q4. If any part of the HGMU currently experiences a high degree of disturbance, what is its cause? *(Recording Sheet 1)*

a. Harvesting of herbaceous vegetation *(C.3.4.1b (answer c))*
b. Ploughing of land *(C.3.5.1b.(answer c))*
c. Woodland management and forestry *(C.3.6.1b (answer c))*
d. Trampling by livestock *(C.3.7.1b (answer c))*
e. Recreational activities *(C.3.15.1 answer c)*
f. None

Q5. If there are trees in the HGMU, are they wind-cropped? *(Recording Sheet 3, A.5)*

3.2.E.2 *Physical Remobilisation of Trace Elements Appraisal Sheet*

HGMU code: ______________	**AA name:** ______________

CV1 Trace element input/concentration in soil					
Q1	a	b	c	d	e
Q2	a	b	c	d	

CV2 Hydrological regime				
Q1	Y	N	U/NA	
Q2	Y	N	U/NA	
Q3	Y	N	U/NA	
Q4	a	b	c	d
Q5	Y	N	U/NA	
Q6	Y	N	U/NA	

CV3 Landform			
Q1	Y	N	U/NA
Q2	Y	N	U/NA
Q3	Y	N	U/NA
Q4	a	b	c
Q5	Y	N	U/NA
Q6	a	b	c
Q7	a	b	c

CV4 Soil/sediment erosion						
Q1	Y	N	U/NA			
Q2	Y	N	U/NA			
Q3	Y	N	U/NA			
Q4	a	b	c	d	e	f
Q5	Y	N	U/NA			

HGMU process assessment code: ______________________________

3.2.E.2 Look-up Table 1

<table>
<tr><th>CV1</th><th>CV2</th><th>CV3</th><th>CV4</th><th>Rationale</th><th>Code</th></tr>
<tr><td>1a|1b|2a</td><td>1</td><td>(1|2|3)
(4a|4b|5)</td><td>(1|2)|
((1|2)
(3|4a–e))</td><td>The HGMU receives a regular input of trace elements or there is a highly or moderately elevated concentration of trace elements in the soil. Trace elements are likely to be exported to nearby rivers or water bodies in run-off as particulate matter. There is evidence that erosion occurs within the HGMU.
NB: The HGMU acts as a source of trace elements and therefore supplies them to water bodies receiving water from the wetland.</td><td>1</td></tr>
<tr><td>1a|1b|2a</td><td>2 3</td><td>(1|2|3)
(4a|4b|5)</td><td>(1|2)|
((1|2)
(3|4a–e))</td><td>The HGMU receives a regular input of trace elements or there is a highly or moderately elevated concentration of trace elements in the soil. Trace elements are likely to be exported into the river or water body as particulate matter in retreating floodwater. There is evidence that erosion occurs within the HGMU.
NB: The wetland acts as a source of trace elements and therefore supplies them to water bodies receiving water from the wetland.</td><td>1</td></tr>
<tr><td>1a|1b|2a</td><td>4a 6</td><td>6c 7c</td><td>2|3|
4a–e|5</td><td>The HGMU receives a regular input of trace elements or there is a highly or moderately elevated concentration of trace elements in the soil. Conditions conducive to wind erosion are present. The HGMU is exposed and wind velocity is high. There is evidence that erosion occurs within the HGMU. Particulate trace elements are likely to be exported from the HGMU via wind erosion.</td><td>1</td></tr>
</table>

3.2.E.2 Look-up Table 2

<table>
<tr><th>CV1</th><th>CV2</th><th>CV3</th><th>CV4</th><th>Rationale</th><th>Code</th></tr>
<tr><td>1c|2b</td><td>*</td><td>*</td><td>*</td><td>Trace element concentrations in the soil are only slightly elevated and/or trace element inputs are rare or insignificant. Therefore no significant quantities of trace elements can be physically remobilised.</td><td>2</td></tr>
</table>

3.2.E.2 Look-up Table 3

<table>
<tr><th>CV1</th><th>CV2</th><th>CV3</th><th>CV4</th><th>Rationale</th><th>Code</th></tr>
<tr><td>1e 2c</td><td>*</td><td>*</td><td>*</td><td>It is not certain whether the HGMU receives trace elements or if trace element concentrations are elevated in the soil.</td><td>1/x</td></tr>
<tr><td>*</td><td>1|(2 3)</td><td>#(1 2 3)|
#(4a 4b 5)</td><td>#(1 2)</td><td>Evidence for erosion is not very strong.</td><td>1/x</td></tr>
<tr><td>*</td><td>2 #3</td><td>*</td><td>*</td><td>Flow velocity is not fast enough to cause erosion.</td><td>1/x</td></tr>
<tr><td>*</td><td>4a 6</td><td>6b|7b</td><td>*</td><td>Conditions for wind erosion are not optimal.</td><td>1/x</td></tr>
</table>

3.2.E.2 Look-up Table 4

<table>
<tr><th>CV1</th><th>CV2</th><th>CV3</th><th>CV4</th><th>Rationale</th><th>Code</th></tr>
<tr><td>1d 2d</td><td>*</td><td>*</td><td>*</td><td>There are no trace element inputs to the HGMU and soil trace element concentrations are not elevated. Therefore it is not possible for the process of physical remobilisation of trace elements to be performed.</td><td>4</td></tr>
<tr><td>*</td><td>#(1 2 3
4a 6)</td><td>*</td><td>*</td><td>There is no apparent hydrological pathway for trace elements to be exported.</td><td>4</td></tr>
<tr><td>*</td><td>#(1 2 3)
4a 6</td><td>6a|7a</td><td>*</td><td>There is no floodwater or run-off, which can transport eroded trace elements. Wind is another possible mechanism for erosion, but low wind velocities, or a sheltering of the HGMU, mean that wind erosion is not possible.</td><td>4</td></tr>
<tr><td>*</td><td>*</td><td>*</td><td>#(1 2 3
4a–e 5) 4f</td><td>There is no evidence of erosion.</td><td>4</td></tr>
</table>

Assessment of the process

Compare the results recorded for each HGMU on the appraisal sheet with the possible outcomes below, and assign an appropriate assessment code. Record this on the appraisal sheet. The syntax codes used are described in the introduction to Section 3.

Look-up tables

1. The process is definitely being performed
If answers to the questions for this process agree with any of the combinations in Look-up Table 1, then the HGMU is definitely performing the process of physical remobilisation of trace elements. It is difficult to quantify the process, because it is highly variable both temporally and spatially. Therefore only a qualitative assessment is given (the code consists only of a number, the quantitative information being excluded).

2. The process is not (significantly) being performed
If answers to the questions for this process agree with any of the combinations in Look-up Table 2, then the HGMU is not significantly performing the process of physical remobilisation of trace elements.

3. The process is being performed, but there are constraining factors or uncertainties
If neither Look-up Table 1 nor Look-up Table 2 leads to an assessment, the process of physical remobilisation of trace elements is performed, but there are constraining factors or uncertainties. In Look-up Table 3 you can find out how many constraining factors influence the process, or which factors cause the uncertainty.

4. The process is definitely not being performed
If answers to the questions for this process agree with any of the combinations in Look-up Table 4, then the HGMU is definitely not performing the process of physical remobilisation of trace elements.

This completes the functional assessment for the process of physical remobilisation of trace elements. The output codes should be compared with those in Table 3.2.8 on page 485 and the appropriate information entered in the Functional Assessment Table in Section 4.0.

3.2.E.3 *Biogeochemical remobilisation of trace elements (process)*

Further reading and references

Caetano, M., Madureira, M.J. and Vale, C. (2003). Metal remobilisation during resuspension of anoxic contaminated sediment: short-term laboratory study, *Water, Air, and Soil Pollution* **143**(1–4), 23–40.

Chen, W., Tan, S.K. and Tay, J.H. (1996). Distribution, fractional composition and release of sediment-bound heavy metals in tropical reservoirs, *Water, Air, and Soil Pollution* **92**(3–4), 273–287.

Martino, M., Turner, A., Nimmo, A. and Millward, G.E. (2002). Resuspension, reactivity and recycling of trace metals in the Mersey Estuary, UK, *Marine Chemistry* **77**(2–3), 171–186.

Mitsch, W.J. and Gosselink, J.G. (2007). *Wetlands*, 4th edn, Wiley, New York.

Petersen, W., Willer, E. and Willamowski, C. (1997). Remobilization of trace elements from polluted anoxic sediments after resuspension in oxic water, *Water, Air, and Soil Pollution* **99**, 515–522.

Sahuquillo, A., Rigol, A. and Rauret, G. (2002). Comparison of leaching tests for the study of trace metals remobilisation in soils and sediments, *Journal of Environmental Monitoring* **4**(6), 1003–1009.

Wallschläger, D., Desai, M.V.M. and Wilken, R.-D. (1996). The role of humic substances in the aqueous mobilization of mercury from contaminated floodplain soils, *Water, Air, and Soil Pollution* **90**, 507–520.

Introduction

Process description:

- The process evaluates the risk of the remobilisation of soil-bound trace elements by biogeochemical processes.
- Remobilisation occurs because of the elimination of trace element binding sites through processes such as the reduction of oxides following soil waterlogging, the degradation of soil organic matter or changes in land-use or vegetation characteristics.
- Further biogeochemical remobilisation can occur via the movement of water within soils caused by the particular hydrological regime or through semi-natural acidification processes, which alter the solubility of trace elements.

Significance – rationale for assessment: The biogeochemical remobilisation of trace elements increases their potential for transfer to and pollution of surface water bodies.

Controlling variables

CV1 Biogeochemical retention of trace elements
CV2 Hydrological regime
CV3 Land-use/vegetation

Answer all the questions below for *each* HGMU; unless otherwise instructed continue from one question to the next. On the appraisal sheet, circle the correct answer. The location of the relevant information is given in brackets after each question. Use a new appraisal sheet for each HGMU.

Note: Before this process can be assessed it is necessary first to carry out an assessment of the biogeochemical retention of trace elements (process 3.2.D.2).

CV1 *Biogeochemical retention of trace elements*

Q1 What is the assessment code for the process of biogeochemical retention of trace elements? *(Functional Assessment 3.2.D.2)*
a. **1/1–1/46**
b. **2**
c. **1/x**
d. **4**

CV2 *Hydrological regime (see Fig. 3.2.15)*

Q1. Does overbank flooding occur at least once a year in the HGMU? *(Recording Sheet 1, C.3.19.8b (answer a))*
a. Yes
b. No
c. Unknown

Q2. Does groundwater discharge occur or is the mean annual groundwater depth < 0.5 m below the soil surface? *(Functional Assessment 3.1.C, Code 1/n or Recording Sheet 1, C.3.22.1c)*
a. Yes
b. No
c. Unknown

Q3. Do long continuous periods (> 90 days) occur when soils are saturated or inundated? *(Recording Sheet 3, B.4.22)*
a. Yes
b. No
c. Unknown

3.2.15 A depression on the floodplain of the 'Middle' Elbe (Germany): (a) during dry conditions; (b) the same depression during wet conditions (photo: M. Scholz).

CV3 *Land-use/vegetation*

Q1. How often is any land within the HGMU ploughed? *(Recording Sheet 1, C.3.5.1b)*

a. No history of ploughing in the last 20 years
b. Not currently ploughed but previously ploughed within last 20 years or ploughed but less than once a year
c. Currently ploughed (every year)
d. Unknown

Q2. Is at least one of the following vegetation types present within the HGMU (*Recording Sheet 3, D*)?

a. Forest and woodland (codes **Wx**) or scrub and shrubland (codes **Sx**)
b. Wetland vegetation: humid grasslands (codes **Gx**), mires (codes **Mx**), or aquatic (codes **Ax**)

Assessment of the process

Compare the results recorded for each HGMU on the appraisal sheet with the possible outcomes below, and assign an appropriate assessment code. Record this on the appraisal sheet. The syntax codes used are described in the introduction to Section 3.

3.2.E.3 *Biogeochemical Remobilisation of Trace Elements Appraisal Sheet*

HGMU code: ______________________	**AA name:** ______________________

CV1 Biogeochemical retention of trace elements				
Q1	a	b	c	d

CV3 Land-use/vegetation				
Q1	a	b	c	d
Q2	a		b	

CV2 Hydrological regime			
Q1	a	b	c
Q2	a	b	c
Q3	a	b	c

HGMU process assessment code: ______________________________________

3.2.E.3 Look-up Table 1

<table>
<tr><th>CV1</th><th>CV2</th><th>CV3</th><th>Rationale</th><th>Code</th></tr>
<tr><td>1a</td><td>1a|2a|3a</td><td>*</td><td>Trace elements are biogeochemically retained within the soil in the HGMU. It is likely that pedogenic oxide-bound trace elements become remobilised during periods of anaerobicity caused by periodic waterlogging of the soil.</td><td>1/1</td></tr>
<tr><td>1a</td><td>*</td><td>1c</td><td>Trace elements are biogeochemically retained within the soil in the HGMU. It is likely that trace elements become remobilised following ploughing as a result of consequent degradation of soil organic matter to which trace elements will be bound.</td><td>1/2</td></tr>
<tr><td>1a</td><td>*</td><td>2a</td><td>Trace elements are biogeochemically retained within the soil in the HGMU. It is likely that trace elements become remobilised as a result of semi-natural acidification processes associated with the vegetation communities present in the HGMU.</td><td>1/3</td></tr>
</table>

3.2.E.3 Look-up Table 2

CV1	CV2	CV3	Rationale	Code
1b	*	*	There is no significant retention of trace elements within the HGMU, therefore it is not possible for the process of remobilisation to occur to a significant degree.	**2**

3.2.E.3 Look-up Table 3

CV1	CV2	CV3	Rationale	Code
1c	*	*	The extent to which trace elements are biogeochemically retained within the soil in the HGMU is unknown or uncertain. Therefore it is not possible to assess whether trace elements are significantly remobilised within the HGMU.	**1/x**
*	1c 2c 3c	1d	The hydrological regime of the HGMU is uncertain and it is not known if the land is ploughed regularly. Therefore it is uncertain if the HGMU performs the process of biogeochemical remobilisation of trace elements.	**1/x**

3.2.E.3 Look-up Table 4

<table>
<tr><th>CV1</th><th>CV2</th><th>CV3</th><th>Rationale</th><th>Code</th></tr>
<tr><td>1d</td><td>*</td><td>*</td><td>The HGMU does not perform the process of retention of trace elements and therefore cannot perform the process of remobilisation of trace elements.</td><td>4</td></tr>
<tr><td>*</td><td>#(1 2 3)</td><td>1a|2b</td><td>There are no mechanisms occurring in the HGMU that are likely to result in the biogeochemical remobilisation of any trace elements which may be bound in the soil.</td><td>4</td></tr>
</table>

Look-up tables

1. The process is definitely being performed
If answers to the questions for this process agree with any of the combinations in Look-up Table 1, then the HGMU is definitely performing the process of biogeochemical remobilisation of trace elements.

2. The process is not (significantly) being performed
If answers to the questions for this process agree with any of the combinations in Look-up Table 2, then the HGMU is not significantly performing the process of biogeochemical remobilisation of trace elements.

3. The process is being performed, but there are constraining factors or uncertainties
If answers to the questions for this process agree with any of the combinations in Look-up Table 3 or for any combination of answers not featured in any of the other tables the process is probably being performed but the extent to which it is occurring is unknown because of constraining factors or uncertainties.

4. The process is definitely not being performed
If answers to the questions for this process agree with any of the combinations in Look-up Table 4, then the HGMU is definitely not performing the process of biogeochemical remobilisation of trace elements.
This completes the functional assessment for the process of biogeochemical remobilisation of trace elements. The output codes should be compared with those in Table 3.2.8 on page 485 and the appropriate information entered in the Functional Assessment Table in Section 4.0.

3.2.E.4 Functional assessment: Trace element export

To determine the overall performance of the trace element export function, the assessment must consider the operation of the component processes:

- export of trace elements through vegetation management (E1)
- physical remobilisation of trace elements (E2)
- biogeochemical remobilisation of trace elements (E3)

Trace element export is definitely being performed (high performance or intermediate performance)

The function 'trace element export' is definitely being performed by the HGMU if one or more of the component processes (E1–3) of this function is occurring, i.e. if:

Any of the process assessment codes = **1/n** or **1/x** (results from Look-up Tables 1 and/or 3 in Section 3.2.E)

The statement 'trace element export is definitely being performed' is a broad statement. The statement does not include any quantitative or semi-quantitative information. Not all processes contribute equally to the function. Therefore, if only one process with a low maximum export rate is contributing to the function, less of the trace metals will be exported than if several of these low export processes, or if one process with a high maximum export rate, contribute to the function. To estimate the performance of the function, all the processes contributing to the function have to be calculated. This is done by comparing the combination of assessment outcome codes determined for each process with the list in Table 3.2.8.

Trace element export is not (significantly) being performed (low performance)

The function 'trace element export' is not being significantly performed by an HGMU if none of the component processes of this function are occurring, i.e. if all process assessment codes = **2**

Trace element export is definitely not being performed (not performed)

The function 'trace element export' is definitely not being performed by an HGMU if none of the component processes of this function are occurring, i.e.

all process assessment codes = **4**

This completes the functional analysis for trace element export. The output codes can be recorded on the Functional Analysis Summary Table in Section 4.0, and used to compare functions among HGMUs across the wetland. If you wish to assess all functions, move on to the next step of the FAPs: Organic Carbon Concentration Control.

When all the HGMUs in the AA have been assessed, a general pattern of trace element export for this metal across the whole wetland will be evident. By manipulation of your answers to the controlling variable (CV) questions, you have the capacity to explore the probable outcomes of any management options under consideration and relate them to specific functions. The electronic version of the FAPs can do this more quickly and easily.

Table 3.2.8 Functional assessment of trace element export. Summing of trace element export processes to determine a function through semi-quantification. Compare the combinations of process output codes (e.g. **1**, **1/x**, **2**, **4**) and read off the semi-quantitative score to gain an overall assessment. **1/n** = 1/any number

Trace element export – assessment outcome codes				**Trace element export**	
Process	**Process**	**Process**	**Score**	**Performance**	**Degree of performance**
1/n	**1/n**	**1/n**	1	Definitely performed	High performance
1/n	**1/n**	**1/x**	0.9		
1/n	**1/n**	**2**	0.9		
1/n	**1/n**	**4**	0.9		
1/n	**1/x**	**1/x**	0.8		
1/n	**1/x**	**2**	0.8		
1/n	**1/x**	**4**	0.8		
1/n	**2**	**2**	0.8		
1/n	**2**	**4**	0.8		
1/n	**4**	**4**	0.8		
1/x	**1/x**	**1/x**	0.5	Definitely performed but with constraining factors or uncertainties	Intermediate performance
1/x	**1/x**	**2**	0.4		
1/x	**1/x**	**4**	0.4		
1/x	**2**	**2**	0.3		
1/x	**2**	**4**	0.3		
1/x	**4**	**4**	0.2		
2	**2**	**2**	0.1	Not significantly performed	Low performance
4	**4**	**4**	0	Definitely not performed	Not performed

3.2.F Organic carbon concentration control (function)

3.2.F.1 *Organic carbon input into surface waters (through leaf fall) (process)*

Further reading and references

Hongve, D., Riise, G. and Kristiansen, J.F. (2004). Increased colour and organic acid concentrations in Norwegian forest lakes and drinking water – a result of increased precipitation? *Aquatic Science* **66**(2), 231–238.

Kadlec, R.H. and Knight, R.L. (1996). *Treatment Wetlands*, CRC, Boca Raton, FL.

Mitsch, W.J. and Gosselink, J.G. (2007). *Wetlands*, 4th edn, Wiley, New York.

Moore, T.R. (2003). Dissolved organic carbon in a northern boreal landscape, *Global Biogeochemical Cycles* **17**(4), 1109.

Nieminen, M. (2004). Export of dissolved organic carbon, nitrogen and phosphorus following clear-cutting of three Norway spruce forests growing on drained peatlands in southern Finland, *Silva Fenn* **38**(2), 123–132.

Tate, C.M. and Meyer, J.L. (1983). The influence of hydrologic conditions and successional state on dissolved organic-carbon export from forested watersheds, *Ecology* **64**(1), 25–32.

Vannote, R.L., Minshall, G.W., Cummins, K.W., Sedell, J.R. and Gushing, E. (1980). The river continuum concept, *Canadian Journal of Fisheries and Aquatic Sciences* **37**, 130–137.

Introduction

Process description: Leaf fall into the water body is a source of particulate organic carbon (POC). Insects play a major role in breaking down the litter into particles, and microbial activity further assists the eventual transformation into dissolved organic carbon (DOC). The occurrence and transformation of POC is one of the key components of what is referred to as the *river continuum concept* in which the physical environment (in this case riparian vegetation that provides the source of DOC) forms a template for the aquatic ecosystem (see Fig. 3.2.16). Disturbance of peatlands (ploughing, draining) and climate change (precipitation amount, precipitation timing and air temperature) can also affect DOC export.

Significance – rationale for assessment:

- POC and DOC of terrestrial origin can be a significant source of energy in aquatic ecosystems, and the dominant source of energy in brown-water lakes.
- DOC effects the light regime in surface waters.
- DOC is a source of acidity that can influence surface water pH.
- DOC is a strong complexing agent, which alters the bioavailability of trace metals and organic pollutants in surface waters by increasing the solubility of trace metals. DOC may determine the water transport capacity of trace metals.

Controlling variables

CV1 Vegetation type
CV2 Landform

Answer all the questions below for *each* HGMU; unless otherwise instructed, continue from one question to the next. Circle the correct answer on the appraisal sheet. The location of the relevant information is given in brackets after each question. Use a new appraisal sheet for each HGMU.

CV1 *Vegetation type*

Q1. Is the vegetation classed as a mire type, code **Mx**? *(Recording Sheet 3, D)*

Q2. Is the vegetation classed as one of the following types indicative of the presence of bog moss (*Sphagnum* spp.), Codes **M1–M7**, **M10–M12**, **A1**, **W4**, *(Recording Sheet 3, D)*

Q3. Is the vegetation classed as one of the following codes indicative of humid grassland, codes **G2–G4**, **G6–G9**, **G11**, *(Recording Sheet 3, D)*

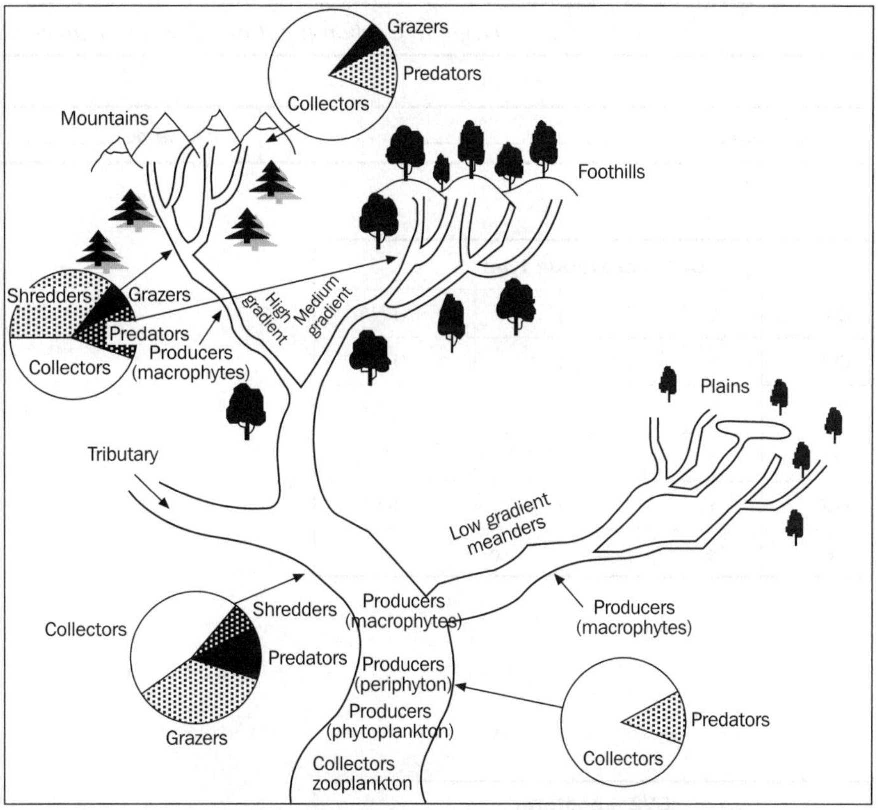

3.2.16 The river continuum concept. The functional role of particulate organic matter in water courses was one of the examples that was used to introduce the *river continuum concept* to the ecological community (Vannote *et al*., 1980). The terms in the pie charts represent invertebrate consumers of organic matter. 'Grazers' scrape surfaces for algae and bacteria, 'shredders' shred large pieces of organic mater, making it more suitable for consumption by grazers and 'collectors', which sieve water to extract food or pick it up from surfaces, and 'predators' hunt other species. The changes in relative proportions of these groups are dictated by proportions of different organic materials remaining in the water column, e.g. the proportion of shredders reduces and that of collectors increases as stream order increases, i.e. downstream.

Q4. Is the vegetation classed as a snow patch community, code **S1**? *(Recording Sheet 3, D)*

Q5. Is the vegetation classed as one of the following codes indicative of woodland on perennially waterlogged soil or sites subjected to flooding codes **W4–W8**, **W11**, **W12**? (*Recording Sheet 3, D)*

Q6. Is leafy deciduous vegetation present on the stream banks?

a. Yes, it currently overhangs the stream channel *(Recording Sheet 3, C.4 ddi)*
b. Yes, it has the potential to overhang the stream channel *(Recording Sheet 3, C.4 ddii)*
c. No
d. Unknown

CV2 *Landform*

Q1. Is it possible that water from the (main) water body (river, stream, ditch or lake) has contact with the HGMU? *(Recording Sheet 1, C.3.19.7b)*

a. No never/not applicable
b. Yes, the HGMU borders the (main) water body and horizontal exchange (either via surface or via subsurface flow) can occur freely between the HGMU and the water body
c. Yes, although the HGMU does not border the water body, the HGMU has contact with the water from the water body during flood events

3.2.F.1 *Organic Carbon input into Surface Waters (through leaf fall) Appraisal Sheet*

HGMU code: ____________________	**AA name:** ____________________

CV1 Vegetation type				
Q1	Y	N	U/NA	
Q2	Y	N	U/NA	
Q3	Y	N	U/NA	
Q4	Y	N	U/NA	
Q5	Y	N	U/NA	
Q6	a	b	c	d

CV2 Landform						
Q1	a	b	c	d	e	f

HGMU process assessment code: ____

3.2.F.1 Look-up Table 1

CV1	CV2	Rationale	Code
6a	1b|1c	The trees in this riparian zone are a seasonal source of particulate organic carbon when large amounts of leaves fall into the watercourse. Invertebrates play a major role in dividing the litter fall into particles.	**1/1**
6b	1b|1c	The trees in this riparian zone are a seasonal source of particulate organic carbon only when the riparian zone is flooded, or mobilisation of leaves by wind may lead to performance of this function. If the trees grow to overhang the stream, this function will occur to a greater extent.	**1/2**
5	1b|1c	This HGMU may perform this function during periods of flooding that flush particulate organic matter from the forest floor.	**1/2**

3.2.F.1 Look-up Table 2

CV1	CV2	Rationale	Code
1|2|3|4	*	There is not a significant source of leaves that could become a particulate organic matter input to the watercourse. Note: Any of the vegetation types in ***CV1*** are unlikely to be performing the function.	**2**

3.2.F.1 Look-up Table 4

CV1	CV2	Rationale	Code
*	1a	There is no hydrological connection to the water body, so the site cannot be a source of particulate organic carbon.	**4**
6c	*	There is no riparian vegetation and thus no source of particulate organic carbon.	**4**

d. Contact between water in the water body and the HGMU does take place once or more than once a year, but in a way other than mentioned in answers b and c (e.g. indirectly, via drainage ditches)
e. Contact between water in the water body and the HGMU does take place, but in a way other than mentioned in answers b and c (e.g. indirectly, via drainage ditches) and less than once a year
f. Unknown

Assessment of the process

Compare the results recorded for each HGMU on the appraisal sheet with the possible outcomes below, and assign an appropriate assessment code. Record this on the appraisal sheet. The syntax codes used are described in the introduction to Section 3.

Look-up tables

1. The process is definitely being performed
If answers to the questions for this function agree with any of the combinations in Look-up Table 1, then the HGMU is definitely performing the process of organic carbon input into surface waters through leaf fall. The codes comprise two numbers based on the certainty of the performance of the process (first part of the code, e.g. '**1**/1', *definitely* being performed) and the strength of evidence supporting the rationale statement (second part of the code, e.g. 1/**1**), **1** being the strongest evidence.

2. The process is not (significantly) being performed
If answers to the questions for this function agree with any of the combinations in Look-up Table 2, then the HGMU is not significantly performing the process of organic carbon input into surface waters through leaf fall.

3. The process may be performed, but there are constraining factors or uncertainties
If the two tables above and the table below do not lead to an assessment, the process of organic carbon input into surface waters through leaf fall is probably being performed, but it is unclear to what degree or the reasons why. Assign code '**1/x**' in these cases.

4. The process is definitely not being performed
If answers to the questions for this process agree with any of the combinations in Look-up Table 4, then the process of organic carbon input into surface waters through leaf fall is definitely not occurring.

This completes the functional assessment for the process of organic carbon input into surface waters through leaf fall. The output codes and quantification values can be recorded on the Functional Assessment Table at the end of Section 3.2.F.3.

3.2.F.2 Organic carbon input into surface waters (through soil) (process)

Further reading and references

Freeman, C., Fenner, N., Ostle, N.J., Kang, H., Dowrick, D.J., Reynolds, B., Lock, M.A., Sleep, D., Hughes, S. and Hudson, J. (2004). Export of dissolved organic carbon from peatlands under elevated carbon dioxide levels, *Nature* **430**(6996), 195–198.

Grunewald, K., Korth, A., Scheithauer, J., Schmidt, W. (2003). Verstärkte Huminstoffeinträge in Trinkwasserspeicher zentraleuropäischer Mittelgebirge, *Wasser & Boden* **55**(4), 47–51.

Hope, D., Billet, M.F., Cresser, M.S. (1994). A review of the export of carbon in river water: fluxes and processes, *Environmental Pollution* **84**, 301–324.

Ingram, H.A.P. (1983) Hydrology. In: Gore, A.J.P. (ed.), *Ecosystems of the World Volume 4A – Mires: Swamp, Bog, Fen, and Moor*, Elsevier, Amsterdam, The Netherlands, 67–168.

Kullberg, A., Bishop, K.H., Hargeby, A., Jansson, M. and Petersen, R.C. (1993). The biological effects of aquatic humus, *Ambio* **22**(5), 331–337.
Laudon, H., Westling, O., Löfgren, S. and Bishop, K. (2001). Modelling pre-industrial ANC and pH during spring flood in northern Sweden, *Biogeochemistry* **54**, 171–195.
Thurman, E.M. (1985). *Organic Geochemistry of Natural Waters*, Martin Nijhoff/Dr W Junk Publishers, Dordrecht, The Netherlands.

Introduction

Process description: Dissolved organic carbon (DOC) may be transported in significant quantities from soils to surface waters when riparian or lake marginal flow pathways traverse organic-rich soils. This process is often positively correlated with run-off rates. Disturbance of peatlands (ploughing, draining) and climate change, (precipitation amount, precipitation timing and air temperature) can also affect DOC export.

Significance – rationale for assessment:

- DOC of terrestrial origin can be a significant source of energy in aquatic ecosystems, and the dominant source of energy in brown-water lakes.
- DOC affects the light regime in surface waters.
- DOC is a source of acidity that can influence surface water pH (Fig. 3.2.17).
- DOC is a strong complexing agent, which alters the bioavailability of trace metals and organic pollutants in surface waters by increasing the solubility of trace metals. DOC may determine the water transport capacity of trace metals.

Even though the HGMU closest to surface water is most likely to be the source of DOC export, it is possible for the actual source of DOC to exist at some distance from the bank or shore. In such cases, the HGMU that is the source of the DOC should be assigned that function.

Controlling variables

CV1 Presence of peat
CV2 Soil water regime
CV3 Climatic regime
CV4 Water quality

Answer all the questions below for *each* HGMU; unless otherwise instructed, continue from one question to the next. Circle the correct answer on the appraisal

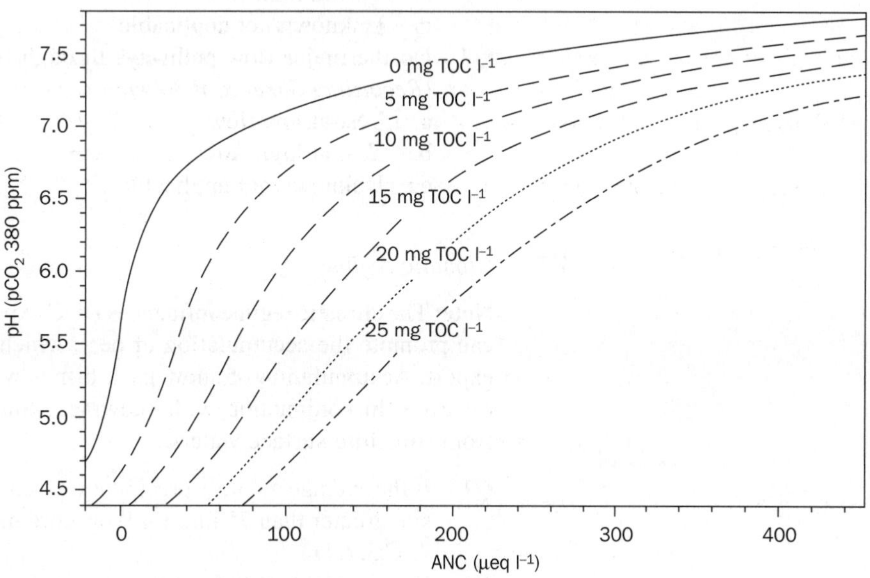

3.2.17 Relationship between pH and DOC (ANC = acid neutralising capacity).

sheet. The location of the relevant information is given in brackets after each question. Use a new appraisal sheet for each HGMU.

CV1 *Presence of peat*

Q1. Is the soil described as an organic soil? *(Recording Sheet 2a, C.3.6)* [*If No, continue with **CV2***]

Q2. How often is any land within the HGMU ploughed? *(Recording Sheet 1, C.3.5.1b)*
- a. No history of ploughing in the last 20 years
- b. Not currently ploughed but previously ploughed within last 20 years or ploughed but less than once a year
- c. Currently ploughed (every year)
- d. Unknown

Q3. How frequently is peat cut and removed from the HGMU? *(Recording Sheet 1, C.3.10.1b)*
- a. No peat cutting or extraction occurs
- b. Hand cutting or extraction of peat occurs at a local or traditional scale
- c. Mechanised cutting or extraction of peat occurs at an industrial or commercial scale
- d. Unknown

CV2 *Soil water regime*

Q1. Is the moisture of the top 20 cm of the soil described as either flooded or wet? *(Recording Sheet 2a, C.3.2)*

Q2. Has the HGMU been assigned a hydrological code indicating groundwater discharge, i.e. **H1**, **H2**, **H3**, **H4**, **H11**, **H12**, **H14** or **H15**? *(Recording Sheet 2a, F1)*

Q3. Has the HGMU been assigned a hydrological code indicating run-off input, i.e. **H2**, **H3**, **H4**, **H5**, **H6**, **H7**, **H13** or **H14**? *(Recording Sheet 2a, F2)*

Q4. Has the HGMU been assigned a hydrological code indicating overbank flooding, i.e. **H4**, **H7**, **H9**, **H10**, **H12**, **H13**, **H14** or **H15**? *(Recording Sheet 2a, F4)*

Q5. Are any areas within the AA drained through the use of surface drainage measures? (*Recording Sheet 1, C.3.9.1)*

Q6. Do the major flow pathways through the soil exit directly into a river or lake? *(Recording Sheet 3, B.3.16)*
- a. Yes, at low flow
- b. Yes, at high flow
- c. Unknown/not applicable

Q7. Do the major flow pathways through this soil traverse organic rich soils? *(Recording Sheet 3, B.3.17)*
- a. Yes, at low flow
- b. Yes, at high flow
- c. Unknown/not applicable

CV3 *Climatic regime*

Note: The climatic regime influences DOC export in a number of ways. Wet climates can promote the accumulation of peat, which then are good sources of DOC for export. Accumulation of snow pack brings with it the likelihood of a major run-off event in conjunction with snowmelt. Such events can flush organic carbon from soils into surface waters.

Q1. Is the average monthly precipitation from the nearest meteorological recording site greater than 75 mm for 9 or more months of the year? *(Recording Sheet 1, C.3.1.1)*

Q2. Does more than 20 % of the mean annual precipitation accumulate as snow? *(Recording Sheet 1, C.3.1.11 (answer b))*

CV4 *Water quality*

Q1. Is there more than 5 mg l^{-1} of DOC in the surface water of the AA (i.e. is the water brown-coloured after filtering with a 0.45 μm filter) during base flow? *(Recording Sheet 1, C.3.20.10)*

Q2. Is there more than 5 mg l^{-1} of DOC in the surface water of the AA (i.e. is the water brown-coloured after filtering with a 0.45 μm filter) during periods of high flow? *(Recording Sheet 1, C.3.20.11)*

Q3. Is there more than 5 mg l^{-1} of DOC in any nearby streams or lakes in the AA or CA (i.e. is the water brown-coloured after filtering with a 0.45 μm filter) during periods of high flow or baseflow? *(Recording Sheet 1, C.3.20.12i)*

Q4. Is there less than 5 mg l^{-1} of DOC in the AA at high flow and low flow, but more than 5 mg l^{-1} in any nearby streams or lakes (i.e. is the water brown-coloured after filtering with a 0.45 μm filter) during periods of high flow or base flow? *(Recording Sheet 1, C.3.20.12ii)*

In the coloured waters of Fenno-Scandia, dissolved organic carbon (DOC), much of which originates from the soil, is a defining feature of surface water quality and aquatic ecology. Wetlands, and changes in flow paths through wetlands associated with human influences such as drainage, are an important influence on the amount of DOC reaching streams (Ingram, 1983). Changes in the amount of DOC are of concern not only for the biota, but also drinking water supplies. Increases in DOC, such as have been observed in parts of Europe during the 1990s, can result in the need for expensive treatment measures (Grunewald *et al.*, 2003). The importance of DOC for water quality and aquatic ecology stems from its complex biochemistry. One example of this is the acidity of DOC. Levels of DOC commonly found in the Nordic environment, change the pH of water, especially in weakly-buffered waters (Fig. 3.2.17). These pH effects influence the speciation of trace elements, and thereby both their bioavailability and toxicity (Kullberg *et al.*, 1993).

Assessment of the process

Compare the results recorded for each HGMU on the appraisal sheet with the possible outcomes below, and assign an appropriate assessment code. Record this on the appraisal sheet. The syntax codes used are described in the introduction to Section 3.

Look-up tables

1. The process is definitely being performed
If the answers to the questions for this function agree with any of the combinations in Look-up Table 1, then the HGMU is definitely performing the process of organic carbon input into surface waters through soil. The codes comprise two numbers based on the certainty of the performance of the process (first part of the code, e.g. '**1**/1', *definitely* being performed) and the strength of evidence supporting the rationale statement (second part of the code, e.g. 1/**1**), **1** being the strongest supporting evidence.

2. The process is not (significantly) being performed
If the answers to the questions for this process agree with any of the combinations in Look-up Table 2, then the HGMU is not significantly performing the process of process of DOC input into surface waters (through soil).

3. The process may be performed, but there are uncertainties and constraining factors
If the two tables above and the table below do not lead to an assessment, the process of dissolved organic carbon input into surface waters (through soil) is probably being performed but there are uncertainties as to the degree to which it is being performed. Assign code **'1/x'** in these cases.

3.2.F.2 *Organic Carbon Input into Surface Waters (through soil) Appraisal Sheet*

HGMU code: ____________	**AA name:** ____________

	CV1 Presence of peat			
Q1	Y		N	U/NA
Q2	a	b	c	d
Q3	a	b	c	d

	CV3 Climatic regime		
Q1	Y	N	U/NA
Q2	Y	N	U/NA

	CV4 Water quality		
Q1	Y	N	U/NA
Q2	Y	N	U/NA
Q3	Y	N	U/NA
Q4	Y	N	U/NA

	CV2 Soil water regime		
Q1	Y	N	U/NA
Q2	Y	N	U/NA
Q3	Y	N	U/NA
Q4	Y	N	U/NA
Q5	Y	N	U/NA
Q6	a	b	c
Q7	a	b	c

HGMU process assessment code: ______________________________

3.2.F.2 Look-up Table 1

CV1	CV2	CV3	CV4	Rationale	Code
1	6a 6b 7a 7b	*	1 2 3	The last soils traversed by run-off before it becomes surface water are rich in organic material. The pronounced colour of the stream or river water is evidence that these flow paths are an important source of DOC at both low flow and high flows.	**1/1**
1	6b 7b	*	2 3	The last soils traversed by run-off before it becomes surface water during periods of peak flow are rich in organic material. During low flow, the flow paths do not go through source areas for DOC export. The pronounced increase in the colour of the stream or river water during high flow is evidence that there is a source of DOC at high flow.	**1/1**
1	6b 7b	2	2 3	The last soils traversed by run-off before it becomes surface water during periods of snowmelt are rich in organic material. During low flow, the flow paths do not go through source areas for DOC export. The pronounced increase in the colour of the stream or river water during high flow is evidence that there is a source of DOC at high flow.	**1/1**
1	5 6b 7b	*	2 3	This riparian wetland was likely to have been a source of DOC at both high flow and low flow. Alteration of flow paths to improve drainage has taken away the DOC output at low flow. The last soils now traversed by run-off before it becomes surface water during periods of peak flow are rich in organic material. During low flow, the flow paths do not currently go through source areas for DOC export. Prior to human intervention though, even low flow pathways were rich in DOC available for export. The pronounced increase in the colour of the stream or river water during high flow is evidence that there is a source of DOC at high flow.	**1/2**

3.2.F.2 Look-up Table 2

CV1	CV2	CV3	CV4	Rationale	Code
*	5	*	4	This riparian wetland may have been a source of DOC, and there is significant DOC in nearby surface waters. Alteration of flow paths to improve drainage has taken away the potential for this DOC output function.	**2**

3.2.F.2 Look-up Table 4

CV1	CV2	CV3	CV4	Rationale	Code
#1	*	*	#(1 2 3)	The lack of organic matter in the soil and the lack of significant dissolved organic matter in run-off mean that this HGMU is not a source of DOC at either low flow or high flow. It is unlikely that this function was ever a natural feature of this HGMU.	**4**

4. The process is definitely not being performed
If the answers to the questions for this function agree with any of the combinations in Look-up Table 4, then the function is definitely not being performed in the HGMU.

This completes the functional assessment for the process of organic carbon input into surface waters through soil. The output codes and quantification values can be recorded on the Functional Assessment Table in Section 3.2.F.3.

3.2.F.3 *Functional assessment: Organic carbon concentration control*

In the Functional Assessment Table below, record the functional outcome for each HGMU within your AA. By examining the results in relation to the extent of individual HGMUs within the AA, it is possible to give indications of the probable effectiveness of the wetland to control the concentration of organic carbon in the adjacent water body.

If the outcome for *either* 3.2.F.1 or 3.2.F.2 is code '**1/1**' or '**1/2**', assign the maximum value of 1 to the HGMU (the function is definitely being performed). If neither process outcome is **1/1** or **1/2** but one is **1/x** (the function is definitely being performed, but there are constraining factors or uncertainties), assign code **1/x** to the HGMU. If both 3.2.F.1 and 3.2.F.2 process outcomes are the same, e.g. 'code **2**', assign that code (e.g. 'code **2**') to the HGMU. If the two processes have different outcomes, then assign the greater value to the HGMU. For example, if 'code **2**' and 'code **4**' are the outcomes, then assign 'code **2**' to the HGMU.

This completes the functional analysis for organic carbon concentration control. The output codes can be recorded on the Functional Analysis Summary Table in Section 4, and used to compare functions among HGMUs across the wetland. If you wish to assess all functions, move on to the next step of the FAPs: Ecological functions.

When all the HGMUs in the assessment area have been assessed, a general pattern of organic carbon concentration control across the whole wetland will be evident. By manipulation of your answers to the controlling variable (CV) questions, you have the capacity to explore the probable outcomes of any management options under consideration and relate them to specific functions. The electronic version of the FAPs can do this more quickly and easily.

Functional Assessment Table: 3.2.F Organic carbon concentration control

HGMU code	Organic carbon concentration control

AA name: ____________________

3.3 Ecological functions

Introduction

Wetlands support often extensive food chains and a rich biodiversity, including species which span both terrestrial and aquatic conditions. However, the range of wetland community types is as variable as the wetlands themselves and high biodiversity is not universal. Wetlands have a particular ecological significance because of their diverse and often highly dynamic hydrological characteristics, owing to their transitional locations in the landscape between terrestrial and aquatic ecosystems, together with a seasonal cycle of rising and falling water levels throughout the year. The highly dynamic nature of water table movements and periodic flooding events results in extreme or short-term variations in the environments of both plants and animals. Mountford and Manchester (2005) review present knowledge on European wetland ecology, and particularly ecological functioning and the interaction of wetland biota with the physico-chemical environment and anthropogenic impacts. Ecological significance is expressed in terms of overall biodiversity, the presence of rare and endangered species and the role of habitat and primary productivity in supporting food chains including those well beyond the boundaries of the wetland itself.

Wetland ecosystems are disproportionately important in supporting biodiversity and include a significant number of highly specialised rare or endangered species. The abundance of rare and endangered species is in part the result of the historic loss, fragmentation and degradation of wetland habitats.

Wetlands can differ from other ecosystems in a number of key respects, which relate to:

1. hydrological dynamics (the rise and fall of the water table together with the impact of any flood pulses);
2. species adapted to flooding or waterlogging together with associated anaerobic conditions;
3. characteristic variations in habitat structure and ecological functions.

Some wetlands are among those ecosystems with the highest levels of primary productivity, such as reedbeds at eutrophic lake margins, whilst others have extremely low levels of production, such as shore weed lawns of oligotrophic lake margins and oligotrophic peat-based ecosystems.

High levels of productivity are often the result of the hydrological dynamics, which may provide a constant supply of new nutrients. Physical disturbance, which removes senescent or dead tissue, can help to maintain more vigorous and young growth as well as maintaining water supplies. However, accumulation of debris can be important in maintaining the structure of floodplain systems.

Some wetlands, such as floodplain forests, have high structural diversity, whilst others can be dominated by a single species, such as reedmace (*Typha* spp.) stands or reedbeds (*Phragmites australis*), sometimes as a result of eutrophication or deliberate management. Structural diversity is a key to the support of biodiversity through the provision of various sites suitable for species with different ecological niche requirements. Vegetation structure provides important places for niche occupancy by fauna for feeding and breeding. These requirements are inextricably linked to hydrological and nutrient dynamics. This is illustrated in Table 3.3.1, which indicates the variation in hydrological conditions tolerated by two major groups of wetland plants (Newbold and Mountford, 1997). The three columns show

Table 3.3.1 Water level lolerance requirements of selected aquatic plants. English common names are given in brackets. Negative values represent maximum depth (cm) of the water table (below soil surface); positive values represent the depth (cm) of standing water above the soil surface (after Newbold and Mountford, 1997) values.

	Dry (cm)	Preferred (cm)	Wet (cm)
Large tolerance (generalist species):			
Alisma plantago aquatica (Water plantain)	–60	–5 +15	+30
Butomus umbellatus (Flowering rush)	–30	–10 +20	+100
Carex curta (Pale sedge)	–100	–40 +10	+30
Glyceria fluitans (Floating sweetgrass)	–50	–10 +10	+50
Iris pseudacorus (Yellow flag)	–60	–10 +10	+60
Littorella uniflora (Shoreweed)	–100	0 +100	+400
Phragmites australis (Common reed)	–100	–20 –0	+50
Typha latifolia (Reedmace)	–20	+10 +75	+100
Narrow tolerance (specialist species):			
Eleogiton fluitans (Floating mud-rush)	–5	+5 +10	+20
Eriophorum angustifolium (Common cotton-grass)	–50	–30 –0	+10
Hydrocotyle vulgaris (Marsh pennywort)	–20	–4 +2	+4
Juncus bulbosus (Bulbous rush)	–30	0	+5
Oenanthe fistulosa (Tubular water-dropwort)	–30	–20 –0	+10
Pinguicula vulgaris (Common butterwort)	–30	–10 –2	0
Rhynchospora alba (White beak-sedge)	–20	–10 –0	+10
Veronica scutellata (Marsh speedwell)	–10	0	+5

the driest recorded situation, 'dry' (often well below soil surface), the wettest recorded situation, 'wet' (often with surface water) and the range where the growth of the plant is most vigorous ('preferred').

Wheeler *et al.* (2004) have produced guidelines on the eco-hydrological requirements of selected freshwater wetland communities in the lowlands of East Anglia, UK. These communities have been identified as having European importance under the Habitats Directive (EC, 1992) and include wet grasslands, fens, ditches and swamps.

Wetland plants and animals have a wide range of adaptations, which allow them to tolerate, thrive and out-compete other species under the particular stresses of this environment. In the case of plants, these include specialised tissue (notably aerenchyma), which enables oxygen to move easily from the atmosphere to plant parts that are submerged in anaerobic, waterlogged substrates, and physiological adaptations such as the ability to tolerate the toxic effects of the products of microbial anaerobic respiration (e.g. as organic acids, hydrogen sulphide, methane and ferrous iron). Some plants can also convert nutrients into cellular material without the benefit of oxygen, a chemical process which, in some cases, can result in the accumulation of significant quantities of ethanol. Sometimes in wetlands, the chemical building blocks for proteins can be lacking, and one response strategy to this, developed by a number of highly specialised, carnivorous plants, has been the ability to trap and digest insects and small animals as a means of making up deficiencies in nitrogen and phosphorus. These plants have special adaptations such as leaves with sticky secretions, e.g. sundews *(Drosera* spp.) or traps, e.g. bladderworts (*Utricularia* spp.), which ensnare prey.

Animals are often dependent on particular features of vegetation and hydrology but they also have a range of adaptations to cope with particular conditions. Amphibians generally tolerate, and some may actually require, the periodic drying out of their wetland habitat. The life cycle of many wetland animals is well adjusted to the more or less cyclical variation in the habitat between wet and dry conditions. Many fish species migrate to the flooded wetland margins of rivers and lakes to spawn. Such areas are frequently the nursery sites for juveniles before they return to the main channel or water body as water levels decline. This also emphasises the importance of connectivity in the landscape, of which wetlands are a key part, and the existence of complementary habitats within the landscape complex. Figure 3.3.1 illustrates the importance of habitat connectivity for fish. This distinguishes five guilds (groups of species that exploit the same type of environmental resource in the same way) according to fish life cycle ranges, preferred adult habitat (circles) and spawning and nursery sites (black arrows) (Schiemer, 2000).

The primary production of biomass and its export downstream or through harvesting are vital to the support of food webs (Fig. 3.3.2), both within individual wetlands and beyond their boundaries. Ecological studies have aimed to characterise sites in terms of plant communities and to identify ways in which plants and selected faunal groups can act as bioindicators of the performance of ecological functions and the 'general health' of the wetland ecosystem (Maltby *et al.*, 1996).

Unlike many other ecosystems, the effect of significant changes in the hydrology of wetlands is often to prevent the maintenance of a single ecological equilibrium. Instead, the system may change within wide limits of species composition and structure. Also, diversity may change in different ways in response to changing levels of connectivity to other systems, especially aquatic ecosystems. This section of the functional assessment procedures (FAPs) focuses on two key aspects of ecological functioning: ecosystem maintenance and food-web support. Both are of importance to the overall conservation and biodiversity value of wetlands.

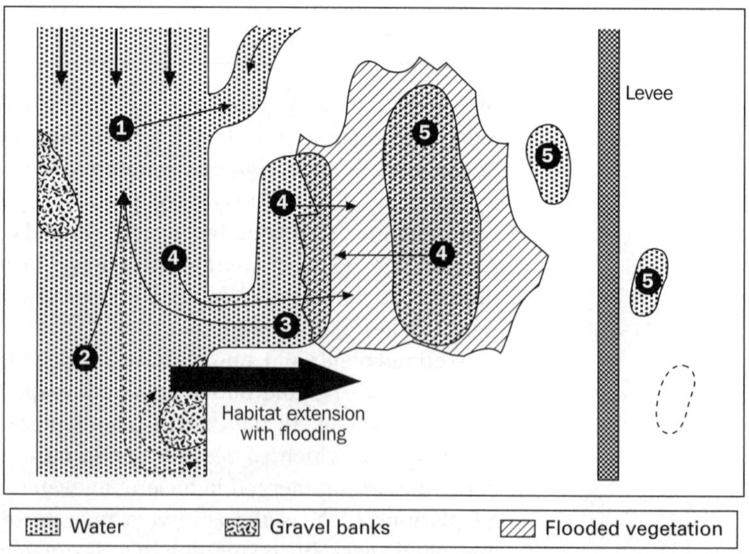

3.3.1 Ecological integrity of European regulated floodplain rivers. Habitat connectivity for fish (after Schiemer, 2000). Numbers indicate habitat types used by five fish guilds. Arrows indicate spawning and nursery habitats.

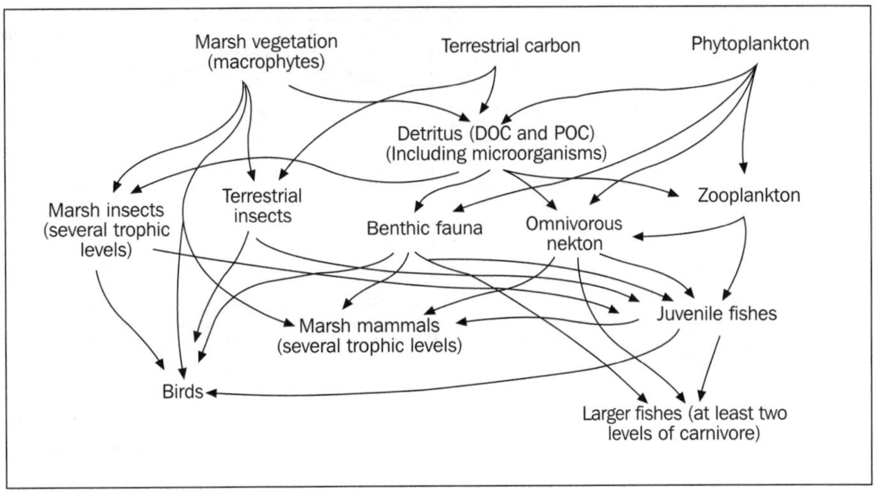

3.3.2 Energy flow diagram for a freshwater marsh showing the major groups of organisms and energy pathways (after Odum *et al.*, 1984).

Ecosystem maintenance

Wetlands can be important habitats supporting high biodiversity. One of the key factors in determining biodiversity is the provision of habitat diversity. Spatial heterogeneity within a habitat will support more species, as they provide a variety of microhabitats, microclimates, sources of food and refugia (Begon *et al.*, 200). In common with many other ecosystems, wetlands can provide a variety of habitats within a small area, owing to their spatial and temporal variability.

This function assesses the ability of a wetland to support biodiversity through an assessment of habitat diversity and quality. A number of processes are examined relating to overall habitat structural diversity, which provides habitat micro-sites for a range of animal and plant species.

Disturbance is an important factor in determining habitat diversity. In oligotrophic systems, the highest level of diversity occurs at intermediate levels of disturbance; however, if disturbance sources are removed, a relatively high diversity is maintained because of low levels of competition between plant species. High intensity of disturbance leads to a pronounced decrease in plant species diversity in oligotrophic systems. In comparison, diversity in eutrophicated systems is generally less important at low disturbance levels because of the effects of higher levels of competition and, conversely, diversity increases with elevated levels of disturbance (Clément and Maltby, 1996). The two curves shown in Fig. 3.3.3 correspond to the intermediate disturbance hypothesis (IDH). The IDH predicts that the highest species diversity can be found at intermediate disturbance levels, where 'levels' can refer to the intensity, frequency, size or time since disturbance (Palmer, 1994). The difference between oligotrophic and eutrophic habitats is concerned with the availability of N and P in the ecosystem. Trophic levels can be characterised by plant assemblages or plant productivity. For example, most wet dwarf grasslands are comparatively unproductive (< 300 g m^{-2} dry weight), and are classified as oligotrophic, whereas reedbeds are mostly highly productive (> 1000 g m^{-2}), and are classified as eutrophic.

Four classes of increasing disturbance intensity are recognised as follows:

1. NoD: *no disturbance* following land abandonment (A);
2. LD: *low disturbance* corresponding to mowing (M) practices that suppress the dominance and negative effects of litter accumulation;
3. MD: *medium disturbance*, e.g. mowing + extensive grazing (E-G);
4. HD: *high disturbance* with intensive grazing (I-G) which includes heavy trampling (poached soils) and some degree of fertilisation in oligotrophic habitats.

Disturbance is considered as a major source of spatial heterogeneity, which is a main factor in controlling species diversity (e.g. Connell, 1978; White and Pickett,

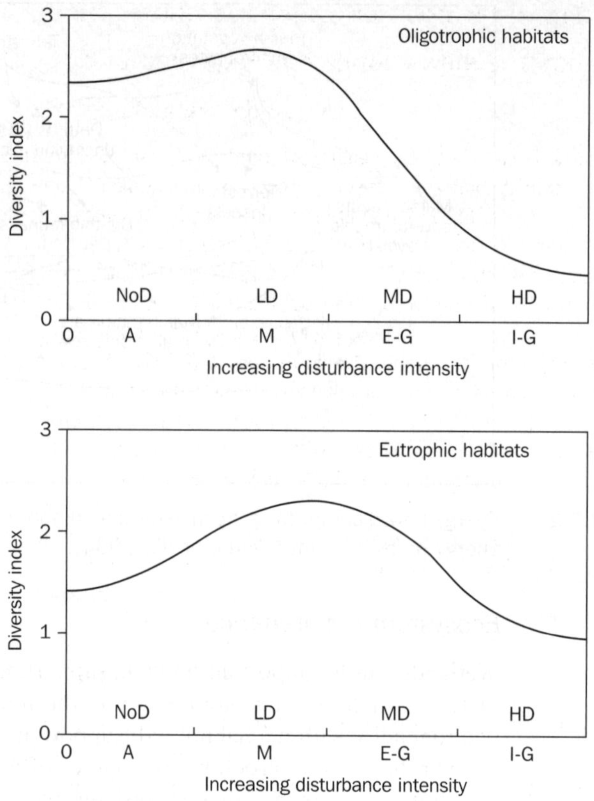

3.3.3 Response of grasslands and other herbaceous ecosystems to disturbance.

1985; Palmer, 1994). Environments with high spatial heterogeneity can be expected to create a greater variety of habitats and resources as well as refuge from predators, and therefore can accommodate greater species diversity. Plant structural diversity also needs to be considered, with trees supporting more insects than shrubs which, in turn, support more species than do herbs (Strong and Levin, 1979).

Birds are one of the most prominent groups to benefit from wetlands, and large numbers of migratory birds can be seen in wetlands throughout Europe (Fig. 3.3.4, see colour section). The importance of wetlands for birds has been one of the key factors driving wetland protection in recent decades. For example the conservation of water birds was the main original aim of the Ramsar Convention (www.ramsar.org) and, even though the Convention now has a much wider rationale for wetland protection, two out of the eight criteria relating to the selection of Ramsar sites relate solely to birds.

Amphibians are another group closely associated with wetlands. They share their life between terrestrial and aquatic environments with a need to return to the aquatic phase for their reproduction, specifically the laying of eggs and larval development. The majority of riverine lake and pond wetlands are consequently remarkable habitats which support a great potential diversity of amphibians, whose life cycles are often synchronized with the hydrological cycle. Therefore, human impacts on wetland hydrology can have a severe impact on amphibians. Most adult amphibians are carnivores and prefer emergent plant habitats, where the invertebrate macrofauna is best provided. The riverine terrestrial zone offers a refuge against predators and protection from low water temperatures. Hibernation under leaves or litter is commonplace for a number of amphibians. Therefore, amphibians will thrive in areas where a variety of microhabitats exist. Figure 3.3.5 compares the normal migrations of the common, pool and agile frogs.

In contrast to amphibians, most reptiles are not directly dependent on an aquatic environment, but there are exceptions such as the European Pond Turtle (*Emys orbicularis*) and the viperine snake (*Natrix maura*). However, some species (*Lacerta vivipara*, *Natrix natrix*, *Vipera berus*) benefit from proximity to wetland habitats, which can provide plentiful food in the form of amphibians, fish and small

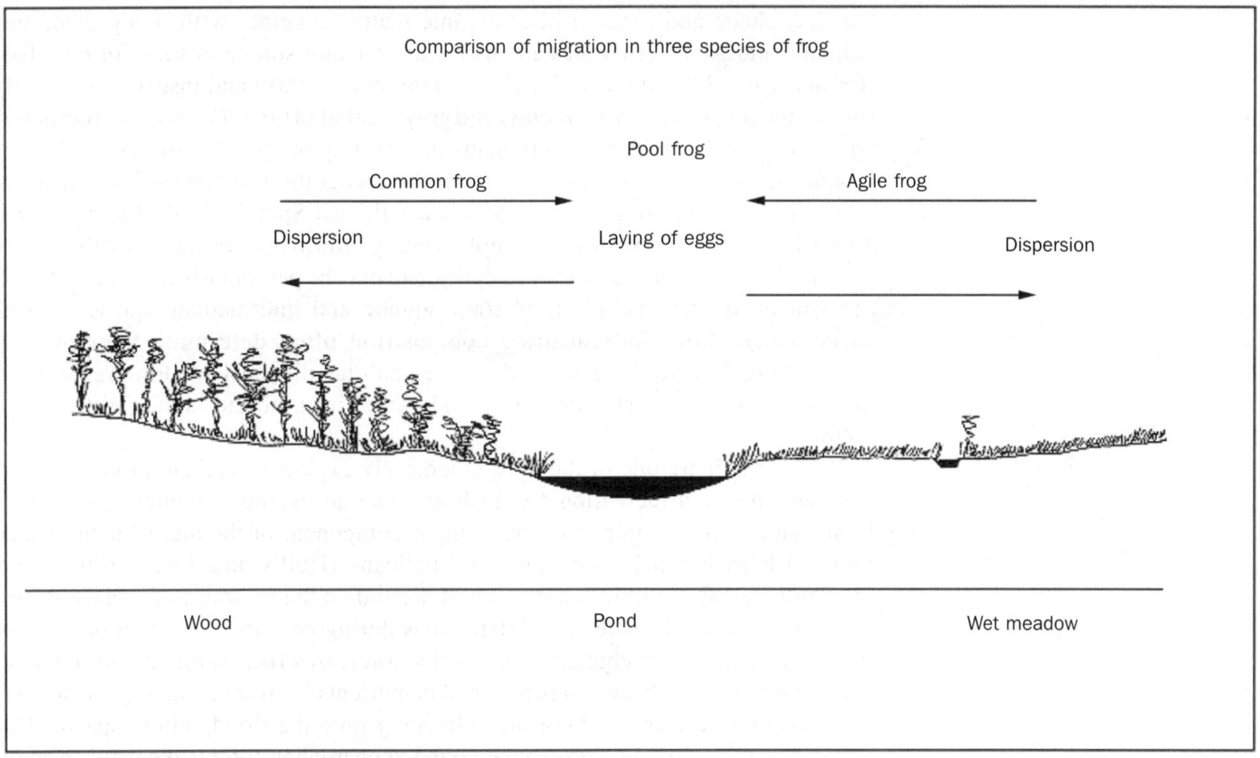

3.3.5 Migration of frog species (Bernard Le Garff, personal communication).

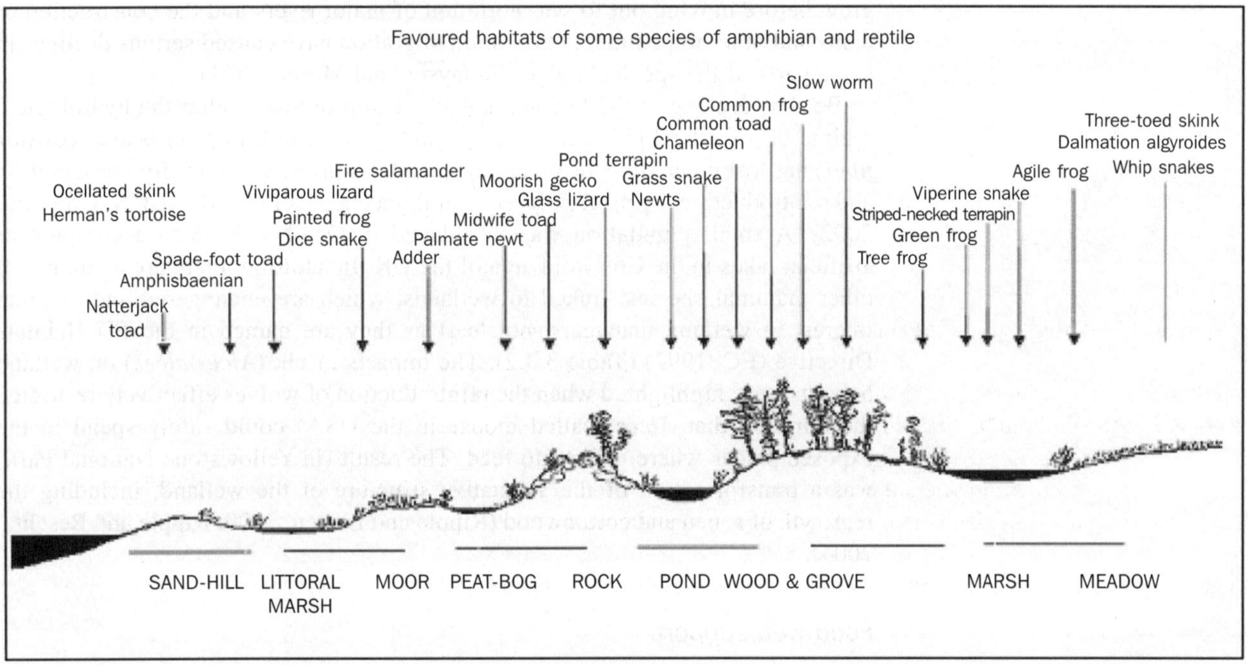

3.3.6 Preferred habitats for some amphibian and reptile species (adapted from Bernard Le Garff, personal communication).

mammals, and suitable micro-sites for thermoregulation through basking (e.g. grass tussocks) as well as refuges from predators. Approximate habitat preferences for some amphibian and reptile species are shown in Fig. 3.3.6.

Macroinvertebrates are frequently used as biological indicators of environmental quality (e.g. Wright *et al.*, 1993). Apart from being used as indicators, macroinvertebrates perform a range of functions in the wetland environment such as influencing nutrient cycling by moving nutrients from one part of the ecosystem to another (Kitchell *et al.*, 1979) and changing, through their feeding activities, the size, shape and form of dead organic matter, together with living plant and animal material. Invertebrates are also a direct major source of food for both fish (Crisp *et al.*, 1978; Mann 1974, 1982; Mann *et al.*, 1989) and insectivorous birds such as the dipper (*Cinclus cinclus*) and grey wagtail (*Motacilla cinerea*) (Ormerod and Tyler, 1987). Habitat requirements of a variety of specific groups (molluscs, ground beetles and hoverflies) underpin the recognition of functional indicators within wetlands (Castella *et al.*, 1994; Castella and Speight 1996; Maltby *et al.*, 1996). In addition, they provide complementary information because of differences in their life cycles and patterns of mobility, and may be particularly useful in wetland assessment as they include terrestrial, aquatic and intermediate species. Local variations in flow and substrate composition often determine their precise distribution. The maintenance of natural habitat diversity within wetlands is therefore essential to support a rich and varied invertebrate fauna (Ward *et al.*, 1994).

Wetland fish are one of the most extensively exploited wetland products, both commercially and recreationally. Fish are also an essential element in wetland food chains; for example, they are a major component of the diet of many water birds such as herons, cormorants and pelicans (Hollis and Jones, 1991). The distribution of fish within a wetland will depend on the presence of surface water, but fish may exploit wider floodplain areas during periods of inundation. As the rising water floods up channels and creeks, it releases fish confined within ponds and swampy areas. This water is enriched in nutrients from decaying organic matter. Many fish migrate upstream or move laterally onto the floodplain to spawn. The eggs hatch within a few days, so the young appear when food is plentiful (Keddy, 2000). Other, anadromous, species such as the salmon (*Salmo salar*), mature at sea but migrate to rivers to spawn in fresh water, where the immature fish initially grow before moving out to sea. Pollution of major rivers and the construction of dams and other impediments to salmon migration have caused serious declines in some parts of the species' range (Finlayson and Moser, 1991).

Beavers are one of the few animals that can profoundly alter the hydrological regime of a wetland or create new wetland areas. The European beaver (*Castor fiber*) has lost much of its original range, and is a priority species for conservation and reintroduction, being classed as 'near threatened (NT)' by the IUCN (Gippoliti, 2002). A small population was introduced in October 2005 to a complex of artificial lakes in the Cotswold area of the UK. In addition, there are a number of other mammal species, linked to wetlands, which are endangered and of great interest to wetland managers, not least as they are named in the EU Habitats Directive (EC, 1992) (Table 3.3.2). The impacts of elk (*Alces alces*) on wetland habitats were highlighted when the reintroduction of wolves effectively restricted the duration that *Alces* (called moose in the USA) could safely spend in the exposed places where it liked to feed. The result (in Yellowstone National Park) was a transformation of the vegetative structure of the wetland, including the regrowth of aspen and cottonwood (Ripple and Larsen, 2000; Ripple and Beschta, 2004).

Food-web support

A food web is an interlinked network of predator and prey organisms (Fig. 3.3.2). At its base is some form of input of biomass, which is consumed by the lowest tier of organisms. This, in turn, is consumed by animals from the next tier and so on. The biomass can be produced on site, through plant growth, or can be transported to the site from elsewhere in the form of dissolved matter, detritus or mobile organisms. Similarly, biomass can be removed from a site by harvesting, natural processes or the movement of organisms. The biomass available for food-web support in a wetland is a balance of production and these import and export processes. However, in wetlands, as decomposition of organic material may be restricted by waterlogging, there may be an accumulation of organic material. Moore (1990) expressed this in the following way:

$$\mathrm{GPP} + \mathrm{IMP} = \mathrm{B} + \mathrm{ER} + \mathrm{EXP} + \mathrm{P}$$

Table 3.3.2 European wetland mammal species included in Annex II of the EU Habitats Directive and/or in the IUCN Red List (EN: Endangered, VU: Vulnerable, NT: Nearly threatened, LR: Lower risk, LR/cd: Lower risk conservation dependent, LR/lc: Lower risk least concern)

Species name	Common name	Associated wetland	EU Habitats Directive Annex II	IUCN Red List
Galemys pyrenaicus	Pyrenean desman	Rivers, streams, creeks, lakes, bogs, marshes, swamps, fens, peatlands	Yes	VU
Myotis capaccinii	Long-fingered bat	Open freshwater areas	Yes	VU
Myotis dasycneme	Pond bat	Open freshwater areas	Yes	VU
Castor fiber	Eurasian beaver	Streams, rivers, lakes	Yes	NT
Alopex lagopus	Arctic fox	Tundra wetlands	Yes	–
Ursus arctos	Brown bear	Tundra wetlands	Yes	LR
Ursus maritimus	Polar bear	Arctic coastal zone	No	LR/cd
Gulo gulo	Wolverine	Tundra wetlands	Yes	VU
Lutra lutra	Otter	Streams, rivers, creeks, ponds, lakes, bogs, marshes, swamps, fens, peatlands, tundra wetlands, alpine wetlands, deltas, estuaries	Yes	NT
Mustela lutreola	European mink	Marshes, rivers, lakes	Yes	EN
Phoca vitulina	Common seal	Shallow coastal waters, estuaries	Yes	LR/lc
Pusa hispida	Ringed seal	Shallow coastal waters, estuaries, freshwater lakes	Yes	LR/lc
Alces alces	European elk	Floodplain and lakeshore wetlands	No	LR/lc
Rangifer tarandus	Reindeer	Tundra wetlands	Yes	LR/lc

where GPP is gross primary productivity, IMP is imported organic material, B is change in wetland biomass, ER is respiration of the whole system, EXP is exported organic material and P is the change in soil organic matter storage. A positive imbalance in the equation produces an annual surplus of organic matter, ultimately resulting in the accumulation of peat.

Hydrology has an impact on primary productivity, with through-flow conditions encouraging productivity. Productivity is high where floodwater brings nutrients to a wetland (Mitsch and Gosselink, 2007), for example in riparian wetlands. Riparian wetlands also benefit from a more oxygenated root zone than is found in stagnant conditions. Physical processes, such as the movement of material by wind or water, can result in the import and export of organic material. Hydrology can impact on the availability of organic material, not only by being a major factor in determining the nature of plant communities, but also by controlling the transport of water-borne organic material (Hughes and Johnes, 1995). As a result, riparian wetlands can export relatively large amounts of organic carbon via the river channel, while wetlands which are hydrologically isolated from flowing waters are unlikely to have much input or export of water-borne organic material. Nutrients and contaminants adsorbed onto sediment particles require wetlands to have hydrological inflows to enable their removal from the water column by deposition. Alternatively, nutrient export by chemical transformation (e.g. denitrification) can take place with or without the presence of surface water flows.

Further reading and references

Begon, M., Harper, J.L. and Townsend, C.R. (2006). *Ecology: Individuals, Populations and Communities*, 4th edn, Blackwell, Oxford, UK.

Castella, E. and Speight, M.C.D. (1996). Knowledge representation using fuzzy coded variables: example based on the use of Syrphidae (Insecta, Diptera) in the assessment of riverine wetlands, *Ecological Modelling* **85**(1), 13–25.

Castella, E., Speight, M.C.D., Obrdlik, P., Schneider, E. and Lavery, T.A. (1994). A methodological approach to the use of terrestrial invertebrates for the assessment of alluvial wetlands, *Wetlands Ecology and Management* **3**, 17–36.

Clément, B. and Maltby, E. (1996). Quelques facteurs de la biodiversité végétale dans les prairies humides des corridors fluviaux, *Acta Botanica Gallica* **143**(4/5), 309–316.

Connell, J.H. (1978). Diversity in tropical rainforests and coral reefs, *Science* **199**, 1302–1310.

Crisp, D.T., Mann, R.H.K. and McCormack, J.C. (1978). The effects of impoundment and regulation upon the stomach contents of fish at Cow Green, Upper Teesdale, *Journal of Fish Biology* **12**, 287–301.

EC (1992). Council Directive 92/43/EEC of 21 May 1992 on the conservation of natural habitats and of wild fauna and flora, *OJ* **L206**, 22.7.1992, 7–50.

Finlayson, M. and Moser, M. (1991), *Wetlands*, International Waterfowl and Wetlands Research Bureau (IWRB), Facts on File Limited, Oxford, UK.

Gippoliti, S. (2002). *Castor fiber*. In: IUCN 2007, *2007 IUCN Red List of Threatened Species*, www.redlist.org

Hollis, G.E. and Jones, T.A. (1991). Europe and the Mediterranean Basin. In: Finlayson, M. and Moser, M. (eds), *Wetlands*, International Waterfowl and Wetlands Research Bureau (IWRB), Facts on File Limited, Oxford, UK.

Hughes, J.M.R. and Johnes, P.J. (1995). Overview of the Ecology and Management of British Wetlands. In: Hughes, J.M.R. and Heathwaite, A.L. (eds), *Hydrology and Hydrochemistry of British Wetlands*, Wiley, Chichester, UK.

Keddy, P.A. (2000). *Wetland Ecology Principles and Conservation*, Cambridge University Press, Cambridge, UK.

Kitchell, J.F., O'Neil, R.V., Webb, D., Gallepp, G.W., Bartell, S.M., Koonee, J.F. and Ausmus, B.S. (1979). Consumer regulation of nutrient cycling, *Bioscience* **29**, 28–34.

Maltby, E., Hogan, D.V. and McInnes, R.J. (eds) (1996). *Functional Analysis of European Wetland Ecosystems – Phase I FAEWE*, Ecosystems Research Report 18, Office for Official Publications of the European Communities, Luxembourg.

Mann, R.H.K. (1974). Observations on the age, growth, reproduction and food of the dace *Leuciscus leuciscus* (L.) in two rivers in southern England, *Journal of Fish Biology* **6**, 237–253.

Mann, R.H.K. (1982). The annual food consumption and prey preference of pike (*Esox lucius*) in the River Frome, Dorset, *Journal of Animal Ecology* **51**, 81–90.

Mann, R.H.K., Blackburn, J.H. and Beaumont, W.R.C. (1989). The ecology of brown trout *Salmo trutta* in English chalk streams, *Freshwater Biology* **21**, 57–70.

Mitsch, W.J. and Gosselink, J.G. (2007). *Wetlands*, 4th edn, Wiley, New York.

Moore, P.D. (1990). Soils and ecology: temperate wetlands. In: Williams, M. (ed.) *Wetlands: a threatened landscape*, Blackwell, Oxford, UK 95–114.

Mountford, J.O and Manchester, S.J. (2005). Ecological functions of wetlands. In: Négrel, Ph., Petelete-Giraud, E., Sgouridis, F. and Maltby, E. (eds), *EUROWET, Integration of European Wetland research in sustainable management of the water cycle. Final Report – Volume 2, Synthesis of the current research in Europe*, EU contract GOCE-CT-2003-505586, Royal Holloway Institute for Environmental Research, University of London, 63–160.

Newbold, C. and Mountford, J.O. (1997). *Water Level Requirements of Wetland Plants and Animals*, English Nature Freshwater Series No. 5. English Nature, Peterborough, UK.

Odum, W.E., Smith III, T.J., Hoover, J.K. and McIvor, C.C. (1984). *The Ecology of Tidal Freshwater Marshes of the United States East Coast: a Community Profile*, FWS/OBS-83/17, Fish and Wildlife Service, Washington, DC.

Ormerod, F.J. and Tyler, S.J. (1987). Dippers (*Cinclus cinclus*) and grey wagtails (*Motacilla cinerea*) as indicators of stream acidity in upland Wales. In: Diamond, A.W. and Filion, F. (eds) *The Use of Birds*, ICBP Technical Publication No 6, International Council for Bird Preservation, Cambridge, UK, 191–209.

Palmer, M.W. (1994). Variation in species richness: towards a unification of hypotheses, *Geobotanica et Phytotaxonomica* **29**, 511–530.

Ripple, W.J. and Beschta, R.L. (2004). Wolves and the ecology of fear: Can predation risk structure ecosystems? *Bioscience* **54**(8), 755–766.

Ripple, W.J. and Larsen, E.J. (2000). Historic aspen recruitment, elk, and wolves in northern Yellowstone National Park, USA, *Biological Conservation* **95**, 361–370.

Schiemer, F. (2000). Fish as indicators for the assessment of the ecological integrity of large rivers, *Hydrobiologia* **422/423**, 271–278.

Strong, D.R. Jr and Levin, D.A. (1979). Species richness of plant parasites and growth form of their hosts, *American Naturalist* **114**, 1–22.

van der Valk, A. (2006). *The Biology of Freshwater Wetlands*, Oxford University Press, Oxford, UK.

Van Eerden, M.R., Drent, R.H., Stahl, J. and Bakker, J.P. (2005). Connecting seas: western Palaearctic continental flyway for water birds in the perspective of changing land use and climate, *Global Change Biology* **11**(6), 894–908.

Ward, D., Holmes, N. and José, P. (1994). *The New Rivers and Wildlife Handbook*, RSPB, NRA and RSNC.

Wheeler, B.D., Gowing, D.J.G., Shaw, S.C., Mountford, J.O. and Money, R.P. (2004). *Ecohydrological Guidelines for Lowland Wetland Plant Communities* (eds Brooks, A.W., José, P.V. and Whiteman, M.I.), Environment Agency (Anglian Region).

White, P.S. and Pickett, S.T.A. (1985). Natural disturbance and patch dynamics: an introduction. In: Pickett, S.T.A. and White, P.S. (eds), *The Ecology of Natural Disturbance and Patch Dynamics*, Academic Press, New York, 3–13.

Wright, J.F., Furse, M.T., Armitage, P.D. and Moss, D. (1993). New procedures for identifying running water sites subject to environmental stress and for evaluating sites for conservation, based on the macroinvertebrate fauna, *Verh. International Verein Limnology* **127**, 319–326.

Ecological functions and processes assessed in these procedures

3.3.A *Ecosystem maintenance (function)*

3.3.A.1 Provision of overall habitat structural diversity (process)
3.3.A.2.1 Provision of micro-sites for macroinvertebrates (process)
3.3.A.2.2 Provision of micro-sites for fish (process)
3.3.A.2.3 Provision of micro-sites for herptiles (process)
3.3.A.2.4 Provision of micro-sites for birds (process)
3.3.A.2.5 Provision of micro-sites for mammals (process)
3.3.A.3 Provision of plant and habitat diversity (process)
3.3.A.4 Functional assessment

3.3.B *Food-web support (function)*

3.3.B.1 Productivity/biomass production (process)
3.3.B.2 Biomass import via physical processes (accumulation of processes)
3.3.B.2.1 Biomass import via watercourses (process)
3.3.B.2.2 Biomass import via overland flow (process)
3.3.B.2.3 Biomass import via wind transport (process)
3.3.B.3 Biomass import via biological processes (accumulation of processes)
3.3.B.4 Biomass export via physical processes (accumulation of processes)
3.3.B.4.1 Biomass export via watercourses (process)
3.3.B.4.2 Biomass export via overland flow (process)
3.3.B.4.3 Biomass export via wind transport (process)
3.3.B.5 Biomass export via biological processes (accumulation of processes)
3.3.B.5.1 Biomass export via fauna (process)
3.3.B.5.2 Biomass export via anthropogenic means (process)
3.3.B.6 Functional assessment

3.3.A Ecosystem maintenance (function)

In the processes that provide micro-sites for faunal groups (Sections 3.3.A.2.1–5), the controlling variable of protected and/or red list status (***CV1***) is assessed on the basis of data applying to the entire assessment area (AA). In assessing these processes for individual hydrogeomorphic units (HGMUs) the FAPs assume that the presence of key species recorded for the AA as a whole applies also to individual component HGMUs. In consequence, it is recommended that, at some stage, a local expert is consulted to confirm the validity of that assumption, and that where this is considered to be unlikely, the final assessment outcome is suitably qualified.

3.3.A.1 Provision of overall habitat structural diversity (process)

Further reading and references

Albrecht, T. (2004). Edge effect in wetland–arable land boundary determines nesting success of scarlet rosefinches (*Carpodacus erythrinus*) in the Czech Republic, *The Auk* **121**(2), 361–371.

Begon, M., Harper, J.L. and Townsend, C.R. (2006). *Ecology: Individuals, Populations and Communities*, 4th edn, Blackwell, Oxford, UK.

Chan, S.S., Larson, D. and Anderson, P.D. (2004). *Microclimate Patterns Associated with Density Management and Riparian Buffers*, An Interim Report on the Riparian Buffer Component of the Density Management Studies, USDA Forest Service, Pacific Northwest Research Station, Corvallis, OR, available from: http://ocid.nacse.org/nbii/density/pdfFiles/BLMDMSInterimReport2July2004.pdf

Chesson, P. (2000). Mechanisms of maintenance of species diversity, *Annual Review of Ecology and Systematics* **31**, 343–366.

Corbacho, C., Sánchez, J.M. and Costillo, E. (2003). Patterns of structural complexity and human disturbance of riparian vegetation in agricultural landscapes of a Mediterranean area, *Agriculture, Ecosystems and Environment*, **95**(2/3), 495–507.

Heilmann-Clausen, J., Aude, E. and Christensen, M. (2005). Cryptogam communities on decaying deciduous wood – does tree species diversity matter? *Biodiversity and Conservation* **14**(9), 2061–2078.

Kotowski, W., van Andel, J., van Diggelen, R. and Hogendorf, J. (2004). Responses of fen plant species to groundwater level and light intensity, *Plant Ecology* **155**(2), 147–156.

Menge, B.A. and Sutherland, J.P. (1976). Species diversity gradients: synthesis of the roles of predation, competition and temporal heterogeneity, *American Naturalist* **110**(973), 351–369.

Paine, R.T. (1966). Food web complexity and species diversity, *American Naturalist* **100**(910), 65–75.

Uuttera, J., Maltamo, M. and Hotanen, J-P. (1997). The structure of forest stands in virgin and managed peatlands: a comparison between Finnish and Russian Karelia, *Forest Ecology and Management* **96**(1/2), 125–138.

van der Valk A. (2006). *The Biology of Freshwater Wetlands*, Oxford University Press, Oxford, UK.

van Wirdum, G. (1993). An ecosystems approach to base-rich freshwater wetlands, with special reference to fenlands, *Hydrobiologia* **265**(1–3), 129–153.

Verhoeven, J.T.A. (1980). The ecology of *Ruppia*-dominated communities in Western Europe. II. Synecological classification. Structure and dynamics of the macroflora and macrofauna communities, *Aquatic Botany* **8**, 1–85.

Vivian-Smith, G. (1997). Microtopographic heterogeneity and floristic diversity in experimental wetland communities, *Journal of Ecology* **85**(1), 71–82.

Wassen, M.J. and Grootjans, A.P. (2004). Ecohydrology: an interdisciplinary approach for wetland management and restoration, *Plant Ecology* **126**(1), 1–4.

Whittaker, R.H. (1972). Evolution and measurement of species diversity, *Taxon* **23**(2–3), 213–251.

Wilby, A., Shachak, M. and Boeken, B. (2001). Integration of ecosystem engineering and trophic effects of herbivores, *Oikos* **92**(3), 436–444.

Worrall, P., Peberdy, K.J. and Millett, M.C. (1997). Constructed wetlands and nature conservation, *Water Science and Technology* **35**(5), 205–213.

Introduction

Process description: Provision of diverse conditions with the potential to support a variety of organisms. Structural diversity is critical to the support of biodiversity, through the provision of various micro-sites suitable for species with different ecological niche requirements. Vegetation structure provides important micro-sites for niche occupancy by fauna for feeding and breeding.

Potential significance – rationale for assessment:

- A diverse system is likely to be of conservation importance.
- A diverse system is likely to be more resilient to impact.
- A diverse system is likely to perform a variety of other functions.

Controlling variables

CV1 Degree of perturbation
CV2 Architectural variety
CV3 Microtopography
CV4 Nature of litter layer
CV5 Nature of soil organic matter

Answer all the questions on *each* of the controlling variables for all HGMUs;

unless otherwise instructed continue from one question to the next. Circle the correct answer on the appraisal sheet. Use a new appraisal sheet for each HGMU. The location of the relevant information is given in brackets after each question.

CV1 *Degree of perturbation*

Q1. Has any distinct area within the HGMU been managed by burning? *(Recording Sheet 1, C.3.3.1b; AA Map 2)*

a. No evidence of a history of burning in the last 20 years
b. No evidence of burning within the last 5 years but evidence of previous burning between 5 and 20 years ago
c. Evidence of current burning once or more than once in the last 5 years
d. Unknown

Q2. Has any distinct area within the HGMU been subjected to harvesting of herbaceous vegetation? *(Recording Sheet 1, C.3.4.1b, C.3.4.2b; AA Map 2)*

a. No history of harvesting of herbaceous vegetation within the last 20 years
b. Not currently harvested but previously harvested within the last 20 years *or* harvested less than once a year *or* yearly harvested using predominately traditional, manual practices *(C.3.4.1b (answer b))* or *(C.3.4.1b (answer c))* and *C3.4.2b (answer b))*
c. Currently harvested once or more than once a year using fully mechanised practices *(C.3.4.1b (answer c))* and *C.3.4.2b (answer a))*
d. Unknown

Q3. Has any distinct area within the HGMU been subjected to ploughing? *(Recording Sheet 1, C.3.5.1b; AA Map 2)*

a. No history of ploughing within the last 20 years
b. Not currently ploughed but previously ploughed within the last 20 years or ploughed but less than once a year
c. Currently ploughed once or more than once a year
d. Unknown

Q4. Has any distinct area within the HGMU been subjected to forestry and woodland management (felling and removal)? *(Recording Sheet 1, C.3.6.1b; AA Map 2)*

a. Never in the last 20 years
b. No current felling and removal but previously practised within the last 20 years, or felling and removal occurs but less than once in 5 years
c. Currently practised or previously practised within the last 5 years
d. Unknown

Q5. Are livestock kept in the HGMU at least once during the year at densities greater than the following? *(Recording Sheet 1, C.3.7.1b; AA Map 2)*

- Cattle or horses or donkeys 1 ha^{-1}
- Pigs 3 ha^{-1}
- Sheep or goats 10 ha^{-1}
- Turkeys or geese 30 ha^{-1}
- Chickens or ducks 100 ha^{-1}

a. No history of such a livestock density in the last 20 years
b. None currently kept but previously kept within the last 20 years, or currently kept but at a lower density
c. Livestock currently kept at least at this density
d. Unknown

Q6. Are any areas within the HGMU subjected to peat cutting or peat mining and subsequent drying and export? *(Recording Sheet 1, C.3.10.1b; AA Map 2)*

a. No peat cutting or extraction occurs except at a local or traditional scale *(C.3.10.1b (answer a or b))*
b. Mechanised cutting or mining of peat occurs at an industrial or commercial scale
c. Unknown

Q7. Do any areas within the AA possess some form of direct recreation use? *(Recording Sheet 1, C.3.15.1; AA Map 2)*
   a. No direct or only light recreational use *(C.3.15.1 (answer a or b))*
   b. Heavy recreational use *(C.3.15.1 (answer c))*
   c. Unknown

Q8. Are any areas within the HGMU currently undergoing any form of mining, quarrying or extraction (other than peat mining)? *(Recording Sheet 1, C.3.11.1b; AA Map 2)*
   a. No mining, quarrying, or extraction of any form, or currently inactive *(C.3.11.1b (answer a or b))*
   b. Active areas of mining, quarrying, or extraction, currently in use *(C.3.11.1b (answer c))*
   c. Unknown

Q9. How often is industrial inorganic waste applied to the HGMU? *(Recording Sheet 1, C.3.12.1b; AA Map 2)*
   a. No history of being applied within the last 20 years *(C.3.12.1b (answer a or b))*
   b. Currently applied once or more than once a year *(C.3.12.1b (answer c))*
   c. Unknown

Q10. Are any areas within the HGMU currently undergoing any form of construction work? *(Recording Sheet 1, C.3.13.1b; AA Map 2)*

CV2 *Architectural variety*

Q1. How many layers of vegetation are described in the HGMU? *(Recording Sheet 3, C.1)*
   a. One or two
   b. Three or more
   c. Not applicable

CV3 *Microtopography*

Q1. Which of the following best describes the microtopographic variation within the HGMU? *(Recording Sheet 3, A.3)*
   a. Negligible
   b. Significant

CV4 *Nature of the litter layer*

Q1. Which of the following best describes the litter layer? *(Recording Sheet 3, C.2)*
   a. Negligible
   b. Herbaceous or forest debris litter layer
   c. Mixed litter layer
   d. Unknown or not applicable

CV5 *Nature of soil organic matter*

Q1. Which of the following best describes the soil organic matter? *(Recording Sheet 2a, C.3.8)*
   a. Poorly defined or well mixed in a mineral horizon
   b. Amorphous peat or humus (lacking visible plant remains)
   c. Fibrous or semi-fibrous peat
   d. Unknown or not applicable

Assessment of the process

Compare the results recorded for each HGMU on the appraisal sheet with the possible outcomes below, and assign an appropriate assessment code. Record this

3.3.A.1 *Provision of Overall Habitat Structural Diversity Appraisal Sheet*

HGMU code: ____________________	**AA name:** ____________________

CV1 Degree of perturbation				
Q1	a	b	c	d
Q2	a	b	c	d
Q3	a	b	c	d
Q4	a	b	c	d
Q5	a	b	c	d
Q6	a	b	c	
Q7	a	b	c	
Q8	a	b	c	
Q9	a	b	c	
Q10	Y	N	U/NA	

CV2 Architectural variety			
Q1	a	b	c

CV3 Microtopography		
Q1	a	b

CV4 Nature of litter layer				
Q1	a	b	c	d

CV5 Nature of soil organic matter				
Q1	a	b	c	d

HGMU process assessment code: ____

3.3.A.1 Look-up Table 1

<table>
<tr><th>CV1</th><th>CV2</th><th>CV3</th><th>CV4</th><th>CV5</th><th>Rationale</th><th>Code</th></tr>
<tr><td>Any 8 or more of (1a|1b)|(2a|2b)|(3a|3b)|(4a|4b)|(5a|5b)|6a|7a|8a|9a</td><td>1b</td><td>1b</td><td>1c</td><td>1c</td><td>Light land-use and management practices are contributing towards the production and maintenance of highly diversified habitats. The architectural diversity of plants within the habitat is high. Small-scale microtopographical variations, the composite nature of the litter layer and soil organic matter contribute towards maintaining a great number of hydro-chemical niches available for plants (and invertebrates).</td><td>1/1</td></tr>
<tr><td>Any 8 or more of (1a|1b)|(2a|2b)|(3a|3b)|(4a|4b)|(5a|5b)|6a|7a|8a|9a</td><td>1b</td><td>1b</td><td>1b</td><td>1c</td><td>Light land-use and management practices are contributing towards the production and maintenance of highly diversified habitats. The architectural diversity of plants within the habitat is high. Small-scale microtopographical variations and soil organic matter contribute towards maintaining a great number of ecological niches available for plants (and invertebrates).</td><td>1/1</td></tr>
<tr><td>Any 8 or more of (1a|1b)|(2a|2b)|(3a|3b)|(4a|4b)|(5a|5b)|6a|7a|8a|9a</td><td>1b</td><td>1b</td><td>1c</td><td>1b</td><td>Light land-use and management practices are contributing towards the production and maintenance of highly diversified habitats. The architectural diversity of plants within the habitat is high. Small-scale microtopographical variations and the composite nature of the litter layer contribute towards maintaining a great number of ecological niches available for plants (and invertebrates).</td><td>1/1</td></tr>
<tr><td>Any 8 or more of (1a|1b)|(2a|2b)|(3a|3b)|(4a|4b)|(5a|5b)|6a|7a|8a|9a</td><td>1b</td><td>1b</td><td>1b</td><td>1b</td><td>Light land-use and management practices are contributing towards the production and maintenance of highly diversified habitats. The architectural diversity of plants within the habitat is high. Small-scale microtopographical variations contribute towards maintaining a number of ecological niches available for plants.</td><td>1/1</td></tr>
<tr><td>Any 8 or more of (1a|1b)|(2a|2b)|(3a|3b)|(4a|4b)|(5a|5b)|6a|7a|8a|9a</td><td>1a</td><td>1b</td><td>1c</td><td>1c</td><td>Light land-use and management practices are contributing towards the production and maintenance of highly diversified habitats. Small-scale microtopographical variations, the composite nature of the litter layer and soil organic matter contribute towards maintaining a great number of ecological niches available for plants (and invertebrates).</td><td>1/1</td></tr>
<tr><td>Any 8 or more of (1a|1b)|(2a|2b)|(3a|3b)|(4a|4b)|(5a|5b)|6a|7a|8a|9a</td><td>1a</td><td>1b</td><td>1b</td><td>1c</td><td>Light land-use and management practices are contributing towards the production and maintenance of highly diversified habitats. Small-scale microtopographical variations and soil organic matter contribute towards maintaining a great number of ecological niches available for plants (and invertebrates).</td><td>1/1</td></tr>
<tr><td>Any 8 or more of (1a|1b)|(2a|2b)|(3a|3b)|(4a|4b)|(5a|5b)|6a|7a|8a|9a</td><td>1a</td><td>1b</td><td>1c</td><td>1b</td><td>Light land use and management practices are contributing towards the production and maintenance of highly diversified habitats. Small-scale microtopographical variations, and the composite nature of the litter layer contribute towards maintaining a great number of ecological niches available for plants (and invertebrates).</td><td>1/1</td></tr>
</table>

<table>
<tr><td>Any 8 or more of (1a|1b)|(2a|2b)|(3a|3b)|(4a|4b)|(5a|5b)|6a|7a|8a|9a</td><td>1a</td><td>1b</td><td>1b</td><td>1b</td><td>Light land-use and management practices are contributing towards the production and maintenance of highly diversified habitats. Small-scale microtopographical variations contribute towards maintaining a number of ecological niches available for plants.</td><td>1/1</td></tr>
<tr><td>Any 5 or more of (1a|1b)|(2a|2b)|(3a|3b)|(4a|4b)|(5a|5b)|6a|7a|8a|9a</td><td>1b</td><td>1b</td><td>1c</td><td>1c</td><td>Current or historic management practices are contributing towards the production and maintenance of diverse habitats. The architectural diversity of plants within the habitat is high. Small-scale microtopographical variations, the composite nature of the litter layer and soil organic matter contributes towards maintaining a great number of ecological niches available for plants (and invertebrates).</td><td>1/2</td></tr>
<tr><td>Any 5 or more of (1a|1b)|(2a|2b)|(3a|3b)|(4a|4b)|(5a|5b)|6a|7a|8a|9a</td><td>1b</td><td>1b</td><td>1b</td><td>1c</td><td>Current or historic management practices are contributing towards the production and maintenance of diverse habitats. The architectural diversity of plants within the habitat is high. Small-scale microtopographical variations and soil organic matter contribute towards maintaining a great number of ecological niches available for plants (and invertebrates).</td><td>1/2</td></tr>
<tr><td>Any 5 or more of (1a|1b)|(2a|2b)|(3a|3b)|(4a|4b)|(5a|5b)|6a|7a|8a|9a</td><td>1b</td><td>1b</td><td>1c</td><td>1b</td><td>Current or historic management practices are contributing towards the production and maintenance of diverse habitats. The architectural diversity of plants within the habitat is high. Small-scale microtopographical variations and the composite nature of the litter layer contribute towards maintaining a great number of ecological niches available for plants (and invertebrates).</td><td>1/2</td></tr>
<tr><td>Any 5 or more of (1a|1b)|(2a|2b)|(3a|3b)|(4a|4b)|(5a|5b)|6a|7a|8a|9a</td><td>1b</td><td>1b</td><td>1b</td><td>1b</td><td>Current or historic management practices are contributing towards the production and maintenance of diverse habitats. The architectural diversity of plants within the habitat is high. Small-scale microtopographical variations contribute towards maintaining number of ecological niches available for plants.</td><td>1/2</td></tr>
<tr><td>Any 5 or more of (1a|1b)|(2a|2b)|(3a|3b)|(4a|4b)|(5a|5b)|6a|7a|8a|9a</td><td>1a</td><td>1b</td><td>1c</td><td>1c</td><td>Current or historic management practices are contributing towards the production and maintenance of diverse habitats. Small-scale microtopographical variations, the composite nature of the litter layer and soil organic matter contributes towards maintaining a great number of ecological niches available for plants (and invertebrates).</td><td>1/2</td></tr>
<tr><td>Any 5 or more of (1a|1b)|(2a|2b)|(3a|3b)|(4a|4b)|(5a|5b)|6a|7a|8a|9a</td><td>1a</td><td>1b</td><td>1b</td><td>1c</td><td>Current or historic management practices are contributing towards the production and maintenance of diverse habitats. Small-scale microtopographical variations and soil organic matter contribute towards maintaining a great number of ecological niches available for plants (and invertebrates).</td><td>1/2</td></tr>
<tr><td>Any 5 or more of (1a|1b)|(2a|2b)|(3a|3b)|(4a|4b)|(5a|5b)|6a|7a|8a|9a</td><td>1a</td><td>1b</td><td>1c</td><td>1b</td><td>Current or historic management practices are contributing towards the production and maintenance of diverse habitats. Small-scale microtopographical variations and the composite nature of the litter layer contribute towards maintaining a great number of ecological niches available for plants (and invertebrates).</td><td>1/2</td></tr>
</table>

3.3.A.1 Look-up Table 1 cont'd

CV1	CV2	CV3	CV4	CV5	Rationale	Code
Any 5 or more of (1a|1b)|(2a|2b)|(3a|3b)|(4a|4b)|(5a|5b)|6a|7a|8a|9a	1a	1b	1b	1b	Current or historic management practices are contributing towards the production and maintenance of diverse habitats. Small-scale microtopographical variations contribute towards maintaining number of ecological niches available for plants.	**1/2**
Any 4 or more of (1a|1b)|(2a|2b)|(3a|3b)|(4a|4b)|(5a|5b)|6a|7a|8a|9a	1b	1b	1b	1c	Current or historic management practices are contributing towards low diversity of habitats. The architectural diversity of plants within the habitat is *high*. Small-scale microtopographical variations and soil organic matter contribute towards maintaining a great number of ecological niches available for plants (and invertebrates).	**1/3**
Any 4 or more of (1a|1b)|(2a|2b)|(3a|3b)|(4a|4b)|(2a|2b)|(3a|3b)|(4a|4b)|(5a|5b)|6a|7a|8a|9a	1b	1b	1c	1b	Current or historic management practices are contributing towards low diversity of habitats. The architectural diversity of plants within the habitat is high. Small-scale microtopographical variations, and the composite nature of the litter layer contribute towards maintaining a great number of ecological niches available for plants (and invertebrates).	**1/3**
Any 4 or more of (1a|1b)|(2a|2b)|(3a|3b)|(4a|4b)|(5a|5b)|6a|7a|8a|9a	1b	1b	1b	1b	Current or historic management practices are contributing towards low diversity of habitats. The architectural diversity of plants within the habitat is high. Small-scale microtopographical variations contribute towards maintaining number of ecological niches available for plants.	**1/3**
Any 4 or more of (1a|1b)|(2a|2b)|(3a|3b)|(4a|4b)|(5a|5b)|6a|7a|8a|9a	1a	1b	1c	1c	Current or historic management practices are contributing towards low diversity of habitats. Small-scale microtopographical variations, the composite nature of the litter layer and soil organic matter contribute towards maintaining a great number of ecological niches available for plants (and invertebrates).	**1/3**
Any 4 or more of (1a|1b)|(2a|2b)|(3a|3b)|(4a|4b)|(5a|5b)|6a|7a|8a|9a	1a	1b	1b	1c	Current or historic management practices are contributing towards low diversity of habitats. Small-scale microtopographical variations and soil organic matter contributes towards maintaining a great number of ecological niches available for plants (and invertebrates).	**1/3**
Any 4 or more of (1a|1b)|(2a|2b)|	1a	1b	1c	1b	Current or historic management practices are contributing towards low diversity of habitats. Small-scale microtopographical variations and the composite nature of the litter layer contribute	**1/3**

(3a|3b)|(4a|4b)|(5a|5b)|6a|7a|8a|9a					towards maintaining a great number of ecological niches available for plants (and invertebrates).	
Any 4 or more of (1a|1b)|(2a|2b)|(3a|3b)|(4a|4b)|(5a|5b)|6a|7a|8a|9a	1a	1b	1b	1b	Current or historic management practices are contributing towards low diversity of habitats. Small-scale microtopographical variations contribute towards maintaining number of ecological niches available for plants.	**1/3**

3.3.A.1 Look-up Table 2

CV1	CV2	CV3	CV4	CV5	Rationale	Code
1c|2c|3c|4c|5c|6b|7b|8b|9b|10	1a	1a	1a	1a	Land-use and management practices have caused excessive disturbance. The architectural diversity of the habitat is low. The diversity provided by microtopography, litter cover and soil organic matter is negligible.	**2**
1c|2c|3c|4c|5c|6b|7b|8b|9b|10	1a	1a	1a	*	Land-use and management practices have caused excessive disturbance. The architectural diversity of the habitat is low. The diversity provided by microtopography, litter cover is negligible.	**2**
1c|2c|3c|4c|5c|6b|7b|8b|9b|10	1a	1a	*	*	Land-use and management practices have caused excessive disturbance. The architectural diversity of the habitat is low.	**2**
1c|2c|3c|4c|5c|6b|7b|8b|9b|10	1a	*	*	*	Land-use and management practices have caused excessive disturbance. The architectural diversity of the habitat is low.	**2**

on the appraisal sheet. The syntax codes used are described in the introduction to Section 3.

Look-up tables

1. The process is definitely being performed

If answers to the questions for this process agree with any of the combinations in Look-up Table 1, then the HGMU is definitely performing the process of provision of overall habitat structural diversity. * Indicates where the answer to questions on the controlling variable can be either YES or NO. Answers are given in order according to both the certainty of occurrence of the process (first part of the code, e.g. '**1**/1') and the degree to which it occurs (second part of the code, e.g. '1/**1**'), based on the strength of evidence resulting from the combination of controlling variables: for code **1/1** the process is performed at the highest rate.

2. The process is not (significantly) being performed

If answers to the questions for this process agree with any of the combinations in Look-up Table 2, then the HGMU is not (significantly) performing the process of overall habitat structural diversity.

3. The process is probably being performed, but there are constraining factors or uncertainties

For any other combinations of answers not given in Look-up Tables 1 and 2, the HGMU is performing the process of overall habitat structural diversity, but only to a small degree, and there are constraining factors or uncertainties in determining the function's performance, which can be identified from the answers given. Assign assessment code **1/x** to these HGMUs.

4. The process is definitely not being performed

This statement is not applicable for this function.

This completes the functional assessment for the process of provision of overall habitat structural diversity. This process comprises sub-function 1 of the function of Ecosystem maintenance. The output codes and quantification values can be recorded on the Functional Assessment Table at the end of Section 3.3.A.4. The following five processes together form sub-function 2 of the function.

3.3.A.2.1 *Provision of micro-sites for macroinvertebrates (process)*

Further reading and references

Botch, M.S. and Masing, V.V. (1983). Mire ecosystems in the USSR. In: Gore, A.J.P. (ed.), *Ecosystems of the World, Volume 4B, Mires: Swamp, Bog, Fen and Moor*, Elsevier, New York, 95–152.

Collinson, N.H., Biggs, J., Corfield, A., Hodson, M.J., Walker, D., Whitfield, M. and Williams, P.J. (1995). Temporary and permanent ponds: An assessment of the effects of drying out on the conservation value of aquatic macroinvertebrate communities, *Biological Conservation* **74**(2), 125–133.

Convention on the Conservation of European Wildlife and Natural Habitats (Bern Convention), Council Decision 82/72/EEC, entry into force 03/12/1981 (http://europa.eu/scadplus/leg/en/(vb/128050.htm)

Convention on the Conservation of Migratory Species of Wild Animals (Bonn Convention); Council Decision 82/461/EEC, entry into force 24.06.1982 (www.cms.int)

Cragg, J.B. (1961). Some aspects of the ecology of moorland animals, *Journal of Animal Ecology* **30**, 205–234.

Damman, A.W.H. and French, T.W. (1987). *The ecology of peat bogs of the glaciated Northeast United States: A community profile*, Biological Report 85 (7.16), US Fish and Wildlife Service, Washington, DC.

EC (1979). Council Directive 79/409/EEC of 2 April 1979 on the conservation of wild birds, *OJ* **L103**, 25.4.1979, 1.

EC (1992). Council Directive 92/43/EEC of 21 May 1992 on the conservation of natural habitats and of wild fauna and flora, *OJ* **L206**, 22/07/1992, 7–50.

Haberman, H. (1959). Eesti madalsoode mesofaunast. In: *Entomoloogiline koguwik* 1, Tartu, 7–28 (with Russian and German extracts), cited in: Botch, M.S. and Masing V.V. (1983) op.cit.

Hooper, D.U., Solan, M., Symstad, A., Diaz, S., Gessner, M.O., Buchmann, N., Degrange, V., Grime, P., Hulot, F., Mermillod-Blondin, F., Roy, J., Spehn, E. and van Peer, L. (2002). Species diversity, functional diversity and ecosystem functioning. In: Loreau, M., Naeem, S. and Inchausti, P. (eds), *Biodiversity and Ecosystem Functioning, Synthesis and Perspectives*, Oxford University Press, Oxford, UK, 195–208.

IUCN (2001). *IUCN Red List Categories and Criteria: Version 3.1*, IUCN Species Survival Commission, IUCN, Gland, Switzerland and Cambridge, UK.

Kaminski, R.M. and Prince, H.H. (1981). Dabbling duck and aquatic macroinvertebrate responses to manipulated wetland habitat, *Journal of Wildlife Management* **45**, 1–15.

Keddy, P.A. (2000). *Wetland Ecology: Principles and Conservation*, Cambridge University Press, Cambridge, UK.

Kuzovkina, Y.A. and Quigley, M.F. (2005). Willows beyond wetlands: uses of Salix L. species for environmental projects, *Water, Air, and Soil Pollution* **162**(1–4), 183–204.

Lakly, M.B. and McArthur, J.V. (2000). Macroinvertebrate recovery of a post-thermal stream: habitat structure and biotic function, *Ecological Engineering* **15** (Supplement 1), S87–S100.

Lemly, A.D. (1982). Modification of benthic insect communities in polluted streams: Combined effects of sedimentation and nutrient enrichment, *Hydrobiologia* **87**, 229–245.

McLaughlin, D.B. and Harris, H.J. (1990). Aquatic insect emergence in two Great Lakes marshes, *Wetlands Ecological Management* **1**, 111–121.

Mitsch, W.J. and Gosselink, J.G. (2007). *Wetlands*, 4th edn, Wiley, New York.

Moore, P.D. and Bellamy, D.J. (1974). *Peatlands*, Springer, New York.

Nicolet, P., Biggs, J., Fox, G., Hodson, M.J., Reynolds, C., Whitfield, M. and Williams, P. (2004). The wetland plant and macroinvertebrate assemblages of temporary ponds in England and Wales, *Biological Conservation* **120**(2), 261–278.

Rinne, J.N. (1990). The utility of stream habitat and biota for identifying potential conflicting forest land uses: Montane riparian areas, *Forest Ecology and Management* **33** and **34**(1–4), 363–383.

Rogers, D.L. and Montalvo, A.M. (2004). *Genetically Appropriate Choices for Plant Materials to Maintain Biological Diversity*, University of California, USDA Forest Service, Rocky Mountain Region, Lakewood, CO, available at: http://www.fs.fed.us/r2/publications/botany/plantgenetics.pdf

Spieles, D.J. and Mitsch, W.J. (2003). A model of macroinvertebrate trophic structure and oxygen demand in freshwater wetlands, *Ecological Modelling* **161**(3), 181–192.

Steedman, R.J. (1988). Modification and assessment of an index of biotic integrity to quantify stream quality in southern Ontario, *Canadian Journal of Fisheries and Aquatic Sciences* **45**, 492–501.

United Nations Educational, Scientific and Cultural Organization (UNESCO), Convention on Wetlands of International Importance especially as Waterfowl Habitat, Ramsar, Iran, 2.2.1971 (www.ramsar.org/)

van der Valk, A. (2006). *The Biology of Freshwater Wetlands*, Oxford University Press, Oxford, UK.

Voigts, D.K. (1976). Aquatic invertebrate abundance in relation to changing marsh vegetation, *American Midland Naturalist* **95**, 312–322.

Weller, M.W. (1994). *Freshwater Marshes: Ecology and Wildlife Management*, 3rd edn, University of Minnesota Press, Mineapolis, MN.

Introduction

Process description: The provision of different micro-sites (small areas with distinct physical and ecological characteristics) for animals to inhabit within the wetland AA. Different groups of animals have different physical and ecological requirements. This section refers to macroinvertebrates. An example of the diversity of macroinvertebrates found in fens is given in Table 3.3.3.

Significance – rationale for assessment:

- The provision of specific microsites is important for certain rare species, which are likely to be important for biodiversity.
- A diverse fauna is likely to be of conservation value.

Table 3.3.3 Numbers of individuals of the main groups of invertebrates in fens as % of total; + = small larvae present but not counted (from Haberman, 1959)

	Field layer			
Group of invertebrates	**Moss layer**	**Rich fen**	**Poor fen**	**Shrub layer**
Oligochaeta	7.1	–	–	–
Mollusca	18.3	2.2	–	2.1
Araneae	19.9	14.9	8.1	6.8
Homoptera, Cicadina	4.7	37.1	16.6	17.6
Psyllina	–	–	–	10.8
Hemiptera	3.7	5.1	9.4	3.4
Thysanoptera	–	7.5	2.1	–
Coleoptera (larvae + adults)	19.5	4.3	7.3	20.7
Lepidoptera (larvae)	–	0.9	1.3	2.3
Hymenoptera	–	3.9	7.7	6.2
Terebrantia	11.2	1.2	1.6	2.1
Diptera, Brachycera	+	10.7	18.1	17.1
Nematocera	3.8	5.2	24.0	8.3
Mean no. of individuals	95.5	222	315	240

Controlling variables

CV1 Protected or Red List status
CV2 Macro-habitat type
CV3 Architectural variety

Answer all the questions below for *each* HGMU; unless otherwise instructed, continue from one question to the next. Circle the correct answer on the appraisal sheet. Use a new appraisal sheet for each HGMU. The location of the relevant information is given in brackets after each question.

CV1 *Protected and/or Red List status*

Q1. Have recent surveys of the AA (i.e. within the last 15 years) described the presence of macroinvertebrates of international importance? *(Recording Sheet 1, C.3.17.3 (answer a)* (Protected Species in the Bern Convention or the Bonn Convention, the European Habitat Directive (annex IV and annex V) or the Red List of threatened species database (UNEP-WCMC)) [*If YES, move on to the Look-up table*]

Q2. Have recent surveys of the AA (i.e. within the last 15 years) described the presence of protected macroinvertebrates of national or sub-national (local) importance? *(Recording Sheet 1, C.3.17.4 (answer a), C.3.17.5 (answer a)* (Protected macroinvertebrates listed through national legislation or Red Lists of threatened species) *[If YES, move on to the Look-up table].*

Q3. Are any areas within the AA formally protected or considered worthy of protection, to maintain their international conservation value according to Ramsar Convention, EU Habitat Directive (annex I and annex II) or EU Bird Directive? *(Recording Sheet 1, C.3.16.1)*

Q4. Are any areas within the AA formally protected or considered worthy of protection, to maintain their conservation value according to national or sub-national (local) legislation? *(Recording Sheet 1, C.3.16.2, C.3.16.3)*

CV2 *Macro-habitat type*

Q1. Has the area within the HGMU been assigned one of the following?
Vegetation codes: *(Recording Sheet 3, D)*
grassland – **G9, G17**
mire – **M9, M11–M14**
woodland – **W4, W7, W8, W12**

Geomorphic codes *(Recording Sheet 2a, C.2)*
BS, CO, CV

CV3 *Architectural variety*

Q1. How many layers are present for the vegetation described in the HGMU? *(Recording Sheet 3, C.1)*

a. One layer
b. Two or more layers
c. Not applicable

Q2. With regard to woody vegetation within the HGMU, how many of the following are represented? *(Recording Sheet 3, C.2 C.4 & C.5)*

(i) Canopy-forming trees *(C.4 (answer a))*
(ii) Understorey trees *(C.4 (answer b))*
(iii) Shrubs or bushes *(C.4 (answer c))*
(iv) Very old, mature trees *(C.5 (answer a))*
(v) Standing dead trees with hardly any or no bark *(C.5 (answer b))*
(vi) Standing dead trees with a substantial covering of bark *(C.5 (answer c))*
(vii) Fallen trees with little or no bark *(C.5 (answer d))*
(viii) Fallen trees with a substantial covering of bark *(C.5 (answer e))*
(ix) Tree stumps *(C.5 (answer f))*
(x) Tree wounds or sap runs *(C.5 (answer g))*
(xi) Rotten holes in trees *(C.5 (answer h))*
(xii) Forest or combination litter *(C.2 (answers c or d))*

a. none or not applicable
b. ≥ 3
c. 1–2

Q3. With regard to herbaceous vegetation within the HGMU, how many of the following are represented? *(Recording Sheet 3, C.2, C.4 & C.5)*

(i) Tall herbaceous vegetation *(C.4 (answer d))*
(ii) Medium height herbaceous vegetation *(C.4 (answer e))*
(iii) Low herbaceous vegetation *(C.4 (answer f))*
(iv) Dense ground vegetation *(C.4 (answer g))*
(v) Sparse ground vegetation *(C.4 (answer h))*
(vi) Grass tussocks *(C.5 (answer n))*
(vii) Herbaceous litter *(C.2 (answer b))*

a. None or not applicable
b. ≥ 3
c. 1–2

Q4. With regard to areas with little or no vegetation within the HGMU, how many of the following surface composition categories are represented? *(Recording Sheet 3, A.2, B.3 and C.5)*

(i) Bare boulders *(A.2 (answer a))*
(ii) Bare gravel *(A.2 (answer b))*
(iii) Bare sand *(A.2 (answer c))*
(iv) Bare silt or mud *(A.2 (answer d))*
(v) Bare peat *(A.2 (answer e))*
(vii) Strand-line (flood) debris *(B.4.9)*
(viii) Herbivorous animal dung *(C.5 (answer l))*
(ix) Ant nest mounds *(C.5 (answer m))*

a. None or not applicable
b. ≥ 3
c. 1–2

Q5. With regard to any areas of water within the HGMU, how many of the following are represented? *(Recording Sheet 3, B.2, B.4; Recording Sheet 2a, F1, F3)*

(i) Temporary running water *(B.2.5)*
(ii) Permanent standing water *(B.4.3)*

3.3.A.2.1 Provision of Micro-sites for Macroinvertebrates Appraisal Sheet

HGMU code: ____________	**AA name:** ____________

CV1 Protected and/or Red List status			
Q1	Y	N	U/NA
Q2	Y	N	U/NA
Q3	Y	N	U/NA
Q4	Y	N	U/NA

CV2 Macro-habitat type			
Q1	Y	N	U/NA

CV3 Architectural variety			
Q1	a	b	c
Q2	a	b	c
Q3	a	b	c
Q4	a	b	c
Q5	a	b	c

HGMU process assessment code: ____________________________

3.3.A.2.1 Look-up Table 1

<table>
<tr><th>CV1</th><th>CV2</th><th>CV3</th><th>Rationale</th><th>Code</th></tr>
<tr><td>1</td><td>*</td><td>*</td><td>The AA is of international importance with regards to macroinvertebrates and is already protected or considered worthy of protection.</td><td>1/1</td></tr>
<tr><td>2</td><td>*</td><td>*</td><td>The AA is of national or local importance with regards to macroinvertebrates and is already protected or considered worthy of protection.</td><td>1/1</td></tr>
<tr><td>3|4</td><td>1</td><td>3 or more of 1b|2b|3b|4b|5b</td><td>The AA is of international, national or local conservation importance, and is already protected or considered worthy of protection. The HGMU represents a typical wetland habitat of the Atlantic or central European zone, and has a very wide-ranging diversity of micro-sites that are available for occupancy by macroinvertebrates.</td><td>1/1</td></tr>
<tr><td>3|4</td><td>#1</td><td>3 or more of 1b|2b|3b|4b|5b</td><td>The AA is of international, national or local conservation importance, and is already protected or considered worthy of protection. The HGMU represents a non-typical wetland habitat of the Atlantic or central European zone, and has a very wide-ranging diversity of micro-sites that are available for occupancy by macroinvertebrates.</td><td>1/2</td></tr>
<tr><td>3|4</td><td>1</td><td>1 or 2 of 1b|2b|3b|4b|5b</td><td>The AA is of international, national or local conservation importance, and is already protected or considered worthy of protection. The HGMU represents a typical wetland habitat of the Atlantic or central European zone, but has a limited ranging diversity of micro-sites that are available for occupancy by macroinvertebrates.</td><td>1/2</td></tr>
<tr><td>#(1|2|3|4)</td><td>#1</td><td>3 or more of 1b|2b|3b|4b|5b</td><td>The HGMU represents a non-typical wetland habitat of the Atlantic or central European zone, and has a very wide ranging diversity of micro-sites that are available for occupancy by macroinvertebrates.</td><td>1/2</td></tr>
<tr><td>3|4</td><td>#1</td><td>1 or 2 of 1b|2b|3b|4b|5b</td><td>The AA is of international, national or local conservation importance, and is already protected or considered worthy of protection. The HGMU represents a non-typical wetland habitat of the Atlantic or central European zone, and has a limited diversity of micro-sites that are available for occupancy by macroinvertebrates.</td><td>1/3</td></tr>
</table>

3.3.A.2.1 Look-up Table 2

<table>
<tr><th>CV1</th><th>CV2</th><th>CV3</th><th>Rationale</th><th>Code</th></tr>
<tr><td>#(1 2 3 4)</td><td>*</td><td>#(3 or more of 1b|2b|3b|4b|5b)</td><td>The AA is not of conservation importance with regards to macroinvertebrates. The HGMU has a very low diversity of micro-sites that are available for occupancy by Carabidae, Syrphidae or Molluscs.</td><td>2</td></tr>
</table>

(iii) Drainage ditches *(B.2.3, B.2.6)*
(iv) Springs *(F1)*
(v) Running water *(F3)*

a. None or not applicable
b. ≥ 3
c. 1–2

Assessment of the process

Compare the results recorded for each HGMU on the appraisal sheet with the possible outcomes below, and assign an appropriate assessment code. Record this on the appraisal sheet. The syntax codes used are described in the introduction to Section 3.

Look-up tables

1. The process is definitely being performed
If answers to the questions for this process agree with any of the combinations in Look-up Table 1, then the HGMU is definitely performing the process of provision of micro-sites for invertebrates. * Indicates where the answer to questions on the controlling variable can be either YES or NO. Answers are given in order according to both the certainty of occurrence of the process (first part of the code, e.g. '**1**/1') and the degree to which it occurs (second part of the code, e.g. '1/**1**'), based on the strength of evidence resulting from the combination of controlling variables: for code **1/1** the process is performed at the highest rate.

2. The process is not (significantly) being performed
If answers to the questions for this process agree with any of the combinations in Look-up Table 2, then the HGMU is not significantly performing the process of provision of micro-sites for invertebrates.

3. The process is being performed, but there are constraining factors or uncertainty
For any other combinations of answers not given in Look-up Tables 1 and 2, the HGMU is performing the process of provision of micro-sites for invertebrates, but there are constraining factors or uncertainty which can be identified from the answers given. Assign assessment code **1/x** to the HGMU.

4. The process is definitely not being performed
This statement is not applicable for this process.

This completes the functional assessment for the process of provision of micro-sites for macroinvertebrates. The output codes and quantification values can be recorded on the Functional Assessment Table at the end of Section 3.3.A.4.

3.3.A.2.2 Provision of micro-sites for fish (process)

Further reading and references

Barbour, C.D. and Brown, J.H. (1974). Fish species diversity in lakes, *The American Naturalist* **108**, 473–489.

Carpentier, A., Paillisson, J-M., Marion, L., Feunteun, E., Baisez, A. and Rigaud, C. (2003). Trends of a bitterling (*Rhodeus sericeus*) population in a man-made ditch network, *Comptes Rendus Biologies* **326** (Supp 1), 166–173.

Cote, I.M., Vinyoles, D., Reynolds, J-D., Doadrio, I. and Perdices, A. (1999). Potential impacts of gravel extraction on Spanish populations of river blennies (*Salaria fluviatilis*) (Pisces, Blenniidae), *Biological Conservation* **87**(3), 359–367.

Crivelli, A.J. and Mestre, D. (1988). Life history traits of pumpkinseed, *Lepomis gibbosus*, introduced into the Camargue, a Mediterranean wetland, *Archiv fur Hydrobiologie* **111**(3), 449–466.

Darlington, P.J. (1957). *Zoogeography: The Geographical Distribution of Animals*, Wiley, New York.

Derksen, A.J. (1989). Autumn movements of under yearling northern pike, *Esox lucius*, from a large Manitoba marsh, *Canadian Field Naturalist* **103**(3), 429–431.
Goulding, M. (1980). *The Fishes and the Forest: Exploration in Amazonian Natural History*, University of California Press, Berkeley, CA.
Hampel, H., Cattrijsse, A. and Vincx, M. (2003). Habitat value of a developing estuarine brackish marsh for fish and macrocrustaceans, *ICES Journal of Marine Science* **60**(2), 278–289.
Keddy, P.A (2000). *Wetland Ecology: Principles and Conservation*, Cambridge University Press, Cambridge, UK.
King, D.R and Hunt, G.S. (1967). Effect of carp on vegetation in Lake Erie marsh, *Journal of Wildlife Management* **31**(1), 181–188.
Lewin, W-C., Okun, N. and Mehner, T. (2004). Determinants of the distribution of juvenile fish in the littoral area of a shallow lake, *Freshwater Biology* **49**(4), 410–424.
Lowe-McConnell, R.H. (1975). *Fish Communities in Tropical Freshwaters: Their Distribution, Ecology and Evolution*, Longman, London, UK.
Maitland, P.S. (1995). The conservation of freshwater fish: past and present experience, *Biological Conservation* **72**(2), 259–270.
Minns, C.K., Kelso, J.R.M. and Randall, R.G. (1996). Detecting the response of fish to habitat alterations in freshwater ecosystems, *Canadian Journal of Fisheries and Aquatic Sciences* **53**(S1), 403–414.
Mitsch, W.J. and Gosselink, J.G. (2007). *Wetlands*, 4th edn, Wiley, New York.
Peters, D.S., Ahrenholz, D.W. and Rice, T.R. (1979). Harvest and value of wetlands associated fish and shellfish, In: Greeson, P.E., Clark, J.R. and Clark, J.E. (eds) *Wetland Functions and Values: The State of our Understanding*, American Water Resources Association, Minneapolis, 606–617.
Poizat, G. and Corivelli, A.J. (1979). Use of seasonally flooded marshes by fish in a Mediterranean wetland: timing and demographic consequences, *Journal of Fish Biology* **51**(1), 106–119.
Pokorný, J. and Hauser, V. (2002). The restoration of fish ponds in agricultural landscapes, *Ecological Engineering* **18**(5), 555–574.
Stephenson, T.D. (1990). Fish reproductive utilization of coastal marshes of Lake Ontario near Toronto, *Great Lakes Research* **16**(1), 71–81.
Tanner, D.K., Brazner, J.C., Brady, V.J. and Regal, R.R. (2004). Habitat associations of larval fish in a Lake Superior coastal wetland, *Great Lakes Research* **30**(3) 349–359.
Tockner, K. and Schiemer, F. (1997). Ecological aspects of the restoration strategy for a river-floodplain system on the Danube River in Austria, *Global Ecology and Biogeography Letters* **6**, 321–329.
van der Valk, A. (2006). *The Biology of Freshwater Wetlands*, Oxford University Press, Oxford, UK.
Welcomme, R.L. (1979). *Fisheries Ecology of Floodplain Rivers*, Longman, London, UK.
Welcomme, R.L. (1986). Fish of the Niger system. In: Davis, B.R. and Walker, K.F. (eds), *The Ecology of River Systems*, Junk, Dordrecht, The Netherlands, 25–48.
Whillans, T.H. (1989). Wetlands and fisheries in Central Ontario lakes, In: Bardecki, M.J and Patterson, N. (eds), *Ontario Wetlands: Inertia or Momentum*, Proceedings of conference, Ryerson Polytechnical Institute, Toronto, Canada, Oct 21–22, 1988.
Willby, N.J. and Eaton, J.W. (2004). Backwater habitats and their role in nature conservation on navigable waterways, *Hydrobiologia* **340**(1–3), 333–338.
World Conservation Monitoring Centre (1992). *Global Biodiversity: Status of the Earth's Living Resources*, Chapman and Hall, London, UK.

Introduction

Process description: The provision of different micro-sites (small areas with distinct physical and ecological characteristics) for animals to inhabit within the wetland AA. Different groups of animals have different physical and ecological requirements. This section refers to fish.

Potential significance – rationale for assessment:

- Due to their rarity, certain species are important with regard to biodiversity.
- A diverse fauna is likely to be of conservation value.
- Fish are an important economic good supported by wetlands (see Table 3.3.4).
- Some species may have indicator value for the ecological conditions of the wetland.

Table 3.3.4 Dominant commercial and recreational wetland-associated fish and shellfish (after Peters *et al.*, 1979)

Habitat	Fish variety
Headwaters	Dace, Loaches, Bullhead
Gravelly streams and rivers	Salmonids, Gudgeon, Dace, Barbel, Sturgeon, Stickleback
Shallow, muddy streams and marginal areas	Lampreys, Stickleback, Roach, Eel
Stagnant pools and unvegetated lakes	Mud minnow, Carp, Tench, Loaches, Eel, Zander, Stickleback
Shallow vegetated lakes	Pike, Bream, Roach, Rudd, Orfe, Minnow, Eel, Perch, Zander, Ruffe, Stickleback
Deeper lakes	Trout, Wels

Answer all the questions on *each* of the controlling variables for the AA as a whole. Unless otherwise instructed, continue from one question to the next. Circle the correct answer on the appraisal sheet. Use a new appraisal sheet for each HGMU. The location of the relevant information is given in brackets after each question.

Controlling variables

CV1 Protected and/or Red List status
CV2 Habitat requirements

CV1 *Protected and/or Red List status*

Q1. Have recent surveys of the AA (i.e. within the last 15 years) described the presence of fish of international importance? *(Recording Sheet 1, C.3.17.3 (answer b)* Protected Species in the Bern Convention or the Bonn Convention (appendix II), the European Habitat Directive (annex IV and annex V) or the Red List of threatened species database (UNEP-WCMC))
[If YES, move on to the Look-up table]

Q2. Have recent surveys of the AA (*i.e.* within the last fifteen years) described the presence of protected fish of national or sub-national (local) importance? *(Recording Sheet 1, C.3.17.4 (answer b) C.3.17.5 (answer b)* Protected fish listed through national legislation or red lists of threatened species)
[If YES, move on to the Look-up table]

Q3. Are any areas within the AA formally protected or considered worthy of protection, to maintain their international conservation value according to Ramsar Convention, EU Habitat Directive (annex I and annex II) or EU Bird Directive? *(Recording Sheet 1, C.3.16.1)*

Q4. Are any areas within the AA formally protected or considered worthy of protection, to maintain their conservation value according to national or sub-national (local) legislation? *(Recording Sheet 1, C.3.16.2, C.3.16.3)*

CV2 *Habitat requirements*

Q1. Are there any records of migratory fish species regularly frequenting the water body, or tributaries adjacent to the HGMU? *(Recording Sheet 1, C.3.18.2)*

Q2. Does the AA contain any HGMUs that have been assigned geomorphic codes **TB**, **CO**, **DT**, **BS**, **OX**, **LK**, **CN**? *(Recording Sheet 2a, C.2)*

Q3. Are there prolonged periods of flooding (>1 month)? *(Recording Sheet 1, C.3.19.5)*

Q4. Does the HGMU contain submerged banks of gravel? *(Recording Sheet 3, C.5 (answer k))*

3.3.A.2.2 *Provision of Micro-sites for Fish Appraisal Sheet*

HGMU code: ____________________	**AA name:** ____________________

CV1 Protected and/or Red List status			
Q1	Y	N	U/NA
Q2	Y	N	U/NA
Q3	Y	N	U/NA
Q4	Y	N	U/NA

CV2 Habitat requirements			
Q1	Y	N	U/NA
Q2	Y	N	U/NA
Q3	Y	N	U/NA
Q4	Y	N	U/NA
Q5	Y	N	U/NA
Q6	Y	N	U/NA

HGMU process assessment code: ______________________________

3.3.A.2.2 Look-up Table 1

<table>
<tr><th>CV1</th><th>CV2</th><th>Rationale</th><th>Code</th></tr>
<tr><td>1</td><td>*</td><td>The AA is of international importance with regards to fish and is already protected or considered worthy of protection.</td><td>**1/1**</td></tr>
<tr><td>2</td><td>*</td><td>The AA is of national or local importance with regards to fish and is already protected or considered worthy of protection.</td><td>**1/1**</td></tr>
<tr><td>3|4</td><td>At least 5 of (1–6)</td><td>The AA is of international, national or local conservation importance, and is already protected or considered worthy of protection. The site has a high potential to support fish.</td><td>**1/1**</td></tr>
<tr><td>3|4</td><td>3 or 4 of (1–6)</td><td>The AA is of international, national or local conservation importance, and is already protected or considered worthy of protection. The site has the potential to support fish.</td><td>**1/2**</td></tr>
<tr><td>#(1|2|3|4)</td><td>At least 5 of (1–6)</td><td>The AA has high potential to support fish.</td><td>**1/2**</td></tr>
<tr><td>#(1|2|3|4)</td><td>3 or 4 of (1–6)</td><td>The AA has potential to support fish.</td><td>**1/3**</td></tr>
</table>

3.3.A.2.2 Look-up Table 2

<table>
<tr><th>CV1</th><th>CV2</th><th>Rationale</th><th>Code</th></tr>
<tr><td>#1 #2 #3 #4</td><td>#1 #2 #3 #4 #5 #6</td><td>The AA is not of conservation importance with regards to fish. The site is not protected or considered worthy of protection for fish. The site does not have the potential to support fish.</td><td>**2**</td></tr>
</table>

Q5. Does the HGMU contain submerged or emergent plants?
Vegetation codes **A1–A10**, **M13**, **M15** (*Recording Sheet 3, D)*

Q6. Does the HGMU have a vegetation code of **U17** or **U18**? *(Recording Sheet 3, D)*

Assessment of the process

Compare the results recorded for the AA on the appraisal sheet with the possible outcomes below, and assign the appropriate assessment code to the overall AA across all the HGMUs. Record this on the appraisal sheet. The syntax codes used are described in the introduction to Section 3.

Look-up tables

1. The process is definitely being performed
If answers to the questions for this process agree with any of the combinations in Look-up Table 1, then the AA is definitely performing the process of provision of micro-sites for fish. * Indicates where the answer to questions on the controlling variable can be either YES or NO. Answers are given in order according to both the certainty of occurrence of the process (first part of the code, e.g. '**1**/1') and the degree to which it occurs (second part of the code, e.g. '1/**1**'), based on the strength of evidence resulting from the combination of controlling variables: for code **1/1** the process is performed at the highest rate.

2. The process is not (significantly) being performed
If answers to the questions for this process agree with any of the combinations in Look-up Table 2, then the AA is not significantly performing the process of provision of micro-sites for fish.

3. The process is being performed, but there are constraining factors or uncertainties
For any other combinations of answers not given in Look-up Tables 1 and 2, the AA is performing the process of provision of micro-sites for fish, but there are constraining factors or uncertainty, which can be identified from the answers given. Assign assessment code **1/x** to the AA.

4. The process is definitely not being performed
This statement is not applicable for this process.

This completes the functional assessment for the process of provision of micro-sites for fish. The output codes and quantification values can be recorded on the Functional Assessment Table at the end of Section 3.3.A.4.

3.3.A.2.3 Provision of micro-sites for herptiles (process)

Further reading and references

Beebee, T.J.C. (1997). Changes in dewpond numbers and amphibian diversity over 20 years on chalk downland in Sussex, England, *Biological Conservation* **81**(3), 215–219.

Beja, P. and Alcazar, R. (2003). Conservation of Mediterranean temporary ponds under agricultural intensification: an evaluation using amphibians, *Biological Conservation* **114**(3), 317–326.

Benayas, J.M.R and de la Montaña, E. (2003). Identifying areas of high-value vertebrate diversity for strengthening conservation, *Biological Conservation* **114**(3), 357–370.

Brinson, M.M., Bradshaw, H.D., Holmes, R.N. and Elkins, J.B. Jr (1980). Litterfall, stemflow and throughfall nutrient fluxes in an alluvial swamp forest, *Ecology* **61**, 827–835.

Collins, J.P. and Storfer, A. (2003). Global amphibian declines: sorting the hypotheses, *Diversity & Distributions* **9**(2), 89–98.

Convention on the Conservation of European Wildlife and Natural Habitats (Bern Convention), Council Decision 82/72/EEC, entry into force 03/12/1981 (http://europa.eu/scadplus/leg/en/lvb/128050.htm)

Convention on the Conservation of Migratory Species of Wild Animals (Bonn Convention), Council Decision 82/461/EEC, entry into force 24.06.1982 (www.cms.int)

Denk, T., Frotzler, N. and Davitashvili, N. (2001). Vegetational patterns and distribution of relict taxa in humid temperate forests and wetlands of Georgia (Transcaucasia), *Biological Journal of the Linnean Society* **72**(2), 287–332.

de Nooij, R.J.W., Verberk, W.C.E.P., Lenders, H.J.R., Leuven, R.S.E.W. and Nienhuis, P.H. (2006). The importance of hydrodynamics for protected and endangered biodiversity of lowland rivers, *Hydrobiologia* **565**(1), 153–162.

Dodd, C.K. and Cade, B.S. (1998). Movement patterns and the conservation of amphibians breeding in small, temporary wetlands, *Conservation Biology* **12**(2), 331–339.

EC (1979). Council Directive 79/409/EEC of 2 April 1979 on the conservation of wild birds, *OJ* **L103**, 25.4.1979, 1.

EC (1992). Council Directive 92/43/EEC of 21 May 1992 on the conservation of natural habitats and of wild fauna and flora, *OJ* **L206**, 22.07.1992, 7–50.

Faber, P.A., Keller, E., Sands, A. and Masser, B.M. (1989). *The Ecology of Riparian Habitats of the Southern Californian Coastal Region: A Community Profile*, Biological Report 85 (7.27), US Fish and Wildlife Service, Washington, DC.

Ficetola, G.F. and De Bernardi, F. (2004). Amphibians in a human-dominated landscape: the community structure is related to habitat features and isolation, *Biological Conservation* **119**(2), 219–230.

Fredrickson, L.H. (1979). Lowland hardwood wetlands: Current status and habitat values for wildlife. In: Greeson, P.E, Clark, J.R. and Clark, J.E. (eds), *Wetland Functions and Values: The State of our Understanding*, American Water Resources Association, Minneapolis, MN, 296–306.

Hall, H.D. and Lambou, V.W. (1990). The ecological significance to fisheries of bottomland hardwood ecosystems: values, detrimental impacts, and assessment: the report of the fisheries workgroup. In: Gosselink, J.G., Lee, L.C. and Muir, T.A. (eds), *Ecological Processes and Cumulative Impacts: Illustrated by Bottomland Hardwood Wetland Ecosystems*, Lewis, Chelsea, MI, 481–531.

IUCN (2001). *IUCN Red List Categories and Criteria: Version 3.1*. IUCN Species Survival Commission. IUCN, Gland, Switzerland and Cambridge, UK.

Knutson, M.G., Sauer, J.R., Olsen, D.A., Mossman, M.J., Hemesath, L.M. and Lannoo, M.J. (1999). Effects of landscape composition and wetland fragmentation on frog and toad abundance and species richness in Iowa and Wisconsin, U.S.A, *Conservation Biology* **13**(6), 1437–1446.

Kolozsvary, M.B. and Swihart, R.K. (1999). Habitat fragmentation and the distribution of amphibians: patch and landscape correlates in farmland, *Canadian Journal of Zoology* **77**(8), 1288–1299.

Mazerolle, M.J. (2003). Detrimental effects of peat mining on amphibian abundance and species richness in bogs, *Biological Conservation* **113**, 215–223.

Mitsch, W.J. and Gosselink, J.G. (2007). *Wetlands* 4th edn, Wiley, New York.

Paton, P.W.C. (2005). A review of vertebrate community composition in seasonal forest pools of the northeastern United States, *Journal of Wetlands Ecology and Management* **13**(3), 235–246.

Roe, J.H., Kingsbury, B.A. and Herbert, N.R. (2004). Comparative water snake ecology: conservation of mobile animals that use temporally dynamic resources, *Biological Conservation* **118**, 79–89.

Rossi, E. and Kuitunen, M. (1996). Ranking of habitats for the assessment of ecological impact in land use planning, *Biological Conservation* **77**(2–3), 227–234.

Semlitsch, R.D. and Bodie, J.R. (2003). Biological criteria for buffer zones around wetlands and riparian habitats for amphibians and reptiles, *Conservation Biology* **17**(5), 1219–1228.

United Nations Educational, Scientific and Cultural Organization (UNESCO), Convention on Wetlands of International Importance especially as Waterfowl Habitat, Ramsar, Iran, 2.2.1971 (www.ramsar.org)

van der Valk, A. (2006). *The Biology of Freshwater Wetlands*, Oxford University Press, Oxford, UK.

Welch, N.E. and MacMahon, J.A. (2005). Identifying habitat variables important to the rare columbia spotted frog in Utah (U.S.A.): an information-theoretic approach, *Conservation Biology* **19**(2), 473–481.

Welcomme, R.L. (1979). *Fisheries Ecology of Floodplain Rivers*, Longman, London, UK.

Whitfield Gibbons, J. (2003). Terrestrial habitat: a vital component for herpetofauna of isolated wetlands, *Wetlands* **23**(3), 630–635.

Introduction

Process description: The provision of different micro-sites (small areas with distinct physical and ecological characteristics) for animals to inhabit within the wetland AA. Different groups of animals have different physical and ecological requirements. This section refers to herptiles (reptiles and amphibians). An example of the distribution of different species of herptiles in different habitats is given in Fig. 3.3.6.

Potential significance – rationale for assessment:

- Due to their rarity, certain species are important with regard to biodiversity.
- A diverse fauna is likely to be of conservation value.
- Some species may have indicator value on the ecological conditions of the wetland.

Answer all the questions on *each* of the controlling variables for all HGMUs. Unless otherwise instructed, continue from one question to the next. Circle the correct answer on the appraisal sheet. Use a new appraisal sheet for each HGMU. The location of the relevant information is given in brackets after each question.

Controlling variables

CV1 Protected and/or Red List status
CV2 Habitat requirements

CV1 *Protected and/or Red List status*

Q1. Have recent surveys of the AA (i.e. within the last 15 years) described the presence of herptiles of international importance? *(Recording Sheet 1, C.3.17.3 (answer c))* (Protected Species in the Bern Convention or the Bonn Convention (appendix II), the European Habitat Directive (annex IV and annex V) or the Red List of threatened species database (UNEP-WCMC)) *[If YES, move on to the Look-up table]*

Q2. Have recent surveys of the AA (i.e. within the last 15 years) described the presence of protected herptiles of national or sub-national (local) importance? *(Recording Sheet 1, C.3.17.4 (answer c), C.3.17.5 (answer c))* (Protected herptiles listed through national legislation or Red Lists of threatened species) *[If YES, move on to the Look-up table]*

Q3. Are any areas within the AA formally protected or considered worthy of protection, to maintain their international conservation value according to Ramsar Convention, EU Habitat Directive (annex I and annex II) or EU Bird Directive? *(Recording Sheet 1, C.3.16.1)*

Q4. Are any areas within the AA formally protected or considered worthy of protection, to maintain their conservation value according to national or sub-national (local) legislation? *(Recording Sheet 1, C.3.16.2, C.3.16.3)*

CV2 *Habitat requirements*

Q1. Within the AA are there any areas that have a south-easterly aspect and could be used as basking sites? *(Recording Sheet 3, C.5 (answer p))*

Q2. Within the AA are there any habitat features that are shady and damp? *(Recording Sheet 3, C.5, answer q)*

Q3. Within the AA are there any HGMUs that have surface water with emergent or submerged vegetation?
Vegetation codes: **A1–A10, M13, M15** *(Recording Sheet 3, D)*

Q4. Within the AA are there any HGMUs that are wet heath?
Vegetation code **M5** *(Recording Sheet 3, D)*

Q5. Within the AA are there any HGMUs that are sand dunes?
Vegetation codes **U2, U4** *(Recording Sheet 3, D)*

Q6. Within the AA are there any HGMUs that are dry heaths?

3.3.A.2.3 *Provision of Micro-sites for Herptiles Appraisal Sheet*

HGMU code: ______________________	AA name: ______________________

CV1 Protected and/or Red List status			
Q1	Y	N	U/NA
Q2	Y	N	U/NA
Q3	Y	N	U/NA
Q4	Y	N	U/NA

CV2 Habitat requirements			
Q1	Y	N	U/NA
Q2	Y	N	U/NA
Q3	Y	N	U/NA
Q4	Y	N	U/NA
Q5	Y	N	U/NA
Q6	Y	N	U/NA
Q7	Y	N	U/NA
Q8	Y	N	U/NA
Q9	Y	N	U/NA

HGMU process assessment code: ____

3.3.A.2.3 Look-up Table 1

<table>
<tr><th>CV1</th><th>CV2</th><th>Rationale</th><th>Code</th></tr>
<tr><td>1</td><td>*</td><td>The AA is of international importance with regard to herptiles and is already protected or considered worthy of protection.</td><td>1/1</td></tr>
<tr><td>2</td><td>*</td><td>The AA is of national or local importance with regards to herptiles and is already protected or considered worthy of protection.</td><td>1/1</td></tr>
<tr><td>3|4</td><td>6 or more of (1|2|3|4|5|6|7|8|9)</td><td>The AA is of international, national or local conservation importance, and is already protected or considered worthy of protection. The site has high potential to support herptiles.</td><td>1/1</td></tr>
<tr><td>3|4</td><td>3 to 5 of (1|2|3|4|5|6|7|8|9)</td><td>The AA is of international, national or local conservation importance, and is already protected or considered worthy of protection. The site has the potential to support herptiles.</td><td>1/2</td></tr>
<tr><td>#(1|2|3|4)</td><td>3 or more of (1|2|3|4|5|6|7|8|9)</td><td>The site has the potential to support herptiles.</td><td>1/3</td></tr>
</table>

3.3.A.2.3 Look-up Table 2

<table>
<tr><th>CV1</th><th>CV2</th><th>Rationale</th><th>Code</th></tr>
<tr><td>#1 #2 #3 #4</td><td>#1 #2 #3 #4 #5 #6 #7 #8 #9</td><td>The AA is not of conservation importance with regards to herptiles. The site is not protected or considered worthy of protection. The site does not have the potential to support herptiles.</td><td>2</td></tr>
</table>

Vegetation code **S8** *(Recording Sheet 3, D)*

Q7. Within the AA are there any HGMUs that are upland grass-heath? Vegetation codes **S9**, **G12**, **G15** *(Recording Sheet 3, D)*

Q8. Within the AA are there any HGMUs that are dry grasslands? Vegetation codes **G18–G32** *(Recording Sheet 3, D)*

Q9. Within the AA are there any HGMUs that are dry xeric scrub? Vegetation codes **S5–7**, **S10–S18** *(Recording Sheet 3, D)*

Assessment of the process

Compare the results recorded for each HGMU on the appraisal sheet with the possible outcomes below, and assign the appropriate assessment code. Record this on the appraisal sheet. The syntax codes used are described in the introduction to Section 3.

Look up tables

1. The process is definitely being performed

If answers to the questions for this process agree with any of the combinations in Look-up Table 1, then the HGMU is definitely performing the process of provision of micro-sites for herptiles. * Indicates where the answer to questions on the controlling variable can be either YES or NO. Answers are given in order according to both the certainty of occurrence of the process (first part of the code, e.g. '**1**/1') and the degree to which it occurs (second part of the code, e.g. '1/**1**'), based on the strength of evidence resulting from the combination of controlling variables: for code **1/1** the process is performed at the highest rate.

2. The process is not (significantly) being performed

If answers to the questions for this process agree with any of the combinations in Look-up Table 2, then the HGMU is not significantly performing the process of provision of micro-sites for herptiles.

3. The process is being performed, but there are constraining factors or uncertainties

For any other combinations of answers not given in Look-up Tables 1 and 2, the HGMU is performing the process of provision of micro-sites for herptiles, but only to a small degree because of constraining factors or uncertainty, which can be identified from the answers given. Assign assessment code **1/x** to these HGMUs.

4. The process is definitely not being performed

This statement is not applicable for this process.

This completes the functional assessment for the process of provision of micro-sites for herptiles. The output codes and quantification values can be recorded on the Functional Assessment Table at the end of Section 3.3.A.4.

3.3.A.2.4 Provision of micro-sites for birds (process)

Further reading and references

Adamus, P.R. (1992). Choices in monitoring wetlands. In: McKenzie, D.H., Hyatt, D.E. and McDonald, V.J. (eds), *Ecological Indicators*, Elsevier, London, UK, 571–592.

Arvidsson, B.L., Bostroem, U., Dahlen, B., Jong, A. de, Kolmodin, U. and Nilsson, S.G. (1992). The importance of mires as breeding habitat for wetland birds in Sweden, *Ornis Svecica* **2**(2), 67–76.

Ball, P.J. and Nudds, T.D. (1989). Mallard habitat selection: an experiment and implications for management. In: Sharitz, R.R. and Gibbons, J.W. (eds), *Freshwater Wetlands and Wildlife*, US Department of Energy, Office of Scientific and Technical Information, Oak Ridge, TN, 659–671.

Bethke, R.W. and Nudds, T.D. (1993). Variation in the diversity of ducks along a gradient of environmental variability, *Oecologia* **93**, 242–250.

Camphuysen, C.J., Berrevoets, C.M., Cremers, H.J.W.M., Dekinga, A., Dekker, R., Ens, B.J., van der Have, T.M., Kats, R.K.H., Kuiken, T., Leopold, M.F., Meer, J. and Piersma, T. (2002). Mass mortality of common eiders (*Somateria mollissima*) in the Dutch Wadden Sea, winter 1999/2000: starvation in a commercially exploited wetland of international importance, *Biological Conservation* **106**(3), 303–317.

Convention on the Conservation of European Wildlife and Natural Habitats (Bern Convention), Council Decision 82/72/EEC, entry into force 03/12/1981 http://europa.eu/scadplus/leg/en/lvb/128050.htm)

Convention on the Conservation of Migratory Species of Wild Animals (Bonn Convention), Council Decision 82/461/EEC, entry into force 24.06.1982 (www.cms.int)

Crivelli, A.J., Grillas, P., Jerrentrup, H. and Nazirides, T. (1995). Effects of fisheries and waterbirds of raising water levels at Kerkini Reservoir, a Ramsar site in northern Greece, *Environmental Management* **19**(3), 431–443.

Darlington, P.J. (1957). *Zoogeography: The Geographical Distribution of Animals*, Wiley, New York.

EC (1979). Council Directive 79/409/EEC of 2 April 1979 on the conservation of wild birds, *OJ* **L103**, 25.4.1979, 1

EC (1992). Council Directive 92/43/EEC of 21 May 1992 on the conservation of natural habitats and of wild fauna and flora, *OJ* **L206**, 22.07.1992, 7–50.

Hanowski, J.M., Niemi, G.J., Lima, A.R. and Regal, R.R. (1997). Response of breeding birds to mosquito control treatments of wetlands, *Wetlands* **17**(4), 485–492.

Huston, M.A. (1994). *Biological Diversity: The Coexistence of Species on Changing Landscapes*, Cambridge University Press, Cambridge, UK.

IUCN (2001). *IUCN Red List Categories and Criteria: Version 3.1*. IUCN Species Survival Commission, IUCN, Gland, Switzerland and Cambridge, UK.

Jackson, S.F., Kershaw, M. and Gaston, K.J. (2004). The buffer effect and the selection of protected areas for waterbirds in Britain, *Biological Conservation* **120**(1), 137–143.

Jackson, S.F., Kershaw, M. and Gaston, K.J. (2004). The performance of procedures for selecting conservation areas: waterbirds in the UK, *Biological Conservation* **118**(2), 261–270.

Kaminski, R.M. and Prince, H.H. (1981). Dabbling duck and aquatic macroinvertebrate responses to manipulated wetland habitat, *Journal of Wildlife Management* **45**, 1–15.

Keddy, P.A. (2000). *Wetland Ecology: Principles and Conservation*, Cambridge University Press, Cambridge, UK.

Kershaw, M. and Cranswick, P.A. (2003). Numbers of wintering waterbirds in Great Britain, 1994/1995–1998/1999: I. Wildfowl and selected waterbirds, *Biological Conservation* **111**(1), 91–104.

Lurz, P.W.W., Shirley, M.D.F., Rushton, S.P. and Sanderson, R.A. (2002). Modelling the consequences of duck migration patterns on the genetic diversity of aquatic organisms: a first step towards a predictive tool for wetland management, *Acta Oecologica* **23**(3), 205–212.

MacArthur, R.H. and MacArthur, J. (1961). On bird species diversity, *Ecology* **42**, 594–598.

Mauchamp, A., Chauvelon, P. and Grillas, P. (2002). Restoration of floodplain wetlands: Opening polders along a coastal river in Mediterranean France, Vistre marshes, *Ecological Engineering* **18**(5), 619–632.

Mitsch, W.J. and Gosselink, J.G. (2007). *Wetlands*, 4th edn, Wiley, New York.

Petts, G.E., Bickerton, M.A., Crawford, C., Lerner, D.N. and Evans, D. (1999). Flow management to sustain groundwater-dominated stream ecosystems, *Hydrological Processes* **13**(3), 497–513.

Pyrovetsi, M. and Crivelli, A. (1988). Habitat use by water-birds in Prespa National Park, Greece, *Biological Conservation* **45**(2), 135–153.

Stroud, D.A., Teed, T.M., Pienkowski, M.W. and Lindsay, R.A. (1987). *Birds, Bogs and Forestry. The Peatlands of Caithness and Sutherland*, Nature Conservancy Council, Peterborough, UK.

Tamisier, A., Grillas, P. (1994). A review of habitat changes in the Camargue: an assessment of the effects of the loss of biological diversity on the wintering waterfowl community, *Biological Conservation* **70**(1), 39–47.

Toivonen, H. and Nybom, C. (1989). Aquatic vegetation and its recent succession in the waterfowl wetland Koijaervi, S. Finland, *Annales Botanici Fennici* **26**(1), 1–14.

United Nations Educational, Scientific and Cultural Organization (UNESCO), Convention on Wetlands of International Importance especially as Waterfowl Habitat, Ramsar, Iran, 2.2.1971 (www.ramsar.org)

van der Valk, A. (2006). *The Biology of Freshwater Wetlands*, Oxford University Press, Oxford, UK.
Weller, M.W. (1994). Bird habitat relationships in a Texas estuarine marsh during summer, *Wetlands* **14**, 293–300.

Introduction

Process description: The provision of different micro-sites (small areas with distinct physical and ecological characteristics) for animals to inhabit within the wetland AA. Different groups of animals have different physical and ecological requirements. This section refers to birds. An example of the distributions of different bird species in a floodplain hardwood forest is given in Fig. 3.3.7.

Potential significance – rationale for assessment:

- Due to their rarity, certain species are important with regard to biodiversity.
- A diverse fauna is likely to be of conservation value.
- Wetlands form important staging posts in bird migration.
- Wetlands are important for the reproduction of many bird species.
- Some species have indicator value for the ecological conditions of the wetland.

Answer all the questions on *each* of the controlling variables for the AA as a whole. Unless otherwise instructed, continue from one question to the next. Circle the correct answer on the appraisal sheet. Use *one* appraisal sheet for the *whole AA*. The location of the relevant information is given in brackets after each question.

Controlling variables

CV1 Protected and/or Red List status
CV2 Migratory birds and bird population density
CV3 Habitat requirements

CV1 *Protected and/or Red List status*

Q1. Have recent surveys of the AA (i.e. within the last 15 years) described the presence of birds of international importance? *(Recording Sheet 1, C.3.17.3 (answer d))*

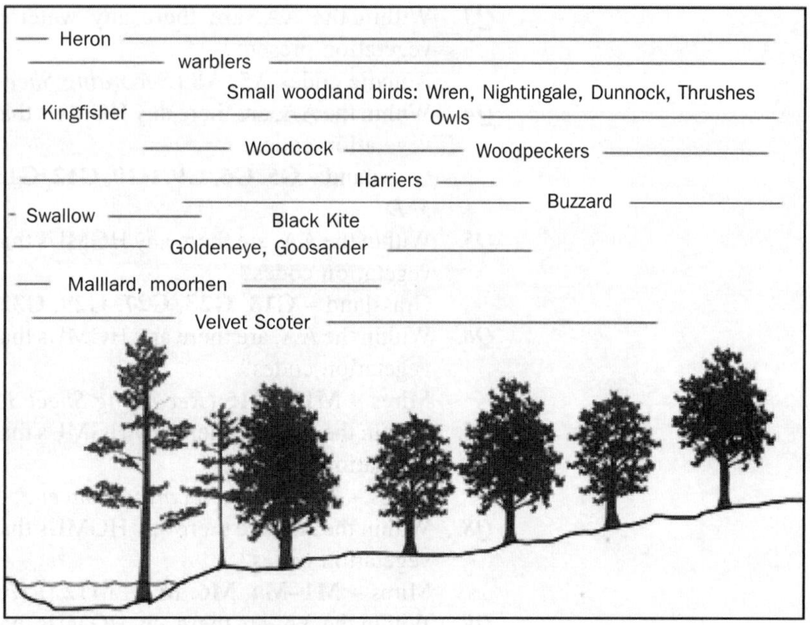

3.3.7 Distributions of preferred breeding/feeding habitats of selected bird species in a floodplain forest (adapted from Mitsch and Gosselink, 2007).

(Protected Species in the Bern Convention or the Bonn Convention (appendix II), the European Habitat Directive (annex IV and annex V) or the Red List of threatened species database (UNEP-WCMC)).
[If YES, move on to the Look-up table]

Q2. Have recent surveys of the AA (i.e. within the last 15 years) described the presence of protected birds of national or sub-national (local) importance? *(Recording Sheet 1, C.3.17.4 (answer d), C.3.17.5 (answer d))*
(Protected Species listed through national legislation or red lists of threatened species)
[If YES, move on to the Look-up table]

Q3. Are any areas within the AA formally protected, or considered worthy of protection, to maintain their international conservation value according to Ramsar Convention, EU Habitat Directive (annex I and annex II) or EU Bird Directive? *(Recording Sheet 1, C.3.16.1)*

Q4. Are any areas within the AA formally protected or considered worthy of protection, to maintain their conservation value according to national or sub-national (local) legislation? *(Recording Sheet 1, C.3.16.2, C.3.16.3)*

CV2 *Migratory birds and bird population density*

Q1. Are there any records of migratory bird species regularly frequenting the AA? *(Recording Sheet 1, C.3.18.3)*

Q2. What is the approximate density of birds in the AA? *(Recording Sheet 1, C.3.8.3a)*

a. 1000–2000 birds ha^{-1}
b. > 2000 but < 5000 birds ha^{-1}
c. 5000–10 000 birds ha^{-1}
d. > 10 000 birds ha^{-1}
e. < 1000 birds ha^{-1}

CV3 *Habitat requirements*

Q1. Within the AA, are there any water bodies with submerged aquatic vegetation present?
Aquatic codes **A2**, **A3**, **A4**, **A7**, **A9** *(Recording Sheet 3, D)*

Q2. Within the AA, are there any water bodies with floating vegetation present?
Aquatic codes **A1**, **A6**, **A10** *(Recording Sheet 3, D)*

Q3. Within the AA, are there any water bodies with emergent macrophyte vegetation present?
Aquatic codes **A5**, **A8** *(Recording Sheet 3, D)*

Q4. Within the AA, are there any HGMUs that have been designated the following vegetation codes?
Grassland – **G5**, **G6**, **G9**, **G10**, **G12**, **G13**, **G16**, **G17**, **M15** *(Recording Sheet 3, D)*

Q5. Within the AA, are there any HGMUs that have been designated the following vegetation codes?
Grassland – **G18**, **G23**, **G27**, **G29**, **G32**, **G33** *(Recording Sheet 3, D)*

Q6. Within the AA, are there any HGMUs that have been designated the following vegetation codes?
Mires – **M13–M15** *(Recording Sheet 3, D)*

Q7. Within the AA, are there any HGMUs that have been designated the following vegetation codes?
Mires – **M9**, **M10** *(Recording Sheet 3, D)*

Q8. Within the AA, are there any HGMUs that have been designated the following vegetation codes?
Mires – **M1–M4**, **M6**, **M11**, **M12** *(Recording Sheet 3, D)*

Q9. Within the AA, are there any HGMUs that have been designated the following vegetation codes?
Woodland – **W26**, **W31** *(Recording Sheet 3, D)*

Q10. Within the AA, are there any HGMUs that have been designated the following vegetation codes?
Woodland – **W7**, **W11** *(Recording Sheet 3, D)*

Q11. Within the CA, are there any areas that have been designated the following land-use code?
OG: including non-agricultural grassland, rough pasture or amenity grassland *(CA Map 1)*

Q12. Within the CA, are there any areas that have been designated the following land-use code?
CW: coniferous woodland *(CA Map 1)*

Q13. Within the CA, are there any areas that have been designated the following land-use code?
IW: inland non-marine water *(CA Map 1)*

Q14. Within the CA, are there any areas that have been designated the following land-use code?
BG: bog or FMS (fen, marsh and swamp) *(CA Map 1)*

Q15. Within the AA, are there any HGMUs that have been designated the following geomorphic codes?
TB, **DT**, **CN**: flowing or standing open water *(Recording Sheet 2a, C.2)*

Q16. Within the AA, are there any HGMUs that have been designated the following geomorphic codes? *(Recording Sheet 2a, C.2)*
LK, **OX**, **CO**: standing open water *(Recording Sheet 2a, C.2)*

Q17. Within the AA, are there any HGMUs that have been designated the following geomorphic code?
PB: point bar *(Recording Sheet 2a, C.2)*

Q18. Within the AA, are there any islands? *(AA Map 3)*

Q19. Within the AA, are there any water bodies with abundant algae present? *(Recording Sheet 3, C.5 (answer r))*

Q20. Within the AA are there any steep sandy riverbanks or lake cliffs? *(Recording Sheet 3, C.5 (answer s))*

Q21. Within the AA are there any bridges, aqueducts, or viaducts? *(Recording Sheet 3, C.5 (answer w))*

Q22. Within the AA are there boulders or bare rocks within or adjacent to the river channel or lake? *(Recording Sheet 3, C.5 (answer bb))*

Q23. Is the water quality in the main river channel or lake of sufficient quality to pass all the described EC water quality directives? *(Recording Sheet 1, C.3.20.1, C.3.20.2, C.3.20.3, C.3.20.4, C.3.20.5)*

Q24. For the overall AA could the degree of human disturbance caused by all of the following activities be described as low?

a. Hunting, shooting or fishing *(Recording Sheet 1, C.3.14.1 (answers a or b))*
b. Recreational activities *(Recording Sheet 1, C.3.15.1 (answers a or b))*
c. Construction activity *(Recording Sheet 1, C.3.13.1a; AA Map 2 (answers 'no'))*
d. Mining or extraction *(Recording Sheet 1, C.3.11.1a (answers a or b); AA Map 2)*
e. Agricultural activity (burning, harvesting, ploughing, forestry, grazing) *(Recording Sheet 1, C.3.3.1a (answers a or b), C.3.4.1a (answers a or b), C.3.5.1a (answers a or b), C.3.6.1a (answers a or b), C.3.7.1a (answers a or b); AA Map 2)*

Assessment of the process

Compare the results recorded for the AA on the appraisal sheet with the possible outcomes below, and assign the appropriate assessment code to the overall AA across all the HGMUs. Record this on the appraisal sheet. The syntax codes used are described in the introduction to Section 3.

3.3.A.2.4 *Provision of Micro-sites for Birds Appraisal Sheet*

HGMU code: ______________________	**AA name:** ______________________

CV1 Protected and/or Red List status			
Q1	Y	N	U/NA
Q2	Y	N	U/NA
Q3	Y	N	U/NA
Q4	Y	N	U/NA

CV2 Migratory birds and bird population density					
Q1	Y		N		U/NA
Q2	a	b	c	d	e

CV3 Habitat requirements			
Q1	Y	N	U/NA
Q2	Y	N	U/NA
Q3	Y	N	U/NA
Q4	Y	N	U/NA
Q5	Y	N	U/NA
Q6	Y	N	U/NA
Q7	Y	N	U/NA
Q8	Y	N	U/NA
Q9	Y	N	U/NA
Q10	Y	N	U/NA
Q11	Y	N	U/NA
Q12	Y	N	U/NA
Q13	Y	N	U/NA
Q14	Y	N	U/NA
Q15	Y	N	U/NA
Q16	Y	N	U/NA
Q17	Y	N	U/NA
Q18	Y	N	U/NA
Q19	Y	N	U/NA
Q20	Y	N	U/NA
Q21	Y	N	U/NA
Q22	Y	N	U/NA
Q23	Y	N	U/NA
Q24	Y	N	U/NA

HGMU process assessment code: ____

3.3.A.2.4 Look-up Table 1

CV1	CV2	CV3	Rationale	Code
1	*	*	The AA is of international importance with regards to birds. It is an offence to destroy this habitat.	**1/1**
2	*	*	The AA is of national importance with regards to birds.	**1/1**
3|4	1|(2c|2d)	16 or more of (1–24)	The AA is of conservation importance, and is already protected or considered worthy of protection. Migratory birds, or very large populations of birds, regularly frequent the site. The range of environmental conditions, in terms of macro-habitats, micro-sites and environmental quality present within the AA provide the potential to support a wide range of bird species.	**1/1**
3|4	1|2a–d	*	The AA is of conservation importance, and is already protected or considered worthy of protection. Migratory birds, or large populations of birds, regularly frequent the site.	**1/2**
3|4	*	16 or more of (1–24)	The AA is of conservation importance, and is already protected or considered worthy of protection. An extremely diverse range of macro-habitats, micro-sites and environmental conditions are present within the AA, providing the potential conditions to support a wide range of bird species.	**1/2**
3|4	1|(2a|2b)	16 or more of (1–24)	The AA is of conservation importance, and is already protected or considered worthy of protection. Migratory birds, or large populations of birds, regularly frequent the site. The range of environmental conditions, in terms of macro-habitats, micro-sites and environmental quality present within the AA provide the potential to support a wide range of bird species.	**1/2**
3|4	1|(2e)	16 or more of (1–24)	The AA is of conservation importance, and is already protected or considered worthy of protection. Migratory birds, or populations of birds regularly frequent the site. The range of environmental conditions, in terms of macro-habitats, micro-sites and environmental quality, present within the AA provide the potential to support a wide range of bird species.	**1/2**
3|4	1|(2c|2d)	5 to 15 of (1–24)	The AA is of conservation importance, and is already protected or considered worthy of protection. Migratory birds, or very large populations of birds regularly frequent the site. The range of environmental conditions, in terms of macro-habitats, micro-sites and environmental quality, present within the AA provide a low potential to support a wide range of bird species.	**1/3**
3|4	1|(2a|2b)	5 to 15 of (1–24)	The AA is of conservation importance, and is already protected or considered worthy of protection. Migratory birds or large populations of birds regularly frequent the site. The range of environmental conditions, in terms of macro-habitats, micro-sites and environmental quality, present within the AA provide a low potential to support a wide range of bird species.	**1/3**
3|4	1|(2e)	5 to 15 of (1–24)	The AA is of conservation importance, and is already protected or considered worthy of protection. Migratory birds or populations of birds regularly frequent the site. The range of environmental conditions, in terms of macro-habitats, micro-sites and environmental quality, present within the AA provide a low potential to support a wide range of bird species.	**1/3**
#(1 2 3 4)	1|2a–d	*	Migratory birds or large populations of birds regularly frequent the site.	**1/3**
#(1 2 3 4)	*	16 or more of (1–24)	An extremely diverse range of macro-habitats, micro-sites and environmental conditions are present within the AA, providing the potential conditions to support a wide range of bird species.	**1/3**

3.3.A.2.4 Look-up Table 2

CV1	CV2	CV3	Rationale	Code
*	*	1 4 5 6 8 11 13 14 16 18 24	The range of environmental conditions, in terms of macro-habitats, micro-sites and environmental quality, present within the AA provide the potential to support a wide range of *grazing bird* species.	**grazing bird**
*	*	2 3 4 6 7 8 10 13 14 15 16 22 23 24	The range of environmental conditions, in terms of macro-habitats, micro-sites and environmental quality, present within the AA provide the potential to support a wide range of *diving omnivorous* bird species.	**diving omnivorous**
*	*	2 3 4 7 8 10 13 14 15 16 19 23 24	The range of environmental conditions, in terms of macro-habitats, micro-sites and environmental quality, present within the AA provide the potential to support a wide range of *surface feeding omnivorous* bird species.	**surface feeding omnivorous**
*	*	2 3 6 13 15 16 23 24	The range of environmental conditions, in terms of macro-habitats, micro-sites and environmental quality, present within the AA provide the potential to support a wide range of *diving seed and weed eating* bird species.	**diving seed and weed eating**
*	*	3 4 6 7 8 10 13 14 15 16 23	A wide range of environmental conditions are present that will provide the potential to support a wide range of *wetland* bird species.	**wetland**
*	*	3 6 7 10 13 15 16 17 22 23 24	The range of environmental conditions, in terms of macro-habitats, micro-sites and environmental quality, present within the AA provide the potential to support a wide range of *diving carnivorous* bird species.	**diving carnivorous**
*	*	4 5 6 7 8 9 10 11 12 14 15 20 21 23 24	The range of environmental conditions, in terms of macro-habitats, micro-sites and environmental quality, present within the AA provide the potential to support a wide range of *aerial piscivorous* bird species.	**aerial piscivorous**
*	*	4 6 8 13 14 15 16 17 23 24	The range of environmental conditions, in terms of macro-habitats, micro-sites and environmental quality, present within the AA provide the potential to support a wide range of *wading carnivorous or omnivorous* bird species.	**wading carnivorous or omnivorous**
*	*	6 7 10 15 24	The range of environmental conditions, in terms of macro-habitats, micro-sites and environmental quality, present within the AA provide the potential to support a wide range of *aerial insectivorous* bird species.	**aerial insectivorous**

3.3.A.2.4 Look-up Table 3

CV1	CV2	CV3	Rationale	Code
#(1 2 3 4)	*	#(16 or more of 1–23)	The AA is not of conservation importance for birds. The AA does not provide sufficient environmental conditions to provide the feeding conditions for most wetland bird species.	**2**
#(1 2 3 4)	*	#(3 4 6 7 8 10 13 14 15 16 23 24)	The AA is not of conservation importance for birds. The AA does not provide the key habitats, such as open water, reed or sedge beds, wet grasslands and riparian woodland necessary to support most wetland bird species. Human disturbance of the site is great and the water quality in the river is poor, producing conditions that are not conducive to supporting a diverse range of bird species.	**2**
#(1 2 3 4)	*	#(3 4 6 7 8 10 13 14 15 16 23)	The AA is not of conservation importance for birds. The AA does not provide the key habitats, such as open water, reed or sedge beds, wet grasslands and riparian woodland necessary to support most wetland bird species. The water quality in the river is poor, producing conditions that are not conducive to supporting a diverse range of bird species.	**2**
#(1 2 3 4)	*	#(3 4 6 7 8 10 13 14 15 16 24)	The AA is not of conservation importance for birds. The AA does not provide the key habitats, such as open water, reed or sedge beds, wet grasslands and riparian woodland necessary to support most wetland bird species. Human disturbance of the site is great, producing conditions that are not conducive to supporting a diverse range of bird species.	**2**
#(1 2 3 4)	*	#(3 4 6 7 8 10 13 14 15 16)	The AA is not of conservation importance for birds. The AA does not provide the key habitats, such as open water, reed or sedge beds, wet grasslands and riparian woodland necessary to support most wetland bird species.	**2**

Look-up tables

1. The process is definitely being performed

1.1. Importance: If answers to the questions for this process agree with any of the combinations in Look-up Table 1, then the AA is definitely performing the process of provision of micro-sites for birds. * Indicates where the answer to questions on the controlling variable can be either YES or NO. Answers are given in order according to both the certainty of occurrence of the process (first part of the code, e.g. '**1**/1') and the degree to which it occurs (second part of the code, e.g. '1/**1**'), based on the strength of evidence resulting from the combination of controlling variables: for code **1/1** the process is performed at the highest rate.

1.2. Assessment of specific groups of birds (feeding groups): The specific groups of birds that are likely to be supported by the potential habitats provided are shown in Look-up Table 2.

2. The process is not significantly being performed
If answers to the questions for this process agree with any of the combinations in Look-up Table 3 below, then the AA is not significantly performing the process of provision of micro-sites for birds.

3. The process is being performed, but there are constraining factors or uncertainty
For any other combinations of answers not given in Look-up Tables 1 or 2, the AA is performing the process of provision of micro-sites for birds, but only to a small degree because of constraining factors or uncertainty, which can be identified from the answers given. Assign assessment code **1/x** to the AA.

4. The process is definitely not being performed
This statement is not applicable for this process.

This completes the functional assessment for the process of provision of micro-sites for birds. The output codes and quantification values can be recorded on the Functional Assessment Table at the end of Section 3.3.A.4.

3.3.A.2.5 *Provision of micro-sites for mammals (process)*

Further reading and references

Billington, G. (2002). *Radio Tracking Study of Greater Horseshoe Bats at Chudleigh Caves and Woods Site of Special Scientific Interest*, English Nature Research Reports No 496, English Nature, Peterborough, UK.

Convention on the Conservation of Migratory Species of Wild Animals (Bonn Convention); Council Decision 82/461/EEC, entry into force 24.06.1982 (www.cms.int)

Convention on the Conservation of European Wildlife and Natural Habitats (Bern Convention), Council Decision 82/72/EEC, entry into force 03/12/1981 (http://europa.eu/scadplus/leg/en/lvb/128050.htm)

EC (1979). Council Directive 79/409/EEC of 2 April 1979 on the conservation of wild birds, *OJ* **L103**, 25.4.1979, 1.

EC (1992). Council Directive 92/43/EEC of 21 May 1992 on the conservation of natural habitats and of wild fauna and flora, *OJ* **L206**, 22.07.1992, 7–50.

Faber, P.A., Keller, E., Sands, A. and Masser, B.M. (1989). *The Ecology of Riparian Habitats of the Southern Californian Coastal Region: A Community Profile*, Biological Report 85 (7.27), US Fish and Wildlife Service, Washington DC.

Feierabend, J.S. (1989). Wetlands: the lifeblood of wildlife. In: Hammer, D.A. (ed.), *Constructed Wetlands for Wastewater Treatment: Municipal, Industrial and Agricultural*, Lewis, Chelsea, MI, 107–118.

Fredrickson, L.H. (1979). Lowland hardwood wetlands: Current status and habitat values for wildlife. In: Greeson, P.E, Clark, J.R and Clark, J.E (eds), *Wetland Functions and Values: The State of our Understanding*, American Water Resources Association, Minneapolis, MN, 296–306.

Glaser, P.H. (1987). *The Ecology of Patterned Boreal Peatlands of Northern Minnesota: A Community Profile*, Biological Report 12, US Fish and Wildlife Service, Washington, DC.

Gordon, I.J., Duncan, P., Grillas, P. and Lecomte, T. (1980). The use of domestic herbivores in the conservation of the biological richness of European wetlands, *Bulletin d'Ecologie* **21**(3), 49–60.

Hall, H.D. and Lambou, V.W. (1990). The ecological significance to fisheries of bottomland hardwood ecosystems: values, detrimental impacts, and assessment: the report of the fisheries workgroup. In: Gosselink, J.G., Lee, L.C. and Muir, T.A. (eds), *Ecological Processes and Cumulative Impacts: Illustrated by Bottomland Hardwood Wetland Ecosystems*, Lewis, Chelsea, MI, 481–531.

IUCN (2001). *IUCN Red List Categories and Criteria: Version 3.1.* IUCN Species Survival Commission. IUCN, Gland, Switzerland and Cambridge, UK.

Jobin, B., Bélanger, L., Boutin, C. and Maisonneuve, C. (2004). Conservation value of agricultural riparian strips in the Boyer River watershed, Québec (Canada), *Agriculture, Ecosystems and Environment* **103**(3), 413–423.

Lanszki, J. (2003). Feeding habits of stone martens in a Hungarian village and its surroundings, *Folia Zoologica* **52**(4), 367–377.

Lode, T. (1993). Diet composition and habitat use of sympatric polecat and American mink in western France, *Acta Therologica* **38**, 161–166.

Lode, T. (1994). Environmental factors influencing habitat exploitation by the polecat *Mustela putorius* in western France, *Journal of Zoology* **234**(1), 75–88.

Maisonneuve, C. and Rioux, S. (2001). Importance of riparian habitats for small mammal and herpetofaunal communities in agricultural landscapes of southern Québec, *Agriculture, Ecosystems and Environment* **83**(1–2), 165–175.

McBrien, M. and Hayes, D. (1998). Habitat development and wastewater treatment wetlands: Natural combination or recipe for failure? In: Hayes, D. (ed.), *Engineering Approaches to Ecosystem Restoration (Proceedings of the 1998 ASCE Wetlands Engineering and River Restoration Conference)*, Denver, CO, March 22–27 (CD-ROM).

Minnesota DNR (1984). *Recommendations for the Protection of Ecologically Significant Peatlands in Minnesota*, Division of Lands and Minerals, Minnesota Department of Natural Resources, St Paul, MN.

Minshall, G.W., Jensen, W.S.E. and Platts, W.S. (1989). *The Ecology of Stream and Riparian Habitats of the Great Basin Region: A Community Profile*, Biological Report 85 (7.24), US Fish and Wildlife Service, Washington, DC.

Mitsch, W.J. and Gosselink, J.G. (2007). *Wetlands*, 4th edn, Wiley, New York.

Nordquist, G.E. and Birney, E.C. (1980). *The Importance of Peatland Habitats to Small Mammals in Minnesota*, Unpublished Final Report, Minnesota Museum of Natural History, University of Minnesota Occasional paper No. 6.

Peintinger, M., Bergamini, A. and Schmid, B. (2003). Species-area relationships and nestedness of four taxonomic groups in fragmented wetlands, *Basic and Applied Ecology* **4**(5), 385–394.

Prigioni, C. and de Marnis, A.M. (1995). Diet of the polecat Mustela putorzus L. in riverine habitats (northern Italy), *Hystrix* (n.s.) **7**(1–2), 69–72, Proceedings of 2nd Italian Symposium on Carnivores.

Rossi, E. and Kuitunen, M. (1996). Ranking of habitats for the assessment of ecological impact in land use planning, *Biological Conservation* **77**(2–3), 227–234.

United Nations Educational, Scientific and Cultural Organization (UNESCO), Convention on Wetlands of International Importance especially as Waterfowl Habitat, Ramsar, Iran, 2.2.1971, (www.ramsar.org)

Sterzynska, M. and Lesinski, G. (2001). Resource assessment and a strategy for the protection of wetland fauna. In: Andersson, L., Marciau, R., Paltto, H., Tardy, B. and Read, H. (eds), *Tools for Preserving Biodiversity in the Nemoral and Boreonemoral Biomes of Europe*, Textbook 1, NACONEX programme, September 2001.

van der Valk, A. (2006). *The Biology of Freshwater Wetlands*, Oxford University Press, Oxford, UK.

Wharton, C.H., Kitchens, W.M., Pendleton, E.C and Sipe, T.W (1982). *The Ecology of Bottomland Hardwood Swamps of the Southeast: A Community Profile*, Biological Services Program FWS/OBS-81/37, US Fish and Wildlife Service, Washington, DC.

Wike, L.D., Martin, F.D., Hanlin, H.G. and Paddock, L.S. (2000). Small mammal populations in a restored stream corridor, *Ecological Engineering* **15**(1), S121–S129.

Introduction

Process description: The provision of different micro-sites (small areas with distinct physical and ecological characteristics) for animals to inhabit within the wetland AA. Different groups of animals have different physical and ecological requirements. This section refers to mammals. Table 3.3.5 gives examples of mammalian species that can be found in European wetlands.

Potential significance – rationale for assessment:

- Due to their rarity, certain species are important with regard to biodiversity.
- A diverse fauna is likely to be of conservation value.
- Mammal species can be important links in food webs.
- Impacts of mammals can be important agents for ecosystem structuring.
- Hunting can provide income for wetland conservation.

Answer all the questions on *each* of the controlling variables for the AA as a whole, unless otherwise instructed. Continue from one question to the next. Circle the correct answer on the appraisal sheet. Use a new appraisal sheet for each HGMU. The location of the relevant information is given in brackets after each question.

Note: River and lake marginal wetlands have the potential to support both those mammals whose specific habitat requirements include wetlands and those with a broader niche that can include wetland habitat.

Controlling variables

CV1 Protected and/or Red List status
CV2 Migratory and/or frequentation records
CV3 Habitat requirements

CV1 *Protected and/or Red List status*

Q1. Have recent surveys of the AA (i.e. within the last 15 years) described the presence of mammals of international importance? *(Recording Sheet 1, C.3.17.3 (answer e))*
(Protected Species in the Bern Convention or the Bonn Convention (appendix II), the European Habitat Directive (annex IV and annex V) or the Red List of threatened species database (UNEP-WCMC))
[If YES, move on to the Look-up table]

Q2. Have recent surveys of the AA (i.e. within the last 15 years) described the presence of protected mammals of national or sub-national (local) importance? *(Recording Sheet 1, C.3.17.4 (answer e), C.3.17.5 (answer e))*

Table 3.3.5 Mammalian groups and species found in European wetlands

Bats	Daubentons Bat (*Myotis daubentonii*), Felten's Myotis Bat (*Myotis punicus*), Pond Bat (*Myotis dasycneme*), Pipistrelle (*Pipistrellus pipistrellus*), Natterer's Bat (*Myotis nattereri*), Long-fingered Bat (*Myotis capaccinii*)
Insectivores	Pygmy Shrew (*Sorex minutus*), Water Shrew (*Sorex palustris*), Pyrenean Desman (*Galemys pyrenaicus*)
Rodents	Root Vole (*Microtus oeconomus*), Water Vole (*Arvicola terrestris*), Northern Birch Mouse (*Sicista betulina*), Algerian Mouse (*Mus spretus*), Harvest Mouse (*Micromys minutus*), Muskrat (*Ondatra zibethica*)
Weasels	Stoat (*Mustela erminea*), Weasel (*Mustela nivalis*), Polecat (*Mustela putorius*), European Mink (*Mustela lutreola*)
Larger mammals	Fox (*Vulpes vulpes*), Wolf (*Canis lupus*), Raccoon-dog (*Nyctereutes procyonoides*), Wild Cat (*Felis sylvestris grampia*), Deer, Elk (*Cervus elaphus*), Brown Bear (*Ursus arctos*), Coypu (*Myocastor coypus*)

(Protected mammals listed through national legislation or red lists of threatened species)
[If YES, move on to the Look-up table]

Q3. Are any areas within the AA formally protected or considered worthy of protection, to maintain their international conservation value according to Ramsar Convention, EU Habitat Directive (annex I and annex II) or EU Bird Directive? *(Recording Sheet 1, C.3.16.1)*

Q4. Are any areas within the AA formally protected or considered worthy of protection, to maintain their conservation value according to national or sub-national (local) legislation? *(Recording Sheet 1, C.3.16.2, C.3.16.3)*

CV2 *Migratory and/or frequentation records*

Q1. Are there any records of migratory mammal species frequenting the AA? *(Recording Sheet 1, C.3.18.4)*

Q2. Within the AA, are there any records of burrows, dens, lairs, holts, holes or other domiciles of mammals? *(Recording Sheet 3, C.5 (answer y))*

Q3. Are there any tunnels or runways through the vegetation? *(Recording Sheet 3, C.5 (answer z))*

Q4. Are there any records of beaver dams present within the AA or adjoining river channel? *(Recording Sheet 3, C.5 (answer aa))*

CV3 *Habitat requirements*

Q1. Is the river a fast flowing mountain stream with the majority of the water originating from glacial or snow melt? *(CA Map 2)*

Q2. Within the AA, are there any HGMUs that have been designated a grassland (**Gx**) vegetation code? *(Recording Sheet 3, D)*

Q3. Within the AA, are there any HGMUs that have been designated the following vegetation codes? *(Recording Sheet 3, D)*
G9, **G11**, **G1**

Q4. Within the AA, are there any HGMUs that have been designated a woodland (**Wx**) vegetation code?

Q5. Within the AA, are there any riparian HGMUs that have been designated the following woodland vegetation codes?
woodland – **W16**, **W17**, **W42** *(Recording Sheet 3, D)*

Q6. Within the AA, are there any areas designated as coniferous woodland?
vegetation codes **W4**, **W21–W24**, **W26**, **W29–W31** *(Recording Sheet 3, D)*

Q7. Within the AA, are there any HGMUs that have been designated the following vegetation codes?
mires – **M13–M15** *(Recording Sheet 3, D)*

Q8. Within the AA, are there any HGMUs that have been designated the following vegetation codes?
mires – **M9–M10** *(Recording Sheet 3, D)*

Q9. Within the AA, are there any riparian HGMUs that have been designated a scrub and dwarf shrub (**Sx**) vegetation code? *(Recording Sheet 3, D)*

Q10. Within the AA, is there any flowing open water (vegetation code **U17**, possibly also **A3–A10**)? *(Recording Sheet 3, D)*

Q11. Within the AA, is there any standing open water (vegetation code **U18**, possibly also **A1–A10**)? *(Recording Sheet 3, D)*

Q12. Within the AA, are the HGMUs that border the river channel predominantly vegetated all the way down to the water's edge? *(Recording Sheet 3, D)*

Q13. Within the AA, are there any areas that are inland cliffs or screes (vegetation code **U5–U8**, **U10**, **U12–U14**)? *(Recording Sheet 3, D)*

Q14. Within the AA, are there any of the following?

a. Very old, mature trees *(Recording Sheet 3, C.5 (answer a))*
b. Standing dead trees with little or no bark *(Recording Sheet 3, C.5 (answer b))*
c. Standing dead trees with a substantial covering of bark *(Recording Sheet 3, C.5 (answer c))*

Q15. Within the AA, are there any river or lake cliffs? *(Recording Sheet 3, C.5 (answer s))*

Q16. Within the AA, are there any of the following?

a. Rotten holes in trees *(Recording Sheet 3, C.5 (answer h))*
b. Hollow trees *(Recording Sheet 3, C.5 (answer i))*
c. Bat boxes *(Recording Sheet 3, C.5 (answer x))*

Q17. Within the AA, are there any of the following?

a. Derelict or uninhabited buildings *(Recording Sheet 3, C.5 (answer u))*
b. Caves, disused tunnels and mines *(Recording Sheet 3, C.5 (answer v))*
c. Old bridges, aqueducts or viaducts *(Recording Sheet 3, C.5 (answer w))*

Q18. Within the AA, are there any boulders or rocks within or adjacent to the river channel? *(Recording Sheet 3, C.5 (answer bb))*

Q19. Is the water quality in the main river or lake of sufficient quality to pass all the described EC water quality directives? *(Recording Sheet 1, C.3.20.1, C.3.20.2, C.3.20.3, C.3.20.4, C.3.20.5)*

Q20. For the overall AA, could the degree of human disturbance caused by all of the following activities be described as low?

a. Hunting, shooting or fishing *(Recording Sheet 1, C.3.14.1 (answers a or b))*
b. Recreational activities *(Recording Sheet 1, C.3.15.1 (answers a or b))*
c. Construction activity*(Recording Sheet 1, C.3.13,1a, answers 'no'; AA Map 2)*
d. Mining or extraction *(Recording Sheet 1, C.3.11.1a (answers a or b); AA Map 2)*
e. Agricultural activity (burning, harvesting, ploughing, forestry, grazing) *(Recording Sheet 1, C.3.3.1a (answers a or b), C.3.4.1a (answers a or b), C.3.5.1a (answers a or b), C.3.6.1a (answers a or b), C.3.7.1a (answers a or b); AA Map 2)*

Assessment of the process

Compare the results recorded for the AA on the appraisal sheet with the possible outcomes below, and assign the appropriate assessment code to the overall AA across all the HGMUs. Record this on the appraisal sheet. The syntax codes used are described in the introduction to Section 3.

Look-up tables

1. The process is definitely being performed

If answers to the questions for this process agree with any of the combinations in Look-up Table 1, then the AA is definitely performing the process of provision of micro-sites for mammals. * Indicates where the answer to questions on the controlling variable can be either YES or NO. Answers are given in order according to both the certainty of occurrence of the process (first part of the code, e.g. '**1**/1') and the degree to which it occurs (second part of the code, e.g. '1/**1**'), based on the strength of evidence resulting from the combination of controlling variables: for code **1/1** the process is performed at the highest rate.

2. The process is not (significantly) being performed

If answers to the questions for this process agree with any of the combinations in Look-up Table 2 below, then the AA is not significantly performing the process of micro-sites for mammals.

3. The process is being performed, but there are constraining factors or uncertainties

For any other combinations of answers not given in Look-up Tables 1 and 2, the AA is performing the process of provision of micro-sites for mammals, but only to a small degree because of constraining factors or uncertainty which can be identified from the answers given. Assign assessment code **1/x** to the AA.

3.3.A.2.5 Provision of Micro-sites for Mammals Appraisal Sheet

HGMU code: ______________	**AA name:** ______________

CV1 Protected and/or Red List status			
Q1	Y	N	U/NA
Q2	Y	N	U/NA
Q3	Y	N	U/NA
Q4	Y	N	U/NA

CV2 Migratory and/or frequentation records			
Q1	Y	N	U/NA
Q2	Y	N	U/NA
Q3	Y	N	U/NA
Q4	Y	N	U/NA

CV3 Habitat requirements			
Q1	Y	N	U/NA
Q2	Y	N	U/NA
Q3	Y	N	U/NA
Q4	Y	N	U/NA
Q5	Y	N	U/NA
Q6	Y	N	U/NA
Q7	Y	N	U/NA
Q8	Y	N	U/NA
Q9	Y	N	U/NA
Q10	Y	N	U/NA
Q11	Y	N	U/NA
Q12	Y	N	U/NA
Q13	Y	N	U/NA
Q14	Y	N	U/NA
Q15	Y	N	U/NA
Q16	Y	N	U/NA
Q17	Y	N	U/NA
Q18	Y	N	U/NA
Q19	Y	N	U/NA
Q20	Y	N	U/NA

HGMU process assessment code: ______________________________

3.3.A.2.5 Look-up Table 1

CV1	CV2	CV3	Rationale	Code
1	*		The AA is of international importance with regards to mammals. It is an offence to destroy this habitat.	**1/1**
2	*		The AA is of national importance with regards to mammals.	**1/1**
3|4	(1|2|3|4)	4 or more of (1–19) 20	The AA is of conservation importance, and is already protected or considered worthy of protection. There are records of mammals frequenting the AA. The AA has high potential to support mammals, and does not suffer from human interference.	**1/1**
3|4	(1|2|3|4)	4 or more of (1–19)	The AA is of conservation importance, and is already protected or considered worthy of protection. There are records of mammals frequenting the AA. The AA has potential to support mammals, but suffers from human interference.	**1/2**
3|4	(1|2|3|4)	2 or 3 of (1–19) 20	The AA is of conservation importance, and is already protected or considered worthy of protection. There are records of mammals frequenting the AA. The AA has potential to support mammals, and does not suffer from human interference.	**1/2**
3|4	*	2 or 3 of (1–19) 20	The AA is of conservation importance, and is already protected or considered worthy of protection. There are no clear records of mammals frequenting the AA. The AA has high potential to support mammals, and does not suffer from human interference.	**1/2**
3|4	(1|2|3|4)	20	The AA is of conservation importance, and is already protected or considered worthy of protection. There are records of mammals frequenting the AA. The AA does not suffer from human interference.	**1/2**
3|4	(1|2|3|4)	2 or 3 of (1–19)	The AA is of conservation importance, and is already protected or considered worthy of protection. There are records of mammals frequenting the AA. The AA has low potential to support mammals, and suffers from human interference.	**1/3**
3|4		2 or more of (1–19)	The AA is of conservation importance, and is already protected or considered worthy of protection. There are no clear records of mammals frequenting the AA. The AA has potential to support mammals, but suffers from human interference.	**1/3**
# (1 2 3 4)	*	4 or more of (1–19) 20	There are records of mammals frequenting the AA. The AA has the potential to support mammals, and does not suffer from human interference.	**1/3**

Please ignore blank cellls.

3.3.A.2.5 Look-up Table 2

CV1	CV2	CV3	Rationale	Code
#(1 2 3 4)	#(1|2|3|4)	#(1–19)	The AA is not of conservation importance with regards to mammals. The site is not protected or considered worthy of protection. There is no evidence of mammals frequenting the site. The site does not have the potential to support mammals.	**2**
#(1 2 3 4)	#(1|2|3|4)	#20	The AA is not of conservation importance with regards to mammals. There is no evidence of mammals frequenting the AA. A high level of human disturbance and pollution dominates the AA.	**2**

4. The process is definitely not being performed
This statement is not applicable for this process.

This completes the functional assessment for the process of provision of micro-sites for mammals. This process comprises sub-function 2 of the function of ecosystem maintenance. The output codes and quantification values can be recorded on the Functional Assessment Table at the end of section 3.3.A.4. The next process forms sub-fuction 3 of the function.

3.3.A.3 *Provision of plant and habitat diversity (process)*

Further reading and references

Alvarez-Cobelas, M., Cirujano, S. and Sánchez-Carrillo, S. (2001). Hydrological and botanical man-made changes in the Spanish wetland of Las Tablas de Daimiel, *Biological Conservation* **97**(1), 89–98.

Barrat-Segretain, M-H. (2001). Biomass allocation in three macrophyte species in relation to the disturbance level of their habitat, *Freshwater Biology* **46**(7), 935–945.

Blom, C.W.P.M. and Voesenek, L.A.C.J. (1996). Flooding: the survival strategies of plants, *Trends in Ecology & Evolution* **11**(7), 290–295.

Braendle, R. and Crawford, R.M.M. (1987). *Rhizome Anoxia Tolerance and Habitat Specialization in Wetland Plants*, Special publication of the British Ecological Society, London, UK.

Bragg, O.M. (2002). Hydrology of peat-forming wetlands in Scotland, *The Science of The Total Environment* **294**(1–3), 111–129.

Brix, H., Sorrell, B.K. and Lorenzen, B. (2001). Are *Phragmites*-dominated wetlands a net source or net sink of greenhouse gases? *Aquatic Botany* **69**(2–4), 313–324.

Burgess, N.D., Evans, C.E. and Thomas, G.J. (1990). Vegetation change on the Ouse Washes Wetland, England, 1972-88 and effects on their conservation importance, *Biological Conservation* **53**(3), 173–189.

Convention on the Conservation of Migratory Species of Wild Animals (Bonn Convention); Council Decision 82/461/EEC, entry into force 24.06.1982 (www.cms.int)

Convention on the Conservation of European Wildlife and Natural Habitats (Bern Convention), Council Decision 82/72/EEC, entry into force 03/12/1981 (http://europa.eu/scadplus/leg/lvb/128050.htm)

Copp, G.H. (1991). Typology of aquatic habitats in the Great Ouse, a small regulated lowland river, *Regulated Rivers Research & Management* **6**(2), 125–134.

Crivelli, A.J., Grillas, P. and Lacaze, B. (1995). Responses of vegetation to a rise in water level at Kerkini Reservoir (1982–1991), a Ramsar site in northern Greece, *Environmental Management* **19**(3), 417–430.

EC (1979). Council Directive 79/409/EEC of 2 April 1979 on the conservation of wild birds, *OJ* **L103**, 25.4.1979, 1.

EC (1992). Council Directive 92/43/EEC of 21 May 1992 on the conservation of natural habitats and of wild fauna and flora, *OJ* **L206**, 22.07.1992, 7–50.

Gerdol, R. and Tomaselli, M. (1991). The vegetation of wetlands in the northern Apennines (Italy), *Phytocoenologia* **21**(4), 421–469.

Girel, J. and Manneville, O. (1998). Present species richness of plant communities in alpine stream corridors in relation to historical river management, *Biological Conservation* **85**(1–2), 21–33.

Hroudová, Z., Krahulcová, A., Zákravsky, P. and Jarolímová, V. (1996). The biology of *Butomus umbellatus* in shallow waters with fluctuating water level, *Hydrobiologia* **340**(1–3), 27–30.

IUCN (2001). *IUCN Red List Categories and Criteria: Version 3.1*, IUCN Species Survival Commission, IUCN, Gland, Switzerland and Cambridge, UK.

Keddy, P.A. (2000). *Wetland Ecology: Principles and Conservation*, Cambridge University Press, Cambridge, UK.

Kennedy, M.P. and Murphy, K.J. (2003). Hydrological and hydrochemical conditions characterising *Carex chordorrhiza L. fil.* (String Sedge) habitat in a Scottish riverine floodplain wetland, *Aquatic Botany* **77**(4), 243–255.

Kirkman, L.K., Drew, M.B., West, L.T. and Blood, E.R. (1998). Ecotone characterization between upland longleaf pine/wiregrass stands and seasonally-ponded isolated wetlands, *Wetlands* **18**(3), 346–364.

Lienert, J., Fischer, M. and Diemer, M. (2002). Local extinctions of the wetland specialist *Swertia perennis* L. (*Gentianaceae*) in Switzerland: a revisitation study based on herbarium records, *Biological Conservation* **103**(1), 65–76.

Linz, G.M., Blixt, D.C., Bergman, D.L. and Bleier, W.J. (1996). Response of ducks to glyphosate-induced habitat alterations in wetlands, *Wetlands* **16**(1), 38–44.

Mander, U., Kuusemets, V. and Ivask, M. (1995). Nutrient dynamics of riparian ecotones: A case study from the Porijõgi River catchment, Estonia, *Landscape and Urban Planning* **31**(1–3), 333–348.

Mountford, J.O. and Chapman, J.M. (1993). Water regime requirements of British wetland vegetation: using the moisture classification of Ellenberg and Londo, *Journal of Environmental Management* **38**(4), 275–288.

Murphy, K.J., Hudson, K.D. and Mitchell, J. (1994). Freshwater and wetland plant communities of Loch Lomond, *Hydrobiologia* **290**(1–3), 63–74.

Nicolet, P., Biggs, J., Fox, G., Hodson, M.J., Reynolds, C., Whitfield, M. and Williams, P. (2004). The wetland plant and macroinvertebrate assemblages of temporary ponds in England and Wales, *Biological Conservation* **120**(2), 261–278.

Petts, G.E., Bickerton, M.A., Crawford, C., Lerner, D.N. and Evans, D. (1999). Flow management to sustain groundwater-dominated stream ecosystems, *Hydrological Processes* **13**(3), 497–513.

Putten, W.H.V.D. (1997). Die-back of *Phragmites australis* in European wetlands: an overview of the European Research Programme on Reed Die-back and Progression (1993–1994), *Aquatic Botany* **59**(3–4), 263–275.

Söderström, B., Svensson, B., Vessby, K. and Glimskär, A. (2001). Plants, insects and birds in semi-natural pastures in relation to local habitat and landscape factors, *Biodiversity and Conservation* **10**(11), 1839–1863.

United Nations Educational, Scientific and Cultural Organization (UNESCO), Convention on Wetlands of International Importance especially as Waterfowl Habitat, Ramsar, Iran, 2.2.1971, (www.ramsar.org)

van der Valk, A. (2006). *The Biology of Freshwater Wetlands*, Oxford University Press, Oxford, UK.

Wheeler, B.D. and Shaw, S.C. Above-ground crop mass and species richness of the principal types of herbaceous rich-fen vegetation of lowland England and Wales, *Journal of Ecology* **79**(2), 285–301.

Introduction

Process description:

- The provision of conditions, which support a variety of specialised wetland plants. Figure 3.3.8 gives examples of the relative abundance of plant species in four different wetland sites.
- The location of the wetland in relation to other types of area is important as these areas may have important biological and physicochemical interactions.

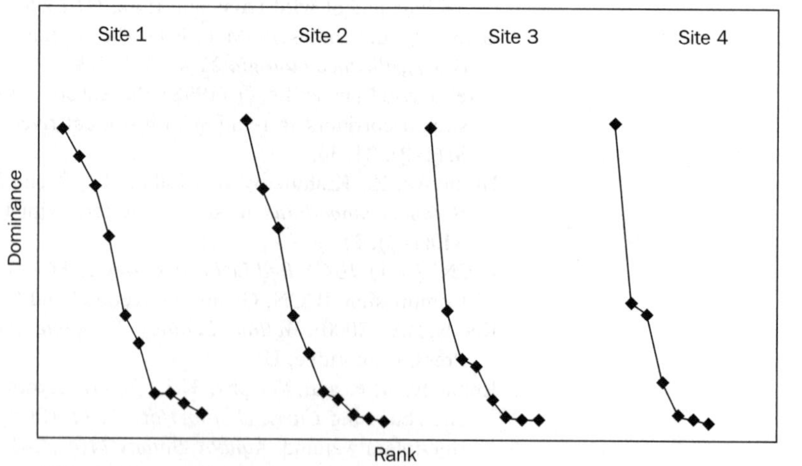

3.3.8 Dominance diversity curves for herbaceous species present in transects of four wetland sites. The 'dominance' axis shows relative abundance. Sites 1 and 2 contain several common species, while species in sites 3 and 4 show strong dominance by few species and a high proportion of rarer species (unpublished data).

Potential significance – rationale for assessment:

- Due to their rarity, certain species are important with regard to biodiversity.
- A diverse flora is likely to be of conservation value.
- A large wetland, well connected to other similar sites, is likely to support larger populations of organisms.
- Diverse habitats are likely to support a wide range of animal species.

Controlling variables

CV1 Protected and/or Red List status
CV2 Vegetation type
CV3 Natural or semi-natural habitat diversity and extent

Answer the questions on *each* of the controlling variables for the whole AA. Unless otherwise instructed, continue from one question to the next. Circle the correct answers on the appraisal sheet. Use a new appraisal sheet for each HGMU. The location of the relevant information is given in brackets at the end of each question.

CV1 *Protected plant species and/or Red List status*

Q1. Have recent surveys of the AA (i.e. within the last 15 years) described the presence of protected plant species of international importance? *(Recording Sheet 1, C.3.17.3 (answer f))*
(Protected Species in the Bern Convention (appendix I), the annex II of the European Habitat directive or the Red List of threatened species database (UNEP-WCMC))
[If YES, move on to the Look-up table]

Q2. Have recent surveys of the AA (i.e. within the last 15 years) described the presence of protected plant species of national or sub-national (local) importance? *(Recording Sheet 1, C.3.17.4 (answer f), C.3.17.5 (answer f))*
(Protected plant species listed through national legislation or Red Lists of threatened species)
[If YES, move on to the Look-up table]

CV2 *Vegetation type*

Q1. Does the AA contain any areas given one of the following vegetation codes? *(Recording Sheet 3, D)*
A2, A4, A5, A7
G3, G6, G7, G11, G13
M1, M2, M3, M4, M5, M7, M10, M11, M12
S2, S4,
U18,
W4, W5, W7, W9, W11

Q2. Does the AA contain any areas given one of the following vegetation codes? *(Recording Sheet 3, D)*
G5, G9, G12, G16, G17, G18, G23, G26, G27, G28
S3, S6, S7, S8, S9, S11, S12, S12, S18
U4, U5, U6, U7, U8, U11, U12, U13, U15, U16
W6, W10, W12, W13, W14, W15, W16, W18, W20, W21, W23, W24, W25, W26, W28, W29, W30, W31, W32, W33, W34, W36, W37, W38, W41, W42, W43, W44

Q3. Are any areas within the AA formally protected or considered worthy of protection, to maintain their international conservation value according to the Ramsar Convention or EU Habitats Directive? *(Recording Sheet 1, C.3.16.1)*

Q4. Are any areas within the AA formally protected or considered worthy of protection, to maintain their international conservation value according to the EU Bird Directive? *(Recording Sheet 1, C.3.16.1)*

3.3.A.3 Provision of Plant and Habitat Diversity Appraisal Sheet

HGMU code: ______________________	**AA name:** ______________________

CV1 Protected and/or Red List status			
Q1	Y	N	U/NA
Q2	Y	N	U/NA

CV3 Natural or semi-natural habitat diversity and extent			
Q1	a	b	c
Q2	a	b	c
Q3	a		b
Q4	Y	N	U/NA

CV2 Vegetation type			
Q1	Y	N	U/NA
Q2	Y	N	U/NA
Q3	Y	N	U/NA
Q4	Y	N	U/NA
Q5	Y	N	U/NA

HGMU process assessment code: ____

Q5. Are any areas within the AA formally protected or considered worthy of protection, to maintain their conservation value according to national or sub-national (local) legislation? *(Recording Sheet 1, C.3.16.2, C.3.16.3)*

CV3 *Natural or semi-natural habitat diversity and extent*

Q1. How many different habitat/vegetation codes were assessed within the AA? *(Recording Sheet 3, C.4 and C.5; AA Map 2)*

a. More than 15
b. 5 to 15
c. Less than 5

Q2. Is the area of the largest tract of natural or semi-natural vegetation within the AA: *(Recording Sheet 3, G; AA Map 2)*

a. Greater than 50 ha
b. 5–50 ha
c. Less than 5 ha

Q3. Is the natural or semi-natural vegetation within the AA: *(CA Map 1)*

a. Continuous with or in close proximity (< 10 km) to similar areas outside the AA
b. Distant (10 km or more) from similar areas outside the assessment area?

Q4. Is the AA surrounded by areas largely free from intensive agriculture, development or buildings (vegetation codes other than **G8**, **G32**, **G34**, **U1**)? *(CA Map 1)*

Assessment of the process

Compare the results recorded for the AA on the appraisal sheet with the possible outcomes below, and assign the appropriate assessment code to the overall AA across all the HGMUs. Record this on the appraisal sheet. The syntax codes used are described in the introduction to Section 3.

Look-up tables

1. The process is definitely being performed
If answers to the questions for this process agree with any of the combinations in Look-up Table 1, then the AA is definitely performing the process of provision of plant and habitat diversity. * Indicates where the answer to questions on the controlling variable can be either YES or NO. Answers are given in order according to both the certainty of occurrence of the process (first part of the code, e.g. '**1**/1') and the degree to which it occurs (second part of the code, e.g. '1/**1**'), based on the strength of evidence resulting from the combination of controlling variables: for Code **1/1** the process is performed at the highest rate.

2. The process is not (significantly) being performed
If answers to the questions for this process agree with any of the combinations in Look-up Table 2, then the AA is not significantly performing the process of provision of plant and habitat diversity.

3. The process is being performed, but there are constraining factors or uncertainty
For any other combinations of answers not given in Look-up Tables 1 and 2, the AA is performing the process of provision of plant and habitat diversity, but only to a small degree because of constraining factors or uncertainty, which can be identified from the answers given. Assign assessment code **1/x** to the AA.

3.3.A.3 Look-up Table 1

<table>
<tr><th>CV1</th><th>CV2</th><th>CV3</th><th>Rationale</th><th>Code</th></tr>
<tr><td>1|2</td><td>*</td><td>*</td><td>The AA is of international, national or local importance with regards to plant species. It is an offence to destroy this habitat.</td><td>1/1</td></tr>
<tr><td>*</td><td>1|3</td><td>*</td><td>The AA is of conservation importance, and is already protected or considered worthy of protection. The AA probably contains an area of a wetland habitat type described by the Habitats Directive or the Ramsar Convention as of Community Interest, and therefore is likely to be of conservation importance.</td><td>1/1</td></tr>
<tr><td>*</td><td>2</td><td>*</td><td>The AA probably contains an area of a non-wetland habitat type described by the Habitat Directive as of Community Interest, and therefore is likely to be of conservation importance.</td><td>1/1</td></tr>
<tr><td>*</td><td>4</td><td>*</td><td>The AA is of international conservation importance, and is already protected or considered worthy of protection. The AA probably contains a habitat described by the EU Bird Directive as of Community Interest, and therefore is possibly of conservation importance.</td><td>1/1</td></tr>
<tr><td>*</td><td>5</td><td>*</td><td>The AA is of national or local conservation importance, and is already protected or considered worthy of protection. The AA contains an area of a vegetation type described by the national or local legislation as of conservation importance.</td><td>1/1</td></tr>
<tr><td>*</td><td>*</td><td>1a</td><td>The AA presents a high diversity in terms of habitats or vegetation types.</td><td>1/1</td></tr>
<tr><td>*</td><td>*</td><td>1b 2a</td><td>The AA presents medium-diversified habitats or vegetation types. The natural or semi-natural vegetation is of a very large area, and therefore likely to support a high diversity of organisms.</td><td>1/2</td></tr>
<tr><td>*</td><td>*</td><td>1b 2b 3a 4</td><td>The AA presents medium-diversified habitats or vegetation types. The natural or semi-natural vegetation is of a medium-sized area. The AA forms part of a larger tract of habitat.</td><td>1/2</td></tr>
<tr><td>*</td><td>*</td><td>1b 2b 3a</td><td>The AA presents medium-diversified habitats or vegetation types. The natural or semi-natural vegetation is of a medium-sized area. The AA forms part of a larger tract of habitat but surrounded by areas of intensive agriculture, development or buildings.</td><td>1/3</td></tr>
<tr><td>*</td><td>*</td><td>1c 2a-b</td><td>The AA presents a low diversity in terms of habitats or vegetation types. The natural or semi-natural vegetation is of a relatively large area.</td><td>1/3</td></tr>
<tr><td>*</td><td>*</td><td>1c 2b</td><td>The AA presents a low diversity in terms of habitats or vegetation types.</td><td>1/3</td></tr>
</table>

3.3.A.3 Look-up Table 2

CV1	CV2	CV3	Rationale	Code
#1#2	#(1 2 3 4 5)	#(1a 1b 2a 2b 3a 3b)	The AA is not of conservation importance with regards to plant species. The site is not protected or considered worthy of protection. The site is small and isolated.	**2**

4. The process is definitely not being performed
This statement is not applicable for this process.

This completes the functional assessment for the process of provision of plant and habitat diversity. This process comprises sub-function 3 of the function of Ecosystem maintenance. The output codes and quantification values can be recorded on the Functional Assessment Table at the end of Section 3.3.A.4.

3.3.A.4 *Functional assessment: Ecosystem maintenance*

The purpose of this section is to determine whether the Ecosystem maintenance function is occurring and, if so, to what degree. Follow the method below for *each* HGMU. In the Functional Assessment Table at the end of this section, record the functional outcome for each HGMU within your AA.

Step 1: Separate the function into sub-functions
There are seven processes linked to three sub-functions within the main ecological function of Ecosystem maintenance.
These sub-functions are:

- provision of habitat diversity (one process)
- provision of faunal diversity (five processes)
- provision of floral diversity (one process)

Sub-functions	Processes
1 Provision of habitat diversity	Provision of overall habitat structural diversity
2 Provision of faunal diversity	Provision of micro-sites for macroinvertebrates Provision of micro-sites for fish Provision of micro-sites for herptiles Provision of micro-sites for birds Provision of micro-sites for mammals
3 Provision of floral diversity	Provision of plant and habitat diversity

Step 2: Derive a score for each process
Functional assessment of the ecological function of Ecosystem maintenance is based on qualitative indicators. For each process, outputs are scaled from 0 to 1 to give a homogenous and semi-quantitative evaluation, as shown in the table below.

Process output code	Process score for functional evaluation	Functional performance
1/1	1	Definitely performed
1/2	0.75	Definitely performed
1/3	0.5	Definitely performed
1/x	0.25	Performed but with constraining factors or uncertainties
2	0	Not (significantly) performed

Step 3: Convert the scores from processes to sub-functions
For sub-functions 1 and 3, scores are those derived from the single supporting process. For sub-function 2, if one of the five processes has a code **1/1** then the sub-function is given the maximum score of 1. If no process is coded with this maximum score, then a mean value is taken of all five process scores (e.g. scores of 0.5, 0.5, 0, 0.75 and 0.25 lead to a score of 0.4 for the sub-function).

Functional Assessment Table: 3.3.A Ecosystem maintenance

HGMU code	Ecosystem maintenance		
	Sub-function habitat diversity	Sub-function faunal diversity	Sub-function floral diversity

AA name: ______________________

Step 4: Functional assessment

To assess the performance of the overall 'ecosystem maintenance' function, add the individual scores obtained for each sub-function, and compare the total with the scale in the table below, which also indicates the functional performance for each range of scores. For example, if the total score is 0.4, then the function would be performed with constraining factors or uncertainties. Note that if any of the sub-functions scores is 1, then the overall 'ecosystem maintenance' function is definitely being performed.

Functional score	Functional performance
1	Definitely performed
0.75–< 1	Definitely performed
0.5–< 0.75	Definitely performed
0.25–< 0.5	Performed but with constraining factors or uncertainties
< 0.25	Not (significantly) performed

This completes the functional analysis for ecosystem maintenance. The output codes and quantification values can be recorded on the Functional Analysis Summary Table in Section 4.0, and used to compare functions among HGMUs across the wetland. If you wish to assess all functions, move on to the next step of the FAPs: Food-web Support.

When all the HGMUs in the AA have been assessed, a general pattern of the provision of ecosystem maintenance across the whole wetland will be evident. By manipulation of your answers to the controlling variable (CV) questions, you have the capacity to explore the probable outcomes of any management options under consideration and relate them to specific functions. The electronic version of the FAPs can do this more quickly and easily.

3.3.B Food-web support (function)

3.3.B.1 Productivity/biomass production (process)

Further reading and references

Barrat-Segretain, M-H. (2001). Biomass allocation in three macrophyte species in relation to the disturbance level of their habitat, *Freshwater Biology* **46**(7), 935–945.

Bernadez, F.G., Rey Benayas, J.M. and Martinez, A. (1993). Ecological impact of groundwater extraction on wetlands (Douro Basin, Spain), *Journal of Hydrology* **141**(1–4), 219–238.

Breber, P. (1993). The misguided policy of draining estuarine wetlands for agriculture: the case of Goro, Po Delta, Italy, *Landscape and Urban Planning* **27**(1), 1–5.

Ennabili, A., Ater, M. and Radoux, M. (1998). Biomass production and NPK retention in macrophytes from wetlands of the Tingitan Peninsula, *Aquatic Botany* **62**(1), 45–56.

Greulich S. and Bornette G. (2003). Being evergreen in an aquatic habitat with attenuated seasonal contrasts – a major competitive advantage? *Plant Ecology* **167**(1), 9–18.

Guesewell, S. and Edwards, P. (1999). Shading by *Phragmites australis*: a threat for species-rich fen meadows? *Applied Vegetation Science* **2**, 61–70.

Guesewell, S., Zorzi, A. and Gigon, A. (2000). Mowing in early summer as a remedy to eutrophication in Swiss fen meadows: are really more nutrients removed? *Bulletin of the Geobotanical Institute ETH* **66**, 11–24.

Hofmann, M. and Isselstein, J. (2005). Species enrichment in an agriculturally improved grassland and its effects on botanical composition, yield and forage quality, *Grass & Forage Science* **60**(2), 136–145.

Kang, H., Freeman, C. and Ashendon, T.W. (2001). Effects of elevated CO_2 on fen peat biogeochemistry, *The Science of The Total Environment* **279**(1–3), 45–50.

Keddy, P.A (2000). *Wetland Ecology: Principles and Conservation*, Cambridge University Press, Cambridge, UK.

Liira J. and Zobel K. (2000). The species richness–biomass relationship in herbaceous plant communities: what difference does the incorporation of root biomass data make? *Oikos* **91**(1), 109–114.

Mitsch, W.J. and Gosselink, J.G. (2007). *Wetlands*, 4th edn, Wiley, New York.
Moreau, S., Bosseno, R., Gu, X.F. and Baret, F. (2003). Assessing the biomass dynamics of Andean bofedal and totora high-protein wetland grasses from NOAA/AVHRR, *Remote Sensing of Environment* **85**(4), 516–529.
Njøs, A. (1994). Future land utilization and management for sustainable crop production, *Soil and Tillage Research* **30**(2–4), 345–357.
Olff, H. and Bakker, J.P. (1991). Long-term dynamics of standing crop and species composition after the cessation of fertilizer application to mown grassland, *Journal of Applied Ecology* **28**(3), 1040–1052.
Pauli, D., Peintinger, M. and Schmid, B. (2002). Nutrient enrichment in calcareous fens: effects on plant species and community structure, *Basic and Applied Ecology* **3**(3), 255–266.
Reeder, B.C. and Binion, B.M. (2001). Comparison of methods to assess water column primary production in wetlands, *Ecological Engineering* **17**(4), 445–449.
Silvius, M.J., Oneka, M. and Verhagen, A. (2000). Wetlands: lifeline for people at the edge, *Physics and Chemistry of the Earth, Part B: Hydrology, Oceans and Atmosphere* **25**(7–8), 645–652.
Stockey, A. and Hunt, R. (1994). Predicting secondary succession in wetland mesocosms on the basis of autecological information on seeds and seedlings, *Journal of Applied Ecology* **31**(3), 543–559.
Summerfield, R.J. (1973). The growth and productivity of *Narthecium Ossifragum* on British mires, *Journal of Ecology* **61**(3), 717–727.
van der Valk, A. (2006). *The Biology of Freshwater Wetlands*, Oxford University Press, Oxford, UK.
Wheeler, B.D. and Shaw, S.C. (1991). Above-ground crop mass and species richness of the principal types of herbaceous rich-fen vegetation of lowland England and Wales, *Journal of Ecology* **79**(2), 285–301.

Introduction

Process description:

- The production of plant biomass from light and chemicals; different plants do this at different rates; rates vary according to wetland and other environmental conditions.
- The biomass produced is available for harvesting or for animals to feed on.

Table 3.3.6 shows the comparison of primary productivity in some major wetland types in relation to N availability.

Significance – rationale for assessment:

- High rates of productivity can support larger populations of animals.
- High productivity can store significant quantities of nutrients, contaminants and carbon, at least temporarily.

Controlling variables

CV1 Vegetation type
CV2 Degree of disturbance or stress

Table 3.3.6 Approximate primary productivities and their relationships to nitrogen dynamics in major freshwater wetland types. Only a small proportion of sediment N is available for plant growth (data from Mitsch and Gosselink, 2007)

Wetland type	Net primary production (kg N ha^{-1} a^{-1})	Total biomass (kg ha^{-1})	Soil N store (kg N ha^{-1} a^{-1})	N loading (kg N ha^{-1} a^{-1})	Plant N uptake (kg N ha^{-1} a^{1})
Bog	5 600	530 000	5 000	8	90
Inland fresh marsh	19 800	460 000	16 000	220	480
Tidal fresh marsh	13 700	460 000	13 400	750	540
Swamp forest	8 700	520 000	13 000	9 000	140
Riparian forest	10 400	370 000	9 000	100 000	170

Answer all the questions below for *each* HGMU; unless otherwise instructed continue from one question to the next. On the appraisal sheet circle the correct answer. The location of the relevant information is given in brackets after each question. Use a new appraisal sheet for each HGMU.

CV1 *Vegetation type*

Note: Vegetation categories a to c in ***Q1*** and ***Q2*** are considered to be at their optimal or maximum density with little or no litter accumulation.

Q1. Has the vegetation been assigned one of the following grassland (**Gx**) codes? *(Recording Sheet 3, D)*

a. **G4, G11**
b. **G7–10, G17**
c. **G6, G12–13, G16, G22, G27**
d. **G1, G3, G5, G18, G20–21, G23**
e. **G2, G14, G15, G19, G30–32**

[If YES go to ***CV2****]*

Q2. Has the vegetation been assigned one of the following mire (**Mx**) codes? *(Recording Sheet 3, D)*

a. **M13**
b. **M9–10, M14–15**
c. **M4, M6-8, M11**
d. **M1–3, M5, M12**

[If YES go to ***CV2****]*

Q3. Has the vegetation been assigned one of the following aquatic (**Ax**) codes? *(Recording Sheet 3, D)*

a. **A7**
b. **A1, A4, A6, A9–10**
c. **A3**
d. **A2, A8**

[If YES go to ***CV2****]*

Q4. Has the vegetation been assigned one of the scrub (**Sx**) or woodland (**Wx**) codes? *(Recording Sheet 3, D)*

[If YES go to ***CV2****]*

Q5. Has the vegetation been assigned an unvegetated or sparsely vegetated lands (**Ux**) codes? *(Recording Sheet 3, D)*

CV2 *Degree of disturbance or stress*

Q1. What is the duration of waterlogging or flooding? *(Recording Sheet 3, B.4.22 (or Recording Sheet 1, C.3.19.6 for flooding information))*

a. One month or less
b. More than one month to 6 months, until March–April
c. More than one month to 6 months, until May or later
d. More than 6 months

Q2. Is the HGMU subject to harvesting of herbaceous vegetation at least once a year? *(Recording Sheet 1, C.3.4.1b (answer c); AA Map 2)*

Q3. Is any livestock kept in the HGMU, once or more than once a year, at densities greater than the following? *(Recording Sheet 1, C.3.7.1b (answer c)); AA Map 2)*

- Cattle or horses or donkeys 1 ha^{-1}
- Pigs 3 ha^{-1}
- Sheep or goats 10 ha^{-1}
- Turkeys or geese 30 ha^{-1}
- Chickens or ducks 100 ha^{-1}

Q4. Is agricultural fertiliser (nitrogen or phosphate) applied, at least once a year, to the HGMU *(Recording Sheet 1, C.3.2.1b (answer c))* or to upslope land linked hydrologically with the HGMU through surface run-off? *(Recording Sheet 1, C.3.2.1b (answer c))*

Q5. Are there any heavy recreational uses or current activities such as mining, extraction and construction occurring that strongly disturb the vegetation or surface of the HGMU? *(Recording Sheet 1, C.3.11.1b (answer b); AA Map 2, and/or Recording Sheet 1, C.3.13.1a; AA Map 2, and/or Recording Sheet 1, C.3.15.1a (answer c); AA Map 2)*

Assessment of the process

Compare the results recorded for each HGMU on the appraisal sheet with the possible outcomes below, and assign an appropriate assessment code. Record this on the appraisal sheet. The syntax codes used are described in the introduction to Section 3.

Look-up tables

1. The process is definitely being performed
If answers to the questions for this process agree with any of the combinations in Look-up Table 1, then the HGMU is definitely performing the process of biomass production and an estimated amount of productivity in dry mass (DM) per year is given as follows (10^3 kg DM ha^{-1} a^{-1}):

For wet grasslands:
- > 12 for **1/a** code
- 6–16 for **1/b** code
- 2–8 for **1/c** code
- < 4 for **1/d** code

For fens, swamps and aquatics:
- > 16 for **1/a** code
- 8–24 for **1/b** code
- 4–12 for **1/c** code
- < 6 for **1/d** code

* Indicates where answers to questions of the controlling variable can be either YES or NO. Answers are given in order according to both the certainty of occurrence of the process (first part of the code, e.g. '**1**/a') and the range of quantification assigned (second part of the code, e.g. '1/**a**'), based on the strength of evidence resulting from the combination of controlling variables: for code **1/a** the process is performed at the highest rate.

2. The process is not (significantly) being performed
If answers to the questions for this process agree with any of the combinations in Look-up Table 2, then the HGMU is definitely not performing the process of biomass production. * Indicates where the answers to questions of the controlling variable can be either YES or NO.

3. The process is being performed, but there are constraining factors or uncertainty
The constraining factors or uncertainty could be as, follows:

- The vegetation type had been recognised, but quantification is difficult because of spatial heterogeneity or temporal variability (***CV1***: **1e**, **3d**)
- The vegetation type was not recognised at the current state of structure or composition (***CV1***: ***Q1–3***, all answers U/NA)
- The vegetation structure (***CV1***: **4**) is complex (e.g. more than one vegetation layer). If the undergrowth vegetation (under-storey) is recognised as grassland (**Gx**) or mire (**Mx**) vegetation type, then productivity may be assessed according to the specific (**Gx/Mx**) procedure.

For any other combinations of answers, either YES or NO, the HGMU is performing the process, but only to a small degree. Assign assessment code **1/x** to these HGMUs.

4. The process is definitely not being performed
This statement is not applicable for this process.

This completes the functional assessment for the process of biomass production. The output codes and quantification values can be recorded on the Functional Assessment Table at the end of Section 3.3.B.6.

3.3.B.1 *Productivity Appraisal Sheet*

HGMU code: ______________________	**AA name:** ______________________

CV1 Vegetation type						
Q1	a	b	c	d	e	U/NA
Q2	a	b	c	d	U/NA	
Q3	a	b	c	d	U/NA	
Q4	Y	N	U/NA			
Q5	Y	N	U/NA			

CV2 Degree of disturbance or stress				
Q1	a	b	c	d
Q2	**Y**	N	U/NA	
Q3	**Y**	N	U/NA	
Q4	**Y**	N	U/NA	
Q5	**Y**	N	U/NA	

HGMU process assessment code: ____

3.3.B.1 Look-up Table 1

<table>
<tr><th>CV1</th><th>CV2</th><th>Rationale</th><th>Code</th><th>Quantification</th></tr>
<tr><td>1a</td><td>1a 2 3 4|1a 3|
(1a|1b|1c)|
(1b|1c) 3 4|
1c 2|1c 4</td><td>Wet grassland with very high productivity is present. Flooding or waterlogging duration is less than 6 months. Exploitation (grazing or nutrient addition) is likely to maintain an optimal productivity level. The HGMU is subjected to a low degree of disturbance from recreational activities, mining or construction.</td><td>1/a</td><td rowspan="2">> 12×10^3 kg DM ha^{-1} a^{-1}</td></tr>
<tr><td>1a|1b</td><td>1a 3 4|
1a 2| (1a|1b)
2* 4| 1b 2 3 4|
1c 2 4</td><td>Wet grassland with very high or high productivity is present. Soil is flooded or waterlogged for less than 6 months. Exploitation (mowing, grazing or nutrient addition) is likely to increase productivity. The HGMU is subjected to a low degree of disturbance from recreational activities, mining or construction.</td><td>1/a</td></tr>
<tr><td>1a</td><td>1a 2 3|1b 2* 3|
1c 2 3 4*|1c 3</td><td>Wet grassland with very high productivity is present. Soil is flooded or waterlogged for less than 6 months. Exploitation (mowing or grazing, trampling by livestock) is likely to decrease productivity. The HGMU is subjected to a low degree of disturbance from recreational activities, mining or construction.</td><td>1/b</td><td rowspan="3">6–16 $\times 10^3$ kg DM ha^{-1} a^{-1}</td></tr>
<tr><td>1b</td><td>1a 2 3 4|
1a 3*| (1b/1c)
3 4| (1a|1b|1c)
|1c 4| 1c 2*</td><td>Wet grassland with high productivity is present. Flooding or waterlogging duration is less than 6 months. Exploitation (grazing or nutrient addition) is likely to maintain an optimal productivity level. The HGMU is subjected to a low degree of disturbance from recreational activities, mining or construction.</td><td>1/b</td></tr>
<tr><td>1c</td><td>1a 3* 4| 1a 2 4*|
1b 2 3 4| 1b 2* 4|
1b 2|1c 2 4</td><td>Wet grassland with low productivity is present. Soil is flooded or waterlogged for less than 6 months. Exploitation (mowing, grazing or nutrient addition) is likely to increase productivity. The HGMU is subjected to a low degree of disturbance from recreational activities, mining or construction.</td><td>1/b</td></tr>
<tr><td>1b</td><td>1a 2 3| 1b 2* 3|
1c 2 3 4*| 1c 3</td><td>Wet grassland with high productivity is present. Soil is flooded or waterlogged for less than 6 months. Exploitation (mowing or grazing, trampling by livestock) is likely to decrease productivity. The HGMU is subjected to a low degree of disturbance from recreational activities, mining or construction.</td><td>1/c</td><td rowspan="3">2–8 $\times 10^3$ kg DM ha^{-1} a^{-1}</td></tr>
<tr><td>1c</td><td>1c 2*| (1a|1b|1c)</td><td>Wet grassland with low productivity is present. Flooding or waterlogging duration is less than 6 months. Exploitation (grazing or nutrient addition) is likely to maintain the productivity level. The HGMU is subjected to a low degree of disturbance from recreational activities, mining or construction.</td><td>1/c</td></tr>
<tr><td>1d</td><td>1a 3* 4| 1a 2 4*|
1b 2 3 4| 1b 4|
1b 2 4*| 1c 2 4</td><td>Wet grassland with very low productivity level is present. Soil is flooded or waterlogged for less than 6 months. Exploitation (mowing, grazing or nutrient addition) is likely to increase productivity. The HGMU is subjected to a low degree of disturbance from recreational activities, mining or construction.</td><td>1/c</td></tr>
<tr><td>1d</td><td>1a 2 3 4| 1a 3*|
1b 3 4|
(1a|1b|1c)|
1c 3* 4| 1c 2*</td><td>Wet grassland with very low productivity is present. Flooding or waterlogging duration is less than 6 months. Exploitation (grazing or nutrient addition) is likely to maintain the productivity level. The HGMU is subjected to a low degree of disturbance from recreational activities, mining or construction.</td><td>1/d</td><td rowspan="2">< 4×10^3 kg DM ha^{-1} a^{-1}</td></tr>
<tr><td>1c|1d</td><td>1a 2 3| 1b 2* 3|
1c 2 3 4*| 1c 3</td><td>Wet grassland with low or very low productivity is present. Soil is flooded or waterlogged for less than 6 months. Mowing, grazing and/or nutrient addition are likely to maintain or decrease productivity. The HGMU is subjected to a low degree of disturbance from recreational activities, mining or construction.</td><td>1/d</td></tr>
</table>

2a|2b	1d 2 4	Reedbeds or fens with very high or high productivity are present. Soil is flooded or waterlogged for more than 6 months. Mowing and/or nutrient addition are likely to maintain or increase productivity level. The HGMU is subjected to a low degree of disturbance from recreational activities, mining or construction.	**1/a**	> 16×10^3 kg DM ha^{-1} a^{-1}
2a	1d 2*	Reedbeds with very high or high productivity are present. Soil is flooded or waterlogged for more than 6 months. The HGMU is subjected to a low degree of disturbance from recreational activities, mining or construction.	**1/a**	
2a|2b|3a	1d 4	Reedbeds, fens or aquatic vegetation with high productivity are present. Flooding or waterlogging duration is more than 6 months. Nutrient addition is likely to increase the productivity level. The HGMU is subjected to a low degree of disturbance from recreational activities, mining or construction.	**1/a**	
2b	1d 2	Fens with high productivity are present. Soil is flooded or waterlogged for more than 6 months. Harvesting of herbaceous vegetation is likely to maintain productivity level. The HGMU is subjected to a low degree of disturbance.	**1/b**	8–24 × 10^3 kg DM ha^{-1} a^{-1}
2c	1d 2 4	Fens or swamps with very high or high productivity are present. Soil is flooded or waterlogged for more than 6 months. Mowing and/or nutrient addition are likely to increase productivity. The HGMU is subjected to a low degree of disturbance from recreational activities, mining or construction.	**1/b**	
2b|3a	1d	Fens or aquatic vegetation with high productivity are present. Flooding or waterlogging duration is more than 6 months. The HGMU is subjected to a low degree of disturbance.	**1/b**	
2c|3b	1d 4	Fens, swamps or aquatic vegetation are present. Flooding or waterlogging duration is more than 6 months. Nutrients are added to the HGMU. The HGMU is subjected to a low degree of disturbance.	**1/b**	
2c	1d 2	Fens or swamps with low productivity are present. Soil is subjected to flooding or waterlogging for more than 6 months. Harvesting of herbaceous vegetation is likely to maintain productivity level. The HGMU is subjected to a low degree of disturbance.	**1/c**	4–12 × 10^3 kg DM ha^{-1} a^{-1}
2d	1d 2 4	Bogs or fens with very low productivity are present. Soil is flooded or waterlogged for more than 6 months. Mowing and/or nutrient addition are likely to increase productivity. The HGMU is subjected to a low degree of disturbance.	**1/c**	
2c|3b	1d	Fens, swamps or aquatic vegetation are present. Flooding or waterlogging duration is more than 6 months. The HGMU is subjected to a low degree of disturbance.	**1/c**	
2d|3c	1d 4	Bogs, fens or aquatic vegetation are present. Flooding or waterlogging duration is more than 6 months. Nutrients are added to the HGMU. The HGMU is subjected to a low degree of disturbance.	**1/c**	
2d	1d 2	Bogs or fens with very low productivity are present. Soil is subjected to flooding or waterlogging for more than 6 months. Harvesting of herbaceous vegetation is likely to maintain the productivity level. The HGMU is subjected to a low degree of disturbance.	**1/d**	< 6×10^3 kg DM ha^{-1} a^{-1}
2d|3c	1d	Bogs, fens or aquatic vegetation are present. Flooding or waterlogging duration is more than 6 months. The HGMU is subjected to a low degree of disturbance.	**1/d**	

3.3.B.1 Look-up Table 2

CV1	CV2	Rationale	Code
5	1* 4* 5*	There is little or no vegetation present.	2
*	1* 2* 3* 4* 5	The HGMU suffers from a high degree of disturbance from recreational activities, mining or construction.	2

3.3.B.2 Biomass import via physical processes (process)

3.3.B.2.1 *Biomass import via watercourses (process)*

Further reading and references

Åström, M., Aaltonen, E-K. and Koivusaari, J. (2001). Impact of ditching in a small forested catchment on concentrations of suspended material, organic carbon, hydrogen ions and metals in stream water, *Aquatic Geochemistry* **7**(1), 57–73.

Balla, D. and Kalettka, T. (2005). Sulphate retention capacity of the German wetland region Spreewald. In: *Integrated Land and Water Resources Management: Towards Sustainable Rural Development*, 15–19 May, Frankfurt (Oder), Germany and Slubice, Poland, Proceedings 21st European Regional Conference of the International Commission on Irrigation and Drainage, available at: http://www.zalf.de/icid/ICID_ERC2005/HTML/ERC2005PDF/Topic_1/Balla.pdf

Blankenberg, A-G.B., Braskerud, B. and Haarstad, K. (2006). Pesticide retention in two small constructed wetlands: treating non-point source pollution from agriculture runoff, *International Journal of Environmental & Analytical Chemistry* **86**(3–4), 225–231.

Čížková-Končalová, H., Květ, J. and Lukavská, J. (1996). Response of *Phragmites australis*, *Glyceria maxima*, and *Typha latifolia* to additions of piggery sewage in a flooded sand culture, *Wetlands Ecology and Management* **4**(1), 43–50.

Engelhardt, C., Krüger, A., Karrasch, B. and Baborowski, M. (1999). Input–output balances of nutrients and plankton in a flooded area of the lower Odra, *Acta Hydrochimica et Hydrobiologica* **27**(5), 325–330.

Hardej, M. and Ozimek, T. (2002). The effect of sewage sludge flooding on growth and morphometric parameters of *Phragmites australis* (Cav.) Trin. ex Steudel, *Ecological Engineering* **18**(3), 343–350.

Heim, S., Schwarzbauer, J., Kronimus, A., Littke, R. and Hembrock-Heger, A. (2003). Organic pollutants in riparian wetlands of the Lippe river (Germany), *Environmental Chemistry Letters* **1**(3), 169–173.

Keddy, P.A. (2000). *Wetland Ecology: Principles and Conservation*, Cambridge University Press, Cambridge, UK.

Kimura, M., Murase, J. and Lu, Y. (2004). Carbon cycling in rice field ecosystems in the context of input, decomposition and translocation of organic materials and the fates of their end products (CO_2 and CH_4), *Soil Biology and Biochemistry* **28**, 1545–1547.

Loiselle, S., Bracchini, L., Bonechi, C. and Rossi, C. (2001). Modelling energy fluxes in remote wetland ecosystems with the help of remote sensing, *Ecological Modelling* **145**(2–3), 243–261.

Lung, W.S. and Light, R.N. (1996). Modelling copper removal in wetland ecosystems, *Ecological Modelling* **93**(1–3), 89–100.

Mitsch, W.J. and Gosselink, J.G. (2007). *Wetlands*, 4th edn Wiley, New York.

Moreau, S., Bosseno, R., Gu, X.F. and Baret, F. (2003). Assessing the biomass dynamics of Andean bofedal and totora high-protein wetland grasses from NOAA/AVHRR, *Remote Sensing of Environment* **85**(4), 516–529.

Pomogyi, P. (1993). Nutrient retention by the Kis-Balaton Water Protection System. *Hydrobiologia* **251**(1–3), 309–320.

Puchalski, W. (2003). The importance of accumulated organic matter in nutrient transport and storage in floodplains, *Towards Natural Flood Reduction Strategies*, Warsaw, 6–13 Sept, available at: http://levis.sggw.waw.pl/ecoflood/contents/articles/S1/html/1_15L.pdf

Small, L.F., McIntire, C.D., MacDonald, K.B., Lara-Lara, J.B., Frey, B.E., Amspoker, M.C. and Winfield, T. (1990). Primary production, plant and detrital biomass, and particle transport in the Columbia River Estuary, *Progress In Oceanography* **25**(1–4), 175–210.

Styczen, M. and Storm, B. (1993). Modelling of N-movements on catchment scale – a tool for analysis and decision making, *Nutrient Cycling in Agroecosystems* **36**(1), 7–17.

Sutula, M.A., Perez, B.C., Reyes, E., Childers, D.E., Davis, S., Day J.W. Jr, Rudnick, D. and Sklar, F. (2003). Factors affecting spatial and temporal variability in material exchange between the Southern Everglades wetlands and Florida Bay (USA), *Estuarine, Coastal and Shelf Science* **57**(5–6), 757–781.

Ulbrich, K., Marsula, R., Jeltsch, F., Hofmann, H. and Wissel, C. (1997). Modelling the ecological impact of contaminated river sediments on wetlands, *Ecological Modelling* **94**(2–3), 221–230.

Wassen, M.J. (1995). Hydrology, water chemistry and nutrient accumulation in the Biebrza fens and floodplains (Poland), *Wetlands Ecology and Management* **3**(2), 125–137.

Xiong, S., Nilsson, C., Johansson, M.E. and Jansson, R. (2001). Responses of riparian plants to accumulation of silt and plant litter: the importance of plant traits, *Journal of Vegetation Science* **12**, 481–490.

Introduction

Process description:

- The transport of biomass into the wetland by water transport processes. The biomass may then be available to support the wetland food web.
- Transport of biomass can occur during normal flow conditions if there is a direct removal from the river channel, but can also occur during flood events when river waters leave the main channel and flow over the floodplain.

Potential significance – rationale for assessment: Areas outside the wetland may be contributing biomass to the biological processes that occur in the wetland.

Controlling variables

CV1 Landform
CV2 Susceptibility to flooding
CV3 Evidence of biomass movement

Answer all the questions on *each* of the controlling variables for all HGMUs. Unless otherwise instructed continue from one question to the next. Undertake the assessment also for B.2.2, B.2.3 and B.3. On the recording sheet circle the correct answer. The location of the relevant information is given in brackets after each question. Use a new appraisal sheet for each HGMU.

CV1 *Landform*

Q1. Does the HGMU share a boundary with the main water body? *(AA Map 3)*

CV2 *Susceptibility to flooding*

Q1. Is the HGMU within an area of floodplain known to flood? *(AA Map 2)*
Q2. Is there any evidence of recent deposition of sediment, sand, gravel or organic material at the surface of the HGMU? *(Recording Sheet 3, B.5.4, B.5.5)*
Q3. Does the HGMU receive overbank flooding? *(Recording Sheet 2a, F4; Recording Sheet 3, B.4.4)*
Q4. Has a strand-line of flood debris been identified within the HGMU, presumably derived from a nearby main water body, which is normally at or below the HGMU level? *(Refer to the neighbouring HGMUs: Recording Sheet 3, B.4.9; AA Map 3)*

If the answer to all the questions in ***CV1*** and ***CV2*** is NO assign a score of **2** to the HGMU and go to Section 3.3.B.2.2 Biomass import via overland flow.

CV3 *Evidence of biomass movement*

Q1. Is there any flood debris deposited in the HGMU? *(Recording Sheet 3, B.4.9)*
Q2. Is there evidence that the litter has originated from a different vegetation type than that present in the HGMU? *(Recording Sheet 3, C.3)*

Assessment of the process

Compare the results recorded for each HGMU on the appraisal sheet with the possible outcomes below, and assign an appropriate assessment code. Record this on the appraisal sheet. The syntax codes used are described in the introduction to Section 3.

3.3.B.2.1 Biomass Import via Watercourses Appraisal Sheet

HGMU code: ____________________	**AA name:** ____________________

CV1 Landform			
Q1	Y	N	U/NA

CV2 Susceptibility to flooding			
Q1	Y	N	U/NA
Q2	Y	N	U/NA
Q3	Y	N	U/NA
Q4	Y	N	U/NA

CV3 Evidence of biomass movement			
Q1	Y	N	U/NA
Q2	Y	N	U/NA

HGMU process assessment code: ______________________________

3.3.B.2.1 Look-up Table 1

CV1	CV2	CV3	Rationale	Code
1	2 or more of (1–4)	1 2	The HGMU is in a position where direct removal of biomass from the watercourse is likely. There is evidence that water flows pass over the HGMU. There is evidence of biomass movement via watercourses.	**1/1**
1	2 or more of (1–4)	1	The HGMU is in a position where direct removal of biomass from the watercourse is likely. There is evidence that water flows pass over the HGMU. There is evidence of biomass movement via watercourses.	**1/2**
1	2 or more of (1–4)	2	The HGMU is in a position where direct removal of biomass from the watercourse is likely. There is evidence that water flows pass over the HGMU. There is evidence of biomass movement.	**1/2**
1	*	1 2	The HGMU is in a position where direct removal of biomass from the watercourse is likely. There is evidence of biomass movement via water courses.	**1/2**
1	*	1	The HGMU is in a position where direct removal of biomass from the watercourse is likely. There is evidence of biomass movement via water courses.	**1/2**
*	2 or more of (1–4)	1	There is evidence that water flows pass over the HGMU. There is evidence of biomass movement via watercourses.	**1/2**
*	2 or more of (1–4)	2	There is evidence that water flows pass over the HGMU. There is evidence of biomass movement.	**1/3**
1		2	The HGMU is in a position where direct removal of biomass from the watercourse is likely. There is evidence of biomass movement.	**1/3**

Note: Please ignore blank cells.

3.3.B.2.1 Look-up Table 2

CV1	CV2	CV3	Rationale	Code
#1	#(1 2 3 4)	*	The HGMU is not in a position where direct removal of biomass from the watercourse is likely. There is no evidence that water flows pass over the HGMU.	**2**
#1	#(1 3 4)	#(1 2)	The HGMU is not in a position where direct removal of biomass from the watercourse is likely. There is no evidence that water flows pass over the HGMU. There is no evidence of biomass movement via watercourses.	**2**

Look-up tables

1. The process is definitely being performed

If answers to the questions for this process agree with any of the combinations in Look-up Table 1, then the HGMU is definitely performing the process of biomass import via watercourses. Answers are given in order according to both the certainty of occurrence of the process (first part of the code, e.g. '**1**/1') and the degree to which it is being performed (second part of the code, e.g. '1/**1**'), based on the strength of evidence resulting from the combination of controlling variables: for code **1/1** the process is performed at the highest rate.

2. The process is not (significantly) being performed

If answers to the questions for this process agree with any of the combinations in Look-up Table 2, then the HGMU is not significantly performing the process of biomass import via watercourses. * Indicates where the answer to questions on the controlling variable can be either YES or NO.

3. The process is being performed, but there are constraining factors or uncertainty

For any other combinations of answers, either YES or NO, the HGMU is performing the process of biomass import via watercourses, but only to a small degree because of constraining factors or uncertainty which can be identified from the answers given. Assign assessment code **1/x** to these HGMUs.

4. The process is definitely not being performed

This statement is not applicable for this process.

This completes the functional assessment for the process of biomass import via water courses. The output codes and quantification values can be recorded on the Functional Assessment Table at the end of Section 3.3.B.6.

3.3.B.2.2 *Biomass import via overland flow (process)*

Further reading and references

Andersson, L., Arheimer, B., Larsson, M., Olsson, J., Pers, C., Rosberg, J., Tonderski, K. and Ulén, B. (2003). Integrated modelling of phosphorus fluxes at the catchment scale, *Diffuse Pollution Conference*, Dublin, Eire, 17–22 Aug. (CD-ROM).

Collins, R. (2003). Relationships between streamwater E.coli concentrations and environmental factors in New Zealand, *Diffuse Pollution Conference*, Dublin, Eire, 17–22 Aug. (CD-ROM).

Engelhardt, C., Krüger, A., Karrasch, B. and Baborowski, M. (1999). Input-output balances of nutrients and plankton in a flooded area of the lower Odra, *Acta Hydrochimica et Hydrobiologica* **27**(5), 325–330.

Gessner, M.O., Schieferstein, B., Müller, U., Barkmann, S. and Lenfers, U.A. (1996). A partial budget of primary organic carbon flows in the littoral zone of a hardwater lake, *Aquatic Botany* **55**(2), 93–105.

Hammer, D.E. and Kadlec, R.H. (1986). Model for wetland surface water dynamics, *Water Resources Research* **22**(13), 1951–1958.

Hardej, M. and Ozimek, T. (2002). The effect of sewage sludge flooding on growth and morphometric parameters of *Phragmites australis* (Cav.) Trin. ex Steudel, *Ecological Engineering* **18**(3), 343–350.

Hengeveld, H. and De Vocht, C. (1982). Role of water in urban ecology, *Urban Ecology* **6**(1–4), 5–347.

Iritz, L., Johansson, B. and Lundin, L. (1994). Impacts of forest drainage on floods, *Hydrological Sciences Journal* **39**(6), 637–662.

Lidén, R. (1999). A new approach for estimation suspended sediment yield, *Hydrology and Earth System Sciences* **3**, 285–294.

Keddy, P.A. (2000). *Wetland Ecology: Principles and Conservation*, Cambridge University Press, Cambridge, UK.

Matgen, P., El Idrissi, A., Henry, J.-B., Tholey, N., Hoffmann, L., de Fraipont, P. and Pfister, L. (2005). Patterns of remotely sensed floodplain saturation and its use in runoff predictions, *Hydrological Processes* **20**(8), 1805–1825.

Mitsch, W.J. and Gosselink, J.G. (2007). *Wetlands*, 4th edn, Wiley, New York.

Moreau, S., Bosseno, R., Gu, X.F. and Baret, F. (2003). Assessing the biomass dynamics of Andean bofedal and totora high-protein wetland grasses from NOAA/AVHRR, *Remote Sensing of Environment* **85**(4), 516–529.

Reddy, K.R., Flaig, E.G. and Graetz, D.A. (1996). Phosphorus storage capacity of uplands, wetlands and streams of the Lake Okeechobee Watershed, Florida, *Agriculture, Ecosystems & Environment* **59**(3), 203–216.

Scarton, F., Day, J.W. Jr, Rismondo, A., Cecconi, G. and Are, D. (2000). Effects of an intertidal sediment fence on sediment elevation and vegetation distribution in a Venice (Italy) lagoon salt marsh, *Ecological Engineering* **16**(2), 223–233.

Small, L.F., McIntire, C.D., MacDonald, K.B., Lara-Lara, J.B., Frey, B.E., Amspoker, M.C. and Winfield, T. (1990). Primary production, plant and detrital biomass, and particle transport in the Columbia River Estuary, *Progress In Oceanography*, **25**(1–4), 175–210.

Stephan, U., Hengl, M. and Schmid, B.H. (2005). Sediment retention in constructed wetland ponds–a laboratory study, *Journal of Environmental Science and Health, Part A: Toxic/Hazardous Substances & Environmental Engineering* **40**(6–7), 1415–1430.

Tao, S., Lin, B., Liu, X. and Cao, J. (2000). Release kinetics of water soluble organic carbon (WSOC) from river sediment and wetland soil, *Water, Air, and Soil Pollution* **118**, 407–418.

Tockner, K. and Schiemer, F. 1997. Ecological aspects of the restoration strategy for a river-floodplain system on the Danube River in Austria, *Global Ecology and Biogeography Letters* **6**(3/4), 321–329.

Tyrrel, S.F., Leeds-Harrison, P.B. and Harrison, K.S. (2002). Removal of ammoniacal nitrogen from landfill leachate by irrigation onto vegetated treatment planes, *Water Research* **36**(1), 291–299.

Ulen, B., Johansson, G. and Kyllmar, K. (2001). Model predictions and long-term trends in phosphorus transport from arable lands in Sweden, *Agricultural Water Management* **49**(3), 197–210.

Vompersky, S.E. and Sirin, A.A. (1988). Runoff formation and regime in drained peatland forests, *Proceedings of the International Symposium on the Hydrology of Wetlands in Temperate and Cold Regions*, Joensuu, Finland, 6–8 June, Volume 1, The Academy of Finland, Helsinki, Finland.

Walton, R., Chapman, R.S. and Davis, J.E. (1996). Development and application of the Wetlands Dynamic Water Budget Model, *Wetlands*, **16**(3), 347–357.

White, D.L., Porter, D.E. and Lewitus, A.J. (2004). Spatial and temporal analyses of water quality and phytoplankton biomass in an urbanized versus a relatively pristine salt marsh estuary, *Journal of Experimental Marine Biology and Ecology* **298**(2), 255–273.

Introduction

Process description:

- The transport of biomass into the wetland by overland flow transport processes. This biomass may then be available to support the wetland food web.
- These flows arise when water is unable to penetrate the soil, and are most likely to occur during storm events. Biomass imported in such a way is likely to have originated from a nearby, upslope area.

Potential significance – rationale for assessment: Areas outside the wetland may be subsidising the biomass available to support decomposition and biogeochemical processes that occur within the wetland.

Controlling variables

CV1 Surface morphology
CV2 Soil water regime
CV3 Evidence of biomass movement

Answer all the questions below for *each* HGMU; unless otherwise instructed continue from one question to the next. On the recording sheet circle the correct answer. The location of the relevant information is given in brackets after each question. Use a new appraisal sheet for each HGMU.

CV1 *Surface morphology*

Q1. Is there a clear flow path, without any obstacles or barriers, which would impede water flow from the neighbouring area overland into the HGMU? *(AA Map 3; Recording Sheet 3, B.2.1; B.2.4)*
*[If NO assign a score of **2** to the HGMU and go to Section 3.3.B.2.3 Biomass import via wind transport].*

Q2. Has a neighbouring upslope area (inside or outside the AA) a gradient greater than 11°? (*Gradient code: **G3** in the neighbouring HGMUs Recording Sheet 2a, C.2)*

CV2 *Soil water regime*

Note: All questions are referring to upslope area either inside (neighbouring HGMUs) or outside the AA.

Q1. Is the moisture status of the topsoil upslope from the HGMU described as either flooded or wet? *(Recording Sheet 2a, C.3.2)*

Q2. Are there any springs or seepage zones present upslope from the HGMU? *(AA Map 2; Recording Sheet 3, B.3.1, B.3.2)*

Q3. If there is an HGMU upslope in the AA, do field indicators suggest that it is an area of groundwater discharge? *(AA Map 3; Recording Sheet 2a, F1)*

Q4. Is the topsoil (0–10 cm) of the area upslope to the HGMU predominantly composed of fine mineral material (clay)? *(AA Map 3; Recording Sheet 2a, C.3.7)*

Q5. Within the soil profile of HGMUs upslope, is there a distinct horizon of hard indurate material (a hard pan) within 50 cm of the surface that could prevent surface water draining through the soil? *(AA Map 3; Recording Sheet 2a, C.3 General comments)*

CV3 *Evidence of biomass movement*

Q1. Is there evidence that the litter has originated from a different vegetation type than that present in the HGMU, but does not form part of an overbank flood strand line (i.e. not from flooding)? *(Recording Sheet 3, C.3, B.4.9)*

Assessment of the process

Compare the results recorded for each HGMU on the appraisal sheet with the possible outcomes below, and assign an appropriate assessment code. Record this on the appraisal sheet. The syntax codes used are described in the introduction to Section 3.

Look-up tables

1. The process is definitely being performed
If answers to the questions for this process agree with any of the combinations in Look-up Table 1, then the HGMU is definitely performing the process of biomass import via overland flow. Answers are given in order according to both the certainty of occurrence of the process (first part of the code, e.g. '**1**/1') and the degree to which it occurs (second part of the code, e.g. '1/**1**'), based on the strength of evidence resulting from the combination of controlling variables: for code **1/1** the process is performed at the highest rate.

2. The process is not significantly being performed
If answers to the questions for this process agree with any of the combinations in Look-up Table 2, then the HGMU is not significantly performing the process of biomass import via overland flow. * Indicates where the answer to questions on the controlling variable can be either YES or NO.

3.3.B.2.2 *Biomass Import via Overland Flow Appraisal Sheet*

HGMU code: ____________________	**AA name:** ____________________

CV1 Surface morphology			
Q1	Y	N	U/NA
Q2	Y	N	U/NA

	CV2 Soil water regime		
Q1	Y	N	U/NA
Q2	Y	N	U/NA
Q3	Y	N	U/NA
Q4	Y	N	U/NA
Q5	Y	N	U/NA

CV3 Evidence of biomass movement			
Q1	Y	N	U/NA

HGMU process assessment code: ________________________________

3.3.B.2.2 Look-up Table 1

CV1	CV2	CV3	Rationale	Code
1 2	3 or more of (1–5)	1	The HGMU is in a position where overland water flow is highly likely. Surface water flows are likely to carry biomass into the HGMU. There is evidence of biomass movement.	**1/1**
1 2	3 or more of (1–5)	*	The HGMU is in a position where overland water flow is highly likely. Surface water flows are likely to carry biomass into the HGMU.	**1/2**
1	3 or more of (1–5)	1	The HGMU is in a position where overland water flow is likely. Surface water flows are likely to carry biomass into the HGMU. There is evidence of biomass movement.	**1/2**
1	3 or more of (1–5)	*	The HGMU is in a position where overland water flow is likely. Surface water flows are likely to carry biomass into the HGMU.	**1/2**
1	*	1	The HGMU is in a position where overland water flow is likely. There is evidence of biomass movement.	**1/3**
1	*	*	The HGMU is in a position where overland water flow is likely.	**1/3**

3.3.B.2.2 Look-up Table 2

CV1	CV2	CV3	Rationale	Code
#1	*	*	Surface water flows are unlikely in neighbouring upslope areas.	**2**
#2	#(1 2 3 4 5)	#1	Surface water flows are unlikely in neighbouring upslope areas. Surface water flows are not likely to bring biomass into the HGMU.	**2**

3. The process is being performed, but there are constraining factors or uncertainty

For any other combinations of answers, either YES or NO, the HGMU is performing the process of biomass import via overland flow, but only to a small degree because of constraining factors or uncertainty which can be identified from the answers given. Assign assessment code **1/x** to these HGMUs.

4. The process is definitely not being performed

This statement is not applicable for this process

This completes the functional assessment for the process of biomass import via overland flow. The output codes and quantification values can be recorded on the Functional Assessment Table at the end of Section 3.3.B.6.

3.3.B.2.3 *Biomass import via wind transport (process)*

Further reading and references

Armstrong, J. Armstrong, W. and Beckett, P. (1992). *Phragmites australis*: Venturi- and humidity-induced pressure flows enhance rhizome aeration and rhizosphere oxidation, *New Phytologist* **120**(2), 197–207.

Buet, A., Roche, H., Habert, H., Caquet, T. and Ramade, F. (1998). Evaluation of a contamination level by organic micropolluants in fishes of the Camargue protected zone. Proposition of an experimental plan to validate pertinent biomarkers, *Ichtyophysiologica acta* **21**, 61–76.

Ellis, S. (1996). The inadequacy of passive, wind-driven traps in assessing the aerial insects of Delta Marsh, *UFS (Delta Marsh) Annual Report*, Volume 31, available at: http://www.umanitoba.ca/faculties/science/delta_marsh/reports/1996/ellis.pdf

Emeis, K., Christiansen, C., Edelvang, K., Jähmlich, S., Kozuch, J., Laima, M., Leipe, T., Löffler, A., Lund-Hansen, L.C., Miltner, A., Pazdro, K., Pempkowiak, J., Shimmield, G., Shimmield, T., Smith, J., Voss, M. and Witt, G. (2002). Material transport from the near shore to the basinal environment in the southern Baltic Sea: II: Synthesis of data on origin and properties of material, *Journal of Marine Systems* **35**(3–4), 151–168.

Green, M.B. (1996). Experience with establishment and operation of reed bed treatment for small communities in the UK, *Wetlands Ecology and Management* **4**(3), 147–158.

Hodson, M. and Threlkeld, S. (2002). *Local and Regional Influence on Taxonomic and Ecological Composition of Herbaceous Species in Headwater Wetlands*, ESA Annual Meeting, Ecological Society of America, Tucson, AZ, abstract available at: http://abstracts.co.allenpress.com/pweb/esa2002/document/?ID=5575

Jankauskas, B. (2003). Soil erosion monitoring metadata for the modeling of landscape change. In: Müller, F., Kepner, W. and Caesar, K. (eds), *Ecosys: Landscape Sciences for Environmental Assessment*, Proceedings from a CCMS Pilot Study Meeting at Salzau Castle, Germany, (Bd. 10), Verein zur Forderung der Okosystemforschung zu Kiel e.V., 88–97.

Jankauskas, B. and Jankauskiene, G. (2003). Erosion-preventive crop rotations for landscape ecological stability in upland regions of Lithuania, *Agriculture, Ecosystems & Environment* **95**(1), 129–142.

Keddy, P.A. (2000). *Wetland Ecology: Principles and Conservation*, Cambridge University Press, Cambridge, UK.

Loiselle, S., Bracchini, L., Bonechi, C. and Rossi, C. (2001). Modelling energy fluxes in remote wetland ecosystems with the help of remote sensing, *Ecological Modelling* **145**(2–3), 243–261.

Mitsch, W.J. and Gosselink, J.G. (2007). *Wetlands*, 4th edn, Wiley, New York.

Scarton, F., Day, J.W. Jr, Rismondo, A., Cecconi, G. and Are, D. (2000). Effects of an intertidal sediment fence on sediment elevation and vegetation distribution in a Venice (Italy) lagoon salt marsh, *Ecological Engineering* **16**(2), 223–233.

Small, L.F., McIntire, C.D., MacDonald, K.B., Lara-Lara, J.B., Frey, B.E., Amspoker, M.C. and Winfield, T. (1990). Primary production, plant and detrital biomass, and particle transport in the Columbia River Estuary, *Progress In Oceanography* **25**(1–4), 175–210.

Sutula, M.A., Perez, B.C., Reyes, E., Childers, D.E., Davis, S., Day, J.W. Jr, Rudnick, D. and Sklar, F. (2003). Factors affecting spatial and temporal variability in material exchange between the Southern Everglades wetlands and Florida Bay (USA), *Estuarine, Coastal and Shelf Science* **57**(5–6), 757–781.

Introduction

Process description:

- The transfer of biomass into the wetland by wind transport processes. This biomass may then be available to support the wetland food web.
- This is most likely to occur where high winds are disrupted, or where there are sudden changes in wind velocity.

Potential significance – rationale for assessment: Areas outside the wetland may be contributing to the biological processes that occur in the wetland.

Controlling variables

CV1 Exposure
CV2 Vegetation type
CV3 Evidence of biomass movement

Answer all the questions below for *each* HGMU; unless otherwise instructed continue from one question to the next. On the recording sheet circle the correct answer. The location of the relevant information is given in brackets after each question. Use a new appraisal sheet for each HGMU.

CV1 *Exposure*

Q1. What is the topographic exposure of the HGMU? *(Recording Sheet 3, A.4)*
  a. 3 (Sheltered)
  b. 2 (Moderate)
  c. 1 (Exposed)

Q2. What is the average wind speed? *(Recording Sheet 1, C.3.1.4)*
  a. Below 4.8 m s^{-1}
  b. Between 4.8 and 6.6 m s^{-1}
  c. Above 6.6 m s^{-1}

CV2 *Vegetation type*

Q1. Are there any solitary trees, windbreaks (*Recording Sheet 3, C.4 (answer i)*), or hedgerows (***W2*** *Recording Sheet 3, D)* within the HGMU?

Q2. Is the vegetation of the HGMU taller than surrounding vegetation? *(Recording Sheet 3, C.1)*

CV3 *Evidence of biomass movement*

Q1. Is there evidence that the litter has originated from a different vegetation type than that present in the HGMU? *(Recording Sheet 3, C.3)*

Assessment of the process

Compare the results recorded for each HGMU on the appraisal sheet with the possible outcomes below, and assign an appropriate assessment code. Record this on the appraisal sheet. The syntax codes used are described in the introduction to Section 3.

Look-up tables

1. The process is definitely being performed
If answers to the questions for this process agree with any of the combinations in Look-up Table 1, then the HGMU is definitely performing the process of biomass import via wind transport. Answers are given in order according to both the certainty of occurrence of the process (first part of the code, e.g. '**1**/1') and the degree to which it occurs (second part of the code, e.g. '1/**1**'), based on the strength

3.3.B.2.3 Biomass Import via Wind Transport Appraisal Sheet

HGMU code: ______________	**AA name:** ______________

CV1 Exposure			
Q1	a	b	c
Q2	a	b	c

CV2 Vegetation type			
Q1	Y	N	U/NA
Q2	Y	N	U/NA

CV3 Evidence of biomass movement			
Q1	Y	N	U/NA

HGMU process assessment code: ______________________________

3.3.B.2.3 Look-up Table 1

CV1	CV2	CV3	Rationale	Code
1c 2c	1 2	1	The site is exposed to wind. The vegetation is likely to disrupt wind flows. There is evidence of biomass movement.	**1/1**
1c 2c	1|2	1	The site is exposed to wind. The vegetation is likely to disrupt wind flows. There is evidence of biomass movement.	**1/2**
1c 2c	*	1	The site is exposed to wind. There is evidence of biomass movement.	**1/2**
1c	1 2	1	The site is exposed to wind. The vegetation is likely to disrupt wind flows. There is evidence of biomass movement.	**1/2**
1c	1|2	1	The site is exposed to wind. The vegetation is likely to disrupt wind flows. There is evidence of biomass movement.	**1/3**
1c	*	1	The site is exposed to wind. There is evidence of biomass movement.	**1/3**

3.3.B.2.3 Look-up Table 2

CV1	CV2	CV3	Rationale	Code
#(1c 1b 2c 2b)	#1 #2	#1	The site is sheltered from wind. The vegetation is unlikely to disrupt wind flows. There is no evidence of biomass movement.	**2**
#(1c 1b 2c 2b)	#1	#1	The site is sheltered from wind. The vegetation is unlikely to disrupt wind flows. There is no evidence of biomass movement.	**2**
#(1c 1b 2c 2b)	#2	#1	The site is sheltered from wind. The vegetation is unlikely to disrupt wind flows. There is no evidence of biomass movement.	**2**
#(1c 1b 2c 2b)	*	#1	The site is sheltered from wind. There is no evidence of biomass movement.	**2**

of evidence resulting from the combination of controlling variables: for code **1/1** the process is performed at the highest rate.

2. The process is not significantly being performed

If answers to the questions for this process agree with any of the combinations in Look-up Table 2, then the HGMU is not significantly performing the process of biomass import via wind transport. * Indicates where the answer to questions on the controlling variable can be either YES or NO.

3. The process is being performed, but there are constraining factors or uncertainty

For any other combinations of answers, either YES or NO, the HGMU is performing the process of biomass import via wind transport, but only to a small degree because of constraining factors or uncertainty which can be identified from the answers given. Assign assessment code **1/x** to these HGMUs.

4. The process is definitely not being performed

This statement is not applicable for this process.

This completes the functional assessment for the process of biomass import via wind transport. The output codes and quantification values can be recorded on the Functional Assessment Table at the end of Section 3.3.B.6.

3.3.B.3 Biomass import via biological processes (process)

3.3.B.3.1 Biomass import via animal movements (process)

Further reading and references

Batty, L.C. (2003). Wetland plants – more than just a pretty face? *Land Contamination and Reclamation* **11**(2), 173–180.

Boros, E., Bánfi, S. and Forró, L. (2006). Anostracans and microcrustaceans as potential food sources of waterbirds on sodic pans of the Hungarian plain, *Hydrobiologia* **567**(1), 341–349.

Dwyer, T.J. (1975). Time budget of breeding gadwalls, *The Wilson Bulletin* **18**(3), 335–344.

Fox, A.D. and Bell, M.C. (1994). Breeding bird communities and environmental variable correlates of Scottish peatland wetlands, *Hydrobiologia* **279–280**(1), 297–307.

Fox, A.D., Glahder, C.M. and Walsh, A.J. (2003). Spring migration routes and timing of Greenland white-fronted geese – results from satellite telemetry, *Oikos* **103**(2), 415–425.

Gessner, M.O., Schieferstein, B., Müller, U., Barkmann, S. and Lenfers, U.A. (1996). A partial budget of primary organic carbon flows in the littoral zone of a hardwater lake, *Aquatic Botany* **55**(2), 93–105.

Grellmann, D. (2002). Plant responses to fertilization and exclusion of grazers on an arctic tundra heath, *Oikos* **98**(2), 190–204.

Heath, M.R. (1995). Size spectrum dynamics and the planktonic ecosystem of Loch Linnhe, *ICES Journal of Marine Science* **52**(3–4), 627–642.

Hengeveld, H. and De Vocht, C. (1982). Role of water in urban ecology, *Urban Ecology* **6**(1–4), 5–347.

Holland, M.M., Whigham, D.F. and Gopal, B. (1990). The characteristics of wetland ecotones. In: Naiman, R.J. and Décamps, H. (eds), *The Ecology and Management of Aquatic-Terrestrial Ecotones*, Unesco, Paris, Parthenon Publishing Group, Carnforth, UK, 171–198.

Keddy, P.A. (2000). *Wetland Ecology: Principles and Conservation*, Cambridge University Press, Cambridge, UK.

Lugo, A.E., Brown, S. and Brinson, M.M. (1990). Concepts in wetland ecology. In: *Forested Wetlands. Ecosystems of the World, Volume 15*, Elsevier, New York, 53–85.

Mander, U., Kull, A., Kuusemets, V. and Tamm, T. (2000). Nutrient runoff dynamics in a rural catchment: Influence of land-use changes, climatic fluctuations and ecotechnological measures, *Ecological Engineering* **14**(4), 405–417.

Mitsch, W.J. and Gosselink, J.G. (2007). *Wetlands*, 4th, edn, Wiley, New York.

Nolet, B.A. and Klaassen, M. (2005). Time and energy constraints in demanding phases of the annual cycle: an example of time limitation in refuelling migratory swans, *Oikos* **111**(2), 302–310.

Nyström, B., Van Slooten, K.B., Bérard, A., Grandjean, D., Druart, J.C. and Leboulanger, C. (2002). Toxic effects of Irgarol 1051 on phytoplankton and macrophytes in Lake Geneva, *Water Research* **36**(8), 2020–2028.

Olofson, J. (2006). Short- and long-term effects of changes in reindeer grazing pressure on tundra heath vegetation, *Journal of Ecology* **94**(2), 431–440.

Romare, P. and Hansson, L-A. (2003). A behavioral cascade: top-predator induced behavioral shifts in planktivorous fish and zooplankton, *Limnology and Oceanography* **48**(5), 1956–1964.

Salvado, V., Quintana, X.D. and Hidalgo, M. (2006). Monitoring of nutrients, pesticides, and metals in waters, sediments, and fish of a wetland, *Archives of Environmental Contamination and Toxicology* **51**(3), 377–386.

Simenstad, C., Tanner, C., Crandell, C., White, J. and Cordell, J. (2005). Challenges of habitat restoration in a heavily urbanized estuary: evaluating the investment, *Journal of Coastal Research* **40**, 6–23.

Small, L.F., McIntire, C.D., MacDonald, K.B., Lara-Lara, J.B., Frey, B.E., Amspoker, M.C. and Winfield, T. (1990). Primary production, plant and detrital biomass, and particle transport in the Columbia River Estuary, *Progress In Oceanography* **25**(1–4), 175–210.

Stachowicz, K., Czernoch, M. and Dubiel, E. (1994). Field pond as a sink for nutrients migrating from agrocenoses to freshwaters, *Aquatic Sciences* **56**(4), 363–375.

Van den Brick, F.W.B., Van der Velde, G., Buijse, A.D. and Klink, A.G. (1996). Biodiversity in the lower Rhine and Meuse River-floodplains: its significance for ecological river management, *Netherlands Journal of Aquatic Ecology* **30**(2–3), 129–149.

Vermeer, H.J.G. (1986). The effect of nutrient addition and lowering of the water table on shoot biomass and species composition of a wet grassland community (Cirsio-Molinietum) Siss. et de Vries, 1942, *Acta Oecologica* **7**(2), 145–155.

Wetzel, R.G. (1992). Wetlands as metabolic gates, *Great Lakes Research* **18**(4), 529–532.

Zhao, Y.Q., Sun, G. and Allen, S.J. (2004). Anti-sized reed bed system for animal wastewater treatment: a comparative study, *Water Research* **38**(12), 2907–2917.

Introduction

Process description:

- The transport of biomass produced outside the AA to the AA as a result of the activities of animals, which may then be available for other animals to feed on.
- Migration, feeding, harvesting, defecation and senescence can enhance this process.

Potential significance – rationale for assessment: Areas outside the wetland may be contributing to the biological processes that occur in the wetland.

Answer all the questions below for *each* HGMU; unless otherwise instructed continue from one question to the next. On the recording sheet circle the correct answer. The location of the relevant information is given in brackets after each question. Use a new appraisal sheet for each HGMU.

Controlling variables

CV1 Animal population migration/movement
CV2 Provision of micro-sites

CV1 *Animal population migration/movement*

Q1. Are there any records of macroinvertebrate species, which undergo migrations in large numbers, visiting the AA? *(Recording Sheet 1, C.3.18.1)*

Q2. Are there any records of macroinvertebrate species that move between land and water as they pass through their life cycle? *(Recording Sheet 1, C.3.17.1, C.3.17.2)*

Q3. Are there any records of migratory fish species regularly frequenting the water body or tributaries? *(Recording Sheet 1, C.3.18.2)*

Q4. Are there any records of migratory bird species frequenting the AA? *(Recording Sheet 1, C.3.18.3)*

3.3.B.3 *Biomass Import via Biological Processes Appraisal Sheet*

HGMU code: ____________________	**AA name:** ____________________

CV1 Animal population migration/movement			
Q1	Y	N	U/NA
Q2	Y	N	U/NA
Q3	Y	N	U/NA
Q4	Y	N	U/NA
Q5	Y	N	U/NA
Q6	Y	N	U/NA

CV2 Provision of micro-sites			
Q1a	Y	N	U/NA
Q1b	Y	N	U/NA
Q1c	Y	N	U/NA
Q1d	Y	N	U/NA

HGMU process assessment code: ____

3.3.B.3 Look-up Table 1

<table>
<tr><th>CV1</th><th>CV2</th><th>Rationale</th><th>Code</th></tr>
<tr><td>3 or more of (1–6)</td><td>3 or more of (1a|1b|1c|1d)</td><td>Very large numbers of animals move between the AA and other areas. Environmental conditions provide the potential to support micro-sites for a wide range of animal species. As a result of the movement of animals, significant quantities of biomass will be transported.</td><td>1/1</td></tr>
<tr><td>3 or more of (1–6)</td><td>1 or 2 of (1a|1b|1c|1d)</td><td>Very large numbers of animals move between the AA and other areas. As a result, significant quantities of biomass are likely to be transported.</td><td>1/2</td></tr>
<tr><td>1 or 2 of (1–5)</td><td>3 or more of (1a|1b|1c|1d)</td><td>Large numbers of animals move between the AA and other areas. Environmental conditions provide the potential to support micro-sites for a wide range of animal species. As a result of the movement of animals, significant quantities of biomass will be transported.</td><td>1/2</td></tr>
</table>

3.3.B.3 Look-up Table 2

<table>
<tr><th>CV1</th><th>CV2</th><th>Rationale</th><th>Code</th></tr>
<tr><td>#(1 2 3 4 5 6)</td><td>#(3 or more of (1a|1b|1c|1d)</td><td>There is no evidence of significant animal movement between the AA and other areas. Environmental conditions provide low potential to support micro-sites for a wide range of animal species. Importation of significant quantities of biomass is unlikely.</td><td>2</td></tr>
</table>

Q5. Are there any records of migratory mammal species frequenting the AA? *(Recording Sheet 1, C.3.18.4)*

Q6. Within the AA, are there any dens, holts, lairs or other domiciles of mammals with ranges extending far outside the AA? *(Recording Sheet 3, C.5 (answer y); Recording Sheet 1, C.3.17.1, C.3.17.2)*

CV2 *Provision of micro-sites*

Q1. Has the AA been allocated a code of **1/1 or 1/2** for provision of micro-sites for:

a. Macroinvertebrates *(process 3.3.A.2.1)*
b. Fish *(process 3.3.A.2.2)*
c. Birds *(process 3.3.A.2.4)*
d. Mammals *(process 3.3.A.2.5)*

Assessment of the process

Compare the results recorded for each HGMU on the appraisal sheet with possible outcomes below, and assign an appropriate assessment code. Record this on the appraisal sheet. The syntax codes used are described in the introduction to Section 3.

Look-up tables

1. The process is definitely being performed
If answers to the questions for this process agree with any of the combinations in Look-up Table 1, then the AA is definitely performing the process of biomass import via animal movements. Answers are given in order according to both the certainty of occurrence of the process (first part of the code, e.g. '**1**/1') and the degree to which it occurs (second part of the code, e.g. '1/**1**'), based on the strength of evidence resulting from the combination of controlling variables: for code **1/1** the process is performed at the highest rate.

2. The process is not significantly being performed
If answers to the questions for this process agree with any of the combinations in Look-up Table 2, then the AA is not significantly performing the process of biomass import via biological processes.

3. The process is being performed, but there are constraining factors or uncertainty
For any other combinations of answers, either YES or NO, the AA is performing the process of biomass import via biological processes, but only to a small degree because of constraining factors or uncertainty which can be identified from the answers given. Assign assessment code **1/x** to the AA.

4. The process is definitely not being performed
This statement is not applicable for this process.

This completes the functional assessment for the process of biomass import by animal movements. The output codes and quantification values can be recorded on the Functional Assessment Table at the end of Section 3.3.B.6.

3.3.B.4 Biomass export via physical processes (accumulation of processes)

3.3.B.4.1 Biomass export via watercourses (process)

Further reading references

Åström, M., Aaltonen, E-K. and Koivusaari, J. (2001). Impact of ditching in a small forested catchment on concentrations of suspended material, organic carbon, hydrogen ions and metals in stream water, *Aquatic Geochemistry* **7**(1), 57–73.

Bouchard, V. and Lefeuvre, J.C. (2000). Primary production and macro-detritus dynamics in a European salt marsh: carbon and nitrogen budgets, *Aquatic Botany* **67**(1), 23–42.
Brylinsky, M. and Daborn, G.R. (1987). Community structure and productivity of the Cornwallis Estuary, Minas Basin, *Continental Shelf Research* **7**(11–12), 1417–1420.
Buzzelli, C.P., Childers, D.L., Dong, Q. and Jones, R.D. (2000). Simulation of periphyton phosphorus dynamics in Everglades National Park, *Ecological Modelling* **134**(1), 103–115.
Fisher, J. and Acreman, M.C. (2004). Wetland nutrient removal: a review of the evidence, *Hydrology and Earth System Sciences* **8**(4), 673–685.
Gessner, M.O., Schieferstein, B., Müller, U., Barkmann, S. and Lenfers, U.A. (1996). A partial budget of primary organic carbon flows in the littoral zone of a hardwater lake, *Aquatic Botany* **55**(2), 93–105.
Gupta, M.K., Shrivastava, P. and Singhal, P.K. (1996). Decomposition of young water hyacinth leaves in lake water, *Hydrobiologia* **335**(1), 33–41.
Hein, T., Baranyi, C., Reckendorfer, W. and Schiemer, F. (2004). The impact of surface water exchange on the nutrient and particle dynamics in side-arms along the River Danube, Austria, *Science of The Total Environment* **328**(1–3), 207–218.
Keddy, P.A. (2000). *Wetland Ecology: Principles and Conservation*, Cambridge University Press, Cambridge, UK.
Kowalczewski, A. (1977). Importance of a bordering wetland for chemical properties of lake water, *Internationale Vereinigung fur Theoretische und Angewandte Limnologie*, **20**(4), 2182–2185.
Kufel, I. and Kufel, L. (1989). In situ decomposition of *Phragmites australis* Trin. ex Steudel and *Typha angustifolia* L., *Ekologia Polska* **36**(3–4), 459–470.
Mitsch, W.J. (1992). Landscape design and the role of created, restored, and natural riparian wetlands in controlling nonpoint source pollution, *Ecological Engineering* **1**(1–2), 27–47.
Mitsch, W.J. and Gosselink, J.G. (2007). Wetlands, 4th edn, Wiley, New York.
Mourad, D. and van der Perk, M. (2004). Modelling nutrients fluxes from diffuse and point emissions to river loads: the Estonian part of the transboundary Lake Peipsi/Chudskoe drainage basin (Russia/Estonia/Latvia), *Water Science & Technology* **49**(3), 21–28.
Nieminen, M., Ahti, E., Nousiainen, H., Joensuu, S. and Vuollekoski, M. (2005). Capacity of riparian buffer zones to reduce sediment concentrations in discharge from peatlands drained for forestry, *Silva Fennica* **39**(3), 331–339.
Overbeck, G., Kiehl, K. and Abs, C. (2003). Seedling recruitment of *Succisella inflexa* in fen meadows: importance of seed and microsite availability, *Applied Vegetation Science* **6**(1), 97–104.
Peverly, J.H. (1982). Stream transport of nutrients through a wetland, *Journal of Environmental Quality* **11**(1), 38–43.
Rodrigues, L.C., Train, S., Roberto, M.C. and Pagioro, T.A. (2002). Seasonal fluctuation of some limnological variables on a floodplain lake (Patos lagoon) of the Upper Paraná River, Mato Grosso do Sul State, Brazil, *Brazilian Archives of Biology and Technology* **45**(4), 499–513.
Serrano, L. and Guisande, C. (1990). Effects of polyphenolic compounds on phytoplankton, *Internationale Vereinigung fuer Theoretische und Angewandte Limnologie* **24**(1), 282–288.
Small, L.F., McIntire, C.D., MacDonald, K.B., Lara-Lara, J.B., Frey, B.E., Amspoker, M.C. and Winfield, T. (1990). Primary production, plant and detrital biomass, and particle transport in the Columbia River Estuary, *Progress In Oceanography* **25**(1–4), 175–210.
Tao, S., Lin, B., Liu, X. and Cao, J. (2000). Release kinetics of water soluble organic carbon (WSOC) from river sediment and wetland soil, *Water, Air, and Soil Pollution* **118**, 407–418.
Weisner, S.E.B., Eriksson, P.G., Graneli, W. and Leonardson, L. (1994). Influence of macrophytes on nitrate removal in wetlands, *Ambio* **23**(6), 363–366.

Introduction

Process description:

- The transport of biomass produced within the wetland to areas outside of the wetland as a result of water movement; this may then be available for animals in areas outside the AA to feed on.
- The export of particulate organic matter (POM) can occur during normal flow conditions if there is a direct input to watercourse. It can also occur during

flood events when the stream or river water leaves the main channel and flows over the surrounding soils of the floodplain.
- The export of dissolved organic carbon (DOC) into the streams depends highly on the riparian flow paths, which vary markedly depending on flow rates. The export of DOC is often positively correlated with run-off rates.

Potential significance – rationale for assessment:

- The wetland may be contributing to the biological processes that occur in areas outside the wetland.
- Allochthonous carbon (POC and DOC) are energy sources at the base of the aquatic food web. Downstream lakes with > 8 mg l^{-1} DOC are normally considered to be heterotrophic ecosystems. The availability of POC and DOC has significant ramifications for the character of the entire food web.

Controlling variables

CV1 Landform
CV2 Susceptibility to flooding
CV3 Vegetation type
CV4 Nature of litter layer
CV5 Degree of disturbance

Answer all the questions below for *each* HGMU; unless otherwise instructed continue from one question to the next. On the recording sheet circle the correct answer. The location of the relevant information is given in brackets after each question. Use a new appraisal sheet for each HGMU.

CV1 *Landform*

Q1. Does the HGMU share a boundary with the watercourse or has it been assigned one of the following geomorphic codes? *(AA Map; Recording Sheet 2a, C.2)*
TB, **FC**, **DT**, **LK**, **OX**, **BS**, **CN**
[If NO move on to ***CV2****]*

Q2. If the answer to ***Q1*** was YES, is there a clear flow path without any obstacles or barriers, which would otherwise prevent water draining into the main channel? *(Recording Sheet 3, B.2.2 or B.2.3; AA Map 3)*

CV2 *Susceptibility to flooding*

Q1. Is the HGMU within an area of floodplain known to flood? *(Recording Sheet 3, B.4.1; AA Map 2)*?

Q2. Is there evidence of recent deposition of sediment, sand, gravel, or organic material at the surface of the HGMU? *(Recording Sheet 3, B.5.4 or B.5.5)*

Q3. Does the HGMU receive overbank flooding? *(Recording Sheet 2a, F4; Recording Sheet 3, B.4.4)*

Q4. Has a strand-line of flood debris been identified within the HGMU, presumably derived from a nearby main water body, which is normally at or below the HGMU level? *(Refer to the neighbouring HGMUs: Recording Sheet 3, B.4.9; AA Map 3)*

If the answers to all the questions in ***CV1*** and ***CV2*** are NO then the process is not (significantly) occurring. In this case, assign a score of **2** to the HGMU and go to Section 3.3.B.4.2 Biomass export via overland flow.

CV3 *Vegetation type*

Q1. Has the HGMU been classed as **1/a** or **1/b** for the assessment of productivity? *(process 3.3.B.1)*

Q2. If the habitat is a watercourse, the vegetation is likely to provide biomass for export. Are the vegetation roots below the surface of the water (i.e. could

it be classed as rooted, floating, emergent or submerged)? (vegetation codes **A2–A5**, **A7–A10**, **M8–M11**, **M13**, **M15**, possibly **W4–W8**) *(Recording Sheet 3,D).*
[*If YES go to* ***CV3***]

Q3. If the vegetation is next to a watercourse, it is more likely to provide biomass for export. Does the vegetation extend to the water edge, or overhang the water? *(Recording Sheet 3, C.5 (answer j))*

Q4. Is the vegetation classed as a woodland (code **WX**)? *(Recording Sheet 3, D)*

CV4 *Nature of litter layer*

Q1. Which of the following best describes the litter layer? *(Recording Sheet 3, C.2)*

a. Negligible
b. Herbaceous litter layer
c. Forest debris litter layer (Fig. 3.3.9).
d. Combination litter layer

CV5 *Degree of disturbance*

Q1. Does any part of the HGMU currently experience a high degree of disturbance from the following?

a. Harvesting of herbaceous vegetation *(Recording Sheet 1, C.3.4.1b (answer c))*
b. Ploughing of land *(Recording Sheet 1, C.3.5.1b (answer c))*
c. Woodland management and forestry *(Recording Sheet 1, C.3.6.1b (answer c))*
d. Trampling by livestock *(Recording Sheet 1, C.3.7.1b (answer c))*

Or in the AA:

e. Recreational activities *(Recording Sheet 1, C.3.15.1 (answer c))*

Q2. If the HGMU shares a boundary with a watercourse, is there any evidence of undercutting of the bank or other erosion processes (Fig. 3.3.10) *(Recording Sheet 3, B.5.2)*

3.3.9 Woody debris entering the river channel and available for transport. As well as constituting a source of biomass, woody debris is valuable habitat for aquatic organisms, particularly fish (photo: E. Maltby).

3.3.10 Undercutting of bank and gravel deposition are clear indications of active erosion and deposition (photo: E. Maltby).

Assessment of the process

Compare the results recorded for each HGMU on the appraisal sheet with possible outcomes below, and assign an appropriate assessment code. Record this on the appraisal sheet. The syntax codes used are described in the introduction to Section 3.

Look-up tables

1. The process is definitely being performed
If answers to the questions for this process agree with any of the combinations in Look-up Table 1, then the HGMU is definitely performing the process of biomass export via watercourses. * Indicates where answer to questions on the controlling variable can be either YES or NO. Answers are given in order according to both the certainty of occurrence of the process (first part of the code, e.g. '**1**/1') and the degree to which it occurs (second part of the code, e.g. '1/**1**'), based on the strength of evidence resulting from the combination of controlling variables: for code **1/1** the process is performed at the highest rate.

2. The process is not significantly being performed
If answers to the questions for this process agree with any of the combinations in Look-up Table 2, then the HGMU is not significantly performing the process of biomass export via watercourses. * Indicates where answer to questions on the controlling variable can be either YES, NO or NA.

3. The process is being performed, but there are constraining factors or uncertainties
For any other combinations of answers, either YES or NO, the HGMU is performing the process of biomass export via water courses, but only to a small degree because of constraining factors or uncertainties which can be identified from the answers given. Assign assessment code **1/x** to these HGMUs.

4. The process is definitely not being performed
This statement is not applicable for this process

This completes the functional assessment for the process of biomass export via water courses. The output codes and quantification values can be recorded on the Functional Assessment Table at the end of Section 3.3.B.6.

3.3.B.4.1 Biomass Export via Watercourses Appraisal Sheet

HGMU code: ____________________	**AA name:** ____________________

CV1 Landform			
Q1	Y	N	U/NA
Q2	Y	N	U/NA

CV4 Nature of litter layer				
Q1	a	b	c	d

CV5 Degree of perturbation disturbance			
Q1a	Y	N	U/NA
Q1b	Y	N	U/NA
Q1c	Y	N	U/NA
Q1d	Y	N	U/NA
Q1e	Y	N	U/NA
Q2	Y	N	U/NA

CV2 Susceptibility to flooding			
Q1	Y	N	U/NA
Q2	Y	N	U/NA
Q3	Y	N	U/NA
Q4	Y	N	U/NA

CV3 Vegetation type			
Q1	Y	N	U/NA
Q2	Y	N	U/NA
Q3	Y	N	U/NA
Q4	Y	N	U/NA

HGMU process assessment code: ____

3.3.B.4.1 Look-up Table 1

CV1	CV2	CV3	CV4	CV5	Rationale	Code
1 2	1|2|3|4	1 2	1b|1c|1d	1a|1b|1c|1d|1e|2	The HGMU is in a position where direct inputs of biomass into the watercourse are likely. There is evidence that water flows pass over the HGMU. The vegetation is productive. The vegetation is in a position where inputs into the watercourse are likely. Disturbance of the vegetation is likely to increase inputs of biomass into the watercourse.	**1/1**
1 2	1|2|3|4	1 3	1b|1c|1d	1a|1b|1c|1d|1e|2	The HGMU is in a position where direct inputs of biomass into the watercourse are likely. There is evidence that water flows pass over the HGMU. The vegetation is productive. The vegetation is in a position where inputs into the watercourse are likely. Disturbance of the vegetation is likely to increase inputs of biomass into the watercourse.	**1/1**
1 2	1|2|3|4	3 4	1b|1c|1d	1a|1b|1c|1d|1e|2	The HGMU is in a position where direct inputs of biomass into the watercourse are likely. There is evidence that water flows pass over the HGMU. The vegetation is in a position where inputs into the watercourse are likely. Disturbance of the vegetation is likely to increase inputs of biomass into the watercourse.	**1/1**
1 2	1|2|3|4	1 2	1b|1c|1d	*	The HGMU is in a position where direct inputs of biomass into the watercourse are likely. There is evidence that water flows pass over the HGMU. The vegetation is productive. The vegetation is in a position where inputs into the watercourse are likely.	**1/2**
1 2	1|2|3|4	1 3	1b|1c|1d	*	The HGMU is in a position where direct inputs of biomass into the watercourse are likely. There is evidence that water flows pass over the HGMU. The vegetation is productive. The vegetation is in a position where inputs into the watercourse are likely.	**1/2**
1 2	2|3|4	1	1b|1c|1d	1a|1b|1c|1d|1e|2	The HGMU is in a position where direct inputs of biomass into the watercourse are likely. There is evidence that water flows pass over the HGMU. The vegetation is productive. The vegetation is in a position where inputs into the watercourse are likely. Disturbance of the vegetation is likely to increase inputs of biomass into the watercourse.	**1/2**
1 2	2|3|4	1	1b|1c|1d	*	The HGMU is in a position where direct inputs of biomass into the watercourse are likely. There is evidence that water flows pass over the HGMU. The vegetation is productive. The vegetation is in a position where inputs into the watercourse are likely.	**1/2**
1 2	1|2|3|4	3 4	1b|1c|1d	*	The HGMU is in a position where direct inputs of biomass into the watercourse are likely. There is evidence that water flows pass over the HGMU. The vegetation is in a position where inputs into the watercourse are likely.	**1/2**
1 2	*	1 2	1b|1c|1d	1a|1b|1c|1d|1e|2	The HGMU is in a position where direct inputs of biomass into the watercourse are likely. The vegetation is productive. The vegetation is in a position where inputs into the watercourse are likely. Disturbance of the vegetation is likely to increase inputs of biomass into the watercourse.	**1/2**
1 2	*	1 3	1b|1c|1d	1a|1b|1c|1d|1e|2	The HGMU is in a position where direct inputs of biomass into the watercourse are likely. The vegetation is productive. The vegetation is in a position where inputs into the watercourse are likely. Disturbance of the vegetation is likely to increase inputs of biomass into the watercourse.	**1/2**

1 2	*	1	1b|1c|1d	1a|1b|	The HGMU is in a position where direct inputs of biomass into the watercourse are likely. The vegetation is productive. The vegetation is in a position where inputs into the watercourse are likely. Disturbance of the vegetation is likely to increase inputs of biomass into the watercourse.	**1/2**
1 2	*	3 4	1b|1c|1d	1a|1b|1c|1d|1e|2	The HGMU is in a position where direct inputs of biomass into the watercourse are likely. The vegetation is in a position where inputs into the watercourse are likely. Disturbance of the vegetation is likely to increase inputs of biomass into the watercourse.	**1/2**
1 2	*	1 2	1b|1c|1d	*	The HGMU is in a position where direct inputs of biomass into the watercourse are likely. The vegetation is productive. The vegetation is in a position where inputs into the watercourse are likely.	**1/3**
1 2	*	1 3	1b|1c|1d	*	The HGMU is in a position where direct inputs of biomass into the watercourse are likely. The vegetation is productive. The vegetation is in a position where inputs into the watercourse are likely.	**1/3**
1 2	*	1	1b|1c|1d	*	The HGMU is in a position where direct inputs of biomass into the watercourse are likely. The vegetation is productive. The vegetation is in a position where inputs into the watercourse are likely.	**1/3**
1 2	*	3 4	1b|1c|1d	*	The HGMU is in a position where direct inputs of biomass into the watercourse are likely. The vegetation is in a position where inputs into the watercourse are likely.	**1/3**
*	1|2|3|4	1	1b|1c|1d	1a|1b|1c|1d|1e|2	There is evidence that water flows pass over the HGMU. The vegetation is productive. Disturbance of the vegetation is likely to increase inputs of biomass into the watercourse.	**1/3**
*	1|2|3|4	1	1b|1c|1d	*	There is evidence that water flows pass over the HGMU. The vegetation is productive.	**1/3**

3.3.B.4.1 Look-up Table 2

<table>
<tr><th>CV1</th><th>CV2</th><th>CV3</th><th>CV4</th><th>CV5</th><th>Rationale</th><th>Code</th></tr>
<tr><td>#(1 2)</td><td>#(1 2 3 4)</td><td>#(1 2 3 4)</td><td>#(1b|1c|1d)</td><td>#(1a 1b 1c 1d 1e 2)</td><td>The HGMU is not in a position where direct inputs of biomass into the watercourse are likely. There is no evidence that water flows pass over the HGMU. The vegetation is not productive. The vegetation is not in a position where inputs into the watercourse are likely. There is no disturbance to increase inputs of biomass into the watercourse.</td><td>2</td></tr>
<tr><td>#(1 2)</td><td>#(1 2 3 4)</td><td>#(1 2 3 4)</td><td>#(1b|1c|1d)</td><td>*</td><td>The HGMU is not in a position where direct inputs of biomass into the watercourse are likely. There is no evidence that water flows pass over the HGMU. The vegetation is not productive. The vegetation is not in a position where inputs into the watercourse are likely.</td><td>2</td></tr>
<tr><td>#(1 2)</td><td>#(1 2 3 4)</td><td>*</td><td>*</td><td>*</td><td>The HGMU is not in a position where direct inputs of biomass into the watercourse are likely. There is no evidence that water flows pass over the HGMU.</td><td>2</td></tr>
</table>

3.3.B.4.2 *Biomass export via overland flow (process)*

Further reading and references

Burkert, U., Ginzel, G., Babenzien, H.D. and Koschel, R. (2005). The hydrogeology of a catchment area and an artificially divided dystrophic lake – consequences for the limnology of Lake Fuchskuhle, *Biogeochemistry* **71**(2), 225–246.

Butera, M.K., Browder, J.A. and Frick, A.L. (1984). Preliminary report on the assessment of wetland productive capacity from a remote-sensing-based model–A NASA/NMFS joint research project, *IEEE Transactions on Geoscience and Remote Sensing* **GE-22**(6), 502–511.

Buzzelli, C.P., Childers, D.L., Dong, Q. and Jones, R.D. (2000). Simulation of periphyton phosphorus dynamics in Everglades National Park, *Ecological Modelling* **134**(1), 103–115.

Clair, T.A., Arp, P., Moore, T.R., Dalva, M. and Meng, F.-R. (2002). Gaseous carbon dioxide and methane, as well as dissolved organic carbon losses from a small temperate wetland under a changing climate, *Environmental Pollution* **116**(Suppl 1), S143–S148.

Collier, K.J., Jackson, R.J. and Winterbourn, M.J. (1989). Dissolved organic carbon dynamics of developed and undeveloped wetland catchments in Westland, New Zealand, *Archiv fur Hydrobiologie* **117**(1), 21–38.

Gessner, M.O., Schieferstein, B., Müller, U., Barkmann, S. and Lenfers, U.A. (1996). A partial budget of primary organic carbon flows in the littoral zone of a hardwater lake, *Aquatic Botany* **55**(2), 93–105.

Groffman, P.M., Gold, A.J. and Simmons, R.C. (1992). Nitrate dynamics in riparian forests: microbial studies, *Journal of Environmental Quality*, **21**, 666–671.

Hein, T., Baranyi, C., Reckendorfer, W. and Schiemer, F. (2004). The impact of surface water exchange on the nutrient and particle dynamics in side-arms along the River Danube, Austria, *Science of The Total Environment* **328**(1–3), 207–218.

Hengeveld, H. and de Vocht, C. (eds), (1982). *Role of Water in Urban Ecology* (Developments in Landscape Management and Urban Planning, Vol. 5) (reprinted from *Urban Ecology*, **6**), Elsevier Scientific, Amsterdam, Oxford, New York.

Keddy, P.A. (2000). *Wetland Ecology: Principles and Conservation*, Cambridge University Press, Cambridge, UK.

Mitsch, W.J. and Gosselink, J.G. (2007). *Wetlands*, 4th edn, Wiley, New York.

Mwanuzi, F., Aalderink, H. and Mdamo, L. (2003). Simulation of pollution buffering capacity of wetlands fringing the Lake Victoria, *Environment International* **29**(1), 95–103.

Odum, E.P. (1981). A functional classification of wetlands, *Proceedings, U.S. Fish and Wildlife Service Workshop on Coastal Ecosystems of the Southeastern United States*, Big Pine Key, FL, 18–22 Feb 1980, Fish and Wildlife Service, Office of Biological Services, Report FWS/OBS-80/79.

Post, J., Krysanova, V. and Suckow, F. (2004). *Simulation of Water and Carbon Fluxes in Agro- and Forest Ecosystems at the Regional Scale*, Potsdam Institute for Climate Impact Research, Potsdam, Germany, available at: http://www.pik-potsdam.de/~post/iEMSs_paper.pdf.pdf

Sachse, A., Babenzien, D., Ginzel, G., Gelbrecht, J. and Steinberg, C.E.W. (2001). Characterization of dissolved organic carbon (DOC) in a dystrophic lake and an adjacent fen, *Biogeochemistry* **54**(3), 279–296.

Small, L.F., McIntire, C.D., MacDonald, K.B., Lara-Lara, J.B., Frey, B.E., Amspoker, M.C. and Winfield, T. (1990). Primary production, plant and detrital biomass, and particle transport in the Columbia River Estuary, *Progress In Oceanography* **25**(1–4), 175–210.

van der Peijl, M.J. and Verhoeven, J.T.A. (2000). Carbon, nitrogen and phosphorus cycling in river marginal wetlands; a model examination of landscape geochemical flows, *Biogeochemistry* **50**(1), 45–71.

Walbridge, M.R. and Lockaby, B.G. (1994). Effects of forest management on biogeochemical functions in southern forested wetlands, *Wetlands* **14**(1), 10–17.

Introduction

Process description:

- The transport of biomass produced within the wetland to areas outside the wetland as a result of overland flow; this may then be available for animals in areas outside the AA to feed on.
- These flows arise when water is unable to penetrate the soil, and are most

likely to occur during storm events. Biomass is exported in such a way that it is likely to be deposited in the river channel before final removal from the wetland.

Potential significance – rationale for assessment: The wetland may contribute to the biological processes that occur in areas outside the wetland.

Controlling variables

CV1 Surface morphology
CV2 Soil water regime
CV3 Vegetation type
CV4 Nature of litter layer
CV5 Degree of disturbance

Answer all the questions below for *each* HGMU; unless otherwise instructed continue from one question to the next. On the recording sheet circle the correct answer. The location of the relevant information is given in brackets after each question. Use a new appraisal sheet for each HGMU.

CV1 *Surface morphology*

Q1. Is there a clear flow path (without any obstacles or barriers which would prevent water flowing) into neighbouring HGMUs and subsequently to the water body? *(AA Map 3 Recording Sheet 3, B.2.2, B.2.3)*

If the answer to ***Q1*** is NO assign a score of **2** to the HGMU and go to Section 3.3.B.4.3 Biomass export via wind transport.

Q2. Has the HGMU been classed as having a gradient greater than 11° (gradient code **G3**)? *(Recording Sheet 2a, C.2)*

CV2 *Soil water regime*

Q1. Is the moisture status of the topsoil described as either flooded, or wet? *(Recording Sheet 2a, C.3.2)*
Q2. Are there any springs or seepage zones present within the HGMU? *(Recording Sheet 3, B.3.1, B.3.2)*
Q3. Do the field indicators suggest that there is a high degree of confidence that the HGMU is in an area of ground water discharge? *(Recording Sheet 2a, F1)*
Q4. Is the topsoil (0–10 cm) of the HGMU predominantly composed of fine mineral material or clay? *(Recording Sheet 2a, C.3.7)*
Q5. Within the soil profile, is there a distinct horizon (0–10 cm) of indurated material (such as a hard pan) that could prevent surface water draining through the soil? *(Recording Sheet 2a, C.3 General comments)*

CV3 *Vegetation type*

Q1. Has the HGMU been classed as **1/a** or **1/b** for the assessment of productivity? *(Process 3.3.B.1)*
Q2. Is the vegetation classed as a woodland (coded **W**)? *(Recording Sheet 3, D)*

CV4 *Nature of litter layer*

Q1. Which of the following best describes the litter layer? *(Recording Sheet 3, C.2)*
a. Negligible
b. Herbaceous litter layer
c. Forest debris litter layer
d. Combination litter layer

3.3.B.4.2 Biomass Export via Overland Flow Appraisal Sheet

HGMU code: ______________________	**AA name:** ______________________

CV1 Surface morphology			
Q1	Y	N	U/NA
Q2	Y	N	U/NA

CV2 Soil water regime			
Q1	Y	N	U/NA
Q2	Y	N	U/NA
Q3	Y	N	U/NA
Q4	Y	N	U/NA
Q5	Y	N	U/NA

CV3 Vegetation type			
Q1	Y	N	U/NA
Q2	Y	N	U/NA

CV4 Nature of litter layer				
Q1	a	b	c	d

CV5 Degree of disturbance			
Q1 a	Y	N	U/NA
Q1 b	Y	N	U/NA
Q1 c	Y	N	U/NA
Q1 d	Y	N	U/NA
Q1 e	Y	N	U/NA
Q2	Y	N	U/NA

HGMU process assessment code: ____

3.3.B.4.2 Look-up Table 1

<table>
<tr><th>CV1</th><th>CV2</th><th>CV3</th><th>CV4</th><th>CV5</th><th>Rationale</th><th>Code</th></tr>
<tr><td>1</td><td>3 or more of (1–5)</td><td>1 2</td><td>1b|1c|1d</td><td>1a|1b|1c|1d|1e|2</td><td>Surface water flows are likely within the HGMU. Surface water flows are likely to carry biomass. The vegetation is productive. Disturbance of the vegetation is likely to increase biomass availability for export.</td><td>1/1</td></tr>
<tr><td>1</td><td>3 or more of (1–5)</td><td>1</td><td>1b|1c|1d</td><td>1a|1b|1c|1d|1e|2</td><td>Surface water flows are likely within the HGMU. Surface water flows are likely to carry biomass. The vegetation is productive. Disturbance of the vegetation is likely to increase biomass available for import and export.</td><td>1/2</td></tr>
<tr><td>1</td><td>3 or more of (1–5)</td><td>2</td><td>1b|1c|1d</td><td>1a|1b|1c|1d|1e|2</td><td>Surface water flows are likely within the HGMU. Surface water flows are likely to carry biomass. The vegetation is productive. Disturbance of the vegetation is likely to increase biomass available for import and export.</td><td>1/2</td></tr>
<tr><td>1</td><td>3 or more of (1–5)</td><td>*</td><td>1b|1c|1d</td><td>1a|1b|1c|1d|1e|2</td><td>Surface water flows are likely within the HGMU. Surface water flows are likely to carry biomass. Disturbance of the vegetation is likely to increase biomass available for import and export.</td><td>1/2</td></tr>
<tr><td>1</td><td>*</td><td>1 2</td><td>1b|1c|1d</td><td>1a|1b|1c|1d|1e|2</td><td>Surface water flows are likely within the HGMU. The vegetation is productive. Disturbance of the vegetation is likely to increase biomass available for import and export.</td><td>1/2</td></tr>
<tr><td>1</td><td>*</td><td>1</td><td>1b|1c|1d</td><td>1a|1b|1c|1d|1e|2</td><td>Surface water flows are likely within the HGMU. The vegetation is productive. Disturbance of the vegetation is likely to increase biomass available for import and export.</td><td>1/2</td></tr>
<tr><td>1</td><td>*</td><td>2</td><td>1b|1c|1d</td><td>1a|1b|1c|1d|1e|2</td><td>Surface water flows are likely within the HGMU. The vegetation is productive. Disturbance of the vegetation is likely to increase biomass available for import and export.</td><td>1/3</td></tr>
<tr><td>1</td><td>*</td><td>*</td><td>1b|1c|1d</td><td>1a|1b|1c|1d|1e|2</td><td>Surface water flows are likely within the HGMU. Disturbance of the vegetation is likely to increase biomass available for import and export.</td><td>1/3</td></tr>
<tr><td>1</td><td>*</td><td>*</td><td>1b|1c|1d</td><td></td><td>Surface water flows are likely within the HGMU.</td><td>1/3</td></tr>
</table>

Please ignore blank cells.

3.3.B.4.2 Look-up Table 2

CV1	CV2	CV3	CV4	CV5	Rationale	Code
#1	*	*	*	*	Surface water flows are not likely to remove biomass from the HGMU.	**2**
#2	#(1 2 3 4 5)	*	*	*	Surface water flows are unlikely.	**2**

CV5 *Degree of disturbance*

Q1. Does any part of the HGMU currently suffer a high degree of disturbance from the following:

a. Harvesting of herbaceous vegetation *(Recording Sheet 1,C.3.4.1b (answer c))*

b. Ploughing of land *(Recording Sheet 1, C.3.5.1b (answer c))*

c. Woodland management and forestry *(Recording Sheet 1, C.3.6.1b (answer c))*

d. Trampling by livestock *(Recording Sheet 1, C.3.7.1b (answer c))*

Or in the AA:

e. Recreational activities *(Recording Sheet 1,C.3.15.1 (answer c))*

Q2. Is there any evidence of substantial patches of bare or eroded land within the HGMU? *(Recording Sheet 3, B.5.2)*

Assessment of the process

Compare the results recorded for each HGMU on the appraisal sheet with the possible outcomes below, and assign an appropriate assessment code. Record this on the appraisal sheet. The syntax codes used are described in the introduction to Section 3.

Look-up tables

1. The process is definitely being performed.
If answers to the questions for this process agree with any of the combinations in Look-up Table 1, then the HGMU is definitely performing the process of biomass export via overland flow. * Indicates where answer to questions on the controlling variable can be either YES or NO. Answers are given in order according to both the certainty of occurrence of the process (first part of the code, e.g. '**1**/1') and the degree to which it occurs (second part of the code, e.g. '1/**1**'), based on the strength of evidence resulting from the combination of controlling variables: for code **1/1** the process is performed at the highest rate.

2. The process is not significantly being performed
If answers to the questions for this process agree with any of the combinations in Look-up Table 2, then the HGMU is not significantly performing the process of biomass export via overland flow. * Indicates where answer to questions on the controlling variable can be either YES, NO, or NA.

3. The process is being performed, but there are constraining factors or uncertainties
For any other combinations of answers, either YES or NO, the HGMU is performing the process of biomass export via overland flow, but only to a small degree because of constraining factors or uncertainty which can be identified from the answers given. Assign assessment code **1/x** to these HGMUs.

4. The process is definitely not being performed
This statement is not applicable for this process

This completes the functional assessment for the process of biomass export via overland flow. The output codes and quantification values can be recorded on the Functional Assessment Table at the end of Section 3.3.B.6.

3.3.B.4.3 *Biomass export via wind transport (process)*

Further reading and references

Chambers, F.M. (1999). Comment on D. M. Wilkinson (1997). Plant colonization: are wind dispersed seeds really dispersed by birds at larger spatial and temporal scales? *Biogeography* **26**(2), 425–427.

Clair, T.A., Arp, P., Moore, T.R., Dalva, M. and Meng, F.-R. (2002). Gaseous carbon dioxide and methane, as well as dissolved organic carbon losses from a small temperate wetland under a changing climate, *Environmental Pollution* **116** (Suppl 1), S143–S148.
Coops, H. and van der Velde, G. (1996). Impact of hydrodynamic changes on the zonation of helophytes, *Netherlands Journal of Aquatic Ecology* **30**(2–3), 165–173.
Ellis, S. (1996). *The Inadequacy of Passive, Wind-driven Traps in Assessing the Aerial Insects of Delta Marsh*, UFS (Delta Marsh) Annual Report, Volume 31, available at: http://www.umanitoba.ca/faculties/science/delta_marsh/reports/1996/ellis.pdf
Haukos, D.A. and Smith, L.M. (1994). Composition of seed banks along an elevational gradient in playa wetlands, *Wetlands* **14**(4), 301–307.
Hölzel, N. and Otte, A. (2003). Restoration of a species-rich flood meadow by topsoil removal and diaspore transfer with plant material, *Applied Vegetation Science* **6**(2), 131–140.
Keddy, P.A. (2000). *Wetland Ecology: Principles and Conservation*, Cambridge University Press, Cambridge, UK.
Mitsch, W.J. and Gosselink, J.G. (2007). *Wetlands*, 4th edn, Wiley, New York.
Muller, S., Dutoit, T., Alard, D. and Grévilliot, F. (1998). Restoration and rehabilitation of species-rich grassland ecosystems in France: a review, *Restoration Ecology* **6**(1), 94–101.
Overbeck, G., Kiehl, K. and Abs, C. (2003). Seedling recruitment of *Succisella inflexa* in fen meadows: Importance of seed and microsite availability, *Applied Vegetation Science* **6**(1), 97–104.
Patzelt, A., Wild, U. and Pfadenhauer, J. (2001). Restoration of wet fen meadows by topsoil removal: vegetation development and germination biology of fen species, *Restoration Ecology* **9**(2), 127–136.
Small, L.F., McIntire, C.D., MacDonald, K.B., Lara-Lara, J.B., Frey, B.E., Amspoker, M.C. and Winfield, T. (1990). Primary production, plant and detrital biomass, and particle transport in the Columbia River Estuary, *Progress In Oceanography* **25**(1–4), 175–210.
Smith, R.S. and Rushton, S.P. (1994). The effects of grazing management on the vegetation of mesotrophic (meadow) grassland in Northern England, *Journal of Applied Ecology* **31**, 13–24.
Saura-Mas, S. and Lloret, F. (2005). Wind effects on dispersal patterns of the invasive alien *Cortaderia selloana* in Mediterranean wetlands, *Acta Oecologica* **27**(2), 129–133.
Watts, C.H. and Didham, R.K. (2006), Rapid recovery of an insect–plant interaction following habitat loss and experimental wetland restoration, *Oecologia* **148**(1), 61–69.
Wilkinson, D.M. (1997). Plant colonization: are wind dispersed seeds really dispersed by birds at larger spatial and temporal scales? *Biogeography* **24**, 61–65.

Introduction

Process description:

- The transport of biomass produced within the wetland to areas outside the wetland by wind; this may then be available for animals in areas outside the AA to feed upon.
- This is most likely to occur where land is exposed, winds are high or wind erosion occurs.

Potential significance – rationale for assessment: The wetland may be contributing to the biological processes that occur in areas outside the wetland.

Controlling variables

CV1 Exposure
CV2 Soil moisture
CV3 Nature of litter layer
CV4 Vegetation type
CV5 Degree of disturbance

Answer all the questions below for *each* HGMU; unless otherwise instructed continue from one question to the next. On the recording sheet circle the correct answer. The location of the relevant information is given in brackets after each question. Use a new appraisal sheet for each HGMU.

CV1 *Exposure*

Q1. What is the topographic exposure of the HGMU? *(Recording Sheet 3, A.4)*
- a. 3 (Sheltered)
- b. 2 (Moderate)
- c. 1 (Exposed)

Q2. What is the average wind speed? *(Recording Sheet 1, C.3.1.4)*
- a. < 4.8 m s^{-1}
- b. 4.8–6.6 m s^{-1}
- c. > 6.6 m s^{-1}

If the answer to both the ***Q1*** and ***Q2*** is a. the process is not (significantly) being performed. In this case assign code **2** to the HGMU and go to Section 3.3.B.5 Biomass export via biological processes.

CV2 *Soil moisture*

Q1. Is the total annual precipitation below 500 mm? *(Recording Sheet 1, C.3.1.1ii)*

Q2. Is the topsoil moisture status described as: *(Recording Sheet 2a, C.3.2)*
- a. Dry
- b. Moist
- c. Wet
- d. Flooded

Q3. Is the topsoil particle size described as: *(Recording Sheet 2a, C.3.7)*
- a. Fine
- b. Medium
- c. Coarse

CV3 *Nature of litter layer*

Q1. Which of the following best describes the litter layer? *(Recording Sheet 3, C.2)*
- a. Negligible
- b. Herbaceous litter layer
- c. Forest debris litter layer
- d. Combination litter layer

CV4 *Vegetation type*

Q1. If there are trees in the HGMU, are they wind-cropped? *(Recording Sheet 3, A.5)*

Q2. Is the habitat or the vegetation classed as a sand dune type (**U2**, **U4**, **G26**) or an eroded peatland (**M4**)? *(Recording Sheet 3, D)*

Q3. Is the vegetation classed as a woodland (code $\mathbf{W_X}$) a plantation (**W1**), or thicket (**S2**, **S3**, **S18**)? *(Recording Sheet 3, D)*

Q4. Is the vegetation classed as a productive grassland type (codes **G4**, **G7–G11**, **G17**)? *Recording Sheet 3, D and 1/a or 1/b for productivity, Process 3.3.B.1)*

Q5. Are there any solitary trees, tree lines, windbreaks or hedgerows (**W2**) within the HGMU? *(Recording Sheet 3, D)*

CV5 *Degree of disturbance*

Q1. Does any part of the HGMU currently experience a high degree of disturbance from the following?
- a. Harvesting of herbaceous vegetation *(Recording Sheet 1,C.3.4.1b (answer c))*
- b. Ploughing of land *(Recording Sheet 1, C.3.5.1b (answer c))*
- c. Woodland management and forestry *(Recording Sheet 1, C.3.6.1b (answer c))*
- d. Trampling by livestock *(Recording Sheet 1, C.3.7.1b (answer c))*

3.3.B.4.3 Biomass Export via Wind Transport Appraisal Sheet

HGMU code: ____________________	**AA name:** ____________________

CV1 Exposure			
Q1	a	b	c
Q2	a	b	c

CV2 Soil moisture				
Q1	Y		N	U/NA
Q2	a	b	c	d
Q3	a		b	c

CV3 Nature of litter layer				
Q1	a	b	c	d

CV4 Vegetation type			
Q1	Y	N	U/NA
Q2	Y	N	U/NA
Q3	Y	N	U/NA
Q4	Y	N	U/NA
Q5	Y	N	U/NA

CV5 Degree of disturbance			
Q1 a	Y	N	U/NA
Q1 b	Y	N	U/NA
Q1 c	Y	N	U/NA
Q1 d	Y	N	U/NA
Q1 e	Y	N	U/NA
Q2	Y	N	U/NA

HGMU process assessment code: ____________________

3.3.B.4.3 Look-up Table 1

<table>
<tr><th>CV1</th><th>CV2</th><th>CV3</th><th>CV4</th><th>CV5</th><th>Rationale</th><th>Code</th></tr>
<tr><td>1c 2c</td><td>1 2a 3c</td><td>(1b|1c|1d)</td><td>1
(2|3|4|5)</td><td>1a|1b|
1c|1d|
1e|2</td><td>The site is exposed to wind. The soil is prone to wind erosion. There is biomass available for movement. There is evidence of biomass movement. The vegetation appears to be affected by wind. Disturbance of the vegetation is likely to increase biomass available for import and export.</td><td>1/1</td></tr>
<tr><td>1c</td><td>1 2a 3c</td><td>(1b|1c|1d)</td><td>1
(2|3|4|5)</td><td>1a|1b|
1c|1d|
1e|2</td><td>The site is exposed to wind. The soil is prone to wind erosion. There is biomass available for movement. There is evidence of biomass movement. The vegetation appears to be affected by wind. Disturbance of the vegetation is likely to increase biomass available for import and export.</td><td>1/2</td></tr>
<tr><td>1c 2c</td><td>1 2a 3c</td><td>(1b|1c|1d)</td><td>1
(2|3|4|5)</td><td>*</td><td>The site is exposed to wind. The soil is prone to wind erosion. There is biomass available for movement. There is evidence of biomass movement. The vegetation appears to be affected by wind.</td><td>1/2</td></tr>
<tr><td>1c</td><td>1 2a 3c</td><td>(1b|1c|1d)</td><td>1
(2|3|4|5)</td><td>*</td><td>The site is exposed to wind. The soil is prone to wind erosion. There is biomass available for movement. There is evidence of biomass movement. The vegetation appears to be affected by wind.</td><td>1/2</td></tr>
<tr><td>1c 2c</td><td>1 2a</td><td>(1b|1c|1d)</td><td>1
(2|3|4|5)</td><td>1a|1b|
1c|1d|
1e|2</td><td>The site is exposed to wind. The soil is prone to wind erosion. There is biomass available for movement. There is evidence of biomass movement. The vegetation appears to be affected by wind. Disturbance of the vegetation is likely to increase biomass available for import and export.</td><td>1/2</td></tr>
<tr><td>1c</td><td>1 2a</td><td>(1b|1c|1d)</td><td>1
(2|3|4|5)</td><td>1a|1b|
1c|1d|
1e|2</td><td>The site is exposed to wind. The soil is prone to wind erosion. There is biomass available for movement. There is evidence of biomass movement. The vegetation appears to be affected by wind. Disturbance of the vegetation is likely to increase biomass available for import and export.</td><td>1/2</td></tr>
<tr><td>1c 2c</td><td>1 2a</td><td>(1b|1c|1d)</td><td>1
(2|3|4|5)</td><td>*</td><td>The site is exposed to wind. The soil is prone to wind erosion. There is biomass available for movement. There is evidence of biomass movement.</td><td>1/2</td></tr>
<tr><td>1c</td><td>1 2a</td><td>(1b|1c|1d)</td><td>1
(2|3|4|5)</td><td>*</td><td>The site is exposed to wind. The soil is prone to wind erosion. There is biomass available for movement. There is evidence of biomass movement. The vegetation appears to be affected by wind.</td><td>1/3</td></tr>
<tr><td>1c 2c</td><td>2a</td><td>(1b|1c|1d)</td><td>1
(2|3|4|5)</td><td>1a|1b|
1c|1d|
1e|2</td><td>The site is exposed to wind. The soil is prone to wind erosion. There is biomass available for movement. There is evidence of biomass movement. The vegetation appears to be affected by wind. Disturbance of the vegetation is likely to increase biomass available for import and export.</td><td>1/2</td></tr>
<tr><td>1c</td><td>2a</td><td>(1b|1c|1d)</td><td>1
(2|3|4|5)</td><td>1a|1b|
1c|1d|
1e|2</td><td>The site is exposed to wind. The soil is prone to wind erosion. There is biomass available for movement. There is evidence of biomass movement. The vegetation appears to be affected by wind. Disturbance of the vegetation is likely to increase biomass available for import and export.</td><td>1/3</td></tr>
<tr><td>1c 2c</td><td>2a</td><td>(1b|1c|1d)</td><td>1
(2|3|4|5)</td><td>*</td><td>The site is exposed to wind. The soil is prone to wind erosion. There is biomass available for movement. There is evidence of biomass movement.</td><td>1/3</td></tr>
<tr><td>1c</td><td>2a</td><td>(1b|1c|1d)</td><td>1
(2|3|4|5)</td><td>*</td><td>The site is exposed to wind. The soil is prone to wind erosion. There is biomass available for movement. There is evidence of biomass movement. The vegetation appears to be affected by wind.</td><td>1/3</td></tr>
</table>

3.3.B.4.3 Look-up Table 1 cont'd

<table>
<tr><th>CV1</th><th>CV2</th><th>CV3</th><th>CV4</th><th>CV5</th><th>Rationale</th><th>Code</th></tr>
<tr><td>1c 2c</td><td>*</td><td>(1b|1c|1d)</td><td>1
(2|3|4|5)</td><td>1a|1b|
1c|1d|
1e|2</td><td>The site is exposed to wind. There is biomass available for movement. There is evidence of biomass movement. The vegetation appears to be affected by wind. Disturbance of the vegetation is likely to increase biomass available for import and export.</td><td>1/3</td></tr>
<tr><td>1c</td><td>*</td><td>(1b|1c|1d)</td><td>1
(2|3|4|5)</td><td>1a|1b|
1c|1d|
1e|2</td><td>The site is exposed to wind. There is biomass available for movement. There is evidence of biomass movement. The vegetation appears to be affected by wind. Disturbance of the vegetation is likely to increase biomass available for import and export.</td><td>1/3</td></tr>
<tr><td>1c 2c</td><td>*</td><td>(1b|1c|1d)</td><td>1
(2|3|4|5)</td><td>*</td><td>The site is exposed to wind. There is biomass available for movement. There is evidence of biomass movement.</td><td>1/3</td></tr>
<tr><td>1c</td><td>*</td><td>(1b|1c|1d)</td><td>1
(2|3|4|5)</td><td>*</td><td>The site is exposed to wind. There is biomass available for movement. There is evidence of biomass movement. The vegetation appears to be affected by wind.</td><td>1/3</td></tr>
<tr><td>1c 2c</td><td>*</td><td>(1b|1c|1d)</td><td>1</td><td>1a|1b|
1c|1d|
1e|2</td><td>The site is exposed to wind. There is biomass available for movement. There is evidence of biomass movement. The vegetation appears to be affected by wind. Disturbance of the vegetation is likely to increase biomass available for import and export.</td><td>1/3</td></tr>
<tr><td>1c</td><td>*</td><td>(1b|1c|1d)</td><td>1</td><td>1a|1b|
1c|1d|
1e|2</td><td>The site is exposed to wind. There is biomass available for movement. There is evidence of biomass movement. The vegetation appears to be affected by wind. Disturbance of the vegetation is likely to increase biomass available for import and export.</td><td>1/3</td></tr>
<tr><td>1c 2c</td><td>*</td><td>(1b|1c|1d)</td><td>1</td><td>*</td><td>The site is exposed to wind. There is biomass available for movement. There is evidence of biomass movement. The vegetation appears to be affected by wind.</td><td>1/3</td></tr>
<tr><td>1c</td><td>*</td><td>(1b|1c|1d)</td><td>1</td><td>*</td><td>The site is exposed to wind. There is biomass available for movement. There is evidence of biomass movement. The vegetation appears to be affected by wind.</td><td>1/3</td></tr>
<tr><td>1c 2c</td><td>*</td><td>(1b|1c|1d)</td><td>*</td><td>*</td><td>The site is exposed to wind. There is biomass available for movement. There is evidence of biomass movement.</td><td>1/3</td></tr>
<tr><td>1c</td><td></td><td>(1b|1c|1d)</td><td></td><td></td><td>The site is exposed to wind. There is biomass available for movement. There is evidence of biomass movement.</td><td>1/3</td></tr>
<tr><td>1c 2c</td><td></td><td></td><td></td><td></td><td>The site is exposed to wind.</td><td>1/3</td></tr>
<tr><td>1c</td><td></td><td></td><td></td><td></td><td>The site is exposed to wind.</td><td>1/3</td></tr>
</table>

Please ignore blank cells.

3.3.B.4.3 Look-up Table 2

CV1	CV2	CV3	CV4	CV5	Rationale	Code
#(1c 2b 2c)	*	*	*	*	The site is sheltered from wind.	**2**
#(1b 1c)	#(1 2a 2b 2c 3c)	#(1a 1b 1c)	#(1 2 3 4 5)	#(1 2)	The site is sheltered from wind. The soil is not prone to wind erosion. There is no biomass available for movement. There is no evidence of biomass movement. The vegetation does not appear to be affected by wind. There is no disturbance of the vegetation.	**2**
#(1b 1c)	#(1 2a 2b 2c 3c)	*	*	*	The site is sheltered from wind. The soil is not prone to wind erosion	**2**

Or in the AA:

e. Recreational activities *(Recording Sheet 1, C.3.15.1 (answer c))*

Q2. Is there any evidence of substantial patches of bare or eroded land within the HGMU? *(Recording Sheet 3, B.5.2)*

Assessment of the process

Compare the results recorded for each HGMU on the appraisal sheet with the possible outcomes below, and assign an appropriate assessment code. Record this on the appraisal sheet. The syntax codes used are described in the introduction to Section 3.

Look-up tables

1. The process is definitely being performed
If answers to the questions for this process agree with any of the combinations in Look-up Table 1, then the HGMU is definitely performing the process of biomass export via wind transport. The answers are given in order according to both the certainty of occurrence of the process (first part of the code, e.g. '**1**/1') and the degree to which it occurs (second part of the code, e.g. '1/**1**'), based on the strength of evidence resulting from the combination of controlling variables: for code **1/1** the process is performed at the highest rate.

2. The process is not significantly being performed
If answers to the questions for this process agree with any of the combinations in Look-up Table 2, then the HGMU is not significantly performing the process of biomass export via wind transport. * Indicates where answer to questions on the controlling variable can be either YES or NO.

3. The process is being performed, but there are constraining factors or uncertainty
For any other combinations of answers, either YES or NO, the HGMU is performing the process of biomass export via wind transport, but only to a small degree because of constraining factors or uncertainty which can be identified from the answers given. Assign assessment code **1/x** to these HGMUs.

4. The process is definitely not being performed
This statement is not applicable for this process.

This completes the functional assessment for the process of biomass export via wind transport. The output codes and quantification values can be recorded on the Functional Assessment Table at the end of Section 3.3.B.6.

3.3.B.5 Biomass export via biological processes (accumulation of processes)

3.3.B.5.1 Biomass export via fauna (process)

Further reading and references

Ausden, M., Hall, M., Pearson, P. and Strudwick, T. (2005). The effects of cattle grazing on tall-herb fen vegetation and molluscs, *Biological Conservation* **122**(2), 317–326.

Blancher, P.J. and McNicol, D.K. (1991). Tree swallow diet in relation to wetland acidity, *Canadian Journal of Zoology* **69**(10), 2629–2637.

Bouchard, V. and Lefeuvre, J.C. (2000). Primary production and macro-detritus dynamics in a European salt marsh: carbon and nitrogen budgets, *Aquatic Botany* **67**(1), 23–42.

Gardarsson, A. and Einarsson, A. (1994). Responses of breeding duck populations to changes in food supply, *Hydrobiologia* **279–280**(1), 15–27.

Hanowski, J.M., Niemi, G.J., Lima, A.R. and Regal, R.R. (1997). Response of breeding birds to mosquito control treatments of wetlands, *Wetlands* **17**(4), 485–492.

Keddy, P.A. (2000). *Wetland Ecology: Principles and Conservation*, Cambridge University Press, Cambridge, UK.

Kutrup, B., Çakir, E. and Yilmaz, N. (2005). Food of the banded newt, *Triturus vittatus ophryticus* (Berthold, 1846) at different sites in Trabzon, *Turkish Journal of Zoology* **29**, 83–89.

Lanszki, J. and Molnar, T. (2003). Diet of otters living in three different habitats in Hungary, *Folia Zoologia* **52**(4), 378–388.

Lurz, P.W.W., Shirley, M.D.F., Rushton, S.P. and Sanderson, R.A. (2002). Modelling the consequences of duck migration patterns on the genetic diversity of aquatic organisms: a first step towards a predictive tool for wetland management, *Acta Oecologica* **23**(3), 205–212.

Mathevet, R., Bousquet, F., Le Page, C. and Antona, M. (2003). Agent-based simulations of interactions between duck population, farming decisions and leasing of hunting rights in the Camargue (Southern France), *Ecological Modelling* **165**(2–3), 107–126.

Mitsch, W.J. and Gosselink, J.G. (2007). *Wetlands*, 4th edn, Wiley and, New York.

Paillisson, J-M., Carpentier, A., Le Gentil, J. and Marion, L. (2004). Space utilization by a cormorant (*Phalacrocorax carbo* L.) colony in a multi-wetland complex in relation to feeding strategies, *Comptes Rendus Biologies* **327**(5), 493–500.

Schmidt, M.H., Lefebvre, G., Poulin, B. and Tscharntke, T. (2005). Reed cutting affects arthropod communities, potentially reducing food for passerine birds. *Biological Conservation* **121**(2), 157–166.

Watson, A.M. and Ormerod, S.J. (2004). The distribution of three uncommon freshwater gastropods in the drainage ditches of British grazing marshes, *Biological Conservation* **118**(4), 455–466.

Introduction

Process description:

- The transport of biomass produced within the wetland to areas outside the wetland as a result of the activities of animals; this may then be available for animals in other areas to feed on.
- This can involve migration, feeding or harvesting.

Potential significance – rationale for assessment:

- The wetland may be contributing to the biological processes that occur in areas outside the wetland.
- The wetland may be supporting economically important activities, such as the harvesting of wild populations or agriculture.

Answer all the questions below for *each* HGMU; unless otherwise instructed continue from one question to the next. On the recording sheet circle the correct answer. The location of the relevant information is given in brackets after each question. Use a new appraisal sheet for each HGMU.

Controlling variables

CV1 Evidence and condition of animal migrations

Note: The process of biomass export via fauna can be used to determine the effects of individual groups of animals as required by following only the relevant controlling variable.

CV1 *Faunal migrations and visitors*

Q1. Are there any records of invertebrates that migrate in large numbers frequenting the AA? *(Recording Sheet 1,C.3.18.1)*

Q2. Are there any records of migratory fish species regularly feeding in the water body or tributaries? *(Recording Sheet 1, C.3.18.2)*

Q3. Does the AA contain any HGMUs that have been assigned **TB**, **DT**, **BS**, **OX**, **LK** geomorphic codes and are connected to the water body by continuous water? *(Recording Sheet 2a, C.2; AA Map 3)*

Q4. Is the AA susceptible to prolonged periods of flooding (> 2 months)? *(Recording Sheet 1, C.3.19.5 (c); AA Map 2)*

3.3.B.5.1 Biomass Export via Fauna Appraisal Sheet

HGMU code: ____________	**AA name:** ____________

CV1 Evidence and condition of animal migrations			
Q1	Y	N	U/NA
Q2	Y	N	U/NA
Q3	Y	N	U/NA
Q4	Y	N	U/NA
Q5	Y	N	U/NA
Q6	Y	N	U/NA
Q7	Y	N	U/NA
Q8	Y	N	U/NA

HGMU process assessment code: ____________

3.3.B.5.1 Look-up Table 1

<table>
<tr><th>CV1</th><th>Rationale</th><th>Code</th></tr>
<tr><td>1 2 (3|4) 5 6 7 8</td><td>Large numbers of invertebrates, fish, birds and mammals move between the AA and other areas, with the result that large quantities of biomass are likely to be exported.</td><td>1/1</td></tr>
<tr><td>5 or 6 of (1|2|3|4|5|6|7|8)</td><td>Clear evidence exists of animal movements between the AA and other areas and as a result, significant quantities of biomass are likely to be exported.</td><td>1/2</td></tr>
<tr><td>3 or 4 of (1|2|3|4|5|6|7|8)</td><td>Evidence is limited for the movement of animals between the AA and other areas, with the result that low quantity of biomass are likely to be exported.</td><td>1/3</td></tr>
</table>

3.3.B.5.1 Look-up Table 2

<table>
<tr><th>CV1</th><th>Rationale</th><th>Code</th></tr>
<tr><td>#(1 2 (3|4) 5 6 7 8)</td><td>There is no evidence of significant animal movement between the AA and other areas. Export of significant quantity of biomass is unlikely.</td><td>2</td></tr>
</table>

Q5. Are there any records of migratory bird species (e.g. geese) frequenting the AA? *(Recording Sheet 1, C.3.18.3)*

Q6. Are there any records of birds that roost elsewhere visiting the AA in large numbers? *(Recording Sheet 1, C.3.8.1a)*

Q7. Are there any records of migratory mammal species frequenting the AA? *(Recording Sheet 1, C.3.18.4)*

Q8. Is there any evidence for, or records of, mammals feeding in the AA, and which have domiciles outside the AA? (Include domestic livestock removed from the AA on a daily or periodic basis) *(Recording Sheet 3, C.5 (answer y); Recording Sheet 1, C.3.17.1, C.3.17.2)*

Assessment of the process

Compare the results recorded for the AA on the appraisal sheet with the possible outcomes below, and assign an appropriate assessment code. Record this on the appraisal sheet. The syntax codes used are described in the introduction to Section 3.

Look-up tables

1. The process is definitely being performed
If answers to the questions for this process agree with any of the combinations in Look-up Table 1, then the AA is definitely performing the process of biomass import and export via biological processes. Answers are given in order according to both the certainty of occurrence of the process (first part of the code, e.g. '**1**/1') and the degree to which it occurs (second part of the code, e.g. '1/**1**'), based on the strength of evidence resulting from the combination of controlling variables: for code **1/1** the process is performed at the highest rate.

2. The process is not significantly being performed
If answers to the questions for this process agree with any of the combinations in Look-up Table 2, then the AA is not significantly performing the process of biomass export via fauna.

3. The process is being performed, but there are constraining factors or uncertainties
For any other combinations of answers, either YES or NO, the AA is performing the process of biomass export via fauna, but only to a small degree because of constraining factors or uncertainty which can be identified from the answers given. Assign assessment code **1/x** to the AA.

4. The process is definitely not being performed
This statement is not applicable for this process.

This completes the functional assessment for the process of biomass export via fauna. The output codes and quantification values can be recorded on the Functional Assessment Table at the end of Section 3.3.B.6.

3.3.B.5.2 *Biomass export via anthropogenic means (process)*

Further reading and references

Cattin, M.F., Blandenier, G., Banašek-Richter, C. and Bersier, L-F. (2003). The impact of mowing as a management strategy for wet meadows on spider (Araneae) communities, *Biological Conservation* **113**(2), 179–188.

Cofer, W.R.III, Levine, J., Winstead, E.L. and Stocks, B. (1991). Trace gas and particulate emissions from biomass burning in temperate ecosystems. In: Levine, J.S. (ed.), *Global Biomass Burning*, MIT Press, Cambridge, MA, 203–208.

Connolly, R.M. (2003). Differences in trophodynamics of commercially important fish between artificial waterways and natural coastal wetlands, *Estuarine, Coastal and Shelf Science* **58**(4), 929–936.

Croonquist, M.J. and Brooks, R.P. (1991). Use of avian and mammalian guilds as indicators of cumulative impacts in riparian-wetland areas, *Environmental Management* **15**(5), 701–714.

Delmas, R. (1994). An overview of present knowledge on methane emission from biomass burning, *Nutrient Cycling in Agroecosystems* **37**(3), 181–190.

Evans, D.M. and Day, K.R. (2002). Hunting disturbance on a large shallow lake: the effectiveness of waterfowl refuges, *Ibis* **144**(1), 2–8.

Gale, M.R., McLaughlin, J.W., Jurgensen, M.F., Trettin, C.C., Soelsepp, T. and Lydon, P.O. (1998). Plant community responses to harvesting and post-harvest manipulations in a Picea–Larix–Pinus wetland with a mineral substrate, *Wetlands* **18**(1), 150–159.

Keddy, P.A. (2000). *Wetland Ecology: Principles and Conservation*, Cambridge University Press, Cambridge, UK.

Kilgore, M.A. and Blinn, C.R. (2004). Policy tools to encourage the application of sustainable timber harvesting practices in the United States and Canada, *Forest Policy and Economics* **6**(2), 111–127.

Kim, S.Y. and Geary, P.M. (2001). The impact of biomass harvesting on phosphorus uptake by wetland plants, *Water Science Technology* **44**(11–12), 61–67.

Kuusemets, V. and Lohmus, K. (2005). Nitrogen and phosphorus accumulation and biomass production by scirpus sylvaticus and phragmites australis in a horizontal subsurface flow constructed wetland, *Journal of Environmental Science and Health, Part A: Toxic/Hazardous Substances & Environmental Engineering* **40**(6–7), 1167–1175.

Mathevet, R., Bousquet, F., Le Page, C. and Antona, M. (2003). Agent-based simulations of interactions between duck population, farming decisions and leasing of hunting rights in the Camargue (Southern France), *Ecological Modelling* **165**(2–3), 107–126.

Maurin, H., Bergot, F. and Vigneux, E. (1997). Man and mammals of metropolitan France: historical evolution and introductions of species in wetlands and aquatic systems, *Bulletin Francais de la Peche et de la Pisciculture* **344–345**, 117–132.

Mitsch, W.J. and Gosselink, J.G. (2007). Wetlands, 4th edn, Wiley, New York.

Musgrove, A.J., Clark, N.A., Gill, J. and Ravenscroft, N.O.M. (2001). *A Review of Wildfowling on the Stour Estuary*. Research Report, British Trust for Ornithology, Thetford, UK.

Ostendorp, W. (1995). Impact of winter reed harvesting and burning on the nutrient economy of reed beds, *Wetlands Ecology and Management* **3**(4), 233–248.

Pothier, D., Prévost, M. and Auger, I. (2003). Using the shelterwood method to mitigate water table rise after forest harvesting, *Forest Ecology and Management* **179**(1–3), 573–583.

Richardson, C.J. (1994). Ecological functions and human values in wetlands: a framework for assessing forestry impacts, *Wetlands* **14**(1), 1–9.

Swetnam, R.D., Mountford, J.O., Manchester, S.J. and Broughton, R.K. (2004). Agri-environmental schemes: their role in reversing floral decline in the Brue floodplain, Somerset, UK, *Journal of Environmental Management* **71**(1), 79–93.

Tolvanen, P., Eskelin, T., Aarvak, T., Eichhorn, G., Øien, I.J. and Gurtovaya, E. (1999). *Monitoring the Autumn Staging of Lesser White-fronted Geese in Kazakstan, October 1999,* Fennoscandian Lesser White-fronted Goose Conservation Project – Annual report.

Zalidis, G.C., Mantzavelas, A.L. and Gourvelou, E. (1997). Environmental impacts on Greek wetlands, *Wetlands* **17**(3), 339–345.

Zetterberg, L., Uppenberg, S. and Åhman, M. (2004). Climate impact from peat utilisation in Sweden, *Mitigation and Adaptation Strategies for Global Change* **9**(1), 37–76.

Introduction

Process description: The transport of biomass produced within the wetland to areas outside wetland as a result of the activities of humans; this may then be available for animals in other areas to feed on.

Potential significance – rationale for assessment: The wetland may be supporting economically important activities, such as the harvesting of grass, conserved as hay for the feeding of stock off site.

Answer all the questions below for *each* HGMU; unless otherwise instructed continue from one question to the next. On the recording sheet circle the correct answer. The location of the relevant information is given in brackets after each question. Use a new appraisal sheet for each HGMU.

Controlling variables

CV1 Productivity
CV2 Land-use and management

CV1 *Productivity*

Q1. Has the HGMU been classed as **1/a or 1/b** for the assessment of productivity? *(process 3.3.B.1)*?

CV2 *Land-use and management*

Q1. Is the vegetation in the HGMU subjected to harvesting of herbaceous vegetation? *(Recording Sheet 1, C.3.4.1b)*
- a. No history of harvesting of herbaceous vegetation in the last 20 years
- b. Not currently harvested but previously harvested within 20 years, or harvested less than once a year
- c. Currently harvested once or more than once a year
- d. Unknown

Q2. If the answer to ***Q1*** is c, which of the following would best describe the harvesting practices? *(Recording Sheet 1, C.3.4.2b)*
- a. Fully mechanised practices
- b. Traditional, labour-intensive manual practices

Q3. Is the AA used for hunting, shooting, or fishing (*Recording Sheet 1, C.3.14.1)?*
- a. Not used for hunting, shooting, or fishing
- b. Used for hunting, shooting, or fishing at a recreational scale
- c. Used for hunting, shooting, or fishing at a commercial scale
- d. Unknown

Q4. Is the vegetation of the HGMU subjected to management by burning? *(Recording Sheet 1, C.3.3.1b)*
- a. No history of burning in the last 20 years
- b. Not burnt within the last 5 years but previously burnt between 5 and 20 years ago
- c. Currently burnt once or more than once in the last 5 years
- d. Unknown

Q5. Is the vegetation in the HGMU subjected to forestry and woodland management? *(Recording Sheet 1, C.3.6.1b)*
- a. Never
- b. No current felling and removal but previously practised within 20 years, or currently less than once in 5 years
- c. Current felling and removal, or practised within the last 5 years
- d. Unknown

Q6. Are any livestock kept for the purpose of meat, milk or wool production in the HGMU? *(Recording Sheet 1, C.3.7.2b)*

Q7. If the answer to ***Q6*** was YES, are any of the following forms of livestock currently kept at least once a year at a density greater than the figure indicated alongside each category? *(Recording Sheet 1, C.3.7.1b (answer c))*
- Cattle or horses or donkeys $> 1\ ha^{-1}$
- Pigs $> 3\ ha^{-1}$
- Sheep or goats $> 10\ ha^{-1}$
- Turkeys or geese $> 30\ ha^{-1}$
- Chickens or ducks $> 100\ ha^{-1}$

Assessment of the process

Compare the results recorded for each HGMU on the appraisal sheet with the possible outcomes below, and assign an appropriate assessment code. Record this on the appraisal sheet. The syntax codes used are described in the introduction to Section 3.

3.3.B.5.2 Biomass export via Anthropogenic Means Appraisal Sheet

HGMU code: ______________	**AA name:** ______________

CV1 Productivity			
Q1	Y	N	U/NA

CV2 Land-use and management				
Q1	a	b	c	d
Q2	a		b	
Q3	a	b	c	d
Q4	a	b	c	d
Q5	a	b	c	d
Q6	Y		N	U/NA
Q7	Y		N	U/NA

HGMU process assessment code: ______________________________

3.3.B.5.2 Look-up Table 1

<table>
<tr><th>CV1</th><th>CV2</th><th>Rationale</th><th>Code</th></tr>
<tr><td>1</td><td>2a 3c
(4c|5c|7)</td><td>The vegetation is productive. Human activities remove large quantities of biomass from the HGMU.</td><td>1/1</td></tr>
<tr><td>1</td><td>(4c|5c|7)</td><td>The vegetation is productive. Human activities remove large quantities of biomass from the HGMU.</td><td>1/1</td></tr>
<tr><td>1</td><td>2a 3c</td><td>The vegetation is productive. Human activities remove large quantities of biomass from the HGMU through harvesting.</td><td>1/1</td></tr>
<tr><td>1</td><td>2a</td><td>The vegetation is productive. Human activities are likely to remove significant quantity of biomass from the HGMU through harvesting.</td><td>1/1</td></tr>
<tr><td></td><td>3c
(4c|5c|7)</td><td>Human activities remove very large quantities of biomass from the HGMU through harvesting.</td><td>1/1</td></tr>
<tr><td></td><td>(4c|5c|7)</td><td>Human activities remove very large quantities of biomass from the HGMU.</td><td>1/2</td></tr>
<tr><td></td><td>2a 3c</td><td>Human activities are likely to remove significant quantity of biomass from the HGMU by harvesting.</td><td>1/2</td></tr>
<tr><td></td><td>2a|3c</td><td>Human activities are likely to remove significant quantity of biomass from the HGMU by harvesting.</td><td>1/2</td></tr>
<tr><td>1</td><td>(4c|5c|7)</td><td>The vegetation is productive. Human activities remove large quantities of biomass from the HGMU.</td><td>1/2</td></tr>
<tr><td>1</td><td>2a
(4c|5c|7)</td><td>The vegetation is productive. Human activities remove large quantities of biomass from the HGMU.</td><td>1/2</td></tr>
<tr><td></td><td>2a 3c
(4c|5c|7)</td><td>Human activities remove very large quantities of biomass from the HGMU through harvesting.</td><td>1/2</td></tr>
<tr><td></td><td>2a
(4c|5c|7)</td><td>Human activities remove very large quantities of biomass from the HGMU through harvesting.</td><td>1/3</td></tr>
<tr><td></td><td>(4c|5c|7)</td><td>Human activities are likely to remove significant quantity of biomass from the HGMU.</td><td>1/3</td></tr>
<tr><td></td><td>2a 3c</td><td>Human activities are likely to remove significant quantity of biomass from the HGMU by harvesting.</td><td>1/3</td></tr>
</table>

Please ignore blank cells.

3.3.B.5.2 Look-up Table 2

<table>
<tr><th>CV1</th><th>CV2</th><th>Rationale</th><th>Code</th></tr>
<tr><td>#1</td><td>#(2a 3c
4c 5c 7)</td><td>Human activities do not remove large quantities of biomass from the AA.</td><td>2</td></tr>
</table>

Look-up tables

1. The process is definitely being performed
If answers to the questions for this process agree with any of the combinations in Look-up Table 1, then the HGMU is definitely performing the process of biomass export via anthropogenic means. * Indicates where the answer to questions on the controlling variable can be either YES or NO.

2. The process is not significantly being performed
If answers to the questions for this process agree with any of the combinations in Look-up Table 2, then the HGMU is not significantly performing the process of biomass export via biological processes.

3. The process is being performed, but there are constraining factors or uncertainty
For any other combinations of answers, either YES or NO, the HGMU is performing the process of biomass export via anthropogenic means, but only to a small degree because of constraining factors or uncertainty which can be identified from the answers given. Assign assessment code **1/x** to these HGMUs.

4. The process is definitely not being performed
This statement is not applicable for this process.

This completes the functional assessment for the process of biomass export via anthropogenic means. The output codes and quantification values can be recorded on the functional assessment table at the end of Section 3.3.B.6.

3.3.B.6 *Functional assessment: Food-web support*

The purpose of this section is to determine whether the function Food-web support is occurring and if so, to what degree. Follow the method below for *each* HGMU. In the Functional Assessment Table at the end of this section, record the functional outcome for each HGMU within your AA.

Step 1: Identify all the processes which contribute to the Food-web support function
A single function (Food-web support) includes the 10 processes:

- productivity/biomass production
- biomass import via water courses
- biomass import via wind transport
- biomass import via overland flow
- biomass import via animal movements
- biomass export via water courses
- biomass export via overland flow
- biomass export via wind transport
- biomass export via fauna
- biomass export via anthropogenic means

The change in biomass within an HGMU (i.e. the material available for food-web support) is related to the amount produced and imported minus the amount exported and this is the basis for the calculations determining the function of food-web support.

Step 2: Derive a score and a weighting coefficient for each process
Two main steps are involved in deriving a functional outcome from process assessments:

1. Process assessment output codes are translated into scores (S1–S10), as shown in Table 3.3.7.
2. The relative contribution of each process to the function should be determined.

Table 3.3.7 Food-web support scoring and weighting

1. Productivity and biomass W1 = 0.6	
Output code(s)	**Score (S1)**
1/a	1
1/b	0.75
1/c	0.5
1/x, 1/d	0.25
2	0
2. Import via water courses W2 = 0.1	
Output code(s)	**Score (S2)**
1/1	1
1/2	0.67
1/x, 1/3	0.33
2	0
3. Import via overland flow W3 = 0.1	
Output code(s)	**Score (S3)**
1/1	1
1/2	0.67
1/x, 1/3	0.33
2	0
4. Import via wind transport W4 = 0.1	
Output code(s)	**Score (S4)**
1/1	1
1/2	0.67
1/x, 1/3	0.33
2	0
5. Import via animal movements W5 = 0.1	
Output code(s)	**Score (S5)**
1/1	1
1/2	0.67
1/x	0.33
2	0
6. Export via water courses W6 = 0.1	
Output code(s)	**Score (S6)**
1/1	1
1/2	0.67
1/x, 1/3	0.33
2	0

Table 3.3.7 cont'd

7. Export via overland flow W7 = 0.1	
Output code(s)	**Score (S7)**
1/1	1
1/2	0.67
1/x, 1/3	0.33
2	0
8. Import via wind transport W8=0.1	
Output code(s)	**Score (S8)**
1/1	1
1/2	0.67
1/x, 1/3	0.33
2	0
9. Export via fauna W9 = 0.1	
Output code(s)	**Score (S9)**
1/1	1
1/2	0.67
1/x	0.33
2	0
10. Export via anthropogenic means W10 = 0.6	
Output code(s)	**Score (S10)**
1/a	1
1/b	0.75
1/c	0.5
1/x, 1/d	0.25
2	0

A weighting coefficient (W1–W10) is assigned to each process as indicated in Table 3.3.7. These weightings could be changed by the user to reflect special conditions.

Step 3: Calculate the functional assessment outcome

Linking the processes to the function is expressed as the sum of the available biomass through primary production and biomass import minus the exported biomass, assuming that the export does not exceed the total available biomass in the HGMU.

The following calculation should be used:

$$(1+ (S1 * W1 + S2 * W2 + S3 * W3 + S4 * W4 + S5 * W5) - (S6 * W6 + S7 * W7 + S8 * W8 + S9 * W9 + S10 * W10))/2$$

For example, if the output codes for an HGMU were as follows:

3.3.B.1	1/b
3.3.B.2.1	1/2
3.3.B.2.2	1/1

3.3.B.2.3	1/3
3.3.B.3	1/2
3.3.B.4.1	1/1
3.3.B.4.2	1/x
3.3.B.4.3	1/2
3.3.B.5.1	2
3.3.B.5.2	1/c

Reference to Table 3.3.7 would result in the following calculation:

$$(1+((0.75 \times 0.6) + (0.67 \times 0.1) + (1 \times 0.1) + (0.33 \times 0.1) + (0.67 \times 0.1)) - ((1 \times 0.1) + (0.33 \times 0.1) + (0.67 \times 0.1) + (0 \times 0.1) + (0.5 \times 0.6))) \div 2$$

which is equivalent to:

$$(1 + (0.45 + 0.067 + 0.01 + 0.033 + 0.067) - (0.01 + 0.033 + 0.067 + 0 + 0.3) \div 2$$

which is equivalent to:

$$(1 + (0.627 - 0.41)) \div 2$$

which is equivalent to: $(1 + 0.217) \div 2$

which equals 0.6085

The outcome of this calculation should be compared with the following table and the functional code may be determined.

Calculation outcome	Functional code
0.76–1	1/1
0.51–0.75	1/2
0.26–0.5	1/3
0.1–0.25	1/4
0	4

This completes the functional analysis for food-web support. The output codes and quantification values can be recorded on the Functional Analysis Summary Table in Section 4.0, and used to compare functions among HGMUs across the wetland.

When all the HGMUs in the assessment area have been assessed, a general pattern of the provision of food-web support across the whole wetland will be evident. By manipulation of your answers to the controlling variable (CV) questions, you have the capacity to explore the probable outcomes of any management options under consideration and relate them to specific functions. The electronic version of the FAPs can do this more quickly and easily.

Functional Assessment Table: 3.3.B Food-web support

HGMU code	Food-web support

AA name: ______________________________

Section 4

Scoring and appendices

4.0 Functional analysis summary

Functional Analysis Summary Table

A Functional Analysis Summary Table is shown on p. 620. Photocopy this to record the functional assessment outputs for each HGMU. An example of a completed form is shown on p. 621.

When the functional assessments have been carried out

After completion of the FAPs for the first HGMU to be assessed, the procedure is simply repeated for each HGMU in turn until functional assessments are complete for the entire AA. When this is complete, the user will have obtained ranked outputs and descriptions of each *process* studied in each HGMU, and categorised outputs for each *function* studied in each HGMU.

The FAPs were designed to provide answers to management questions that otherwise would require detailed scientific research. There are no pages of data that need to be interpreted by scientists, but the FAPs supply the broad answers to the same questions. They are able to do this because of the extensive research that was conducted in the contributory research programmes (see Section 1.8). The findings of this research were used to match the processes and functions that occur in wetlands with the structure of the questions that comprise the database and assessment sections of the FAPs. It is this underlying research that enables the FAPs to assess wetland functioning effectively in a way that is much quicker and cheaper than site-specific empirical scientific research.

The information obtained from the FAPs enables the user to compile a picture of functional performance across the entire wetland assessment area. It is useful to do this on a map, since it can be used directly for planning any changes in management. The WEDSS computer package, which is a sister publication to the FAPs developed during the EVALUET research programme, can be downloaded directly from the SWIMMER website under 'WEDSS download' (http://www.liv.ac.uk/swimmer/research/past_projects.html). The WEDSS has a mapping facility integral to its structure, and can assist further in management and policy planning.

The FAPs outputs can be used in a number of ways. Among them are:

- to assess current functioning in the wetland generally or for strategic management targets, e.g. to evaluate the wetland for wading bird habitat, or its potential to mitigate the downstream impacts of hydrological events;
- to put the knowledge to work for planning management changes to the wetland;
- to test potential management scenarios for their probable consequences both for the intended target and for non-target functions;
- to compile a broad picture of wetland functioning across a wide area of landscape;
- to contribute to an understanding of ecosystem services provided by the wetland.

Scenario testing is possible by changing the answers given in Sections 2 and 3 to reflect the management options being considered. For example, if the user wishes to explore the potential for extending the period of flooding of an area of wetland by installation of a bund or the operation of sluice gates (perhaps to encourage a species of bird), the answers to certain questions in the appropriate sections (e.g.

Functional Analysis Summary Table: AA name: ______________________________

HGMU code	Floodwater Detention 3.1.A	Groundwater Recharge 3.1.B	Groundwater Discharge 3.1.C	Sediment Retention 3.1.D	Nutrient Retention 3.2.A	Nutrient Export 3.2.B	In-situ Carbon Retention 3.2.C	Trace Element Storage 3.2.D	Trace Element Export 3.2.E	Organic Carbon Conc. Control 3.2.F	Ecosystem Maintenance 3.3.A	Food Web Support 3.3.B

Functional Analysis Summary Table: example AA name: Kismeldon Meadows

HGMU code	Floodwater Detention 3.1.A	Groundwater Recharge 3.1.B	Groundwater Discharge 3.1.C	Sediment Retention 3.1.D	Nutrient Retention 3.2.A	Nutrient Export 3.2.B	In-situ Carbon Retention 3.2.C	Trace Element Storage 3.2.D	Trace Element Export 3.2.E	Organic Carbon Conc. Control 3.2.F	Ecosystem Maintenance 3.3.A	Food Web Support 3.3.B
F/GO/ELV	1/x	2	4	1/x	1/x	2	2	0.8	0.3	0.8	0.5 to < 0.75	1/3
F/GO/DBS	1	2	1/x	1	1/x	1	1	0.2	0.5	1	1	1/4
F/G/DDT	1	2	1/x	1	1	1	1	0.4	0.5	0.2	0.25 to < 0.5	4
S/G1/IUS	4	4	1	4	2	1/x	1/x	0.1	0.7	0.5	0.25 to < 0.5	1/1

hydrological, soils, vegetation) of the FAPs can be altered in line with the proposed changes, and the FAPs outputs will predict the effects of the changes on all functions dependent on them.

This capacity can be taken further to predict functioning in a range of potential management options, and was illustrated in Section 1.10, where a description was given of the ability of the FAPs to facilitate the evaluation of wetlands on a landscape scale. The method relies on matching HGMUs in the AA, which have already been assessed, with other HGMUs delineated in the target landscape that have similar characteristics. There is an inherent assumption that the contributory areas (see Section 2) in the already-assessed area and the target landscape are similar and, where this is obviously not the case, the user should establish the character of the target area where it influences the HGMU, and conduct other functional assessments as appropriate.

Nonetheless, the extrapolation of functioning in assessed HGMUs to similar HGMUs across a broad landscape area can be used to make a comparatively rapid evaluation on a landscape scale, as illustrated in Table 1.10.2. The example in Section 1.10 goes further in that it combines extrapolation of the functions to a wider area with the testing of three alternative potential management options. These were: carrying on as normal ('Degraded'), complying with minimum policy for environmental performance ('Policy Compliance'), and management to maximise the environmental value of the area ('Green'). Evaluation of the proposed management alternatives was then possible for functioning, and the resulting ecosystem services, across the entire landscape area. Further adjustments to the FAPs inputs are, of course, possible in order to explore further management alternatives.

Having reached this far, the user will now be in a position to perform scenario-testing using changes to the inputs. This can, however, be time-consuming, and we recommend the use of the electronic version of the FAPs, which accompanies this manual. The computerised versions make scenario-testing relatively quick and easy, and the results are displayed on maps that show the predicted changes on the ground.

Ecosystem services

Ecosystem services refers to the outcomes of ecosystem functioning as seen from the perspective of human society. All ecosystems perform functions from which human societies gain or, in some cases, lose. In the case of wetlands, these may be economically important and life-sustaining when they result in harvests of food and materials or threatening if they facilitate disease transmission. Additionally, wetlands can be potentially life-saving when floods or droughts are minimised by their presence.

In recent years, ecosystem services have been the subject of intensive evaluation, with a view to giving economic weight to arguments for conserving them (see e.g. Barbier *et al.*, 1997; Turner *et al.*, 2003). If the combined values of wetlands' multiple ecosystem services can be presented economically, they can be set against the predicted economic value of any proposed non-wetland conversion of the area. Commonly, the value to human societies of services provided by ecosystems can exceed that potentially to be gained by their conversion to other uses (Balmford *et al.*, 2002).

Application of the FAPs to a wetland gives a measure of the ecosystem services provided by it. This can be the first stage in the process of a more comprehensive economic valuation of the wetland. At the end of the functional assessment process, the Functional Assessment Procedures CD-ROM contains a visual appraisal that helps the user to gauge the overall performance of the HGMU, allowing the user to gain an at-a-glance impression of functioning and the delivery of ecosystem services (Fig. 4.1). Figure 4.2 shows a process chart from the Functional Assessment Procedures CD-ROM, providing a display of all processes occurring in the HGMU, along with clickable access to the FAPs output (score and rationale statement). This facilitates comparisons among any of the processes assessed.

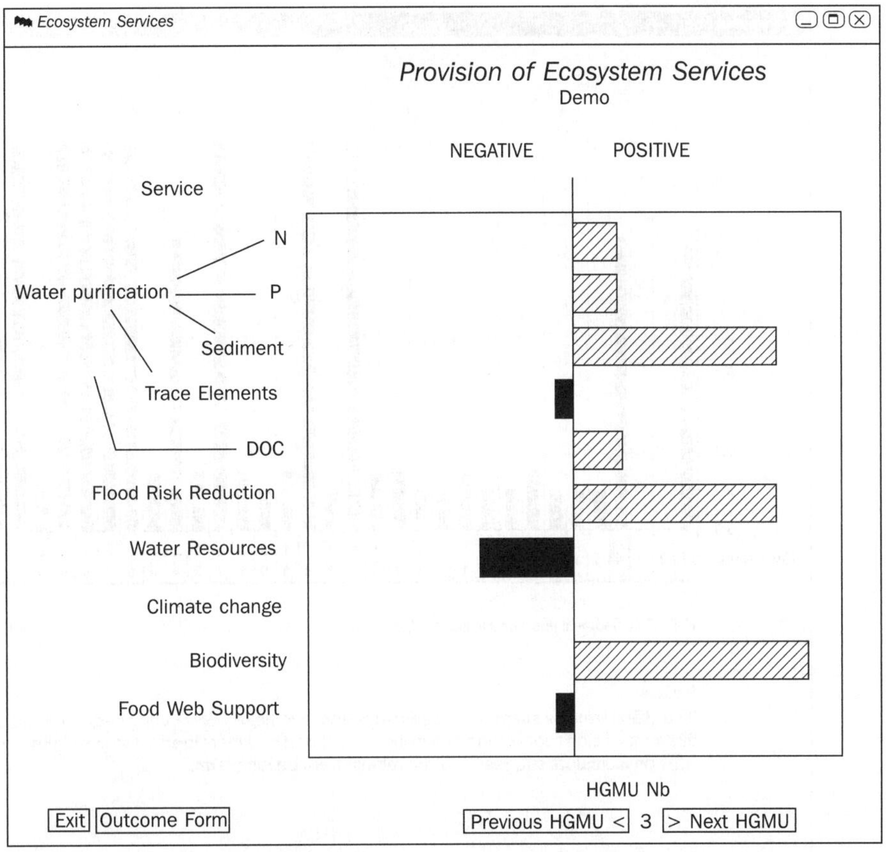

4.1 Part of the output display of the Functional Assessment Procedures CD-ROM, giving a rapid impression of the contribution to selected ecosystem services by a particular HGMU during a wetland assessment. This output shows that management presently favours biodiversity and species health, contributes somewhat to the improvement of water quality, and has little or no effect on water resources.

During development of the FAPs, several assessments were made of wetland areas in different parts of Europe. One example is reported in Janssen *et al.* (2005). Another assessment was made in Hovran, Central Sweden. This examined wetland functions supporting biodiversity, recreation (bird watching), and water quality, especially with regard to phosphorus transport.

The area was assessed under five different management scenarios. These were *business as usual, historical land use, dynamic water, pasture-based* management and best available *agricultural technologies* (buffer-zones, spring ploughing or catch crops). The functional assessment outputs (shown in Figs 4.3–4.5) revealed that the most suitable management options differed according to the function under consideration.

The benefits of all management proposals under consideration were compared using multicriteria analysis (Fig. 4.6), which selected *pasture-based* management as the option providing optimal environmental balance for all of the target criteria combined. *Dynamic water, agricultural technologies, business as usual*, and *historical land-use* ranked lower on performance for the chosen ecosystem services.

References

Balmford, A., Bruner, A., Cooper, P., Costanza, R. and 15 others (2002). Economic reasons for conserving wild nature, *Science* **297**, 950–953.

Barbier, E.B., Acreman, M.C. and Knowler, D. (1997). *Economic Valuation of Wetlands: a Guide for Policy Makers and Planners*, Ramsar Convention Bureau, Gland, Switzerland.

Janssen, R., Herwijnen, M.V. and Beinat, E. (2001). *DEFINITE: Cases and User Manual*, Report R-01/01, Institute for Environmental Studies, Vrije Universiteit Amsterdam, The Netherlands.

Turner, R.K., van den Bergh, J.C.J.M. and Brouwer, R. (2003). *Managing Wetlands: an Ecological Economics Approach*, Edward Elgar Publishing, Cheltenham, UK.

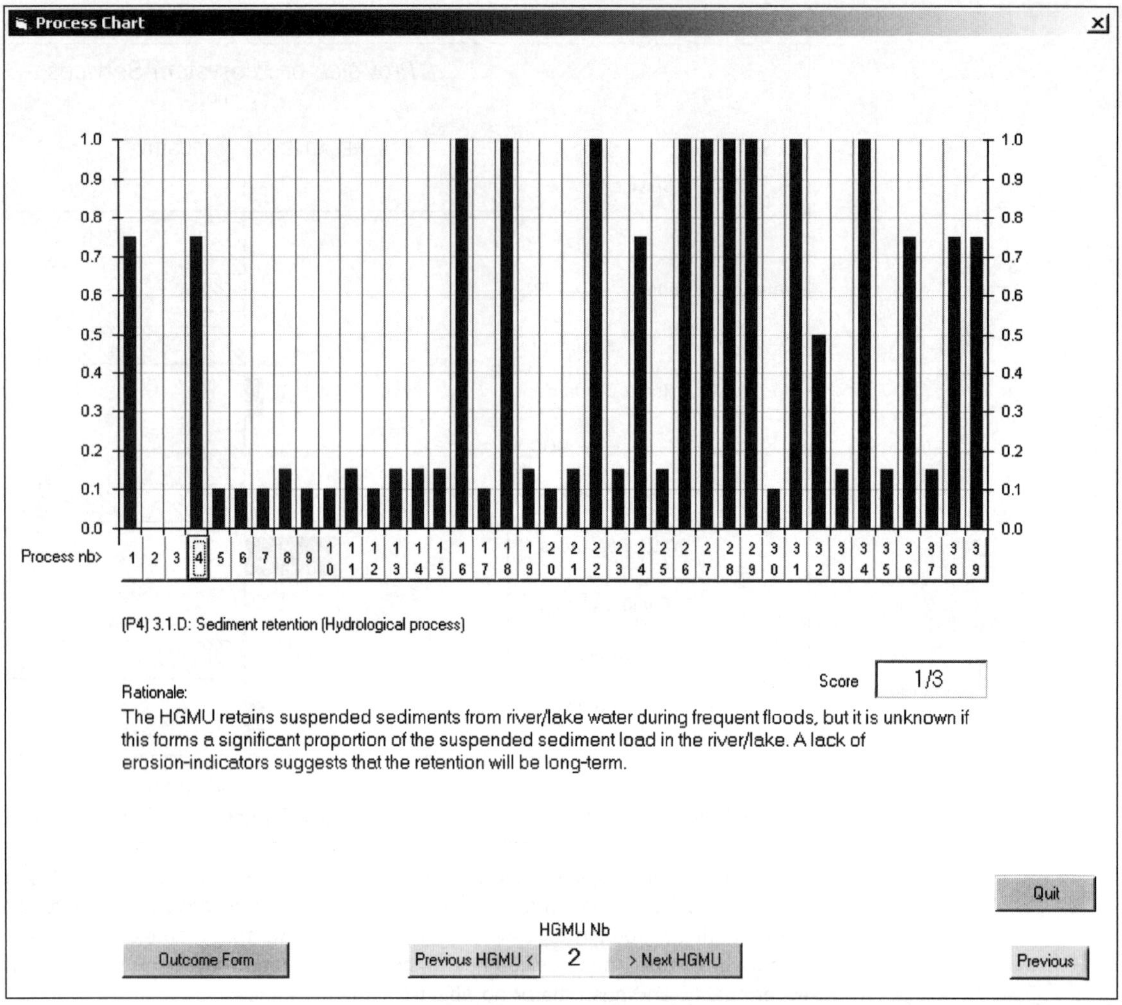

4.2 An example from the Functional Assessment Procedures CD-ROM showing comparative performance of processes in an HGMU. A scale of 0–1 is used to render the outputs comparable. Details about the score and rationale statement for any single process can be accessed by use of the mouse.

Manual scoring

The Functional Assessment Procedures CD-ROM automatically produces scores for the functional outputs of the FAPs, and the answers to the CV questions can also be keyed directly into the WEDSS; however, if you wish to do the scoring manually, a method is described below.

Scoring the functions

The functional analyses made using Sections 3.1 (hydrological), 3.2 (biogeochemical) and 3.3 (ecological) can be used directly from the section outputs in order to evaluate HGMU performance for each process and, when combined, for the resulting functions. The output values for functions, which can be entered in the Functional Analysis Summary Table on p. 620, differ according to the characteristics of each function. The process outputs may, however, be summed separately in a way that gives a numerical score for each function. This makes it easier to compare functional performance before and after a change in management practice, or among different parts of the wetland assessment area. Although the FAPs on the CD-ROM do this automatically, the same process can be achieved manually using the output codes conversion boxes in the following sub-section.

The purpose of this numerical scoring system is to permit comparisons of the same function across HGMUs or in the same HGMU over time. This enables the user to assess the effects of changes made to CVs and to make quick comparisons

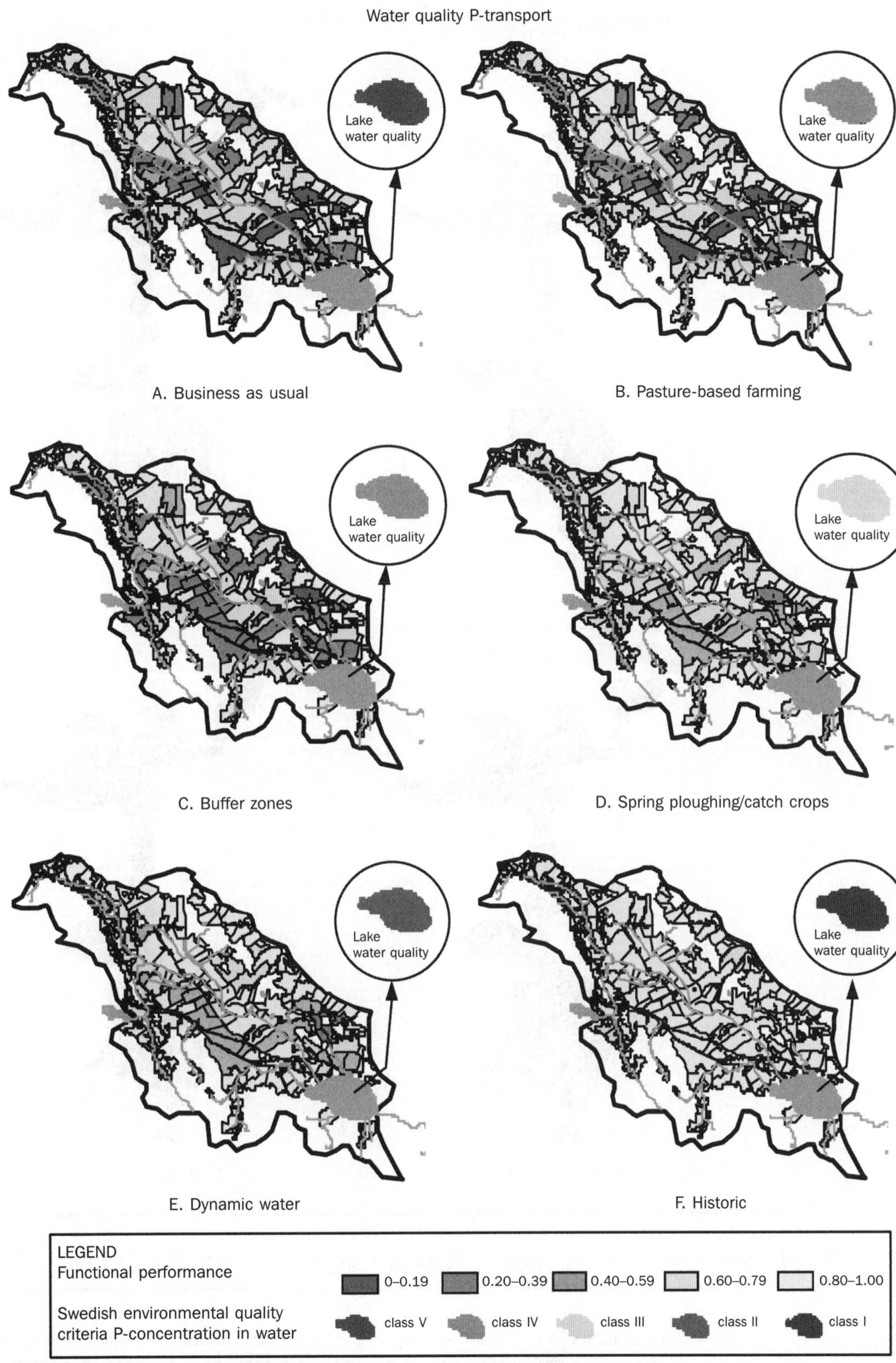

4.3 Outcomes of functional assessments for a lake catchment within the assessment area in Hogvan, Sweden. Output scores were translated into a scale of 0–1. Different management scenarios were tested for their potential effects on in-lake phosphorus concentrations. The result shows clearly the deleterious effects of modern land management in comparison with historical practices (data from Magnus Carlsson).

4.4 Outcomes of functional assessments for a wetland assessment area in Hogvan, Sweden. Output scores were translated into a scale of 0–1. Different management scenarios were tested for their potential effects on floral diversity. Traditional forms of land management were shown to increase the diversity of meadow flora when compared with other forms of management (data from Magnus Carlsson).

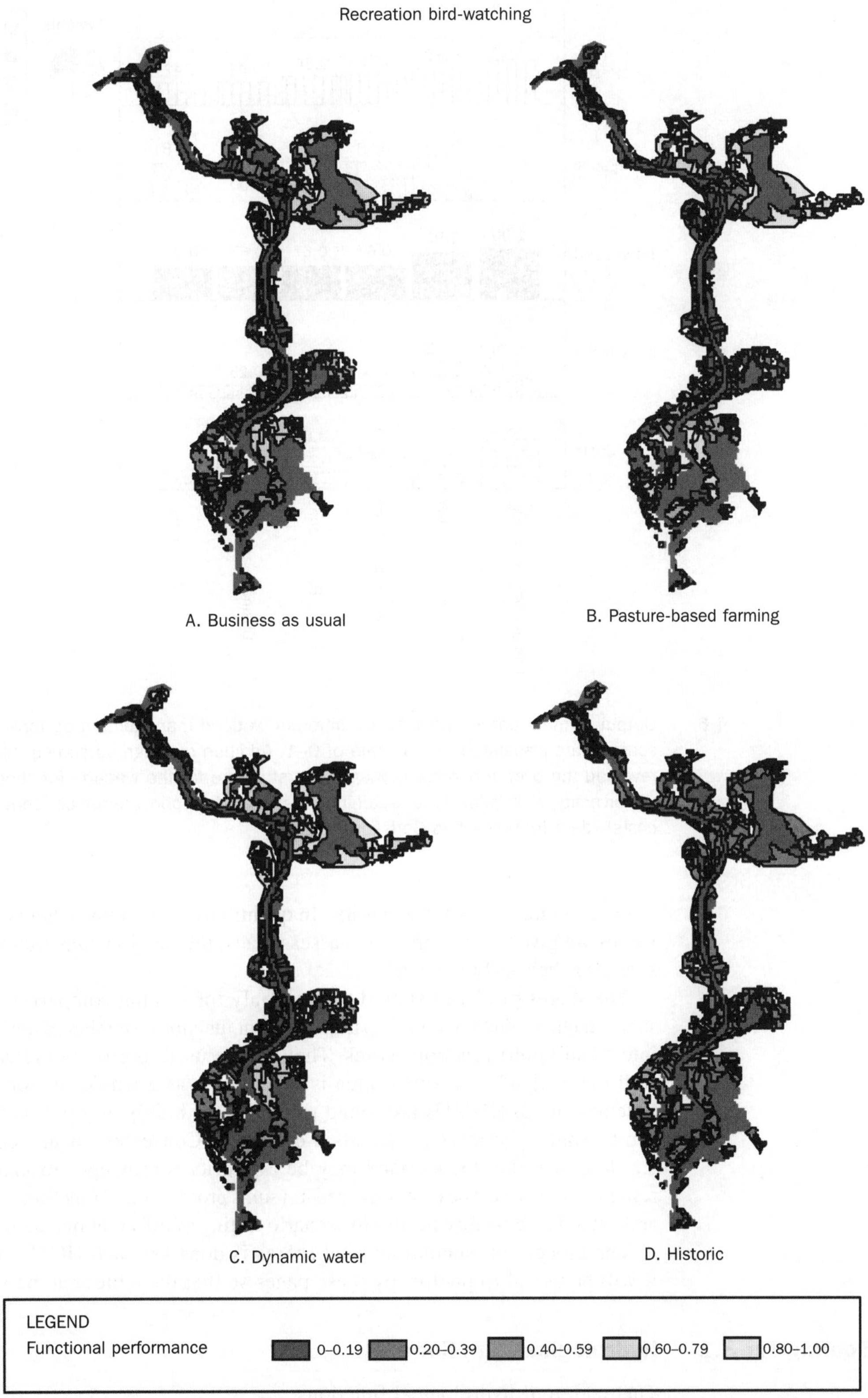

4.5 Outcomes of functional assessments for a wetland area in Hogvan, Sweden, in which different management scenarios were tested for their potential effects on provision of habitat for birds. Output scores were translated into a scale of 0–1. The introduction of a dynamic water regime was found to be the optimum management for birds (data from Magnus Carlsson).

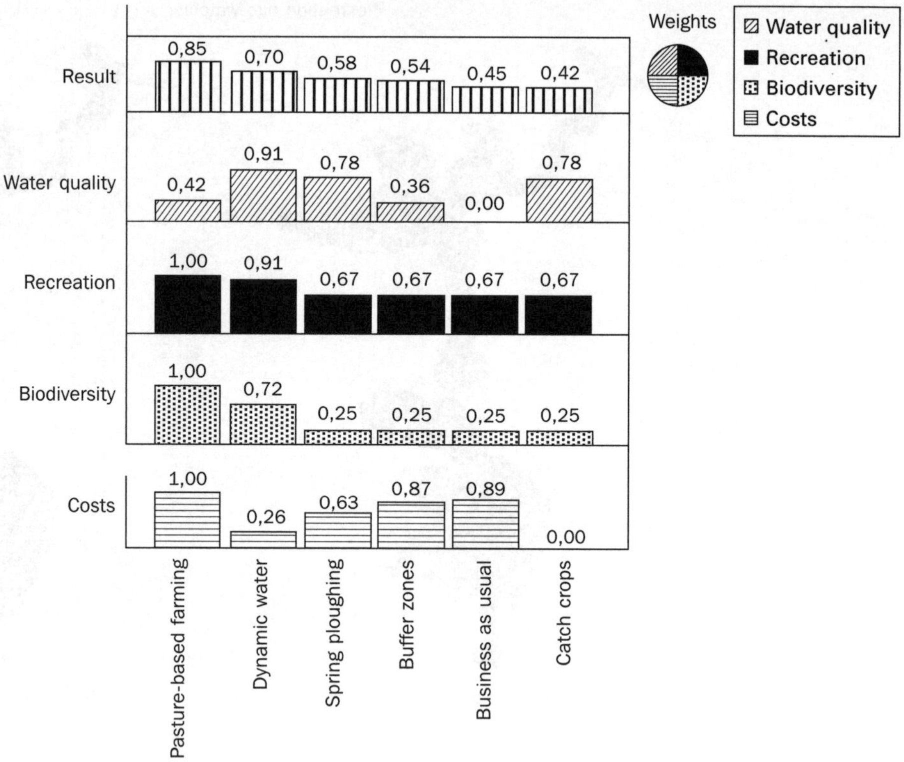

4.6 Output of multi-criteria analysis for different wetland management options. Output scores were translated into a scale of 0–1. Addition of weighted scores (top line) revealed the overall optimal management structure for the wetland functions assessed (biodiversity, recreation (bird watching), water quality (phosphorus concentration) and costs) (data from Magnus Carlsson).

of different management scenarios. In quantitative functions, where concentration values are given in the functional assessments, this section constitutes a shorthand score for their easy comparison.

The scores produced should be used only for making comparisons among the same functions, and not for comparing different functions, and should not be aggregated into 'total' multi-function scores. They can be used to give an indication of how well the wetland assessment area is functioning as a whole. If, for example, all functions in all HGMUs are found to score very highly, say 0.7–1, then the area can be said to be in generally good condition. Conversely, if all scores are low, e.g. less than 0.4, the wetland may be a priority for changes in management or restoration activity. The electronic version sums processes and functions automatically, and extends the facility for use in scenario testing of different management options.

The process of calculating total scores is done for each HGMU individually. It will be useful to photocopy these pages so that the tables can be written on.

Scoring for 3.1 Hydrological functions

There are four hydrological functions:

3.1.A Floodwater detention (comprises 1 process)
3.1.B Groundwater recharge (comprises 1 process)
3.1.C Groundwater discharge (comprises 1 process)
3.1.D Sediment retention (comprises 1 process)

Referring to Section 3, read off the correct scores in the following box, and enter them in the Comparative Score Table (a photocopiable table will be found on p. 636).

Scoring box for 3.1 Hydrological functions

3.1.A	Score	3.1.B	Score	3.1.C	Score	3.1.D	Score
1/1	1	*1/1*	1	*1/1*	1	*1/1*	1
1/2	0.9	*1/2*	0.9	*1/2*	0.9	*1/2*	0.9
1/3–1/5	0.8	*1/3*	0.8	*1/3*	0.8	*1/3–1/6*	0.8
1/6–1/7	0.7	*1/4*	0.7	*1/4–1/5*	0.7	*1/7–1/11*	0.7
1/8–1/12	0.6	*1/5*	0.6	*1/6–1/7*	0.6	*1/12–1/17*	0.6
1/13–1/14	0.5	*1/6*	0.5	*1/8*	0.5	*1/18–1/20*	0.5
2	0.1	*2*	0.1	*2*	0.1	*2*	0.1
4	0	*4*	0	*4*	0	*4*	0

Scoring for 3.2 Biogeochemical functions

There are six biogeochemical functions:

3.2.A Nutrient retention (comprises 5 processes)
3.2.B Nutrient export (comprises 4 processes)
3.2.C In situ carbon retention (comprises 1 process)
3.2.D Trace element storage (comprises 3 processes)
3.2.E Trace element export (comprises 3 processes)
3.2.F Organic carbon concentration control (comprises 2 processes).

Scoring for 3.2.A *Nutrient retention (5 processes)*

The functional analysis output codes for the function (5 processes) are quantitative. The derived values should be entered in the functional analysis summary table above and used for quantitative comparisons among HGMUs.

A non-quantitative score for the function can be derived using the following box.

Scoring box for 3.2.A Nutrient retention

3.2.A.1	Score	3.2.A.2	Score	3.2.A.3	Score	3.2.A.4	Score	3.2.A.5	Score
1/a	1	*1/a*	1	*1*	1	*1*	1	*1*	1
1/b	0.9	*1/b*	0.9	–	–	–	–	–	–
1/c	0.8	–	–	–	–	–	–	–	–
1/x	0.4	*1/x*	0.4	*1/x*	0.4	*1/x*	0.4	*1/x*	0.4
2	0.1	*2*	0.1	*2*	0.1	*2*	0.1	*2*	0.1

A score for the entire function is obtained by taking a simple average of the individual process scores obtained, e.g. if the scores obtained for the processes were as follows: *3.2.A.1* = 0.8; *3.2.A.2* = 1; *3.2.A.3* = 0.4; *3.2.A.4* = 0.1; and *3.2.A.5* = 0.4, then the overall function score would be (0.8 + 1 + 0.4 + 0.1 + 0.4)/5 = 0.54. The following box will be useful.

3.2.A	Process	Process	Process	Process	Process	**Mean**
Score						

Enter the mean score in the Comparative Score Table on p. 636.

Scoring for 3.2.B *Nutrient export (4 processes)*

Use the same procedure to derive a mean score for the function of Nutrient Export.

Scoring box for 3.2.B Nutrient export

3.2.B.1	Score	**3.2.B.2**	Score	**3.2.B.3**	Score	**3.2.B.4**	Score
1/a	1	***1***	1	***1/a***	1	**1**	1
1/b	0.9	–	–	***1/b***	0.9	–	–
1/c	0.8	–	–	***1/c***	0.8	–	–
–	–	–	–	***1/d***	0.7	–	–
1/x	0.25	***1/x***	0.25	***1/x***	0.25	***1/x***	0.25
2	0.1	**2**	0.1	**2**	0.1	**2**	0.1

The following box will be useful.

3.2.B	Process	Process	Process	Process	**Mean**
Score					

Enter the mean score in the Comparative Score Table on p. 636.

Scoring for 3.2.C *In-situ carbon retention (1 process)*

Referring to Section 3, read off the score given in the following box, and enter it in the Comparative Score Table on p. 636.

Scoring box for 3.2.C *In-situ* carbon retention

3.2.C.1	Score
1/a	1
1/b	0.9
–	–
1/x	0.4
2	0.1

Scoring for 3.2.D *Trace element storage (3 processes)*

The assessment for trace elements must be placed in context of the typical values found for the specific trace element of concern in the catchment of the wetland. Please refer to the cautionary note given in the text (Actions 3.2.D.2 and 3.3.E.2). Compare the combinations of process output codes (e.g. **1**, **1/x**, **2**, **4**) derived in Section 3, and read off the semi-quantitative score in the following box to gain an overall cumulative score. Enter it in the Comparative Score Table on p. 636.

Scoring box for 3.2.D Trace element storage

Process	*Process*	*Process*	Score
1	**1**	**1**	1
1	**1**	**1/x**	0.9
1	**1**	**2**	0.9
1	**1**	**4**	0.9
1	**1/x**	**1/x**	0.8
1	**1/x**	**2**	0.8
1	**1/x**	**4**	0.8
1	**2**	**2**	0.8
1	**2**	**4**	0.8
1	**4**	**4**	0.8
1/x	**1/x**	**1/x**	0.5
1/x	**1/x**	**2**	0.4
1/x	**1/x**	**4**	0.4
1/x	**2**	**2**	0.3
1/x	**2**	**4**	0.3
1/x	**4**	**4**	0.2
2	**2**	**2**	0.1
4	**4**	**4**	0

Scoring for 3.2.E *Trace element export (3 processes)*

Compare the combinations of process output codes (e.g. **1**, **1/x**, **2**, **4**) derived in Section 3, and read off the semi-quantitative score in the following box to gain an overall cumulative score. Enter it in the Comparative Score Table on p. 636.

Scoring box for 3.2.E Trace element export

Process	*Process*	*Process*	Score
1	**1**	**1**	1
1	**1**	**1/x**	0.9
1	**1**	**2**	0.9
1	**1**	**4**	0.9
1	**1/x**	**1/x**	0.8
1	**1/x**	**2**	0.8
1	**1/x**	**4**	0.8
1	**2**	**2**	0.8
1	**2**	**4**	0.8
1	**4**	**4**	0.8
1/x	**1/x**	**1/x**	0.5
1/x	**1/x**	**2**	0.4
1/x	**1/x**	**4**	0.4
1/x	**2**	**2**	0.3
1/x	**2**	**4**	0.3
1/x	**4**	**4**	0.2
2	**2**	**2**	0.1
4	**4**	**4**	0

Scoring for 3.2.F *Organic carbon concentration control (2 processes)*

In the following box, select the appropriate row and column for the output score and read off the cumulative score. Enter the result in the Comparative Score Table on p. 636.

Scoring box for 3.2.F Organic carbon concentration control

Output code	*1/1*	*1/2*	*1/x*	*2*	*4*
1/1	1	0.925	0.7	0.55	0.5
1/2		0.85	0.625	0.475	0.425
1/x			0.4	0.25	0.2
2				0.1	0.05
4					0

This score assumes that both processes carry equal weight.

Scoring for 3.3 Ecological functions

There are two ecological functions:

3.3.A Ecosystem maintenance (comprises 7 processes)
3.3.B Food-web support (comprises 9 processes)

This part of the summing is based on the same scoring structure as in Section 3.3, Action A.2, Step 2 but is reached in a simpler way.

Scoring for 3.3.A.1 *Ecosystem maintenance (1 process); sub-function 1*

Refer to Section 3 and read off the code from the box below, then enter the score in the 3.3.A Sub-function Summing Table on p. 634.

Scoring box for 3.3.A.1 Ecosystem maintenance

3.3.A.1	Score
1/1	1
1/2	0.75
1/3	0.5
1/x	0.25
2	0

Scoring for 3.3.A.2 *Provision of micro-sites (5 processes); sub-function 2*

- 3.3.A.2.1 invertebrates
- 3.3.A.2.2 fish
- 3.3.A.2.3 herptiles
- 3.3.A.2.4 birds
- 3.3.A.2.5 mammals

Refer to Section 3 and, for each process output code, read off the appropriate score in the following box, then enter the result in the Mean Score Table on p. 633.

Scoring box for 3.A.2 Provision of micro-sites

Output code	Score
1/1	5
1/2	0.75
1/3	0.5
1/x	0.25
2	0

Mean score table

Process	Score
3.3.A.2.1	
3.3.A.2.2	
3.3.A.2.3	
3.3.A.2.4	
3.3.A.2.5	
Total	
Mean	

The maximum valid score is 1. *Mean* scores above this figure should be disregarded. For example, if the scores for individual processes in sub-function 2 were 0.75, 0.5, 0.25, 5 and 0.75 (total 7.25), the mean would be 1.45 (7.25/5 = 1.45). This should be entered as the maximum score of 1, and the other 0.45 ignored. Next, enter the score in the 3.3.A Sub-function Summing Table on p. 634.

Scoring for 3.3.A.3 *Provision of plant and habitat diversity (1 process); sub-function 3*

Refer to Section 3 and read off the code from the following box, then enter the score in the 3.3.A Sub-function Summing Table on p. 634.

Scoring box for 3.3.A.3 Provision of plant and habitat diversity

3.3.A.3	Score
1/1	1
1/2	0.75
1/3	0.5
1/x	0.25
2	0

Enter the three sub-function scores in the Sub-function Summing Table below (i.e. the scores from sub-functions 1 and 3, and the mean score for sub-function 2), and take a simple average to derive a mean score for the whole function of ecosystem maintenance. This mean score can then be transferred to the Comparative Score Table on p. 636.

Sub-function Summing Table

Sub-function	Score
1	
2	
3	
Total	
Mean	

Where the wetland habitat favours a priority species type, this is reflected in the scoring for sub-function 2. Scores for non-priority species can be used to assess the feasibility of broadening the habitat for increased biodiversity where this does not jeopardise conditions for priority species. Your experience and judgement will be important.

Scoring for 3.3.B *Food-web support (ten processes)*

The individual influences of the ten processes of biomass import and export are not equal in ecological importance, so cannot be compared directly one against the other. Consequently, we have used expert judgement to weight the individual process scores, and so allow comparisons of the relative contribution to food-web support of each process.

Processes that add biomass:
3.3.B.1 Biomass production
3.3.B.2.1 Biomass import by physical processes: watercourses
3.3.B.2.2 Biomass import by physical processes: overland flow
3.3.B.2.3 Biomass import by physical processes: wind transport
3.3.B.3 Biomass import by biological processes: animal movements

Processes that remove biomass:
3.3.B.4.1 Biomass export by physical processes: watercourses
3.3.B.4.2 Biomass export by physical processes: overland flow
3.3.B.4.3 Biomass export by physical processes: wind transport
3.3.B.5.1 Biomass export by physical processes: fauna
3.3.B.5.2 Biomass export by physical processes: anthropogenic means

Refer to Section 3.3.B and use the process output codes to derive a score according to the following boxes, then enter the score in the weighting table below. The scoring is similar to that done in Section 3.3, with the scores given in the following boxes.

Scoring box for 3.3.B.1 Biomass import

3.3.B.1	Score	*3.3.B.2.1*	Score	*3.3.B.2.2*	Score	*3.3.B.2.3*	Score	*3.3.B.3*	Score
1/a	1	*1/1*	1	*1/1*	1	*1/1*	1	*1/1*	1
1/b	0.75	*1/2*	0.67	*1/2*	0.67	*1/2*	0.67	*1/2*	0.67
1/c	0.5	*1/3*	0.33	*1/3*	0.33	*1/3*	0.33	–	–
1/d	0.25	–	–	–	–	–	–	–	–
1/x	0.25	*1/x*	0.33	*1/x*	0.33	*1/x*	0.33	*1/x*	0.33
2	0	*2*	0	*2*	0	*2*	0	*2*	0

Scoring box for 3.3.B.2 Biomass export

3.3.B.4.1	Score	***3.3.B.4.2***	Score	***3.3.B.4.3***	Score	***3.3.B.5.1***	Score	***3.3.B5.2***	Score
1/1	1	***1/1***	1	***1/1***	1	***1/1***	1	***1/a***	1
1/2	0.67	***1/2***	0.67	***1/2***	0.67	***1/2***	0.67	***1/b***	0.75
1/3	0.33	***1/3***	0.33	***1/3***	0.33	***1/3***	0.33	***1/c***	0.5
–	–	–	–	–	–	–	–	***1/d***	0.25
1/x	0.33	***1/x***	0.33	***1/x***	0.33	***1/x***	0.33	***1/x***	0.25
2	0	***2***	0	***2***	0	***2***	0	***2***	0

Transfer the scores to the weighting tables below. Multiply the output score for each process by the weighting factor indicated in the 'weighting' column. Enter the result in the 'weighted scores' column, e.g. $0.75 \times 0.6 = 0.45$; $0.67 \times 0.1 = 0.067$. Then add the weighted scores and enter the total in the bottom right hand corner.

Weighting table (import)

Import process	Process output score	***Weighting***	Weighted scores
3.3.B.1		0.6	
3.3.B.2.1		0.1	
3.3.B.2.2		0.1	
3.3.B.2.3		0.1	
3.3.B.3		0.1	
Total	–	–	

Weighting table (export)

Export process	Process output score	***Weighting***	Weighted scores
3.3.B.4.1		0.1	
3.3.B.4.2		0.1	
3.3.B.4.3		0.1	
3.3.B.5.1		0.1	
3.3.B.5.2		0.6	
Total	–	–	

Finally, calculate an overall score for Food-web support as follows: Subtract the export total from the import total, add 1, then divide the answer by 2, i.e. if the *import* weighted scores were 0.45, 0.067, 0.033, 0.1 and 0.067 (total 0.717), and the *export* weighted scores were 0.033, 0.067, 0.033, 0.067 and 0.15 (total 0.35), then the equation would be $((0.717 - 0.35) + 1)/2 = 0.6835$, which can be rounded to 0.7. The first row of the Food-web Support Final Score Table on p. 636 gives this example as an illustration of the process.

Food-web Support Final Score Table

HGMU I.D.	Import total score	Export total score	+ 1	Sub-total	Divide by 2	=
e.g. **S/G1/IUS**	*0.717*	*0.35*	*1*	*1.367*	*2*	*0.6835*
			1		2	
			1		2	
			1		2	
			1		2	
			1		2	
			1		2	
			1		2	

The Comparative Score Table below can be used to record the numerical scores for performance of each function in each HGMU. If you wish to keep a record of scores for sub-functions use the Sub-function Scoring Table on page 637.

Comparative Score Table

Function	HGMU I.D.	HGMU I.D.	HGMU I.D.	HGMU I.D.	HGMU I.D.	HGMU I.D.
3.1.A						
3.1.B						
3.1.C						
3.1.D						
3.2.A						
3.2.B						
3.2.C						
3.2.D						
3.2.E						
3.2.F						
3.3.A						
3.3.B						

Sub-function Scoring Table

HGMU	Sub-function scores	3.3 Score (component)	3.3 Score (component)	3.3 Score (component)
	3.3.A	(sf-1)	(sf-2)	(sf-3)
	3.3.B	(import)	(export)	–
	3.3.A	(sf-1)	(sf-2)	(sf-3)
	3.3.B	(import)	(export)	–
	3.3.A	(sf-1)	(sf-2)	(sf-3)
	3.3.B	(import)	(export)	–
	3.3.A	(sf-1)	(sf-2)	(sf-3)
	3.3.B	(import)	(export)	–
	3.3.A	(sf-1)	(sf-2)	(sf-3)
	3.3.B	(import)	(export)	–
	3.3.A	(sf-1)	(sf-2)	(sf-3)
	3.3.B	(import)	(export)	–
	3.3.A	(sf-1)	(sf-2)	(sf-3)
	3.3.B	(import)	(export)	–
	3.3.A	(sf-1)	(sf-2)	(sf-3)
	3.3.B	(import)	(export)	–
	3.3.A	(sf-1)	(sf-2)	(sf-3)
	3.3.B	(import)	(export)	–
	3.3.A	(sf-1)	(sf-2)	(sf-3)
	3.3.B	(import)	(export)	–
	3.3.A	(sf-1)	(sf-2)	(sf-3)
	3.3.B	(import)	(export)	–
	3.3.A	(sf-1)	(sf-2)	(sf-3)
	3.3.B	(import)	(export)	–
	3.3.A	(sf-1)	(sf-2)	(sf-3)
	3.3.B	(import)	(export)	–
	3.3.A	(sf-1)	(sf-2)	(sf-3)
	3.3.B	(import)	(export)	–
	3.3.A	(sf-1)	(sf-2)	(sf-3)
	3.3.B	(import)	(export)	–

Appendix 1: Glossary and alternative terminology

It has been necessary to use a variety of scientific and specialised terms within the FAPs. While some of these are in common usage, they are nevertheless included in this Glossary to ensure uniformity of meaning and interpretation. Other terms are more specialised and are included here to aid the non-specialist user in application and understanding of the FAP methodology. Terms followed by code letters in brackets are those features requiring identification during the field study stage of database establishment.

Scientific terms

cm	centimetre
g	gram
ha	hectare [a hectare is 10 000 m^2 (2.471 acres; an acre is 0.405 ha)]
ha^{-1}	per ha [this $^{-1}$ rule generally replaces terms such as m/s]
HGMU	hydrogeomorphic unit
kg^{-1}	per kilogram
l	litre
l^{-1}	per litre
m	metre
ml (or mL)	millilitre
m^2	square metres
m^{-2}	per square metre
m^{-3}	per cubic metre
$m\ s^{-1}$	metres per second
mg	milligram
μg	microgram
m.O.D.	ordnance datum Newlyn (also ODN) [a standard level used in the UK, defined as: the mean level of the sea at Newlyn in Cornwall, England between May 1915 and April 1921]
sp.	of a species where the genus but not the species is identified, e.g. *Typha* sp. means 'a species of *Typha*'
spp.	of more than one species, e.g. *Typha* spp. means more than one species of *Typha* is discussed
a	year (annum)
a^{-1}	per year

Glossary

alluvial fan – sediments derived from erosion processes that fan out in deposits at the base of slopes.

aquifer – a rock formation or sediment that can store and transmit water capable of being abstracted and utilised as a resource.

arterial drainage – main part of a river or artificial drainage system.

artificial depression (AD) – excavated depression on a floodplain, made for a variety of purposes such as mineral extraction or footings for building construction.

artificial ponds – small, man-made water bodies, often lined with an impermeable layer or substrate to help retain water.

backswamp (BS) – permanently waterlogged low-lying land at the back of a floodplain.

bank – the side of a river channel.

biogeochemical – biotic and abiotic compartments of an ecosystem.

biomass – the quantity of living plant and animal material. Biomass production values represent the amount of organic material which has accumulated within an ecosystem over a specified period of time.

braided channel – characterised by successive division and rejoining of a streamflow channel with accompanying islands or braid bars.

canal (CN) – artificial navigational waterway.

cascade – a low, curving fall of water over a rock outcrop or boulders, in which the flow remains in contact with the substrate (see *waterfall*).

catchment – all the land which drains to a single discharge channel. In the context of a wetland, it refers to all land upslope of that wetland which receives rainfall that can enter the wetland as surface run-off. The same as the term '*watershed*' in North American usage.

cave (CV) – includes any natural cave, deep undercut, or disused mine.

channel – distinctive natural or man-made topographical trough-like depression that periodically or continuously conveys moving water through the landscape. River, stream, tributary and drainage are some of the terms used to describe channels.

common birds – all species of birds other than those listed as being threatened at some scale (either locally or globally).

construction – any building or infrastructural development such as housing, offices, factories, roads, railways, bridges, and the disturbance created by all related activities including provision of access roads and embankment construction.

cut-off channel (CO) – former river course that no longer transmits river flow and contains no permanent water (see *oxbow lake*) but is seasonally flooded.

depression – a feature in the landscape that is of lower elevation than that of the surrounding land.

disturbance – any factor which destroys or alters the structure of biomass, including the effects of trampling, grazing and fire damage.

ditch (DT) – man-made channel constructed to convey agricultural drainage; may carry permanent or intermittent flows of water, or water may be penned at a high level for wildlife conservation purposes.

ecosystem – (ecological system) all plants, animals, protists, fungi and micro-organisms and their products within a defined area that interact with each other and with the environment.

embankment (EM) – man-made linear elevated feature, usually constructed parallel to a river channel, either set back from the channel, or built up on top of a natural levee; intended to restrict the extent of inundation by river water at times of high discharge.

entrenched – containment of a river or stream by artificial deepening of the channel.

flat area (FA) – near horizontal area of land situated on a floodplain, where any surface irregularities are less than 0.5 m in height or depth.

flood duration – the length of time an area remains inundated by flood waters during a flooding event.

flood extent – the spatial extent over which an area becomes inundated during a flooding event.

flood frequency – the frequency (or rate of return) at which an area is flooded.

floodplain – the land adjacent to a river channel that receives water from overbank flooding and is composed of alluvial material. Floodplains can be active (still flood) or historic (no longer flood).

former channel (FC) – sinuous depression, seasonally flooded, still connected to the river channel.

function – the net product of naturally-occurring ecosystem processes and structure ('the things that wetlands do').

functional assessment – the evaluation of the capability of a wetland to perform functions; an assessment outcome may be in the form of a qualitative statement, or a value or rate of performance.

geomorphic – refers to the shape of the land and associated landforms.

geomorphology – the study of the land surface and the evolution of landforms.

gradient – slope of the ground surface or of the longitudinal profile of a river or its channel.

groundwater – water that is moving or being stored underground.

growing season – the part of the year with temperatures high enough to allow plant growth. Usually defined as the number of days between the last 'killing frost' of spring and the first one of autumn. In general its duration decreases with distance from the equator, with increasing altitude, and with distance from the sea.

humose – a resilient humus-containing, but not humus-rich, layer of soil, formed from decomposition of organic material. Usually dark brown to black in colour.

humus – organic matter in or on soil that is stable, and unlikely to break down further.

hydrogeomorphic unit (HGMU) – an area of homogeneous geomorphology and hydrology and, under normal conditions, soil. The basic unit upon which the functional assessment procedures are undertaken.

hydrologic – dealing with the topic of hydrology and the movement of water.

hydrological continuity – the ability of water to pass from one medium (e.g. rock type) to another without being impeded.

hydrology – the study of water, especially in relation to its occurrence in streams, rivers, lakes and wetlands.

hydrophytic vegetation – plants that live in water or saturated soil, such as floating and submerged plants.

indurate – as in a soil layer: to harden or grow hard.

karst – limestone or dolomite of high porosity due to dissolution of the rock along joints, fractures or bedding planes. Can contain underground streams and be vulnerable to groundwater contamination. These hydrological systems are now included in the definition of wetlands under the Ramsar Convention.

lake or pond (LK) – standing water bodies including man-made examples such as fish ponds, mill ponds and gravel pits.

lentic – pertaining to standing, or very slow flowing water bodies; includes lakes and reservoirs (*cf* lotic).

levee (LV) – natural elevation resulting from sediment deposition on a floodplain adjacent to a river channel.

livestock unit (LSU) – the unit used for livestock feed budgeting to enable feed energy requirements to be expressed in terms of a single livestock type. Approximate large animal equivalents are one cow, bullock, horse or mule are equal to three pigs and also equal to six sheep or goats.

lotic – pertaining to flowing water bodies, including rivers and streams (*cf* lentic).

meander – the winding of a stream channel.

mole drain – unlined small diameter 'tunnel' artificially created below the ground surface by pulling a bullet-shaped cylinder through the soil; intended to improve soil drainage or, more recently in wetland restoration, used as a technique to rewet land by subsoil irrigation.

non-wetland – areas that are not characterised by soils inundated or saturated at a frequency that allows them to develop the properties associated with waterlogging, or allowing the support of hydrophytic vegetation. Typical non-wetland areas may include naturally well drained (or artificially drained) agricultural land and levees.

ox-bow lake (OX) – a lake formed in a meandering river system, when a crescent-shaped course of the former channel is detached from the main channel as the river cuts across the narrowing neck of the loop, thereby creating a by-pass of the original course in a new channel. These still backwaters can dry up and become cut-off channels.

particle-size – the effective diameter of individual particles within a soil; sand (2 mm–60 μm), silt (2–60 μm) and clay (< 2 μm); the combined contribution of size classes gives the overall particle-size class (texture) of the soil, e.g. sandy loam, silty clay.

peat – unconsolidated, predominantly organic material consisting largely of plant remains accumulated under waterlogged conditions; the degree of preservation of remains varies from well preserved in fibrous peat to highly decomposed in amorphous peat.

peat dome (PD) – smooth convex elevated peat deposit on a floodplain or bordering a back-slope or terrace; can represent the initial stage of raised bog formation.

pipe drains – nowadays trenchless pipe installation uses continuous slotted PVC; formerly round clay pipes (tiles) were placed end-to-end in a trench; installed at a range of depths in the subsoil, depending on soil properties, crop drainage requirements, slope and rainfall.

point bar (PB) – area of sedimentation on the inside of a meander loop, often occurring as concentric ridges and swales; following inundation, sloughs of dead water sometimes form in the swales; can be vegetated or unvegetated.

pool – a distinct river channel feature of deeper water; pools are mainly central or occupy most of the channel width (*cf* riffle).

process – any physical, chemical or ecological interaction that has consequences for functioning and on which functioning depends.

Ramsar Sites – wetland sites that have been designated for conservation under the guidelines laid down in what is commonly known as The Ramsar Convention (*The Convention on Wetlands of International Importance, especially as Waterfowl Habitat*), adopted in 1971 at Ramsar in Iran.

riffle – fast-flowing shallow water with distinctly disturbed surface; spans most of the channel width but may be diagonal in sinuous streams (*cf* pool).

river dune (RD) – sand dunes, between 1 m and 10 m high, found on floodplains.

soil matrix – the major abiotic material, excluding stones, comprising soil horizons.

soil profile – a vertical section of the soil from the surface through all its layers (horizons).

step-pool – a sequential feature of cascade and pool that can occur where bedrock or boulders form a series of steps.

strand lines – elongated piles of floatable debris such as twigs, branches, herbaceous material, litter and sediment, deposited on floodplains at the highest point of inundation, following flood events.

stress – any factor which impairs physiological processes and/or reduces the rate of accumulation of biomass; includes shortage of light, water and nutrients.

sub-surface drainage – artificial drainage measures installed within the subsoil; can include tiles, clay pipes and mole drains.

surface drainage – artificial drainage measures that are open at the surface, such as channels and ditches.

terrace (TC) – a horizontal shelf or bench that is found in a position elevated above the active floodplain but below adjacent valleyside slopes, and is only inundated during exceptional flood events. River terraces can represent the remnants of a former floodplain, isolated at a higher level than the active floodplain, and dissected by renewed down-cutting by a rejuvenated river. They can also be the formations left by freely swinging meanders cutting into a former floodplain.

tile drains – pipes made of clay, concrete or ceramic material, laid with open joints in trenches excavated through the subsoil to collect and carry excess water from the soil.

topsoil – the upper layer of the soil, usually organic-rich and subject to oxidisation and hydration. In cultivated soils, often synonymous with 'the plough layer'.

tributary (TB) – stream or channel traversing a floodplain to enter a larger river and contribute to the discharge of the latter.

unidentified depression (UD) – depressional feature on the floodplain, with an unknown origin.

unidentified elevation (UE) – elevated feature on the floodplain, with an unknown origin.

upslope – any area that lies at higher elevation than a reference point.

waterfall – a feature of bedrock channels, where flow separates from the rock back wall of a distinct vertical feature (see also *cascade*).

wetland – a functional definition: heterogeneous but distinctive ecosystem in which special ecological, biogeochemical and hydrological functions arise from the dominance and particular sources, chemistry and periodicity of inundation or saturation by water; occurs in a wide range of landscapes and may support permanent shallow (generally < 2 m) or temporary standing water; has soils, substrates and biota adapted to flooding and/or waterlogging and associated conditions of restricted aeration.

wetland values – human interpretation and perception of the worth of wetland functions.

Alternative terminology

The following is a list of commonly used terms and phrases which are equivalent in part or total in meaning to the functions addressed by the FAPs.

Function code and name	Commonly used or alternative terms and phrases
3.1.A Floodwater detention	Flood attenuation Flood control Flood defence Flood desynchronisation Flood protection Decreased/reduced flooding downstream of wetland Flood water storage Inundation protection
3.1.B Groundwater recharge	Aquifer recharge Groundwater storage
3.1.C Groundwater discharge	Groundwater seepage Springline discharge
3.1.D Sediment retention	Deposition of sediment Heavy metals removal Sediment trapping Suspended solids removal Toxic substance removal
3.2.A Nutrient retention	Nitrogen storage Phosphorus storage
3.2.A Nutrient retention and 3.2.B Nutrient export	Buffer zone identification Eutrophication protection Fertiliser run-off removal Nitrate or nitrogen removal Nutrient attenuation Nutrient buffer Nutrient cycling Nutrient removal Phosphorus removal Water quality maintenance Water quality improvement
3.2.C *In situ* carbon retention	Biomass accumulation Carbon storage Organic carbon accumulation Peat formation
3.2.D Trace element storage	Trace element removal or retention Heavy metal removal, retention or storage
3.2.E Trace element export	Trace element release or removal Heavy metal export, release or removal
3.2.F Organic carbon concentration control	Dissolved organic carbon export or removal
3.3.A Ecosystem maintenance	Biodiversity maintenance Conservation of habitat, species or wetland ecosystem Ecological integrity Habitat diversity Habitat for plants Habitat for faunal groups (amphibians, birds, fish, invertebrates, mammals) Habitat for wildlife Maintenance of habitat structure Riparian habitat protection
3.3.B Food-web support	Biomass export Biomass production Food chain support Primary production Productivity

Appendix 2: Information sources (United Kingdom)

Some information sources that may prove useful for the collection of data from desk studies are listed below for the United Kingdom. This is not comprehensive and, in particular, web sites are constantly being updated. Many other sources may also be available.

Climate
County records centres
Local libraries
Met Office
http://www.metoffice.com/
http://www.metoffice.gov.uk/climate/uk/
Met information sources (University of Reading):
http://www.met.rdg.ac.uk/~brugge/bi.html
Defra (climate change): http://www.defra.gov.uk/environment/climatechange/
NERC: http://www.nerc.ac.uk
http://www.zetnet.co.uk/sigs/weather/scotclim/scotclim.html
AIRC/defra (Centre for Air Transport and the Environment): www.ace.mmu.ac.uk/
Atmospheric deposition: Defra; Environment Agency

Land use
County records centres
Local libraries
National Land Use Database: http://www.nlud.org.uk
Centre for Land Use and Water Resources Research: http://www.cluwrr.ncl.ac.uk/
Defra statistics: http://defra.gov.uk/environment/statistics/land/index.htm
Countryside Survey 2000: http://www.cs2000.org.uk/index.htm
UK Government statistics: http://www.statistics.gov.uk
UK Clearing House Mechanism for Biodiversity:
http://chm.org.uk/library/ecosys/agric/ag001.htm
UK agriculture statistics: http://www.ukagriculture.com/
Wildfowl faecal deposits: land owner, local wildlife trust
Human, industrial and domestic waste: Environment Agency, Scottish Environmental Protection Agency, River Purification Boards, water companies, county records centres
Growing season: land owner, MAFF (Defra) publications
Agricultural ammonium application: land owner, MAFF(Defra) publications
Agricultural crop burning: land owner, MAFF(Defra) publications
Agricultural faecal deposits: land owner, MAFF(Defra) publications
Agricultural nitrate application: land owner, MAFF(Defra) publications
Association of Rivers Trusts: http://www.associationofriverstrusts.org.uk
Follow links for details of individual rivers trusts.

Management
County records centres
Wetlands and natural resources research groups
Chartered Institute of Water and Environmental Management: http://ciwem.org.uk/events/wetlands.asp
Wise use of wetlands: www.ramsar.org/wurc/wurc_library.htm
Local wildlife trusts

Defra Code of Practice on Conservation, Access and Recreation, 2000: http://www.defra.gov.uk/rural/ruralwp/whitepaper/chapter11.htm

Conservation status

Biodiversity Action Plans, county records centres

The Wildfowl and Wetlands Trust: www.wwt.org.uk

Joint Nature Conservation Committee (JNCC), advisor to Government: http://www.jncc.gov.uk/

Natural England: http://www.naturalengland.org.uk

Wetland conservation: issues and application: http://www.gisdevelopment.net/ application/environment/wetland/index.htm

WWF: http://www.wwf.org.uk/researcher/focalregions/

Flooding information

County records centres

Defra, Flood and Coastal Zone Risk Management
www.defra.gov.uk/environ/fcd/default.htm
Tel: 0845 9881188
info-fm@defra.gsi.gov.uk

Flood event maps: Environment Agency; River Purification Boards; Water Companies

Historical flood event records: Environment Agency, River Purification Boards, Water Companies, county records centres

Historical river stage levels: Environment Agency, River Purification Boards; Water Companies, county records centres

Overbank floodwater inputs of nitrogen: Environment Agency, River Purification Boards, Water Companies

Groundwater information

Environment Agency; Water Companies, River Purification Boards

Depth of water table

Environment Agency; Water Companies, River Purification Boards

Other potential sources of information

1. The British Library
   Tel: +44 (0)870 444 1500)
   Location 1
   The British Library
   St Pancras
   96 Euston Road
   London
   NW1 2DB

   Location 2
   British Library Newspapers
   Colindale Avenue
   London
   NW9 5HE

   Location 3
   The British Library
   Boston Spa
   Wetherby
   West Yorkshire
   LS23 7BQ

   Environmental Information Service (EIS), Tel: +44 (0)20 7412 7477; eis@bl.uk
   Maps, Tel: +44 (0)20 7412 7702; Maps@bl.uk
   http://www.bl.uk/
2. Remote Sensing Internet Resources Directory – sources of data on the World

Wide Web on remote sensing of coastal and marine environments. http://www.ncl.ac.uk/tcmweb/msc_tcm/rs.htm
Local airfields and flying clubs

3. British Waterways (Environment and Heritage)
Tel: 01923 201120
enquiries.hq@britishwaterways.co.uk
http://www.britishwaterways.co.uk/home/index.html
4. The Countryside Agency – see Natural England
5. Department for Environment, Food and Rural Affairs (DEFRA),
Freedom of Information and Data Protection Division
Lord Chancellor's Department
Selborne House
London SW1E 6QW
Tel: +44 (0) 20 7210 2665
fidp@lcdhq.gsi.gov.uk
http://www.defra.gov.uk/
6. English Heritage
Customer Services Department
PO Box 569
Swindon
SN2 2YP
Tel: +44 (0) 870 333 1181
customers@english-heritage.org.uk
http://www.english-heritage.org.uk/
7. English Nature – see Natural England
8. Environment Agency (England and Wales)
Tel: 08708 506 506 (UK only)
+44 1709 389 201 (non-UK)
http://www.environment-agency.gov.uk/
9. Forestry Commission
231 Corstorphine Road
Edinburgh
Scotland
EH12 7AT
Tel: +44 (0) 131 334 0303
enquiries@forestry.gsi.gov.uk
http://www.forestry.gov.uk/
10. Government Gateway (online government information)
http://www.gateway.gov.uk/
11. Joint Nature Conservation Committee
http://www.jncc.gov.uk/
12. Scottish Environment Protection Agency (SEPA)
Erskine Court
Castle Business Park
Stirling
FK9 4TR
Tel: +44 (0) 1786 457700
http://www.sepa.org.uk/
13. National Trust – Conservation Directorate
39 Sheep Street
Cirencester
Glos.
GL7 1RQ
Tel: +44(0) 8706095382
http://www.nationaltrust.org.uk/main/w-chl.htm
14. Natural England
Northminster House
Peterborough
PE1 1UA

Tel: +44 (0)1733 455000
enquiries@english-nature.org.uk
http://www.english-nature.org.uk/
* Natural England is a partnership formed in 2006 from English Nature, the Countryside Agency (Landscape, Access and Recreation Division) and Defra (Rural Development Service)

15. The Wildlife Trusts – Water Policy Team
c/o Wiltshire Wildlife Trust
Elm Tree Court
Long Stree
Devizes
Wiltshire
SN10 1NJ
Tel: +44(0) 1380 725670
http://www.waterpolicyteam.org/General/EntryPage.htm

Appendix 3: Reference list for codes

The procedures use codes in some cases for recording information on AA or CA maps. These codes are used to reduce the amount of written information needed on the maps. The following table provides a quick reference for these codes.

Code	Section in procedures	Interpretation
S/E	2.1.C.2	Sea/estuary
IW	2.1.C.2	Inland (non-marine) waters
B/M/C	2.1.C.2	Beach/mudflat/cliffs
SM	2.1.C.2	Salt marsh
IG	2.1.C.2	Improved grassland
WG	2.1.C.2	Wet grassland
DSH	2.1.C.2	Dwarf shrub heath
BMW	2.1.C.2	Broadleaved/mixed woodland
PC	2.1.C.2	Permanent crops
CW	2.1.C.2	Coniferous woodland
BG	2.1.C.2	Bog
FMS	2.1.C.2	Fen, marsh and swamp
AH	2.1.C.2	Arable/horticultural land
SRD	2.1.C.2	Suburban/rural development
UID	2.1.C.2	Urban/industrial development
BG/R+S	2.1.C.2	Bare ground/rocks and scree
OG	2.1.C.2	Other grassland
FERT **OWASTE** **LIME**	2.1.C.3.2.14–15	The areas within the AA/CA where fertiliser (**FERT**), organic waste (**OWASTE**) or lime (**LIME**) are applied and indicate the type, rate and frequency of fertiliser applied.
BURN	2.1.C.3.3.2	The areas within the AA that are subjected to burning.
HARV	2.1.C.3.4.4	The areas within the AA where harvesting of herbaceous vegetation occurs. Also indicate on the map where different harvesting practices (mechanised or manual) are used.
PLOU	2.1.C.3.5.3–4	The areas within the AA/CA that are ploughed.
FOREST	2.1.C.3.6.2	The woodland areas within the AA where felling and removal of trees, coppicing, or pollarding occurs.

LV	2.1C.3.7.4/6	The areas within the AA/CA where livestock are kept at densities above those listed in C.3.7.1.
LPB	2.1.C.3.8.2	The areas where large populations of birds occur within the AA.
ACTSD	2.1C.3.9.2/4	Any currently active surface drains within the AA/CA.
ACTSSD	2.1.C.3.9.6/8	Any currently active sub-surface drains within the AA/CA.
PEATEX	2.1.C.3.10.3/4	The location and extent (by shading with a pencil) of current cutting of peat within the AA/CA.
MINEX	2.1.C.3.11.3/4	The location and extent (by shading with a pencil) of mining, quarrying or extraction activities within the AA/CA.
IIWASTE	2.1.C.3.12.3/4	The areas within the AA/CA subjected to industrial inorganic waste disposal.
CONSTR	2.1.C.3.13.3/4	The location and extent of construction activities within the AA/CA.
HUNSF	2.1.C.3.14.2	The location and extent of hunting, shooting, or fishing activities within the AA.
RECRE	2.1.C.3.15.2	The location and extent of recreation activities within the AA.
ConII **ConNI** **ConLI**	2.1.C.3.16.4	The location and extent of conservation areas within the AA of: international importance (**ConII**) national importance (**ConNI**) local importance (**ConLI**)

Appendix 4: Publications and theses produced during FAP development

* Indicates where references are likely to be most readily obtainable direct from the author. A list of author contacts is given in the introductory pages.

Abernethy, V.J. (1994). *Functional Ecology of Euhydrophyte Communities in European Riverine Wetlands*, Ph.D Thesis, University of Glasgow, UK.

Abernethy, V.J. (1994). Functional ecology of euhydrophyte communities in European riverine wetland ecosystems using functional vegetation groups, *Freshwater Biology* **49**(3), 346–356.

Aerts, R. and Van der Peijl, M.J. (1993). A simple model to explain the dominance of low-productive perennials in nutrient-poor habitats, *Oikos* **66**, 144–147.

Amiaud, B. and Touzard, B. (2004). The relationship between soil seed bank, above ground vegetation and disturbance in old embanked marshlands of Western France, *Flora* **199**(3), 25–35.

Baker, C.J. and Maltby, E. (1995). Nitrate removal by river marginal wetlands: factors affecting the provision of a suitable denitrification environment. In: Hughes, J.M.R. and Heathwaite, A.L. (eds), *Hydrology and Hydrochemistry of British Wetlands*, Wiley, Chichester, UK, 291–313.

Bishop, K.H. and Lee, Y.H. (1997). Catchments as a source of mercury/methylmercury in boreal surface waters. In: Sigel, H. and Sigel, A. (eds), *Metal Ions in Biological Systems, Volume 34, Mercury and its Effects on Environment and Biology,* Marcel Dekker, New York, 114–130.

Bishop, K.H., Laudon, H. and Köhler, S.J. (2000). Separating the natural and anthropogenic components of spring flood pH decline: A method for areas that are not chronically acidified, *Water Resources Research* **36**(7), 1873–1884.

Blackwell, M.S.A. (1997). *Zones of Enhanced Denitrification in River Marginal Wetlands*, PhD Thesis, Department of Geography, Royal Holloway University of London, UK.

Blackwell, M.S.A. and Maltby, E. (1998). Identification and field testing of zones of enhanced denitrification (ZEDs). In: McComb, A.J. and Davis, J.A. (eds), *Wetlands for the Future*, contributions from INTECOL's V International Wetlands Conference, Gleneagles Publishing, Adelaide, Australia, 443–456.

Brandt, O. (2003). *Eintrags- und Wirkungspfade von Schwermetallen und Arsen in Flußaue-Systemen am Beispiel der Mulde zwischen Bitterfeld/Wolfen und Dessau, Sachsen-Anhalt (Input/Output and Dynamics of Heavy Metals and Arsenic in River Wetlands – a Case Study in the Wetlands of River Mulde between Bitterfeld/Wolfen and Dessau, Saxony-Anhalt)*, Ph.D Thesis, Technical University of Berlin, Germany, *http://edocs.tu-berlin.de/diss/2003/brandt_oliver.htm*

Brandt, O., Geyer, S., *Henle, K. and Matheis, G. (2000). Verteilung und Verhalten von Spurenelementen im Sicker- und Grundwasser der Muldeaue bei Bitterfeld/Wolfen. In: Friese, K., Witter, B., Miehlich, G., Rode, M. (Hrsg.), *Stoffhaushalt von Auenökosystemen. Böden und Hydrologie*, Schadstoffe, Bewertungen, Springer, Berlin, Heidelberg, New York, 327–336 (in German).

Brandt, O., Geyer, S., Matheis, G. and *Henle, K. (1999). Verteilung und Verhalten von anorganischen Spurenelementen im Sicker- und Grundwasser im Bereich Bitterfeld/Wolfen, *Muldeaue Workshop Stoffhaushalt von Auenökosystemen der Elbe und ihrer Nebenflüsse*, Magdeburg, 1–3.2.1999. UFZ-Bericht 1/1999, UFZ Leipzig-Halle GmbH, Leipzig, 95–98 (in German).

Brandt, O., Geyer, S., Matheis, G. and Henle, K. (1999). Verteilung von Spurenelementen in Sedimenten der Muldeaue bei Bitterfeld/Wolfen als Indikator räumlich und zeitlich differenzierbarer Eintragspfade, *Leipziger Geowissenschaften* **11**, 153–160 (in German).

Castella, E. and Speight, M.C.D (1996). Knowledge representation using fuzzy coded variables: example based on the use of Syrphidae (Insecta, Diptera) in the assessment of riverine wetlands, *Ecological Modelling* **85**(1), 13–25.

Castella, E., Speight, M.C.D., Obrdlik, P., Schneider, E. and Lavery, T.A. (1994). A methodological approach to the use of terrestrial invertebrates for the assessment of alluvial wetlands, *Wetlands Ecology and Management* **3**, 17–36.

*Clément, B. (1994). Synecology of plant communities within river marginal wetlands, *Proceedings of the 4th workshop of the FAEWE EC-STEP project*, Madrid, Spain, Feb.

Clément, B. and Maltby, E. (1996). Quelques facteurs de la biodiversité végétale dans les prairies humides des corridors fluviaux, *Acta Botanica Gallica* **143**, 309–316 (in French).

Clément, B., Maltby, E., Hogan, D.V. and McInnes, R.J. (1996). Relationships between vegetation, hydrology and soil properties in river marginal wetlands of the river Torridge basin. In: Merot, P. and Jigorel, A. (eds), *Hydrologie dans les pays celtiques*, INRA Editions, No. 79, Paris, 305–314.

David, S. (2000). *Analyses Descriptives et Fonctionnelles Comparées de Zones Humides de Lac et d'Estuaire*, PhD Thesis, University of Rennes 1, France.

De la Hera, A. (1999). *Analisis hidrologico de los humedales de la Mancha Humeda y plan de restauracion de un humedal ribereño degradado*, Unpublished PhD Thesis, Department of Geodynamics, Faculty of Geological Sciences. Complutense University of Madrid, Spain (in Spanish).

De la Hera, A. (2002). Upper Guadiana Basin characterisation. In: Coleto, C., Cortina, L.M. and Llamas, M.R. (eds), *Conflictos entre el desarrollo de las aguas subterraneas y la conservación de los humedales: la cuenca alta del Guadiana*, Marcelino Botín Foundation, Madrid, Spain, 165–196 (in Spanish).

Florin, M., Montes, C. and Rueda, F. (1993). Origin, hydrologic functioning and morphometric characteristics of small, shallow, semi-arid lakes (lagunas) in La Mancha, Central Spain, *Wetlands* **13**(4), 247–259.

Flynn, N., Gardner, P.J. and Maltby, E. (1999). The measurement and analysis of denitrification rates obtained using soil columns from river marginal wetlands, *Soil Use and Management* **15**, 150–156.

Flynn, N., Maltby, E. and Gardner, P.J. (1998). Nitrogen interactions between floodwater and floodplain soils. In: McComb, A.J. and Davis, J.A) (eds), *Wetlands for the Future*, contributions from INTECOL's V International Wetlands Conference, Gleneagles Publishing, Adelaide, Australia, 433–442.

Fornés, J.M. (1994). *Hidrología de algunas lagunas de Castilla-La Mancha*, unpublished PhD Thesis, Facultad de Ciencias Geológicas, Servicio de Publicaciones, Vicerrectorado de Extensión Universitaria, Universidad Complutense de Madrid (in Spanish).

Fornés, J.M., de la Hera, A., Cruces, J. and Llamas, M.R. (1995). Funcionamiento hidrológico de un humedal ribereño, El Masegar, en el conjunto de los humedales de la Mancha Húmeda de la cuenca alta del río Cigüela, *Estudios Geológicos* **51**(5–6), 259–276 (in Spanish).

Fornés, J.M., de la Hera, A. and Llamas, M.R. (1999). Landscape changes and ecological impacts caused by groundwater abstraction in the Upper Guadiana basin, Spain. Gambling with groundwater–physical, chemical and biological aspects of aquifer–stream relations. In: Brahana, J.V. *et al.* (eds) *Proceedings of the Conference of the International Association of Hydrogeologists and the American Institute of Hydrology*, 137–142.

Garcia, M. and Llamas, M.R. (1993). Impacto de los regadios en La Llanura Manchego, *Ecosistemas* **5**, 41–43 (in Spanish).

Hills, J.M., Murphy, K.J., Pulford, I.D. and Flowers, T.M. (1994). A method for classifying European riverine wetland ecosystems using functional vegetation groups, *Functional Ecology* **8**, 242–252.

Gardner, P.J., Flynn, N. and Maltby, E. (2000). A temperature correction procedure for field-based soil denitrtification experiments, *Soil Use and Management* **16**, 260–263.

Gardner, P.J., Flynn, N. and Maltby, E. (2001). A simple method for the determination of ionic diffusion coefficients in flooded soils, *Hydrological Processes* **15**, 511–518.

Hills, J.M. (1994). *Functional Analysis of European Wetland Vegetation*, Ph.D Thesis, University of Glasgow, UK.

Hills, J.M. and Murphy, K.J. (1996). Evidence for consistent functional groups of wetland vegetation across a broad geographical range of Europe, *Wetlands Ecology and Management* **4**, 51–63.

Hills, J.M., Murphy, K.J., Pulford, I.D. and Flowers, T.H. (1994). A method for classifying European riverine wetland ecosystems using functional vegetation groups, *Functional Ecology* **8**, 242–252.

Hooijer, A. (1996). *Floodplain hydrology. An Ecologically-orientated Study of the Shannon Callows, Ireland*, PhD Thesis, Free University Amsterdam, The Netherlands.

Hruška, J., Köhler, S.J. and Bishop, K.H. (1999). Buffering processes in a boreal stream during experimental acidification, *Environmental Pollution* **106**, 55–65.

Janssen, R., Goosen, H., Verhoeven, M.L., Verhoeven, J.T.A., Omtzigt, A.Q.A. and Maltby,

E. (2005). Decision support for integrated wetland management, *Environmental Modelling & Software* **20**, 215–229.

Köhler, S.J. (1999). *Quantifying the Role of Natural Organic Acids on pH and Buffering in Swedish Surface Waters*, PhD Thesis, Swedish University of Agricultural Sciences, Department of Forest Ecology SE-90183 Umeå, Sweden.

Köhler, S.J., Hruška, J. and Bishop, K.H. (1999). Influence of organic acid site density on pH modelling of Swedish lakes, *Canadian Journal of Fishery and Aquatic Sciences* **56**(8), 1461–1470.

Köhler, S., Laudon, H., Wilander, A. and Bishop, K.H. (2000). Estimating organic acids dissociation in natural surface water using total alkalinity and TOC, *Water Research* **34**(5), 1425–1434.

Laudon, H. and Bishop, K.H. (1999). Quantifying sources of ANC depression during spring flood episodes in Northern Sweden, *Environmental Pollution* **105**, 427–455.

Laudon, H., Köhler, S.J. and Bishop, K.H. (1999). Natural acidity or anthropogenic acidification in spring flood of Northern Sweden, *Science of the Total Environment* **234**, 63–73.

Laudon, H., Westling, O. and Bishop, K.H. (2000). Variability of spring flood hydrochemistry in Northern Sweden, *Canadian Journal of Fisheries and Aquatic Sciences* **57**, 1888–1900.

Lloyd, J.W. and Tellam, J.H. (1995). Groundwater fed wetlands in the UK. In: Hughes, J.M.R. and Heathwaite, A.L. (eds), *Hydrology and Hydrochemistry of British Wetlands*, Wiley, Chichester, UK, 39–62.

Maltby, E. (1998). Wetlands within catchments – some issues of scale and location in linking processes and functions. In: McComb, A.J. and Davis, J.A. (eds), *Wetlands for the Future*, contributions from INTECOL's V International Wetlands Conference, Gleneagles Publishing, Adelaide, Australia, 383–392.

Maltby, E., Hogan, D.V., Immirzi, C.P., Tellam, J.H. & van der Peijl, M.J. (1994). Building a new approach to the investigation and assessment of wetland ecosystem functioning. In: Mitsch, W.J. (ed.), *Global Wetlands: Old World and New*, Elsevier, Amsterdam, The Netherlands, 637–658.

Maltby, E., Hogan, D.V. and McInnes, R.J. (1996). *Functional Analysis of European Wetland Ecosystems. Phase I (FAEWE)*, Ecosystems Research Report No 18, European Commission Directorate General Science, Research & Development, Brussels.

Maltby, E., Hogan, D.V. and Oliver, G.A. (1995). Wetland soil hydrology and ecosystem functioning. In: Hughes, J.M.R. and Heathwaite, A.L. (eds), *Hydrology and Hydrochemistry of British Wetlands*, Wiley, Chichester, UK, 325–362.

Maltby, E., Jewkes, E.C., Baker, C.J., Mockler, N.J., Flynn, N. and Gardner, P.J. (1998). A simple method for estimating nitrate removal from floodwaters in river marginal wetlands, *Soil Use and Management* **14**, 90–95.

McInnes, R.J., Maltby, E., Neuber, M.S. and Rostron, C.P. (1998). In: McComb, A.J. and Davis, J.A. (eds), *Wetlands for the Future*, contributions from INTECOL's V International Wetlands Conference, Gleneagles Publishing, Adelaide, Australia, 407–432.

*Murphy, K.J. (1995). Predicting the impacts of changed hydrological regime on European riverine wetland plant communities, *Abs. 80th Ecological Society of America Meeting*, Utah, USA.

Murphy, K.J., Castella, E., Clement, B., Hills, J.M., Obrdlik, P., Pulford, I.D., Schneider, E., Speight, M.C.D. (1994). Biotic indicators of riverine wetland ecosystem functioning. In: Mitsch, W.J. (ed.), *Global Wetlands–Old World and New*, Elsevier, Amsterdam, The Netherlands, 659–682.

Papatolios, K.T., *Tellam, J.H. and Lloyd, J.W. (1997). Hydrogeological assessment of river marginal wetlands in north Devon, UK. In: Marinos, P., Koukis, G., Tsiambos, G. and Stournaras, G. (eds), *Proceedings of the International Conference on Engineering Geology and the Environment*, Athens, June 23–27, Balkema, Rotterdam, The Netherlands, 3003–3008.

*Petry, D., Lutosch, I. and Scholz, M. (2003). *Auenschutz und Wasserrahmenrichtlinie – eine Herausforderung für Naturschutz und Landschaftsplanung im Europäischen Kontext (Floodplain ecosystems and the Water Framework Directive – a new challenge for nature conservation)*, UVP-report 17/3+4, 129–133. (*contact K. Henle) (in German).

*Petry, D., Scholz, M. and Lutosch, I. (eds) (2002). *Relevanz der EU-Wasserrahmenrichtlinie für den Naturschutz in Auen*, UFZ-Bericht 22/2002, Leipzig, Halle, http://www.ufz.de/data/ufzbericht22-02543.pdf (*contact K. Henle) (in German).

Russell, M.A. and Maltby, E. (1995). The role of hydrologic regime on phosphorus, dynamics in a seasonally waterlogged soil. In: Hughes, J.M.R. and Heathwaite, A.L. (eds), *Hydrology and Hydrochemistry of British Wetlands*, Wiley, Chichester, UK, 245–260.

Scholz, M., Horlitz, T., Lutosch, I. and *Henle, K. (2002). Neue Wege in der Analyse und Bewertung von Landschaftsökologie und Sozio-Ökonomie in Auen am Beispiel der Elbe in Hinsicht auf die Umsetzung der EU-WRRL. In: Geller, W. *et al.* (eds), *Elbe – neue Horizonte des Flussgebietsmanagements. 10. Magdeburger Gewässerschutzseminar*, B.G. Teubner, Stuttgart, Leipzig, Wiesbaden, Germany, 133–136 (in German).

*Scholz, M., Schulz, C. and Horlitz, T. (2004). Analyse und Bewertung ökologischer und sozioökonomischer Auenfunktionen. In: Möltgen, J. and Petry, D. (Hrsg.), *Interdisziplinäre Methoden des Flussgebietsmanagements*, Münster, Germany, Workshopbeiträge 15./16. März, IfGI prints **21**, 205–212. (* contact K. Henle) (in German)

*Schulz, C., Scholz, M., Horlitz, T. and Lutosch, I. (2003). Funktionale Analyse und Bewertung von Auen am Beispiel der Mittleren Elbe. In: *International Association for Landscape Ecology Deutschland, Tagungsband zur 4. Jahrestagung der IALE –Region Deutschland: Landnutzungswandel: Ist-Zustand & Rahmenbedingungen – Szenarien – Bewertung*, Eberswalde, 80. (* contact K. Henle) (in German)

Simpson, M. (2000). *A Functional Classification of European Wetland Ecosystems*, unpublished PhD Thesis, Royal Holloway University of London.

Sirin, A., Köhler, S.J. and Bishop, K.H. (1998). Resolving flow pathways and geochemistry in a headwater forested wetland with multiple tracers. Kovar, K., Tappeiner, U., Peters, N.E. and Craig, R.G. (eds), In: *Hydrology, Water Resources and Ecology in Headwaters*, IAHS Publ. No. 248, Wallingford, UK, 337–342.

Speight, M.C.D. and Castella, E. (1995). Bugs in the system: Relationships between distribution data, habitat and site evaluation in development of an environmental assessment procedure based on invertebrates. In: Valovirta, J., Harding, P.T. and Kime, D. (eds), *Proceedings of the 9th International Colloquium of the European Invertebrate Survey, Helsinki, 3-4 September 1993*, number 7 in WWF Finland Report, 1–9.

*Spink, A. (1994). Phytometric estimation of fertility. In: *Proceedings of the 4th Workshop of the FAEWE-EC STEP Project*, Madrid, Spain. (*contact J.T. A. Verhoeven)

Touzard, B. (1999). *Plant Diversity Dynamics in Wet Herbaceous Systems After the Release of Agricultural Practices*, PhD Thesis, University of Rennes, France.

Touzard, B., Amiaud, B., Langlois, E., Lemauviel, S. and Clément, B (2002). The relationship between soil seed bank, above ground vegetation and disturbance in a eutrophic alluvial wetland of Western France, *Flora* **197**(3), 175–185.

Touzard, B. and Clément, B. (2001). Effets de microperturbations experimentales sur la dynamique de la diversité végétale d'une roselière eutrophe, *Botanica Helvetica* **111**, 45–58 (in French).

Touzard, B., Clément, B. and Lavorel, S. (2002). Successional patterns in a eutrophic alluvial wetland of western France, *Wetlands* **22**, 111–125.

Vadineanu, A., Cristofor, S., Ignat, G., Romanca, G., Ciubuc, C., Botnariuc, N. and Sarbu, A. (1998). Ecological processes, functions and sustainable management in the Danube Delta, In: McComb, A.J. and Davis, J.A. (eds), *Wetlands for the Future*, contributions from INTECOL's V International Wetlands Conference, Gleneagles Publishing, Adelaide, Australia, 373–382.

Van den Wyngaert, I.J.J. (2001). *Grazing of Extensive Reed Beds by Moulting Greylag Geese: Effects on Nutrient Dynamics and Growth of The Phragmites Australis Vegetation and Consequenes for the Lake Ecosystem*, PhD Thesis, Utrecht University, The Netherlands.

Van den Wyngaert, I.J.J., Wienk, L.D., Sollie, S., Bobbink, R. and Verhoeven, J.T.A. (2003). Long-term effects of yearly grazing by moulting greylag geese (*Anser anser*) on reed (*Phragmites australis*) growth and nutrient dynamics, *Aquatic Botany* **75**, 229–248.

Van der Peijl, M.J. (1997). *Modelling of Biogeochemical Processes and Spatial Patterns in Freshwater Wetlands*, PhD Thesis, Utrecht University, The Netherlands.

Van der Peijl, M.J., Van Oorschot, M.M.P. and Verhoeven, J.T.A. (2000). Simulation of the effects of nutrient enrichment on nutrient and carbon dynamics in a river marginal wetland, *Ecological Modelling* **134**, 169–184.

Van der Peijl, M.J. and Verhoeven, J.T.A. (1999). A model for carbon, nitrogen and phosphorus dynamics and their interactions in river marginal wetlands, *Ecological Modelling* **118**, 95–130.

Van der Peijl, M.J. and Verhoeven, J.T.A. (2000). Carbon, nitrogen and phosphorus cycling in river marginal wetlands; using a dynamic simulation model to examine the importance of landscape geochemical flows, *Biogeochemistry* **50**, 45–71.

Van Oorschot, M.M.P. (1994). Plant production, nutrient uptake and mineralization in river marginal wetlands: the impact of nutrient additions due to former land-use. In: Mitsch, W.J. (ed.), *Global Wetlands: Old World and New*, Elsevier, Amsterdam, The Netherlands, 133–150.

Van Oorschot, M.M.P. (1996). *Effects of the Vegetation on Carbon, Nitrogen and Phosphorus*

Dynamics in English and French Riverine Grasslands, PhD Thesis, Utrecht University, The Netherlands.

Van Oorschot, M.M.P., Hayes, C. and Van Strien, I. (1998). The influence of soil desiccation on plant production, nutrient uptake and plant nutrient availability in two French floodplain grasslands, *Regulated Rivers: Research and Management* **14**, 313–327.

Van Oorschot, M.M.P., Robbemont, E., Boerstal, M., Van Strien, M. and Van Kerkhoven-Schmitz, M. (1997). Effects of enhanced nutrient availability on plant and soil nutrient dynamics in two English riverine ecosystems, *Journal of Ecology* **85**, 167–179.

Van Oorschot, M.M.P., Spink, A., Van Gaalen, A.N., Maltby, E., Mockler, N.J. and Verhoeven, J.T.A. (2000). Experimental manipulation of soil water levels in two French riverine grassland soils, *Acta Oecologica* **21**, 49–62.

Verhoeven, J.T.A. (1998). Wetland restoration and creation: consequences for nutrient-related processes. In: McComb, A.J. and Davis, J.A. (eds), *Wetlands for the Future*, Contributions from INTECOL's V International Wetlands Conference, Gleneagles Publishing, Adelaide, Australia, 503–516.

Appendix 5: Conversion of process outcomes to functional scores – Functional Assessment Procedures CD-ROM

1. Hydrological functions

Expert judgement has been used to translate process performance codes into scores that can be compared within related functions. Scores range between 0 (no significant performance of function) and 1 (full performance of function).

(i) Floodwater detention

Process performance code	Score for function
1/1x	1
1/2x	0.9
1/3 or **1/5**	0.8
1/4 or **1/6**	0.7
1/7 or **1/9**	0.6
1/8 or **1/10**	0.5
1/11 or **1/13**	0.4
1/12 or **1/14**	0.3
1/15	0.2
1/16	0.1
2	0
4	0

(ii) Groundwater recharge/Groundwater discharge

Process performance code	Score for function
1/1	1
1/2	0.9
1/3	0.8
1/4	0.7
1/5	0.6
1/6	0.5
1/7	0.4
1/8	0.3
2	0
4	0

(iii) Sediment retention

Process performance code	Score for function
1/1	1
1/3 or **1/10** or **1/11**	0.8
1/2 or **1/4** or **1/12** or **1/14**	0.6
1/5 or **1/6** or **1/7** or **1/8** or **1/9** or **1/13** or **1/20**	0.4
1/15 or **1/16** or **1/17** or **1/18** or **1/19**	0.2
2 or **4**	0

2. Biogeochemical functions

Means are estimated for quantitative functions.

(i) Nutrient retention

Nitrogen		**Plant uptake (N)**	**Storage in soil organic matter (N)**	**Adsorption of N as ammonium**	**Retention of particulate nutrients (N)**
Code 1/a	Min	10* kg N ha^{-1} a^{-1}	10 kg N ha^{-1} a^{-1}		
	Max	80 kg N ha^{-1} a^{-1}	50 kg N ha^{-1} a^{-1}		
	Average	45 kg N ha^{-1} a^{-1}	30 kg N ha^{-1} a^{-1}		
Code 1/b	Min	10* kg N ha^{-1} a^{-1}	5 kg N ha^{-1} a^{-1}		
	Max	50 kg N ha^{-1} a^{-1}	25 kg N ha^{-1} a^{-1}		
	Average	30 kg N ha^{-1} a^{-1}	15 kg N ha^{-1} a^{-1}		
Code 1/c	Min	5* kg N ha^{-1} a^{-1}			
	Max	25 kg N ha^{-1} a^{-1}			
	Average	15 kg N ha^{-1} a^{-1}			
Code 1/d	Min				
	Max				
	Average				
Code 1/x	Estimated average	7.5* kg N ha^{-1} a^{-1}	7.5* kg N ha^{-1} a^{-1}	15* kg N ha^{-1} a^{-1}	25* kg N ha^{-1} a^{-1}
Code 1	Estimated average			30 kg N ha^{-1} a^{-1}	50 kg N ha^{-1} a^{-1}
Code 2	Average	0 kg N ha^{-1} a^{-1}	0 kg N ha^{-1} a^{-1}	0 kg N ha^{-1} a^{-1}	0 kg N ha^{-1} a^{-1}

* estimated value

Phosphorus		Plant uptake (P)	Storage in soil organic matter (P)	Adsorption of Phosphorus	Retention of particulate nutrients (P)
Code 1/a	Min	1* kg P ha^{-1} a^{-1}	2 kg P ha^{-1} a^{-1}		
	Max	5 kg P ha^{-1} a^{-1}	10 kg P ha^{-1} a^{-1}		
	Average	3 kg P ha^{-1} a^{-1}	6 kg P ha^{-1} a^{-1}		
Code 1/b	Min	1* kg P ha^{-1} a^{-1}	1 kg P ha^{-1} a^{-1}		
	Max	3 kg P ha^{-1} a^{-1}	5 kg P ha^{-1} a^{-1}		
	Average	2 kg P ha^{-1} a^{-1}	3 kg P ha^{-1} a^{-1}		
Code 1/c	Min	0.5 * kg P ha^{-1} a^{-1}			
	Max	1.5 kg P ha^{-1} a^{-1}			
	Average	1 kg P ha^{-1} a^{-1}			
Code 1/d	Min				
	Max				
	Average				
Code 1/x	Estimated average	0.5* kg P ha^{-1} a^{-1}	1.5* kg P ha^{-1} a^{-1}	2.5* kg P ha^{-1} a^{-1}	5* kg P ha^{-1} a^{-1}
Code 1	Estimated average			5* kg P ha^{-1} a^{-1}	10* kg P ha^{-1} a^{-1}
Code 2	Average	0 kg P ha^{-1} a^{-1}	0 kg P ha^{-1} a^{-1}	0 kg P ha^{-1} a^{-1}	0 kg P ha^{-1} a^{-1}

* estimated value

Degree of performance	Nitrogen	Phosphorus
High performance → 0.99	≥ 100 ″ 155 kg ha^{-1} a^{-1}	≥ 10 ″ 24 kg ha^{-1} a^{-1}
Intermediate perform → 0.66	≥ 15 < 100 kg ha^{-1} a^{-1}	≥ 2 < 10 kg ha^{-1} a^{-1}
Low performance → 0.33	> 0 < 15 kg ha^{-1} a^{-1}	> 0 < 2 kg ha^{-1} a^{-1}
Not performed → 0	0 kg ha^{-1} a^{-1}	0 kg ha^{-1} a^{-1}

(ii) Nutrient export

Nitrogen		Denitrification	Ammonia volatilisation	Vegetation management (N)
Code 1/a	Average	155 kg N ha^{-1} a^{-1}		225 kg N ha^{-1} a^{-1}
Code 1/b	Average	45 kg N ha^{-1} a^{-1}		115 kg N ha^{-1} a^{-1}
Code 1/c	Average	10 kg N ha^{-1} a^{-1}		60 kg N ha^{-1} a^{-1}
Code 1/d	Average			30 kg N ha^{-1} a^{-1}
Code 1/x	Estimated average	5* kg N ha^{-1} a^{-1}	20* kg N ha^{-1} a^{-1}	15* kg N ha^{-1} a^{-1}
Code 1	Estimated average		40* kg N ha^{-1} a^{-1}	
Code 2	Average	0 kg N ha^{-1} a^{-1}	0 kg N ha^{-1} a^{-1}	0 kg N ha^{-1} a^{-1}

* estimated value

Phosphorus		Vegetation management (P)
Code 1/a	Average	22.5 kg P ha^{-1} a^{-1}
Code 1/b	Average	11.5 kg P ha^{-1} a^{-1}
Code 1/c	Average	6 kg P ha^{-1} a^{-1}
Code 1/d	Average	3 kg P ha^{-1} a^{-1}
Code 1/x	Estimated average	1.5* kg P ha^{-1} a^{-1}
Code 1	Estimated average	
Code 2	Average	0 kg P ha^{-1} a^{-1}

* estimated value

Degree of performance	Nitrogen	Phosphorus
High performance → 0.99	≥ 100 ″ 420 kg ha^{-1} a^{-1}	≥ 10 ″ 22.5 kg ha^{-1} a^{-1}
Intermediate performance → 0.66	≥ 15 < 100 kg ha^{-1} a^{-1}	≥ 2 < 10 kg ha^{-1} a^{-1}
Low performance → 0.33	> 0 < 15 kg ha^{-1} a^{-1}	> 0 < 2 kg ha^{-1} a^{-1}
Not performed → 0	0 kg ha^{-1} a^{-1}	0 kg ha^{-1} a^{-1}

(iii) *In-situ carbon retention (one process in function)*

3.2.C.1	Score
1/a	1
1/b	0.9
–	–
1/x	0.4
2	0.1

(iv) *Trace element storage/export*

Correspondence for storage processes. Compilation of trace element process outcomes are assigned scores, which are summed before translation into scores for the overall function.

Process outcome	Process score
1	1
1/x	10
2	100
4	1000

Sum of scores	Score for function
3	1
12	0.9
102	0.9
1002	0.9
21	0.8
111	0.8
1011	0.8
201	0.8
1101	0.8
2001	0.8
30	0.6
120	0.5
1020	0.5
210	0.4
1110	0.4
2010	0.4
300	0.2
3000	0

(v) *Organic carbon concentration control (syntax)*

If code1(i, 18) = "1/1" Then score(i, 18) = 1
If code1(i, 18) = "1/2" Then score(i, 18) = 0.5
If code1(i, 18) = "2" Then score(i, 18) = 0
If code1(i, 18) = "4" Then score(i, 18) = 0
If code1(i, 19) = "1/1" Then score(i, 19) = 1
If code1(i, 19) = "1/2" Then score(i, 19) = 0.67
If code1(i, 19) = "1/3" Then score(i, 19) = 0.33
If code1(i, 19) = "2" Then score(i, 19) = 0
If code1(i, 19) = "4" Then score(i, 19) = 0
funscore(i, 8) = (score(i, 18) + score(i, 19))/2

3. Ecological functions

The range of ecological processes makes them widely varied in impact. Expert judgement has been used to weight each process for its relative environmental impact on wetland functioning.

(i) *Ecosystem maintenance*

Process outcome	Score for function
1/1	1
1/2	0.75
1/3	0.5
1/x	0.5
2	0.25

(ii) *Food-web support*

Biomass production (weighting = 0.6)

Process outcome	Score for function
1/a	1
1/b	0.75
1/c	0.5
1/d	0.25
1/x	0.25
2	0

Import and export of biomass – except export via biological processes; (weighting = 0.1)

Process outcome	Score for function
1/1	1
1/2	0.67
1/3	0.33
1/x	0.33
2	0

Export of biomass via biological processes (weighting = 0.6)

Process outcome	Score for function
1/a	1
1/b	0.75
1/c	0.5
1/x	0.25
2	0

Index